Oxidative Stress in Applied Basic Research and Clinical Practice

Editor-in-Chief
Donald Armstrong

For other titles published in this series, go to
www.springer.com/series/8145

Note from the Editor-in-Chief

All books in this series illustrate point-of-care testing and critically evaluate the potential of antioxidant supplementation in various medical disorders associated with oxidative stress. Future volumes will be up-dated as warranted by emerging new technology, or from studies reporting clinical trials.

Donald Armstrong
Editor-in-Chief

Toshio Miyata • Kai-Uwe Eckardt
Masaomi Nangaku
Editors

Studies on Renal Disorders

Humana Press

Editors
Toshio Miyata
Division of Molecular Medicine
and Therapy
United Centers for Advanced Research
and Translational Medicine (ART)
Tohoku University Graduate School
of Medicine
2-1 Seiryo-Machi, Aoba-ku
Sendai, Miyagi 980-8575
Japan
miyata@med.tohoku.ac.jp

Kai-Uwe Eckardt
Department of Nephrology
and Hypertension
Friedrich-Alexander-University
Erlangen-Nuremberg
Krankenhausstrasse 10-12 D-91054
Erlangen, Germany
Kai-Uwe.Eckardt@med4.med.
uni-erlangen.de

Masaomi Nangaku
University of Tokyo School of Medicine
7-3-1 Hongo, Bunkyo-ku
Tokyo 113-8655
Japan
mnangaku-tky@umin.ac.jp

ISBN 978-1-4939-5805-4 ISBN 978-1-60761-857-7 (eBook)
DOI 10.1007/978-1-60761-857-7
Springer New York Dordrecht Heidelberg London

Printed on acid-free paper

Humana Press is part of Springer Science+Business Media (www.springer.com)

Preface

Broad derangements of oxygen metabolism, such as oxidative stress and hypoxia, have been implicated in the genesis of kidney disease, independently of hemodynamic and metabolic abnormalities. They further impact various biological reactions linked to oxygen metabolism, such as nitrosative stress, advanced glycation, carbonyl stress, and endoplasmic reticulum stress. This causal role of impaired oxygen metabolism in kidney disease has implications for our understanding of current therapeutic benefits accruing from antihypertensive agents, the control of hyperglycemia/hyperinsulinemia or of hyperlipidemia, and the dietary correction of obesity. The defense mechanisms against oxidative stress (e.g., the Nrf2-Keap1 system) and hypoxia (e.g., the HIF-PHD system) have been recently explored in various cells, including kidney cells, and they include intracellular sensors for oxidative stress and hypoxia. Novel approaches targeting these sensors may offer clinical benefits in several disorders in which oxidative stress or hypoxia is a final, common pathway. Leading basic researchers and clinical scientists have contributed to this book and provide up-to-date, cutting-edge reviews on recent advances in the pathobiology of oxygen metabolism in kidney disease, especially oxidative stress and hypoxia.

Sendai
Erlangen
Tokyo
February 8, 2010

Toshio Miyata, MD, PhD
Kai-Uwe Eckardt, MD
Masaomi Nangaku, MD, PhD

Contents

Part II Clinical Aspects of Oxidative Stress in the Kidney

Part III Current Therapy Targeting Oxidative tress

Contributors

Abolfazl Zarjou
Division of Nephrology, THT 647, University of Alabama at Birmingham, 1900 University Boulevard, Birmingham, Alabama 35294, USA

Adam Whaley-Connell
Division of Nephrology, Department of Internal Medicine, University of Missouri-Columbia School of Medicine, CE416, 5 Hospital Drive, Columbia, MO 65212, USA

Agnes B. Fogo
Division of Renal Pathology, Vanderbilt University School of Medicine, C-3310 Medical Center North Department of Pathology, Vanderbilt University Medical Center, Nashville, TN 37232-2561, USA
agnes.fogo@vanderbilt.edu

Akira Nishiyama
Department of Pharmacology and Hypertension and Kidney Disease Research Center, Kagawa University Medical School, 1750-1 Ikenobe, Miki-cho, Kita-gun, Kagawa Prefecture 761-0793, Japan
akira@kms.ac.jp

Alexander Angerhofer
Division of Renal Diseases and Hypertension, University of Colorado, Aurora, CO 80045 USA

Antonio L. Bartorelli
Centro Cardiologico Monzino, Via Parea 4, 20138 Milan, Italy

Anupam Agarwal
Division of Nephrology, THT 647, University of Alabama at Birmingham, 1900 University Boulevard, Birmingham, Alabama 35294, USA
agarwal@uab.edu

Bernardo Rodriguez-Iturbe
Servicio de Nefrología, Hospital Universitario, Universidad del Zulia and IVIC-Zulia, Maracaibo, Estado Zulia, Venezuela
bernardori@telcel.net.ve

Carsten C. Scholz
Conway Institute of Biomolecular and Biomedical Research,
School of Medicine and Medical Science, College of Life Sciences,
University College Dublin, Belfield, Dublin 4, Ireland

Carsten Willam
Department of Nephrology and Hypertension, Friedrich-Alexander-University
Erlangen, Krankenhausstrasse 12, D-91054 Erlangen, Germany
carsten.willam@uk-erlangen.de

Charles van Ypersele de Strihou
Division of Molecular Medicine and Therapy, United Centers for Advanced
Research and Translational Medicine (ART), Tohoku University Graduate
School of Medicine, 2-1 Seiryo-Machi, Aoba-ku, Sendai, Miyagi 980-8575, Japan

Chih-Wei Yang
Kidney Research Center, Department of Nephrology,
Chang Gung Memorial Hospital, 199, Tun-Hua North Road, Taipei, Taiwan
cwyang@ms1.hinet.net

Christian Rosenberger
Department of Medicine, Hadassah University Hospital,
Mount Scopus, 24035, Jerusalem 91240, Israel

Christopher S. Wilcox
Department of Medical Cell Biology, Uppsala University,
Box 571751 23 Uppsala, Sweden

Colin R. Lenihan
Conway Institute of Biomolecular and Biomedical Research,
School of Medicine and Medical Science, College of Life Sciences,
University College Dublin, Belfield, Dublin 4, Ireland

Cormac T. Taylor
Conway Institute of Biomolecular and Biomedical Research,
School of Medicine and Medical Science, College of Life Sciences,
University College Dublin, Belfield, Dublin 4, Ireland
cormac.taylor@ucd.ie

Danilo Fliser
Department of Internal Medicine – Renal and Hypertensive Diseases
Saarland University Medical Centre, Kirrberger Strasse,
66421 Homburg/Saar, Germany
indfli@uks.eu

David R. Sell
Department of Pathology and Biochemistry, Case Western Reserve University,
Wolstein Research Bldg., 2103 Cornell Road, Cleveland, OH 44106-7288, USA

Eisei Noiri
Department of Medicine, Tokyo University, Tokyo, Japan
noiri-tky@umin.ac.jp

Erminio Sisillo
Centro Cardiologico Monzino, Via Parea 4, 20138 Milan, Italy

Ferdinand H. Bahlmann
Department of Internal Medicine – Renal and Hypertensive Diseases
Saarland University Medical Centre, Kirrberger Strasse,
66421 Homburg/Saar, Germany

Francesco Addabbo
Department of Medicine, Tokyo University, Tokyo, Japan

Fredrik Palm
Department of Medical Cell Biology, Uppsala University,
Box 571751 23 Uppsala, Sweden
Fredrik.Palm@mcb.uu.se

George Henderson
Division of Renal Diseases and Hypertension, University of Colorado,
Aurora, CO 80045 USA

Giancarlo Marenzi
Centro Cardiologico Monzino, I.R.C.C.S. Department of Cardiovascular Sciences,
University of Milan, Milan, Italy
Giancarlo.marenzi@ccfm.it

Gregg L. Semenza
Vascular Program, Institute for Cell Engineering; McKusick-Nathans
Institute of Genetic Medicine; and Departments of Pediatrics, Medicine,
Oncology, Radiation Oncology, and Biological Chemistry,
The Johns Hopkins University School of Medicine, Baltimore, MD 21205, USA
gsemenza@jhmi.edu

Guo-Hua Fong
Department of Cardiology, Kyushu University of Medicine,
3-1-1 Maidashi, Higashi-ku, Fukuoka 812-8582, Japan

György Balla
Division of Nephrology, THT 647, University of Alabama at Birmingham,
1900 University Boulevard, Birmingham, Alabama 35294, USA

Hidenori Koyama
Department of Metabolism, Endocrinology and Molecular Medicine
Osaka City University School of Medicine, 1-4-3, Asahi-machi,
Abeno, Osaka 545-8585, Japan
hidekoyama@med.osaka-cu.ac.jp

Hideyasu Kiyomoto
Department of Pharmacology and Hypertension and Kidney Disease Research Center, Kagawa University, 1750-1 Ikenobe, Miki-cho, Kita-gun, Kagawa Prefecture 761-0793, Japan

Hiroshi Nishi
Division of Nephrology and Endocrinology, University of Tokyo School of Medicine, 7-3-1 Hongo, Bunkyo-ku, Tokyo, Japan
mnangaku-tky@umin.ac.jp

Hiroyuki Kobori
Department of Pharmacology and Hypertension and Kidney Disease Research Center, Kagawa University, 1750-1 Ikenobe, Miki-cho, Kita-gun, Kagawa Prefecture 761-0793, Japan

Ina Nemet
Department of Pathology and Biochemistry, Case Western Reserve University, Wolstein Research Bldg., 2103 Cornell Road, Cleveland, OH 44106-7288, USA

Jean-Marc Odee
Department of Medicine, Hadassah University Hospital, Mountt Scopus, 24035, Jerusalem 91240, Israel

Johannes Jacobi
Department of Nephrology and Hypertension, Friedrich-Alexander-University Erlangen, Krankenhausstrasse 12, D-91054 Erlangen, Germany

Josef Pfeilschifter
Pharmazentrum frankfurt/ZAFES, University Hospital, Goethe University, Theodor-Stern-Kai 7, D-60590, Frankfurt am Main, Germany
pfeilschifter@em.uni-frankfurt.de

Josephine M. Forbes
Diabetes Complications Division, Baker IDI Heart and Diabetes Institute, Glycation and Diabetes, Baker IDI Heart and Diabetes Institute, 75 Commercial Road, Melbourne, VIC 3004, Australia
Josephine.Forbes@bakeridi.edu.au

József Balla
Division of Nephrology, THT 647, University of Alabama at Birmingham, 1900 University Boulevard, Birmingham, Alabama 35294, USA

Kai-Uwe Eckardt
Department of Nephrology and Hypertension, Friedrich-Alexander-University, Erlangen-Nuremberg, Krankenhausstrasse 10-12, D-91054 Erlangen, Germany

Karl-Friedrich Beck
Pharmazentrum frankfurt/ZAFES, University Hospital, Goethe University, Theodor-Stern-Kai 7, D-60590 Frankfurt am Main, Germany

Kotaro Takeda
Department of Cardiology, Kyushu University of Medicine,
3-1-1 Maidashi, Higashi-ku, Fukuoka 812-8582, Japan
ktakeda@cardiol.med.kyushu-u.ac.jp

Kyung Mee Kim
Division of Renal Diseases and Hypertension, University of Colorado,
Aurora, CO 80045 USA

Li-Jun Ma
Division of Renal Pathology, Vanderbilt University School of Medicine,
C-3310 Medical Center North Department of Pathology,
Vanderbilt University Medical Center, Nashville, TN 37232-2561, USA

Lina Nordquist
Department of Medical Cell Biology, Uppsala University,
Box 571751 23 Uppsala, Sweden

Lu-Ping Li
Department of Radiology, Evanston Northwestern Healthcare, Suite G507,
Evanston Hospital, 2650 Ridge Avenue, Evanston, IL 60201, USA

Mark E. Cooper
Diabetes Complications Division, Baker IDI Heart and Diabetes Institute,
Glycation and Diabetes, Baker IDI Heart and Diabetes Institute,
75 Commercial Road, Melbourne, VIC 3004, Australia

Masaomi Nangaku
Division of Nephrology and Endocrinology, University of Tokyo School
of Medicine, 7-3-1 Hongo, Bunkyo-ku, Tokyo 113-8655, Japan
mnangaku-tky@umin.ac.jp

Michael S. Goligorsky
Division of Nephrology and Endocrinology, University of Tokyo School
of Medicine, 7-3-1 Hongo, Bunkyo, Tokyo 113-8655, Japan

Miriam F. Weiss
Department of Pathology and Biochemistry, Case Western Reserve University,
Wolstein Research Bldg., 2103 Cornell Road, Cleveland, OH 44106-7288, USA

Mogher Khamaisi
Department of Medicine, Hadassah University Hospital, Mount Scopus,
24035, Jerusalem 91240, Israel

Motoko Yanagita
Division of Nephrology, Jikei Medical University, 3-25-8, Nishi-Shimbashi,
Minato-ku, Tokyo 105-8461, Japan

Naila Rabbani
Warwick Medical School, Clinical Sciences Research Institute, University of Warwick, University Hospital, Coventry CV2 2DX, UK

Nosratola D. Vaziri
Servicio de Nefrología, Hospital Universitario,
Universidad del Zulia and IVIC-Zulia, Maracaibo, Estado Zulia, Venezuela

Olivier Devuyst
Laboratory of Nephrology, Université catholique, de Louvain Medical School, 54 Avenue Hippocrate, UCL 5450, 1200 Brussels, Belgium
olivier.devuyst@uclouvain.be

Paul J. Thornalley
Warwick Medical School, Clinical Sciences Research Institute, University of Warwick, University Hospital, Coventry CV2 2DX, UK
P.J.Thornalley@warwick.ac.uk

Peter Hansell
Department of Medical Cell Biology, Uppsala University,
Box 571751 23 Uppsala, Sweden

Pottumarthi V. Prasad
Department of Radiology, Evanston Northwestern Healthcare, Suite G507, Evanston Hospital, 2650 Ridge Avenue, Evanston, IL 60201, USA
pprasad@enh.org

Ravi Nistala
Division of Nephrology, Department of Internal Medicine, University of Missouri-Columbia School of Medicine, CE416, 5 Hospital Drive, Columbia, MO 65212, USA
nistalar@health.missouri.edu

Raymond C. Harris
Division of Nephrology, Vanderbilt University, School of Medicine and Nashville Veterans Affairs Hospital, Nashville, TN 37232, USA
ray.harris@vanderbilt.edu

Reiko Inagi
Division of Nephrology and Endocrinology, University of Tokyo School of Medicine, 7-3-1, Hongo Bunkyo-ku, Tokyo 113-8655, Japan
inagi-npr@umin.ac.jp

Richard J. Johnson
Division of Renal Diseases and Hypertension, University of Colorado, Aurora, CO 80045, USA
richard.johnson@ucdenver.edu

Roland Veelken
Department of Nephrology and Hypertension, Friedrich-Alexander-University Erlangen, Krankenhausstrasse 12, D-91054, Germany

Sadayoshi Ito
Division of Nephrology, Endocrinology and Vascular Medicine,
Tohoku University Graduate School of Medicine, 1-1 Seiryocho,
Aoba-ku, Sendai 980-8574, Japan

Samuel N. Heyman
Department of Medicine, Hadassah University Hospital,
Mount Scopus, 24035, Jerusalem 91240, Israel
heyman@cc.huji.ac.il

Sara Terryn
Laboratory of Nephrology, Université catholique, de Louvain Medical School,
54 Avenue Hippocrate, UCL 5450, B-1200, Brussels, Belgium

Seymour Rosen
Department of Medicine, Hadassah University Hospital, Mount Scopus,
24035, Jerusalem 91240, Israel

Takashi Dan
Division of Molecular Medicine and Therapy, United Centers for Advanced Research and Translational Medicine (ART), Tohoku University Graduate School of Medicine, 2-1 Seiryo-Machi, Aoba-ku, Sendai, Miyagi 980-8575, Japan
dantks@m.tains.tohoku.ac.jp

Takashi Yokoo
Division of Nephrology, Jikei Medical University, 3-25-8, Nishi-Shimbashi,
Minato-ku, Tokyo 105-8461, Japan
tyokoo@jikei.ac.jp

Takefumi Mori
Division of Nephrology, Endocrinology and Vascular Medicine,
Tohoku University Graduate School of Medicine, 1-1 Seiryocho,
Aoba-ku, Sendai 980-8574, Japan
tmori@med.tohoku.ac.jp

Tetsuhiro Tanaka
Division of Nephrology and Endocrinology, University of Tokyo School of Medicine, 7-3-1 Hongo, Bunkyo-ku, 113-8655 Tokyo, Japan
tetsu-tky@umin.ac.jp

Tilmann Ditting
Department of Nephrology and Hypertension, Friedrich-Alexander-University Erlangen, Krankenhausstrasse 12, D-91054, Germany

Toshio Miyata
Division of Molecular Medicine and Therapy, United Centers for Advanced Research and Translational Medicine (ART), Tohoku University Graduate School of Medicine, 2-1 Seiryo-Machi, Aoba-ku, Sendai, Miyagi 980-8575, Japan
t-miyata@mail.tains.tohoku.ac.jp

Ulrike Bruning
Conway Institute of Biomolecular and Biomedical Research, School of Medicine and Medical Science, College of Life Sciences, University College Dublin, Belfield, Dublin 4, Ireland

Vincent M. Monnier
Department of Pathology and Biochemistry, Case Western Reserve University, Wolstein Research Bldg., 2103 Cornell Road, Cleveland, OH 44106-7288, USA
vmm3@cwru.edu

Volker H. Haase
Division of Nephrology and Hypertension, Vanderbilt University, C-3119A, MCN, 1161 21st Avenue, Nashville, TN 37232, USA
volker.haase@vanderbilt.edu

Wanja M. Bernhardt
Department of Nephrology and Hypertension, Friedrich-Alexander-University Erlangen, Krankenhausstrasse 10-12, D-91054 Erlangen, Germany
wanja.bernhardt@uk-erlangen.de

Witcha Imaram
Division of Renal Diseases and Hypertension, University of Colorado, Aurora, CO 80045, USA

Yoshiki Nishizawa
Department of Metabolism, Endocrinology and Molecular Medicine
Osaka City University School of Medicine, 1-4-3, Asahi-machi, Abeno, Osaka, 545-8585, Japan

Yuri Y. Sautin
Division of Renal Diseases and Hypertension, University of Colorado, Aurora, CO 80045, USA

Part I
Oxidative Stress

Chapter 1
Oxidative Stress Injury in Glomerular Mesangium

Josef Pfeilschifter and Karl-Friedrich Beck

Abstract The renal mesangium consists of glomerular mesangial cells and their surrounding extracellular matrix. This highly specialized pericapillary tissue constitutes the inner part of the glomerulus that supports the structure of the glomerular capillaries. In the healthy glomerulus, mesangial cells act by synthesis and control of extracellular matrix, by a well-organized production of a series of different growth factors, and by a fine-tuned cross-talk with the neighboring glomerular cells, namely the podocytes and the endothelial cells, as a regulatory device for the maintenance of glomerular structure and function. Besides invading immune cells, resident mesangial cells are considered the key players in modulating inflammatory signaling processes within the glomerulus. After exposure to cytokines, mesangial cells amplify the inflammatory process by synthesizing high amounts of cytokines, reactive oxygen species (ROS), and nitric oxide (NO). Oxidative and nitrosative stress affects surrounding glomerular cells that subsequently may collapse by apoptotic or necrotic mechanisms. We describe the sources and action of oxidative and nitrosative signaling processes in the mesangium and we discuss therapeutic strategies for the treatment of stress-induced cell damage in the course of glomerular diseases.

Keywords Reactive oxygen species · Nitric oxide · Oxidative stress · Nitrosative stress

---

J. Pfeilschifter (✉)
Pharmazentrum frankfurt/ZAFES, University Hospital, Goethe University,
Theodor-Stern-Kai 7, D-60590 Frankfurt am Main, Germany
e-mail: pfeilschifter@em.uni-frankfurt.de

T. Miyata et al. (eds.), *Studies on Renal Disorders*, Oxidative Stress in Applied Basic Research and Clinical Practice, DOI 10.1007/978-1-60761-857-7_1,

Abbreviations

AA	Arachidonic acid
ACE	Angiotensin converting enzyme
Ang II	Angiotensin II
AP-1	Activator protein 1
ELAV	Embryonal lethal abnormal vision
eNOS	Endothelial NO synthase
ET-1	Endothelin-1
FeTSPP	5,10,15,20-Tetrakis (4-sulfonatophenyl) porphyrinato iron III chloride
FP15	Fe(III)Tetrakis-2-(*N*-triethylene glycol monomethyl ether) pyridyl porphyrin
H_2O_2	Hydrogen peroxide
HIF	Hypoxia-inducible factor
IL-1β	Interleukin-1β
IL-8	Interleukin 8
iNOS	Inducible NO synthase
L-NIL	L-N^6-(L-iminoethyl) lysine dihydrochloride
L-NMMA	N^G-Monomethyl-L-arginine
LPS	Lipopolysaccharide
MAPK	Mitogen-activated protein kinase
MCP-1	Monocyte chemoattractant protein1
MIP-2	Macrophage inflammatory protein
MMP-9	Matrix metalloproteinase 9
Mox1	Mitogenic oxidase 1
NADPH	Nicotinamide adenine dinucleotide phosphate
NFκB	Nuclear factor κB
NIK	NFκB-Inducing kinase
nNOS	Neuronal NO synthase
Nox	NADPH oxidase
O_2^-	Superoxide peroxynitrite
PDGF	Platelet-derived growth factor
PDGFRα	PDGF receptor α
PKB	Protein kinase B
PKC	Protein kinase C
PPAR	Peroxisome poliferator activated receptor
PTK	Protein tyrosine kinase
PTP	Protein tyrosine phosphatase
RNS	Reactive nitrogen species
ROS	Reactive oxygen species
RT-PCR	Reverse transcription polymerase chain reaction
sGC	Soluble guanylyl cyclase
SOD	Superoxide dismutase
TIMP-1	Tissue inhibitor or metalloproteinases 1

1 Introduction

The renal mesangium consists of glomerular mesangial cells and their surrounding extracellular matrix. This highly specialized pericapillary tissue constitutes the inner part of the glomerulus that supports the structure of the glomerular capillaries. Mesangial cells are able to contract, and this helps to maintain the structure of capillary loops and to regulate the capillary flow and ultrafiltration surface (for review see [1–4]). However, the major regulation of the single-nephron glomerular filtration rate (SNGFR) is conferred by the efferent and afferent arterioles. Therefore, mesangial cell contraction is currently considered to contribute to the fine tuning of SNGFR. In the healthy glomerulus, mesangial cells act by synthesis and control of extracellular matrix, by a well-organized production of a series of different growth factors and by a fine-tuned cross-talk with the neighboring glomerular cells, namely the podocytes and the endothelial cells, as a regulatory device for the maintenance of glomerular structure and function [4, 5]. However, disease states, caused by diabetes, autoimmune disorders, or inflammatory processes, dramatically disturb the glomerular homeostasis, and this processes may lead to necrosis, apoptosis, excessive production of extracellular matrix, fibrosis, and, subsequently, loss of glomerular function [4]. In an inflammatory setting, polymorphonuclear cells and macrophages, which produce large amounts of inflammatory mediators such as interleukin-1β (IL-1β) and tumor necrosis factor α (TNFα), but also reactive oxygen species and nitric oxide, invade the glomerulus and activate glomerular cells to participate in the production of these mediators. Besides invading immune cells, resident mesangial cells are considered the key players in modulating inflammatory signaling processes within the glomerulus [4, 5]. After exposure to cytokines, mesangial cells amplify the inflammatory process by synthesizing themselves into high amounts of cytokines, reactive oxygen species (ROS), and nitric oxide (NO). The primary ROS synthesized enzymatically is the radicalic superoxide anion O_2^-, which is decomposed to H_2O_2 or the highly reactive hydroxyl radical. NO can react with different ROS or even molecular oxygen to form the most prominent mediators of nitrosative stress, namely peroxynitrite or N_2O_3, respectively. Oxidative and nitrosative stress affects surrounding glomerular cells that subsequently may collapse by apoptotic or necrotic mechanisms. Here we describe the sources and action of oxidative and nitrosative stress in the mesangium and we discuss therapeutic strategies for the treatment of stress-induced cell damage in the course of inflammatory glomerular diseases.

2 Sources of Oxidative and Nitrosative Stress in Mesangial Cells

Generally, all enzymes that use molecular oxygen as a substrate are potentially generators of ROS. In most cases, ROS formation is undesirable and occurs when the respective enzymatic reaction runs under adverse conditions such as substrate or cofactor deficiency. In contrast, nicotinamide adenine dinucleotide phosphate (NADPH) oxidases produce exclusively ROS to kill invading micro-organisms or to trigger mitogenic signaling cascades. Most of ROS-producing enzymatic reactions result first in the generation of the superoxide anion, which is rapidly further processed to more reactive compounds such as hydrogen peroxide, hydroxyl radical, or hypochlorite. The synthesis of NO occurs – with few exceptions in – in a more coordinated manner by the action of three different NO syntheses that are tightly regulated at expression and activity levels and that fulfill specific physiological tasks [6, 7]. Subsequently, reactive nitrogen species (RNS) are formed by the reaction of NO with ROS.

The first evidence that rat mesangial cells are capable of producing ROS was shown in 1983 [8]. Baud et al. demonstrated that phagocytosis of serum-treated zymosan particles by mesangial cells was accompanied with the production of superoxide anion (O_2^-) and hydrogen peroxide (H_2O_2). Intriguingly, ROS production was associated with the activation of lipoxygenase. Inhibition of lipoxygenase activity by three different inhibitors attenuated ROS production, thus identifying lipoxygenase as a source of ROS in mesangial cells. Later, ROS production was also observed in human mesangial cells after stimulation with the cytokines IL-1β and TNFα [9], and an NADPH-dependent oxidase has been characterized to be the responsible enzyme for mesangial ROS production [10]. Importantly, cytokine-induced ROS production is responsible for TNFα-induced apoptosis in rat mesangial cells [11, 12]. Meanwhile, we know that a series of mediators such as high glucose, angiotensin II, and many others induce ROS formation in mesangial cells, indicating these glomerular pericytes as targets in diabetic nephropathy and hypertension (reviewed in [13, 14]). At the molecular level, in particular NADPH oxidases but also mediators derived from the arachidonic acid metabolism are the responsible devices in ROS generating processes. It is worth mentioning that even if NADPH oxidases have been known as the main source of ROS for a long time, the catalytic subunit of the classical phagocytic NADPH oxidase $p91^{phox}$ or Nox2 has never been detected immunologically in mesangial cells. This was amazing, as all accessory subunits such as $p22^{phox}$, $p47^{phox}$, and $p67^{phox}$, which are a prerequisite for a functional NADPH oxidase complex, are expressed in mesangial cells (for review see [15]). This enigma was deciphered in 2003 when Nox4 (formerly renox), a catalytic subunit that can replace gp91phox that was first characterized in 2000 in the renal cortex [16]. Nox4 has been demonstrated to confer angiotensin II-mediated activation of protein kinase B/Akt in mesangial cells [17]. Nox1 (formerly Mox1 for mitogenic oxidase), an NADPH oxidase subunit with high homology to Nox2 and Nox4, was first characterized in vascular smooth muscle cells and different tissues [18]. Nox1 was also found to trigger ROS formation in rat mesangial cells. Interestingly, Nox1 is expressionally controlled at the mRNA and protein level by nitric oxide (NO) via a cyclic GMP (cGMP)-dependent mechanism in rat mesangial cells [19].

The synthesis of NO by the inducible form of NO synthase (iNOS, also referred to as NOSII or macrophage type of NOS) in mesangial cells was first described indirectly by the measurement of cGMP, the main downstream effector of

NO-induced signaling in rat mesangial cells that were stimulated with IL-1β and TNFα [20, 21]. iNOS expression is also induced by cyclic AMP (cAMP) [22] and cytokine or cAMP-induced expression of iNOS is drastically inhibited by growth factors such as platelet-derived growth factor (PDGF), transforming growth factor β (TGF-β), or endothelin-1 (ET-1) [23–26] and glucocorticoids [27, 28]. By contrast, cytokine-induced iNOS expression is upregulated in rat mesangial cells by basic fibroblast growth factor [29] and ROS donors such as the hypoxanthine/xanthine oxidase system or the redox-cycler DMNQ [30]. In human mesangial cells considerable iNOS expression occurs only when combinations of cytokines were administered as a cocktail [31]. Cloning and analysis of the 5'-flanking regions of murine, rat, and human iNOS revealed a high species and cell specificity. In mouse macrophages, induction of iNOS by interferon-γ or lipopolysaccharide (LPS) is mediated by γ-activated sites, IFN-γ-responsive elements, and nuclear factor κB (NFκB) [32–34]. In rat mesangial cells, binding of the NFκB at the appropriate site on the iNOS promoter is essential for the induction of iNOS by IL-1β [35, 36] in rat mesangial cells. In contrast, ET-1 inhibits cytokine-induced iNOS expression without affecting NFκB binding capacity, suggesting additional mechanisms that are essential for cytokine-induced iNOS expression [37]. For the induction of iNOS by cAMP, enhanced binding of the transcription factors CAAT/enhancer-binding protein (C/EBP) and cAMP-responsive element-binding protein (CREB) on the iNOS promoter has been reported [36]. The most important mechanism that triggers human iNOS expression is obviously the IFN-γ-JAK-2-STAT1α pathway [38]. The role of NFκB in the expression of iNOS is possibly restricted to warrant basal expression that is strongly modified by factors that confer posttranscriptional iNOS expression by regulating iNOS mRNA stability [39]. Probably the synthesis of NO by mesangial cells occurs exclusively via the iNOS. Conflicting results exist regarding the role of the so-called constitutive isoforms of NOS, namely the neuronal form (nNOS, also referred to as bNOS or NOSI) and the endothelial form (eNOS, also referred to as ecNOS or NOSIII) in mesangial cells. So far, there is only one report describing the activity of eNOS in mesangial cells [40]. Owada et al. [40] describe the formation of cGMP after activation of the endothelin B receptor. In contrast, another report demonstrates that ET-1 induces NO formation in the glomerulus but not in mesangial cells [41].

Interestingly, NO and ROS have a strong impact on the expression of their own sources in mesangial cells. As exemplified for iNOS, both ROS and NO are able to amplify cytokine-induced iNOS expression and NO synthesis in rat renal mesangial cells [30, 42]. Importantly, in the same cell system cytokine-induced NO inhibits the expression of Nox1 and the formation of ROS [19]. These results further corroborate the hypothesis that ROS and NO – by modulating the expression of their generating enzyme systems – shift the ROS/NO balance to a NO-dominated chemistry in the course of an inflammatory process in mesangial cells [43].

At this point, it is important to note that the description of the sources of mesangial ROS and NO production as mentioned above does not necessarily mean that mesangial cells react exclusively in an autocrine/paracrine manner to ROS or NO produced by their own synthesis machinery. The situation in the glomerulus is much more complex. The main glomerular source of oxidative and nitrosative stress is invading immune cells that produce high amounts of ROS and NO predominantly by the activity of Nox2 and iNOS in the glomerulus. Moreover, the contribution of resident glomerular cells (e.g., endothelial cells by the activity of Nox4 or eNOS) or the macula densa by the activity of nNOS on glomerular ROS/NO production has to be considered.

3 Actions of Oxidative and Nitrosative Stress in Mesangial Cells

Both ROS and NO modulate a series of signaling pathways that finally may alter the susceptibility to cell death by necrosis and apoptosis or that affect other phenotypic changes as cell growth or migration. At the molecular level, ROS and NO determine the constitution of the phosphoproteome by affecting protein kinases or phosphatases, they modify protein activity by oxidation, nitration, or nitrosation, or they directly change the transcription by influencing the activity of several transcription factors [44–46]. Intriguingly, ROS and NO may counteract their specific functions by scavenging each other or they act together forming reactive metabolites such as peroxynitrite or other reactive compounds that exert toxic effects by nitrating proteins or lipids [43]. However, the synthesis of ROS and NO occurs timely and spatially separated within the glomerulus and, as mentioned above, the machineries of NO and ROS synthesis are potently influenced by their own products. Taking these phenomena into account, we developed the hypothesis that, for example, in the inflamed mesangium, the chemistry of oxidative and nitrosative stress is dominated in the early phases by the action of "classical" ROS such as O_2^- and H_2O_2 mainly via the activation of NADPH oxidases, followed by a phase of nitrosative stress after the induction of iNOS with subsequent formation of peroxynitrite and N_2O_3, and finally a resolution phase with suppressed ROS production and with the presence of high amounts of NO. We hypothesized that the activity of iNOS is eventually inhibited by substrate depletion or NO-mediated inhibition of the iNOS enzyme, leading to the resolution of the inflammatory process and retaining a balanced synthesis of small amounts ROS and NO within the glomerulus [43, 47].

3.1 Oxidative and Nitrosative Stress and Apoptosis

Removal of dispensable or injured cells by apoptosis is a prerequisite for the homeostasis in healthy tissue, but also for the resolution of many diseases. A role for apoptosis in removing mesangial cells in areas of mesangial hypercellularity was first demonstrated in 1994 by Baker et al. [48] in a rat model of anti-Thy1.1

mesangioproliferative nephritis. However, whether mesangial apoptosis has beneficial or detrimental effect depends on the biological context (e.g., the presence of growth factors, cytokines, ROS, NO, or the composition of the extracellular matrix) [49]. The participation of NO and ROS in apoptotic or necrotic cell death in cultured mesangial cells has been shown by several reports [11, 12, 50–52]. Remarkably, whereas ROS and NO donors dose-dependently increased apoptotic or necrotic cell death in rat mesangial cells, exposure of mesangial cells to equimolar amounts of NO and ROS had no toxic effects indicative for high susceptibility of mesangial cells to oxidative stress or NO. Otherwise, peroxynitrite, a most prominent anticipated mediator of nitrosative stress that is formed by the reaction of O_2^- with NO, does not obviously exert toxic effects. This reflects the potent intrinsic defense system against RNS in rat mesangial cells [51]. These results were later corroborated by a report that describes enhanced formation of ceramide after stimulation with NO or ROS, indicating a potential role for ceramide as a potential mediator of NO- and ROS-induced cell death [53]. Remarkably, SIN-1, a substance that produces equimolar amounts of NO and O_2^- and that is therefore accepted as a peroxynitrite generator, had no effect on ceramide formation and cell death, and this clearly indicates that peroxynitrite is not toxic to mesangial cells. In contrast, glomerular endothelial cells showed enhanced ceramide formation and cell death after treatment with SIN-1. Obviously, endothelial cells lack a protective system against nitrosative stress comparable to that of mesangial cells

3.2 Effects of ROS and NO on the Phosphoproteome

Phosphorylation and dephosphorylation of serine/threonine or tyrosine residues in target proteins by protein kinases or protein phosphatases determine the final outcome of most important cellular signaling processes. Consequently, the activity of the responsive kinases and phosphatases governs the constitution of the cellular phosphoproteome. Analysis of the human genome revealed the existence of about 500 protein kinases that represent the so-called kinome [54]. Moreover, alternative splicing of many of these genes may result in a high structural diversity of protein kinases [55]. Nearly 200 genes encode for protein phosphatases representing the phosphatome. Notably, around 700 genes control the phosphoproteome of a cell, and this indicates the high complexity of physiologic and pathophysiologic processes triggered by these important groups of enzymes.

Intriguingly, ROS and NO greatly affect the constitution of the cellular phosphoproteome. In a recent review, Chiarugi and Buricchi propose an elegant model for the effects of ROS on protein tyrosine phosphorylation [56]. Protein tyrosine phosphatases (PTPs) are commonly downregulated by oxidation of their sulfhydryl groups by hydrogen peroxide or other ROS [57]. In most cases this process is reversible by the action of antioxidants, and therefore, constitutes a well-defined regulatory mechanism [58]. In contrast, many protein tyrosine kinases (PTKs) are upregulated by oxidation. Mechanistically, as exemplified for the insulin receptor kinase, activation by ROS is achieved by a conformational change of the enzyme after oxidation of several cysteine residues [59]. Since many PTKs regulate their activity by autophosphorylation, ROS-mediated downregulation of PTPs strongly contributes to a ROS-induced activity of PTKs. Importantly, this process is reversible, and a decrease of the intracellular redox potential by a reduced activity of ROS-producing enzymes or by an oxidant-mediated increase of protective mechanisms results in the reconstitution of the phosphoproteome. In this scenario PTPs become hyperactive due to their high phosphorylation state (as a result of high activity of PTKs in the presence of ROS) and the recovery of their activity by reduction of cysteine residues. Simultaneously, cysteine residues of PTKs were reduced and – due to the high activity of PTPs – PTKs become dephosphorylated, leading to an inhibition of PTK activity and subsequently to the abrogation of a series of signaling pathways induced by growth factors and cytokines. Taken together, a fine-tuned redox balance determines induction and resolution of PTK-mediated signaling cascades. As exemplified for PKCγ, a classical calcium- and diacylglycerol-dependent form of protein kinase C, ROS are also able to activate serine/threonine kinases [60].

Nitric oxide and peroxynitrite are also able to modulate kinase or phosphatase activity, leading to changes, for example, in the three main mitogen-activated protein kinase (MAPK) pathways [44, 61, 62]. Changes in the phosphorylation pattern by NO or peroxynitrite are triggered by different mechanisms, affecting, for example, sulfhydryl groups by nitrosation or tyrosine residues by nitration. However, in contrast to ROS, which exert their effects mainly by oxidative processes, NO bears additionally the ability to act via a well-defined physiological receptor, the soluble guanylyl cyclase (sGC). Cyclic GMP produced by the activity of sGC triggers phosphorylation cascades that may counteract the oxidative response. In rat mesangial cells, elevated cGMP levels as evoked by stimulation with atrial natriuretic peptide (ANP) augment the expression of MAP kinase phosphatase in glomerular mesangial cells, thereby affecting phorbol ester-induced mesangial cell proliferation [63] and inhibiting endothelin-1-induced activation of c-Jun, NH_2-terminal kinase (JNK), and p42/p44 (ERK-1/ERK-2) MAPK pathways [64, 65]. In contrast, both NO and ROS are able to activate p42/p44 (ERK-1/ERK-2) MAPK-mediated pathways in mesangial cells. Nitric oxide activates p42/p44 MAPK via cGMP-dependent and -independent mechanisms by inhibition of tyrosine phosphatases [66], whereas ROS formation induced by angiotensin II treatment stimulates phosphorylation of p42/p44 MAPK via the activity of Nox4 [67]. It is worth mentioning that angiotensin II also triggers PKB/Akt-dependent signaling [17], phosphorylation of 3-phosphoinositide-dependent protein kinase-1 (PDK-1) [68], and JNK [69] in a ROS-dependent manner, indicating that angiotensin II-evoked redox signaling triggers nearly all pathways required for mesangial cell matrix formation or proliferation. Augmented PDGF receptor phosphorylation by NO-dependent inhibition of PDGF receptor phosphotyrosine phosphatase by NO may counteract cGMP-mediated anti-proliferative effects in mesangial cells [70].

3.3 *Effects of Oxidative and Nitrosative Stress on the Transcription Pattern of Mesangial Cells*

Besides their ability to potently impact signaling processes that alter the gene expression pattern, both autacoids NO and ROS are able to directly affect gene expression at the transcriptional, posttranscriptional, and posttranslational levels [44–46, 71]. A series of so-called redox-sensitive transcription factors are responsive to nitrosative or oxidative stress, and such effects were best characterized for nuclear factor kappa B (NFκB), activator protein 1 (AP-1), and hypoxia-inducible factor 1 (HIF-1) [45, 71–74]. A series of reports describe the action of ROS or NO on the gene expression pattern in mesangial cells of murine, human, or rat origin. For these investigations, some authors administered ROS or NO exogenously using compounds such as the redox cycler 2,3-dimethoxy-1,4-naphthoquinone (DMNQ) or the xanthine oxidase/(hypo)xanthine system for the production of the superoxide anion, glucose oxidase for the production of H_2O_2, or different NO-producing agents such as diethylenetriamine/NO (DETA/NO) or S-nitrosoglutathione (GSNO). To assess the effects of endogenously produced ROS or NO, mesangial cells were prone to conditions mimicking (a) activation of the renin-angiotensin-aldosterone system (RAAS), (b) diabetic nephropathy by elevated glucose levels, or (c) inflammation by the cytokines IL-1β and TNF-α or aggregated immunoglobulin. The resulting expression patterns were then compared with that of mesangial cells that were additionally treated with suitable inhibitors for NADPH oxidases or inducible NO-synthase such as diphenylene iodonium (DPI) or N^G-monomethyl-L-arginine (L-NMMA), respectively, or with scavengers of ROS such as *N*-acetyl-cysteine.

To our knowledge, the first report that describes the action of endogenously produced ROS on gene expression in mesangial cells was published by Satriano et al. in 1993 [75]. The authors demonstrate that ROS, produced by murine mesangial cells after stimulation with TNF-α and aggregated immunoglobulin, enhanced the expression of monocyte chemoattractant protein 1 (MCP-1) and moncyte colony-stimulating factor 1 (CSF-1). The same group also observed later a ROS-dependent upregulation of the genes for RANTES (regulated upon activation, normal T cell expressed and secreted) and ICAM-1 (intercellular adhesion molecule-1), further corroborating a prominent role for ROS in mesangial chemokine signaling [76]. Intriguingly, originally noninflammatory pathways are able to induce the expression of proinflammatory chemokines. Ha et al. [77] demonstrated that high glucose induced the expression of MCP-1 via the activation of the transcription factor NFκB and the production of ROS, clearly indicating a proinflammatory effect of high glucose conditions. In fact, ROS-induced activation of MAP kinases, followed by activation of the transcription factors NFκB and AP-1, is a key signaling pathway induced by ROS in mesangial cells as exemplified for the expression of the matrix metalloprotease 9 gene [78]. Moreover, a further target that exerts chemokine-like functions, namely osteopontin, has been characterized to be upregulated via aldosterone/cytokine-induced ROS formation in rat mesangial cells [79].

High glucose potently induces generation of ROS via activation of protein kinase C (PKC) in mesangial cells [80]. Besides activation of the transcription factors AP-1 and NFκB by PKC-induced ROS, elevation of TGF-β levels is a main feature of ROS activity in mesangial cells. TGF-β-induced SMAD signaling cascades, in turn led by an upregulation of collagens 1, 3, and 4, fibronectin, and plasminogen activator inhibitor 1 (PAI-1) to an excessive formation of ECM, which caused glomerular fibrosis as a cardinal symptom of diabetic nephropathy not only by a enhanced expression of matrix components [81, 82] but also by a reduced expression of ECM-degrading factors [83]. Importantly, PKC is not only a prerequisite for high glucose-induced ROS formation. Activation of PKC is also regulated by a fine-tuned redox mechanism that may under certain circumstances constitute a positive regulatory loop, further amplifying the actions of ROS on cell death or gene regulation [84, 85].

It is worth mentioning that one of the first cell systems that was used to analyze NO-mediated gene expression were human renal mesangial cells. Brown et al. [86] found that protein and mRNA expression of the inflammatory chemokine IL-8 was drastically upregulated by NO. Meanwhile, a series of genes were found to be under expressional control by NO or ROS in mesangial cells. In our lab, we could demonstrate that certain genes are regulated in a coordinated fashion by NO and ROS [30, 42], whereas others are regulated in an opposite manner by the autacoids NO and ROS [78, 87].

The aim of most studies performed to analyze NO-dependent gene expression was to define a more or less unique and ubiquitous signaling machinery that triggers NO-mediated gene expression. This goal has so far not been attained. By contrast, the analysis of NO-modulated mRNA and protein expression revealed that a variety of signaling cascades are involved in NO-evoked cell responses.

Nevertheless, our current understanding of the biology of NO signaling focuses on three different pathways that are involved in triggering NO-driven responses directly to the transcriptional machinery of the cell or in mediating posttranscriptional or posttranslational mechanisms and, subsequently, to changes in the gene expression pattern. These pathways include (a) the cGMP signaling pathway [88], (b) interference with hypoxia-mediated signaling [89], and (c) interference with redox signaling (e.g., posttranslational changes of proteins by nitration of tyrosine residues or nitrosation of cysteine residues) [47].

During the past 15 years, several so-called differential display methods have been developed or improved to analyze the mRNA and protein expression patterns of cultured cells and tissues. We successfully used the RNA arbitrarily primed reverse transcription–polymerase chain reaction (RAP-PCR) to analyze NO-mediated gene expression on the mRNA level [90]. We also directly focused on NO-driven protein expression using a two-dimensional protein gel electrophoresis protocol. Besides these "arbitrary" approaches, we also investigated the NO-dependent regulation of gene products that are considered as key players in several forms of glomerulonephritis. These experiments revealed a series of genes that were expressionally controlled by NO. As exemplified for the NO-dependent regulation of SPARC (secreted protein acidic and rich in cysteine), SMOC-1 (secreted modular calcium-binding protein-1), biglycan, PDGF receptor α (PDGFRα), MnSOD, and Nox1, relatively small amounts of NO (20–100 nM) are sufficient for the activation of GC, which in turn drives the expression of these genes that in general provide mesangial cells with a protective capacity, namely antifibrotic, antiapoptotic, or antioxidative in nature [19, 91–95]. In contrast, as shown for macrophage inflammatory protein (MIP-2), high amounts of NO are redox-active and mediate proinflammatory signals, targeting redox-sensitive transcription factors such as AP-1 or NFκB [95] in mesangial cells. Obviously, cGMP-independent signaling by high amounts of NO triggers proinflammatory gene expression as shown for iNOS and MIP-2 [42, 95]. Another group found the gene for connective tissue growth factor to be downregulated by NO in a cGMP-dependent manner [96]. Recently, we could demonstrate that this effect may be driven by a cGMP-dependent downregulation of SMOC-1, a matricellular protein that we have found to support TGF-β-mediated signaling devices [91]. cGMP- and redox-dependent effects of NO on gene expression are summarized in Fig. 1. Importantly, we could demonstrate that NO is able to act on gene expression at transcriptional, posttranscriptional, and post-translational levels in mesangial cells. By analyzing the effects of NO on MMP-9 expression in more detail, we found that decreased levels of MMP-9 mRNA are due to a reduced expression of the mRNA-stabilizing protein HuR, which belongs to the group of embryonal lethal abnormal vision (ELAV) proteins [97]. Remarkably, as mentioned above, ROS act on MMP-9 expression in an opposite manner by activating the transcription factors NFκB and AP-1 in the rat MMP-9 promoter [77]. Furthermore, we could demonstrate that NO affects the expression of neutral ceramidase, an enzyme that plays a central role

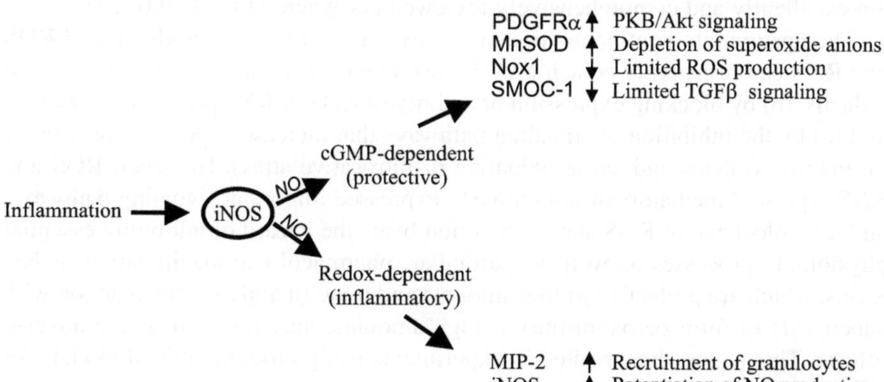

Fig. 1 The role of cGMP- and redox-mediated gene expression by nitric oxide. Low amounts of nitric oxide are sufficient for the activation of sGC and subsequent elevation of cGMP levels. Enhanced cGMP levels regulate gene expression in a protective direction as exemplified PDGF receptor α, MnSOD, Nox1, and SMOC-1, respectively. High levels of nitric oxide induce redox-dependent signaling mechanisms that induce inflammatory genes such as MIP-2 and iNOS independently of changes in cellular cGMP levels. *PDGFRα* PDGF receptor α; *PKB/Akt* protein kinase B; *MIP-2* macrophage inflammatory protein-1; *Nox1* NADPH oxidase 1

for the homeostasis of the important lipid second messenger ceramide by the induction of its proteosomal degradation [98, 99].

Taken together, ROS and NO potently act on gene expression. Both autacoids induce redox-dependent signaling processes acting via protein modifications, in particular on protein phosphatases, which usually results in a diminished activity of the affected protein. Furthermore, ROS and NO trigger gene expression mainly by affecting redox-sensitive transcription factors. Both autacoids are able to scavenge each other and form reactive nitrogen species that have been identified to cause apoptotic or necrotic cell death in many cell types. However, mesangial cells are obviously well protected against nitrosative stress. It is worth noting that NO acts via a physiological receptor, namely sGC, that drives the gene expression pattern in mesangial cells into a protective direction, thereby antagonizing adverse redox actions of ROS and NO.

4 Therapeutic Opportunities

Diseases such as diabetes, hypertension, hyperlipidemia, and several forms of systemic or local inflammation may cause oxidative or nitrosative stress within the renal mesangium, but also in many other tissues. Proper conventional and systemic treatment of these diseases (e.g., with insulin, angiotensin converting enzyme inhibitors, lipid-lowering, or anti-inflammatory agents) is usually sufficient not only to decrease local mesangial but also systemic ROS and NO production. The effects of such treatments on oxidative and nitrosative stress in the mesangium are excellently and comprehensively reviewed elsewhere [14, 74, 100–103].

Therapeutic intervention to target a massive and detrimental production of ROS and RNS can principally be achieved by (a) direct scavenging or neutralization of radicals, (b) by blocking expression or activity of ROS or RNS producing enzymes, and (c) by the inhibition of signaling pathways that increase expression or activity of enzyme systems and cause oxidative or nitrosative stress. However, ROS and RNS represent mediators of ubiquitously expressed important signaling pathways, and total blockage of ROS and RNS action bears the hazard of inhibiting essential physiologic processes as well. In particular, pharmacologic manipulation of NO levels, which are protective in low amounts and toxic (mainly via the reaction with superoxide to form peroxynitrite) in high amounts, may result in severe adverse effects. This is best documented in experiments using different animal models for mesangioproliferative nephritis.

The first report that definitely indicates a role for NO in the progression of glomerulonephritis has been provided by Weinberg et al. [104]. The authors used MRL-lpr/lpr mice, which spontaneously develop a lupus-like disease with the symptoms of arthritis and immune complex glomerulonephritis. Administration of the unspecific NOS inhibitor L-NMMA blocked the development of glomerulonephritis and ameliorated inflammatory arthritis, indicating a pro-inflammatory role of NO in this animal model. Members of the same group reported more recently that the specific iNOS inhibitor L-*N*6-(l-iminoethyl) lysine dihydrochloride (L-NIL) is more effective than L-NMMA to ameliorate glomerulonephritis in MRL-lpr/lpr mice [105]. Similar results were also reported in a rat model of glomerulonephritis. Narita et al. [106] used the anti-Thy.1 model of glomerulonephritis, an acute inflammatory rat model that resembles the human form of mesangioproliferative glomerulonephritis like immunoglobulin A (IgA) nephritis. This model is characterized by a complement-dependent early loss of mesangial cells and a subsequent immense accumulation of matrix. Remarkably, all symptoms of anti-Thy.1-induced glomerulonephritis disappear in a time period of about 6 weeks, resulting in a complete resolution of the complement-induced disease. Blockage of NO synthesis by L-NMMA or a dietary reduction of L-arginine availability reduced both mesangiolysis and matrix deposition, clearly suggesting a detrimental role for NO in this model. The same group further corroborated these data by showing that a high L-arginine diet supports mesangiolysis and matrix formation in the same experimental model [107]. In contrast, the Lewis rat substrain (LEW/Maa) develops a chronic form of glomerulonephritis after the injection of an anti-Thy.1 antibody [108]. Interestingly, pretreatment with the iNOS-specific inhibitor L-NIL enhanced proteinuria and fibronectin deposition by nearly 50%, indicating a protective role of iNOS-derived NO. Most importantly, the use of the iNOS-specific inhibitor L-NIL did not affect blood pressure or the glomerular filtration rate, further indicating that the activity of iNOS is protective at least in this rat model of chronic glomerulonephritis. A protective role of iNOS-derived NO in attenuating hypercellularity during the glomerular repair phase has also been documented [109]. Interestingly, administration of Bay 41-2272, an activator of the soluble guanylyl cyclase (sGC), to nephritic rats slowed progression of anti-Thy.1-induced glomerulosclerosis in LEW/Maa rats, suggesting that the protective action of iNOS is at least in part mediated by sGC [110].

In summary, a series of publications describe protective or deleterious actions of NO in different models of inflammatory glomerular diseases. Whether NO is a beneficial or deleterious mediator in an inflammatory process is still a matter of debate. Nevertheless, it is obvious that specific inhibition of iNOS is superior to a nonspecific inhibition of all NOS isoforms. Particularly, it is necessary to leave the protective role of the eNOS enzyme in regulating vascular homeostasis untouched. Moreover, NO-independent activation of sGC may be sufficient to mimic the protective effect of NO, irrespective of whether it is produced by eNOS or iNOS. This indicates that the protective effects of NO are mainly mediated by sGC [111].

Treatment of diseases accompanied by oxidative stress with classic antioxidants such a vitamins A, C, and E or selenium has been shown to be effective in some animal models, but antioxidant therapy in humans has, with only a few exceptions, no or even adverse effects [14, 112]. Also the use of peroxynitrite decomposition catalysts such as the metal porphyrin complexes 5,10,15,20-tetrakis (4-sulfonatophenyl) porphyrinato iron III chloride (FeTSPP), Fe(III) tetra-mesitylporphyrin octasulfonate, or Fe(III)tetrakis-2-(*N*-triethylene glycol monomethyl ether) pyridyl porphyrin (FP15) to reduce nitrosative stress has been proven only in animal models. Nevertheless, by scavenging peroxynitrite as a main effector of nitrosative stress but not influencing the bioactivity of beneficial NO, such compounds may represent a promising tool to treat conditions accompanied with nitrosative stress.

Another strategy to avoid oxidative and concomitantly nitrosative stress is the inhibition of superoxide synthesis and its toxic metabolites. Due to the various enzyme systems that generate ROS in disease conditions, blocking of the expression or activity of a specific enzyme is not sufficient in most cases to totally block ROS production. Nonetheless, inhibition of NADPH oxidase by apocynin, a methyl-substituted catechol that blocks assembly of the $p47^{phox}$ subunit to the membrane, has importance in treating inflammatory diseases in the traditional medicine of Peruvian Indians [113]. Since NADPH oxidases (Nox2 in invading neutrophils and macrophages, Nox1 and Nox4 in mesangial cells) are the most prominent ROS sources within the mesangium, treatment with apocynin could also help to treat glomerular diseases accompanied with oxidative stress, but one has to consider that Nox4 activity is not inhibited by apocynin, since this isoform does not require $p47^{phox}$ for activity. As for NO, ROS have detrimental effects but they also act as important second messengers. The detailed knowledge of the subcellular distribution and the special tasks of the different isoforms are prerequisites to develop future specific devices to inhibit NADPH oxidase by small molecules or by silencing their expression using appropriate siRNA protocols that target directly the mesangium. To date, we can consider several potent pharmaceuticals that may help to avoid the deleterious effects of oxidative stress in the mesangium upstream from the activity of NADPH oxidases. Since angiotensin II signaling produces high amounts of ROS via activation of NADPH oxidases, angiotensin II receptor 1 (AT_1) blockade or inhibition of angiotensin converting enzyme may be useful to avoid oxidative stress within the mesangium. Moreover, statins could diminish glomerular

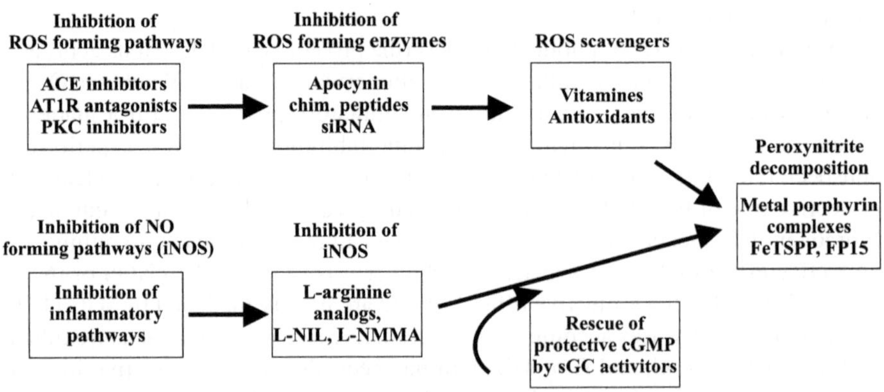

Fig. 2 Therapeutic opportunities for inhibition of ROS and RNS formation and activity. Reduced formation of ROS or NO can be achieved by treatment of the causation of ROS- of NO-forming pathways, by direct inhibition of ROS- or NO-producing enzymes, or by scavenging ROS and RNS. Using total inhibition of iNOS, rescue of protective cGMP-signaling by administration of cGC activators has to be considered. *AT_1R* angiotensin II type 1 receptor; *cGMP* cyclic GMP; *FP15* Fe(III)tetrakis-2-(*N*-triethylene glycol monomethyl ether) pyridyl porphyrin; *FeTSPP* 5,10,15,20-tetrakis (4-sulfonatophenyl) porphyrinato iron III chloride; *sGC* guanylyl cyclase

oxidative stress by inhibiting NADPH oxidase activity in mesangial cells [114–116]. A further potent approach to limit oxidative stress in the mesangium, particularly in diabetic nephropathy, is the inhibition of PKC. Since ROS activate PKC isoforms and, in turn, activation of PKC triggers ROS production, blockage of this vicious cycle may drastically reduce oxidative stress. However, different PKC isoforms fulfill obviously different tasks in renal cells, and the expression of the isoform PKCε is even protective in diabetic nephropathy, resulting in the need for isotype-specific PKC inhibitors. The therapeutic potential to treat glomerular diseases by PKC inhibition is comprehensively reviewed elsewhere [117, 118]. Possible pharmacological interventions of oxidative and nitrosative stress are summarized in Fig. 2.

5 Conclusion

The synthesis and action of reactive oxygen and nitrogen metabolites is a complex and tightly regulated facet of the signaling systems operative in higher organisms. Even if our knowledge about redox regulation processes has grown dramatically within the past two decades, this interesting field of biomedical research is far from being fully understood. In particular, the translation of the immense amount of data obtained from cell culture studies and animal models to pharmacological therapy represents a main challenge for the next decade. Nonetheless, further efforts have to be undertaken to fully characterize all enzyme systems responsible for ROS and NO synthesis to be able to develop specific inhibitors. Moreover, the detailed knowledge of transcriptional or posttranscriptional mechanisms that trigger the expression of enzymes, which produce ROS or NO, could help to develop strategies to manipulate ROS and NO synthesis in a direction that favors the resolution of glomerular diseases accompanied by imbalanced oxidative or nitrosative stress. In addition, we have to learn in more detail which signaling pathways and ROS/NO producing enzymes are responsible for oxidative and nitrosative stress in the mesangium as a prerequisite to successfully use inhibitors in the clinical practice.

Acknowledgments We apologize to those colleagues whose important contributions to this field could not be cited due to space limitations. The work of the authors' lab was supported by the German Research Foundation (SFB 815, FOG 784, GRK 757, GRK 1172, EXC 147, PF361/6-1) and by the EU (grant number LSHM-CT-2004-005033, EICOSANOX).

References

1. Schlöndorff D (1987) The glomerular mesangial cell: an expanding role for a specialized pericyte. FASEB J 1:272–281
2. Schlöndorff D, Banas B (2009) The mesangial cell revisited: no cell is an island. J Am Soc Nephrol 20:1179–1187

3. Stockand JD, Sansom SC (1998) Glomerular mesangial cells: electrophysiology and regulation of contraction. Physiol Rev 78:723–744
4. Pfeilschifter J (1989) Cross-talk between transmembrane signalling systems: a prerequisite for the delicate regulation of glomerular haemodynamics by mesangial cells. Eur J Clin Invest 19:347–361
5. Pfeilschifter J (1994) Mesangial cells orchestrate inflammation in the renal glomerulus. News Physiol Sci 9:271–276
6. Förstermann U, Closs EI, Pollock JS et al. (1994) Nitric oxide synthase isozymes. Characterization, purification, molecular cloning, and functions. Hypertension 23:1121–1131
7. Mungrue IN, Bredt DS, Stewart DJ et al. (2003) From molecules to mammals: what's NOS got to do with it? Acta Physiol Scand 179:123–135
8. Baud L, Hagege J, Sraer J et al. (1983) Reactive oxygen production by cultured rat glomerular mesangial cells during phagocytosis is associated with stimulation of lipoxygenase activity. J Exp Med 158:1836–1852
9. Radeke HH, Meier B, Topley N et al. (1990) Interleukin 1-alpha and tumor necrosis factor-alpha induce oxygen radical production in mesangial cells. Kidney Int 37:767–775
10. Radeke HH, Cross AR, Hancock JT et al. (1991) Functional expression of NADPH oxidase components (alpha- and beta-subunits of cytochrome b558 and 45-kDa flavoprotein) by intrinsic human glomerular mesangial cells. J Biol Chem 266:21025–21029
11. Moreno-Manzano V, Ishikawa Y, Lucio-Cazana J et al. (2000) Selective involvement of superoxide anion, but not downstream compounds hydrogen peroxide and peroxynitrite, in tumor necrosis factor-alpha-induced apoptosis of rat mesangial cells. J Biol Chem 275:12684–12691
12. Böhler T, Waiser J, Hepburn H et al. (2000) TNF-alpha and IL-1alpha induce apoptosis in subconfluent rat mesangial cells. Evidence for the involvement of hydrogen peroxide and lipid peroxidation as second messengers. Cytokine 12:986–991
13. Lee HB, Yu MR, Yang Y et al. (2003) Reactive oxygen species-regulated signaling pathways in diabetic nephropathy. J Am Soc Nephrol 14(Suppl 3):S241–S245
14. Sachse A, Wolf G (2007) Angiotensin II-induced reactive oxygen species and the kidney. J Am Soc Nephrol 18:2439–2446
15. Gill PS, Wilcox CS (2006) NADPH oxidases in the kidney. Antioxid Redox Signal 8:1597–1607
16. Geiszt M, Kopp JB, Várnai P et al. (2000) Identification of renox, an NAD(P)H oxidase in kidney. Proc Natl Acad Sci U S A 97:8010–8014
17. Gorin Y, Ricono JM, Kim NH et al. (2003) Nox4 mediates angiotensin II-induced activation of Akt/protein kinase B in mesangial cells. Am J Physiol Renal Physiol 285:F219–F229
18. Suh YA, Arnold RS, Lassegue B et al. (1999) Cell transformation by the superoxide-generating oxidase Mox1. Nature 401:79–82
19. Plesková M, Beck KF, Behrens MH et al. (2006) Nitric oxide down-regulates the expression of the catalytic NADPH oxidase subunit Nox1 in rat renal mesangial cells. FASEB J 20:139–141
20. Pfeilschifter J, Schwarzenbach H (1990) Interleukin 1 and tumor necrosis factor stimulate cGMP formation in rat renal mesangial cells. FEBS Lett 273:185–187
21. Pfeilschifter J, Rob P, Mülsch A et al. (1992) Interleukin 1 beta and tumour necrosis factor alpha induce a macrophage-type of nitric oxide synthase in rat renal mesangial cells. Eur J Biochem 203:251–255
22. Kunz D, Mühl H, Walker G et al. (1994) Two distinct signaling pathways trigger the expression of inducible nitric oxide synthase in rat renal mesangial cells. Proc Natl Acad Sci USA 91:5387–5391
23. Pfeilschifter J (1991) Platelet-derived growth factor inhibits cytokine induction of nitric oxide synthase in rat renal mesangial cells. Eur J Pharmacol 208:339–340

24. Pfeilschifter J, Vosbeck K (1991) Transforming growth factor beta 2 inhibits interleukin 1 beta- and tumour necrosis factor alpha-induction of nitric oxide synthase in rat renal mesangial cells. Biochem Biophys Res Commun 175:372–379
25. Kunz D, Walker G, Pfeilschifter J (1997) Transforming growth factor-beta 2 inhibits interleukin 1 beta-induced expression of inducible nitric oxide synthase in rat renal mesangial cells. Inflamm Res 46:327–331
26. Beck KF, Mohaupt MG, Sterzel RB (1995) Endothelin-1 inhibits cytokine-stimulated transcription of inducible nitric oxide synthase in glomerular mesangial cells. Kidney Int 48:1893–1899
27. Plüss C, Werner ER, Wachter H et al. (1997) Differential effect of dexamethasone on interleukin 1beta- and cyclic AMP-triggered expression of GTP cyclohydrolase I in rat renal mesangial cells. Br J Pharmacol 122:534–538
28. Pfeilschifter J (1991) Anti-inflammatory steroids inhibit cytokine induction of nitric oxide synthase in rat renal mesangial cells. Eur J Pharmacol 195:179–180
29. Kunz D, Walker G, Eberhardt W et al. (1997) Platelet-derived growth factor and fibroblast growth factor differentially regulate interleukin 1beta- and cAMP-induced nitric oxide synthase expression in rat renal mesangial cells. J Clin Invest 100:2800–2809
30. Beck KF, Eberhardt W, Walpen S et al. (1998) Potentiation of nitric oxide synthase expression by superoxide in interleukin 1 beta-stimulated rat mesangial cells. FEBS Lett 435:35–38
31. Nicolson AG, Haites NE, McKay NG et al. (1993) Induction of nitric oxide synthase in human mesangial cells. Biochem Biophys Res Commun 193:1269–1274
32. Xie QW, Whisnant R, Nathan C (1993) Promoter of the mouse gene encoding calcium-independent nitric oxide synthase confers inducibility by interferon gamma and bacterial lipopolysaccharide. J Exp Med 177:1779–1784
33. Martin E, Nathan C, Xie QW (1994) Role of interferon regulatory factor 1 in induction of nitric oxide synthase. J Exp Med 180:977–984
34. Xie QW, Kashiwabara Y, Nathan C (1994) Role of transcription factor NF-kappa B/Rel in induction of nitric oxide synthase. J Biol Chem 269:4705–4708
35. Eberhardt W, Kunz D, Hummel R et al. (1996) Molecular cloning of the rat inducible nitric oxide synthase gene promoter. Biochem Biophys Res Commun 223:752–756
36. Eberhardt W, Plüss C, Hummel R et al. (1998) Molecular mechanisms of inducible nitric oxide synthase gene expression by IL-1beta and cAMP in rat mesangial cells. J Immunol 160:4961–4969
37. Beck KF, Sterzel RB (1996) Cloning and sequencing of the proximal promoter of the rat iNOS gene: activation of NFkappaB is not sufficient for transcription of the iNOS gene in rat mesangial cells. FEBS Lett 394:263–267
38. Kleinert H, Wallerath T, Fritz G et al. (1998) Cytokine induction of NO synthase II in human DLD-1 cells: roles of the JAK-STAT, AP-1 and NF-kappaB-signaling pathways. Br J Pharmacol 125:193–201
39. Kleinert H, Pautz A, Linker K et al. (2004) Regulation of the expression of inducible nitric oxide synthase. Eur J Pharmacol 500:255–266
40. Owada A, Tomita K, Terada Y et al. (1994) Endothelin (ET)-3 stimulates cyclic guanosine 3',5'-monophosphate production via ETB receptor by producing nitric oxide in isolated rat glomerulus, and in cultured rat mesangial cells. J Clin Invest 93:556–563
41. Tack I, Marin Castano E, Pêcher C et al. (1997) Endothelin increases NO-dependent cGMP production in isolated glomeruli but not in mesangial cells. Am J Physiol 272:F31–F39
42. Mühl H, Pfeilschifter J (1995) Amplification of nitric oxide synthase expression by nitric oxide in interleukin 1 beta-stimulated rat mesangial cells. J Clin Invest 95:1941–1946
43. Pfeilschifter J, Beck KF, Eberhardt W et al. (2002) Changing gears in the course of glomerulonephritis by shifting superoxide to nitric oxide-dominated chemistry. Kidney Int 61:809–815

44. Beck KF, Eberhardt W, Frank S et al. (1999) Inducible NO synthase: role in cellular signalling. J Exp Biol 202:645–653
45. Pfeilschifter J, Eberhardt W, Beck KF (2001) Regulation of gene expression by nitric oxide. Pflugers Arch 442:479–486
46. Sen CK, Packer L (1996) Antioxidant and redox regulation of gene transcription. FASEB J 10:709–720
47. Pfeilschifter J, Eberhardt W, Beck KF et al. (2003) Redox signaling in mesangial cells. Nephron Exp Nephrol 93:e23–e26
48. Baker AJ, Mooney A, Hughes J et al. (1994) Mesangial cell apoptosis: the major mechanism for resolution of glomerular hypercellularity in experimental mesangial proliferative nephritis. J Clin Invest 94:2105–2116
49. Watson S, Cailhier JF, Hughes J et al. (2006) Apoptosis and glomerulonephritis. Curr Dir Autoimmun 9:188–204
50. Mühl H, Sandau K, Brüne B et al. (1996) Nitric oxide donors induce apoptosis in glomerular mesangial cells, epithelial cells and endothelial cells. Eur J Pharmacol 317:137–149
51. Sandau K, Pfeilschifter J, Brüne B (1997) The balance between nitric oxide and superoxide determines apoptotic and necrotic death of rat mesangial cells. J Immunol 158:4938–4946
52. Hruby Z, Beck KF (1997) Cytotoxic effect of autocrine and macrophage-derived nitric oxide on cultured rat mesangial cells. Clin Exp Immunol 107:76–82
53. Pautz A, Franzen R, Dorsch S et al. (2002) Cross-talk between nitric oxide and superoxide determines ceramide formation and apoptosis in glomerular cells. Kidney Int 61:790–796
54. Manning G, Whyte DB, Martinez R et al. (2002) The protein kinase complement of the human genome. Science 298:1912–1934
55. Milanesi L, Petrillo M, Sepe L et al. (2005) Systematic analysis of human kinase genes: a large number of genes and alternative splicing events result in functional and structural diversity. BMC Bioinformatics 6 (Suppl 4):S20
56. Chiarugi P, Buricchi F (2007) Protein tyrosine phosphorylation and reversible oxidation: two cross-talking posttranslation modifications. Antioxid Redox Signal 9:1–24
57. Sullivan SG, Chiu DT, Errasfa M et al. (1994) Effects of H2O2 on protein tyrosine phosphatase activity in HER14 cells. Free Radic Biol Med 16:399–403
58. Cunnick JM, Dorsey JF, Mei L et al. (1998) Reversible regulation of SHP-1 tyrosine phosphatase activity by oxidation. Biochem Mol Biol Int 45:887–894
59. Schmid E, Hotz-Wagenblatt A, Hacj V et al. (1999) Phosphorylation of the insulin receptor kinase by phosphocreatine in combination with hydrogen peroxide: the structural basis of redox priming. FASEB J 13:1491–1500
60. Lin D, Takemoto DJ (2005) Oxidative activation of protein kinase Cgamma through the C1 domain. Effects on gap junctions. J Biol Chem 280:13682–13693
61. Huwiler A, Pfeilschifter J (1999) Nitric oxide stimulates the stress-activated protein kinase p38 in rat renal mesangial cells. J Exp Biol 202:655–660
62. Klotz LO, Schroeder P, Sies H (2002) Peroxynitrite signaling: receptor tyrosine kinases and activation of stress-responsive pathways. Free Radic Biol Med 33:737–743
63. Sugimoto T, Haneda M, Togawa M et al. (1996) Atrial natriuretic peptide induces the expression of MKP-1, a mitogen-activated protein kinase phosphatase, in glomerular mesangial cells. J Biol Chem 271:544–547
64. Sugimoto T, Kikkawa R, Haneda M et al. (1995) Cyclic nucleotides attenuate endothelin-1-induced activation of mitogen-activated protein kinase in cultured rat mesangial cells. J Diabetes Complications 9:249–251
65. Isono M, Haneda M, Maeda S et al. (1998) Atrial natriuretic peptide inhibits endothelin-1-induced activation of JNK in glomerular mesangial cells. Kidney Int 53:1133–1142
66. Callsen D, Pfeilschifter J, Brüne B (1998) Rapid and delayed p42/p44 mitogen-activated protein kinase activation by nitric oxide: the role of cyclic GMP and tyrosine phosphatase inhibition. J Immunol 161:4852–4858

67. Gorin Y, Ricono JM, Wagner B et al. (2004) Angiotensin II-induced ERK1/ERK2 activation and protein synthesis are redox-dependent in glomerular mesangial cells. Biochem J 381:231–239
68. Block K, Eid A, Griendling KK et al. (2008) Nox4 NAD(P)H oxidase mediates Src-dependent tyrosine phosphorylation of PDK-1 in response to angiotensin II: role in mesangial cell hypertrophy and fibronectin expression. J Biol Chem 283:24061–24076
69. Ding G, Zhang A, Huang S et al. (2007) ANG II induces c-Jun NH2-terminal kinase activation and proliferation of human mesangial cells via redox-sensitive transactivation of the EGFR. Am J Physiol Renal Physiol 293:F1889–F1897
70. Callsen D, Sandau KB, Brüne B (1999) Nitric oxide and superoxide inhibit platelet-derived growth factor receptor phosphotyrosine phosphatases. Free Radic Biol Med 26:1544–1553
71. Bogdan C (2001) Nitric oxide and the regulation of gene expression. Trends Cell Biol 11:66–75
72. Genestra M (2007) Oxyl radicals, redox-sensitive signalling cascades and antioxidants. Cell Signal 19:1807–1819
73. Pfeilschifter J, Eberhardt W, Huwiler A (2003) Nitric oxide and mechanisms of redox signaling. J Am Soc Nephrol 14(Suppl 3):S237–S240
74. Valko M, Leibfritz D, Moncol J et al. (2007) Free radicals and antioxidants in normal physiological functions and human disease. Int J Biochem Cell Biol 39:44–84
75. Satriano JA, Shuldiner M, Hora K et al. (1993) Oxygen radicals as second messengers for expression of the monocyte chemoattractant protein, JE/MCP-1, and the monocyte colony-stimulating factor, CSF-1, in response to tumor necrosis factor-alpha and immunoglobulin G. Evidence for involvement of reduced nicotinamide adenine dinucleotide phosphate (NADPH)-dependent oxidase. J Clin Invest 92:1564–1571
76. Satriano JA, Banas B, Luckow B et al. (1997) Regulation of RANTES and ICAM-1 expression in murine mesangial cells. J Am Soc Nephrol 8:596–603
77. Ha H, Yu MR, Choi YJ et al. (2002) Role of high glucose-induced nuclear factor-kappaB activation in monocyte chemoattractant protein-1 expression by mesangial cells. J Am Soc Nephrol 13:894–902
78. Eberhardt W, Huwiler A, Beck KF et al. (2000) Amplification of IL-1 beta-induced matrix metalloproteinase-9 expression by superoxide in rat glomerular mesangial cells is mediated by increased activities of NF-kappa B and activating protein-1 and involves activation of the mitogen-activated protein kinase pathways. J Immunol 165:5788–5797
79. Gauer S, Hauser IA, Obermüller N et al. (2008) Synergistic induction of osteopontin by aldosterone and inflammatory cytokines in mesangial cells. J Cell Biochem 103:615–623
80. Ha H, Lee HB (2000) Reactive oxygen species as glucose signaling molecules in mesangial cells cultured under high glucose. Kidney Int Suppl 77:S19–S25
81. Iglesias-De La Cruz MC, Ruiz-Torres P et al. (2001) Hydrogen peroxide increases extracellular matrix mRNA through TGF-beta in human mesangial cells. Kidney Int 59:87–95
82. Osman B, Doller A, Akool ES et al. (2009) Rapamycin induces the TGFbeta(1)/Smad signaling cascade in renal mesangial cells upstream of mTOR. Cell Signal 21 (12):1806–1817
83. Baricos WH, Cortez SL, Deboisblanc M et al. (1999) Transforming growth factor-beta is a potent inhibitor of extracellular matrix degradation by cultured human mesangial cells. J Am Soc Nephrol 10:790–795
84. Gopalakrishna R, Anderson WB (1991) Reversible oxidative activation and inactivation of protein kinase C by the mitogen/tumor promoter periodate. Arch Biochem Biophys 285:382–387
85. Gopalakrishna R, Jaken S (2000) Protein kinase C signaling and oxidative stress. Free Radic Biol Med 28:1349–1361.
86. Brown Z, Robson RL, Westwick J (1993). L-arginine/nitric oxide pathway: a possible signal transduction mechanism for the regulation of the chemokine IL-8 in human mesangial cells. Adv Exp Med Biol 351:65–75

87. Eberhardt W, Beeg T, Beck KF et al. (2000) Nitric oxide modulates expression of matrix metalloproteinase-9 in rat mesangial cells. Kidney Int 57:59–69
88. Pilz RB, Casteel DE (2003) Regulation of gene expression by cyclic GMP. Circ Res 93:1034–1046
89. Hagen T, Taylor CT, Lam F et al. (2003) Redistribution of intracellular oxygen in hypoxia by nitric oxide: effect on HIF1alpha. Science 302:1975–1978
90. Welsh J, Chada K, Dalal SS et al. (1992) Arbitrarily primed PCR fingerprinting of RNA. Nucleic Acids Res 20:4965–4970
91. Dreieicher E, Beck KF, Lazaroski S et al. (2009) Nitric oxide inhibits glomerular TGF-beta signaling via SMOC-1. J Am Soc Nephrol 20:1963–1974
92. Schaefer L, Beck KF, Raslik I et al. (2003) Biglycan, a nitric oxide-regulated gene, affects adhesion, growth, and survival of mesangial cells. J Biol Chem 278:26227–26237
93. Beck KF, Güder G, Schaefer L et al. (2005) Nitric oxide upregulates induction of PDGF receptor-alpha expression in rat renal mesangial cells and in anti-Thy-1 glomerulonephritis. J Am Soc Nephrol 16:1948–1957
94. Keller T, Plesková M, McDonald MC et al. (2003) Identification of manganese superoxide dismutase as a NO-regulated gene in rat glomerular mesangial cells by 2D gel electrophoresis. Nitric Oxide 9:183–193
95. Walpen S, Beck KF, Schaefer L et al. (2001) Nitric oxide induces MIP-2 transcription in rat renal mesangial cells and in a rat model of glomerulonephritis. FASEB J 15:571–573
96. Keil A, Blom IE, Goldschmeding R et al. (2002) Nitric oxide down-regulates connective tissue growth factor in rat mesangial cells. Kidney Int 62:401–411
97. Akool el-S, Kleinert H, Hamada FM et al. (2003) Nitric oxide increases the decay of matrix metalloproteinase 9 mRNA by inhibiting the expression of mRNA-stabilizing factor HuR. Mol Cell Biol 23:4901–4916
98. Franzen R, Fabbro D, Aschrafi A et al. (2002) Nitric oxide induces degradation of the neutral ceramidase in rat renal mesangial cells and is counterregulated by protein kinase C. J Biol Chem 277:46184–46190
99. Franzen R, Pfeilschifter J, Huwiler A (2002) Nitric oxide induces neutral ceramidase degradation by the ubiquitin/proteasome complex in renal mesangial cell cultures. FEBS Lett 532:441–444
100. Nistala R, Whaley-Connell A, Sowers JR (2008) Redox control of renal function and hypertension. Antioxid Redox Signal 10:2047–2089
101. Kanwar YS, Wada J, Sun L et al. (2008) Diabetic nephropathy: mechanisms of renal disease progression. Exp Biol Med (Maywood) 233:4–11
102. Gao L, Mann GE (2009) Vascular NAD(P)H oxidase activation in diabetes: a double-edged sword in redox signalling. Cardiovasc Res 82:9–20
103. Singh D, Kaur R, Chander V et al. (2006) Antioxidants in the prevention of renal disease. J Med Food 9:443–450
104. Weinberg JB, Granger DL, Pisetsky DS et al. (1994) The role of nitric oxide in the pathogenesis of spontaneous murine autoimmune disease: increased nitric oxide production and nitric oxide synthase expression in MRL-lpr/lpr mice, and reduction of spontaneous glomerulonephritis and arthritis by orally administered NG-monomethyl-L-arginine. J Exp Med 179:651–660
105. Reilly CM, Farrelly LW, Viti D et al. (2002). Modulation of renal disease in MRL/lpr mice by pharmacologic inhibition of inducible nitric oxide synthase. Kidney Int 61:839–846
106. Narita I, Border WA, Ketteler M et al. (1995) Nitric oxide mediates immunologic injury to kidney mesangium in experimental glomerulonephritis. Lab Invest 72:17–24
107. Peters H, Border WA, Noble NA (1999) L-arginine supplementation increases mesangial cell injury and subsequent tissue fibrosis in experimental glomerulonephritis. Kidney Int 55:2264–2273

108. Westenfeld R, Gawlik A, de Heer E et al. (2002) Selective inhibition of inducible nitric oxide synthase enhances intraglomerular coagulation in chronic anti-Thy 1 nephritis. Kidney Int 61:834–838
109. Satriano J, Lortie MJ, Ishizuka S et al. (2005) Inhibition of inducible nitric oxide synthase alters Thy-1 glomeruonephritis in rats. Nephron Physiol 102:17–26
110. Wang Y, Krämer S, Loof T et al. (2005) Stimulation of soluble guanylate cyclase slows progression in anti-thy1-induced chronic glomerulosclerosis. Kidney Int 68:47–61
111. Wang-Rosenke Y, Neumayer HH, Peters H (2008) NO signaling through cGMP in renal tissue fibrosis and beyond: key pathway and novel therapeutic target. Curr Med Chem 15:1396–1406
112. Bjelakovic G, Nikolova D, Gluud LL et al. (2007) Mortality in randomized trials of antioxidant supplements for primary and secondary prevention: systematic review and meta-analysis. JAMA 297:842–857
113. Stolk J, Hiltermann TJ, Dijkman JH et al. (1994) Characteristics of the inhibition of NADPH oxidase activation in neutrophils by apocynin, a methoxy-substituted catechol. Am J Respir Cell Mol Biol 11:95–102
114. Shinozaki K, Nishio Y, Ayajiki K et al. (2007) Pitavastatin restores vascular dysfunction in insulin-resistant state by inhibiting NAD(P)H oxidase activity and uncoupled endothelial nitric oxide synthase-dependent superoxide production. J Cardiovasc Pharmacol 49:122–130
115. Fujii M, Inoguchi T, Maeda Y et al. (2007) Pitavastatin ameliorates albuminuria and renal mesangial expansion by downregulating NOX4 in db/db mice. Kidney Int 72:473–480
116. Yu HY, Inoguchi T, Nakayama M et al. (2005) Statin attenuates high glucose-induced and angiotensin II-induced MAP kinase activity through inhibition of NAD(P)H oxidase activity in cultured mesangial cells. Med Chem 1:461–466
117. Noh H, King GL (2007) The role of protein kinase C activation in diabetic nephropathy. Kidney Int Suppl 106:S49–S53
118. Meier M, Menne J, Haller H (2009) Targeting the protein kinase C family in the diabetic kidney: lessons from analysis of mutant mice. Diabetologia 52:765–775.
119. Walpen S, Beck KF, Eberhardt W et al. (2000) Downregulation of SPARC expression is mediated by nitric oxide in rat mesangial cells and during endotoxemia in the rat. J Am Soc Nephrol 11:468–476

Chapter 2
Transition Metals and Other Forms of Oxidative Protein Damage in Renal Disease

Vincent M. Monnier, Ina Nemet, David R. Sell, and Miriam F. Weiss

Abstract Oxidative and carbonyl stresses are dramatically increased in chronic renal disease, whereby an inverse relationship usually exists between renal clearance and the accumulation of low molecular weight compounds ultimately responsible for the damage to plasma constituents. Damage to proteins results from primary attack to protein residues by reactive oxygen species with or without metal catalyst, or via myeloperoxidase and hypochlorous acid. Secondary, indirect forms of damage result from oxoaldehydes and lipid peroxidation products involved in glycation and glycoxidation reactions with nucleophilic residues. The chemical oxidative pathways responsible for protein damage and its biological and clinical significance are discussed, emphasizing end stage renal disease. Interventions that improve or worsen oxidant stress, such as intravenous iron therapy, are reviewed.

Keywords Oxidative stress · Carbonyl stress · Glycation · Catalytic metals · Myeloperoxidase · Methylglyoxal · Ascorbic acid

1 Introduction

Oxidative modifications of proteins are strongly associated with renal disease. The presence of high levels of protein carbonyl, advanced oxidation protein products (AOPP), metal catalyzed oxidation products, and other oxidative modifications has been understood to be a sign of oxidant stress markers, a vague term that provides little information regarding mechanisms of formation. In addition to metal catalyzed oxidation (MCO), oxidative modifications in renal disease can involve hypochlorous acid, peroxide, peroxinitrite, hydrogen peroxide, and hydroxyl radicals.

V.M. Monnier (✉)
Department of Pathology and Biochemistry, Case Western Reserve University,
Wolstein Research Bldg., 2103 Cornell Road, Cleveland, OH 44106-7288, USA
e-mail: vmm3@cwru.edu

T. Miyata et al. (eds.), *Studies on Renal Disorders*, Oxidative Stress in Applied Basic Research and Clinical Practice, DOI 10.1007/978-1-60761-857-7_2,

To clarify these diverse mechanisms, the first part of this chapter will be devoted to the chemistry of protein oxidation with the ultimate question of which modifications should be measured in order to unequivocally implicate a particular pathway of protein damage. It is useful to distinguish between *primary* damage to proteins (i.e., oxidative modifications of amino acid residues and peptide backbone fragmentation reactions) vs. *secondary* forms of damage involving modifications of nucleophilic amino acids (such as lysine, arginine, and cysteine) by carbonyls from lipoxidation or the Maillard reaction. Thus, *carbonyl* and *oxidative* stress are often tightly linked. However, protein modifications resulting from the latter, and in particular those from the Maillard reaction, will be reviewed elsewhere in this book.

The second part of this chapter will be devoted to clinical, biological, and structural aspects of metal catalyzed oxidation, focusing primarily on advanced oxidation products that are common to MCO and related mechanisms of oxidation.

2 Chemical Pathways of Protein Damage by Oxidation

A number of biomarkers for protein oxidation have been identified in recent years that demonstrate not only the extent of oxidative injury but also provide evidence for the source of the oxidant. Knowledge of the latter is essential for developing adequate interventions to stop or slow down damage. Figure 1 summarizes the major sources of protein oxidants in vivo (i.e., hypochlorous acid [HOCl], peroxynitrite [$ONOO^-$], and free radicals), resulting from a combination of different enzymatic and nonenzymatic reactions.

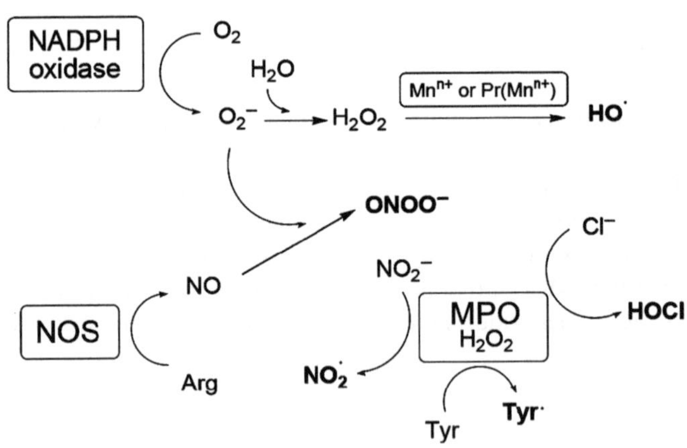

Fig. 1 Major pathways leading to formation of reactive oxidants via the myeloperoxidase (MPO), NADPH oxidase, nitric oxide synthase (NOS), and redox active metal ions — free (Mn^{n+}) and protein bound metals ($Pr(Mn^{n+})$)

Sulfur containing amino acids, Cys and Met, are the major sites of oxidation within proteins. They can be oxidized easily by radical-mediated or two-electron pathways, typically involving HOCl, peroxides, peroxynitrite, or singlet oxygen. The resulting products are primarily disulfides (cystine or mixed disulfides) and methionine sulfoxide (MetSOX), respectively, regardless of the nature of the oxidant. Cysteine reacts with a thiyl radical (RS•) that dimerize with another cysteine or sulfhydryl residue to form the disulfide. The thiyl radical, in a competitive reaction with oxygen, gives a peroxy radical, which usually forms oxyacids (RSO_2H and RSO_3H). The same products are formed in the two-electron mediated oxidation process where the initial adduct species (e.g., RSCl, RSOH, RSNO) form from HOCl, peroxides, and nitric oxide (NO^+), respectively, and then react with another thiol group to produce disulfides or hydrolyze into the oxyacids. These intermediate compounds are usually too labile and complex for quantification in vivo.

Methionine invariably gets oxidized into MetSOX, regardless of whether it is oxidized by free radicals, peroxide, or peroxynitrite [1]. The latter rapidly reacts with Met residues and gives MetSOX, however, in the presence of physiological levels of CO_2 an $ONOOCO_2^-$ adduct is formed that oxidizes Met into MetSOX at a much slower rate than $ONOO^-$, suggesting that the extent of Met oxidation observed in vivo will be modulated to a major extent by local CO_2 levels. MetSOX can be subjected to further oxidation and give of sulfone, however, this reaction occurs to a much lesser extent [1, 2].

In summary, the major end products of Cys and Met oxidation expected in vivo are cystine and methionine sulfoxide. Further oxidation of cystine into cysteic acid requires much more drastic conditions that are not easily achieved in vivo, except by the HOCl system.

2.1 HOCl Mediated Pathways

In vivo HOCl is formed from hydrogen peroxide and chloride ions (Cl^-) in a reaction catalyzed by myeloperoxidase (see Fig. 1). In reactions with protein residues, HOCl generates oxidized (disulfide, MetSOX, carbonyls, and dityrosine) and chlorinated products (chloramines, 3-chlorotyrosine, and 3,5-dichlorotyrosine) (Fig. 2).

Oxidative reactions by HOCl are several orders of magnitude faster than chlorination reactions and therefore more common under physiological conditions. However, the oxidative products can also be generated by other oxidants (Figs. 3 and 4), making chlorinated products specific biomarkers for HOCl mediated protein modifications. The most favored chlorination reaction of HOCl is with amine groups. However, formed chloramines with typical half-lives of about 10 min at 37°C are readily reduced back to parent amines by biological reductants (thiols, ascorbate, or methionine) or spontaneously broken down to aldehydes [3]. Chlorination of tyrosyl residues, although less favored than chlorination of amines,

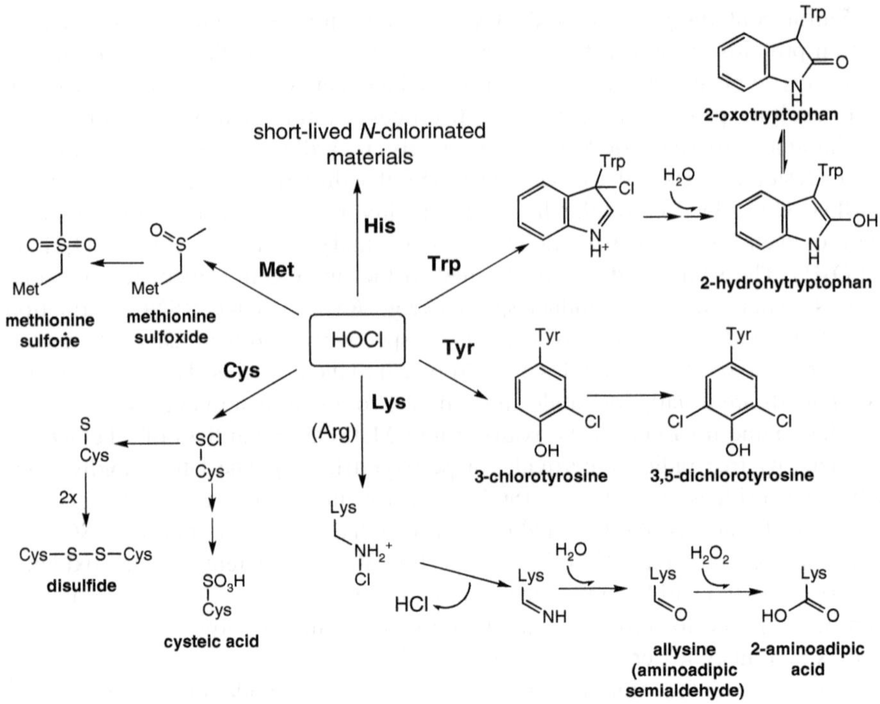

Fig. 2 Products of amino acids residues oxidation by hypochlorous acid (HOCl)

produce stable end products: 3-chlorotyrosine and to a much lesser extent 3,5-dichlorotyrosine, which can be used as biomarkers of HOCl generation [4].

2.2 Free Radicals Oxidation

Highly reactive free radicals, such as HO•, initiate oxidation by abstracting a hydrogen atom from an amino acid residue and generating carbon-centered radicals. These radicals react further with O_2, producing peroxy radicals (see Fig. 3). If the peroxy radical is formed on an α-carbon it undergoes decomposition and forms an amino derivative, which spontaneously hydrolyses and yields two peptide fragments. Peroxy radicals formed on amino-acid side chains rearrange, through incompletely defined reaction pathways, into alcohols or aldehydes. Exposure of tryptophan to HO• results in indole ring opening and kynurenine formation, while histidine generates 2-oxohistidine, asparagine, and aspartate. In a reaction with tyrosyl residues free radicals generate tyrosyl phenoxy radicals. The major fate of these structures is self-dimerization and formation of di-tyrosine crosslinks. Oxidation of phenylalanine, on the other hand, produces *ortho-* (2-hydroxyphenylalanine) and *meta*-tyrosine (3-hydroxyphenylalanine) as well as tyrosine.

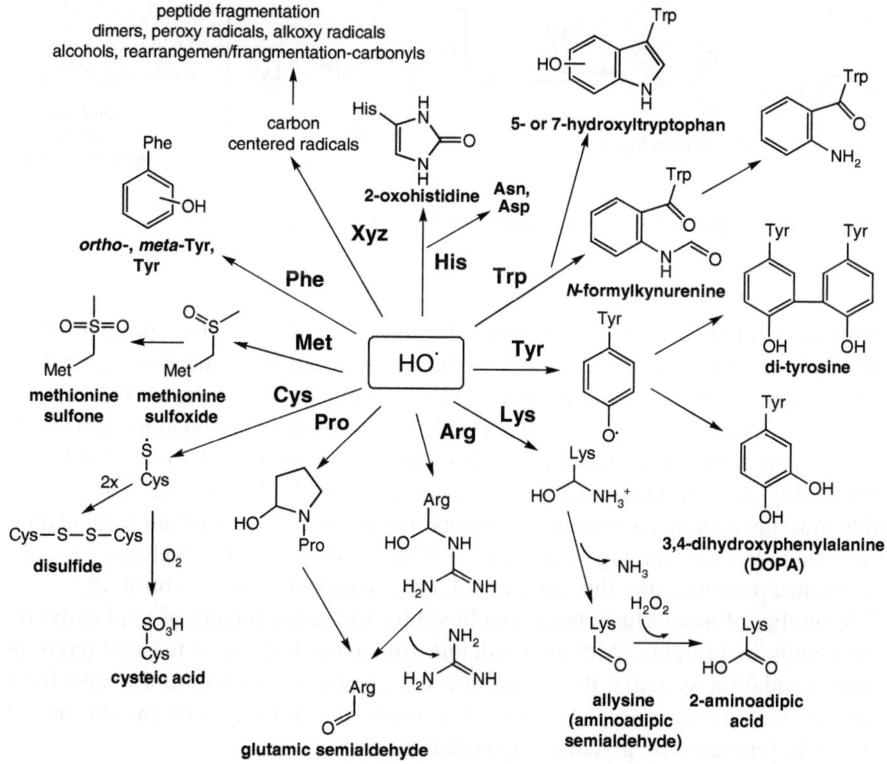

Fig. 3 Products of amino acids residues oxidation by hydroxyl radicals ($HO^{-\bullet}$)

Stable modifications that are commonly measured in vivo include kynurenine, 2-oxohistidine, di-tyrosine, and o-tyrosine.

2.3 Metal-Catalyzed Oxidation (MCO)

Redox-active metal ions, such as Fe(III) and Cu(II), catalyze oxidative damage of proteins in the presence of oxygen and an electron donor or by generating hydroxyl or alkoxyl radicals through Haber-Weiss and Fenton-type chemistry from hydrogen or organic peroxides, respectively. Highly reactive radicals, as described above, nonselectively abstract hydrogen. Basically, this reaction can occur at all available sites, though His, Arg, Lys, Pro, and to a lesser extent Met, and Cys residues are the most common. On the other hand, aromatic amino acid residues, Trp, Tyr, and Phe, are not the major targets for this oxidation system, most probably because these residues are not commonly present at metal-binding sites of proteins [5].

As discussed above, carbonyls can also be generated by Lys and Arg oxidation with HOCl. However, the overall yield of carbonyls through HOCl oxidation

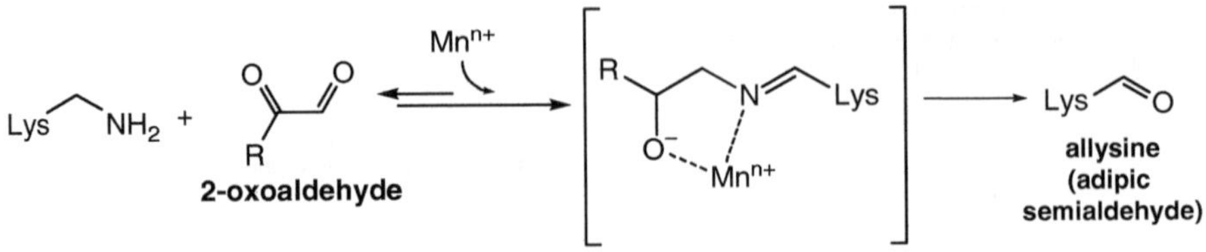

Fig. 4 Suyama's pathway of lysine deamination by α-oxoaldehydes

represents a relatively low percentage of the total [6]. Taken together, protein-bound carbonyls are the major products of oxidation catalyzed by metal ions. Therefore, protein-bound carbonyls can be considered important markers of the MCO, with one notable exception. Allysine can be formed in vivo by lysine oxidation with lysyl-oxidase (particularly in collagen), as well as through α-oxoaldehydes mediated Cu(II) catalyzed lysine deamination (Suyama's pathway) [7] (see Fig. 4). This latter possibility must be recognized in biologic settings characterized by significant participation of dicarbonyls, as in end stage renal disease (ESRD). The extent to which lysyl oxidase can oxidize proteins other that nascent collagen is unclear, if not doubtful [8].

A number of assays have been developed for the measurement of total carbonyl compounds in proteins. Although suitable for quantification of the carbonyls in simple oxidation systems, the assays are not accurate for proteins isolated from complex biological samples because of possible interference with protein-bound carbonyls generated by glycation/lipoxidation.

A better approach includes reduction of the carbonyls with $NaBH_4$ into corresponding alcohols followed by protein hydrolysis and analysis by isotope dilution gas chromatography-mass spectrometry (GC/MS) [9, 10]. However, the carbonyl compounds are not the final oxidation products (AOPPs) since they are prone to further oxidation into corresponding carboxylic acids, either by auto-oxidation or by reaction with H_2O_2 or HOCl. Thus, for the complete oxidation profile, parallel measurements of the carboxylic acid, such as 2-aminoadipic acid, should also be included [10].

2.4 *Oxidation by Peroxynitrite ($ONOO^-$/HONOO)*

The weak oxidants nitric oxide and superoxide, which are produced by several cell types, rapidly combine to form a strong oxidant – peroxynitrite. Due to nitric oxide's longer biological half-life and facile diffusion in comparison with superoxide, peroxynitrite generation will predominantly occur nearer to sites of O_2^- formation [11]. Peroxynitrite anion is relatively stable, but its acid, peroxynitrous acid (HONOO), quickly rearranges to form nitrate. Two major mechanisms of oxidation by peroxynitrite and its acid are accepted. The first mechanism includes homolysis of peroxynitrite and formation of HO• and •NO_2. This pair of radicals can either diffuse apart, giving free radicals that can perform oxidations, or react together to form nitrate or to reform peroxynitrite. Because hydroxyl-radical scavengers do not completely block peroxinitrite reactions, it is thought that the peroxide bond O-O in peroxynitrous acid (HO–O–N=O) can result in a transfer of an HO^- or an oxygen atom by electron donors [12].

Peroxynitrite also promotes nitration of amino acid residues Fig. 5. Most notably, protein tyrosine residues represent the major targets for peroxynitrite-mediated nitrations where 3-nitrotyrosine represents a marker of such modifications. Other mechanisms of biological nitrations exist (such H_2O_2/NO_2^- heme peroxidase, reaction of $^{\bullet}NO_2$ and NO nitration with concomitant oxidation). However, these alternative reactions are more restricted than the nitration reactions mediated by peroxynitrite. Peroxynitrite reacts with carbon dioxide (ubiquitous in biological systems) to form $ONOOCO_2^-$ (nitroso-peroxocarboxylate). The latter homolyses further into a relatively strong one-electron oxidant carbonate radical and nitrogen dioxide – a moderate oxidant and nitrating agent. The net consequence of the reaction of CO_2 on peroxynitrite is to reduce peroxynitrite half-life and distance of diffusion, effects that reduce one-electron oxidation and enhance nitration [11].

From a practical viewpoint, one of the most frequently determined markers of peroxynitrite activity is 3-nitrotyrosine [13]. Not considered here is the broad field of sulfhydryl nitrosylation, which has regulatory effects on protein function as well as stochastic effects [14].

In summary, several oxidative mechanisms lead to protein damage. It is clear that metal catalyzed oxidation cannot be considered entirely separately from other

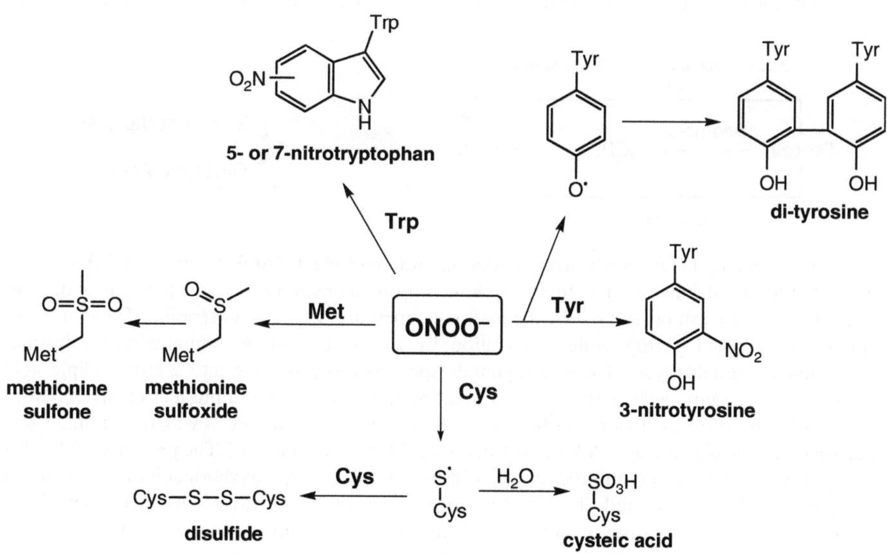

Fig. 5 Products of amino acids residues oxidation by peroxynitrite ($ONOO^-$)

oxidative mechanisms. Below, we present selected human and animal in vivo studies that help shape the "oxidative landscape" in renal disorders.

3 Protein Carbonyls (PCs) and Advanced Oxidation Products (AOPPs) in Renal Disorders

Much of the understanding of the role of AOPPs in renal disease comes from the pioneering work of Witko-Sarsat et al. [15]. The first assay was based on the notion that in renal disease lysines in serum proteins have been oxidized into chloramines by HOCl, which upon reaction with potassium iodide at acidic pH generates yellow tri-iodide, I_3^-. Therefore, a standard curve is produced using chloramine T, a widely used reagent for the radio labeling of proteins with the 125 or 131 isotopes of iodine. Since chloramines are not very stable and tend to decompose into protein carbonyls (see Figs. 2 and 6), it is not exactly clear what the assay measures nor is the exact nature of the color generated known, since multiple secondary reaction products might be expected. In the aggregate, however, yellow color is measured photometrically at 340 nm immediately after the reaction. This assay has been adapted for microtiter plates [15].

The "protein carbonyls" that are assayed in most publications reviewed below are all based on the reaction of protein with 2,4-dinitrophenylhydrazine (DNP). The resulting hydrazone is quantitated spectrophotometrically or using Western blotting with an anti-DNP antibody. Several methods have been developed based on this reaction [17]. Similar to the assay for AOPP, these methods are generally not specific, as they measure carbonyls from both oxidation and glycation. Mass spectrometry methods involving reduction of the protein with sodium borohydride to

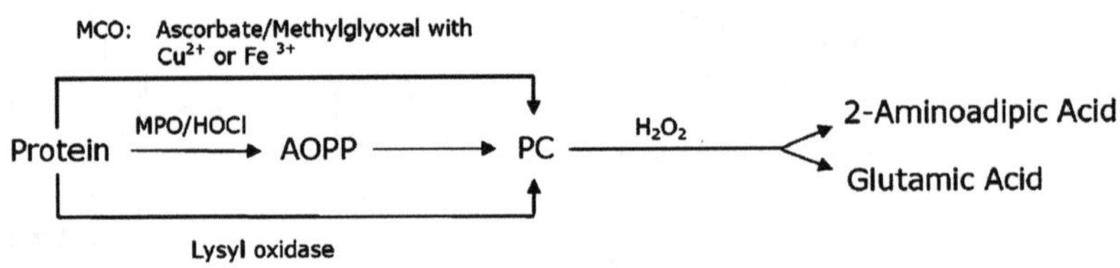

Fig. 6 Relationship between advanced oxidation protein products (AOPPs) and protein carbonyls (PCs). AOPPs are difficult to quantitate by chromatographic methods because they easily decompose into PCs. PCs can be assayed by the hydrazine method, which is not specific. The two major protein PCs formed during protein oxidation by metals are adipic semialdehyde (allysine) and glutamic semialdehyde. These compounds spontaneously oxidize into 2-aminoadipic acid (2-AAA) and glutamic acid, respectively. (Note: glutamic acid is not a product of myeloperoxidase [MPO] or lysyl oxidation.) In highly oxidizing environments, PC levels may remain low because of rapid oxidation to 2-AAA and glutamic acid, respectively [16]. The presence of 2-AAA and the borohydride reduction products of allysine PC (i.e., 6-hydroxynorleucine) are evidence that lysine oxidation occurred. The relative importance of MCO, myeloperoxidase, and lysyl oxidase (if at all) in renal disease and ESRD remains to be determined. In contrast the finding of the borohydride reduction product of glutamyl semialdehyde (i.e., 5-hydroxy-2-aminovaleric acid) is specific for MCO

measure the alcohols of adipic semialdehyde (allysine) and glutamic semialdehyde have greater specificity [9, 10].

In the sections that follow, we will use the terms AOPPs and protein carbonyls (PCs), as they are used by the authors, without going into technical details as to what assay was used and which chemical entity was measured. The relationship between these products is summarized below:

3.1 Clinical Studies

3.1.1 Oxidation Markers in End Stage Renal Failure and Earlier Stages of Renal Disease

Numerous compounds accumulate when kidneys fail, disturbing the function of multiple organ systems [18]. At the top of the list of organs systems affected by uremia is the heart and vasculature. Cardiovascular disease is the leading cause of death in all age groups of patients with ESRD [19]. In uremia, the association between traditional risk factors, such as high cholesterol and accelerated heart disease, is weak. Instead, cardiovascular complications correlate with markers of inflammation and malnutrition [20, 21] and the presence of a markedly increased oxidative burden [22]. In an otherwise puzzling pathogenetic process, oxidative stress has been proposed as the key evidence linking uremia to cardiovascular complications [23].

Direct measurements have been made of increased oxidative activity in the circulation of patients treated by hemodialysis using electron-spin resonance spectroscopy (ESR) and spin-traps. ESR activity is localized to the middle molecular weight range (~3,000 Da) of dialyzable toxins and is reduced significantly by dialysis [24]. Reactive oxidants have short half-lives, and direct measurements are difficult and cumbersome. Therefore, the devastation caused by oxidative stress is evidenced by extensive protein damage (i.e., thiol oxidation, amine oxidation, and carbonyl modification) (Table 1) present on circulating, cellular [30], and tissue proteins as well as lipids [22]. Because reactive oxidants have short half-lives, oxidative stress is better measured clinically as stable oxidized compounds, such as advanced oxidation (AOPPs) and glycation end products (AGEs), which some workers believe represent a form of uremic toxin [31, 32].

Exhaustion of antioxidant defense systems is a distinctive feature of uremia. For example, in patients with ESRD treated by hemodialysis, levels of the antioxidant ascorbate are low, and the ratio of reduced to oxidized ascorbate is several-fold greater than normal [33–35]. In addition to massive oxidant stress there is massive carbonyl stress [36], resulting in a five- to tenfold increase in dicarbonyl adducts to plasma proteins, such as those of glyoxal, methylglyoxal and 3-deoxyglucosone [37, 38]. Thus ascorbate oxidation and accumulation of metabolites such as

Table. 1 Representative plasma and human skin collagen levels of oxidation products

Protein modification	Fluid	Normal	Diabetes	ESRD	Comments	References
AOPP (μmol/L)	Plasma	29.4 ± 4.9	–	137.6 ± 11.1	No correl. w. prot. conc.	[15]
AOPP (μmol/L)	Plasma	0.76 ± 0.51	CRF: 13.73 ± 4.45	16.95 ± 2.62		[25]
3-Chloro-tyrosine μmol/mol tyr.	Plasma	N.D.	–	3.5 ± 0.8		[26]
PC (nmol/mg)	Plasma	2.06 ± 0.33	2.96 ± 0.48 Db + ESRD	2.51 ± 0.18		[27]
PC (nmol/mg)	Plasma	0.093 ± 0.014	0.099 ± 0.014	0.16 ± 0.050	Inversely related with PON1 in ESRD	[28]
PC (nmol/mg)	Plasma	0.35 ± 0.04	–	1.23 ± 0.50		[29]
Alkanals (nM)	Plasma	1,407 ± 286	–	2,088 ± 831		[29]
Alkenals (nM)	Plasma	825 ± 142	–	2,453 ± 821		[29]
4-OH-alkenals (nM)	Plasma	460 ± 102	–	1,732 ± 598		[29]
Free SH nmol/mg	Plasma	4.48 ± 1.12	–	2.69 ± 1.23		[29]
Prot-NH_2 (AU/mg prot)	Plasma	287 ± 163	–	188 ± 106		
Allysine (mmol/mol lys)	Collagen (skin)	0.05 ± 0.3	0.04–0.3	0.8–0.4	Not age-dependent	[10]
2-Aminoadipic acid (mmol/mol lys)	Collagen (skin)	0.2–0.8	0.2–2.0	0.5–7.0	Strongly age-dependent	[10]

dehydroascorbic acid and diketogulonic acid contribute to carbonyl stress and further increase the generation of AGEs [39].

Even patients with mild kidney disease show evidence of increased protein damage, i.e., decreased free sulfhydryl groups on proteins [25], oxidized low-density lipoprotein (LDL) [40], the buildup of advanced oxidation protein products (AOPPs) [41], and AGEs [42–44]. AGE-modified proteins are removed by glomerular filtration followed by tubular reabsorption and degradation [45–47]. Thus impaired renal function itself is an important mechanism of accumulation in renal disease. Mild chronic kidney failure is associated with impaired antioxidant defense systems as well [48]. A strong proof that systemic oxidant stress (and formation of AGEs) is dependent on renal function is that protein carbonyls and other plasma markers of oxidant and carbonyl stress increased following development of chronic allograft nephropathy, but decreased in complication-free transplant patients [49]. Similarly, in the Project to Improve Care in Acute Renal Disease (PICARD) study, plasma PCs and proinflammatory cytokines were increased in patients with acute renal failure (ARF), and protein carbonyls were higher in ARF than in critically ill and ESRD patients and controls [50].

In the presence of renal disease, oxidation reactions form one part of a vicious circle [51]. In turn, modified proteins induce oxidative stress through several mechanisms. At the cellular level, AGEs bind to specific receptors, activate intracellular signals (such as nuclear factor κB [NFκB]) and stimulate cytokine release and inflammation [52, 53]. Neutrophils and other inflammatory cells produce numerous reactive oxygen species, including superoxide anion, hydrogen peroxide, hydroxyl radical, and hypochlorous acid [54]. Thus oxidized proteins and lipids interact with inflammatory cells to generate reactive oxidants, particularly in the vascular endothelium [22].

In renal disease oxidized proteins are both the result of oxidative stress, and one of its major causes, but patterns of protein modifications differ in different disease settings. In a Chinese study of type 2 diabetes, erythrocyte antioxidants stress enzymes and plasma vitamin C were decreased while plasma indices of lipid peroxidation and AOPPs were increased, and were much higher in presence of nephropathy where they correlated with each other [55]. Conversely, while plasma protein carbonyls were found increased in ESRD, they were not in complication-free diabetes (see Table 1) [28]. This suggests that diabetes per se does not increase MCO. However, the more specific MCO marker 2-aminoadipic acid Fig. 6 was elevated in skin collagen from diabetic individuals even in the absence of renal failure, suggesting increased metal catalyzed oxidation in diabetes [10].

Likewise, serum levels of oxidized proteins were elevated in systemic lupus erythematosus (SLE), but not related to renal involvement (i.e., lupus nephritis) [56]. Thus, both in lupus and diabetes there may be an independent systemic effect, one linked to an autoimmune process and hyperglycemia, respectively, the other linked to severity of the renal disease. In further support of a dual mechanism for oxidation, increased serum levels of AOPPs and AGEs were found in patients with amyloid A–related rheumatoid arthritis compared to rheumatoid arthritis patients without amyloidosis, whereby AGE levels were related to kidney function but

AOPPs were not [57]. In another study, protein carbonyl formation was strongly dependent on renal function, but not on essential hypertension alone [58]. This surprising dichotomy may mean that protein carbonyl formation in renal disease proceeds primarily via MCO rather than via MPO and AOPP (Fig. 6). However, when ESRD is reached, both MCO and MPO are expected to contribute to protein PCs and 2-aminoadipic acid, whereby H_2O_2 is a dramatic catalyst for the latter [16].

3.1.2 Relationship Between Oxidation and Complications in ESRD

Patients with ESRD experience an increased risk of cardiac complications, more than 10–20 times the incidence of the general population [59]. Although many patients start dialysis with conditions that predispose to atherosclerosis (such as diabetes mellitus, hypertension, or advanced age), these classic cardiac risk factors are insufficient to explain the extraordinary prevalence of cardiovascular disease and death in uremia [60].

Dialysis patients appear to be chronically in higher inflammatory states, compared to normal subjects, as suggested by high serum levels of C-reactive protein (CRP) [61] and pro-inflammatory cytokines [62, 63]. While inflammation is a source of oxidative stress [64] and high serum CRP levels have been associated with higher mortality in dialysis patients (including those without documented ischemic heart disease) [65], it is unclear which inflammatory mechanisms are involved and how they are related to specific types of clinical cardiovascular disease (CVD) (e.g., coronary artery disease vs. cardiomyopathy).

Specific protein modifications have been associated with the pathophysiology of uremic complications, including arterial calcification [44, 66, 67], carotid atherosclerosis [68], and cardiac hypertrophy [68, 69]. In ESRD, plasma protein carbonyls and oxidant stress markers were elevated in symptomatic chronic heart disease (CHD) patients compared to those without [70]. Carotid artery intima-media thickness (CIMT), serum PCs, and markers of oxidant stress are significantly increased in hemodialysis patients, but only serum thiobarbiturate reactive substances (TBARS) and serum nitrite/nitrate levels actually correlated with CIMT [71]. These observations suggest that MCO in the blood poorly reflects the process in the artery.

Albumin is a major protein target of oxidative activity in uremic plasma as well as a critical defense against oxidative stress [72]. When inflammation or malnutrition leads to low serum albumin levels in renal disease, this protection is impaired or lost [73]. Patients with low albumin levels develop marked increases in the peroxidation of erythrocyte membrane lipids [30]. How sick albumin is in ESRD is exemplified in Fig. 7, which depicts a simple serum protein electrophoresis in which the albumin band is thickened and blurry because of damage by oxidation and glycation. Such albumin from ESRD patients has impaired drug-binding properties [74]. Albumin from hemodialysis (HD) patients was found to have not only decreased antioxidant activity, to be more oxidized, negatively charged, and conformationally altered, but also triggered oxidative burst when incubated with

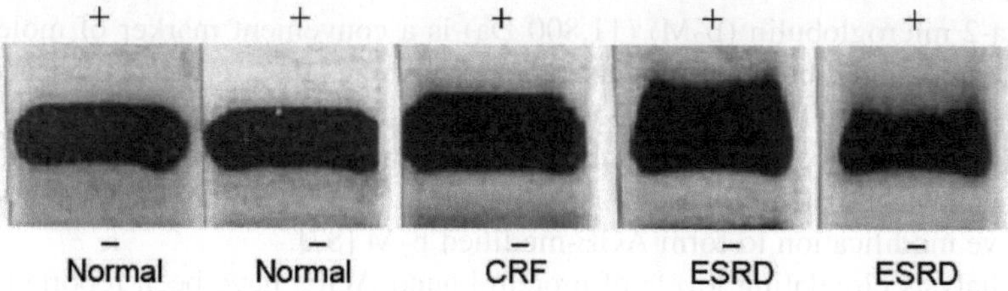

Fig. 7 Appearance of the albumin band from patients with chronic renal failure or end stage renal disease (ESRD) in non-denaturing serum protein electrophoresis in agarose gel. The band is thickened and streaks toward the anode, indicating increased negative charge due to oxidative deamination, CML formation, and loss of arginines from modification by dicarbonyl compounds

human neutrophils [27]. Also, some of the compounds that impair albumin activity appear to be extractable with charcoal [75]. A comprehensive review of the impact of albumin oxidation and other modifications on its function is available elsewhere [76].

From the above, it is not surprising that mortality risk is inversely related to albumin levels [77]. In that regard, plasma PCs and inflammatory stress markers are worse in hypo- than normo-albuminemic patients on HD [78], and it is the combined oxidant and carbonyl stress that is thought to be an important event in the pathogenesis of atherosclerosis [79, 80], the leading cause of morbidity and mortality in uremic patients [81].

Finally, only few efforts have been made to map the target amino acids of oxidative plasma protein damage in patients on hemodialysis. Besides the studies on albumin mentioned above, Alhamdani et al. found elevated levels of plasma alkanals and other lipid peroxidation products [29]. Using GC/MS these authors succeeded in identifying and quantifying up to 11 alkanals, alkenals, and hydroxyl-alkenals in plasma, in addition to the other already known modifications (see Table 1).

3.1.3 Interventions That Decrease Plasma Oxidation Products in Humans

Dialysis Therapy

Multiple compounds accumulate when kidneys fail. Because dialysis is not selective, uremic toxins are classified commonly by their molecular weight [18, 82]. Dialysis removes solutes by diffusion, convection, or adsorption onto the dialyzer membrane surface. Blood levels of markers could be affected directly by any of these mechanisms, with measurable changes occurring during the hemodialysis session, or indirectly by removal of other solutes that are precursors of their metabolic pathways (e.g., glycation pathways) [37, 83]. These indirect effects could occur during the HD session or over longer intervals, depending on the kinetics of chemical reactions and biological events that affect biomarker levels.

Beta-2 microglobulin (β_2M) (11,800 Da) is a convenient marker of molecular size because it can be readily measured and has relevance in the pathogenesis of dialysis-related amyloidosis [84, 85]. Numerous studies over the past decade have shown that high-flux dialysis membranes are associated with higher clearances of β_2M than low-flux membranes [84, 86]. Importantly, β_2M is rendered more toxic by oxidative modification to form AGE-modified β_2M [87].

Predialysis circulating levels of protein-bound AGEs have been reported to be lower in patients treated with high-flux HD than those treated with low-flux HD [37, 88, 89]. In a cross-over study, protein-bound pentosidine (by high-performance liquid chromatography [HPLC]) and AGE-modified β_2M levels were reduced by long-term HD ($\geq$6 months) with high-flux membranes. High flux (F60S, Fresenius) caused a slight to moderate decrease, in contrast to the more pronounced effect of superflux F800S [90]. A recent study used protein-leaking dialyzers (MW cut-off >70 kDa) and demonstrated further decreases in albumin-bound pentosidine levels. Interestingly, these decreases were correlated with significant improvements in levels of inflammatory cytokines (interleukin [IL]-1β, IL-6, tumor necrosis factor [TNF] α), oxidative markers (AOPP, protein carbonyls), and immune-modulating cytokines (IL-10 and interferon [INF] γ) [91].

AGEs are difficult to study in clinical settings because (a) they are chemically and biologically diverse [92, 93], and (b) AGE assays by immunologic methods (e. g., enzyme-linked immunosorbent assay [ELISA]) are not directly comparable to analytic methods (HPLC, liquid chromatography-mass spectrometry [LC/MS]) [88]. In addition, after gastrointestinal digestion, AGEs and glycation adducts present on proteins in food are absorbed as low molecular weight (LMW) or free adducts, which can be removed by HD [94, 95]. Because of the variability in epitope structure and antibody binding characteristics, it is likely that ELISA more readily detects dietary free and LMW AGEs, while underestimating the total AGE content of serum proteins [88]. Thus, while ELISA methods are widely used, analytic chemistry methods are considered the standard of accuracy for detecting AGE proteins [88].

These points are illustrated by two studies of HD patients that yielded apparently contradictory findings. In a Swedish study of 90 patients, pentosidine measured by HPLC positively correlated with inflammatory markers (CRP, fibrinogen, and IL-6), although there was no association between pentosidine and survival [96]. In a German study, baseline AGEs were measured using an ELISA, with the result that higher AGE levels were paradoxically associated with better survival [97]. A potential explanation of the results of the second study is that the intake of highly glycated (better-tasting) food leads to better nutrition and better survival. Food scientists were the first to take interest in the chemistry of AGE formation [98]. In general, foods cooked at high temperatures undergo more "browning" reactions [99], and tasty food has a high content of aromatic, flavorful AGE compounds. Foodstuffs high in AGEs and AGE precursors undergo proteolysis in the intestine and are absorbed as LMW fragments termed glycotoxins [99]. After a meal high in AGEs, an increase in circulating LMW AGEs is readily detected by ELISA [53, 99]. In the absence of kidney clearance, these products accumulate in the circulation

[100]. Thus, patients who have high dietary protein intake would also be expected to have higher circulating levels of LMW AGEs as detected by ELISA [97].

Although dialysis removes filterable low molecular compounds that may have toxicity, dialysis itself can trigger oxidant stress, therefore acting against the desirable effect. In one study protein carbonyls were not decreased immediately after dialysis in contrast to dialyzable molecules such as malonyldialdehyde [101]. Dursun et al. measured protein carbonyls in patients with diabetes and ESRD patients before and after hemodialysis [102]. Baseline plasma protein carbonyls were increased in diabetic patients and further increased in diabetic ESRD patients. Surprisingly, the dialysis procedure contributed to further increase in protein carbonyl levels. The highest levels were observed in patients with diabetes after dialysis. Simultaneously dialysis improved depressed protein-SH levels in patients treated by hemodialysis. Similar observations were made by Ward et al. who found that neutrophil reactive oxygen species (ROS) production was increased in the first 30 min of dialysis with high flux membranes, but then decreased postdialysis [103]. Though protein-SH normalized postdialysis, there was no change in PCs and AOPPs.

These results are best interpreted as follows: protein-SH oxidation is strongly dependent on ROS production and highly reversible. In contrast in a recent study, our laboratory showed that H_2O_2 was not a requirement for protein carbonyl formation via the MCO [16]. However, oxidation of lysine by either MOP/HOCl or metals is quasi-irreversible. Thus, it is likely that dialysis decreases ROS species responsible for SH oxidation, but not those species inducing MCO, including peptide-bound redox active metals.

Antioxidants and Other Drugs

Two small randomized trials respectively suggested the beneficial effect of antioxidants, vitamin E, and *N*-acetylcysteine on CVD outcomes in HD patients [104, 105], in contrast to the lack of benefit shown in large prospective studies in the general population [106, 107]. Thus proponents of antioxidant therapy argue that greater benefit might be found in the setting of abnormal levels of oxidative stress, such as in patients with ESRD [108]. Indeed, protein and DNA oxidation was decreased in hemo- and peritoneal dialysis patients treated with 300 mg vitamin E [109]. Yet, in hemodialysis patients treated with α-tocopherol, circulating levels of anti-inflammatory γ-tocopherol decreased markedly. There were no net changes in oxidative protein modifications (including pentosidine, iso(4)-levuglandin E2 and (E)-4-oxo-2-nonenal) [110].

Further evidence of dissociation between effects of inflammation and oxidation was revealed when γ-tocopherol was administered together with docosahexaenoic acid to hemodialysis patients. Plasma IL-6, white blood cell (WBC), and neutrophil counts decreased. But CRP, F2 isoprostanes, and protein carbonyls remained unchanged. Moreover, hemodialysis per se did not influence plasma levels of protein carbonyls, CRP, IL-6, and IL-10 [111]. Similarly, in patients with advanced diabetic nephropathy, the hypoglycemic agent and peroxisome proliferator-activated receptors (PPAR) agonist pioglitazone improved plasma inflammatory markers, such as TNFα, but not protein carbonyls, suggesting that inflammation and oxidation pathways have independent effects [112].

We conclude that in patients with renal disease, the balance of oxidant/antioxidant systems is highly complex. Both chemical pathways and inflammatory components appear to be involved in the formation of glycoxidation and protein-bound lipid oxidation products [110].

Transplantation

Rapid decreases in PC and F2 isoprostanes and levels of pro-inflammatory proteins IL-6, TNFα, and CRP were observed upon renal transplantation in ESRD patients [113]. Yet surprisingly, in the 24-h period following renal transplantation, plasma protein carbonyls and lipoxidation markers increased due to ischemic stress. At 7 days' posttransplant, lipoxidation markers decreased while the protein carbonyls further increased [114]. Nevertheless, most studies show that renal transplantation is the most efficacious long-term intervention for decreasing AOPPs, AGEs, and proinflammatory markers, especially with normalization of renal function [115–118].

4 Metal Catalyzed Oxidant (MCO) Stress in End Stage Renal Disease and the Impact of Iron Supplementation

What is the source of increased oxidative activity in uremia? As reviewed above, current evidence points to the retention of oxidative compounds as renal function deteriorates. Once oxidative damage is extensive, cause and effect become much more difficult to distinguish from each other. The systemic response to oxidative damage results in accelerated atherosclerosis, and the resultant chronic inflammation leads to malnutrition [22, 23]. When malnutrition leads to low albumin levels, a critical defense against oxidative stress is lost [73].

To complicate the picture, dialysis activates production of reactive oxidants by inflammatory cells directly [119], and it leads to peroxidation of lipids [120]. Capeillere-Blandin et al. investigated whether their original data on AOPPs/protein carbonyl formation could be explained by myeloperoxidase [121]. Incubation of albumin with this enzyme in the presence of H_2O_2/Cl^- lead to chlorination of the protein, suggesting this oxidizing system could explain the data in ESRD patients. Myeloperoxidase, a hemoprotein stored in the azurophilic granules of neutrophils and macrophages, functions to catalyze the conversion of chloride and hydrogen peroxide to hypochlorite. Clinical studies have demonstrated the predictive value of directly measured myeloperoxidase for the presence of coronary artery disease in the general population [122].

Activation of neutrophils to release pro-oxidant molecules is an important concern [123]. However, it is generally accepted that the oxidative burst is transient in dialysis using biocompatible dialysis membranes [124]. Therefore, the net effect of dialysis is to reduce the production of oxidants [103] and to improve reversible biomarkers of oxidation (e.g., high-flux dialysis regenerates free sulfhydryl groups on albumin) [25, 103, 125]. These observations suggest that in the clinical setting dialyzable compounds with oxidative activity are the most important cause of oxidative stress in renal failure.

Oxoaldehydes like methylglyoxal are five- to tenfold elevated in ESRD and can oxidize lysine by the Suyama pathway [126]. This pathway involves complex formation between Fe3+ or Cu2+ and a methylglyoxal-lysine adduct leading to allysine formation. Moreover, the same damage results from the traditional Stadtman pathway of MCO. Previously, in patients with ESRD, we elucidated the mechanisms leading to accelerated formation of two distinct structurally identified AGEs: the glycoxidation products N^{e}(carboxymethyl)lysine (CML) and pentosidine [37]. Though formation of these two modifications proceeds by different pathways, it is markedly enhanced by metal catalyzed oxidation reactions. Conversely, proteins with high AGE content readily bind redox active metal in a dose-dependent fashion [51, 127, 128], taking up three to four times as much iron or copper as in the unglycated state [127]. In the healthy state, metal ions are tightly sequestered in the functional core of hemoglobin and enzymes, or stored and transported on specific binding proteins. In addition, a biologic defense exists against their action as hydroxyl radical generators through the Fenton reaction [129]. In contrast to physiologic iron-binding proteins like transferrin, "glycochelates" are highly oxidative, leading to a vicious circle of increased AGE formation via oxidative pathways [127, 130, 131].

Therapeutic infusion of iron plays an important role in the management of anemia in patients with ESRD. During and immediately after infusion, temporary oversaturation of transferrin is seen, resulting in transient release of redox-active labile iron [132–135]. Increased levels of free iron induce oxidative stress [136], causing oxidation of circulating and cellular lipids [137] and proteins [138], with an increase in plasma MDA, a decrease in nonprotein SH groups [139], and increased albumin protein carbonyl formation and sulfhydryl oxidation [140].

Tovbin et al. similarly reported that intravenous (IV) iron saccharate induced a 37% increase in AOPP level, which correlated positively with prehemodialysis CRP level and was TNFα related [141]. Intravenous iron administration did not affect tetraethylammonium chloride (TEAC), thiol, dityrosine, or TNFα levels. These authors concluded that intravenous iron administration induces an increase in protein oxidation (AOPP levels) that was related to the baseline degree of inflammation. At present the long-term dangers of iron infusion and the safest formulations, dosages, and strategies to minimize iron toxicity remain undefined in ESRD [142–144].

These observations raise the question of how to distinguish between the oxidative toxicity of metal catalyzed and myeloperoxidase pathways of oxidation. For example, in one study, peroxide concentrations significantly increased during HD but were not correlated to IV iron administration and seemed to result from other sources of oxidative stress related to HD [145]. Cavdar et al. [146] observed that IV administered iron in 20 and 100 mg doses did not cause additional deteriorating effect on oxidant stress (plasma MDA) and EDEF (erythrocyte deformability) and was even improved by IV iron. Finally, Driss et al. conducted a study on the effects of intravenous polymaltose iron on oxidant stress and nontransferrin-bound iron (NTBI) in hemodialysis patients [147]. The serum iron and transferrin saturation index increased during iron infusion and NTBI transiently appeared in some patients, but markers of oxidative stress were not significantly affected. They concluded that although ESRD patients have a high prevalence of NTBI in their serum, no correlation could be established between the presence of NTBI and an increased oxidative stress. The slow infusion of maltofer did not promote a significant increase in the plasma concentration of oxidative stress markers.

The apparent quandary may be best explained based on differences in the concentration of infused iron. Michelis et al. determined carbonyls on fibrinogen in hemodialysis patients on maintenance iron [138]. During a dialysis session, carbonyls further increased when 125 mg but not when 62.5 mg or no iron gluconate was administered. The changes in fibrinogen-PC during dialysis showed a significant linear correlation with the calculated values of transferrin saturation and free transferrin.

5 Conclusions and Therapeutic Implications

The first and foremost impression emanating from this review is that plasma protein oxidation remains low if not negligible as long as renal function is not impaired, whether acutely or chronically. There may be some exceptions for diabetes and septicemia and any other massive ubiquitous systemic stress such as SLE and amyloidosis. However, even in these conditions minor impairment of renal function contributes to protein oxidation. It is also well recognized that chronic renal failure and end stage renal disease are associated with massive combined oxidant and carbonyl stress and various specific AGEs, and oxidation products have been shown to be markedly increased in ESRD (see Table 1).

The more difficult question raised by this review is the relative importance of two disparate chemical mechanisms. To what extent is the formation of AOPPs related to MPO/HOCl, and to what extent is the formation of protein carbonyls directly related to metal catalyzed oxidation or to MPO? There is not only a need to confirm the molecular nature of AOPP by mass spectrometric methods, but also to determine the relative roles of myeloperoxidase vs. redox active metals in the formation of PCs. A strong though circumstantial evidence for MCO is the fact that ESRD serum and dialysis fluid specimens easily glycoxidize in vitro [37]. However, myeloperoxidase activity is increased in all renal failure patients, and there is more activation and "priming" of leukocytes in patients treated by hemodialysis [148]. In addition, there is an increase in nicotinamide adenine dinucleotide phosphate (NADPH) oxidase dependent superoxide formation [149].

Some clues might come from the administration of vitamin E. Alpha-tocopherol is a known inhibitor of lipid peroxidation. However, there is a paradoxical effect in that both catalytic metals and lipid peroxides can enhance α-tocopherol oxidation by a similar mechanism, possibly explaining the inability of α-tocopherol to prevent secondary modification of serum proteins by iso[4]-levuglandin E2, 4-hydoxynonenal (HNE), and 4-oxo-2-nonenal (ONE) [110]. As for MPO α-tocopherol was also ineffective against MPO-mediated oxidation and peroxynitrite. Kamgar et al. [150] treated HD patients with a cocktail of vitamins including vitamin E and observed no decrease in PCs. However, since they included 250 mg vitamin C, a classical substrate for MCO, they might have induced rather than suppressed MCO. Nevertheless, the results were similarly negative with γ-tocopherol [151]. In view of the fact that tocopherol is primarily an antioxidant in the lipid phase, while MCO occurs primarily in the aqueous, the most efficient way to differentiate between MPO and MCO will be to carry out interventions with chelating agents, such as desferrioxamine or iron. Although desferrioxamine has been used to treat aluminum overload in ESRD, we did not come across studies on the impact of chelators on PCs in ESRD. Such studies are urgently needed.

Thus far, currently the most efficient therapy to decrease oxidant stress, besides renal transplantation, appears to be the utilization of high flux dialysis membranes. The fact that they decrease levels of methylglyoxal [152], a cofactor for MCO via the Suyama pathway, provides support for this mechanism of protein damage in ESRD. However, current dialysis modalities are altogether still insufficient for the elimination of the AGE, PC, and AOPP burden that is in part responsible for the complications of uremia.

Acknowledgments This work was supported by grants from NIA (AG18436) to VMM, NIDDK/NIH DK57733 and DK45619 to MAF, from the National Institute of Environmental Health Sciences (MAF) grant ES11461, and from the Leonard B. Rosenberg Renal Research Foundation of the Center for dialysis Care, Cleveland Ohio.

References

1. Davies MJ. The oxidative environment and protein damage. Biochim Biophys Acta 2005;1703:93–109.
2. Rees MD, Kennett EC, Whitelock JM, Davies MJ. Oxidative damage to extracellular matrix and its role in human pathologies. Free Radic Biol Med 2008;44:1973–2001.
3. Hazen SL, d'Avignon A, Anderson MM, Hsu FF, Heinecke JW. Human neutrophils employ the myeloperoxidase-hydrogen peroxide-chloride system to oxidize alpha-amino acids to a family of reactive aldehydes. Mechanistic studies identifying labile intermediates along the reaction pathway. J Biol Chem 1998;273:4997–5005.
4. Winterbourn CC, Kettle AJ. Reactions of superoxide with myeloperoxidase and its products. Jpn J Infect Dis 2004;57:S31–3.

5. Stadtman ER. Oxidation of free amino acids and amino acid residues in proteins by radiolysis and by metal-catalyzed reactions. Annu Rev Biochem 1993;62:797–821.
6. Hawkins CL, Pattison DI, Davies MJ. Hypochlorite-induced oxidation of amino acids, peptides and proteins. Amino Acids 2003;25:259–74.
7. Akagawa M, Sasaki T, Suyama K. Oxidative deamination of lysine residue in plasma protein of diabetic rats. Novel mechanism via the Maillard reaction. Eur J Biochem 2002;269:5451–8.
8. Kagan HM, Williams MA, Williamson PR, Anderson JM. Influence of sequence and charge on the specificity of lysyl oxidase toward protein and synthetic peptide substrates. J Biol Chem 1984;259:11203–7.
9. Requena JR, Chao CC, Levine RL, Stadtman ER. Glutamic and aminoadipic semialdehydes are the main carbonyl products of metal-catalyzed oxidation of proteins. Proc Natl Acad Sci U S A 2001;98:69–74.
10. Sell DR, Strauch CM, Shen W, Monnier VM. 2-aminoadipic acid is a marker of protein carbonyl oxidation in the aging human skin: effects of diabetes, renal failure and sepsis. Biochem J 2007;404:269–77.
11. Radi R, Peluffo G, Alvarez MN, Naviliat M, Cayota A. Unraveling peroxynitrite formation in biological systems. Free Radic Biol Med 2001;30:463–88.
12. Pryor WA, Squadrito GL. The chemistry of peroxynitrite: a product from the reaction of nitric oxide with superoxide. Am J Physiol 1995;268:L699–722.
13. Rabbani N, Thornalley PJ. Assay of 3-nitrotyrosine in tissues and body fluids by liquid chromatography with tandem mass spectrometric detection. Meth Enzymol 2008;440:337–59.
14. Martinez-Ruiz A, Lamas S. Nitrosylation of thiols in vascular homeostasis and disease. Curr Atheroscler Rep 2005;7:213–8.
15. Witko-Sarsat V, Friedlander M, Capeillere-Blandin C et al. Advanced oxidation protein products as a novel marker of oxidative stress in uremia. Kidney Int 1996;49:1304–13.
16. Fan X, Zhang J, Theves M et al. Mechanism of lysine oxidation in human lens crystallins during aging and in diabetes. J Biol Chem 2009;284:34618–27.
17. Levine RL, Wehr N, Williams JA, Stadtman ER, Shacter E. Determination of carbonyl groups in oxidized proteins. Methods Mol Biol 2000;99:15–24.
18. Vanholder R, De Smet R, Glorieux G et al. Review on uremic toxins: classification, concentration, and interindividual variability. Kidney Int 2003;63:1934–43.
19. Foley RN, Parfrey PS, Sarnak MJ. Clinical epidemiology of cardiovascular disease in chronic renal disease. Am J Kidney Dis 1998;32(Suppl 3):S112–9.
20. Cheung AK, Sarnak MJ, Yan G et al. Atherosclerotic cardiovascular disease risks in chronic hemodialysis patients. Kidney Int 2000;58:353–62.
21. Stenvinkel P. Malnutrition and chronic inflammation as risk factors for cardiovascular disease in chronic renal failure. Blood Purif 2001;19:143–51.
22. Himmelfarb J, Hakim RM. Oxidative stress in uremia. Curr Opin Nephrol Hypertens 2003;12:593–8.
23. Himmelfarb J, Stenvinkel P, Ikizler TA, Hakim RM. The elephant in uremia: oxidant stress as a unifying concept of cardiovascular disease in uremia. Kidney Int 2002;62:1524–38.
24. Roselaar SE, Nahat NB, Winyard PG, Jones P, Cunningham J, Blake DR. Detection of oxidants in uremic plasma by electron spin resonances spectroscopy. Kidney Int 1995;48:199–206.
25. Himmelfarb J, McMonagle E, McMenamin E. Plasma protein thiol oxidation and carbonyl formation in chronic renal failure. Kidney Int 2000;58:2571–8.
26. Himmelfarb J, McMenamin ME, Loseto G, Heinecke JW. Myeloperoxidase-catalyzed 3-chlorotyrosine formation in dialysis patients. Free Radic Biol Med 2001;31:1163–9.
27. Mera K, Anraku M, Kitamura K et al. Oxidation and carboxy methyl lysine-modification of albumin: possible involvement in the progression of oxidative stress in hemodialysis patients. Hypertens Res 2005;28:973–80.

28. Kalogerakis G, Baker AM, Christov S et al. Oxidative stress and high-density lipoprotein function in Type I diabetes and end-stage renal disease. Clin Sci (Lond) 2005;108:497–506.
29. Alhamdani MS, Al-Kassir AH, Jaleel NA, Hmood AM, Ali HM. Elevated levels of alkanals, alkenals and 4-HO-alkenals in plasma of hemodialysis patients. Am J Nephrol 2006;26:299–303.
30. Soejima A, Matsuzawa N, Miyake N et al. Hypoalbuminemia accelerates erythrocyte membrane lipid peroxidation in chronic hemodialysis patients. Clin Nephrol 1999;51:92–7.
31. Vlassara H. Serum advanced glycosylation end products: a new class of uremic toxins? Blood Purif 1994;12:54–9.
32. Schwenger V, Zeier M, Henle T, Ritz E. Advanced glycation endproducts (AGEs) as uremic toxins. Nahrung 2001;45:172–6.
33. Ponka A, Kuhlback B. Serum ascorbic acid in patients undergoing chronic hemodialysis. Acta Med Scand 1983;213:305–7.
34. Hultqvist M, Hegbrant J, Nilsson-Thorell C et al. Plasma concentrations of vitamin C, vitamin E and/or malondialdehyde as markers of oxygen free radical production during hemodialysis. Clin Nephrol 1997;47:37–46.
35. Hirano H, Tone Y, Otani H et al. Levels of serum ascorbate and its metabolites in hemodialysis patients. Nippon Jinzo Gakkai Shi 2004;46:426–33.
36. Miyata T, Kurokawa K, van Ypersele de Strihou C. Carbonyl stress and long-term uremic complications. Adv Nephrol Necker Hosp 1998;28:311–9.
37. Weiss MF, Erhard P, Kader-Attia FA et al. Mechanisms for the formation of glycoxidation products in end-stage renal disease. Kidney Int 2000;57:2571–85.
38. Thornalley PJ. Advanced glycation end products in renal failure. J Ren Nutr 2006;16:178–84.
39. Miyata T, Horie K, Ueda Y et al. Advanced glycation and lipidoxidation of the peritoneal membrane: respective roles of serum and peritoneal fluid reactive carbonyl compounds. Kidney Int 2000;58:425–35.
40. Maggi E, Bellazzi R, Falaschi F et al. Enhanced LDL oxidation in uremic patients: an additional mechanism for accelerated atherosclerosis? Kidney Int 1994;45:876–83.
41. Witko-Sarsat V, Friedlander M, Nguyen Khoa T et al. Advanced oxidation protein products as novel mediators of inflammation and monocyte activation in chronic renal failure. J Immunol 1998;161:2524–32.
42. Monnier VM, Sell DR, Nagaraj RH et al. Maillard reaction-mediated molecular damage to extracellular matrix and other tissue proteins in diabetes, aging, and uremia. Diabetes 1992;41(Suppl 2):36–41.
43. Beisswenger PJ, Healy JC, Shultz EK. Glycosylated serum proteins and glycosylated hemoglobin in the assessment of glycemic control in insulin-dependent and non-insulin-dependent diabetes mellitus. Metabolism 1993;42:989–92.
44. Makita Z, Yanagisawa K, Kuwajima S et al. Advanced glycation endproducts and diabetic nephropathy. J Diabetes Complicat 1995;9:265–8.
45. Gugliucci A, Bendayan M. Renal fate of circulating advanced glycated end products (AGE): evidence for reabsorption and catabolism of AGE-peptides by renal proximal tubular cells. Diabetologia 1996;39:149–60.
46. Londono I, Bendayan M. Glomerular handling of native albumin in the presence of circulating modified albumins by the normal rat kidney. Am J Physiol Renal Physiol 2005;289: F1201–9.
47. Ozdemir AM, Hopfer U, Rosca MV, Fan XJ, Monnier VM, Weiss MF. Effects of advanced glycation end product modification on proximal tubule epithelial cell processing of albumin. Am J Nephrol 2008;28:14–24.
48. Ceballos-Picot I, Witko-Sarsat V, Merad-Boudia M et al. Glutathione antioxidant system as a marker of oxidative stress in chronic renal failure. Free Radic Biol Med 1996;21:845–53.

49. Oberg BP, McMenamin E, Lucas FL et al. Increased prevalence of oxidant stress and inflammation in patients with moderate to severe chronic kidney disease. Kidney Int 2004;65:1009–16.
50. Himmelfarb J, McMonagle E, Freedman S et al. Oxidative stress is increased in critically ill patients with acute renal failure. J Am Soc Nephrol 2004;15:2449–56.
51. Saxena AK, Saxena P, Wu X, Obrenovich M, Weiss MF, Monnier VM. Protein aging by carboxymethylation of lysines generates sites for divalent metal and redox active copper binding: relevance to diseases of glycoxidative stress. Biochem Biophys Res Commun 1999;260:332–8.
52. Schmidt AM, Hofmann M, Taguchi A, Yan SD, Stern DM. RAGE: a multiligand receptor contributing to the cellular response in diabetic vasculopathy and inflammation. Semin Thromb Hemost 2000;26:485–93.
53. Peppa M, Uribarri J, Cai W, Lu M, Vlassara H. Glycoxidation and inflammation in renal failure patients. Am J Kidney Dis 2004;43:690–5.
54. Rosen H, Crowley JR, Heinecke JW. Human neutrophils use the myeloperoxidase-hydrogen peroxide-chloride system to chlorinate but not nitrate bacterial proteins during phagocytosis. J Biol Chem 2002;277:30463–8.
55. Pan HZ, Zhang L, Guo MY et al. The oxidative stress status in diabetes mellitus and diabetic nephropathy. Acta Diabetol 2009 [Epub ahead of print].
56. Morgan PE, Sturgess AD, Hennessy A, Davies MJ. Serum protein oxidation and apolipoprotein CIII levels in people with systemic lupus erythematosus with and without nephritis. Free Radic Res 2007;41:1301–12.
57. Rysava R, Kalousova M, Zima T, Dostal C, Merta M, Tesar V. Does renal function influence plasma levels of advanced glycation and oxidation protein products in patients with chronic rheumatic diseases complicated by secondary amyloidosis? Kidney Blood Press Res 2007;30:1–7.
58. Agarwal R. Chronic kidney disease is associated with oxidative stress independent of hypertension. Clin Nephrol 2004;61:377–83.
59. Sarnak MJ, Coronado BE, Greene T et al. Cardiovascular disease risk factors in chronic renal insufficiency. Clin Nephrol 2002;57:327–35.
60. Herzog CA, Ma JZ, Collins AJ. Poor long-term survival after acute myocardial infarction among patients on long-term dialysis. N Engl J Med 1998;339:799–805.
61. Libby P. Vascular biology of atherosclerosis: overview and state of the art. Am J Cardiol 2003;91:3A–6.
62. Steinberg JJ, Cajigas A, Brownlee M. Enzymatic shot-gun 5′-phosphorylation and 3′-sister phosphate exchange: a two-dimensional thin-layer chromatographic technique to measure DNA deoxynucleotide modification. J Chromatogr 1992;574:41–55.
63. Kimmel PL, Phillips TM, Simmens SJ et al. Immunologic function and survival in hemodialysis patients. Kidney Int 1998;54:236–44.
64. Hazen SL, Hsu FF, d'Avignon A, Heinecke JW. Human neutrophils employ myeloperoxidase to convert alpha-amino acids to a battery of reactive aldehydes: a pathway for aldehyde generation at sites of inflammation. Biochemistry 1998;37:6864–73.
65. deFilippi C, Wasserman S, Rosanio S et al. Cardiac troponin T and C-reactive protein for predicting prognosis, coronary atherosclerosis, and cardiomyopathy in patients undergoing long-term hemodialysis. JAMA 2003;290:353–9.
66. Sakata N, Imanaga Y, Meng J et al. Increased advanced glycation end products in atherosclerotic lesions of patients with end-stage renal disease. Atherosclerosis 1999;142:67–77.
67. Weiss MF, Scivittaro V, Anderson JM. Oxidative stress and increased expression of growth factors in lesions of failed hemodialysis access. Am J Kidney Dis 2001;37:970–80.
68. Yoshida S, Yamada K, Hamaguchi K et al. Immunohistochemical study of human advanced glycation end-products (AGE) and growth factors in cardiac tissues of patients on maintenance dialysis and with kidney transplantation. Clin Nephrol 1998;49:273–80.

69. Zoccali C, Mallamaci F, Asahia K et al. Pentosidine, carotid atherosclerosis and alterations in left ventricular geometry in hemodialysis patients. J Nephrol 2001;14:293–8.
70. Lim PS, Wu MY, Chien SW et al. Elevated circulating levels of soluble CD-40 ligand in haemodialysis patients with symptomatic coronary heart disease. Nephrology (Carlton) 2008;13:677–83.
71. Dursun B, Dursun E, Suleymanlar G et al. Carotid artery intima-media thickness correlates with oxidative stress in chronic haemodialysis patients with accelerated atherosclerosis. Nephrol Dial Transplant 2008;23:1697–703.
72. Himmelfarb J, McMonagle E. Albumin is the major plasma protein target of oxidant stress in uremia. Kidney Int 2001;60:358–63.
73. Danielski M, Ikizler TA, McMonagle E et al. Linkage of hypoalbuminemia, inflammation, and oxidative stress in patients receiving maintenance hemodialysis therapy. Am J Kidney Dis 2003;42:286–94.
74. Sarnatskaya VV, Ivanov AI, Nikolaev VG et al. Structure and binding properties of serum albumin in uremic patients at different periods of hemodialysis. Artif Organs 1998;22:107–15.
75. Sarnatskaya VV, Lindup WE, Ivanov AI et al. Extraction of uraemic toxins with activated carbon restores the functional properties of albumin. Nephron Physiol 2003;95:10–18.
76. Oettl K, Stauber RE. Physiological and pathological changes in the redox state of human serum albumin critically influence its binding properties. Br J Pharmacol 2007;151:580–90.
77. Phelan PJ, O'Kelly P, Walshe JJ, Conlon PJ. The importance of serum albumin and phosphorous as predictors of mortality in ESRD patients. Ren Fail 2008;30:423–9.
78. Mayer B, Zitta S, Greilberger J et al. Effect of hemodialysis on the antioxidative properties of serum. Biochim Biophys Acta 2003;1638:267–72.
79. Lyons TJ. Glycation, carbonyl stress, EAGLEs, and the vascular complications of diabetes. Semin Vasc Med 2002;2:175–89.
80. Baynes JW, Thorpe SR. Glycoxidation and lipoxidation in atherogenesis. Free Radic Biol Med 2000;28:1708–16.
81. Lindner A, Charra B, Sherrard DJ, Scribner BH. Accelerated atherosclerosis in prolonged maintenance hemodialysis. N Engl J Med 1974;290:697–701.
82. Vanholder R, Argiles A, Baurmeister U et al. Uremic toxicity: present state of the art. Int J Artif Organs 2001;24:695–725.
83. Miyata T, Inagi R, Kurokawa K. Diagnosis, pathogenesis, and treatment of dialysis-related amyloidosis. Miner Electrolyte Metab 1999;25:114–7.
84. Cheung AK, Agodoa LY, Daugirdas JT et al. Effects of hemodialyzer reuse on clearances of urea and beta2-microglobulin. The Hemodialysis (HEMO) Study Group. J Am Soc Nephrol 1999;10:117–27.
85. Leypoldt JK, Cheung AK, Deeter RB. Single compartment models for evaluating beta 2-microglobulin clearance during hemodialysis. ASAIO J 1997;43:904–9.
86. Leypoldt JK, Cheung AK, Deeter RB. Effect of hemodialyzer reuse: dissociation between clearances of small and large solutes. Am J Kidney Dis 1998;32:295–301.
87. Miyata T, Iida Y, Ueda Y et al. Monocyte/macrophage response to beta 2-microglobulin modified with advanced glycation end products. Kidney Int 1996;49:538–50.
88. Henle T, Deppisch R, Beck W, Hergesell O, Hansch GM, Ritz E. Advanced glycated end-products (AGE) during haemodialysis treatment: discrepant results with different methodologies reflecting the heterogeneity of AGE compounds. Nephrol Dial Transplant 1999;14:1968–75.
89. Miyata T, Maeda K, Kurokawa K, van Ypersele de Strihou C. Oxidation conspires with glycation to generate noxious advanced glycation end products in renal failure. Nephrol Dial Transplant 1997;12:255–8.
90. Stein G, Franke S, Mahiout A et al. Influence of dialysis modalities on serum AGE levels in end-stage renal disease patients. Nephrol Dial Transplant 2001;16:999–1008.

91. Galli F, Benedetti S, Floridi A et al. Glycoxidation and inflammatory markers in patients on treatment with PMMA-based protein-leaking dialyzers. Kidney Int 2005;67:750–9.
92. Ahmed N, Thornalley PJ. Quantitative screening of protein biomarkers of early glycation, advanced glycation, oxidation and nitrosation in cellular and extracellular proteins by tandem mass spectrometry multiple reaction monitoring. Biochem Soc Trans 2003;31:1417–22.
93. Valencia JV, Weldon SC, Quinn D et al. Advanced glycation end product ligands for the receptor for advanced glycation end products: biochemical characterization and formation kinetics. Anal Biochem 2004;324:68–78.
94. Faist V, Erbersdobler HF. Metabolic transit and in vivo effects of melanoidins and precursor compounds deriving from the Maillard reaction. Ann Nutr Metab 2001;45:1–12.
95. Uribarri J, Peppa M, Cai W et al. Dietary glycotoxins correlate with circulating advanced glycation end product levels in renal failure patients. Am J Kidney Dis 2003;42:532–8.
96. Suliman ME, Heimburger O, Barany P et al. Plasma pentosidine is associated with inflammation and malnutrition in end-stage renal disease patients starting on dialysis therapy. J Am Soc Nephrol 2003;14:1614–22.
97. Schwedler SB, Metzger T, Schinzel R, Wanner C. Advanced glycation end products and mortality in hemodialysis patients. Kidney Int 2002;62:301–10.
98. Maillard LC. Reaction generale des acides amines sur les sucres: ses consequences biologiques. Presse Med 1912;71:546.
99. Koschinsky T, He CJ, Mitsuhashi T et al. Orally absorbed reactive glycation products (glycotoxins): an environmental risk factor in diabetic nephropathy. Proc Natl Acad Sci U S A 1997;94:6474–9.
100. Miyata T, Fu MX, Kurokawa K, van Ypersele de Strihou C, Thorpe SR, Baynes JW. Autoxidation products of both carbohydrates and lipids are increased in uremic plasma: is there oxidative stress in uremia? Kidney Int 1998;54:1290–5.
101. Raj DS, Dominic EA, Pai A et al. Skeletal muscle, cytokines, and oxidative stress in end-stage renal disease. Kidney Int 2005;68:2338–44.
102. Dursun E, Dursun B, Suleymanlar G, Ozben T. Carbonyl stress in chronic renal failure: the effect of haemodialysis. Ann Clin Biochem 2005;42:64–6.
103. Ward RA, Ouseph R, McLeish KR. Effects of high-flux hemodialysis on oxidant stress. Kidney Int 2003;63:353–9.
104. Tepel M, van der Giet M, Statz M, Jankowski J, Zidek W. The antioxidant acetylcysteine reduces cardiovascular events in patients with end-stage renal failure: a randomized, controlled trial. Circulation 2003;107:992–5.
105. Boaz M, Smetana S, Weinstein T et al. Secondary prevention with antioxidants of cardiovascular disease in endstage renal disease (SPACE): randomised placebo-controlled trial. Lancet 2000;356:1213–8.
106. [No autors listed]. Dietary supplementation with n-3 polyunsaturated fatty acids and vitamin E after myocardial infarction: results of the GISSI-Prevenzione trial. Gruppo Italiano per lo Studio della Sopravvivenza nell'Infarto miocardico. Lancet 1999;354:447–55.
107. Lonn E, Bosch J, Yusuf S et al. Effects of long-term vitamin E supplementation on cardiovascular events and cancer: a randomized controlled trial. JAMA 2005;293:1338–47.
108. Halliwell B. The antioxidant paradox. Lancet 2000;355:1179–80.
109. Domenici FA, Vannucchi MT, Jordao AA, Jr., Meirelles MS, Vannucchi H. DNA oxidative damage in patients with dialysis treatment. Ren Fail 2005;27:689–94.
110. Lu L, Erhard P, Salomon RG, Weiss MF. Serum vitamin E and oxidative protein modification in hemodialysis: a randomized clinical trial. Am J Kidney Dis 2007;50:305–13.
111. Pupim LB, Himmelfarb J, McMonagle E, Shyr Y, Ikizler TA. Influence of initiation of maintenance hemodialysis on biomarkers of inflammation and oxidative stress. Kidney Int 2004;65:2371–9.
112. Agarwal R. Anti-inflammatory effects of short-term pioglitazone therapy in men with advanced diabetic nephropathy. Am J Physiol Renal Physiol 2006;290:F600–5.

113. Simmons EM, Langone A, Sezer MT et al. Effect of renal transplantation on biomarkers of inflammation and oxidative stress in end-stage renal disease patients. Transplantation 2005;79:914–9.
114. Przekwas M, Malgorzewicz S, Zdrojewski Z, Debska-Slizien A, Lysiak-Szydlowska W, Rutkowski B. Influence of predialysis oxidative stress on peroxidation processes after renal transplantation. Transplant Proc 2003;35:2170–3.
115. Hricik DE, Schulak JA, Sell DR, Fogarty JF, Monnier VM. Effects of kidney or kidney-pancreas transplantation on plasma pentosidine. Kidney Int 1993;43:398–403.
116. Hricik DE, Wu YC, Schulak A, Friedlander MA. Disparate changes in plasma and tissue pentosidine levels after kidney and kidney-pancreas transplantation. Clin Transplant 1996;10:568–73.
117. Antolini F, Valente F, Ricciardi D, Baroni M, Fagugli RM. Principal component analysis of some oxidative stress parameters and their relationships in hemodialytic and transplanted patients. Clin Chim Acta 2005;358:87–94.
118. Antolini F, Valente F, Ricciardi D, Fagugli RM. Normalization of oxidative stress parameters after kidney transplant is secondary to full recovery of renal function. Clin Nephrol 2004;62:131–7.
119. Descamps-Latscha B, Goldfarb B, Nguyen AT et al. Establishing the relationship between complement activation and stimulation of phagocyte oxidative metabolism in hemodialyzed patients: a randomized prospective study. Nephron 1991;59:279–85.
120. Himmelfarb J, Ault KA, Holbrook D, Leeber DA, Hakim RM. Intradialytic granulocyte reactive oxygen species production: a prospective, crossover trial. J Am Soc Nephrol 1993;4:178–86.
121. Capeillere-Blandin C, Gausson V, Descamps-Latscha B, Witko-Sarsat V. Biochemical and spectrophotometric significance of advanced oxidized protein products. Biochim Biophys Acta 2004;1689:91–102.
122. Zhang R, Brennan ML, Fu X et al. Association between myeloperoxidase levels and risk of coronary artery disease. JAMA 2001;286:2136–42.
123. Coskun C, Kural A, Doventas Y et al. Hemodialysis and protein oxidation products. Ann N Y Acad Sci 2007;1100:404–8.
124. Ward RA. Phagocytic cell function as an index of biocompatibility. Nephrol Dial Transplant 1994;9(Suppl 2):46–56.
125. Himmelfarb J, McMenamin E, McMonagle E. Plasma aminothiol oxidation in chronic hemodialysis patients. Kidney Int 2002;61:705–16.
126. Akagawa M, Sasaki D, Kurota Y, Suyama K. Formation of alpha-aminoadipic and gamma-glutamic semialdehydes in proteins by the maillard reaction. Ann N Y Acad Sci 2005;1043:129–34.
127. Qian M, Liu M, Eaton JW. Transition metals bind to glycated proteins forming redox active "glycochelates": implications for the pathogenesis of certain diabetic complications. Biochem Biophys Res Commun 1998;250:385–9.
128. Seifert ST, Krause R, Gloe K, Henle T. Metal complexation by the peptide-bound Maillard reaction products N(epsilon)-fructoselysine and N(epsilon)-carboxymethyllysine. J Agric Food Chem 2004;52:2347–50.
129. Halliwell B, Gutteridge JM. Role of free radicals and catalytic metal ions in human disease: an overview. Meth Enzymol 1990;186:1–85.
130. Yim MB, Yim HS, Lee C, Kang SO, Chock PB. Protein glycation: creation of catalytic sites for free radical generation. Ann N Y Acad Sci 2001;928:48–53.
131. Sell DR, Strauch CM, Shen W, Monnier VM. Aging, diabetes, and renal failure catalyze the oxidation of lysyl residues to 2-aminoadipic acid in human skin collagen: evidence for metal-catalyzed oxidation mediated by alpha-dicarbonyls. Ann N Y Acad Sci 2008;1126:205–9.
132. Zanen AL, Adriaansen HJ, van Bommel EF, Posthuma R, Th de Jong GM. "Oversaturation" of transferrin after intravenous ferric gluconate (Ferrlecit(R)) in haemodialysis patients. Nephrol Dial Transplant 1996;11:820–4.

133. Esposito BP, Breuer W, Slotki I, Cabantchik ZI. Labile iron in parenteral iron formulations and its potential for generating plasma nontransferrin-bound iron in dialysis patients. Eur J Clin Invest 2002;32(Suppl 1):42–9.
134. Agarwal R. Transferrin saturation with intravenous irons: an in vitro study. Kidney Int 2004;66:1139–44.
135. Kooistra MP, Kersting S, Gosriwatana I et al. Nontransferrin-bound iron in the plasma of haemodialysis patients after intravenous iron saccharate infusion. Eur J Clin Invest 2002;32 (Suppl 1):36–41.
136. Rooyakkers TM, Stroes ES, Kooistra MP et al. Ferric saccharate induces oxygen radical stress and endothelial dysfunction in vivo. Eur J Clin Invest 2002;32(Suppl 1):9–16.
137. Roob JM, Khoschsorur G, Tiran A, Horina JH, Holzer H, Winklhofer-Roob BM. Vitamin E attenuates oxidative stress induced by intravenous iron in patients on hemodialysis. J Am Soc Nephrol 2000;11:539–49.
138. Michelis R, Gery R, Sela S et al. Carbonyl stress induced by intravenous iron during haemodialysis. Nephrol Dial Transplant 2003;18:924–30.
139. Mimic-Oka J, Savic-Radojevic A, Pljesa-Ercegovac M et al. Evaluation of oxidative stress after repeated intravenous iron supplementation. Ren Fail 2005;27:345–51.
140. Anraku M, Kitamura K, Shinohara A et al. Intravenous iron administration induces oxidation of serum albumin in hemodialysis patients. Kidney Int 2004;66:841–8.
141. Tovbin D, Mazor D, Vorobiov M, Chaimovitz C, Meyerstein N. Induction of protein oxidation by intravenous iron in hemodialysis patients: role of inflammation. Am J Kidney Dis 2002;40:1005–12.
142. Bishu K, Agarwal R. Acute injury with intravenous iron and concerns regarding long-term safety. Clin J Am Soc Nephrol 2006;1(Suppl 1):S19–23.
143. Horl WH. Clinical aspects of iron use in the anemia of kidney disease. J Am Soc Nephrol 2007;18:382–93.
144. Kletzmayr J, Horl WH. Iron overload and cardiovascular complications in dialysis patients. Nephrol Dial Transplant 2002;17(Suppl 2):25–9.
145. Scheiber-Mojdehkar B, Lutzky B, Schaufler R, Sturm B, Goldenberg H. Non-transferrin-bound iron in the serum of hemodialysis patients who receive ferric saccharate: no correlation to peroxide generation. J Am Soc Nephrol 2004;15:1648–55.
146. Cavdar C, Temiz A, Yenicerioglu Y et al. The effects of intravenous iron treatment on oxidant stress and erythrocyte deformability in hemodialysis patients. Scand J Urol Nephrol 2003;37:77–82.
147. Driss F, Vrtovsnik F, Katsahian S et al. Effects of intravenous polymaltose iron on oxidant stress and non-transferrin-bound iron in hemodialysis patients. Nephron Clin Pract 2005;99: c63–7.
148. Sela S, Shurtz-Swirski R, Cohen-Mazor M et al. Primed peripheral polymorphonuclear leukocyte: a culprit underlying chronic low-grade inflammation and systemic oxidative stress in chronic kidney disease. J Am Soc Nephrol 2005;16:2431–8.
149. Fortuno A, Beloqui O, San Jose G, Moreno MU, Zalba G, Diez J. Increased phagocytic nicotinamide adenine dinucleotide phosphate oxidase-dependent superoxide production in patients with early chronic kidney disease. Kidney Int (Suppl 99) 2005:S71–5.
150. Kamgar M, Zaldivar F, Vaziri ND, Pahl MV. Antioxidant therapy does not ameliorate oxidative stress and inflammation in patients with end-stage renal disease. J Natl Med Assoc 2009;101:336–44.
151. Himmelfarb J, Phinney S, Ikizler TA, Kane J, McMonagle E, Miller G. Gamma-tocopherol and docosahexaenoic acid decrease inflammation in dialysis patients. J Ren Nutr 2007;17:296–304.
152. Agalou S, Ahmed N, Dawnay A, Thornalley PJ. Removal of advanced glycation end products in clinical renal failure by peritoneal dialysis and haemodialysis. Biochem Soc Trans 2003;31:1394–6.

Chapter 3
Cyclooxygenase in the Kidney and Oxidative Stress

Raymond C. Harris

Abstract The kidney is a rich source of prostanoids (prostaglandins and thromboxane). In addition to high levels of expression of the constitutive rate-limiting enzyme responsible for prostanoid production, cyclooxygenase-1 (COX-1), the inducible isoform of cyclooxygenase, COX-2, is also constitutively expressed in the kidney and is highly regulated in response to alterations in intravascular volume. COX metabolites have been shown to exert important physiologic functions in maintenance of renal blood flow, mediation of renin release, and regulation of sodium excretion. In addition to physiologic regulation of COX expression in the kidney, increases in COX expression are also seen in a variety of inflammatory renal injuries, and COX metabolites may serve as mediators of inflammatory injury in renal disease. A common factor in many forms of renal disease is increased production of reactive oxygen species (ROS), and prostanoids have been shown to modulate ROS in various renal diseases. In addition, increased cyclooxygenase activity may itself serve as a source of free radical production.

Keywords COX · Free radical · Prostaglandin · Thromboxane

1 Introduction

In the mammalian kidney, prostaglandins are important mediators of physiologic processes, including modulation of vascular tone and salt and water. Prostaglandins arise from enzymatic metabolism of free arachidonic acid, which is cleaved from membrane phospholipids by phospholipase A_2 activity. The cyclooxygenase enzyme system is the major pathway for metabolism of arachidonic acid in the kidney.

R.C. Harris (✉)
Division of Nephrology, Vanderbilt University, School of Medicine
and Nashville Veterans Affairs Hospital, Nashville, TN 37232, USA
e-mail: rayharris@vanderbilt.edu

T. Miyata et al. (eds.), *Studies on Renal Disorders*, Oxidative Stress in Applied Basic Research and Clinical Practice, DOI 10.1007/978-1-60761-857-7_3,

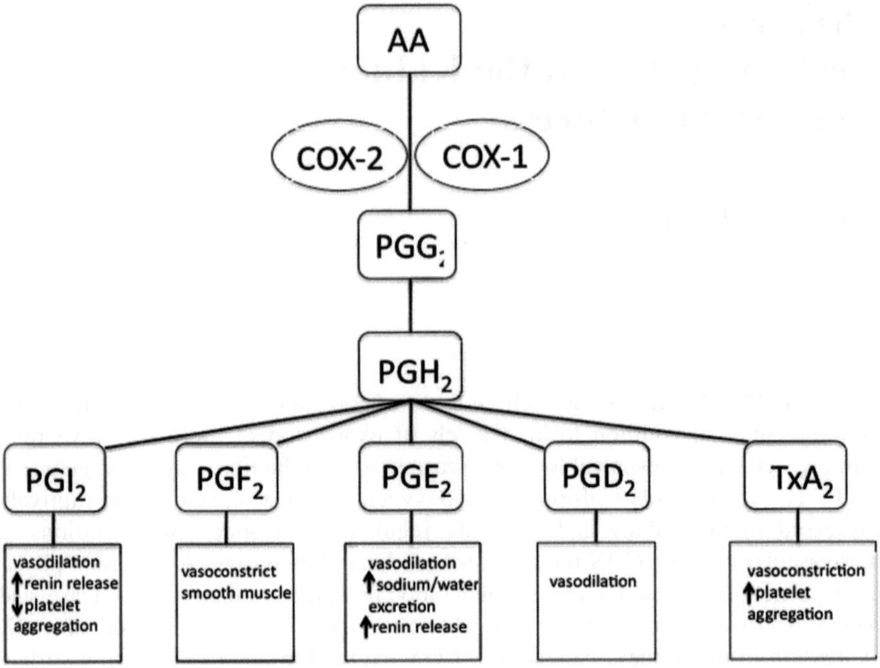

Fig. 1 The arachidonic acid cascade. Either COX-1 or COX-2 convert arachidonic acid (AA) to the common precursor, PGH2, which is then acted upon by specific synthetases to produce the different prostaglandins (PGs) and thromboxane A_2 (TxA_2)

Cyclooxygenases (COX, prostaglandin synthases G_2/H_2) are the enzymes responsible for the initial conversion of arachidonic acid to prostaglandin G_2 and subsequently to prostaglandin H_2, which serves as the precursor for subsequent metabolism by prostaglandin and thromboxane synthases (Fig. 1).

Two COX isoforms exist in mammals: "constitutive" COX-1 and inflammatory-mediated and glucocorticoid-sensitive COX-2. COX-2 expression is highly regulated at the level of transcription, mRNA export from the nucleus, message stability, and translational efficiency [1, 2], thereby controlling the expression of COX-2 in response to many of the same cellular stresses that activate arachidonate release (e.g., cell volume changes, shear stress, hypoxia, oxidative stress), as well as in response to a wide range of pro-inflammatory cytokines and growth factors [3, 4]. In many cases, COX-2 gene transcription is mediated by activation of consensus sequences in the 5' flanking region of the COX-2 gene for nuclear factor κB (NFκB), and nuclear factor interleukin (IL) 6/C-EBP and a cyclic adenosine monophosphate (AMP) response element (CRE) [5]. For example, induction of COX-2 mRNA transcription by endotoxin (lipopolysaccharide) involve CRE sites [6] and NFκB sites [7]. Other pathways for transcriptional regulation of COX-2 have also been described. Suppression of COX-2 expression and prostaglandin synthesis by glucocorticoids represents an additional important mechanism of regulation and appears to be an important mechanism underlying the anti-inflammatory effects of glucocorticoids.

2 Cyclooxygenase Distribution in the Kidney

COX-1 is expressed in normal adult mammalian kidney in vasculature, glomerular mesangial cells, and collecting duct, while COX-2 expression is highest in the macula densa (MD) and adjacent cortical thick ascending limb (cTAL) in the cortex and in lipid-laden medullary interstitial cells in the medulla [8]. Lower levels of expression are also detected under basal conditions in podocytes, mesangial cells, and renal arterioles in some species, including humans [9] (Fig. 2).

In the mammalian kidney, the macula densa is involved in regulating renin release and afferent arteriolar tone (via tubuloglomerular feedback). COX-2 expression in the macula densa increases in response to a low salt diet (Fig. 3), and COX-2 derived prostanoids in the macula densa are mediators of renin expression and release [10–12]. Administration of nonspecific cyclooxygenase inhibitors or selective COX-2 inhibitors prevents the increases in renin release mediated by macula densa sensing of decreases in luminal NaCl. Induction of a high renin state by imposition of a salt deficient diet, ACE inhibition, diuretic administration, or experimental renovascular hypertension all significantly increase macula densa COX-2 expression [13–17].

Macula densa COX-2 expression is modulated by components of the renin-angiotensin system, with both positive and negative feedback mechanisms. In vivo studies indicate that angiotensin II, through AT_1 receptors, inhibits macula densa COX-2 expression. ACE inhibitors or angiotensin receptor blockers (ARBs) increase cortical COX-2 mRNA and immunoreactive protein [14]. In adult wild type mice on a control diet, minimal renal cortical COX-2 immunoreactive protein was detected, while in AT_1R knockout mice (Agtr1a-/-; Agtr1b-/-), abundant COX-2

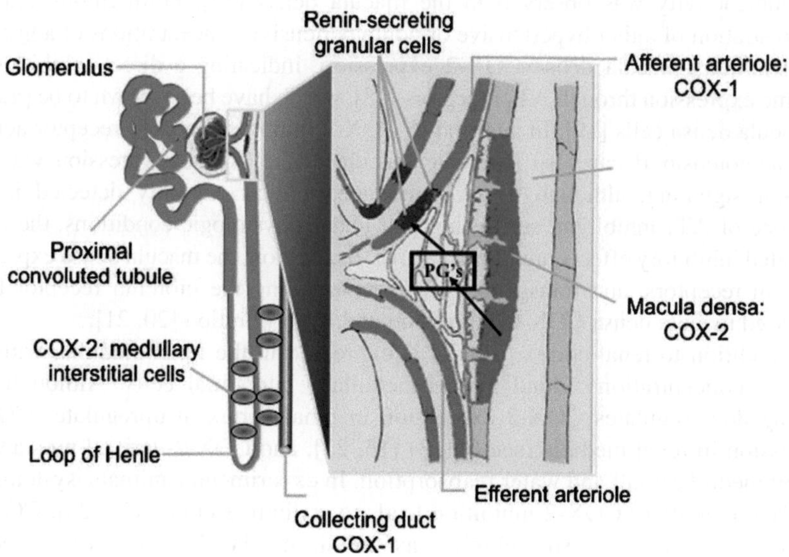

Fig. 2 Localization of expression of cyclooxygenases in the adult mammalian kidney

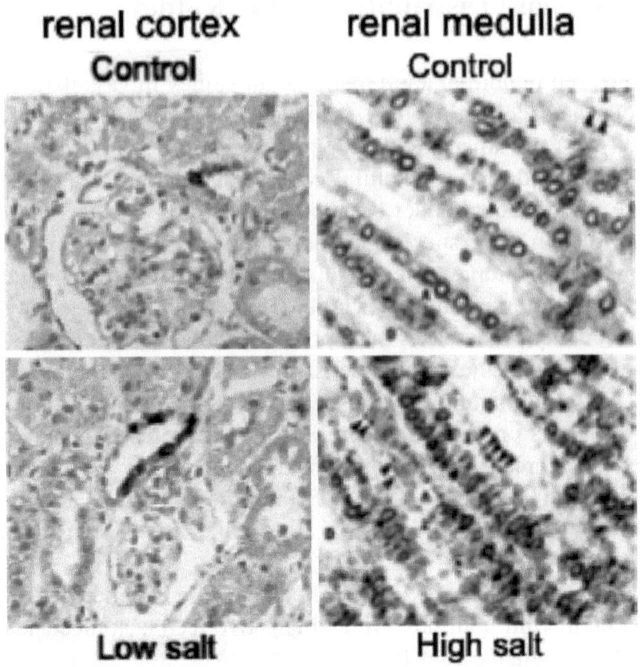

Fig. 3 Regulation of COX-2 expression in the kidney. In the cortex, a sodium deficient diet increases COX-2 expression in the macula densa, while in the inner medulla, high salt increases COX-2 expression in the medullary interstitial cells

immunoreactivity was observed in the macula densa [14]. Furthermore, chronic administration of either hypertensive or nonhypertensive concentrations of angiotensin II inhibits macula densa COX-2 expression, indicating a direct inhibition of enzyme expression through AT_1 receptors [18], which have been shown to be present in macula densa cells [19]. In addition to COX-2 inhibition by AT_1 receptor activation, angiotensin II can also stimulate macula densa COX-2 expression via AT_2 receptor signaling, although the AT_2-mediated effects are only detected in the presence of AT_1 inhibition, suggesting that under physiologic conditions, the AT_1-mediated inhibitory effects predominate [18]. In addition, the macula densa expresses prorenin receptors, and transgenic rats overexpressing the prorenin receptor have increased macula densa COX-2 expression and hyperfiltration [20, 21].

In addition to renal cortex, COX-2 is expressed in the renal medulla, with the greatest concentrations found in the medullary interstitial cells. Although salt loading downregulates COX-2 expression in renal cortex, it upregulates COX-2 expression in renal medulla (see Fig. 3) [13, 22], and COX-2-derived prostanoids inhibit medullary salt and water reabsorption. In experimental animals, systemic or selective medullary COX-2 inhibition leads to sodium retention [23, 24]. COX-2-derived prostanoids also inhibit vasopressin-mediated water reabsorption.

Furthermor, in lithium-induced nephrogenic diabetes insipidus, lithium-induced decreases in GSK-3β activity lead to increased renal COX-2 expression and COX-2-derived urine PGE_2 excretion, and suppression of COX-2 activity blunts lithium-associated polyuria [25]. PGE_2 synthesis is also altered in central diabetes insipidus. Recent studies have indicated that mice with genetic deletion of the PGE_2 receptor subtype EP_1 have a urinary concentrating defect secondary to decreased hypothalamic vasopressin release, indicating that neurally derived PGE_2 may mediate vasopressin synthesis and release in response to water deprivation [26].

3 Potential Roles of Cyclooxygenases in Kidney Injury

COX-2 expression increases in a number of animal models of renal disease, as well as human disease. These include anti-Thy-1 glomerulonephritis [27], Heymann nephritis [28], reflux nephropathy [29], lupus nephritis [30, 31], anti-glomerular basement membrane (GBM)–induced experimental nephritis [32], chronic heart failure [33], and hypertension [33, 34].

COX-derived arachidonic acid metabolites mediate altered renal hemodynamic function in response to inflammatory renal diseases, and studies have indicated that cyclooxygenase metabolites may also be mediators of altered renal function seen in models of diabetes and high protein feeding [35, 36]. Macula densa COX-2 expression is increased in models of hyperfiltration, such as a high protein diet, remnant kidney, and diabetes, and COX-2 inhibition decreases hyperfiltration [35, 36]. The prostanoids involved have not yet been completely characterized, although it is presumed that vasodilatory prostaglandins are involved in mediation of the altered renal hemodynamics.

In addition to hemodynamic effects, COX-2 inhibitors decrease proteinuria and inhibit development of glomerular sclerosis in the remnant kidney model (Fig. 4) and in experimental diabetes [37–39]. COX-2 inhibition was also found to slow progression in the Han rat model of polycystic kidney disease [40].

Increased podocyte COX-2 expression is seen in rats after renal ablation and Thy-1 nephritis and in cultured murine podocytes in response to mechanical stress [38, 41, 42]. Komers et al. detected podocyte COX-2 expression in rats with streptozotocin-induced diabetes [35]. Recent studies in transgenic mice selectively overexpressing COX-2 in podocytes have indicated increased albuminuria and glomerular injury [43]. In cultured podocytes, mechanical stress–induced COX-2 and EP_4 mRNA/protein expression through a p38 MAP kinase pathway and PGE_2 stimulation of stretched cells resulted in a dramatic loss of actin stress fiber organization [44]. When COX-2 activity was inhibited by selective COX-2 inhibitors, proteinuria and progressive renal injury were significantly decreased [45–49].

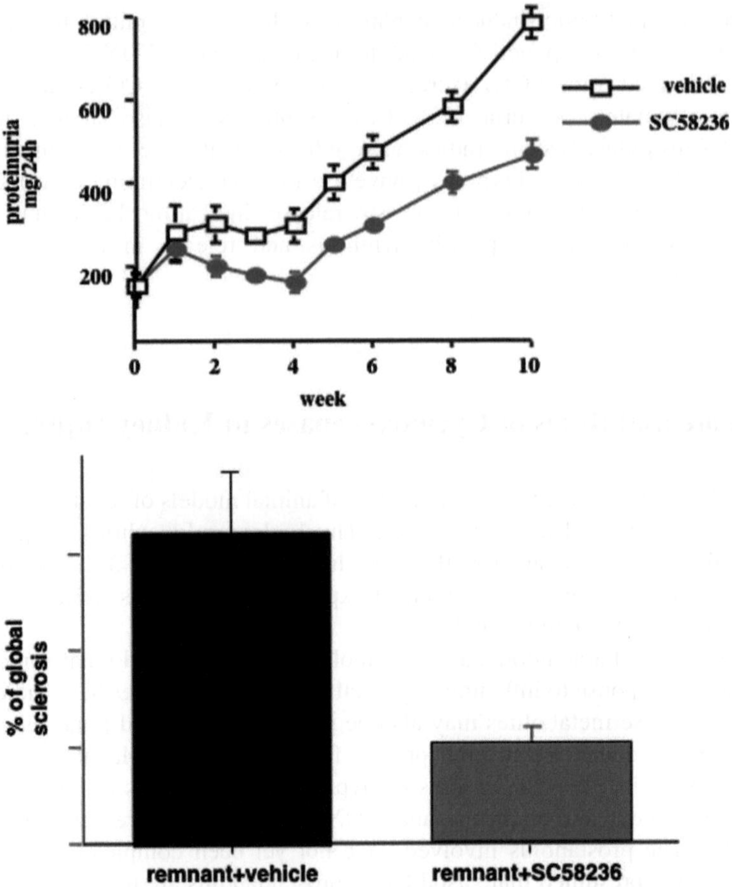

Fig. 4 Effect of COX-2 inhibition on progressive renal damage. In rats subjected to subtotal renal ablation, the highly selective COX-2 inhibitor (SC58236) decreased proteinuria and glomerulosclerosis

4 Glomerular Inflammatory Injury

Cyclooxygenase metabolites have been implicated in functional and structural alterations in glomerular and tubulointerstitial inflammatory diseases [50]. Essential fatty acid deficiency totally prevents the structural and functional consequences of administration of nephrotoxic serum (NTS) to rats, an experimental model of anti-GBM glomerulonephritis [51]. Changes in arteriolar tone during the course of this inflammatory lesion are mediated principally by locally released COX and lipoxygenase (LO) metabolites of arachidonic acid (AA) [51].

Thromboxane A_2 (TxA_2) release appears to play an essential role in mediating the increased renovascular resistance observed during the early phase of this disease [52]. Subsequently, increasing rates of PGE_2 generation may account for progressive dilation of renal arterioles and increases in renal blood flow at later stages of the disease. Consistent with this hypothesis, TxA_2 antagonism ameliorated the drop in renal blood flow (RBF) and glomerular filtration rate (GFR) at 2 h post-NTS administration, but not at 1 day. During the later, heterologous, phase of NTS, COX metabolites mediate both the renal vasodilation as well as the reduction in ultrafiltration coefficient (K_f) that characterize this phase [51]. The net functional result of COX inhibition during this phase of experimental glomerulonephritis, therefore, would depend on the relative importance of renal perfusion vs. the preservation of K_f for the maintenance of GFR. Evidence also indicates that COX metabolites are mediators of pathologic lesions and the accompanying proteinuria in this model [52]. COX-2 expression in the kidney increases in experimental anti-GBM glomerulonephritis [32, 41] and after systemic administration of lipopolysaccharide [53].

A beneficial effect of fish oil diets (enriched in eicosapenataenoic acid), with an accompanying reduction in the generation of COX products, has been demonstrated in genetic murine lupus (MRL-lpr mice). In subsequent studies, enhanced renal TxA_2 and PGE_2 generation was demonstrated in this model, as well as in NZB mice, another genetic model of lupus [52]. In addition, studies in humans demonstrated an inverse relation between TxA_2 biosynthesis and glomerular filtration rate and improvement of renal function following short-term therapy with a thromboxane receptor antagonist in patients with lupus nephritis [52]. More recently, studies have indicated that in humans, as well as NZB mice, COX-2 expression was upregulated in patients with active lupus nephritis, with colocalization to infiltrating monocytes, suggesting that monocytes infiltrating the glomeruli contribute to the exaggerated local synthesis of TxA_2 [30, 31]. COX-2 inhibition selectively decreased thromboxane production, and chronic treatment of NZB mice with a COX-2 inhibitor, and mycophenolate mofetil significantly prolonged survival [30]. Taken together, these data, as well as others from animal and human studies, support a major role for the intrarenal generation of TxA_2 in mediating renal vasoconstriction during inflammatory and lupus-associated glomerular injury.

The demonstration of a functionally significant role for COX metabolites in experimental and human inflammatory glomerular injury has raised the question of the cellular sources of these eicosanoids in the glomerulus. In addition to infiltrating inflammatory cells, resident glomerular macrophages, glomerular mesangial cells, and glomerular epithelial cells represent likely sources for eicosanoid generation. In the anti-Thy1.1 model of mesangioproliferative glomerulonephritis, COX-1 staining was transiently increased in diseased glomeruli at day 6 and was localized mainly to proliferating mesangial cells. COX-2 expression in the macula densa region also transiently increased at day 6 [54, 55]. Glomerular COX-2 expression in this model has been controversial, with one group reporting increased podocyte COX-2 expression [41] and two other groups reporting minimal, if any, glomerular COX-2 expression [54, 55]. However, it is of interest that selective COX-2 inhibitors have been reported to inhibit glomerular repair in the anti-Thy1.1 model [55]. In both anti-Thy1.1 and anti-GBM models of glomerulonephritis, the nonselective COX inhibitor, indomethacin, increased monocyte chemoattractant protein-1 (MCP-1), suggesting that prostaglandins may repress recruitment of monocytes/macrophages in experimental glomerulonephritis [56].

A variety of cytokines have been reported to stimulate PGE_2 synthesis and COX-2 expression in cultured mesangial cells. Furthermore, complementary components, in particular C5b-9, which are known to be involved in the inflammatory models described above, have been implicated in the stimulation of PGE_2 synthesis in glomerular epithelial cells. Cultured glomerular epithelial cells predominantly express COX-1, but exposure to C5b-9 significantly increased COX-2 expression [52].

5 Glomerular Noninflammatory Injury

Studies have suggested that prostanoids may also mediate altered renal function and glomerular damage following subtotal renal ablation, and glomerular prostaglandin production may be altered in such conditions. Glomeruli from remnant kidneys, as well as animals fed a high protein diet, have increased prostanoid production [52]. These studies suggested an increase in cyclooxygenase enzyme activity per se rather than, or in addition to, increased substrate availability, since increases in prostanoid production were noted when excess exogenous arachidonic acid was added.

Following subtotal renal ablation, there are selective increases in renal cortical and glomerular COX-2 mRNA and immunoreactive protein expression, without significant alterations in COX-1 expression [38]. This increased COX-2 expression was most prominent in the macula densa and surrounding cortical thick ascending limb of Henle (cTALH). In addition, COX-2 immunoreactivity was also present in podocytes of remnant glomeruli, and increased prostaglandin production in isolated glomeruli from remnant kidneys was inhibited by a COX-2 selective inhibitor but was not by a COX-1 selective inhibitor [38]. Of interest, Weichert et al. have reported that in the fawn-hooded rat, which develops spontaneous glomerulosclerosis, there is increased cTALH/macula densa COX-2 and neuronal nitric oxide synthase (nNOS) and juxtaglomerular cell renin expression preceding development of sclerotic lesions [34].

When given 24 h after subtotal renal ablation, a nonselective nonsteroidal anti-inflammatory drug (NSAID), indomethacin, normalized increases in renal blood flow and single nephron GFR; similar decreases in hyperfiltration were noted when indomethacin was given acutely to rats 14 days after subtotal nephrectomy, although in this latter study, the increased glomerular capillary pressure (P_{GC}) was not altered because both afferent and efferent arteriolar resistances increased [52]. Previous studies have also suggested that nonselective cyclooxygenase inhibitors may acutely decrease hyperfiltration in diabetes and inhibit proteinuria or structural injury [52]; more recent studies have indicated selective COX-2 inhibitors will decrease the hyperfiltration seen in experimental diabetes or increased dietary protein [35, 57]. Of note, both nonselective NSAIDs and selective

COX-2 inhibitors have also been reported to be effective in reducing proteinuria in patients with nephrotic syndrome [52, 58].

Administration of COX-2 selective inhibitors decreased proteinuria and inhibited development of glomerular sclerosis in rats with reduced functioning renal mass (see Fig. 4) [45, 59]. In addition, COX-2 inhibition decreased mRNA expression of tumor growth factor (TGF)-β1 and types III and IV collagen in the remnant kidney [45]. Similar protection was observed with administration of nitroflurbiprofen (NOF), a nitric oxide (NO)–releasing NSAID without gastrointestinal toxicity [60]. Prior studies have also demonstrated that thromboxane synthase inhibitors retarded progression of glomerulosclerosis, with decreased proteinuria and glomerulosclerosis in rats with remnant kidneys and in diabetic nephropathy, in association with increased renal prostacyclin production and lower systolic blood pressure [61]. Studies in models of type 1 and type 2 diabetes have indicated that COX-2 selective inhibitors retarded progression of diabetic nephropathy [37, 62]. Schmitz et al. confirmed increases in thromboxane excretion in the remnant kidney and correlated decreased arachidonic and linoleic acid levels with increased thromboxane production, since the thromboxane synthase inhibitor U63557A restored fatty acid levels and retarded progressive glomerular destruction [61].

Enhanced glomerular synthesis and urinary excretion of both PGE_2 and TxA_2 have been demonstrated in passive Heymann nephritis (PHN) and adriamycin-induced glomerulopathies in rats. Both COX-1 and COX-2 expression are increased in glomeruli with PHN [63]. Both thromboxane synthase inhibitors and selective COX-2 inhibitors also decreased proteinuria in PHN [52]. As noted, transgenic mice selectively overexpressing COX-2 in podocytes have indicated increased albuminuria and glomerular injury [43], and more recent studies have indicated an important role for thromboxane A_2 to mediate the injury [64]. In addition, PGE_2 has also been implicated in podocyte injury [44], which is of interest since *Stra l3*, a PGE_2-induced gene, has been shown to be a regulator of the redox state of podocytes [65].

In contrast to the putative deleterious effects of TxA_2, the prostacyclin analogue cicaprost retarded renal damage in uninephrectomized dogs fed a high sodium and high protein diet, an effect that was not mediated by amelioration of systemic hypertension [66].

Prostanoids have also been shown to alter extracellular matrix production by mesangial cells in culture. TxA_2 stimulates matrix production by both TGF-β-dependent and -independent pathways [67]. PGE_2 has been reported to decrease steady state mRNA levels of alpha 1(I) and alpha 1(III) procollagens, but not alpha 1(IV) procollagen and fibronectin mRNA, and to reduce secretion of all studied collagen types into the cell culture supernatants. Of interest, this effect did not appear to be mediated by cyclic adenosine monophosphate (cAMP) [68]. PGE_2 has also been reported to increase production of matrix metalloproteinase-2 (MMP-2) and to mediate angiotensin II-induced increases in MMP-2 [69]. Whether vasodilatory prostaglandins mediate decreased fibrillar collagen production and increased matrix degrading activity in glomeruli in vivo has not yet been studied; however, there is compelling evidence in nonrenal cells that prostanoids may either mediate or modulate matrix production [70]. PGE_2 derived from COX-2 has also been shown to decrease platelet-derived growth factor (PDGF)-mediated mesangial cell proliferation by inhibiting production of reactive oxygen species [71].

6 COX-2 in Hypertension

COX-2-derived PGE_2 can decrease angiotensin II-mediated hypertension by inhibiting nicotinamide adenine dinucleotide phosphate (NADPH)-mediated production of ROS [72], while in a high salt/elevated aldosterone model of hypertension, COX-2 inhibitors exacerbate hypertension [73]. COX-2 inhibition elevates blood pressure in both SHR and WKY rats on a high salt diet, but not in rats on low salt intake, indicating that the hypertension induced by COX-2 inhibition can occur independently of a genetic predisposition to hypertension and can be prevented by salt deprivation [74]. In animals on a high salt diet, selective medullary infusion of COX-2 inhibitors will also induce hypertension [24].

The mechanisms underlying upregulation of medullary COX-2 expression in response to volume expansion are probably mutifactorial. There is clear evidence that increased medullary tonicity increases medullary COX-2 expression. Different studies have indicated that a role for NFκB [75], epidermal growth factor receptor (EGFR) transactivation [76], and mitochondrial-generated ROS [77]. Whether these represent parallel pathways or are all interrelated is not yet clear; however, it should be noted that the described EGFR transactivation is mediated by cleavage of the EGFR ligand, TGF-α by ADAM17 (TACE), which is known to be activated by src, which can be activated by ROS. In addition to medullary COX-2, cortical COX-2 expression increases in salt sensitive hypertension, especially in the glomerulus and is inhibited by either the superoxide dismutase mimetic, 4-hydroxy-2,2,6,6-tetramethylpiperidine-1-oxyl (TEMPOL), or an ARB [78].

7 Acute Kidney Injury (AKI)

Acute kidney injury in response to sepsis or accompanying acute administration of endotoxin is characterized by progressive reductions in RBF and GFR, even in the absence of hypotension. Renal histology in such animals is normal, but cortical generation of COX metabolites is markedly elevated. Both medullary and cortical COX-2 expression have been shown to increase in a model of sepsis. This increase appeared to be mediated by activation of Toll-like receptor 4 (TLR4), and in TLR4-/- mice, juxtaglomerular apparatus renin expression was absent [79]. A number of reports have provided evidence for a role for TxA_2-induced renal vasoconstriction in this model of renal dysfunction [80].

In contrast, COX-2 expression decreases in the kidney in response to acute ischemic injury [81]. There is some controversy about the role of cyclooxygenase products in ischemia-reperfusion injury. Roles for prostaglandins and TxA_2 in modulating or mediating renal injury have been suggested in ischemia/reperfusion [82] and in models of toxin-mediated acute tubular injury, including those induced by uranyl nitrate [83], amphotericin B [84], aminoglycosides [85], and glycerol [86]. Furthermore, fibrosis resulting from prolonged ischemic injury has been shown to be ameliorated by nonspecific COX inhibition [87]. In contrast, renal injury in response to ischemia reperfusion is worsened by COX-2 selective inhibitors or in COX-2-/- mice [88], and administration of vasodilator prostaglandins has been shown to ameliorate injury [89], possibly through a PPARδ-dependent mechanism [90].

8 Urinary Tract Obstruction

Following induction of chronic (more than 24 h) ureteral obstruction, renal prostaglandin and TxA_2 synthesis is markedly enhanced, particularly in response to stimuli such as endotoxin or bradykinin. Enhanced prostanoid synthesis likely arises from infiltrating mononuclear cells, proliferating fibroblast-like cells, interstitial macrophages, and interstitial medullary cells. Considerable evidence, derived from studies utilizing specific enzyme inhibitors, suggests a causal relationship between increased renal generation of TxA_2 and the intense vasoconstriction that characterizes the hydronephrotic or postobstructed kidney [50]. In this sense, therefore, hydronephrotic injury can be regarded as a form of subacute inflammatory insult in which intrarenal eicosanoid generation from infiltrating leukocytes contributes to the pathophysiologic process. Finally, TxA_2 has been implicated in the resetting of the tubuloglomerular feedback mechanism observed in hydronephrotic kidneys [91]. Recent studies have also suggested that selective COX-2 inhibitors may prevent renal damage in response to unilateral ureteral obstruction [92, 93].

9 Diabetes Mellitus

In the streptozotocin-induced model of diabetes in rats, COX-2 expression is increased in the macula densa region [35, 37]. COX-2 immunoreactivity has also been detected in the macula densa region in human diabetic nephropathy [33]. Studies suggest that vasodilator prostanoids, PGI_2 and PGE_2, play an important role in the hyperfiltration seen early in diabetes mellitus [94]. In streptozotocin-induced diabetes in rats, previous studies indicated that nonselective cyclooxygenase inhibitors acutely decrease hyperfiltration in diabetes and inhibit proteinuria or structural injury [95], and more recent studies have also indicated that acute administration of a selective COX-2 inhibitor decreased hyperfiltration [35]. The increased COX-2 expression appears to be mediated at least in part by increased

ROS production in diabetes, since the superoxide dismutase analogue TEMPOL blocks the increased expression [96]. At least in cultured mesangial cells, the ROS that mediate increased COX-2 expression in response to high glucose are derived from the mitochondria [97].

Chronic administration of a selective COX-2 inhibitor significantly decreased proteinuria and reduced extracellular matrix deposition, as indicated by decreases in immunoreactive fibronectin expression and in mesangial matrix expansion. In addition, COX-2 inhibition reduced expression of TGF-β, PAI-1, and vascular endothelial growth factor in the kidneys of the diabetic hypertensive animals [37]. TxA_2 may play a role in the development of albuminuria and basement membrane changes with diabetic nephropathy. In addition, administration of a selective PGE_2 EP_1 receptor antagonist prevented development of experimental diabetic nephropathy [98]. In contrast to the proposed detrimental effects of these "vasoconstrictor" prostanoids, administration of a prostacyclin analogue decreased hyperfiltration and reduced macrophage infiltration in early diabetic nephropathy by increasing eNOS expression in afferent arterioles and glomerular capillaries [99].

10 Role of Reactive Oxygen Species as Mediators of COX-2 Actions

In addition to NADPH oxidase, nitric oxide synthase, and xanthine oxidase, COX-2 can also be a source of oxygen radicals [100]. COX-2 enzymatic activity is commonly accompanied by associated oxidative mechanisms (co-oxidation) and free radical production [101]. The catalytic activity of cyclooxygenase consists of a series of radical reactions that use molecular oxygen and generate intermediate ROS [102]. Elevated levels of COX-2 protein are associated with increased ROS production and apoptosis in cultured renal cortical cells [103] and human mesangial cells [97]. It has been recently suggested that COX-2–mediated lipid peroxidation, rather than prostaglandins, can induce DNA damage via adduct formation [104]. A COX-2 specific inhibitor, NS-398, was able to reduce the oxidative activity, with prevention of oxidant stress [105].

In addition to ROS generated by cyclooxygenase per se, prostanoids may also activate intracellular pathways that generate ROS. Locally generated ROS may damage cell membranes, leading to lipid peroxidation and release of arachidonic acid. Prostanoids released during inflammatory reactions cause rapid degenerative changes in some cultured cells, and their potential cytotoxic effect has been suggested to occur by accelerating intracellular oxidative stress [106]. Thromboxane [107] and PGE_2 acting through the EP_1 receptor [108] have been reported to induce NADPH oxidase and ROS production. Of interest, PGE_2 acting through the EP_4 receptor inhibits macrophage oxidase activity [72, 109]. As mentioned previously, there is also evidence for cross-talk between COX-2 and ROS, such that ROS may induce COX-2 expression [104]. Interestingly, during aging there is ROS-mediated NFκB expression, which increases COX-2 expression in the kidney [110]. Furthermore, this appears to induce a vicious cycle, since COX-2 then serves as a source of ROS. The amount of renal ROS resulting from COX activity increases with age, such that up to 25% of total kidney ROS production in aged rat kidneys is inhibited by NSAID administration.

11 Nonenzymatic Metabolism of Arachidonic Acid

It has long been recognized that oxidant injury can result in peroxidation of lipids. In 1990, Morrow and Roberts reported that a series of prostaglandin-like compounds could be produced by free radical catalyzed peroxidation of arachidonic acid that is independent of cyclooxygenase activity [111]. These compounds, which are termed "isoprostanes," are increasingly utilized as a sensitive marker of oxidant injury in vitro and in vivo [112]. In addition, at least two of these compounds, 8-iso-PGF_{2a} (15-F_2-isoprostane) and 8-iso-PGE_2 (15-E_2-isoprostane), are potent vasoconstrictors when administered exogenously [113]. It has been shown that 8-iso-PGF_{2a} will constrict the renal microvasculature and decrease GFR, an effect that is prevented by thromboxane receptor antagonism [114]. However, the role of endogenous isoprostanes as mediators of biologic responses remains unclear.

12 Conclusion

In summary, cyclooxygenase-derived prostanoids play important physiologic roles in the normal kidney but may also be mediators of inflammatory injury in renal disease. A common factor in many forms of renal disease is increased production of reactive oxygen species, and prostanoids have been shown to modulate ROS in various renal diseases. In addition, increased cyclooxygenase activity may itself serve as a source of free radical production.

References

1. Jang BC, Munoz-Najar U, Paik JH, Claffey K, Yoshida M, Hla T. Leptomycin B, an inhibitor of the nuclear export receptor CRM1, inhibits COX-2 expression. J Biol Chem 2002; 278: 2773–2776.
2. Dixon DA, Tolley ND, King PH, Nabors LB, McIntyre TM, Zimmerman GA, Prescott SM. Altered expression of the mRNA stability factor HuR promotes cyclooxygenase-2 expression in colon cancer cells. J Clin Invest 2001; 108: 1657–1665.
3. Inoue H, Taba Y, Miwa Y, Yokota C, Miyagi M, Sasaguri T. Transcriptional and posttranscriptional regulation of cyclooxygenase-2 expression by fluid shear stress in vascular endothelial cells. Arterioscler Thromb Vasc Biol 2002; 22: 1415–1420.

4. Bonazzi A, Mastyugin V, Mieyal PA, Dunn MW, Laniado-Schwartzman M. Regulation of cyclooxygenase-2 by hypoxia and peroxisome proliferators in the corneal epithelium. J Biol Chem 2000; 275: 2837–2844.
5. Hla T, Bishop-Bailey D, Liu CH, Schaefers HJ, Trifan OC. Cyclooxygenase-1 and -2 isoenzymes. Int J Biochem Cell Biol 1999; 31: 551–557.
6. Mestre JR, Mackrell PJ, Rivadeneira DE, Stapleton PP, Tanabe T, Daly JM. Redundancy in the signaling pathways and promoter elements regulating cyclooxygenase-2 gene expression in endotoxin-treated macrophage/monocytic cells. J Biol Chem 2001; 276: 3977–3982.
7. Tanabe T, Tohnai N. Cyclooxygenase isozymes and their gene structures and expression. Prostaglandins Other Lipid Mediat 2002; 68–69: 95–114.
8. Harris RC, McKanna JA, Akai Y, Jacobson HR, Dubois RN, Breyer MD. Cyclooxygenase-2 is associated with the macula densa of rat kidney and increases with salt restriction. J Clin Invest 1994; 94: 2504–2510.
9. Harris RC. COX-2 and the Kidney. J Cardiovasc Pharmacol 2006; 47 (Suppl 1): S37–S42.
10. Peti-Peterdi J, Komlosi P, Fuson AL, Guan Y, Schneider A, Qi Z, Redha R, Rosivall L, Breyer MD, Bell PD. Luminal NaCl delivery regulates basolateral PGE2 release from macula densa cells. J Clin Invest 2003; 112: 76–82.
11. Kim SM, Chen L, Mizel D, Huang YG, Briggs JP, Schnermann J. Low plasma renin and reduced renin secretory responses to acute stimuli in conscious COX-2-deficient mice. Am J Physiol Renal Physiol 2007; 292: F415–F422.
12. Matzdorf C, Kurtz A, Hocherl K. COX-2 activity determines the level of renin expression but is dispensable for acute upregulation of renin expression in rat kidneys. Am J Physiol Renal Physiol 2007; 292: F1782–F1790.
13. Yang T, Singh I, Pham H, Sun D, Smart A, Schnermann JB, Briggs JP. Regulation of cyclooxygenase expression in the kidney by dietary salt intake. Am J Physiol 1998; 274: F481–F489.
14. Cheng HF, Wang JL, Zhang MZ, Miyazaki Y, Ichikawa I, McKanna JA, Harris RC. Angiotensin II attenuates renal cortical cyclooxygenase-2 expression. J Clin Invest 1999; 103: 953–961.
15. Mann B, Hartner A, Jensen BL, Hilgers KF, Hocherl K, Kramer BK, Kurtz A. Acute upregulation of COX-2 by renal artery stenosis. Am J Physiol Renal Physiol 2001; 280: F119–F125.
16. Harding P, Carretero OA, Beierwaltes WH. Chronic cyclooxygenase-2 inhibition blunts low sodium-stimulated renin without changing renal haemodynamics. J Hypertens 2000; 18: 1107–1113.
17. Wang JL, Cheng HF, Harris RC. Cyclooxygenase-2 inhibition decreases renin content and lowers blood pressure in a model of renovascular hypertension. Hypertension 1999; 34: 96–101.
18. Zhang MZ, Yao B, Cheng HF, Wang SW, Inagami T, Harris RC. Renal cortical cyclooxygenase 2 expression is differentially regulated by angiotensin II AT(1) and AT(2) receptors. Proc Natl Acad Sci USA 2006; 103: 16045–16050.
19. Kovacs G, Peti-Peterdi J, Rosivall L, Bell PD. Angiotensin II directly stimulates macula densa Na-2Cl-K cotransport via apical AT(1) receptors. Am J Physiol Renal Physiol 2002; 282: F301–F306.
20. Kaneshiro Y, Ichihara A, Takemitsu T, Sakoda M, Suzuki F, Nakagawa T, Hayashi M, Inagami T. Increased expression of cyclooxygenase-2 in the renal cortex of human prorenin receptor gene-transgenic rats. Kidney Int 2006; 70: 641–646.
21. Nguyen G. Increased cyclooxygenase-2, hyperfiltration, glomerulosclerosis, and diabetic nephropathy: put the blame on the (pro)renin receptor? Kidney Int 2006; 70: 618–620.
22. Ye W, Zhang H, Hillas E, Kohan DE, Miller RL, Nelson RD, Honeggar M, Yang T. Expression and function of COX isoforms in renal medulla: evidence for regulation of salt sensitivity and blood pressure. Am J Physiol Renal Physiol 2006; 290: F542–F549.

23. Qi Z, Hao CM, Langenbach RI, Breyer RM, Redha R, Morrow JD, Breyer MD. Opposite effects of cyclooxygenase-1 and -2 activity on the pressor response to angiotensin II. J Clin Invest 2002; 110: 61–69.
24. Zewde T, Mattson DL. Inhibition of cyclooxygenase-2 in the rat renal medulla leads to sodium-sensitive hypertension. Hypertension 2004; 44: 424–428.
25. Rao R, Zhang MZ, Zhao M, Cai H, Harris RC, Breyer MD, Hao CM. Lithium treatment inhibits renal GSK-3 activity and promotes cyclooxygenase 2-dependent polyuria. Am J Physiol Renal Physiol 2005; 288: F642–F649.
26. Kennedy CR, Xiong H, Rahal S, Vanderluit J, Slack RS, Zhang Y, Guan Y, Breyer MD, Hebert RL. Urine concentrating defect in prostaglandin EP1-deficient mice. Am J Physiol Renal Physiol 2007; 292: F868–F875.
27. Blume C, Heise G, Muhlfeld A, Bach D, Schror K, Gerhardz CD, Grabensee B, Heering P. Effect of flosulide, a selective cyclooxygenase 2 inhibitor, on passive Heymann nephritis in the rat. Kidney Int 1999; 56: 1770–1778.
28. Shinmura K, Tang XL, Wang Y, Xuan YT, Liu SQ, Takano H, Bhatnagar A, Bolli R. Cyclooxygenase-2 mediates the cardioprotective effects of the late phase of ischemic preconditioning in conscious rabbits. Proc Natl Acad Sci USA 2000; 97: 10197–10202.
29. Solari V, Piotrowska AP, Cascio S, Unemoto K, Chertin B, Puri P. Cyclooxygenase-2 up-regulation in reflux nephropathy. J Urol 2003; 170: 1624–1627.
30. Zoja C, Benigni A, Noris M, Corna D, Casiraghi F, Pagnoncelli M, Rottoli D, Abbate M, Remuzzi G. Mycophenolate mofetil combined with a cyclooxygenase-2 inhibitor ameliorates murine lupus nephritis. Kidney Int 2001; 60: 653–663.
31. Tomasoni S, Noris M, Zappella S, Gotti E, Casiraghi F, Bonazzola S, Benigni A, Remuzzi G. Upregulation of renal and systemic cyclooxygenase-2 in patients with active lupus nephritis. J Am Soc Nephrol 1998; 9: 1202–1212.
32. Chanmugam P, Feng L, Liou S, Jang BC, Boudreau M, Yu G, Lee JH, Kwon HJ, Beppu T, Yoshida M, et al. Radicicol, a protein tyrosine kinase inhibitor, suppresses the expression of mitogen-inducible cyclooxygenase in macrophages stimulated with lipopolysaccharide and in experimental glomerulonephritis. J Biol Chem 1995; 270: 5418–5426.
33. Khan KN, Stanfield KM, Harris RK, Baron DA. Expression of cyclooxygenase-2 in the macula densa of human kidney in hypertension, congestive heart failure, and diabetic nephropathy. Ren Fail 2001; 23: 321–330.
34. Weichert W, Paliege A, Provoost AP, Bachmann S. Upregulation of juxtaglomerular NOS1 and COX-2 precedes glomerulosclerosis in fawn-hooded hypertensive rats. Am J Physiol Renal Physiol 2001; 280: F706–F714.
35. Komers R, Lindsley JN, Oyama TT, Schutzer WE, Reed JF, Mader SL, Anderson S. Immunohistochemical and functional correlations of renal cyclooxygenase-2 in experimental diabetes. J Clin Invest 2001; 107: 889–898.
36. Yao B, Xu J, Qi Z, Harris RC, Zhang MZ. Role of renal cortical cyclooxygenase-2 expression in hyperfiltration in rats with high-protein intake. Am J Physiol Renal Physiol 2006; 291: F368–F374.
37. Cheng HF, Wang CJ, Moeckel GW, Zhang MZ, McKanna JA, Harris RC. Cyclooxygenase-2 inhibitor blocks expression of mediators of renal injury in a model of diabetes and hypertension. Kidney Int 2002; 62: 929–939.
38. Wang JL, Cheng HF, Zhang MZ, McKanna JA, Harris RC. Selective increase of cyclooxygenase-2 expression in a model of renal ablation. Am J Physiol 1998; 275: F613–F622.
39. Komers R, Lindsley JN, Oyama TT, Anderson S. Cyclo-oxygenase-2 inhibition attenuates the progression of nephropathy in uninephrectomized diabetic rats. Clin Exp Pharmacol Physiol 2007; 34: 36–41.
40. Sankaran D, Bankovic-Calic N, Ogborn MR, Crow G, Aukema HM. Selective COX-2 inhibition markedly slows disease progression and attenuates altered prostanoid production in Han:SPRD-cy rats with inherited kidney disease. Am J Physiol Renal Physiol 2007; 293: F821–F830.

41. Hirose S, Yamamoto T, Feng L, Yaoita E, Kawasaki K, Goto S, Fujinaka H, Wilson CB, Arakawa M, Kihara I. Expression and localization of cyclooxygenase isoforms and cytosolic phospholipase A2 in anti-Thy-1 glomerulonephritis. J Am Soc Nephrol 1998; 9: 408–416.
42. Martineau LC, McVeigh LI, Jasmin BJ, Kennedy CR. p38 MAP kinase mediates mechanically induced COX-2 and PG EP4 receptor expression in podocytes: implications for the actin cytoskeleton. Am J Physiol Renal Physiol 2004; 286: F693–F701.
43. Cheng H, Wang S, Jo YI, Hao CM, Zhang M, Fan X, Kennedy C, Breyer MD, Moeckel GW, Harris RC. Overexpression of cyclooxygenase-2 predisposes to podocyte injury. J Am Soc Nephrol 2007; 18: 551–559.
44. Martineau LM, McVeigh LI, Jasmin BJ, Kennedy CR. p38 MAP kinase mediates mechanically-induced cyclooxygenase-2 and prostaglandin EP4 receptor expression in podocytes; implications for the actin cytoskeleton. Am J Physiol Renal Physiol 2003.
45. Wang JL, Cheng HF, Shappell S, Harris RC. A selective cyclooxygenase-2 inhibitor decreases proteinuria and retards progressive renal injury in rats. Kidney Int 2000; 57: 2334–2342.
46. Sanchez PL, Salgado LM, Ferreri NR, Escalante B. Effect of cyclooxygenase-2 inhibition on renal function after renal ablation. Hypertension 1999; 34: 848–853.
47. Horiba N, Kumano E, Watanabe T, Shinkura H, Sugimoto T, Inoue M. Subtotal nephrectomy stimulates cyclooxygenase 2 expression and prostacyclin synthesis in the rat remnant kidney. Nephron 2002; 91: 134–141.
48. Kong WX, Ma J, Gu Y, Yang HC, Zuo YQ, Lin SY. Renal protective effects of specific cyclooxygenase-2 inhibitor in rats with subtotal renal ablation. Zhonghua Nei Ke Za Zhi 2003; 42: 186–190.
49. Kang DH, Nakagawa T, Feng L, Watanabe S, Han L, Mazzali M, Truong L, Harris R, Johnson RJ. A role for uric acid in the progression of renal disease. J Am Soc Nephrol 2002; 13: 2888–2897.
50. Klahr S, Morrissey JJ. The role of growth factors, cytokines, and vasoactive compounds in obstructive nephropathy. Semin Nephrol 1998; 18: 622–632.
51. Takahashi K, Kato T, Schreiner GF, Ebert J, Badr KF. Essential fatty acid deficiency normalizes function and histology in rat nephrotoxic nephritis. Kidney Int 1992; 41: 1245–1253.
52. Harris RC, Breyer MD. Arachidonic acid metabolites and the kidney. *The Kidney*, BM Brenner, ed. 2004: 727–776.
53. Yang T, Sun D, Huang YG, Smart A, Briggs JP, Schnermann JB. Differential regulation of COX-2 expression in the kidney by lipopolysaccharide: role of CD14. Am J Physiol 1999; 277: F10–F16.
54. Hartner A, Pahl A, Brune K, Goppelt-Struebe M. Upregulation of cyclooxygenase-1 and the PGE2 receptor EP2 in rat and human mesangioproliferative glomerulonephritis. Inflamm Res 2000; 49: 345–354.
55. Kitahara M, Eitner F, Ostendorf T, Kunter U, Janssen U, Westenfeld R, Matsui K, Kerjaschki D, Floege J. Selective cyclooxygenase-2 inhibition impairs glomerular capillary healing in experimental glomerulonephritis. J Am Soc Nephrol 2002; 13: 1261–1270.
56. Schneider A, Harendza S, Zahner G, Jocks T, Wenzel U, Wolf G, Thaiss F, Helmchen U, Stahl RA. Cyclooxygenase metabolites mediate glomerular monocyte chemoattractant protein-1 formation and monocyte recruitment in experimental glomerulonephritis. Kidney Int 1999; 55: 430–441.
57. Bing Y, Xu J, Qi Z, Harris RC, Zhang MZ. The role of renal cortical cyclooxygenase-2 (COX-2) expression in hyperfiltration in rats with high protein intake. Am J Physiol Renal Physiol 2006; 291(2):F368–F374.
58. Vogt L, de Zeeuw D, Woittiez AJ, Navis G. Selective cyclooxygenase-2 (COX-2) inhibition reduces proteinuria in renal patients. Nephrol Dial Transplant 2009; 24: 1182–1189.
59. Goncalves AR, Fujihara CK, Mattar AL, Malheiros DM, Noronha Ide L, de Nucci G, Zatz R. Renal expression of COX-2, ANG II, and AT1 receptor in remnant kidney: strong

renoprotection by therapy with losartan and a nonsteroidal anti-inflammatory. Am J Physiol Renal Physiol 2004; 286: F945–F954.
60. Fujihara CK, Malheiros DM, Donato JL, Poli A, De Nucci G, Zatz R. Nitroflurbiprofen, a new nonsteroidal anti-inflammatory, ameliorates structural injury in the remnant kidney. Am J Physiol 1998; 274: F573–F579.
61. Schmitz PG, Krupa SM, Lane PH, Reddington JC, Salinas-Madrigal L. Acquired essential fatty acid depletion in the remnant kidney: amelioration with U-63557A. Kidney Int 1994; 46: 1184–1191.
62. Dey A, Maric C, Kaesemeyer WH, Zaharis CZ, Stewart J, Pollock JS, Imig JD. Rofecoxib decreases renal injury in obese Zucker rats. Clin Sci (Lond) 2004; 107: 561–570.
63. Takano T, Cybulsky AV. Complement C5b-9-mediated arachidonic acid metabolism in glomerular epithelial cells : role of cyclooxygenase-1 and -2. Am J Pathol 2000; 156: 2091–2101.
64. Cheng H, Fan X, Guan Y, Moeckel GW, Zent R, Harris RC. Distinct Roles for Basal and Induced COX-2 in Podocyte Injury. J Am Soc Nephrol 2009; 20(9): 1953–1962.
65. Bek MJ, Wahle S, Muller B, Benzing T, Huber TB, Kretzler M, Cohen C, Busse-Grawitz A, Pavenstadt H. Stra13, a prostaglandin E2-induced gene, regulates the cellular redox state of podocytes. FASEB J 2003; 17: 682–684.
66. Villa E, Martinez J, Ruilope L, Mampaso F, Sancho J, Robles R. Cicaprost, a prostacyclin analog, protects renal function in uninephrectomized dogs in the absence of changes in blood pressure. Am J Hypertension 1992; 6: 253–257.
67. Studer R, Negrete H, Craven P, DeRubertis F. Protein kinase C signals thromboxane induced increases in fibronectin synthesis and TGF-beta bioactivity in mesangial cells. Kidney Int 1995; 48: 422–430.
68. Zahner G, Disser M, Thaiss F, Wolf G, Schoeppe W, Stahl R. The effect of prostaglandin E2 on mRNA expression and secretion of collagens I, III, and IV and fibronectin in cultured rat mesangial cells. JASN 1994; 4: 1778–1785.
69. Singhal P, Sagar S, Garg P, Bansal V. Vasoactive agents modulate matrix metalloproteinase-2 activity by mesangial cells. Am J Med Sciences 1995; 310: 235–241.
70. Varga J, Diaz-Perez A, Rosenbloom J, Jimenez SA. PGE_2 causes a coordinate decrease in the steady state levels of fibronectin and types I and III procollagen mRNAs in normal human dermal fibroblasts. Biochem Biophysical Res Comm 1987; 147: 1282–1288.
71. Zahner G, Wolf G, Schroeder S, Stahl RA. Inhibition of platelet-derived growth factor-induced mesangial cell proliferation by cyclooxygenase-2 overexpression is abolished through reactive oxygen species. FEBS Lett 2006; 580: 2523–2528.
72. Jia Z, Guo X, Zhang H, Wang MH, Dong Z, Yang T. Microsomal prostaglandin synthase-1-derived prostaglandin E2 protects against angiotensin II-induced hypertension via inhibition of oxidative stress. Hypertension 2008; 52: 952–959.
73. Yao B, Harris RC, Zhang MZ. Interactions between 11beta-hydroxysteroid dehydrogenase and COX-2 in kidney. Am J Physiol Regul Integr Comp Physiol 2005; 288: R1767–R1773.
74. Hocherl K, Endemann D, Kammerl MC, Grobecker HF, Kurtz A. Cyclo-oxygenase-2 inhibition increases blood pressure in rats. Br J Pharmacol 2002; 136: 1117–1126.
75. Hao CM, Yull F, Blackwell T, Komhoff M, Davis LS, Breyer MD. Dehydration activates an NF-kappaB-driven, COX2-dependent survival mechanism in renal medullary interstitial cells. J Clin Invest 2000; 106: 973–982.
76. Kuper C, Bartels H, Fraek ML, Beck FX, Neuhofer W. Ectodomain shedding of pro-TGF-alpha is required for COX-2 induction and cell survival in renal medullary cells exposed to osmotic stress. Am J Physiol Cell Physiol 2007; 293: C1971–C1982.
77. Yang T, Zhang A, Honeggar M, Kohan DE, Mizel D, Sanders K, Hoidal JR, Briggs JP, Schnermann JB. Hypertonic induction of COX-2 in collecting duct cells by reactive oxygen species of mitochondrial origin. J Biol Chem 2005; 280: 34966–34973.

78. Jaimes EA, Zhou MS, Pearse DD, Puzis L, Raij L. Upregulation of cortical COX-2 in salt-sensitive hypertension: role of angiotensin II and reactive oxygen species. Am J Physiol Renal Physiol 2008; 294: F385–F392.
79. El-Achkar TM, Plotkin Z, Marcic B, Dagher PC. Sepsis induces an increase in thick ascending limb Cox-2 that is TLR4 dependent. Am J Physiol Renal Physiol 2007; 293: F1187–F1196.
80. Wise WC, Cook JA, Tempel GE, Reines HD, Halushka PV. The rat in sepsis and endotoxic shock. Prog Clin Biol Res 1989; 299: 243–252.
81. Villanueva S, Cespedes C, Gonzalez AA, Vio CP, Velarde V. Effect of ischemic acute renal damage on the expression of COX-2 and oxidative stress-related elements in rat kidney. Am J Physiol Renal Physiol 2007; 292: F1364–F1371.
82. Ruschitzka F, Shaw S, Noll G, Barton M, Schulz E, Muller GA, Luscher TF. Endothelial vasoconstrictor prostanoids, vascular reactivity, and acute renal failure. Kidney Int Suppl 1998; 67: S199–S201.
83. Chaudhari A, Kirschenbaum MA. Altered glomerular eicosanoid biosynthesis in uranyl nitrate-induced acute renal failure. Biochim Biophys Acta 1984; 792: 135–140.
84. Hardie WD, Ebert J, Frazer M, Takahashi K, Badr K. The effect of thromboxane A2 receptor antagonism on amphotericin B–induced renal vasoconstriction in the rat. Prostaglandins 1993; 45: 47–56.
85. Higa EM, Schor N, Boim MA, Ajzen H, Ramos OL. Role of the prostaglandin and kallikrein-kinin systems in aminoglycoside-induced acute renal failure. Braz J Med Biol Res 1985; 18: 355–365.
86. Papanicolaou N, Hatziantoniou C, Bariety J. Selective inhibition of thromboxane synthesis partially protected while inhibition of angiotensin II formation did not protect rats against acute renal failure induced with glycerol. Prostaglandins Leukot Med 1986; 21: 29–35.
87. Feitoza CQ, Goncalves GM, Semedo P, Cenedeze MA, Pinheiro HS, Beraldo FC, dos Santos OF, Teixeira Vde P, dos Reis MA, Mazzali M, Pacheco-Silva A, Camara NO. Inhibition of COX 1 and 2 prior to renal ischemia/reperfusion injury decreases the development of fibrosis. Mol Med 2008; 14: 724–730.
88. Patel NS, Cuzzocrea S, Collino M, Chaterjee PK, Mazzon E, Britti D, Yaqoob MM, Thiemermann C. The role of cycloxygenase-2 in the rodent kidney following ischaemia/reperfusion injury in vivo. Eur J Pharmacol 2007; 562: 148–154.
89. Vargas AV, Krishnamurthi V, Masih R, Robinson AV, Schulak JA. Prostaglandin E1 attenuation of ischemic renal reperfusion injury in the rat. J Am Coll Surg 1995; 180: 713–717.
90. Hsu YH, Chen CH, Hou CC, Sue YM, Cheng CY, Cheng TH, Lin H, Tsai WL, Chan P, Chen TH. Prostacyclin protects renal tubular cells from gentamicin-induced apoptosis via a PPARalpha-dependent pathway. Kidney Int 2008; 73: 578–587.
91. Morsing P, Stenberg A, Persson AE. Effect of thromboxane inhibition on tubuloglomerular feedback in hydronephrotic kidneys. Kidney Int 1989; 36: 447–452.
92. Miyajima A, Ito K, Asano T, Seta K, Ueda A, Hayakawa M. Does cyclooxygenase-2 inhibitor prevent renal tissue damage in unilateral ureteral obstruction? J Urol 2001; 166: 1124–1129.
93. Ozturk H, Ozdemir E, Otcu S, Buyukbayram H, Ihsan Dokucu A. Renal effects on a solitary kidney of specific inhibition of cyclooxygenease-2 after 24 h of complete ureteric obstruction in rats. Urol Res 2002; 30: 223–226.
94. DeRubertis FR, Craven PA. Eicosanoids in the pathogenesis of the functional and structural alterations of the kidney in diabetes. Am J Kidney Dis 1993; 22: 727–735.
95. Hommel E, Mathiesen E, Arnold-Larsen S, Edsberg B, Olsen UB, Parving HH. Effects of indomethacin on kidney function in type 1 (insulin-dependent) diabetic patients with nephropathy. Diabetologia 1987; 30: 78–81.
96. Li J, Chen YJ, Quilley J. Effect of TEMPOL on renal cyclooxygenase expression and activity in experimental diabetes in the rat. J Pharmacol Exp Ther 2005; 314: 818–824.

97. Kiritoshi S, Nishikawa T, Sonoda K, Kukidome D, Senokuchi T, Matsuo T, Matsumura T, Tokunaga H, Brownlee M, Araki E. Reactive oxygen species from mitochondria induce cyclooxygenase-2 gene expression in human mesangial cells: potential role in diabetic nephropathy. Diabetes 2003; 52: 2570–2577.
98. Makino H, Tanaka I, Mukoyama M, Sugawara A, Mori K, Muro S, Suganami T, Yahata K, Ishibashi R, Ohuchida S, Maruyama T, Narumiya S, Nakao K. Prevention of diabetic nephropathy in rats by prostaglandin E receptor EP1-selective antagonist. J Am Soc Nephrol 2002; 13: 1757–1765.
99. Yamashita T, Shikata K, Matsuda M, Okada S, Ogawa D, Sugimoto H, Wada J, Makino H. Beraprost sodium, prostacyclin analogue, attenuates glomerular hyperfiltration and glomerular macrophage infiltration by modulating ecNOS expression in diabetic rats. Diabetes Res Clin Pract 2002; 57: 149–161.
100. Virdis A, Colucci R, Fornai M, Blandizzi C, Duranti E, Pinto S, Bernardini N, Segnani C, Antonioli L, Taddei S, Salvetti A, Del Tacca M. Cyclooxygenase-2 inhibition improves vascular endothelial dysfunction in a rat model of endotoxic shock: role of inducible nitric-oxide synthase and oxidative stress. J Pharmacol Exp Ther 2005; 312: 945–953.
101. Xu Z, Choudhary S, Voznesensky O, Mehrotra M, Woodard M, Hansen M, Herschman H, Pilbeam C. Overexpression of COX-2 in human osteosarcoma cells decreases proliferation and increases apoptosis. Cancer Res 2006; 66: 6657–6664.
102. Marnett LJ, Rowlinson SW, Goodwin DC, Kalgutkar AS, Lanzo CA. Arachidonic acid oxygenation by COX-1 and COX-2. Mechanisms of catalysis and inhibition. J Biol Chem 1999; 274: 22903–22906.
103. Rockwell P, Martinez J, Papa L, Gomes E. Redox regulates COX-2 upregulation and cell death in the neuronal response to cadmium. Cell Signal 2004; 16: 343–353.
104. Lee SH, Williams MV, Dubois RN, Blair IA. Cyclooxygenase-2-mediated DNA damage. J Biol Chem 2005; 280: 28337–28346.
105. Mouithys-Mickalad A, Deby-Dupont G, Dogne JM, de Leval X, Kohnen S, Navet R, Sluse F, Hoebeke M, Pirotte B, Lamy M. Effects of COX-2 inhibitors on ROS produced by Chlamydia pneumoniae-primed human promonocytic cells (THP-1). Biochem Biophys Res Commun 2004; 325: 1122–1130.
106. Prasad KN, Hovland AR, La Rosa FG, Hovland PG. Prostaglandins as putative neurotoxins in Alzheimer's disease. Proc Soc Exp Biol Med 1998; 219: 120–125.
107. Wilcox CS. Oxidative stress and nitric oxide deficiency in the kidney: a critical link to hypertension? Am J Physiol Regul Integr Comp Physiol 2005; 289: R913–R935.
108. Jaimes EA, Tian RX, Pearse D, Raij L. Up-regulation of glomerular COX-2 by angiotensin II: role of reactive oxygen species. Kidney Int 2005; 68: 2143–2153.
109. Serezani CH, Chung J, Ballinger MN, Moore BB, Aronoff DM, Peters-Golden M. Prostaglandin E2 suppresses bacterial killing in alveolar macrophages by inhibiting NADPH oxidase. Am J Respir Cell Mol Biol 2007; 37: 562–570.
110. Kim HJ, Kim KW, Yu BP, Chung HY. The effect of age on cyclooxygenase-2 gene expression: NF-kappaB activation and IkappaBalpha degradation. Free Radic Biol Med 2000; 28: 683–692.
111. Morrow J, Harris TM, Roberts LJ 2nd. Noncyclooxygenase oxidative formation of a series of novel prostaglandins: analytical ramifications for measurement of eicosanoids. Anal Biochem 1990; 186: 184–185.
112. Roberts L, Morrow J. Products of the isoprostane pathway: unique bioactive compounds and markers of lipid peroxidation. Cell Mol Life Sci 2002; 59: 808–820.
113. Morrow JD, Roberts LJ. The isoprostanes: unique bioactive products of lipid peroxidation. Prog Lipid Res 1997; 36: 1–21.
114. Takahashi K, Nammour T, Fukunaga M, Ebert J, Morrow JD, Roberts LJ, 2nd, Hoover RL, Badr KF. Glomerular actions of a free radical-generated novel prostaglandin, 8-epi-prostaglandin F2 alpha, in the rat. Evidence for interaction with thromboxane A2 receptors. J Clin Invest 1992; 90: 136–141.

Chapter 4
Renin-Angiotensin System in the Kidney and Oxidative Stress: Local Renin-Angiotensin-Aldosterone System and NADPH Oxidase-Dependent Oxidative Stress in the Kidney

Hideyasu Kiyomoto, Hiroyuki Kobori, and Akira Nishiyama

Abstract The renin-angiotensin system (RAS) plays an important role in the pathogenesis of renal diseases. All of the RAS components can be found in the kidney, and intrarenal angiotensin II (AngII) is formed by multiple independent mechanisms. There is also increasing evidence implicating aldosterone as an important factor besides AngII in the pathogenesis of renal diseases. Therefore, we should include aldosterone as one of the key components of the renin-angiotensin-aldosterone system (RAAS) when trying to understand how renal diseases are induced and augmented and, in turn, how they can be suppressed. Although multiple pathways are involved in the RAAS-dependent renal injury, there is increasing evidence supporting the roles of reactive oxygen species (ROS), which contribute to renal functional aberrations and pro-inflammatory or pro-fibrotic tissue damages. Animal studies have shown that AngII- and aldosterone-dependent renal injury is associated with increased ROS production and nicotinamide adenine dinucleotide phosphate (NADPH) oxidase activity. Furthermore, preclinical and clinical studies have indicated that the reno-protective effects of RAAS inhibitors are associated with their antioxidative effects. In this chapter, we will briefly summarize our current understanding of the role of ROS in mediating RAAS-dependent renal injury with special emphasis on the role of NADPH oxidase.

Keywords Reactive oxygen species · NADPH oxidase · RAAS blockade

A. Nishiyama (✉)
Department of Pharmacology and Hypertension and Kidney Disease
Research Center, Kagawa University Medical School, 1750-1 Ikenobe, Miki-cho, Kita-gun, Kagawa Prefecture 761-0701, Japan
e-mail: akira@kms.ac.jp

T. Miyata et al. (eds.), *Studies on Renal Disorders*, Oxidative Stress in Applied Basic Research and Clinical Practice, DOI 10.1007/978-1-60761-857-7_4,

1 Introduction

Angiotensin II (AngII) is the most powerful and biologically active product of the renin-angiotensin system (RAS), which regulates renal hemodynamics and functions [1, 2]. AngII exerts substantial direct effects on the renal microvasculature and glomerular mesangium by constricting vascular smooth muscle cells and mesangial cells [3–6]. Intrarenal AngII also regulates renal hemodynamics by modulating the sensitivity of the tubuloglomerular feedback mechanisms [7–9]. The roles of AngII in regulating sodium transport by renal epithelial cells are also well documented [10–14]. High intrarenal AngII levels contribute to salt and water retention through direct actions on renal tubular transport function when it is inappropriately stimulated [12, 15–17]. AngII also stimulates the zona glomerulosa of the adrenal cortex to produce the sodium-retaining hormone aldosterone. Aldosterone stimulates ionic transport in the principal cells by increasing the number of open sodium and potassium channels in the luminal membrane and the activity of the Na^+/K^+-ATPase pump in the basolateral membrane [18–20]. These effects of aldosterone are mediated through mineralocorticoid receptors (MRs). Thus, aldosterone promotes sodium chloride reabsorption and potassium secretion in the principal cells of the cortical collecting tubular segment of the nephron via activation of MR. Arima et al. showed that aldosterone constricts afferent and efferent arterioles, eliciting greater vasoconstriction in the efferent arteriole than in the afferent arteriole [21, 22]. However, there is still no evidence for the specific role of aldosterone in the regulation of renal hemodynamics.

In addition to their physiological roles, AngII and aldosterone induce tissue inflammation, cell growth, migration, apoptosis, and differentiation by regulating the gene expression of many bioactive substances and activating multiple intracellular signaling pathways [23, 24]. Among these pathways, reactive oxygen species (ROS) act as important mediators for AngII- and aldosterone-induced renal injury [1, 25–30]. In the kidney, ROS are necessary for normal physiologic processes and signal transduction systems, which include the development of nephrons, erythropoiesis, and the regulation of tubular transport and hemodynamics [1, 2]. However, the loss of redox homeostasis forms excessive ROS, which activate pro-inflammatory and pro-fibrotic pathways, leading to the impaired renal function and renal cell injury [1, 25, 27, 29]. In addition to their direct actions, ROS also induce the above effects by impairing nitric oxide (NO) bioactivity [23, 31–33].

In this chapter, we will briefly review the recent findings related to the regulation of the intrarenal renin-angiotensin-aldosterone system (RAAS) and the role of ROS in mediating RAAS-dependent renal injury, with particular emphasis on the role of nicotinamide adenine dinucleotide phosphate (NADPH) oxidase. Blockade of the RAAS to target ROS will be reviewed elsewhere in this book and will not be discussed in detail in this chapter.

2 Regulation of Circulating RAAS and Its Role

The "classic" RAAS was discovered as a unique endocrine system and is predominantly regulated by the activity of renin secreted from the juxtaglomerular apparatus (JGA) on the afferent arterioles of the kidney [1, 2]. On release into the circulation, renin cleaves angiotensinogen (AGT) at the N-terminus to form the decapeptide angiotensin I (AngI) [1]. AGT is mainly formed and constitutively secreted by hepatic cells. As shown in Fig. 1, the circulating concentrations of AGT is very high, more than 1,000 times greater than the plasma levels of AngI and AngII in anesthetized rats [34–36]. Thus, although some species variation exists, changes in plasma renin activity mainly determine the rate of AngI formation in the plasma from the huge stores of circulating AGT. AngI is easily converted to AngII, not only via the circulating angiotensin converting enzyme (ACE), but also because of the ubiquitous expression of ACE in endothelial cells in most vascular beds, including the lung [37]. Circulating AngII causes peripheral vasoconstriction and tubular sodium reabsorption, and it stimulates aldosterone production in the adrenal gland [1, 2].

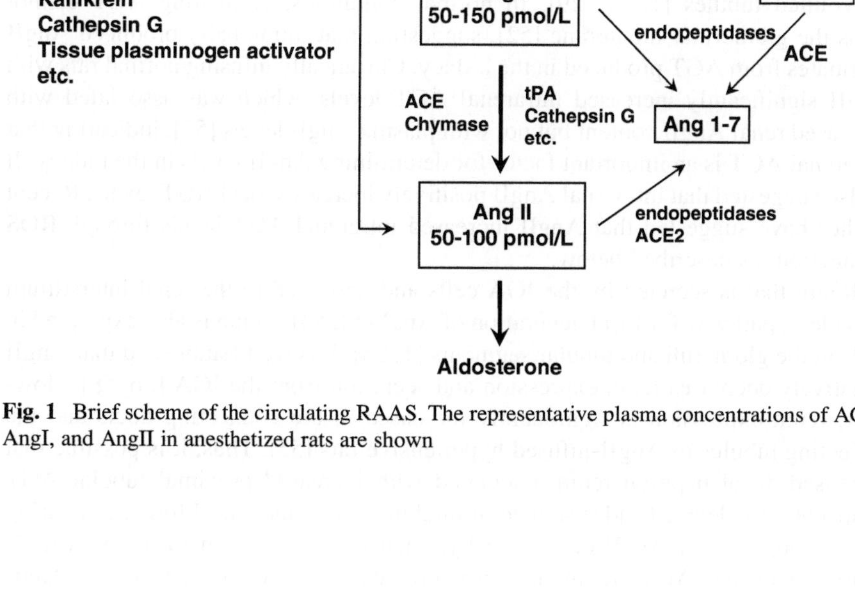

Fig. 1 Brief scheme of the circulating RAAS. The representative plasma concentrations of AGT, AngI, and AngII in anesthetized rats are shown

3 Intrarenal RAAS

The RAS and RAAS have been suggested to be endocrine, paracrine, autocrine, and intracrine systems [38]. Substantial evidence has demonstrated that the local formation of AngII is of major significance in the regulation of AngII levels in the kidney [39–41]. It has been suggested that aldosterone is also produced in renal cells [42–44]. Thus, the local RAAS plays an important role in regulating renal function, but contributes to deranged renal function and tissue injury when it is inappropriately activated [31, 45].

3.1 Regulation of Intrarenal AngII and Aldosterone Levels

Locally produced AngII in the kidney plays an important role in the regulation of renal function, and when inappropriately activated, it contributes to the progression of renal injury by inducting cell activities such as inflammation, cell growth, mitogenesis, apoptosis, migration, and differentiation by activating multiple intracellular signaling pathways [23, 24, 46, 47]. Renal tissue AngII levels are much greater than can be explained by the concentrations delivered by the arterial blood flow [35, 36, 48]. The kidney has all of the RAS components with compartmentalization in the tubular and interstitial networks and intracellular accumulation. Although most of the circulating AGT is produced and secreted by the liver, AGT is also produced in the kidney [41, 49–51]. In the kidney, AGT is predominantly localized in proximal tubules, but it is also expressed in other renal cells, including glomerular endothelial cells, mesangial cells, podocytes, and distal convoluted tubules [1, 25, 29]. In normal conditions, circulating AGT cannot cross the glomerular membrane [52], suggesting that intrarenally produced AngII originates from AGT produced in the kidney. Chronically infusing normal rats with AngII significantly increased intrarenal AGT levels, which was associated with increased renal AngII content but not with plasma AngII levels [52], indicating that intrarenal AGT is an important factor for determining AngII levels in the kidney. It is also suggested that intrarenal AngII positively increases local AGT levels. Recent studies have suggested that AngII increased intrarenal AGT levels through ROS production, as described below.

Renin that is secreted by the JGA cells and delivered to the renal interstitium provides a pathway for local generation of AngI [53, 54]. Renin is also expressed in cells in the glomeruli and tubular segments [1, 55]. It is well established that AngII negatively decreases renin expression and secretion from the JGA [56–58]. However, renin expression is significantly increased in the connecting ducts and the collecting tubules of AngII-infused hypertensive rats [55]. Thus, it is possible that increased distal nephron renin associated with increased proximal tubular AGT contributes to elevated and sustained intratubular AngI and AngII formation during the development of AngII-dependent hypertension. In addition to the enzymatic action to cleave AGT, renin and prorenin also exert direct effects on AngII generation by binding to the (pro)renin receptor [59]. The existence of the (pro) renin receptor in the kidney suggests the potential for local generation of AngII [60, 61]. Ichihara et al. demonstrated that blocking the (pro)renin receptor with handle region peptide prevents the augmentation of internal AngII levels and renal injury in diabetic rats [62]. These data suggest that (pro)renin receptor-dependent intrarenal AngII generation contributes to the development of diabetic nephropathy. Interestingly, the stimulation of the (pro)renin receptor by renin and prorenin directly activates intracellular signal transduction through AngII-independent mechanisms [59, 60, 63]. However, it is still unclear how intimately the AngII-dependent and -independent effects of (pro)renin receptors are involved in the pathogenesis of renal injury.

In the rodent kidney, ACE is abundantly expressed on vascular endothelial cells and on various renal cells, including the brush border membranes of proximal tubules, and rapidly converts AngI into AngII [64–67]. ACE also converts Ang1-9 to Ang1-7 (see Fig. 1). However, vascular endothelial cells in normal human kidneys express very low levels of ACE [68]. AngI is also converted into AngII by chymase, a chymotrypsin-like serine protease [69]. Chymase is inactivated in plasma because abundant chymase inhibitors, including serine protease inhibitors, are mostly contained in the blood, indicating that chymase is only enzymatically active in local tissues [70]. Since rodents have less chymase activity than humans, there are fewer experimental data on the role of chymase in AngII formation within the kidney [71]. Another carboxypeptidase, ACE2, converts AngI into Ang1-9 and AngII into Ang1-7 [1] (see Fig. 1). Therefore, it is possible that ACE2 negatively regulates ACE-dependent AngII formation by stimulating an alternative pathway for AngI degradation. It has been reported that the administration of Ang1-7 decreased NADPH oxidase activity and ameliorated proteinuria in diabetic and hypertensive rats [1].

3.2 *AngII Receptors and MR in the Kidney*

Intrarenal AngII levels are much higher than can be explained by the circulating concentrations of AngII, which is not distributed in a homogeneous manner but is compartmentalized in tubular fluid, interstitial fluid, and intracellular compartments [1, 17, 35, 36, 51, 72]. Some AngII is internalized via AngII receptors type 1 (AT_1R)-mediated endocytosis [73, 74]. Studies have indicated that internalized AngII is relatively protected from peptidase-induced degradation and activates intracellular signal transduction through intracellular AT_1R [47, 75]. Furthermore, intracellular AngII also binds to nuclear receptors [15, 74, 75]. Because AngII stimulates AGT and renin expression in proximal tubules and distal nephrons, respectively [54, 76], it is possible that intracellular AngII may exert genomic effects to regulate AGT or renin mRNA expression. It has also been suggested that intracellular AngII is recycled and secreted by binding to AngII receptors on the cell membrane to mediate its subsequent effects [14, 51, 77].

Two major categories of AngII receptors, AT_1R and type 2 (AT_2R), have been pharmacologically characterized and cloned [78–80]. The AT_1R is widely expressed in proximal tubules, the thick ascending limb of the loop of Henle, glomeruli, arterial vasculature, vasa recta, juxtaglomerular cells, distal tubules, collecting ducts, macula densa cells, and interstitial cells, for example [81]. On the other hand, the AT_2R is highly expressed in humans and rodents in the renal mesenchyme during fetal life, and its expression decreases dramatically after birth [82]. The AT_2R is localized to the glomerular epithelial cells, proximal tubules, collecting ducts, and renal vasculature [83]. Most of the actions of AngII on renal function and injury are due to activation of the AT_1R [83, 84]. Although the role of the AT_2R remains uncertain, it has been suggested that AT_2R activation counteracts the effects of the AT_1R [85]. The regulation of intrarenal AT_1R and AT_2R in pathophysiological conditions is extremely complex and is not fully clarified.

Aldosterone is a steroid hormone that is also secreted from the adrenal gland and regulates body fluid homeostasis by activating the MR in the distal nephron. However, recent studies have indicated that aldosterone is also produced in the kidney [42, 43]. Aldosterone is produced and secreted in response to AngII in cultured human mesangial cells [42]. In diabetic rats, the renal aldosterone contents is significantly increased, and this is associated with increased glomerular expression of CYP11B2, a steroidogenic enzyme [42–44, 86, 87]. However, the amount of intrarenally produced aldosterone and its pathophysiological roles are not clear. Our preliminary experiments using renal microdialysis in normotensive rats showed that the aldosterone concentrations in the renal interstitial fluid are below the detectable range (data not shown). Thus, it is possible that intrarenal aldosterone is mainly derived from circulating aldosterone, whereas the intrarenal generation of aldosterone is negligible in physiological conditions. However, intrarenal production of aldosterone may be augmented in some pathological conditions.

Accumulating evidence suggests that aldosterone is a key factor in mediating cardiovascular injury [31, 88]. Chronic administration of aldosterone with salt loading elicits proteinuria, glomerular mesangial injury, and tubulointerstitial fibrosis in rats [27, 89, 90]. Since these aldosterone-induced renal injuries were prevented by treatment with MR antagonists, it is possible that aldosterone directly induces renal tissue injury by activating locally expressed MR. Based on these observations in aldosterone-infused rats [90, 91], we investigated the existence of MR in cultured rat mesangial cells [92, 93] and renal fibroblasts [94] by Western blotting. An MR-specific antibody detected significant MR protein expression in mesangial cells and renal fibroblasts [92–94]. In fluoro-labeling experiments using confocal microscopy, MR protein expression was observed in both the cytoplasm and the nuclear fraction, with barely detectable levels in the membrane fraction. Administration of aldosterone induces nuclear translocation of cytosolic MR (Fig. 2). MR expression was also detected in podocytes [92–94] and renal tubular cells [95–97]. Taken together, MR is localized not only in distal tubular cells but also in other renal cells to mediate aldosterone-induced renal tissue injury. On the other hand, aldosterone-independent MR activation has also been reported.

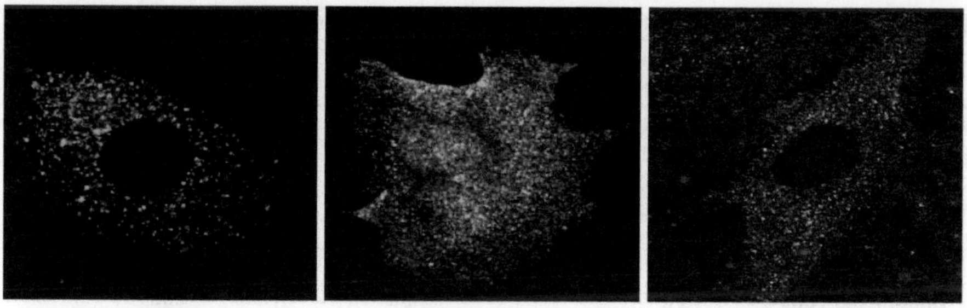

Fig. 2 Effects of aldosterone and eplerenone on nuclear translocation of mineralocorticoid receptors (MR) in cultured rat mesangial cells. Administration of aldosterone induces nuclear translocation of cytosolic MR (*green*). Eplerenone inhibits aldosterone-induced nuclear translocation of cytosolic MR. Original magnification, ×1,600

Shibata et al. demonstrated that constitutively activated Rac1 directly activates MR-dependent activity and induces glomerular injury [98]. Nevertheless, the precise molecular mechanisms responsible for aldosterone/MR-dependent renal injury remain unclear.

4 NADPH Oxidase-Dependent ROS Production in Renal Injury

The mechanisms of ROS production will be described in detail later in this chapter. RAAS activation leads to the activation of ROS-generating enzymes, including the principal metabolic oxygenases such as NADPH oxidase [23, 31, 45]. In terms of quantity, although the mitochondrial respiratory chain system can generate much more superoxide anions, the NADPH oxidase complex is known to be the principal source of abnormal cellular signaling [23, 99]. All of the components necessary for the NADPH oxidase complex to function in the kidney are expressed in the various kidney segments (Table 1). Enhanced ROS production via NADPH oxidase was known to be closely associated with various kidney diseases, including glomerulonephritis in human and experimental animals [100]. All of the components of NADPH oxidase are required for generation of superoxide anions in the kidney. The composition of the membranous subunits are expressed differently in the various kidney cell types [99]. For example, Nox4 is expressed at high levels throughout the kidney. By contrast, other members of the Nox family, including Nox1 and Nox2, are expressed at lower but quantitatively significant levels [101]. Therefore, Nox enzymes have become attractive candidates for the source of ROS generation in the kidney [102, 103].

At least, three isoforms of NADPH oxidase have been discovered in membranous fraction from the kidney, including Nox1, gp91phox, and Nox4 [104]. The other NADPH oxidase components, p22phox, p47phox, and p67phox, are

Table 1 The local expression of NADPH oxidase components in the kidney

	Membranous components	Cytosolic components	References
Whole kidney	Nox1, gp91phox, Nox4	p47phox	[101, 105]
Cortical tissues	Nox1, Nox2, Nox4, p22phox	p67phox, p47phox, Rac1	[32, 106, 110, 117]
Medullar tissues	Nox4, p22pho	p47	[106]
Glomeruli	Nox2, Nox4	p47phox, p67phox, Rac1	[45]
Endothelial cells	n/a	n/a	n/a
Mesangial cells	Nox1, Nox4, p22phox	Rac1, p47, p67	[105, 107, 108, 165]
Epithelial cells (podocytes)	n/a	Rac1	[98]
Tubules			
PCT	p22phox, Nox1, Nox2, Nox4	p47, p67	[105, 111–115]
DTL/ATL/TAL	n/a	n/a	n/a
DCT	Nox4, p22phox	p47, p67	[105]
DCN	n/a	n/a	n/a
CCD	gp91phox, Nox2, p22phox	p47, p67	[105, 166, 167]
MCD	Nox4	n/a	[103, 109]
Macula densa cells	p22phox	p47, p67	[105]

PTC, proximal convoluted tubules; DTL, descending thin limb; ATL, ascending thin limb; TAL, thick ascending limb; DCT, distal convoluted tubules; DCN, distal connecting tubules; CCD, cortical collecting duct; MCD, medullary collecting duct.

expressed in cultured human mesangial cells [105]. It seems that no consensus has been reached for the corticomedullary differences between the different subunits, as the literature to date contains results for different animals undergoing different treatments. Few studies have demonstrated none or very low expression of the regulatory subunits in the medulla. Schluter et al. [106] reported that the differential expression of NADPH oxidase components between renal cortex and medulla in spontaneously hypertensive rats (SHR), in terms of Nox1, Nox2, and p67phox, is higher in the renal medulla than in cortex, and the expression of Nox4 and p22phox is higher in the cortex than in the medulla. On the other hand, the expression of p47phox is ubiquitous in the renal cortex and medulla. In glomeruli, the mesangial cells are unique in that they express Nox1, Nox4, p22phox, p47phox, and p67phox, but not Nox2 [92, 105, 107, 108]. These findings obtained in mesangial cells may be similar to those in vascular smooth muscle cells, which are also deficient in Nox2 or at least express Nox2 at very low levels. It is very important to consider the relationships between ROS production systems and other bioactive substances such as NO. In mesangial cells, there is a putative crosstalk between NO and superoxide production systems because NO inhibits the expression of Nox1 [109]. The antioxidative effect of AngII AT_1R blocker (ARB) ameliorated glomerular injury through suppressing mRNA expressions of p22phox and p47phox in the renal cortex in obese rats [110].

The NADPH oxidase components, p22phox, p47phox, and p67phox, are all expressed in the proximal convoluted tubules (PCT), distal convoluted tubules (DCT), cortical collecting duct (CCD), and the macula densa cells [111–115]. Expression of the most major component Nox1 is detected in PCTs, which shows differential regulation of redox control within the cortex. The gene expression of Nox2 in PCTs and CCDs has been confirmed. Nox4 was first reported to be expressed in the PCTs and DCTs. However, the existence and localization of Nox4 in the kidney has remained controversial. Nox4 mRNA was initially detected by in situ hybridization in the renal cortex, but its expression was much lower in the renal medulla [103, 109]. However, immunohistochemical studies for the human kidney revealed the existence of Nox4 in the distal nephron, and specifically, Nox4 mRNA was found in the medullary collecting ducts (MCD) [103, 109]. Nox4, one of the membranous components of NADPH oxidase, is unique because it enables cells to use NADPH as a substrate for superoxide generation [116]. The gene expression of Nox4 was increased in the cortex and the outer medulla. These regions may predominantly use NADPH as a substrate for generating superoxide, which might be relevant to the kidneys [117].

In addition, renal Nox4 plays a pivotal role in the renal dysfunction in hypertension and insulin resistance [45, 118]. The accumulation of 8-hydroxy-deoxyguanosine (8-OHdG), a representative marker for ROS-dependent DNA damage, occurred concomitantly with increased gene expression of Nox4 and p22phox in the kidney of streptozotocin-induced diabetic rats, and those abnormalities were reversed by daily injection of insulin for 2 weeks [119]. Another group that used the same animal model showed that infusing antisense oligonucleotides targeting Nox4 mRNA for 2 weeks was effective in decreasing ROS in cortical and glomerular homogenates, organelle hypertrophy, and fibronectin expression, which are characteristic markers of diabetic kidney disease [120]. Nistala et al. reported that perivascular fibrosis decreased concomitantly with reduced oxidative stress, as assessed by 3-nitrotyrosine in glomeruli and 8-OHdG in urine [31, 99]. The enzymatic activity of NADPH oxidase is dependent on the translocation of cytosolic enzyme subunits (i.e., Rac1 translocation to the plasma membrane), as confirmed by Western blotting and immunohistochemistry [121].

There is also some evidence for relatively high levels of hydrogen peroxide observed in urine samples [109, 122]. Although infiltrated polymorphonuclear leukocytes and monocytes should be considered an important source of ROS production in renal injury, the resident kidney cells, including mesangial cells, podocytes, and endothelial cells, also release oxidants in response to various substances against glomerulus via immunological and nonimmunological stimuli, including excess ROS associated with RAAS [45, 123, 124].

Welch and Wilcox have demonstrated that ROS impair renal function and NO generation in the kidney [125]. ROS can convert NO to the ROS derivative, peroxynitrite ($ONOO^-$). The reduction in NO decreases tubuloglomerular feedback (TGF), which controls the retention of sodium until the body extracellular fluid is restored. On the other hand, peroxynitrite acting on the JGA promotes the TGF, which leads to the retention of sodium by the tubules, even if the body is replete with sodium. Those reactions involved in the TGF contribute to the development of hypertension and volume overload. Importantly, blocking the AT_1R with an ARB ameliorated ROS-induced increases in TGF, which links RAAS activation to ROS-mediated redox changes [125]. It has been suggested that the Rho/Rho-kinase pathway might be involved in the induction of Nox4 in the kidney because the administration of Fasudil, a Rho-kinase inhibitor, and statins could inhibit this phenomenon [126].

5 Renal ROS Production Associated with Activation of the RAAS

5.1 AngII

There is much evidence that AngII plays a key contribution to progression of renal injury, such as the development of kidney hypertrophy, excess production of oxidative stress, and incidental inflammation, and thus promotes fibrosis in the kidney. Many of the downstream signaling molecules involve AT_1R pathways that are invoked in the vasculature in the kidney. For example, the calcium-dependent protein kinase C (PKC), G-protein/Rac1, mitogen-activated protein kinase (MAPK)/transforming growth factor (TGF)-β1, insulin growth factor-1/insulin receptor substrate-1, and NADPH oxidase/ROS signaling pathways are involved in the response to excess RAAS activity . Even though the agonistic stimuli are widespread throughout the renal cells, the signaling crosstalk among these molecules is complicated and is dependent on the disease condition and the type of cells whose receptors are expressed predominantly.

Studies in streptozotocin-induced diabetic rats, SHRs, and Zucker obese rats have indicated that ARBs ameliorate glomerulosclerosis, probably by suppressing ROS production [99]. The Ren2 transgenic rat is a unique model for investigating the influence of kidney AngII and the RAAS on redox control of the kidney [45, 99, 124]. The Ren2 rat overexpresses the mouse renin transgene and exhibits increased tissue AngII, which activates the local renal tissue RAAS. The activated tissue RAAS was shown to increase oxidative stress in these animals, which contributes to insulin resistance, hypertension, and proteinuria. Whaley-Connell et al. reported that AngII increased Nox4 expression in the kidney, particularly in podocytes, and that this was associated with proteinuria and hypertension in Ren2 rats [45, 124]. Moreover, the levels of Nox2, p47phox, p67phox, and Rac1, in addition to Nox4, are increased as a result of activated NADPH oxidase activity in glomeruli. Electron microscopic morphological examination in Ren2 transgenic rat revealed podocyte abnormalities, such as increased podocyte effacement, basement membrane thickening, slit pore diaphragm widening, and decreased slit pore numbers. These defects were improved by treatment with a superoxide dismutase mimetic, TEMPOL (hydroxy-2,2,6,6-tetramethylpiperidine-1-oxyl), and ARBs [45, 124]. Thus, these data suggest that the dual blockade of ROS production and RAAS ameliorates proteinuria, supporting a potential role of the RAAS-mediated redox imbalance pathway [45]. The dual blockade of RAAS and ROS also abrogated changes of nephrosclerosis at the structural level. Importantly, these experiments indicate that modulation of the RAAS had favorable effects on renal function independent of a blood pressure lowering effect.

5.2 *Aldosterone*

Chronically administering aldosterone and salt to rats caused severe glomerular injury and tubulointerstitial fibrosis [26, 27, 89, 92, 127, 128]. Increases in multiple factors, including MAPKs [90, 94], Rho-kinase [129], plasminogen activator inhibitor-1 [27, 130], TGF-β1 [131], connective tissue growth factor [91], proinflammatory cytokines such as osteopontin [132] and monocyte chemoattractant protein-1 [132], and ROS [90, 128, 130], have been observed in the renal tissues of aldosterone-infused rats. Among those factors, ROS are thought to be essential for mediating aldosterone-induced renal injury [89, 91, 95, 133]. It was shown that administration of aldosterone with high salt loading induces severe glomerulosclerosis and that this renal injury was associated with increased ROS production concomitant with the upregulation of mRNA of NADPH oxidase components, p22phox, Nox4, and gp91phox in renal cortical tissues [90, 91]. Treatment with eplerenone, a selective MR antagonist, or TEMPOL, a superoxide dismutase mimetic, suppressed aldosterone-induced ROS production via NADPH oxidase [90]. Other studies have reported that TEMPOL or eplerenone markedly attenuated salt-induced podocyte injury and proteinuria in animal models of salt-sensitive hypertension and metabolic syndrome [128]. These findings suggest that aldosterone-induced podocyte injury contributes to the pathogenesis of proteinuria, possibly through local MR activation and ROS production [89, 95, 133–135].

The observations in the animal models described above prompted us to investigate whether aldosterone and MR have direct effects on NADPH oxidase activation and superoxide anion generation in cultured rat mesangial cells. We found that aldosterone directly stimulates superoxide anion generation in rat mesangial cells. Aldosterone-induced superoxide anion production was accompanied by increased NADPH oxidase activity and translocation of p47phox and p67phox, cytosolic components of NADPH oxidase, to the rat mesangial cell membrane [92].

6 ROS-Dependent RAAS Activation in the Kidney

It is well known that ROS pathways are located downstream of the AT_1R [136, 137]. On the other hand, several previous studies have implied that AGT expression is augmented by ROS activation via several pathways. Rao showed that hydrogen peroxide induces ERK1/2 in vascular smooth muscle cells [138]. Yoshizumi et al. showed that hydrogen peroxide activates JNK via c-Src-dependent mechanisms in vascular smooth muscle cells [139]. Suzaki et al. showed that hydrogen peroxide activates c-Src-mediated BMK1 in PC12 cells [140]. They also demonstrated c-Src-dependent BMK1 activation in glomeruli of diabetic rats and in glomerular mesangial cells by high glucose conditions [141]. It was also shown by Perona et al. that Rho activates nuclear factor κB (NFκB) in 3T3 cells [142]. Schreck et al. showed that hydrogen peroxide and oxygen radicals activate NF-κB in a human T cell line [143]. Interestingly, all three of these mediators (MAPK, ROS, and NFκB) were reported to activate AGT expression. Zhang et al. showed that AGT gene expression is stimulated via a p38 kinase pathway in immortalized proximal tubular cells of rat kidney [144]. Hsieh et al. showed that AGT gene expression is activated via ROS in a proximal tubular cell line [145]. AGT gene expression is activated by the NFκB p65 transcription factor in hepatocytes [146]. Moreover, recent studies have suggested a link between MAPK activation and NFκB pathways [147, 148]. These data suggest that the augmentation of AGT expression by ROS activation may involve several pathways. Interestingly, using a cell line of rat immortalized renal proximal tubular cells, Chan et al. demonstrated that high glucose augments AGT gene expression [144, 145, 149–152]. Using the same cell line, they also reported that rat AGT gene has a putative insulin-responsive element in its promoter region [153, 154]. These data suggest that high glucose regulates AGT gene expression in vitro. However, there is a lack of evidence in vivo to demonstrate a link between ROS and AGT in the kidneys.

Importantly, it was recently demonstrated that ROS-dependent activation of intrarenal AGT plays an important role in hypertensive rats [49, 155]. Temporary blockade of the RAS at the prediabetic stage was shown to attenuate renal injury in type 2 diabetic rats later in life [135, 156, 157]. Moreover, it was recently demonstrated that elevated ROS and the ROS-associated augmentation of intrarenal AGT initiate the development of diabetic nephropathy in type 2 diabetic rats [158, 159]. The critical role of ROS-dependent AGT/RAS activation was also reported in immunoglobulin-A (IgA) nephropathy patients [41, 160] and in IgA nephropathy mice [161–163]. These data suggest that ROS pathways are located upstream of the AGT/AT_1R axis.

To demonstrate that the administration of an ARB blocks the vicious high glucose-ROS-AGT-Ang II-AT_1R-ROS cycle by suppressing ROS and inflammation, and thus suppresses diabetic nephropathy, 13 hypertensive diabetic nephropathy patients who received ARBs were recruited and the following parameters were evaluated before and 16 weeks after the administration of ARBs: urinary AGT, albumin, 8-OHdG, 8-epi-prostaglandin F2α, monocyte chemoattractant protein-1 (MCP-1), and interleukin (IL)-6 and IL-10 levels [164]. ARB treatment reduced the blood pressure and urinary levels of AGT, albumin, 8-OHdG, 8-epi-prostaglandin F2α, MCP-1, and IL-6, but increased urinary IL-10 levels. The rate of reduction of urinary AGT was correlated with the rate of reduction of blood pressure and urinary levels of albumin, 8-OHdG, 8-epi-prostaglandin F2α, MCP-1, and IL-6 and the increase in urinary IL-10 levels. These results suggest that the mechanisms by which ARBs exert their renoprotective effect may involve the suppression of intrarenal AGT levels in association with their anti-inflammatory and antioxidant properties in patients with type 2 diabetes [164]. The roles of ROS in the regulation of other RAAS components have not been well investigated.

7 Conclusion

In this review, we briefly summarized the evidence supporting the roles of the RAAS and NADPH oxidase-dependent ROS in the pathogenesis of renal dysfunction and disease. In terms of the preclinical and clinical evidence, it seems clear that RAAS blockade offers a potential therapeutic target for controlling renal injury through inhibiting ROS. On the other hand, ROS regulate the renal RAAS activity. Thus, a vicious cycle of activation of the RAAS and ROS plays a pivotal role in the progression of renal injury.

Acknowledgments This work was partially supported by grants-in-aid for scientific research from the Ministry of Education, Culture, Sports, Science, and Technology of Japan (17590830, 20590253 and 2159105) and by a grant from the National Institute of Diabetes and Digestive and Kidney Diseases (R01DK072408). The authors appreciate Drs. Kumiko Moriwaki, Hirofumi Hitomi, and Daisuke Nakano (Kagawa University) for helpful discussion. The authors also thank Ms. Natsuko Ishii for her proofreading of the manuscript.

References

1. Azizi M, Webb R, Nussberger J, Hollenberg NK. Renin inhibition with aliskiren: where are we now, and where are we going? J Hypertens. 2006;24:243–256.
2. Paul M, Poyan Mehr A, Kreutz R. Physiology of local renin-angiotensin systems. Physiol Rev. 2006;86:747–803.
3. Arendshorst WJ, Brannstrom K, Ruan X. Actions of angiotensin II on the renal microvasculature. J Am Soc Nephrol. 1999;10 (Suppl 11):S149–S161.
4. Kimura K, Inokuchi S, Sugaya T, et al. Location and action of angiotensin II type 1 receptor in the renal microcirculation. Kidney Int Suppl. 1997;63:S201–S204.
5. Navar LG, Inscho EW, Imig JD, Mitchell KD. Heterogeneous activation mechanisms in the renal microvasculature. Kidney Int Suppl. 1998;67:S17–S21.
6. Park S, Bivona BJ, Harrison-Bernard LM. Compromised renal microvascular reactivity of angiotensin type 1 double null mice. Am J Physiol Renal Physiol. 2007;293:F60–F67.
7. Mitchell KD, Mullins JJ. ANG II dependence of tubuloglomerular feedback responsiveness in hypertensive ren-2 transgenic rats. Am J Physiol. 1995;268:F821–F828.
8. Navar LG, Saccomani G, Mitchell KD. Synergistic intrarenal actions of angiotensin on tubular reabsorption and renal hemodynamics. Am J Hypertens. 1991;4:90–96.
9. Seeliger E, Wronski T, Ladwig M, et al. The renin-angiotensin system and the third mechanism of renal blood flow autoregulation. Am J Physiol Renal Physiol. 2009;296: F1334–F1345.

10. Cinelli AR, Efendiev R, Pedemonte CH. Trafficking of Na-K-ATPase and dopamine receptor molecules induced by changes in intracellular sodium concentration of renal epithelial cells. Am J Physiol Renal Physiol. 2008;295:F1117–F1125.
11. Efendiev R, Budu CE, Cinelli AR, et al. Intracellular Na+ regulates dopamine and angiotensin II receptors availability at the plasma membrane and their cellular responses in renal epithelia. J Biol Chem. 2003;278:28719–28726.
12. Kolb RJ, Woost PG, Hopfer U. Membrane trafficking of angiotensin receptor type-1 and mechanochemical signal transduction in proximal tubule cells. Hypertension. 2004;44:352–359.
13. Komlosi P, Fuson AL, Fintha A, et al. Angiotensin I conversion to angiotensin II stimulates cortical collecting duct sodium transport. Hypertension. 2003;42:195–199.
14. Li XC, Carretero OA, Navar LG, Zhuo JL. AT1 receptor-mediated accumulation of extracellular angiotensin II in proximal tubule cells: role of cytoskeleton microtubules and tyrosine phosphatases. Am J Physiol Renal Physiol. 2006;291:F375–F383.
15. Li XC, Zhuo JL. Intracellular ANG II directly induces in vitro transcription of TGF-beta1, MCP-1, and NHE-3 mRNAs in isolated rat renal cortical nuclei via activation of nuclear AT1a receptors. Am J Physiol. 2008;294:C1034–C1045.
16. Navar LG, Harrison-Bernard LM, Imig JD, et al. Renal responses to AT1 receptor blockade. Am J Hypertens. 2000;13:45S–54S.
17. Navar LG, Harrison-Bernard LM, Wang CT, et al. Concentrations and actions of intraluminal angiotensin II. J Am Soc Nephrol. 1999;10 (Suppl 11):S189–S195.
18. Kim SW, Wang W, Nielsen J, et al. Increased expression and apical targeting of renal ENaC subunits in puromycin aminonucleoside-induced nephrotic syndrome in rats. Am J Physiol Renal Physiol. 2004;286:F922–F935.
19. Michlig S, Mercier A, Doucet A, et al. ERK1/2 controls Na,K-ATPase activity and transepithelial sodium transport in the principal cell of the cortical collecting duct of the mouse kidney. J Biol Chem. 2004;279:51002–51012.
20. Vinciguerra M, Mordasini D, Vandewalle A, Feraille E. Hormonal and nonhormonal mechanisms of regulation of the NA,K-pump in collecting duct principal cells. Semin Nephrol. 2005;25:312–321.
21. Arima S, Kohagura K, Xu HL, et al. Nongenomic vascular action of aldosterone in the glomerular microcirculation. J Am Soc Nephrol. 2003;14:2255–2263.
22. Arima S, Kohagura K, Xu HL, et al. Endothelium-derived nitric oxide modulates vascular action of aldosterone in renal arteriole. Hypertension. 2004;43:352–357.
23. Sachse A, Wolf G. Angiotensin II-induced reactive oxygen species and the kidney. J Am Soc Nephrol. 2007;18:2439–2446.
24. Wang W, Soltero L, Zhang P, et al. Renal inflammation is modulated by potassium in chronic kidney disease: possible role of Smad7. Am J Physiol Renal Physiol. 2007;293: F1123–F1130.
25. Izuhara Y, Nangaku M, Inagi R, et al. Renoprotective properties of angiotensin receptor blockers beyond blood pressure lowering. J Am Soc Nephrol. 2005;16:3631–3641.
26. Nagase M, Fujita T. Mineralocorticoid receptor activation in obesity hypertension. Hypertens Res. 2009;32:649–657.
27. Nagase M, Matsui H, Shibata S, et al. Salt-induced nephropathy in obese spontaneously hypertensive rats via paradoxical activation of the mineralocorticoid receptor: role of oxidative stress. Hypertension. 2007;50:877–883.
28. Otani H, Otsuka F, Inagaki K, et al. Aldosterone breakthrough caused by chronic blockage of angiotensin II type 1 receptors in human adrenocortical cells: possible involvement of bone morphogenetic protein-6 actions. Endocrinology. 2008;149:2816–2825.
29. Sofue T, Kiyomoto H. Angiotensin II receptor blocker is a renoprotective remedy for metabolic syndrome. Hypertens Res. 2009;32:735–737.

30. Yoneda T, Takeda Y, Usukura M, et al. Aldosterone breakthrough during angiotensin II receptor blockade in hypertensive patients with diabetes mellitus. Am J Hypertens. 2007;20:1329–1333.
31. Nistala R, Wei Y, Sowers JR, Whaley-Connell A. Renin-angiotensin-aldosterone system-mediated redox effects in chronic kidney disease. Transl Res. 2009;153:102–113.
32. Whaley-Connell A, Habibi J, Johnson M, et al. Nebivolol reduces proteinuria and renal NADPH oxidase-generated reactive oxygen species in the transgenic Ren2 rat. Am J Nephrol. 2009;30:354–360.
33. Whaley-Connell A, Habibi J, Wei Y, et al. Mineralocorticoid receptor antagonism attenuates glomerular filtration barrier remodeling in the transgenic Ren2 rat. Am J Physiol Renal Physiol. 2009;296:F1013–F1022.
34. Mitchell KD, Jacinto SM, Mullins JJ. Proximal tubular fluid, kidney, and plasma levels of angiotensin II in hypertensive Ren-2 transgenic rats. Am J Physiol. 1997;273:F246–F253.
35. Nishiyama A, Seth DM, Navar LG. Renal interstitial fluid concentrations of angiotensins I and II in anesthetized rats. Hypertension. 2002;39:129–134.
36. Nishiyama A, Seth DM, Navar LG. Renal interstitial fluid angiotensin I and angiotensin II concentrations during local angiotensin-converting enzyme inhibition. J Am Soc Nephrol. 2002;13:2207–2212.
37. Santos CF, Caprio MA, Oliveira EB, et al. Functional role, cellular source, and tissue distribution of rat elastase-2, an angiotensin II-forming enzyme. Am J Physiol Heart Circ Physiol. 2003;285:H775–H783.
38. Weir MR, Dzau VJ. The renin-angiotensin-aldosterone system: a specific target for hypertension management. Am J Hypertens. 1999;12:205S–213S.
39. Hilgers KF, Veelken R, Muller DN, et al. Renin uptake by the endothelium mediates vascular angiotensin formation. Hypertension. 2001;38:243–248.
40. Kobori H, Harrison-Bernard LM, Navar LG. Expression of angiotensinogen mRNA and protein in angiotensin II-dependent hypertension. J Am Soc Nephrol. 2001;12:431–439.
41. Takamatsu M, Urushihara M, Kondo S, et al. Glomerular angiotensinogen protein is enhanced in pediatric IgA nephropathy. Pediatr Nephrol. 2008;23:1257–1267.
42. Nishikawa T, Suematsu S, Saito J, et al. Human renal mesangial cells produce aldosterone in response to low-density lipoprotein (LDL). J Steroid Biochem Mol Biol. 2005;96:309–316.
43. Siragy HM, Xue C. Local renal aldosterone production induces inflammation and matrix formation in kidneys of diabetic rats. Exp Physiol. 2008;93:817–824.
44. Xue C, Siragy HM. Local renal aldosterone system and its regulation by salt, diabetes, and angiotensin II type 1 receptor. Hypertension. 2005;46:584–590.
45. Whaley-Connell AT, Chowdhury NA, Hayden MR, et al. Oxidative stress and glomerular filtration barrier injury: role of the renin-angiotensin system in the Ren2 transgenic rat. Am J Physiol Renal Physiol. 2006;291:F1308–F1314.
46. Muller DN, Klanke B, Feldt S, et al. (Pro)renin receptor peptide inhibitor "handle-region" peptide does not affect hypertensive nephrosclerosis in Goldblatt rats. Hypertension. 2008;51:676–681.
47. Nishiyama A, Yoshizumi M, Rahman M, et al. Effects of AT1 receptor blockade on renal injury and mitogen-activated protein activity in Dahl salt-sensitive rats. Kidney Int. 2004;65:972–981.
48. Wang CT, Navar LG, Mitchell KD. Proximal tubular fluid angiotensin II levels in angiotensin II-induced hypertensive rats. J Hypertens. 2003;21:353–360.
49. Kobori H, Nishiyama A. Effects of tempol on renal angiotensinogen production in Dahl salt-sensitive rats. Biochem Biophys Res Commun. 2004;315:746–750.
50. Kobori H, Ozawa Y, Suzaki Y, Nishiyama A. Enhanced intrarenal angiotensinogen contribute to early renal injury in spontaneously hypertensive rats. J Am Soc Nephrol. 2005;16:2073–2080.
51. Navar LG, Harrison-Bernard LM, Nishiyama A, Kobori H. Regulation of intrarenal angiotensin II in hypertension. Hypertension. 2002;39:316–322.

52. Kobori H, Nishiyama A, Harrison-Bernard LM, Navar LG. Urinary angiotensinogen as an indicator of intrarenal Angiotensin status in hypertension. Hypertension. 2003;41:42–49.
53. Vargas SL, Toma I, Kang JJ, et al. Activation of the succinate receptor GPR91 in macula densa cells causes renin release. J Am Soc Nephrol. 2009;20:1002–1011.
54. Kihara M, Umemura S, Sugaya T, et al. Expression of neuronal type nitric oxide synthase and renin in the juxtaglomerular apparatus of angiotensin type-1a receptor gene-knockout mice. Kidney Int. 1998;53:1585–1593.
55. Prieto-Carrasquero MC, Harrison-Bernard LM, Kobori H, et al. Enhancement of collecting duct renin in angiotensin II-dependent hypertensive rats. Hypertension. 2004;44:223–229.
56. Kurtz L, Gerl M, Kriz W, et al. Replacement of connexin 40 by connexin 45 causes ectopic localization of renin-producing cells in the kidney but maintains in vivo control of renin gene expression. Am J Physiol Renal Physiol. 2009;297:F403–F409.
57. Kurtz L, Schweda F, de Wit C, et al. Lack of connexin 40 causes displacement of renin-producing cells from afferent arterioles to the extraglomerular mesangium. J Am Soc Nephrol. 2007;18:1103–1111.
58. Rosivall L, Mirzahosseini S, Toma I, et al. Fluid flow in the juxtaglomerular interstitium visualized in vivo. Am J Physiol Renal Physiol. 2006;291:F1241–F1247.
59. Nguyen G, Muller DN. The biology of the (pro)renin receptor. J Am Soc Nephrol. 2010;21:18–23.
60. Ichihara A, Sakoda M, Kurauchi-Mito A, et al. Renin, prorenin and the kidney: a new chapter in an old saga. J Nephrol. 2009;22:306–311.
61. Nguyen G, Contrepas A. Physiology and pharmacology of the (pro)renin receptor. Curr Opin Pharmacol. 2008;8:127–132.
62. Ichihara A, Hayashi M, Kaneshiro Y, et al. Inhibition of diabetic nephropathy by a decoy peptide corresponding to the "handle" region for nonproteolytic activation of prorenin. J Clin Invest. 2004;114:1128–1135.
63. Nguyen G, Delarue F, Burckle C, et al. Pivotal role of the renin/prorenin receptor in angiotensin II production and cellular responses to renin. J Clin Invest. 2002;109:1417–1427.
64. Mezzano S, Droguett A, Burgos ME, et al. Renin-angiotensin system activation and interstitial inflammation in human diabetic nephropathy. Kidney Int Suppl. 2003;86:S64–S70.
65. Mezzano SA, Aros CA, Droguett A, et al. Renal angiotensin II up-regulation and myofibroblast activation in human membranous nephropathy. Kidney Int Suppl. 2003;86:S39–S45.
66. Michel B, Grima M, Coquard C, et al. Effects of dietary protein and uninephrectomy on renal angiotensin converting enzyme activity in the rat. Kidney Int. 1994;45:1587–1592.
67. Schulz WW, Hagler HK, Buja LM, Erdos EG. Ultrastructural localization of angiotensin I-converting enzyme (EC 3.4.15.1) and neutral metalloendopeptidase (EC 3.4.24.11) in the proximal tubule of the human kidney. Lab Invest. 1988;59:789–797.
68. Metzger R, Bohle RM, Pauls K, et al. Angiotensin-converting enzyme in non-neoplastic kidney diseases. Kidney Int. 1999;56:1442–1454.
69. Bacani C, Frishman WH. Chymase: a new pharmacologic target in cardiovascular disease. Cardiol Rev. 2006;14:187–193.
70. Miyazaki M, Takai S. Tissue angiotensin II generating system by angiotensin-converting enzyme and chymase. J Pharmacol Sci. 2006;100:391–397.
71. Miyazaki M, Takai S, Jin D, Muramatsu M. Pathological roles of angiotensin II produced by mast cell chymase and the effects of chymase inhibition in animal models. Pharmacol Ther. 2006;112:668–676.
72. Kobori H, Nangaku M, Navar LG, Nishiyama A. The intrarenal renin-angiotensin system: from physiology to the pathobiology of hypertension and kidney disease. Pharmacol Rev. 2007;59:251–287.
73. Li XC, Hopfer U, Zhuo JL. AT1 receptor-mediated uptake of angiotensin II and NHE-3 expression in proximal tubule cells through a microtubule-dependent endocytic pathway. Am J Physiol Renal Physiol. 2009;297:F1342–F1352.

74. Zhuo JL, Li XC. Novel roles of intracrine angiotensin II and signalling mechanisms in kidney cells. J Renin Angiotensin Aldosterone Syst. 2007;8:23–33.
75. Pendergrass KD, Averill DB, Ferrario CM, et al. Differential expression of nuclear AT1 receptors and angiotensin II within the kidney of the male congenic mRen2. Lewis rat. Am J Physiol Renal Physiol. 2006;290:F1497–F1506.
76. Ingelfinger JR, Jung F, Diamant D, et al. Rat proximal tubule cell line transformed with origin-defective SV40 DNA: autocrine ANG II feedback. Am J Physiol. 1999;276: F218–F227.
77. Zhuo JL, Imig JD, Hammond TG, et al. Ang II accumulation in rat renal endosomes during Ang II-induced hypertension: role of AT(1) receptor. Hypertension. 2002;39:116–121.
78. Murphy TJ, Alexander RW, Griendling KK, et al. Isolation of a cDNA encoding the vascular type-1 angiotensin II receptor. Nature. 1991;351:233–236.
79. Nakajima M, Mukoyama M, Pratt RE, et al. Cloning of cDNA and analysis of the gene for mouse angiotensin II type 2 receptor. Biochem Biophys Res Commun. 1993;197:393–399.
80. Sasamura H, Hein L, Krieger JE, et al. Cloning, characterization, and expression of two angiotensin receptor (AT-1) isoforms from the mouse genome. Biochem Biophys Res Commun. 1992;185:253–259.
81. Tufro-McReddie A, Harrison JK, Everett AD, Gomez RA. Ontogeny of type 1 angiotensin II receptor gene expression in the rat. J Clin Invest. 1993;91:530–537.
82. Norwood VF, Garmey M, Wolford J, et al. Novel expression and regulation of the renin-angiotensin system in metanephric organ culture. Am J Physiol Regul Integr Comp Physiol. 2000;279:R522–R530.
83. Miyata N, Park F, Li XF, Cowley AW, Jr. Distribution of angiotensin AT1 and AT2 receptor subtypes in the rat kidney. Am J Physiol. 1999;277:F437–F446.
84. Bagby SP, LeBard LS, Luo Z, et al. ANG II AT(1) and AT(2) receptors in developing kidney of normal microswine. Am J Physiol Renal Physiol. 2002;283:F755–F764.
85. Siragy HM, Carey RM. Protective role of the angiotensin AT2 receptor in a renal wrap hypertension model. Hypertension. 1999;33:1237–1242.
86. Lee SH, Yoo TH, Nam BY, et al. Activation of local aldosterone system within podocytes is involved in apoptosis under diabetic conditions. Am J Physiol Renal Physiol. 2009;297: F1381–F1390.
87. Taira M, Toba H, Murakami M, et al. Spironolactone exhibits direct renoprotective effects and inhibits renal renin-angiotensin-aldosterone system in diabetic rats. Eur J Pharmacol. 2008;589:264–271.
88. Sato A, Saruta T, Funder JW. Combination therapy with aldosterone blockade and renin-angiotensin inhibitors confers organ protection. Hypertens Res. 2006;29:211–216.
89. Kiyomoto H, Rafiq K, Mostofa M, Nishiyama A. Possible underlying mechanisms responsible for aldosterone and mineralocorticoid receptor-dependent renal injury. J Pharmacol Sci. 2008;108:399–405.
90. Nishiyama A, Yao L, Nagai Y, et al. Possible contributions of reactive oxygen species and mitogen-activated protein kinase to renal injury in aldosterone/salt-induced hypertensive rats. Hypertension. 2004;43:841–848.
91. Fan YY, Kohno M, Nakano D, et al. Inhibitory effects of a dihydropyridine calcium channel blocker on renal injury in aldosterone-infused rats. J Hypertens. 2009;27:1855–1862.
92. Miyata K, Rahman M, Shokoji T, et al. Aldosterone stimulates reactive oxygen species production through activation of NADPH oxidase in rat mesangial cells. J Am Soc Nephrol. 2005;16:2906–2912.
93. Nishiyama A, Yao L, Fan Y, et al. Involvement of aldosterone and mineralocorticoid receptors in rat mesangial cell proliferation and deformability. Hypertension. 2005;45:710–716.
94. Nagai Y, Miyata K, Sun GP, et al. Aldosterone stimulates collagen gene expression and synthesis via activation of ERK1/2 in rat renal fibroblasts. Hypertension. 2005;46:1039–1045.

95. Nishiyama A, Hitomi H, Rahman A, Kiyomoto H. Drug discovery for overcoming chronic kidney disease (CKD): pharmacological effects of mineralocorticoid-receptor blockers. J Pharmacol Sci. 2009;109:1–6.
96. Patni H, Mathew JT, Luan L, et al. Aldosterone promotes proximal tubular cell apoptosis: role of oxidative stress. Am J Physiol Renal Physiol. 2007;293:F1065–F1071.
97. Zhang A, Jia Z, Guo X, Yang T. Aldosterone induces epithelial-mesenchymal transition via ROS of mitochondrial origin. Am J Physiol Renal Physiol. 2007;293:F723–F731.
98. Shibata S, Nagase M, Yoshida S, et al. Modification of mineralocorticoid receptor function by Rac1 GTPase: implication in proteinuric kidney disease. Nat Med. 2008;14:1370–1376.
99. Nistala R, Whaley-Connell A, Sowers JR. Redox control of renal function and hypertension. Antioxid Redox Signal. 2008;10:2047–2089.
100. Budisavljevic MN, Hodge L, Barber K, et al. Oxidative stress in the pathogenesis of experimental mesangial proliferative glomerulonephritis. Am J Physiol Renal Physiol. 2003;285:F1138–F1148.
101. Geiszt M, Kopp JB, Varnai P, Leto TL. Identification of renox, an NAD(P)H oxidase in kidney. Proc Natl Acad Sci USA. 2000;97:8010–8014.
102. Radeke HH, Cross AR, Hancock JT, et al. Functional expression of NADPH oxidase components (alpha- and beta-subunits of cytochrome b558 and 45-kDa flavoprotein) by intrinsic human glomerular mesangial cells. J Biol Chem. 1991;266:21025–21029.
103. Shiose A, Kuroda J, Tsuruya K, et al. A novel superoxide-producing NAD(P)H oxidase in kidney. J Biol Chem. 2001;276:1417–1423.
104. Chabrashvili T, Tojo A, Onozato ML, et al. Expression and cellular localization of classic NADPH oxidase subunits in the spontaneously hypertensive rat kidney. Hypertension. 2002;39:269–274.
105. Jones SA, Hancock JT, Jones OT, et al. The expression of NADPH oxidase components in human glomerular mesangial cells: detection of protein and mRNA for p47phox, p67phox, and p22phox. J Am Soc Nephrol. 1995;5:1483–1491.
106. Schluter T, Grimm R, Steinbach A, et al. Neonatal sympathectomy reduces NADPH oxidase activity and vascular resistance in spontaneously hypertensive rat kidneys. Am J Physiol Regul Integr Comp Physiol. 2006;291:R391–R399.
107. Gorin Y, Ricono JM, Kim NH, et al. Nox4 mediates angiotensin II-induced activation of Akt/protein kinase B in mesangial cells. Am J Physiol Renal Physiol. 2003;285:F219–F229.
108. Moriwaki K, Kiyomoto H, Hitomi H, et al. Interferon-gamma enhances superoxide production in human mesangial cells via the JAK-STAT pathway. Kidney Int. 2006;70:788–793.
109. Orient A, Donko A, Szabo A, et al. Novel sources of reactive oxygen species in the human body. Nephrol Dial Transplant. 2007;22:1281–1288.
110. Namikoshi T, Tomita N, Satoh M, et al. Olmesartan ameliorates renovascular injury and oxidative stress in Zucker obese rats enhanced by dietary protein. Am J Hypertens. 2007;20:1085–1091.
111. Bek MJ, Reinhardt HC, Fischer KG, et al. Up-regulation of early growth response gene-1 via the CXCR3 receptor induces reactive oxygen species and inhibits Na+/K+-ATPase activity in an immortalized human proximal tubule cell line. J Immunol. 2003;170:931–940.
112. Fujii S, Zhang L, Kosaka H. Albuminuria, expression of nicotinamide adenine dinucleotide phosphate oxidase and monocyte chemoattractant protein-1 in the renal tubules of hypertensive Dahl salt-sensitive rats. Hypertens Res. 2007;30:991–998.
113. Kagawa T, Takao T, Horino T, et al. Angiotensin II receptor blocker inhibits tumour necrosis factor-alpha-induced cell damage in human renal proximal tubular epithelial cells. Nephrology (Carlton) 2008;13:309–315.
114. Vetter M, Chen ZJ, Chang GD, et al. Cyclosporin A disrupts bradykinin signaling through superoxide. Hypertension. 2003;41:1136–1142.
115. Wolf G, Hannken T, Schroeder R, et al. Antioxidant treatment induces transcription and expression of transforming growth factor beta in cultured renal proximal tubular cells. FEBS Lett. 2001;488:154–159.

116. Lassegue B, Clempus RE. Vascular NAD(P)H oxidases: specific features, expression, and regulation. Am J Physiol Regul Integr Comp Physiol. 2003;285:R277–R297.
117. Zou AP, Li N, Cowley AW, Jr. Production and actions of superoxide in the renal medulla. Hypertension. 2001;37:547–553.
118. Zou LX, Imig JD, Hymel A, Navar LG. Renal uptake of circulating angiotensin II in Val5-angiotensin II infused rats is mediated by AT1 receptor. Am J Hypertens. 1998;11:570–578.
119. Etoh T, Inoguchi T, Kakimoto M, et al. Increased expression of NAD(P)H oxidase subunits, NOX4 and p22phox, in the kidney of streptozotocin-induced diabetic rats and its reversibility by interventive insulin treatment. Diabetologia. 2003;46:1428–1437.
120. Gorin Y, Block K, Hernandez J, et al. Nox4 NAD(P)H oxidase mediates hypertrophy and fibronectin expression in the diabetic kidney. J Biol Chem. 2005;280:39616–39626.
121. Wassmann S, Laufs U, Muller K, et al. Cellular antioxidant effects of atorvastatin in vitro and in vivo. Arterioscler Thromb Vasc Biol. 2002;22:300–305.
122. Asaba K, Tojo A, Onozato ML, et al. Effects of NADPH oxidase inhibitor in diabetic nephropathy. Kidney Int. 2005;67:1890–1898.
123. Tojo A, Asaba K, Onozato ML. Suppressing renal NADPH oxidase to treat diabetic nephropathy. Expert Opin Ther Targets. 2007;11:1011–1018.
124. Whaley-Connell A, Habibi J, Nistala R, et al. Attenuation of NADPH oxidase activation and glomerular filtration barrier remodeling with statin treatment. Hypertension. 2008;51:474–480.
125. Welch WJ, Wilcox CS. AT1 receptor antagonist combats oxidative stress and restores nitric oxide signaling in the SHR. Kidney Int. 2001;59:1257–1263.
126. Gojo A, Utsunomiya K, Taniguchi K, et al. The Rho-kinase inhibitor, Fasudil, attenuates diabetic nephropathy in streptozotocin-induced diabetic rats. Eur J Pharmacol. 2007;568:242–247.
127. Du J, Fan YY, Hitomi H, et al. Mineralocorticoid receptor blockade and calcium channel blockade have different renoprotective effects on glomerular and interstitial injury in rats. Am J Physiol Renal Physiol. 2009;297:F802–F808.
128. Shibata S, Nagase M, Yoshida S, et al. Podocyte as the target for aldosterone: roles of oxidative stress and Sgk1. Hypertension. 2007;49:355–364.
129. Sun GP, Kohno M, Guo P, et al. Involvements of Rho-kinase and TGF-beta pathways in aldosterone-induced renal injury. J Am Soc Nephrol. 2006;17:2193–2201.
130. Ma J, Weisberg A, Griffin JP, et al. Plasminogen activator inhibitor-1 deficiency protects against aldosterone-induced glomerular injury. Kidney Int. 2006;69:1064–1072.
131. Saito K, Ishizaka N, Aizawa T, et al. Role of aberrant iron homeostasis in the upregulation of transforming growth factor-beta1 in the kidney of angiotensin II-induced hypertensive rats. Hypertens Res. 2004;27:599–607.
132. Blasi ER, Rocha R, Rudolph AE, et al. Aldosterone/salt induces renal inflammation and fibrosis in hypertensive rats. Kidney Int. 2003;63:1791–1800.
133. Nishiyama A, Abe Y. Molecular mechanisms and therapeutic strategies of chronic renal injury: renoprotective effects of aldosterone blockade. J Pharmacol Sci. 2006;100:9–16.
134. Nagase M, Fujita T. Aldosterone and glomerular podocyte injury. Clin Exp Nephrol. 2008;12:233–242.
135. Nishiyama A, Kobori H, Konishi Y, et al. Mineralocorticoid receptor blockade enhances the anti-proteinuric effect of an angiotensin II blocker through inhibiting podocyte injury in type 2 diabetic rats. J Pharmacol Exp Ther. 2010;332:1072–1080.
136. Griendling KK, Minieri CA, Ollerenshaw JD, Alexander RW. Angiotensin II stimulates NADH and NADPH oxidase activity in cultured vascular smooth muscle cells. Circ Res. 1994;74:1141–1148.
137. Ruiz-Ortega M, Ruperez M, Esteban V, Egido J. Molecular mechanisms of angiotensin II-induced vascular injury. Curr Hypertens Rep. 2003;5:73–79.

138. Rao GN. Hydrogen peroxide induces complex formation of SHC-Grb2-SOS with receptor tyrosine kinase and activates Ras and extracellular signal-regulated protein kinases group of mitogen-activated protein kinases. Oncogene. 1996;13:713–719.
139. Yoshizumi M, Abe J, Haendeler J, et al. Src and Cas mediate JNK activation but not ERK1/2 and p38 kinases by reactive oxygen species. J Biol Chem. 2000;275:11706–11712.
140. Suzaki Y, Yoshizumi M, Kagami S, et al. Hydrogen peroxide stimulates c-Src-mediated big mitogen-activated protein kinase 1 (BMK1) and the MEF2C signaling pathway in PC12 cells: potential role in cell survival following oxidative insults. J Biol Chem. 2002;277:9614–9621.
141. Suzaki Y, Yoshizumi M, Kagami S, et al. BMK1 is activated in glomeruli of diabetic rats and in mesangial cells by high glucose conditions. Kidney Int. 2004;65:1749–1760.
142. Perona R, Montaner S, Saniger L, et al. Activation of the nuclear factor-kappaB by Rho, CDC42, and Rac-1 proteins. Genes Dev. 1997;11:463–475.
143. Schreck R, Rieber P, Baeuerle PA. Reactive oxygen intermediates as apparently widely used messengers in the activation of the NF-kappa B transcription factor and HIV-1. EMBO J. 1991;10:2247–2258.
144. Zhang SL, Tang SS, Chen X, et al. High levels of glucose stimulate angiotensinogen gene expression via the p38 mitogen-activated protein kinase pathway in rat kidney proximal tubular cells. Endocrinology. 2000;141:4637–4646.
145. Hsieh TJ, Zhang SL, Filep JG, et al. High glucose stimulates angiotensinogen gene expression via reactive oxygen species generation in rat kidney proximal tubular cells. Endocrinology. 2002;143:2975–2985.
146. Li J, Brasier AR. Angiotensinogen gene activation by angiotensin II is mediated by the rel A (nuclear factor-kappaB p65) transcription factor: one mechanism for the renin angiotensin system positive feedback loop in hepatocytes. Mol Endocrinol. 1996;10:252–264.
147. Brinkmann MM, Glenn M, Rainbow L, et al. Activation of mitogen-activated protein kinase and NF-kappaB pathways by a Kaposi's sarcoma-associated herpesvirus K15 membrane protein. J Virol. 2003;77:9346–9358.
148. Hoffmann E, Thiefes A, Buhrow D, et al. MEK1-dependent delayed expression of Fos-related antigen-1 counteracts c-Fos and p65 NF-kappaB-mediated interleukin-8 transcription in response to cytokines or growth factors. J Biol Chem. 2005;280:9706–9718.
149. Zhang SL, Chen X, Hsieh TJ, et al. Hyperglycemia induces insulin resistance on angiotensinogen gene expression in diabetic rat kidney proximal tubular cells. J Endocrinol. 2002;172:333–344.
150. Zhang SL, Chen X, Wei CC, et al. Insulin inhibits dexamethasone effect on angiotensinogen gene expression and induction of hypertrophy in rat kidney proximal tubular cells in high glucose. Endocrinology. 2002;143:4627–4635.
151. Zhang SL, Filep JG, Hohman TC, et al. Molecular mechanisms of glucose action on angiotensinogen gene expression in rat proximal tubular cells. Kidney Int. 1999;55:454–464.
152. Zhang SL, To C, Chen X, et al. Essential role(s) of the intrarenal renin-angiotensin system in transforming growth factor-beta1 gene expression and induction of hypertrophy of rat kidney proximal tubular cells in high glucose. J Am Soc Nephrol. 2002;13:302–312.
153. Chen X, Zhang SL, Pang L, et al. Characterization of a putative insulin-responsive element and its binding protein(s) in rat angiotensinogen gene promoter: regulation by glucose and insulin. Endocrinology. 2001;142:2577–2585.
154. Hsieh TJ, Fustier P, Wei CC, et al. Reactive oxygen species blockade and action of insulin on expression of angiotensinogen gene in proximal tubular cells. J Endocrinol. 2004;183:535–550.
155. Kobori H, Nishiyama A, Abe Y, Navar LG. Enhancement of intrarenal angiotensinogen in Dahl salt-sensitive rats on high salt diet. Hypertension. 2003;41:592–597.
156. Nagai Y, Yao L, Kobori H, et al. Temporary angiotensin II blockade at the prediabetic stage attenuates the development of renal injury in type 2 diabetic rats. J Am Soc Nephrol. 2005;16:703–711.

157. Nishiyama A, Nakagawa T, Kobori H, et al. Strict angiotensin blockade prevents the augmentation of intrarenal angiotensin II and podocyte abnormalities in type 2 diabetic rats with microalbuminuria. J Hypertens. 2008;26:1849–1859.
158. Miyata K, Ohashi N, Suzaki Y, et al. Sequential activation of the reactive oxygen species/angiotensinogen/renin-angiotensin system axis in renal injury of type 2 diabetic rats. Clin Exp Pharmacol Physiol. 2008;35:922–927.
159. Suzaki Y, Ozawa Y, Kobori H. Intrarenal oxidative stress and augmented angiotensinogen are precedent to renal injury in Zucker diabetic fatty rats. Int J Biol Sci. 2007;3:40–46.
160. Kobori H, Katsurada A, Ozawa Y, et al. Enhanced intrarenal oxidative stress and angiotensinogen in IgA nephropathy patients. Biochem Biophys Res Commun. 2007;358:156–163.
161. Ohashi N, Katsurada A, Miyata K, et al. Activation of reactive oxygen species and the renin-angiotensin system in IgA nephropathy model mice. Clin Exp Pharmacol Physiol. 2009;36:509–515.
162. Ohashi N, Katsurada A, Miyata K, et al. Role of activated intrarenal reactive oxygen species and renin-angiotensin system in IgA nephropathy model mice. Clin Exp Pharmacol Physiol. 2009;36:750–755.
163. Ohashi N, Urushihara M, Kobori H. Activated intrarenal reactive oxygen species and renin angiotensin system in IgA nephropathy. Minerva urologica e nefrologica. Ital J Urol Nephrol. 2009;61:55–66.
164. Ogawa S, Kobori H, Ohashi N, et al. Angiotensin II Type 1 receptor blockers reduce urinary angiotensinogen excretion and the levels of urinary markers of oxidative stress and inflammation in patients with type 2 diabetic nephropathy. Biomark Insights. 2009;4:97–102.
165. Terada Y, Kobayashi T, Kuwana H, et al. Aldosterone stimulates proliferation of mesangial cells by activating mitogen-activated protein kinase 1/2, cyclin D1, and cyclin A. J Am Soc Nephrol. 2005;16:2296–2305.
166. Babilonia E, Lin D, Zhang Y, et al. Role of gp91phox -containing NADPH oxidase in mediating the effect of K restriction on ROMK channels and renal K excretion. J Am Soc Nephrol. 2007;18:2037–2045.
167. Zhang Y, Lin DH, Wang ZJ, et al. K restriction inhibits protein phosphatase 2B (PP2B) and suppression of PP2B decreases ROMK channel activity in the CCD. Am J Physiol. 2008;294:C765–773.

Chapter 5
Thiamine in Diabetic Renal Disease: Dietary Insufficiency, Renal Washout, Antistress Gene Response, Therapeutic Supplements, Risk Predictor, and Link to Genetic Susceptibility

Paul J. Thornalley and Naila Rabbani

Abstract Thiamine (vitamin B_1) has a key role in protection against renal disease by providing the thiamine pyrophosphate cofactor for transketolase and thereby maintaining oxidative and reductive pentosephosphate pathway activities key in countering oxidative and metabolic stress. The link to renal disease is amplified by increased washout and deficiency in renal disease, which in diabetic nephropathy may be linked to tissue-specific downregulation of thiamine transporters. Transketolase is part of the antistress gene response coordinated by transcription factor NF-E2–related factor-2 (nrf2) and activated by dietary bioactive and synthetic activators. Such activators have been found to be beneficial in the treatment of diabetic nephropathy wherein increased transketolase activity may have a key role. High dose thiamine supplements prevented the development of nephropathy in experimental diabetes and in a recent pilot scale trial reversed early stage nephropathy in patients with type 2 diabetes. Transketolase gene *TKT* variability and increased fractional excretion of thiamine were linked to susceptibility and progression of diabetic nephropathy. A definitive, large-scale trial of thiamine supplements for treatment of early stage diabetic nephropathy is now desirable.

Keywords Thiamine · Transketolase · Diabetic nephropathy · NF-E2-related factor-2 (nrf2) · Diabetic nephropathy

---

P.J. Thornalley (✉)
Warwick Medical School, Clinical Sciences Research Institute, University of Warwick, University Hospital, Coventry CV2 2DX, UK
e-mail: P.J.Thornalley@warwick.ac.uk

T. Miyata et al. (eds.), *Studies on Renal Disorders*, Oxidative Stress in Applied Basic Research and Clinical Practice, DOI 10.1007/978-1-60761-857-7_5,

1 Introduction: Role of Thiamine in Metabolism and Nutritional Sufficiency in Renal Disease

Thiamine (vitamin B_1) is a water-soluble vitamin and essential micronutrient. Thiamine is absorbed from the gastrointestinal tract and taken up into tissues and converted to thiamine pyrophosphate (TPP) catalysed by thiamine pyrophosphokinase. Thiamine pyrophosphate plays a major role in carbohydrate metabolism as the cofactor of pyruvate dehydrogenase (PDH), α-ketoglutarate dehydrogenase (KGDH), and transketolase (TKT). As a cofactor for PDH and KGDH, it is vital for pyruvate entry into and maintenance of the citric acid cycle and aerobic glycolysis. As a cofactor for TKT in the nonoxidative branch of the pentose phosphate pathway, it has a key role in supporting the oxidative branch of the pentosephosphate pathway forming nicotinamide adenine dinucleotide phosphate (NADPH) for biosynthetic and enzymatic maintenance of antioxidant and detoxification capacity – via glutathione reductase, aldoketo reductases, and others – and also driving the reductive branch to divert excess fructose-6-phosphate and triosephosphates to ribose-5-phosphate. The saturation of red blood cell TKT with TPP cofactor has conventionally been used as an indicator of thiamine status. The assay measures the increase of TKT activity in red blood cells lysates after addition of a saturating amount of TPP [1]. The increase has also been called erythrocyte-transketolase TPP effect or "thiamine effect." A thiamine effect less than 15% indicates thiamine sufficiency; higher values have been defined as a sign of thiamine deficiency (>15%, moderate thiamine deficiency; >25%, severe thiamine deficiency). Recent investigations, however, suggest that TKT saturation in red blood cell and thiamine excretion may not be adequate parameters to assess thiamine availability in diabetes [2]. It is also an inappropriate assessment in tissues with significant proteasomal proteolysis as apo-TKT has increased susceptibility to proteolysis [3]. Nutritional sufficiency of thiamine has also been defined on the basis of a threshold urinary excretion of thiamine greater than 0.20 μmol/24 h (>60 μg/24 h) [4]. Nutritional intake of thiamine relates to the urinary excretion of thiamine [5].

Dietary reference intake (DRI) of thiamine amounts to 1.3 mg/day in adult men and 1.1 mg/day in adult women. The European Population Reference Intake (PRI) for thiamine is 1.2 and 0.9 mg/day for adult males and females, respectively. In most other countries recommended intake is 1.0–1.4 mg/day for adult males and 0.8–1.1 mg/day for adult females. Estimates based on food intake indicate that reported mean intake of vitamin B_1 in some European countries varied from 1.1 to 2.3 mg/day. In Europe high level intake (95th percentile) varied from 1.9 to 6.4 mg/day. In the United States, the median daily intake of thiamine from food is approximately 2 mg/day and the 95th percentile of intake from both food and supplements was approximately 6.1 mg/day [6]. On this basis of urinary excretion of thiamine, diabetic patients in two recent studies in the United Kingdom and Pakistan had adequate nutritional intake of thiamine [2, 7]. There is concern, however, for inadequate thiamine intake of patients on renal replacement therapy [8] and renal transplant patients [9], including in some cases producing thiamine-responsive Wernicke syndrome [10].

2 Thiamine Transporters and Impaired Renal Handling of Thiamine in Diabetic Nephropathy and Renal Failure

Thiamine is an organic cation and utilizes high affinity organic cation transporters to cross cell membranes at normal physiological concentrations. At high concentrations, thiamine may also cross cell membranes by passive diffusion of the thiazolium ring-opened, unionized form. Genes for thiamine transporters are members of the solute carrier *SLC* family. Four members of this family have been identified: *SLC19A2* and *SLC19A3* – encoding thiamine transporters THTR1 [11] and THTR2 [12]; *SLC19A1* – encoding the reduced folate transporter RFC-1, which transports folic acid and also thiamine monophosphate (TMP) and TPP across cell plasma membranes [13–15]; and *SLC25A19* – encoding the mitochondrial TPP transporter (mTHTR) [16], previously thought to be a mitochondrial deoxynucleotide carrier [17].

Thiamine is absorbed mainly in the proximal part of the small intestine with some absorption also in the stomach and colon by THTR2 on the luminal surface of gastrointestinal epithelial cells and THTR1 on the basolateral surface [18]. It is then distributed throughout the body and taken into tissues by THTR1 and THTR2 transporters: THTR1 has particular high expression in skeletal muscle, placenta, heart, liver, and kidney [11, 19], and THTR2 has high expression in placenta, kidney, and liver [20]. The K_M for thiamine transport by THTR1 is 2.5 μM [11] and by THTR2 is 27 nM [18]. RFC-1 is widely expressed in human tissues, including in the mitochondrial membrane [14, 21]. It has affinities for TMP and TPP of 26 and 32 μM, respectively [14, 15]. TMP and TPP in blood plasma and cerebrospinal fluid arise from cellular efflux probably by RFC-1 [22–24]. mTHTR is expressed and localized to mitochondria of all human tissues, with particularly high levels in the kidney [25]. Thiamine in the glomerular filtrate is reabsorbed by renal brush border membrane high affinity transporters where influx is increased by an outward directed H^+ gradient [26]. RFC-1 is expressed on the apical and basolateral surface of proximal tubular epithelial cells [27] and likely mediates reuptake of TMP and TPP. Proton antiport membrane transport may operate in both intestinal and renal proximal tubular thiamine uptake [28].

Mutations in *SLC19A2* cause malfunction of THTR1, thiamine deficiency, and thiamine-responsive megaloblastic anaemia syndrome [29–32]. High-dose thiamine therapy overcomes this deficiency, consistent with passive diffusion membrane transport of the ring-opened, unionized form of thiamine at high concentrations [31]. Mutations in *SLC25A19* cause Amish lethal microcephaly, which markedly retards brain development and leads to α-ketoglutaric aciduria and premature death, usually by 6 months of age [33]. Knockout of *SLC25A19* in mice produced a similar phenotype with low mitochondrial TPP [16].

Thiamine and TMP transport may be abnormal in diabetes. Human tubular epithelial cells in vitro cultured with high glucose concentration show decreased levels of THTR1 protein [34]. In experimental diabetes, there was decreased intestinal absorption of thiamine and TMP [35] and an eightfold increase in renal clearance associated with decreased tubular reuptake of thiamine [22, 36]. High-dose supplementation of thiamine prevented increased clearance of thiamine and also the development of incipient diabetic nephropathy [36]. Experimental diabetes has recently also been associated with decreased expression of RFC-1 [37]. In clinical diabetes we also found increased renal clearance of thiamine in type 1 and type 2 diabetic patients. Renal mishandling of thiamine precedes the development of microalbuminuria [2]. Nutritional thiamine deficiency usually leads to increased THTR1 expression to scavenge available thiamine [38]. This effect was found in red blood cells of diabetic patients, presumably relating to reticulocytes and erythroblasts sensing a thiamine-deficient environment [2] and also mononuclear leukocytes of diabetic patients [39]. There is a dichotomy of response of THTR1 expression in cell types in diabetes; therefore, renal cells appeared unable to upregulate THTR1

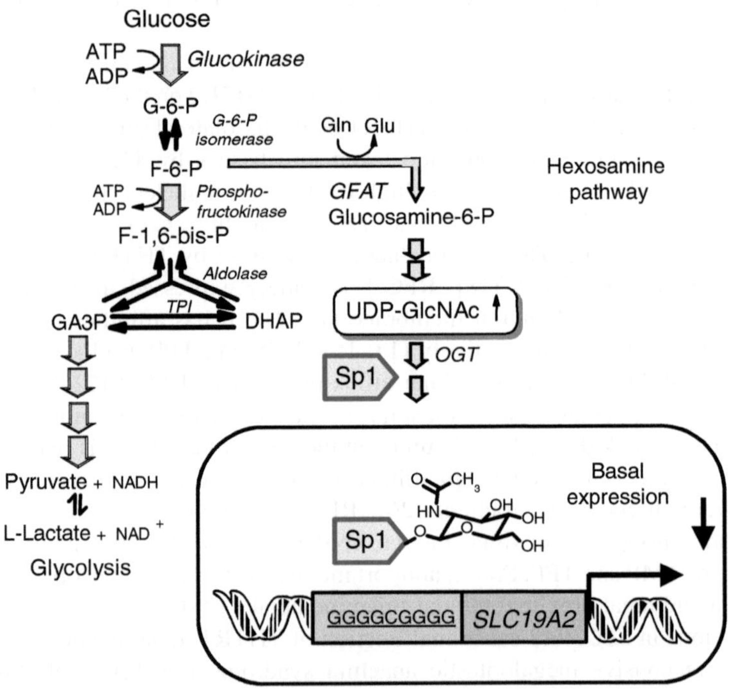

Fig. 1 Putative mechanism for the metabolic signaling via the hexosamine pathway for downregulation of basal expression of thiamine transporter THTR1. *DHAP* dihydroxyacetonephosphate; *F-1,6-bis-P* fructose-1,6-bisphosphate; *F-6-P* fructose-6-phosphate; *GA3P* glyceraldehyde-3-phosphate; *GFAT* glutamine:fructose-6-phosphate amidotransferase; *G-3-P* glycerol-3-phopshate; *G-6-P* glucose-6-phosphate; *OGT* UDP-*N*-acetylglucosaminyl transferase; *UDP-GlcNAc* UDP-*N*-acetylglucosamine

expression in response to increased renal washout of thiamine. This is likely linked to dysfunctional signaling of tubular epithelial cells in hyperglycemia. Basal expression of THTR1, THTR2, and RFC-1 transporters is dependent on transcription factor SP1. In diabetes, cells suffering high cytosolic glucose concentrations activate the hexosamine pathway, leading to increased glycosylation and inactivation of SP1. This is now implicated in tissue-specific decrease in expression of THTR1, THTR2, and RFC-1 transporters in diabetes, including glomerular endothelial cells, mesangial cells and podocytes, and tubular epithelial cells. RFC-1 is also downregulated by phosphorylation by protein kinase C (PKC) [40]. A similar downregulation of THTR1 expression has now been discovered by THTR1 proteins containing PKC phosphorylation sites [11] (Fig. 1).

Decline of glomerular filtrate rate (GFR) in renal disease may decrease washout of thiamine, whereas dialysis renal replacement therapy may exacerbate it. Hemodialysis with low and high flux membranes and peritoneal dialysis (PD) produced a washout and decreased plasma levels of thiamine [41, 42]. Hence, thiamine status is a general concern and supplements are considered for patients receiving renal replacement therapy. Washout of thiamine is thought to contribute to the increased risk of thiamine deficiency and Wernicke encephalopathy in dialysis patients [10].

3 Transketolase: Countering Oxidative and Metabolic Stress and the Antistress Gene Response in Diabetic Nephropathy

Transketolase has a pivotal role in glycolytic metabolism in sustaining both the oxidative and reductive pentosephosphate pathways. Maintenance of NADPH production in the oxidative branch sustains reductase activities for antioxidant capacity, metabolism of carbonyl compounds formed from lipid peroxidation, and the degradation of glucose and glycolytic intermediates (e.g., the metabolism of 5-hydroxynonenal, glyoxal, methylglyoxal, and 3-deoxyglucosone) [43–45]. By this function, TKT has a critical indirect role in enzymatic antioxidant defenses and the enzymatic defense against glycation, particularly by endogenous dicarbonyls ("dicarbonyl stress") and glucose degradation products in PD. The reductive pentosephosphate pathway counters the accumulation of fructose-6-phosphate and triosephosphates and thereby prevents activation of hexosamine and protein kinase C pathways, mitochondrial dysfunction driven by the glycerophosphate shuttle, and increased formation of methylglyoxal and related advanced glycation end products (AGEs) [46]. By this function, TKT has a critical indirect role in countering metabolic stress (Fig. 2).

In thiamine deficiency, TKT and PDH are considered to be the TPP-dependent enzymes most susceptible to rapid loss of TPP cofactor. Thiamine deficiency in vivo induced decreased activity of TKT and PDH [47]. Apo-TKT has a short half-life (approximately 25 min) [3] and so decreased tissue concentrations of TPP are expected to lead to a prompt decrease in TKT protein.

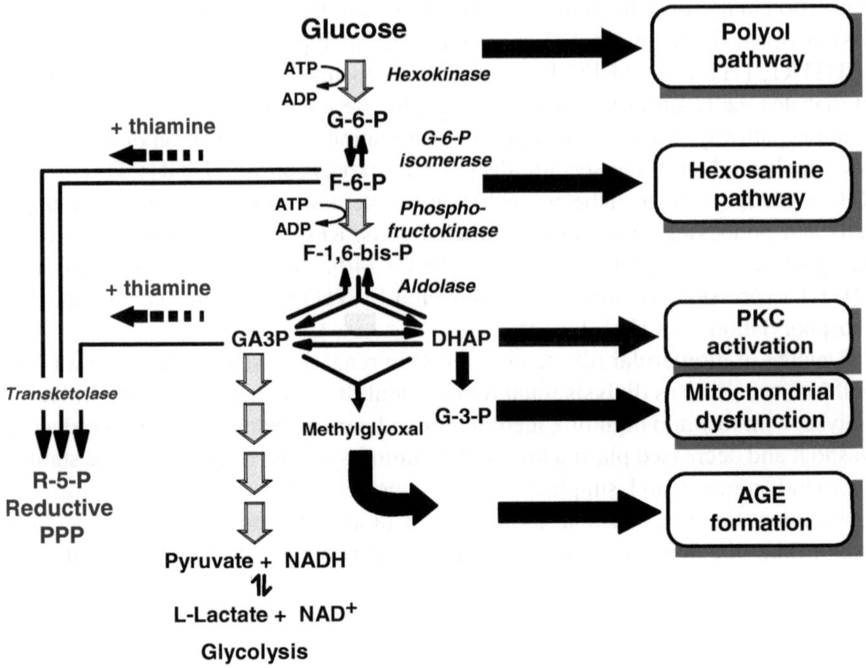

Fig. 2 Metabolic dysfunction linked to the development of diabetic nephropathy. Reversal by high dose thiamine therapy. *DHAP* dihydroxyacetonephosphate; *F-1,6-bis-P* fructose-1,6-bisphosphate; *F-6-P* fructose-6-phosphate; *GA3P* glyceraldehyde-3-phosphate; *GFAT* glutamine:fructose-6-phosphate amidotransferase; *G-3-P* glycerol-3-phopshate; *G-6-P* glucose-6-phosphate; *OGT* UDP-*N*-acetylglucosaminyl transferase; *UDP-GlcNAc* UDP-*N*-acetylglucosamine

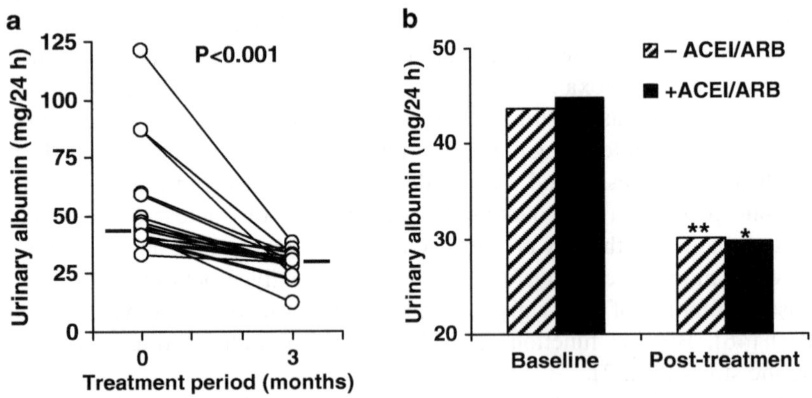

Fig. 3 Effect of high dose thiamine on urinary albumin excretion in type 2 diabetic patients with microalbuminuria. (**a**) Individual changes of urinary albumin excretion in the thiamine treatment group. (**b**) Effect of thiamine treatment on urinary albumin excretion in patients with and without ACEI or ARB therapy. Median values are shown with significance: * and **, $P < 0.05$ and $P < 0.01$, respectively. Data from Refs. [7, 62]

Transketolase and the partner enzyme of the pentosephosphate pathway, transaldolase, both have antioxidant response elements in their gene promoter sequences and have expression inducible by activation of transcription factor NF-E2-related factor-2 (nrf2) [48]. Transcription factor nrf2 coordinates the antistress gene response associated with inducible increased defense against oxidative stress, glycation, metabolic stress, dyslipidemia, and protein damage [48, 49]. It is activated by a wide range of dietary bioactive compounds with health beneficial effects: glucosinolate-derived dietary isothiocyanates [50] and indoles [51], thioethers and disulfides [52], polyphenols [53], flavonoids [54], and carotenoids [55], and omega-3 fatty acids via formation of J_3-isoprostanes [56]; and also novel synthetic compounds such as methyl-2-cyano-3,12-dioxooleana-1,9(11)dien-28-oate (CCDO methyl ester or Bardoxolone) [57]. Both bioactive and synthetic nrf2 activators have recently been found to prevent and reverse renal disease, particularly diabetic nephropathy [58–61]. Increased expression of TKT was a critical factor in prevention of hyperglycemia-induced metabolic dysfunction in microvascular endothelial cells [44] and likely contributes to the emerging benefit of nrf2 activators in renal disease. It is likely that induction of TK expression and activity contributed to these effects. Inducible TK expression cannot have a manifest benefit, however, without access to its obligatory cofactor TPP.

4 Thiamine in Early Stage Diabetic Nephropathy: Therapeutic Supplements, Risk Predictor, and Link to Genetic Susceptibility

From the above considerations it may be concluded that a novel strategy to counter biochemical dysfunction linked to the development of diabetic nephropathy is high-dose thiamine therapy. We tested this hypothesis and found that thiamine therapy prevented the development of diabetic nephropathy in experimental diabetes without improvement of glycemic control [36]. Other remarkable characteristics of the intervention were correction of increased diuresis and glucosuria [22], features also associated with activators of nrf2 [60]. Thiamine supplements corrected decreased expression and activity of TKT in glomeruli, activated the reductive pentosephosphate pathway, and countered activation of PKC and increased AGE formation [36]. In a recent clinical intervention, we were able to translate this therapeutic effect successfully to clinical diabetes.

In a pilot-scale study, we evaluated the effect of oral high dose supplements of thiamine on urinary albumin excretion (UAE), a marker of early stage diabetic nephropathy, in type 2 diabetic patients with microalbuminuria. Patients were given three 100 mg capsules of thiamine or placebo per day for 3 months with a 2-month follow-up washout period. The primary endpoint was change in UAE. Other markers of renal and vascular dysfunction and plasma concentrations of thiamine were determined. Urinary albumin excretion was decreased 41% from baseline in patients receiving thiamine therapy for 3 months with respect to baseline; all patents showed a decrease in UAE after treatment. There was no significant decrease in UAE in patients receiving placebo. There was no effect of thiamine treatment on glycemic control, dyslipidemia, or blood pressure [7]. The treatment of patients with angiotensin converting enzyme inhibitors (ACEIs) and angiotensin receptor blockers (ARBs) in this study was distributed equally between placebo and thiamine treatment groups. After 3 months' treatment with thiamine, UAE was decreased significantly in patients with and without ACEI/ARB therapy [62]. This suggests that the benefits of thiamine supplementation on the renal function of patients with type 2 diabetes with microalbuminuria are available to both those receiving and not receiving ACEI/ARB therapy, supporting the importance of pursuing further thiamine supplementation therapy for diabetic nephropathy. Addition of thiamine supplementation therapy may further improve best current therapy. Thiamine supplements at high dose may provide improved therapy for early stage diabetic nephropathy (Fig. 3).

In extension of these studies, we investigated the link of renal handling of thiamine to decline in renal function in diabetic nephropathy. In collaboration with the Joslin Kidney Study, where patients with type 1 diabetes were identified at the point of development of microalbuminuria and had over 12 years of follow-up study, we found that patients with microalbuminia and early decline in GFR had higher fractional excretion of thiamine compared to patients with stable renal function. This suggested that abnormal handling of thiamine by the kidney may be a risk predictor of decline of renal function in diabetic nephropathy [63].

The above studies have suggested that mishandling of thiamine and related transketolase expression and activity may be related to risk of diabetic nephropathy. This has prompted recent search for genetic polymorphism in the *TKT* gene. *TKT* maps to site 3p14.3 [64], close to a locus of susceptibility for decline in GFR in diabetic nephropathy at 3p14.2 [65]. In a study of 231 diabetic subjects, 15 single nucleotide polymorphisms in TKT were genotyped and haplotypes inferred using Bayesian-based algorithm. Plasma thiamine, plasma and red blood cell TPP, and red blood cell activity of TKT differed significantly between diabetics with and without diabetic nephropathy. Haplotype distribution of TKT differed significantly between case vs. control groups. Carrier state of the risk haplotype was associated with significantly accelerated onset of diabetic nephropathy and lower thiamin concentrations but not with TKT activity. This suggests that *TKT* variability and thiamine status modify susceptibility and progression of diabetic nephropathy [66].

5 Future Prospects

Recent research suggests that dietary thiamine intake is linked to maintenance of normal glycemic control in normal healthy subjects [67] and endothelial progenitor cell number in type 2 diabetic patients [68], and thiamine supplements may improve vascular health, as judged by flow-mediated dilatation response [69]. This suggests that dietary and supplementary thiamine may be determinants of vascular health. We now need a definitive study of the effect of thiamine supplements on progression of early stage diabetic nephropathy to build on the encouraging pilot scale trial outcome [7]. It is not clear if TMP pro-drugs for thiamine therapy will be as effective as thiamine; the outcome of recent pilot-scale trials with S-benzoyl-TMP (Benfotiamine) are awaited. A trial with Benfotiamine (600 mg/day for 6 months) in patients with type 2 diabetes and microalbuminuria showed no benefit on urinary albumin excretion and GFR [70]. A recent report of a similar trial of Benfotiamine in patients with type 2 diabetes and microalbuminuria aslo failed to show beneficial effect [71]. Benfotiamine may load tissues with TMP to levels that inhibit thiamine pyrophosphokinase and may thereby be less effective in trapping thiamine in target tissues than thiamine [22].

Further studies are required on the assessment of thiamine status in patients with renal disease as a predictor of risk of decline in renal function and requirement for thiamine supplement therapy. The link of *TKT* to genetic susceptibility is also deserving of further study, as well as the importance of inducible TKT expression in the likely benefit of dietary bioactive and synthetic activators of nrf2 in treatment of diabetic nephropathy and other renal disease.

Acknowledgments We thank Diabetes UK, Juvenile Diabetes Research Foundation, and the Pakistan Higher Education Commission for support for our thiamine-related research.

References

1. Brady JA, Rock CL, Horneffer MR. Thiamin status, diuretic medications, and the management of congestive heart failures. J Am Diet Assoc 1995; 95: 541–544.
2. Thornalley PJ, Babaei-Jadidi R, Al Ali H et al. High prevalence of low plasma thiamine concentration in diabetes linked to marker of vascular disease. Diabetologia 2007; 50: 2164–2170.
3. Soh Y, Song BJ, Jeng JJ et al. Critical role of Arg(433) in rat transketolase activity as probed by site-directed mutagenesis. Biochem J 1998; 333: 367–372.
4. Finglas PM. Thiamin. Int J Vitam Nutr Res 1993; 63: 270–274.
5. Tasevska N, Runswick SA, McTaggart A et al. Twenty-four-hour urinary thiamine as a biomarker for the assessment of thiamine intake. Eur J Clin Nutr 2007; 62: 1139–1147.
6. European Food Safety Authority. Scientific Opinion of the Panel on Food Additives and Nutrient Sources added to Food (ANS) on a request from the Commission on Benfotiamine, thiamine monophosphate chloride and thiamine pyrophosphate chloride, as sources of vitamin B_1. EFSA J 2008; 864: 1–31.
7. Rabbani N, Shahzad Alam S, Riaz S et al. High dose thiamine therapy for patients with type 2 diabetes and microalbuminuria: a pilot randomised, double-blind, placebo-controlled study. Diabetologia 2008; 52: 208–212.
8. Wang AYM, Sea MMM, Ip R et al. Independent effects of residual renal function and dialysis adequacy on dietary micronutrient intakes in patients receiving continuous ambulatory peritoneal dialysis. Am J Clin Nutr 2002; 76: 569–576.

9. Heaf J, Jakobsen U, Tvedegaard E et al. Dietary habits and nutritional status of renal transplant patients. J Renal Nutr 2004; 14: 20–25.
10. Hung SC, Hung SH, Tarng DC et al. Thiamine deficiency and unexplained encephalopathy in hemodialysis and peritoneal dialysis patients. Am J Kidney Dis 2001; 38: 941–947.
11. Dutta B, Huang W, Molero M et al. Cloning of the human thiamine transporter, a member of the folate transporter family. J Biol Chem 1999; 274: 31925–31929.
12. Rajgopal A, Edmondnson A, Goldman ID et al. SLC19A3 encodes a second thiamine transporter ThTr2. Biochim Biophys Acta Mol Basis Dis 2001; 1537: 175–178.
13. Matherly LH. Molecular and cellular biology of the human reduced folate carrier. Prog Nucleic Acid Res Mol Biol 2001; 67: 131–162.
14. Zhao R, Gao F, Goldman ID. Reduced folate carrier transports thiamine monophosphate: an alternative route for thiamine delivery into mammalian cells. Am J Physiol Cell Physiol 2002; 282: C1512–C1517.
15. Zhao RB, Gao F, Wang YH et al. Impact of the reduced folate carrier on the accumulation of active thiamin metabolites in murine leukemia cells. J Biol Chem 2001; 276: 1114–1118.
16. Lindhurst MJ, Fiermonte G, Song S et al. Knockout of Slc25a19 causes mitochondrial thiamine pyrophosphate depletion, embryonic lethality, CNS malformations, and anemia. PNAS 2006; 103: 15927–15932.
17. Kang J, Samuels DC. The evidence that the DNC (SLC25A19) is not the mitochondrial deoxyribonucleotide carrier. Mitochondrion 2008; 8: 103–108.
18. Said HM, Balamurugan K, Subramanian VS et al. Expression and functional contribution of hTHTR-2 in thiamin absorption in human intestine. Am J Physiol Gastrointest Liver Physiol 2004; 286: G491–G498.
19. Reidling JC, Subramanian VS, Dudeja PK et al. Expression and promoter analysis of SLC19A2 in the human intestine. Biochim Biophys Acta 2002; 1561: 180–187.
20. Eudy JD, Spiegelstein O, Barber RC et al. Identification and characterization of the human and mouse SLC19A3 gene: A novel member of the reduced folate family of micronutrient transporter genes. Mol Genet Metab 2000; 71: 581–590.
21. Trippett TM, Garcia S, Manova K et al. Localization of a human reduced folate carrier protein in the mitochondrial as well as the cell membrane of leukemia cells. Cancer Res 2001; 61: 1941–1947.
22. Babaei-Jadidi R, Karachalias N, Kupich C et al. High dose thiamine therapy counters dyslipidaemia in streptozotocin-induced diabetic rats. Diabetologia 2004; 47: 2235–2246.
23. Chin E, Zhou J, Bondy C. Anatomical and developmental patterns of facilitative glucose transporter gene-expression in the rat kidney. J Clin Invest 1993; 91: 1810–1815.
24. Tallaksen CME, Bohmer T, Karlsen J et al. Determination of thiamin and its phosphate esters in human blood, plasma, and urine. Vitam Coenzymes 1997; 279 (Pt I): 67–74.
25. Dolce V, Fiermonte G, Runswick MJ et al. The human mitochondrial deoxynucleotide carrier and its role in the toxicity of nucleoside antivirals. Proc Natl Acad Sci USA 2001; 98: 2284–2288.
26. Gastaldi G, Coya E, Verri A et al. Transport of thiamin in rat renal brush border membrane vesicles. Kidney Int 2002; 57: 2043–2054.
27. Morshed KM, Ross DM, McMartin KE. Folate transport proteins mediate the bidirectional transport of 5-methyltetrahydrofolate in cultured human proximal tubule cells. J Nutr 1997; 127: 1137–1147.
28. Rindi G, Laforenza U. Thiamine intestinal transport and related issues: recent aspects. PSEBM 2000; 224: 246–255.
29. Balamurugan K, Said HM. Functional role of specific amino acid residues in human thiamine transporter SLC19A2: mutational analysis. Am J Physiol Gastrointest Liver Physiol 2002; 283: G37–G43.
30. Fleming JC, Tartaglini E, Steinkamp MP et al. The gene mutated in thiamine-responsive anaemia with diabetes and deafness (TRMA) encodes a functional thiamine transporter. Nat Genet 1999; 22: 305–308.

31. Diaz GA, Banikazemi M, Oishi K et al. Mutations in a new gene encoding a thiamine transporter cause thiamine-responsive megaloblastic anaemia syndrome. Nat Genet 1999; 22: 309–312.
32. Labay V, Raz T, Baron D et al. Mutations in SLC19A2 cause thiamine-responsive megaloblastic anaemia associated with diabetes mellitus and deafness. Nat Genet 1999; 22: 300–304.
33. Rosenberg MJ, Agarwala R, Bouffard G et al. Mutant deoxynucleotide carrier is associated with congenital microcephaly. Nat Genet 2002; 32: 175–179.
34. Larkin JR, Thornalley PJ. High glucose causes a decrease in expression of thiamine transporters in human proximal tubule epithelial cells in vitro. Diabetologia 2008; 51: 219.
35. Patrini C, Laforenza U, Gastaldi G et al. Effects of insulin on thiamine intestinal transport in rat everted jejunal sacs. J Physiol (London) 1996; 493: 100S–101S.
36. Babaei-Jadidi R, Karachalias N, Ahmed N et al. Prevention of incipient diabetic nephropathy by high dose thiamine and Benfotiamine. Diabetes 2003; 52: 2110–2120.
37. Naggar H, Ola MS, Moore P et al. Downregulation of reduced-folate transporter by glucose in cultured RPE cells and in RPE of diabetic mice. Invest Ophthalmol Vis Sci 2002; 43: 556–563.
38. Laforenza U, Patrini C, Alvisi C et al. Thiamine uptake in human intestinal biopsy specimens, including observations from a patient with acute thiamine deficiency. Am J Clin Nutr 1997; 66: 320–326.
39. AntonySunil A, Babaei-Jadidi R, Rabbani N et al. Increased thiamine transporter contents of red blood cells and peripheral blood mononuclear leukocytes in type 1 and type 2 diabetic patients. Diabetes 2007; 56: A609.
40. Halwachs S, Kneuer C, Honscha W. Downregulation of the reduced folate carrier transport activity by phenobarbital-type cytochrome P450 inducers and protein kinase C activators. Biochim Biophys Acta 2007; 1768: 1671–1679.
41. Heinz J, Domrose U, Westphal S et al. Washout of water-soluble vitamins and of homocysteine during haemodialysis: effect of high-flux and low-flux dialyser membranes. Nephrology 2008; 13: 384–389.
42. Chychul LM, Basu TK, Dasgupta MK. Continuous ambulatory peritoneal dialysis and micronutrients: a review. Nutr Res 1991; 11: 513–525.
43. Cheng JZ, Yang YS, Singh SP et al. Two distinct 4-hydroxynonenal metabolizing glutathione S-transferase isozymes are differentially expressed in human tissues. Biochem Biophys Res Commun 2001; 282: 1268–1274.
44. Xue M, Qian Q, Adaikalakoteswari A et al. Activation of NF-E2-related factor-2 reverses biochemical dysfunction of endothelial cells induced by hyperglycemia linked to vascular disease. Diabetes 2008; 57: 2809–2817.
45. Thornalley PJ, Qian Q, Rabbani N. Prevention of dicarbonyl stress induced by hyperglycaemia in microvascular endothelial cells by sulforaphane. Diabetologia 2008; 51: 1301.
46. Thornalley PJ. The potential role of thiamine (vitamin B_1) in diabetic complications. Curr Diabetes Res 2005; 1: 287–298.
47. Terbukhina RV, Ostrovsky YM, Petushok VG et al. Effect of thiamine deprivation on thiamine metabolism in mice. J Nutr 1981; 111: 505–513.
48. Thimmulappa RK, Mai KH, Srisuma S et al. Identification of nrf2-regulated genes induced by the chemopreventive agent sulforaphane by oligonucleotide array. Cancer Res 2002; 62: 5196–5203.
49. Kwak MK, Wakabayashi N, Itoh K et al. Modulation of gene expression by cancer chemopreventive dithiolethiones through the Keap1-Nrf2 pathway. Identification of novel gene clusters for cell survival. J Biol Chem 2003; 278: 8135–8145.
50. Thornalley PJ, IARC Workgroup. Cruciferous vegetables, isothiocyanates and indoles. *IARC handbook on chemoprevention of cancer*. Lyon: IARC Press, 2004.

51. Maiyoh GK, Kuh JE, Casaschi A et al. Cruciferous indole-3-carbinol inhibits apolipoprotein B secretion in HepG2 cells. J Nutr 2007; 137: 2185–2189.
52. Chen C, Pung D, Leong V et al. Induction of detoxifying enzymes by garlic organosulfur compounds through transcription factor nrf2: effect of chemical structure and stress signals. Free Radic Biol Med 2004; 37: 1578–1590.
53. Tanigawa S, Fujii M, Hou DX. Action of Nrf2 and Keap1 in ARE-mediated NQO1 expression by quercetin. Free Radic Biol Med 2007; 42: 1690–1703.
54. Mann GE, Rowlands DJ, Li FYL et al. Activation of endothelial nitric oxide synthase by dietary isoflavones: role of NO in nrf2-mediated antioxidant gene expression. Cardiovasc Res 2007; 75: 261–274.
55. Ben Dor A, Steiner M, Gheber L et al. Carotenoids activate the antioxidant response element transcription system. Mol Cancer Ther 2005; 4: 177–186.
56. Gao L, Wang JK, Sekhar KR et al. Novel n-3 fatty acid oxidation products activate Nrf2 by destabilizing the association between Keap1 and Cullin3. J Biol Chem 2007; 282: 2529–2537.
57. Yates MS, Tauchi M, Katsuoka F et al. Pharmacodynamic characterization of chemopreventive triterpenoids as exceptionally potent inducers of Nrf2-regulated genes. Mol Cancer Ther 2007; 6: 154–162.
58. Fukuchi Y, Kato Y, Okunishi I et al. 6-Methylsulfinylhexyl isothiocyanate, an antioxidant derived from *Wasabia japonica* MATUM, ameliorates diabetic nephropathy in type 2 diabetic mice. Food Sci Technol Res 2004; 10: 290–295.
59. Sharma S, Anjaneyulu M, Kulkarni SK et al. Resveratrol, a polyphenolic phytoalexin, attenuates diabetic nephropathy in rats. Pharmacology 2006; 76: 69–75.
60. Anjaneyulu M, Chopra K. Quercetin, an anti-oxidant bioflavonoid, attenuates diabetic nephropathy in rats. Clin Exp Pharmacol Physiol 2004; 31: 244–248.
61. Schwartz SL, Denham DS, Hurwitz CA et al. Bardoxolone methyl shown to improve renal function in patients with chronic kidney disease and type 2 diabetes mellitus. Diabetes 2009; 58: A30.
62. Rabbani N, Alam S, Riaz S et al. Thiamine in diabetic nephropathy: a novel treatment modality? Reply to Alkhalaf A, Kleefstra N, Groenier KH et al. [letter]. Diabetologia 2009; 52: 1214–1216.
63. AntonySunil A, Perkins B, Krolewski A et al. Thiamine status and risk of early renal function decline in type 1 diabetic patients. Diabetologia 2008; 51: S98.
64. Lapsys NM, Layfield R, Baker E et al. Chromosomal location of the human transketolase gene. Cytogenet Cell Genet 1992; 61: 274–277.
65. Freedman BI, Bowden DW, Rich SS et al. Genome-wide linkage scans for renal function and albuminuria in type 2 diabetes mellitus: the Diabetes Heart Study. Diabet Med 2008; 25: 268–276.
66. Pacal L, Tomandl J, Tanhauserova V et al. Thiamine levels and transketolase genetic variants as modifiers of progression of diabetic nephropathy. Diabetologia 2009; 52: 1079.
67. Bakker SJL, Hoogeveen EK, Nijpels G et al. The association of dietary fibres with glucose is partly explained by concomitant intake of thiamine: the Hoorn study. Diabetologia 1998; 41: 1168–1175.
68. Wong CY, Qiuwaxi JA, Chen H et al. Daily intake of thiamine correlates with the circulating level of endothelial progenitor cells and the endothelial function in patients with type II diabetes. Mol Nutr Food Res 2008; 52: 1421–1427.
69. Arora S, Lidor A, Abularrage CJ et al. Thiamine (vitamin B_1) improves endothelium-dependent vasodilatation in the presence of hyperglycemia. Ann Vasc Surg 2006; 20: 653–658.
70. Jerums G et al. Does Benfotiamine alter microalbuminia and hyperfiltration in patients with type 2 diabetes. In: The Maillard Reaction: Interface Between Aging, Nutrition and Metabolism (Thomas M and Forbes J eds.), in press.
71. Alkhalaf A, Klooster A, van Oeveren W et al. A Double-Blind, Randomized, Placebo-Controlled Clinical Trial on Benfotiamine Treatment in Patients with Diabetic Nephropathy. Diabetes Care 2010; 33: 1598-1601.

Chapter 6
Novel Members of the Globin Family and Their Function Against Oxidative Stress

Hiroshi Nishi and Masaomi Nangaku

Abstract Globin has been a fundamental research target in the study of evolution, biochemistry, structural biology, genetics, and medicine. Conventional knowledge holds that hemoglobin in erythrocytes and myoglobin in skeletal and cardiac myocytes play an essential role in oxygen handling. Recently, however, this idea has been challenged with new findings that hemoglobin and myoglobin are expressed in visceral organs and play a role in nitric oxide metabolism. Further, neuroglobin and cytoglobin were discovered as novel mammalian globin members with unique tissue distribution. In addition, accumulating evidence supports the idea that all these globin members could share a protective role against oxidative stress. In this chapter, we provide a broad overview of the four types of globin, with particular focus on their sites of expression and scavenging properties against reactive oxygen species.

Keywords Hemoglobin · Cytoglobin · Superoxide · Hypoxia

1 Introduction

2 Globins, Hypoxia, and Oxidative Stress

The globin family of proteins are essential proteins that occur in bacteria, protists, fungi, plants, and animals [1]. This molecule has a heme domain that binds to oxygen molecules, dependent on specific physiological conditions. As oxygen is

H. Nishi (✉)
Division of Nephrology and Endocrinology, University of Tokyo School of Medicine, 7-3-1 Hongo, Bunkyo-ku, Tokyo, Japan
e-mail: mnangaku-tky@umin.ac.jp

T. Miyata et al. (eds.), *Studies on Renal Disorders*, Oxidative Stress in Applied Basic Research and Clinical Practice, DOI 10.1007/978-1-60761-857-7_6,

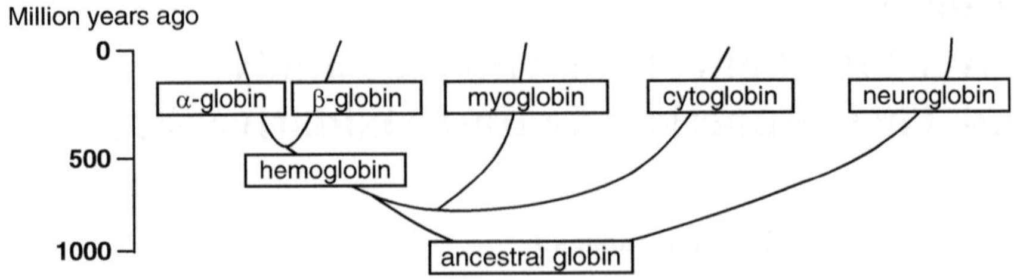

Fig. 1 An evolutionary model of mammalian globins. Hemoglobin α- and β-subunit, myoglobin, cytoglobin, and neuroglobin have the same origin of an ancestral globin. Modified from Burmester et al. [55]

indispensable to the life and activity of aerobic organisms, a shortage in oxygen, or hypoxia, induces in response multiple cellular and molecular protective mechanisms. The globin family is thus a key player in helping supply oxygen molecules to the aerobic metabolism of the respiratory chain [1–5]: as hemoglobin, it transports oxygen in erythrocytes of the circulatory system, and as myoglobin it stores and buffers oxygen in the myocytes of cardiac and skeletal tissues.

In addition to these two conventional globin family members, the two novel globins, neuroglobin and cytoglobin, have recently been discovered as new members of the globin superfamily. From an evolutionary viewpoint, these members have the same origin, named ancestral globin (Fig. 1). In contrast to conventional globins, the molecular function of these two novel globins remains inconclusive and is presently under intensive research. Whatever the outcome, however, the long history of globin biology has resulted in a conventional idea that will be challenged only with great difficulty.

Oxygen is both vital and deleterious [6]; the latter derives from its central role in the generation of reactive oxygen species (ROS), which can participate as benevolent molecules in cell signaling processes but can also induce irreversible cellular damage and death. Numerous findings support the idea that, by their interaction with oxygen, globins play a protective role in the body [4, 5]; the specific working hypothesis is that globins may be beneficial in scavenging ROS induced by severe hypoxia.

3 Hemoglobin

Hemoglobin facilitates the transport of oxygen in erythrocytes of the circulatory system. The globin molecule is an assembly of four globular protein subunits. The most common type in adult humans is hemoglobin A, which normally accounts for more than 95% of the total blood supply. Hemoglobin A consists of two α- and two β-subunits ($\alpha_2\beta_2$). Additionally, δ chain synthesis, which begins late in the third trimester and lasts into adulthood, results in the formation of hemoglobin A_2 ($\alpha_2\delta_2$), with a normal range of 1.5–3.5%. In contrast, hemoglobin F ($\alpha_2\gamma_2$) is the major type in fetuses, while Gower 1 ($\zeta_2\varepsilon_2$), Gower 2 ($\alpha_2\varepsilon_2$), and hemoglobin Portland ($\zeta_2\gamma_2$) are the primary hemoglobins in the embryo [7]. Each subunit has a molecular weight of about 17,000 Da, for a total molecular weight of the tetramer of about 68,000 Da [8]. Each protein chain arranges into a set of α-helix structural segments connected together in a globin fold arrangement, the motif of which is the same arrangement used in other combinations of heme and globin proteins. This folding pattern contains a pocket that strongly binds the heme group, which in turn binds reversibly to oxygen molecules through its iron atom and thereby enables hemoglobin to perform its primary role of picking up oxygen in the lungs and releasing it in the tissues. In 1959, Perutz et al. determined the molecular structure of hemoglobin by X-ray crystallography [9], for which with Kendrew et al. [10] he jointly received the Nobel Prize in Chemistry in 1962.

Accumulating evidence in evolutional biology has revealed that the distribution of hemoglobin in invertebrates and microbes is diverse. It also suggests that some vertebrate globins might have novel functions, such as that of a dioxygenase, oxygen sensor, or terminal oxidase [1, 8, 11–13]. In 1999, Liu et al. challenged the long-standing notion that the expression of mammalian hemoglobin genes was limited to cells of erythroid lineage [14]. They found that treatment of murine peritoneal macrophages with lipopolysaccharide and interferon-γ led to the activation of the β-globin gene in these nonerythroid cells. Seven years later, two groups independently demonstrated that hemoglobin was also expressed in alveolar epithelial type II cells of the lung [15, 16]. In the kidney, several microarray analyses indicated the presence of ectopic hemoglobin expression [17, 18], although these studies lacked detailed analyses and did not exclude the possibility of contamination. To better understand local hemoglobin expression in the normal kidney, we performed reverse transcriptase-polymerase chain reaction for glomeruli obtained by sieving of extensively saline-perfused rat kidneys, and showed expression of α- and β-globin with identical sequences to those of the corresponding globin mRNA [19]. These expressions were confirmed in manually dissected and laser capture microdissected glomeruli samples. Glomerular globin expression was also demonstrated by immunoblotting. Furthermore, in situ hybridization studies showed globin subunit expression in the mesangial region, and immunostaining with two different antibodies and double immunostaining with antibody to OX-7, a marker of rat mesangial cells, revealed mesangial localization of hemoglobin. We also confirmed globin expression in primary cultured rat mesangial cells, but not in cultured rat glomerular endothelial or epithelial cells. The fact that the site of erythropoiesis in fish is the kidney may suggest that ectopic expression of hemoglobin in mammals is a natural consequence of evolution and provide an evolutionary rationale for the role of the kidney in the regulation of bone marrow erythropoiesis in mammals [20]. Several groups subsequently observed ectopic hemoglobin expression in the central nervous system (CNS) [21–24] and endometrium [25].

The biological function of hemoglobin expressed at nonerythroid sites in mammals merits discussion but independent to its role in oxygen transportation in erythrocytes. Several hypotheses about the role of tissue hemoglobin have been proposed. First, as in invertebrates, this molecule may function as a heat transducer by virtue of its oxygenation–deoxygenation cycle, enzymatic activities, and drug interactions [1]. Second, hemoglobin might be involved in interactions of nitric oxide (NO) gas. Stamler et al. found that binding of oxygen to heme irons promoted the rapid and tight binding of NO to Cys93 of β-globin, forming S-nitrosohemoglobin, which contracted blood vessels and decreased cerebral perfusion in the R-structure and relaxed vessels to improve blood flow in the T-structure [12, 26]. Although the question of whether NO-hemoglobin is indispensable to hypoxic erythrocyte-dependent vasodilation remains unanswered [27]. This discovery raises the possibility that nonerythroid hemoglobin might play a role in NO storage and buffering.

Third, hemoglobin might interact with ROS in nonerythroid tissues. Several chemical structural analyses showed that hemoglobin effectively removes hydrogen peroxide [28] and that heme of hemoglobin detoxifies highly oxidizing radicals, yielding the ferric state [29]. With regard to hemoglobin in erythrocytes, those released during hemolysis react with endogenous oxidant hydrogen peroxide to generate transient radicals during the peroxidative consumption of hydrogen peroxide. Thus, the net biologic effect of extracellular hemoglobin in a hydrogen peroxide–rich environment will be determined by the balance of hydrogen peroxide decomposition and radical generation [30]. In contrast, our recent study of nonerythroid hemoglobin clarified that its antioxidant effect was an essential function of hemoglobin expressed by rat kidney mesangial cells [19]. Overexpression of hemoglobin α- and β-subunits in mesangial cells ameliorated oxidative stress induced by the addition of hydrogen peroxide to culture media. In this experiment, we utilized the internal ribosome entry site (IRES) vector, which expresses two different genes in a single cell, on the basis that hemoglobin acts as a tetramer of two different subunits. This finding is interesting because oxidative stress induced by the generation of free radicals is a major inciting mechanism of renal injury through which mesangial cell proliferation, fibrosis, and ultimately glomerulosclerosis may progress [31]. Our results suggest that globin may physiologically act in mesangial cells as an endogenous antioxidant defensive protein, in company with uric acid, ascorbic acid, glutathione, and antioxidant enzymes.

4 Myoglobin

Myoglobin, another member of the conventional globin family, is the most extensively characterized tissue globin. Myoglobin is a small monomeric cytoplasmic tissue globin that has a molecular weight of 16,700 Da [10]. It has eight α-helices and a hydrophobic core and contains a heme prosthetic group in the center, around which the remaining apoprotein folds. Kendrew shared the Nobel Prize with Perutz [9] for identifying the structure. Myoglobin can act as a store of oxygen to help maintain a constant supply during the rapidly fluctuating demands of muscular contraction. This helps to explain why the concentration of myoglobin is highest in the skeletal muscles of diving mammals such as whales and seals.

Vertebrate myoglobin has been thought to occur solely in cardiac and skeletal muscle. However, Fraser et al. reported that the hypoxia-tolerant common carp (*Cyprinus carpio*) has myoglobin not only in muscle but also in other metabolically active tissues, including liver, brain, and gills [32]. They identified two unique myoglobins in carp tissue, Myg-1 and -2. Myg-1 is expressed not only in muscle but also liver, kidney, and gill tissue. Interestingly, the concept of myoglobin expression in nonmuscular tissues gave researchers a key to its novel therapeutic application. Indeed, by expressing myoglobin ectopically in lung tumor cells, hypoxic status in tumor tissues was improved, which in turn delayed tumor engraftment and decreased tumor growth following xenotransplantation into mice [33].

The discovery of Fraser et al. provided another assumption, namely that nonmuscular tissue myoglobin might function in ways not directly related to cellular oxygenation [32]. In fish, Myg-1 expression in the brain was not upregulated under hypoxia, in contrast to that in the liver. Garry et al. reported that myoglobin mutant mice developed cellular and molecular adaptations to promote viability and preserve cardiac function [34], although this was partially explained by the activation of alternative compensatory mechanisms [35]. In addition, the structural analyses led to the hypothesis that myoglobin may play a role in maintaining NO homoeostasis [36, 37]. Another group pointed out that human myoglobin differs from other known mammalian myoglobins in having the presence of Cys110 to regulate NO in vivo [38]. Further, Frauenfelder et al. used mathematical structural analyses to report that, in addition to NO, myoglobin has the capacity to bind with free radicals such as hydrogen peroxide [37]. Thus, this protective role of myoglobin against ROS, separate to its oxygen storage, indicates that the physiological function of myoglobin has yet to be conclusively established.

5 Neuroglobin

Neuroglobin was discovered by Burmester et al. as a third member of the mammalian globin superfamily [39]. These investigators found novel partial globin-like sequences in the Expressed Sequence Tag (EST) database that did not correspond to any known hemoglobin or myoglobin and cloned and sequenced the coding regions of the human and mouse cDNAs. They also demonstrated that this protein is predominantly expressed in the brain, and therefore called it neuroglobin. Neuroglobin is a small monomeric cytoplasmic heme protein (151 amino acids) that is also expressed in rat, puffer fish, and zebra fish brain [40–42]. Although neuroglobin bears only a small sequence similarity to vertebrate myoglobin (<21%) and hemoglobin (<25%), all key determinants of genuine globins are nevertheless conserved: a proximal histidine residue His(F8) links the polypeptide chain to the heme iron, a distal histidine residue His(E7) resides close to the binding site of the diatomic ligand, and a phenylalanine residue Phe(CD1) is involved in π–π stacking interactions with the heme [43]. It is also the first example of a vertebrate globin that is hexacoordinate in the "deoxy" ferrous (Fe^{2+}) form, a feature previously reported only for invertebrate and plant globins [40]. In other words, the distal His(E7) of neuroglobin is coordinated to the heme iron in both the ferrous and ferric forms, in contrast to the great majority of globins, in which the met (Fe^{3+}) state is hexacoordinate with an exogenous ligand at the sixth coordination, whereas the ferrous deoxy (Fe^{2+}) form is generally pentacoordinate. This explains why neuroglobin is conventionally referred to as "unliganded" [44].

Neuroglobin is expressed in focal regions of the brain at low concentrations (less than 0.01% of total protein content) and reversibly binds oxygen, albeit less avidly than myoglobin [39]. The biological function of neuroglobin was therefore initially analyzed with regard to its interaction with oxygen molecules. Indeed, Hankeln et al. described a close correlation between the concentration of neuroglobin and metabolic activity [45], finding that the neuroglobin concentration of approximately 100 μM in mammalian retina, a tissue having among the highest rate of oxygen consumption, was comparable to that of myoglobin in skeletal muscle. Neuroglobin transcript and protein are induced in cultured primary cortical neurons during the acute phases of a hypoxic challenge and in brain tissue in vivo after focal cerebral ischemia [46]. By contrast, Mammen et al. observed no global changes in neuroglobin expression in the brains of mice exposed to chronic hypoxic conditions [47]. These conflicting results may suggest that neuroglobin functions during the acute phases of an ischemic insult.

Next, serial experiments have suggested a potential role for neuroglobin in the detoxification of ROS. In primitive in vitro experiments, human neuroblastoma SH-SY5Y cells transfected with a neuroglobin-expressing vector showed enhanced survival after oxidative insult [48]. In rats, the delivery of neuroglobin antisense into the lateral ventricle, followed by occlusion–reperfusion of the middle cerebral artery, produced a 60% increase in cerebral infarct volume, whereas induction of neuroglobin overexpression via adenoviral vector resulted in a 50% decrease in the size of the cerebral infarct [49]. Further, Kahn et al. engineered transgenic mice that overexpressed murine neuroglobin constitutively [50]: compared with wild-type littermates, cerebral infarct volume in these transgenic mice after arterial occlusion was reduced by 30%, while the volume of myocardial infarcts produced by occlusion of the left anterior descending coronary artery was reduced by 25%.

To understand the molecular mechanism of the protective role of neuroglobin against hypoxic or oxidant injury, Wakasugi et al. used the surface plasmon resonance phenomenon to show that ferric neuroglobin, which is generated spontaneously as a result of rapid autoxidation, binds exclusively to the guanosine diphosphate (GDP)-bound form of the α subunit of heterotrimeric G protein (Gαi) [51]. They reported that ferric neuroglobin behaved as a guanine nucleotide dissociation inhibitor (GDI), inhibiting the rate of exchange of GDP for guanosine 5′-triphosphate (GTP). The interaction of GDP-bound Gαi with ferric neuroglobin liberates Gβγ, leading in turn to protection against ROS-induced neuronal death. Another microscope study indicated that neuroglobin might play a preventive role against actin assembly and death-signaling module polarization in hypoxic neurons [52]. Taken together, these findings indicate that the key role of neuroglobin in the CNS may be ROS scavenging.

The pioneering study of Kahn et al. expanded the known protective mechanisms of neuroglobin beyond the antioxidant effect [53]. They found that cultured cortical neurons isolated from neuroglobin-transgenic mice were resistant to the toxic effects of β-amyloid. In addition, compared with Alzheimer's disease (AD) model mice, double-transgenic (neuroglobin-transgenic × AD mice) mice showed reductions in extracellular β-amyloid deposits and an improved dementia phenotype.

6 Cytoglobin

In 2001, Kawada et al. used proteomic analysis to identify a heme-containing protein upregulated in stellate (Ito) cells of fibrotic liver, which they termed stellate cell-activated protein (STAP) [54]. Human, mouse, and zebra fish homologues were identified in the EST databases by different groups and subsequently named cytoglobin [55, 56]. The size of mammalian (190 amino acids) and fish (174 amino acids) cytoglobin exceeds that of vertebrate myoglobin and hemoglobin, which typically comprise 140–150 residues [55–57]. Sequence conservation among mammalian cytoglobin is extremely high, with mouse and human cytoglobin differing by only 4% of their aminoacid sequences. The structural differences relative to hemoglobin and myoglobin are mainly located in the unusually long N- and C-terminal regions, whereas structural features of the classical globin fold can be recognized in the cytoglobin core region. Further, the key residues important for structure and function in respiratory globins, Phe(CD1), His(E7), and His(F8), are conserved in cytoglobin and neuroglobin. Following neuroglobin, cytoglobin is the second vertebrate hexacoordinated globin identified [55, 56].

Discussion of the sites of expression of cytoglobin has been bumpy and fraught with detours. According to a pioneering paper, the main cell type expressing this fourth globin appeared to be splanchnic fibroblasts [54]. An immunohistochemical analysis suggested that cytoglobin-positive cells are also positive for HSP47, a collagen-specific molecular chaperone, in the normal rat kidney and intestine, and for α-smooth muscle actin (α-SMA), a marker for activated myofibroblast, in the fibrotic rat kidney and pancreas [58]. They reported that the cells are also positive for CD73, a marker for renal cortical fibroblast-like cells in the normal rat kidney interstitium, and for OX-7, a marker for mesangial cells in the normal kidney glomerulus. Recently, our group also used immunostaining with two different antibodies against independent peptide sequences of cytoglobin and observed essentially the same staining pattern in the kidney interstitium, accompanied by merged immunostaining of 5′-ectonuleotidase, a rat kidney fibroblast marker [59], and cytoglobin (Nishi, Nangaku: in submission). In their immunohistochemical studies with four different antibodies, Schmidt et al. noted cytoglobin-positive cells from connective and supportive tissues, represented by not only fibroblasts but also chondrocytes and osteoblasts [60]. Cytoglobin is also reported to be expressed in neurons of the brain [61, 62], like neuroglobin, and in several layers of the retina [63, 64]; and more recently in cardiac myocyte [65], vascular endothelial cell [66], and epithelial cell lineages in a broad range of organs [67]. These controversial results might be attributable to a lack of established immunostaining methods optimized for this protein, extremely low levels of expression in the cells, and possibly also to species differences.

The biological role of cytoglobin has been the subject of intense study. Various functions have been proposed, including oxygen storage [57], radical scavenging [48, 61, 68], collagen synthesis [58, 69, 70], and tumor suppression [71]. Given that its structure indicates the same level of binding affinity for oxygen as myoglobin, an oxygen-storing property is reasonable, suggesting in turn that it may facilitate oxygen diffusion to the mitochondrial respiratory chain in various tissues [56]. Indeed, the marked upregulation of cytoglobin has been reported in hypoxic tissues, including brain, kidney, and connective tissues [60]. Further, Fordel et al. observed that hypoxia-induced upregulation of cytoglobin is attenuated in hypoxia-inducible factor (HIF)-1 heterozygous mutant mice and suggested that cytoglobin is regulated by HIF and its responsive element pathway [72]. On this basis, upregulation of cytoglobin in hypoxic tissue may be an adaptive response to increase the efficiency of mitochondrial respiration under oxygen deprivation states.

Second, given that hypoxic tissue is a major source of ROS, recent studies have suggested that cytoglobin may not only sense oxygen concentration but also act as a regulatory protein that protects cells from ROS. In the kidney, the main target of hypoxic injury is the tubulointerstitial area [31]. Given that oxidative stress mediates many of the deleterious effects of ischemia-reperfusion (IR) injury, we investigated expression of cytoglobin in a kidney IR injury model. Results showed an increased number of cytoglobin-positive cells in tubulointerstitium subjected to IR injury (Nishi, Nangaku: in submission). In addition, several interventional studies support an antioxidant effect of cytoglobin. Fordel et al. demonstrated that overexpression of cytoglobin was protective in human neuroblastoma SH-SY5Y cells exposed to oxidant insults [48], while Li et al. demonstrated that cytoglobin-siRNA transfection exacerbated cell death induced by hydrogen peroxide in N2a neuroblastoma cells [68]. Our experiments with cytoglobin-overexpressing rats more clearly demonstrated its ROS-scavenging properties in vivo (Nishi, Nangaku: in submission): while no pathological phenotypes were observed under normal conditions, overexpression of cytoglobin prevented IR injury of the kidney in experimental rats, accompanied by the decreased deposition of oxidative stress markers such as 4-hydroxy-2-nonenal and nitrotyrosine in the kidney interstitium.

Third, a cloning study provided the initial suggestion of a pro-fibrotic effect of cytoglobin [54]. Tateaki et al. analyzed the immunohistochemistry of fibrotic liver and found that fibrotic change is characterized by a dramatic increase in the number of cytoglobin-positive but fibulin-2-negative cells [70], while Nakatani et al. reported that forced overexpression of cytoglobin in NIH3T3 fibroblasts attenuated cell migration and upregulated collagen $\alpha 1(I)$ [58]. In contrast, Xu et al. demonstrated using a recombinant adeno-associated virus (AAV) vector that toxic and cholestatic liver fibrotic injury was prevented by overexpression of exogenous cytoglobin [69]. Although apparently conflicting, these study results suggested a cause-effect association of cytoglobin with the progression of fibrosis, regardless of pro- or antifibrotic property. Further, we recently found that cytoglobin-transgenic rats showed a decrease in chronic kidney fibrosis after 5/6 nephrectomy (Mimura, Nishi, Nangaku: in submission). The milder histological findings in both studies were accompanied by lower deposition of nitrotyrosine in the liver and kidney, implying that the improvement in tissue fibrosis may be mediated by antioxidant properties of cytoglobin.

Finally, the role of cytoglobin as a tumor suppressor is a fascinating possibility, although evidence is scarce. This idea was originally derived from an observational epigenetic study in a human cohort with oral cancer, which reported the enhancement of promoter methylation of cytoglobin genes in tumors compared to control tissue [73]. Another clinical study of lung cancer patients also noted the hypermethylation of cytoglobin promoter and downregulation of cytoglobin gene in tumor tissues [74]. The subsequent gain-of-function and loss-of-function study in vitro with several cultured cancer cell lines suggested that cytoglobin affects colony formation [71].

7 Conclusion

Globin biology is a traditional category of basic and clinical science, and knowledge is expanding. Recent intensive studies have suggested nontraditional functions for hemoglobin and myoglobin and new functions for neuroglobin and cytoglobin, such as their role in redox regulation, nitric oxide metabolism, and regulation of vascular tone and protection against oxidative stress. These findings also provoke interesting speculation with regard to physiology and pathology and will certainly stimulate further investigation of the globin and oxygen biology of microbes for humans.

References

1. Hardison R (1998) Hemoglobins from bacteria to man: evolution of different patterns of gene expression. *J Exp Biol* 201, 1099–1117.
2. Wittenberg BA & Wittenberg JB (1989) Transport of oxygen in muscle. *Annu Rev Physiol* 51, 857–878.
3. Perutz MF (1990) Mechanisms regulating the reactions of human hemoglobin with oxygen and carbon monoxide. *Annu Rev Physiol* 52, 1–25.
4. Brunori M, Giuffre A, Sarti P, Stubauer G, & Wilson MT (1999) Nitric oxide and cellular respiration. *Cell Mol Life Sci* 56, 549–557.
5. Merx MW, Flogel U, Stumpe T, Godecke A, Decking UK, & Schrader J (2001) Myoglobin facilitates oxygen diffusion. *FASEB J* 15, 1077–1079.

6. Davies KJ (1995) Oxidative stress: the paradox of aerobic life. *Biochem Soc Symp* 61, 1–31.
7. Maniatis T, Fritsch EF, Lauer J, & Lawn RM (1980) The molecular genetics of human hemoglobins. *Annu Rev Genet* 14, 145–178.
8. Hardison RC (1996) A brief history of hemoglobins: plant, animal, protist, and bacteria. *Proc Natl Acad Sci U S A* 93, 5675–5679.
9. Perutz MF, Rossmann MG, Cullis AF, Muirhead H, Will G, & North AC (1960) Structure of haemoglobin: a three-dimensional Fourier synthesis at 5.5-A. resolution, obtained by X-ray analysis. *Nature* 185, 416–422.
10. Kendrew JC, Bodo G, Dintzis HM, Parrish RG, Wyckoff H, & Phillips DC (1958) A three-dimensional model of the myoglobin molecule obtained by x-ray analysis. *Nature* 181, 662–666.
11. Gardner PR, Gardner AM, Martin LA, & Salzman AL (1998) Nitric oxide dioxygenase: an enzymic function for flavohemoglobin. *Proc Natl Acad Sci U S A* 95, 10378–10383.
12. Gow AJ & Stamler JS (1998) Reactions between nitric oxide and haemoglobin under physiological conditions. *Nature* 391, 169–173.
13. Hausladen A, Gow AJ, & Stamler JS (1998) Nitrosative stress: metabolic pathway involving the flavohemoglobin. *Proc Natl Acad Sci U S A* 95, 14100–14105.
14. Liu L, Zeng M, & Stamler JS (1999) Hemoglobin induction in mouse macrophages. *Proc Natl Acad Sci U S A* 96, 6643–6647.
15. Newton DA, Rao KM, Dluhy RA, & Baatz JE (2006) Hemoglobin is expressed by alveolar epithelial cells. *J Biol Chem* 281, 5668–5676.
16. Bhaskaran M, Chen H, Chen Z, & Liu L (2005) Hemoglobin is expressed in alveolar epithelial type II cells. *Biochem Biophys Res Commun* 333, 1348–1352.
17. Hsieh TJ, Chen R, Zhang SL, Liu F, Brezniceanu ML, Whiteside CI, Fantus IG, Ingelfinger JR, Hamet P, & Chan JS (2006) Upregulation of osteopontin gene expression in diabetic rat proximal tubular cells revealed by microarray profiling. *Kidney Int* 69, 1005–1015.
18. Eikmans M, Roos-van Groningen MC, Sijpkens YW, Ehrchen J, Roth J, Baelde HJ, Bajema IM, de Fijter JW, de Heer E, & Bruijn JA (2005) Expression of surfactant protein-C, S100A8, S100A9, and B cell markers in renal allografts: investigation of the prognostic value. *J Am Soc Nephrol* 16, 3771–3786.
19. Nishi H, Inagi R, Kato H, Tanemoto M, Kojima I, Son D, Fujita T, & Nangaku M (2008) Hemoglobin is expressed by mesangial cells and reduces oxidant stress. *J Am Soc Nephrol* 19, 1500–1508.
20. Wickramasinghe SN (1993) Erythropoietin and the human kidney: evidence for an evolutionary link from studies of Salmo gairdneri. *Comp Biochem Physiol Comp Physiol* 104, 63–65.
21. Leffers H, Navarro VM, Nielsen JE, Mayen A, Pinilla L, Dalgaard M, Malagon MM, Castano JP, Skakkebaek NE, Aguilar E, *et al.* (2006) Increased expression of alpha- and beta-globin mRNAs at the pituitary following exposure to estrogen during the critical period of neonatal sex differentiation in the rat. *J Steroid Biochem Mol Biol* 99, 33–43.
22. Schelshorn DW, Schneider A, Kuschinsky W, Weber D, Kruger C, Dittgen T, Burgers HF, Sabouri F, Gassler N, Bach A, *et al.* (2009) Expression of hemoglobin in rodent neurons. *J Cereb Blood Flow Metab* 29, 585–595.
23. Richter F, Meurers BH, Zhu C, Medvedeva VP, & Chesselet MF (2009) Neurons express hemoglobin alpha- and beta-chains in rat and human brains. *J Comp Neurol* 515, 538–547.
24. Biagioli M, Pinto M, Cesselli D, Zaninello M, Lazarevic D, Roncaglia P, Simone R, Vlachouli C, Plessy C, Bertin N, *et al.* (2009) Unexpected expression of {alpha}- and {beta}-globin in mesencephalic dopaminergic neurons and glial cells. *Proc Natl Acad Sci USA.* 106:15454–15459.
25. Dassen H, Kamps R, Punyadeera C, Dijcks F, de Goeij A, Ederveen A, Dunselman G, & Groothuis P (2008) Haemoglobin expression in human endometrium. *Hum Reprod* 23, 635–641.

26. Stamler JS, Jia L, Eu JP, McMahon TJ, Demchenko IT, Bonaventura J, Gernert K, & Piantadosi CA (1997) Blood flow regulation by S-nitrosohemoglobin in the physiological oxygen gradient. *Science* 276, 2034–2037.
27. Isbell TS, Sun CW, Wu LC, Teng X, Vitturi DA, Branch BG, Kevil CG, Peng N, Wyss JM, Ambalavanan N, *et al.* (2008) SNO-hemoglobin is not essential for red blood cell-dependent hypoxic vasodilation. *Nat Med* 14, 773–777.
28. Masuoka N, Kodama H, Abe T, Wang DH, & Nakano T (2003) Characterization of hydrogen peroxide removal reaction by hemoglobin in the presence of reduced pyridine nucleotides. *Biochim Biophys Acta* 1637, 46–54.
29. Goldstein S & Samuni A (2005) Intra- and intermolecular oxidation of oxymyoglobin and oxyhemoglobin induced by hydroxyl and carbonate radicals. *Free Radic Biol Med* 39, 511–519.
30. Widmer CC, Pereira CP, Gehrig P, Vallelian F, Schoedon G, Buehler PW, & Schaer DJ (2009) Hemoglobin can attenuate hydrogen peroxide-induced oxidative stress by acting as an antioxidative peroxidase. *Antioxid Redox Signal.*
31. Nangaku M (2006) Chronic hypoxia and tubulointerstitial injury: a final common pathway to end-stage renal failure. *J Am Soc Nephrol* 17, 17–25.
32. Fraser J, de Mello LV, Ward D, Rees HH, Williams DR, Fang Y, Brass A, Gracey AY, & Cossins AR (2006) Hypoxia-inducible myoglobin expression in nonmuscle tissues. *Proc Natl Acad Sci U S A* 103, 2977–2981.
33. Galluzzo M, Pennacchietti S, Rosano S, Comoglio PM, & Michieli P (2009) Prevention of hypoxia by myoglobin expression in human tumor cells promotes differentiation and inhibits metastasis. *J Clin Invest* 119, 865–875.
34. Garry DJ, Ordway GA, Lorenz JN, Radford NB, Chin ER, Grange RW, Bassel-Duby R, & Williams RS (1998) Mice without myoglobin. *Nature* 395, 905–908.
35. Godecke A, Flogel U, Zanger K, Ding Z, Hirchenhain J, Decking UK, & Schrader J (1999) Disruption of myoglobin in mice induces multiple compensatory mechanisms. *Proc Natl Acad Sci USA* 96, 10495–10500.
36. Flogel U, Merx MW, Godecke A, Decking UK, & Schrader J (2001) Myoglobin: a scavenger of bioactive NO. *Proc Natl Acad Sci U S A* 98, 735–740.
37. Frauenfelder H, McMahon BH, Austin RH, Chu K, & Groves JT (2001) The role of structure, energy landscape, dynamics, and allostery in the enzymatic function of myoglobin. *Proc Natl Acad Sci U S A* 98, 2370–2374.
38. Nicolis S, Pennati A, Perani E, Monzani E, Sanangelantoni AM, & Casella L (2006) Easy oxidation and nitration of human myoglobin by nitrite and hydrogen peroxide. *Chemistry* 12, 749–757.
39. Burmester T, Weich B, Reinhardt S, & Hankeln T (2000) A vertebrate globin expressed in the brain. *Nature* 407, 520–523.
40. Couture M, Burmester T, Hankeln T, & Rousseau DL (2001) The heme environment of mouse neuroglobin. Evidence for the presence of two conformations of the heme pocket. *J Biol Chem* 276, 36377–36382.
41. Dewilde S, Kiger L, Burmester T, Hankeln T, Baudin-Creuza V, Aerts T, Marden MC, Caubergs R, & Moens L (2001) Biochemical characterization and ligand binding properties of neuroglobin, a novel member of the globin family. *J Biol Chem* 276, 38949–38955.
42. Zhang C, Wang C, Deng M, Li L, Wang H, Fan M, Xu W, Meng F, Qian L, & He F (2002) Full-length cDNA cloning of human neuroglobin and tissue expression of rat neuroglobin. *Biochem Biophys Res Commun* 290, 1411–1419.
43. Pesce A, De Sanctis D, Nardini M, Dewilde S, Moens L, Hankeln T, Burmester T, Ascenzi P, & Bolognesi M (2004) Reversible hexa- to penta-coordination of the heme Fe atom modulates ligand binding properties of neuroglobin and cytoglobin. *IUBMB Life* 56, 657–664.
44. Sawai H, Kawada N, Yoshizato K, Nakajima H, Aono S, & Shiro Y (2003) Characterization of the heme environmental structure of cytoglobin, a fourth globin in humans. *Biochemistry* 42, 5133–5142.

45. Hankeln T, Wystub S, Laufs T, Schmidt M, Gerlach F, Saaler-Reinhardt S, Reuss S, & Burmester T (2004) The cellular and subcellular localization of neuroglobin and cytoglobin – a clue to their function? *IUBMB Life* 56, 671–679.
46. Sun Y, Jin K, Mao XO, Zhu Y, & Greenberg DA (2001) Neuroglobin is up-regulated by and protects neurons from hypoxic-ischemic injury. *Proc Natl Acad Sci U S A* 98, 15306–15311.
47. Mammen PP, Shelton JM, Goetsch SC, Williams SC, Richardson JA, Garry MG, & Garry DJ (2002) Neuroglobin, a novel member of the globin family, is expressed in focal regions of the brain. *J Histochem Cytochem* 50, 1591–1598.
48. Fordel E, Thijs L, Martinet W, Lenjou M, Laufs T, Van Bockstaele D, Moens L, & Dewilde S (2006) Neuroglobin and cytoglobin overexpression protects human SH-SY5Y neuroblastoma cells against oxidative stress-induced cell death. *Neurosci Lett* 410, 146–151.
49. Sun Y, Jin K, Peel A, Mao XO, Xie L, & Greenberg DA (2003) Neuroglobin protects the brain from experimental stroke in vivo. *Proc Natl Acad Sci U S A* 100, 3497–3500.
50. Khan AA, Wang Y, Sun Y, Mao XO, Xie L, Miles E, Graboski J, Chen S, Ellerby LM, Jin K, *et al.* (2006) Neuroglobin-overexpressing transgenic mice are resistant to cerebral and myocardial ischemia. *Proc Natl Acad Sci U S A* 103, 17944–17948.
51. Wakasugi K, Nakano T, & Morishima I (2003) Oxidized human neuroglobin acts as a heterotrimeric Galpha protein guanine nucleotide dissociation inhibitor. *J Biol Chem* 278, 36505–36512.
52. Khan AA, Ou Mao X, Banwait S, Dermardirossian CM, Bokoch GM, Jin K, & Greenberg DA (2008) Regulation of hypoxic neuronal death signaling by neuroglobin. *FASEB J* 22, 1737.
53. Khan AA, Mao XO, Banwait S, Jin K, & Greenberg DA (2007) Neuroglobin attenuates beta-amyloid neurotoxicity in vitro and transgenic Alzheimer phenotype in vivo. *Proc Natl Acad Sci U S A* 104, 19114–19119.
54. Kawada N, Kristensen DB, Asahina K, Nakatani K, Minamiyama Y, Seki S, & Yoshizato K (2001) Characterization of a stellate cell activation-associated protein (STAP) with peroxidase activity found in rat hepatic stellate cells. *J Biol Chem* 276, 25318–25323.
55. Burmester T, Ebner B, Weich B, & Hankeln T (2002) Cytoglobin: a novel globin type ubiquitously expressed in vertebrate tissues. *Mol Biol Evol* 19, 416–421.
56. Trent JT III & Hargrove MS (2002) A ubiquitously expressed human hexacoordinate hemoglobin. *J Biol Chem* 277, 19538–19545.
57. Pesce A, Bolognesi M, Bocedi A, Ascenzi P, Dewilde S, Moens L, Hankeln T, & Burmester T (2002) Neuroglobin and cytoglobin. Fresh blood for the vertebrate globin family. *EMBO Rep* 3, 1146–1151.
58. Nakatani K, Okuyama H, Shimahara Y, Saeki S, Kim DH, Nakajima Y, Seki S, Kawada N, & Yoshizato K (2004) Cytoglobin/STAP, its unique localization in splanchnic fibroblast-like cells and function in organ fibrogenesis. *Lab Invest* 84, 91–101.
59. Picard N, Baum O, Vogetseder A, Kaissling B, & Le Hir M (2008) Origin of renal myofibroblasts in the model of unilateral ureter obstruction in the rat. *Histochem Cell Biol* 130, 141–155.
60. Schmidt M, Gerlach F, Avivi A, Laufs T, Wystub S, Simpson JC, Nevo E, Saaler-Reinhardt S, Reuss S, Hankeln T, *et al.* (2004) Cytoglobin is a respiratory protein in connective tissue and neurons, which is up-regulated by hypoxia. *J Biol Chem* 279, 8063–8069.
61. Mammen PP, Shelton JM, Ye Q, Kanatous SB, McGrath AJ, Richardson JA, & Garry DJ (2006) Cytoglobin is a stress-responsive hemoprotein expressed in the developing and adult brain. *J Histochem Cytochem* 54, 1349–1361.
62. Li RC, Lee SK, Pouranfar F, Brittian KR, Clair HB, Row BW, Wang Y, & Gozal D (2006) Hypoxia differentially regulates the expression of neuroglobin and cytoglobin in rat brain. *Brain Res* 1096, 173–179.
63. Ostojic J, Sakaguchi DS, de Lathouder Y, Hargrove MS, Trent JT III, Kwon YH, Kardon RH, Kuehn MH, Betts DM, & Grozdanic S (2006) Neuroglobin and cytoglobin: oxygen-binding proteins in retinal neurons. *Invest Ophthalmol Vis Sci* 47, 1016–1023.

64. Schmidt M, Laufs T, Reuss S, Hankeln T, & Burmester T (2005) Divergent distribution of cytoglobin and neuroglobin in the murine eye. *Neurosci Lett* 374, 207–211.
65. Singh S, Manda SM, Sikder D, Birrer MJ, Rothermel BA, Garry DJ, & Mammen PP (2009) Calcineurin activates cytoglobin transcription in hypoxic myocytes. *J Biol Chem* 284, 10409–10421.
66. Halligan KE, Jourd'heuil FL, & Jourd'heuil D (2009) Cytoglobin is expressed in the vasculature and regulates cell respiration and proliferation via nitric oxide dioxygenation. *J Biol Chem* 284, 8539–8547.
67. Shigematsu A, Adachi Y, Matsubara J, Mukaide H, Koike-Kiriyama N, Minamino K, Shi M, Yanai S, Imamura M, Taketani S, *et al.* (2008) Analyses of expression of cytoglobin by immunohistochemical studies in human tissues. *Hemoglobin* 32, 287–296.
68. Li D, Chen XQ, Li WJ, Yang YH, Wang JZ, & Yu AC (2007) Cytoglobin up-regulated by hydrogen peroxide plays a protective role in oxidative stress. *Neurochem Res* 32, 1375–1380.
69. Xu R, Harrison PM, Chen M, Li L, Tsui TY, Fung PC, Cheung PT, Wang G, Li H, Diao Y, *et al.* (2006) Cytoglobin overexpression protects against damage-induced fibrosis. *Mol Ther* 13, 1093–1100.
70. Tateaki Y, Ogawa T, Kawada N, Kohashi T, Arihiro K, Tateno C, Obara M, & Yoshizato K (2004) Typing of hepatic nonparenchymal cells using fibulin-2 and cytoglobin/STAP as liver fibrogenesis-related markers. *Histochem Cell Biol* 122, 41–49.
71. Shivapurkar N, Stastny V, Okumura N, Girard L, Xie Y, Prinsen C, Thunnissen FB, Wistuba, II, Czerniak B, Frenkel E, *et al.* (2008) Cytoglobin, the newest member of the globin family, functions as a tumor suppressor gene. *Cancer Res* 68, 7448–7456.
72. Fordel E, Geuens E, Dewilde S, Rottiers P, Carmeliet P, Grooten J, & Moens L (2004) Cytoglobin expression is upregulated in all tissues upon hypoxia: an in vitro and in vivo study by quantitative real-time PCR. *Biochem Biophys Res Commun* 319, 342–348.
73. Shaw RJ, Liloglou T, Rogers SN, Brown JS, Vaughan ED, Lowe D, Field JK, & Risk JM (2006) Promoter methylation of P16, RARbeta, E-cadherin, cyclin A1 and cytoglobin in oral cancer: quantitative evaluation using pyrosequencing. *Br J Cancer* 94, 561–568.
74. Xinarianos G, McRonald FE, Risk JM, Bowers NL, Nikolaidis G, Field JK, & Liloglou T (2006) Frequent genetic and epigenetic abnormalities contribute to the deregulation of cytoglobin in non-small cell lung cancer. *Hum Mol Genet* 15, 2038–2044.

Part II
Clinical Aspects of Oxidative Stress in the Kidney

Chapter 7
Hypertension

Bernardo Rodriguez-Iturbe and Nosratola D. Vaziri

Abstract Studies in animals have shown that oxidative stress precedes hypertension, and a variety of antioxidant treatments prevent or ameliorate genetic and acquired hypertension. Human studies have demonstrated deficiencies in several antioxidant mechanisms in patients with essential hypertension, but significant difficulties exist in designing conclusive therapeutic trials and a causal relationship between oxidative stress, and hypertension remains unproven. Increased intravascular pressure, shear, and oscillatory stress induce oxidative stress that results in a reduction in nitric oxide availability and endothelial dysfunction. In addition, oxidative stress activates kinases and phosphatases that induce vascular remodeling. The effects of reactive oxygen species on vascular tone are not uniform, but in the kidney they favor sodium retention. The mechanisms by which a high salt intake increases oxidative stress are reviewed. The interrelation between oxidative stress, inflammation, and angiotensin II activity in the kidney results in a tendency to sodium retention, which is the hallmark of salt-sensitive hypertension. The possible participation of a low-grade autoimmune reactivity in the kidney in the pathogenesis of essential hypertension is discussed.

Keywords Oxidative stress · Nitric oxide · Endothelial dysfunction · Angiotensin II

1 Introduction

Substantial quantities of reactive oxygen species (ROS) are produced in the course of normal metabolism, cell signaling, and innate immunity. ROS play an essential role in regulation of cell growth, gene expression, protein biosynthesis, host defense

B. Rodriguez-Iturbe (✉)
Servicio de Nefrología, Hospital Universitario, Universidad del Zulia and IVIC-Zulia, Maracaibo, Estado Zulia, Venezuela
e-mail: bernardori@telcel.net.ve

T. Miyata et al. (eds.), *Studies on Renal Disorders*, Oxidative Stress in Applied Basic Research and Clinical Practice, DOI 10.1007/978-1-60761-857-7_7,

against microbial infections, fetal development, intercellular signaling and cellular senescence, apoptosis, and necrosis. The primary ROS generated in the body is superoxide (O_2^-), which is formed from single electron reduction of molecular oxygen. The bulk of superoxide produced in the body originates in the mitochondria, representing about 2% of the mitochondrial electron-chain generated by-products [1]. In addition several cytosolic oxidases, principally nicotinamide adenine dinucleotide phosphate (NADPH) oxidase as well as xanthine oxidase (XO), lipoxygenase (LOX), and cyclooxygenase (COX), are involved in generating ROS. Superoxide is converted to H_2O_2 by superoxide dismutase (SOD), and H_2O_2 is in turn converted to O_2 and H_2O by glutathione peroxidase and catalase. However, when present, electron donors (such as iron, copper, or uncontained superoxide) convert H_2O_2 to hydroxyl radical, which is the most reactive/cytotoxic free radical known. In addition, myeloperoxidase, which is a highly abundant enzyme in the phagocytes, converts H_2O_2 to hypochlorous acid ($HOCl^-$), a highly reactive chlorine species. Finally, interaction of superoxide and nitric oxide (NO) leads to formation of peroxynitrite ($OONO^-$), which is an exceedingly reactive nitrogen species. All the primary ROS and their potent derivatives such as $HOCl^-$, OH^- radicals, and $OONO^-$ have the potential of inducing cellular toxicity. Authoritative reviews [2–4] have addressed the participation of redox balance in physiologic process, and previous chapters of this book covered the role played by oxidative stress in renal biology and pathobiology. In this chapter we will review the evidence that associates oxidative stress and hypertension, discuss the mechanisms sustaining this two-way relationship, and examine the interaction between oxidative stress, inflammation, and angiotensin activity within the kidney and the role it plays in the pathogenesis of hypertension.

2 Association Between Oxidative Stress and Hypertension: Experimental and Clinical Evidence

The relationship between hypertension and the disturbances in redox balance that result in oxidative stress have been the subject of intense investigation (Table 1). In experimental animal models it has been amply demonstrated that, first, oxidative stress precedes the development of hypertension and, second, that the reduction of oxidative stress prevents or ameliorates hypertension. Observations that indicate that oxidative stress antecedes hypertension include the finding that genetic deficiency of mitochondrial SOD results in hypertension [6] and the demonstration that increment in renal NADPH oxidase [6], overproduction of O_2^- and endothelial dysfunction [7, 8] in the spontaneously hypertensive rat (SHR) precede hypertension.

Antioxidant treatments have been used to prevent and to correct hypertension. The administration of TEMPOL (hydroxy-2,2,6,6-tetramethylpiperidine-1-oxyl), a SOD mimetic compound to young prehypertensive SHR, prevents the development of hypertension [9] and for hypertensive SHR it ameliorates

Table 1 Oxidative stress and hypertension: experimental and clinical evidence

	References
Experimental evidence	
Oxidative stress precedes hypertension	
Superoxide dismutase deficiency	[6]
Increase in NADPH oxidase	[6, 7]
TEMPOL to normotensive SHR prevents hypertension	[9]
Antioxidant treatment improves/controls hypertension	
Antioxidant diets	[12–15]
Apocynin	[16]
TEMPOL	[10, 11]
Melatonin	[17]
Tetrahydrobiopterin (BH_4)	[20]
Antioxidant treatment does not modify hypertension	
TEMPOL in angiotensin II and phenylephrine infusion	[28]
TEMPOL in SD rats treated with high salt	[29]
Apocynin in SD rats treated with high salt	[29]
Human studies	
Oxidative stress in prehypertension (120–138/80–89 mmHg)	
Lower total antioxidant capacity	[30]
High oxidized lipoproteins	[30]
High plasma carbonyl levels	[31]
Low plasma ascorbic acid levels	[32]
Oxidative stress in hypertension	
Reduced levels of SOD, glutathione peroxidase, catalase	[33–35]
Increased superoxide production (platelets and leukocytes)	[36, 37]
Increased ADMA, increased 13-HODI acid	[38]
Increased ROS production in resistance vessels	[39]
Clinical trials of antioxidants for hypertension[a]	
Vitamin C improved hypertension	[40, 41]
Vitamin C worsened hypertension	[42]
Vitamin E inconclusive effects on hypertension	[13]

ADMA asymmetric dimethylarginine; *13-HODA* 13-hydroxyoctadecadienoic acid
[a]Only trials specifically designed to evaluate effects of antioxidant vitamins on blood pressure

hypertension [10, 11]. Other antioxidant strategies that have been found to ameliorate hypertension include antioxidant-rich diets [12–15], apocynin (inhibitor of NADPH oxidase) [16], melatonin (free radical scavenger) [17], L-arginine supplementation [18], and the administration of the tetrahydrobiopterin (BH_4), a cofactor that would prevent or minimize the O_2^- generation by NO synthase (NOS) uncoupling [19, 20]. The role played by NADPH oxidase and its subunits (NOX1, NOX2, NOX4, $p22^{phox}$) in the generation of ROS has been established in angiotensin II-induced hypertension [16, 21–25], the SHR, in stroke-prone SHR, and deoxycorticosterone acetate (DOCA)-salt-induced hypertension [10, 26, 27]. It should be noted that in specific models of hypertension, some antioxidants failed to correct hypertension; for instance, Zhang et al. [28] found that in acute angiotensin II- and phenylephrine-infusion studies, TEMPOL prevented phosphorylation of mitogen-activated protein kinase (MAPK) and reduced oxidative stress but did not lower blood pressure. Likewise Elmarakby et al. [29] reported that in Sprague-Dawley rats treated with a high salt diet, TEMPOL and apocynin prevented endothelin-mediated increase in plasma 8-isoprostane but did not reduce blood pressure. However, the vast majority of investigations cited before have conclusively shown improvement of hypertension with antioxidant treatments.

In human studies the causal relationship between oxidative stress and hypertension remains unproven (see Table 1). It has been shown that individuals with so-called prehypertension (blood pressure 120–138/80–89 mmHg) have 7% lower total antioxidant capacity and 15% higher levels of oxidized low-density lipoproteins [30], higher plasma protein carbonyl levels, decreased erythrocyte catalase content, increased glutathione peroxidase activity (attributed to a compensatory response to oxidative stress) [31], and reduced plasma levels of ascorbic acid, which inversely correlated with systolic blood pressure [32]. In addition, patients with well-defined hypertension have reduced levels of SOD, glutathione peroxidase and catalase [33–35], increased superoxide production in platelets and leukocytes [36, 37], and increased plasma asymmetric dimethylarginine (ADMA) and the lipid peroxidation product of linoleic acid, 13-hydroxyoctadecadienoic acid, both of which are positively correlated with mean blood pressure [38]. Furthermore, vascular smooth muscle cells from resistance arteries of hypertensive patients exhibit increased ROS production and upregulation of NADPH oxidase [39]. Nevertheless, there is no definitive demonstration of causality in these associations. Small therapeutic trials that evaluated the effects of vitamin C have reported improvement [40, 41] as well as worsening of hypertension [42], and vitamin E supplementation has also given inconsistent results [13]. Several large trials have examined the cardiovascular effects of antioxidant vitamins and have uniformly failed to show significant improvement in cardiovascular outcomes [43–47], but these trials were not designed to evaluate effects of these treatments on blood pressure and the patients included in these studies were patients with arteriosclerosis and other cardiovascular risk factors not ideal when examining isolated effects on hypertension.

There are significant difficulties in designing conclusive therapeutic trials aimed at determining if antioxidant treatment improves hypertension in humans [reviewed in 48–50]. Briefly, from a conceptual viewpoint, many oxidant reactions are designed for the defense against infective organisms and by necessity are relatively resistant to physiological antioxidant systems; this important characteristic makes them equally resistant to exogenous antioxidant therapy. In fact, antioxidant effectiveness against O_2^- production requires a combination of cell permeability and catalytic activity, which explains why vitamins C and E have low efficacy [51]. It is also important to realize that antioxidants given in excess can increase oxidative stress by the formation of their oxidized or free radical form, and, therefore, the dose of a given antioxidant is a critical and essentially undefined element in its therapeutic potential. An example of the necessary balance in antioxidant defenses has been highlighted in the renal ablation model in which a ROS-scavenger therapy with melatonin, which acts as an scavenger of free oxygen radicals, ameliorates renal damage [52], while the administration of a SOD mimetic, TEMPOL, reduces superoxide by shifting the reaction toward hydrogen peroxide, but the existing deficiency in catalase and glutathione peroxidase [53] results in the accumulation of the latter and unchanged ROS-induced nephrotoxic damage [54]. Finally, there is the central fact that although some studies demonstrate a deficiency of antioxidant defense systems in hypertensive patients [33–35], increased production of ROS is a critical element in the inflammatory reactivity associated with a variety of conditions, such as arteriosclerosis, chronic kidney disease, obesity, and hyperglycemia, which are frequent companions of hypertensive cardiovascular disease. Focusing only on the increment of the antioxidant defenses, without improving these conditions, is unlikely to have a favorable effect on oxidative stress.

In summary, while the existence of a two-way causal relationship between oxidative stress and hypertension is difficult to prove in humans, there is compelling evidence supporting this relationship in animal studies [55].

3 Mechanisms Involved in the Relationship Between Oxidative Stress and Hypertension

3.1 *Intravascular Pressure, Shear Stress, and Oscillatory Stress*

Redox-sensitive mechanisms are activated by hemodynamic forces within the arterial vascular wall. In isolated arteries, increase in intraluminal pressure has been shown to induce a protein kinase C-mediated activation of NADPH oxidase, resulting in enhanced superoxide production [56]. The effects of raised intraluminal pressure have also been explored in vivo in the experimental model of aortic coarctation, and these studies have also shown upregulation of the expression of NADPH oxidase subunits above the coarctation [57].

Shear stress, resulting from the movements of blood through the vessels, is likely beneficial since it upregulates NO production by inducing an increase in eNOS mRNA transcription triggered by activation in MAPK pathways [58]. Furthermore, shear stress stimulates compensatory increments in cytosolic SOD [59] and intracellular levels of glutathione peroxidase [60]. In contrast, oscillatory stress induces oxidative stress by increasing O_2^- and H_2O_2 production derived from NADPH oxidase, xanthine oxidase, and mitochondrial enzymes [61, 62]. These effects are particularly evident at sites where atherosclerosis lesions are present [63].

3.2 *Reduction of NO Availability and Endothelial Dysfunction*

Oxidative stress reduces the availability of the NO by several mechanisms. First, the affinity of NO and superoxide is very high, and their reactivity and consequent generation of peroxynitrite is only limited by the availability of superoxide; thus oxidative stress causes avid inactivation of NO and generation of reactive nitrogen species. In addition, oxidative stress oxidizes NO cofactor tetrahydrobiopterin (BH_4), which not only impairs generation of NO but also results in further increments in oxidative stress since it causes NOS to generate superoxide instead of NO (NOS uncoupling).

Reduction in NO bioavailability results in increased vascular resistance, increased adrenergic tone due to depressed NO-mediated inhibition of sympathetic outflow, and attenuation of NO-mediated pressure natriuresis. While all these effects are pro-hypertensive [48, 64], the most important effect of oxidative stress relative to blood pressure is the reduction of NO-mediated modulation of vascular tone. NO abundance is critical for both active vasodilatation and for counterbalancing the effects of vasoconstrictors such as angiotensin II and endothelin-1. Therefore, endothelial dysfunction is an expected consequence of oxidative stress and reduced NO bioavailability [65]. Endothelial dysfunction precedes the development of hypertension in the SHR [7], is independently associated with adverse cardiovascular outcomes [66, 67], and is a relevant feature of the abnormal hemodynamic responses in aging rats [68, 69] and elderly individuals [70].

Antioxidant treatment improves endothelial function, which contributes to the amelioration of hypertension observed with TEMPOL [71–73], with the administration of antioxidant-rich diet and vitamins [12, 74–76], and with melatonin [17] in the SHR.

The generation of ROS in the endothelial cells, vascular smooth muscle cells, adventitial fibroblasts, and inflammatory cells is not only involved in reduction of NO availability and endothelial dysfunction, but it is also a stimulus for endothelial activation, which involves changes in endothelial phenotype with cell-to-cell interactions and expression of adhesion molecules [77]. As mentioned earlier, generation of ROS in the endothelium depends in a large measure on NADPH oxidases that may also be involved in modulating the activity of other vascular sources of ROS [78]. A large number of stimuli have been found to increase NADPH oxidase activity, including metabolic factors such as glucose, insulin, advanced glycated end products, and oxidized lipids, as well as angiotensin II, ET-1, and inflammatory cytokines [78], which, in turn, can be stimulated by the generated oxidative stress, thereby closing the self-sustaining cycle of oxidative stress, inflammation, and angiotensin II activity [48, 79].

3.3 Reactive Oxygen Species, Vascular Tone, and Arteriolar Remodeling

The direct effects of different ROS on the vascular tone are not the same and can cause either vasoconstriction or vasodilation [80]. Hydrogen peroxide can activate potassium channels and induce hyperpolarization and vasodilation in the cerebral and coronary vessels [81–84], and since H_2O_2 is generated by flow [85], it likely plays a physiological role controlling vascular tone. In fact, the beneficial effects of TEMPOL, a SOD mimetic that both acutely and chronically has been found to ameliorate hypertension [86], have been ascribed to the generation of H_2O_2 [87]. Within the kidney the vascular effects of ROS are of physiological relevance in hypertension because they favor sodium retention. The effects of H_2O_2 in the renal medullary blood flow are different from the vasorelaxing effects observed in other vascular beds. Intramedullary infusions of H_2O_2 reduce medullary blood flow and sodium excretion in a dose-dependent manner [88]; moreover, as noted in a later section, chronic intramedullary H_2O_2 infusion produces salt-sensitive increments in blood pressure [89].

Superoxide causes vasoconstriction in renal cortical and in medullary circulation. NADPH oxidase is likely the most important source of superoxide in vascular smooth muscle cells, while mitochondrial respiration and other metabolic oxidases are of minor importance [90]. Angiotensin II-induced superoxide formation is mediated by AT1 receptor activation and results from upregulation of NOX1 and activation of protein kinase C [91]. In isolated renal afferent arterioles, superoxide-induced vasoconstriction is inhibited by the SOD mimetic TEMPOL [92], and since TEMPOL increases the medullary, but not the cortical renal blood flow [93], it is likely that superoxide helps to maintain the basal tone in medullary circulation.

The vasoconstriction induced by ROS results from the increment in intracellular calcium in vascular smooth muscle cells [94], but in addition to its direct effects, several additional mechanisms may contribute to superoxide-induced vasoconstriction. Superoxide increases the production of adenosine, which is a potent renal vasoconstrictor [95], and its modification of the prostaglandin system aggravates renal vasoconstriction. Superoxide inhibits prostacyclin synthase and, thereby, impairs the production of vasodilating PGI_2 [96]. ROS reaction with arachidonic acid produces 8-isoprostane PGF2α, and this compound, by activating the thromboxane A2 receptor, causes preglomerular vasoconstriction [97]. Finally, oxidative stress may also have an indirect influence in vascular tone since it modifies the response to a variety of vasoactive agents. For instance, TEMPOL partially blocks the renal vasoconstriction induced by endothelin-1 and norepinephrine and significantly reduces angiotensin-induced vasoconstriction in aorta and mesenteric vessels [98, 99] and in human forearm [100].

Vascular remodeling is a feature of hypertensive states that may contribute to the severity and maintenance of hypertension. Vascular remodeling is induced by MAPK, protein tyrosine (PT) kinases, and phosphatases and metalloproteinases that are activated by ROS. Angiotensin II activation of p38MAPK is dependent on H_2O_2 [101], and the constrictor effect of angiotensin II is mediated by activation of p38MAPK and H_2O_2. Oxidative stress-induced p38MAPK activation results in vascular smooth muscle cell proliferation and collagen synthesis [102]. Hydrogen peroxide production is required for signal transduction of epidermal growth factor receptor and platelet-derived growth factor receptor [103], and oxidation of cysteine residues by H_2O_2 inactivates PT phosphatases in a reversible process that maintains PT phosphatase in both active and inactive forms [104]. In addition, oxidative stress modulates the activity of metalloproteinases capable of degrading extracellular matrix; both MMP2 and MMP9 may be activated by ROS in vascular smooth muscle cells [22]. Therefore, ROS are capable of inducing functional as well as structural modifications of the vasculature, and oxidative stress is critically involved in the vascular remodeling in hypertensive states.

4 Oxidative Stress in the Kidney, Inflammation, Angiotensin Activity, and Salt Retention

4.1 *Salt and Oxidative Stress*

The relationship between salt intake and hypertension has been recently reviewed [105]. Pertinent to the present chapter, a high salt intake increases oxidative stress not only as a consequence of increased blood pressure levels but also by several additional and related mechanisms. A high salt diet stimulates renal NADPH oxidase activity and increases superoxide production [106]. Chronic consumption of a high salt diet results in a marked downregulation of iNOS and nNOS in renal cortex and renal medulla and downregulation of eNOS expression in the renal cortex; these changes were associated with the development of hypertension in normotensive Sprague-Dawley rats [107]. In experimental models of salt sensitive hypertension, a high salt diet results in a complex pattern of pro-hypertension responses that include upregulation of the renal renin-angiotensin system with the generation of pro-inflammatory and pro-fibrotic cytokines and an inability to raise antioxidant enzyme systems [108]. A high salt diet also increases renal endothelin production by the thick ascending limb of the loop of Henle and ET(B) receptor activation and increases thick ascending limb eNOS expression by increasing the outer medullary osmolarity, which may represent a regulatory response to increments in salt intake [109]. In addition, a high salt intake reduces the cationic amino acid transporter, leading to inhibition of L-arginine transport in the renal medulla; however, this effect may not be critical for raising blood pressure because changes in the sodium content of the diet did not modify the hypertension despite reduction of L-arginine uptake [110].

4.2 *Oxidative Stress and Renal Medullary Blood Flow*

Studies by Cowley and his associates [reviewed in 111] have investigated the role of oxidative stress in the modulation of renal medullary blood flow and its relationship with the pressure natriuresis and salt-sensitive hypertension. The vulnerability to oxidative stress of the medulla and papilla of the kidney, which may contribute to tubulointerstitial injury in the early stages of hypertension [112], is the result of oxygen shunting between the descending and ascending vasa recta, which leads to a large pO_2 gradient between these structures and the cortex. Medullary blood flow modulates pressure natriuresis by inducing changes in the renal interstitial hydrostatic pressure. Reduced medullary blood flow has been demonstrated in SHRs and is increased by chronic infusions of captopril in the medullary interstitial space without changes in cortical blood flow. This maneuver is associated with improvement of hypertension and a correction (shift to the left) of the pressure natriuresis relationship [113]. In salt-sensitive Dahl rats, salt administration causes a decrease in medullary blood flow that is corrected with L-arginine as NO precursor [114].

Increased ROS activity in the renal medulla results in a reduction in sodium excretion [111]. Both O_2^- and H_2O_2 generation have an influence in medullary blood flow. Increased O_2^- production induced by the intramedullary infusion of a SOD inhibitor results in a reduction in medullary blood flow and hypertension [115], and similar effects have been demonstrated with H_2O_2 infusion [115, 116].

The role played by NO in the modulation of medullary blood flow is demonstrated by the profound reductions in medullary blood flow observed with the inhibition of NOS with N^G-nitro-L-arginine methyl ester (L-NAME) administration [117]. However, the slope of pressure natriuresis is unchanged with inhibition of NOS, indicating that NO may participate, but is not essential, for a normal pressure–natriuresis relationship [118].

4.3 *Oxidative Stress, Renal Angiotensin Activity, and Inflammation*

It has been amply demonstrated that angiotensin II infusion stimulates ROS production [119, 120], and that the NADPH oxidase system is involved in the production of O_2^- induced by angiotensin II [121]. NADPH in endothelial cells is composed of gp91^{phox} (Nox2), p22^{phox}, p47^{phox}, p67^{phox}, and Rac1, and their contribution to the generation of oxidative stress in the pathogenesis of hypertension has been extensively reviewed [3, 4, 122] and will not be discussed here. Pertinent to the present discussion, angiotensin infusion induces mRNA and protein expression of gp91^{phox} and p22^{phox} [123] and p67^{phox} is activated in angiotensin-induced ROS production in aortic adventitial fibroblasts [124]. The smooth muscle cells have a low or absent NOX2, and in these cells NOX1 is involved in the O_2^- generated by angiotensin II [3]. Within the kidney, the activity of the renal angiotensin system plays a central role in the tendency toward sodium retention, which is the hallmark of salt-sensitive hypertension. All elements of the renin-angiotensin system are present in the kidney [125], and intrarenal angiotensin activity stimulates sodium reabsorption [126]. It should be emphasized that intrarenal angiotensin activity is independent of circulating angiotensin levels. Not only is it unmodified by plasma expansion [127], but it is stimulated by salt ingestion in salt-sensitive hypertension models that present, in contrast, suppressed plasma levels of angiotensin II [128]. Several studies have shown that resident, as well as infiltrating, immune cells in tubulointerstitial areas of the kidney express angiotensin II [12, 129–132], that renal angiotensin II content [132] and interstitial fluid angiotensin II concentration [128] are directly related to the severity of inflammation, and, as will be discussed later, immune cells produce angiotensin II [133].

The activity of the renal angiotensin system in humans cannot be studied directly, but recent studies by Kobori et al. [134] have demonstrated that it is reflected in the urinary angiotensinogen content, and they have provided evidence of enhanced activity in hypertensive states [135, 136].

In the kidney, oxidative stress, angiotensin II activity, and inflammation are inseparably linked, support one another, and are critical in the pathogenesis of hypertension [48, 79]. ROS promote inflammation by stimulating redox-sensitive signal transduction pathways and activating pro-inflammatory transcription factors, including nuclear factor kappa B and activator protein-1 [137–139]. In addition, oxidative stress induces a heat shock protein (HSP) response that results in further induction of pro-inflammatory cytokines and intercellular adhesion molecules [140]. Enhancement of leukocyte migration and activation in turn increases ROS production and amplifies the pro-inflammatory response.

The importance of renal inflammation in the pathogenesis of hypertension is not only due to the local disturbance of the pressure natriuresis mechanism but also because within the kidney, the interrelation between inflammation, oxidative stress, and angiotensin II activity promotes a tendency to sodium retention (Fig. 1) [48, 79]. A universal feature of experimental models of hypertension is the existence of

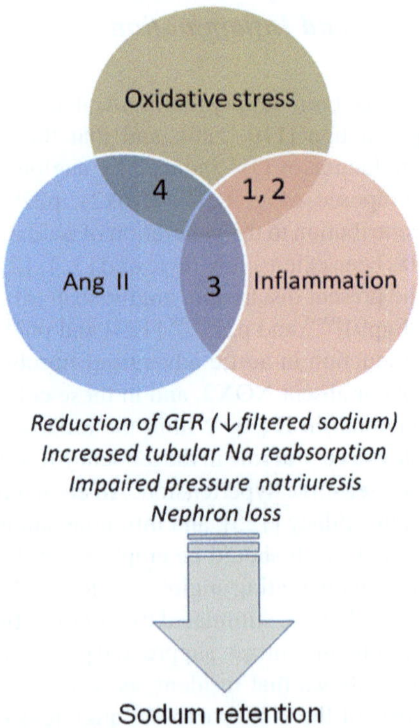

Fig. 1 Interrelation between oxidative stress, inflammation, and angiotensin II activity in the kidney results in a tendency to salt retention and hypertension. *1* Direct O_2^- production by activated inflammatory cells and resident renal cells (mediated by angiotensin II, PDGF and TNF-α). *2* Activation of NFκB, upregulation of MCP-1, MAPK, and intercellular adhesion molecules (E-selectin, ICAM-1, and VCAM-1), and increased expression of heat shock proteins. *3* Activation of NFκB, overexpression of intercellular adhesion molecules and overproduction of MCP-1. *4* Vasoconstriction, ischemia, upregulation of NADPH oxidase

Table 2 Experimental and clinical conditions in which hypertension was ameliorated or prevented with measures directed to reduce renal tubulointerstitial inflammation (updated from references [105, 141])

Experimental/clinical condition	References
Genetic strains of hypertensive rats	
SHR	[12, 17, 131, 143, 152, 165]
Transgenic dTGF rats (double-transgenic rats harboring both human renin and angiotensinogen genes)	[148]
Dahl salt-sensitive rats	[147, 150]
Hypertensive Lyon rats	[153]
Hypertensive NZB mouse	[154]
Prenatally-induced hypertension	[149]
Experimental models of hypertension	
Renal infarct hypertension	[155]
Page (cellophane wrap) hypertension	[132]
Chronic low-dose lead intoxication	[145]
Experimental salt-sensitive hypertension	
DOCA-salt hypertension	[146, 162]
Angiotensin II infusion	[130, 151]
Inhibition of nitric oxide synthase	[142]
Protein overload	[129]
Human studies	
Grade I hypertension	[159]

intrarenal inflammation [105, 141] and a series of studies from our laboratories [12, 17, 129–131, 142–145] and from other research groups [146–156] have demonstrated that reduction of inflammation results in amelioration or prevention of hypertension in genetic and acquired models of hypertension (Table 2). The relationship between immune cells and the renal angiotensin system was suggested in early studies that showed angiotensin II staining in a significant number of infiltrating lymphocytes in experimental models of hypertension [130] and a correlation between tubulointerstitial inflammation and angiotensin II-positive cells in renal biopsies of models of salt-sensitive hypertension and in SHRs [12, 129, 131]. Subsequently, in the Page (cellophane-wrapped) kidney model of hypertension, the renal content of angiotensin II was found to be related to the intensity of the inflammatory infiltration [132]. Franco et al. [128] demonstrated that in salt-sensitive hypertension, the renal interstitial angiotensin II concentration and angiotensin-mediated effects in glomerular hemodynamics may be suppressed by the use of mycophenolate mofetil (MMF) in a manner similar to that observed with angiotensin receptor (ATR) blockade. Recent studies have elegantly demonstrated that T-lymphocytes express angiotensinogen, angiotensin I-converting enzyme, and renin and are equipped to produce physiological levels of angiotensin II; furthermore, inhibition of T-cell angiotensin converting enzyme reduced tumor necrosis factor-alpha (TNF-α) production, indicating that endogenously produced angiotensin II has a regulatory role in this process. Both AT1R and AT2R are involved in this process, and superoxide was a critical mediator of T-cell TNF-α production [133].

The importance of T-lymphocytes in the pathogenesis of hypertension has been recently highlighted by studies of the same investigators showing that mice lacking T-cells are unable to develop angiotensin II-induced hypertension, and adoptive transfer of T-lymphocytes restored blood pressure responsiveness [157].

The abundant experimental evidence that relates renal inflammation and hypertension is in contrast with the few studies that support this association in clinical studies. This is logical since the use of immunosuppressive drugs for essential hypertension is ethically unjustified. In the field of renal transplantation, patients treated with MMF and rapamycin have a lesser incidence of hypertension [158], but these data may not be extrapolated to essential hypertension. Nevertheless, there are now two studies that support the possibility that renal inflammation may be relevant in the pathogenesis of hypertension in man (see Table 2). Herrera et al. [159] studied eight patients with grade 1 essential hypertension and normal renal function who received MMF for the treatment of rheumatoid arthritis or psoriasis. These patients were studied before, during, and after a course of MMF treatment. Although there were no changes in antihypertensive medication or sodium intake, the blood pressure was significantly reduced during MMF and returned to the pretreatment levels after this immunosuppressive therapy was discontinued. Urinary cytokine determinations suggested a reduction in renal inflammatory reactivity during MMF treatment. A role for intrarenal inflammation in the pathogenesis of hypertension is also suggested by data presented by Hughson et al. [160], who examined autopsy kidneys of 107 African Americans and 87 white hypertensive patients, excluding premorbid kidney disease and pathological findings of severe atherosclerosis and diabetic glomerulosclerosis. While not stated by the authors, their data show that hypertensive subjects, irrespective of race, had a significantly higher number of CD68 positive cells infiltrating the kidney (a measure of inflammation) than their normotensive counterparts and, furthermore, the most significant correlation found with mean blood pressure levels in their study was the intensity of renal infiltration of CD68 positive cells.

The importance of inflammation in the pathogenesis of hypertension has been recently recognized by Ruiz-Ortega et al. [161], who suggested that the antihypertensive effects of ATR blockers and angiotensin converting enzyme inhibitors could be partially mediated by blockade of the pro-inflammatory effects of angiotensin II.

4.4 Autoimmune Reactivity as a Potential Cause of Oxidative Stress in Hypertension

More than three decades ago, Svendsen [162] reported that the salt-dependent phase in the DOCA-salt experimental model of hypertension was not observed in the athymic nude mice. Subsequently, a series of investigations reported association between immune dysfunction [163, 164] and hypertension in the SHR, interpreted as an adaptive response to counteract the potentially fatal effects of severe hypertension [165]. Other studies, in contrast, found that immunosuppression with cyclophosphamide and antithymocytic serum ameliorated hypertension [152]. These studies went largely ignored until recently when studies from several laboratories (summarized in Table 2) demonstrated that immune suppression was associated with reduction in intrarenal inflammation, reduction in oxidative stress, and correction or prevention of hypertension. Furthermore, experiments previously cited [133, 157] have demonstrated that the participation of T-lymphocytes is necessary for the development of angiotensin II-induced hypertension.

Based on this collective evidence, we have postulated that persisting renal inflammation, upregulation of intrarenal angiotensin system, and oxidative stress in salt-sensitive hypertension could be the result of low-grade autoimmune reactivity [48, 79, 166]. Since renal overexpression of HSP is a common finding in experimental models of hypertension [167, 168], we raised the possibility that HSP could be involved in the putative local autoimmunity, as shown to occur in experimental models of autoimmune arthritis, multiple sclerosis, and diabetes [169, 170]. Indirect support for this possibility was obtained in experiments that show that splenocytes from animals in several models of salt-sensitive hypertension present a proliferative response, characteristic of delayed hypersensitivity reaction, when challenged with HSP70 [171]. Further experiments are needed to explore the potential participation of autoimmune reactivity in the pathogenesis of salt-sensitive hypertension.

References

1. Sachse A, Wolf G. Angiotensin II-induced reactive oxygen species and the kidney. *J Am Soc Nephrol* 2007; 18: 2439–2446.
2. Dröge W. Free radicals in the physiological control of cell function. *Physiol Rev* 2001; 82: 47–95.
3. Nistala R, Whaley-Connel A, Sowers JR. Redox control of renal function and hypertension. *Antiox Redox Signal* 2008; 10: 2047–2089.
4. Sedeek M, Hébert RL, Kennedy CR, Burns KD, Touyz RM. Molecular mechanisms of hypertension: role of Nox family NADPH oxidases. *Curr Opin Nephrol Hypertens* 2009; 18: 122–127.
5. Rodriguez-Iturbe B, Sepassi L, Quiroz Y, Ni Z, Wallace DC, Vaziri ND. Association of mitochondrial SOD deficiency with salt-sensitive hypertension and accelerated renal senescence. *J Appl Physiol* 2007; 102: 255–260.
6. Chabrashvili T, Tojo A, Onozato ML, Kitiyakara C, Quinn MT, Fujita T, Welch WJ, Wilcox CS. Expression and cellular localization of classic NADPH oxidase subunits in the spontaneously hypertensive rat kidney. *Hypertension* 2002; 39: 269–274.
7. Jameson M, Dai FX, Luscher T, Skopec J, Diederich D, Diederich A. Endothelium-derived contracting factors in resistance arteries of young spontaneously hypertensive rats before development of overt hypertension. *Hypertension* 1993; 21: 280–282.

8. Endemann D, Touyz RM, Li JS, Deng LY, Schiffrin EL. Altered angiotensin II-induced small artery contraction during development of hypertension in spontaneously hypertensive rats. *Am J Hypertens* 1999; 12: 716–723.
9. Nabha L, Garbern JC, Buller CL, Charpie JR. Vascular oxidative stress precedes high blood pressure in spontaneously hypertensive rats. *Clin Exp Hypertens* 2005; 27: 71–82.
10. Park JB, Touyz RM, Chen X, Schiffrin EL. Chronic treatment with a superoxide dismutase mimetic prevents vascular remodeling and progression of hypertension in salt-loaded stroke-prone spontaneously hypertensive rats. *Am J Hypertens* 2002; 15: 78–84.
11. Shokoji T, Nishiyama A, Fujisawa Y, Hitomi H, Kiyomoto H, Takahashi N, Kimura S, Kohno M, Abe Y. Renal sympathetic nerve responses to tempol in spontaneously hypertensive rats. *Hypertension* 2003; 41: 266–273.
12. Rodriguez-Iturbe B, Zhan CD, Quiroz Y, Sindhu RK, Vaziri ND. Antioxidant-rich diet relieves hypertension and reduces renal immune infiltration in spontaneously hypertensive rats. *Hypertension* 2003; 41: 341–346.
13. Houston MC. Nutraceuticals, vitamins, antioxidants, and minerals in the prevention and treatment of hypertension. *Prog Cardiovasc Dis* 2005; 47: 396–449.
14. Chen X, Touyz RM, Park JB, Schiffrin EL. Antioxidant effects of vitamins C and E are associated with altered activation of vascular NADPH oxidase and superoxide dismutase in stroke-prone SHR. *Hypertension* 2001; 38: 606–611.
15. Galley HF, Thornton J, Howdle PD, Walker BE, Webster NR. Combination oral ntioxidant supplementation reduces blood pressure. *Clin Sci (Lond)* 1997; 92: 361–365.
16. Virdis A, Neves MF, Amiri F, Touyz RM, Schiffrin EL. Role of NAD(P)H oxidase on vascular alterations in angiotensin II-infused mice. *J Hypertens* 2004; 22: 535–542.
17. Nava M, Quiroz Y, Vaziri ND, Rodríguez-Iturbe B. Melatonin reduces renal interstitial inflammation and improves hypertension in spontaneously hypertensive rats. *Am J Physiol Renal Physiol* 2003; 284: F447–F454.
18. Fujii S, Zhang L, Igarashi J, Kosaka H. L-arginine reverses p47phox and gp91phox expression induced by high salt in Dahl rats. *Hypertension* 2003; 42: 1014–1020.
19. Landmesser U, Dikalov S, Price SR, McCann L, Fukai T, Holland SM, Mitch WE, Harrison DG. Oxidation of tetrahydrobiopterin leads to uncoupling of endothelial cell nitric oxide synthase in hypertension. *J Clin Invest* 2003; 111: 1201–1209.
20. Hong HJ, Hsiao G, Cheng TH, Yen MH. Supplementation with tetrahydrobiopterin suppresses the development of hypertension in spontaneously hypertensive rats. *Hypertension* 2001; 38: 1044–1048.
21. Fukui T, Ishizaka N, Rajagopalan S, Laursen JB, Capers Q, Taylor WR, Harrison DG, de Leon H, Wilcox JN, and Griendling KK. p22phox mRNA expression and NADPH oxidase activity are increased in aortas from hypertensive rats. *Circ Res* 1997; 80: 45–51.
22. Ragopalan S, Meng XP, Ramasamy S, Harrison DG, Galis ZS. Reactive oxygen species produced by macrophage-derived foam cells regulate the activity of vascular matrix metalloproteinases in vitro Implications for artheriosclerotic plaque stability. *J Clin Invest* 1996; 98: 2572–2579.
23. Landmesser U, Cai H, Dikalov S, McCann L, Hwang J, Jo H, Holland SM, Harrison DG. Role of p47(phox) in vascular oxidative stress and hypertension caused by angiotensin II. *Hypertension* 2002; 40: 511–515.
24. Dikalova A, Clempus R, Lassegue B, Cheng G, McCoy J, Dikalov S, SanMartin A, Lyle A, Weber DS, Weiss D, Taylor WR, Schmidt HH, Owens GK, Lambeth JD, Griendling KK. Nox1 overexpression potentiates angiotensin II-induced hypertension and vascular smooth muscle hypertrophy in transgenic mice. *Circulation* 2005; 112: 2668–2676.
25. Zhang Y, Griendling KK, Dikalova A, Owens GK, Taylor WR. Vascular hypertrophy in angiotensin II-induced hypertension is mediated by vascular smooth muscle cell-derived H_2O_2. *Hypertension* 2005; 46: 732–737.

26. Zalba G, Beaumont FJ, San Jose G, Fortuno A, Fortuno MA, Etayo JC, Diez J. Vascular NADH/NAD(P)H oxidase is involved in enhanced superoxide production in spontaneously hypertensive rats. *Hypertension* 2000; 35: 1055–1061.
27. Akasaki T, Ohya Y, Kuroda J, Eto K, Abe I, Sumimoto H, Iida M. Increased expression of gp91phox homologues of NAD(P)H oxidase in the aortic media during chronic hypertension: involvement of the renin-angiotensin system. *Hypertens Res* 2006; 29: 813–820.
28. Zhang GX, Kimura S, Nishiyama A, Shokoji T, Rahman M, Abe Y. ROS during the acute phase of Ang II hypertension participates in cardiovascular MAPK activation but not vasoconstriction. *Hypertension* 2004; 43: 117–124.
29. Elmarakby AA, Loomis ED, Pollock JS, Pollock DM. NADPH oxidase inhibition attenuates oxidative stress but not hypertension produced by chronic ET-1. *Hypertension* 2005; 45: 283–287.
30. Chrysohoou C, Panagiotakos DB, Pitsavos C, Skoumas J, Economou M, Papadimitriou L, Stefanadis C. The association between prehypertension status and oxidative stress markers related to atherosclerosis disease: the ATTICA study. *Atherosclerosis* 2007; 192: 169–176.
31. Sathiyapriya V, Nandeesha H, Bobby Z, Selvaraj N, Pavithran P. Perturbation of oxidant-antioxidant status in non-obese prehypertensive male subjects. *J Hum Hypertens* 2007; 21: 176–178.
32. Yoshioka M, Matsushita T, Chuman Y. Inverse association of serum ascorbic acid level and blood pressure or rate of hypertension in male subjects age 30-39 years. *Int J Vitam Nutr Res* 1984; 54: 343–347.
33. Redon J, Oliva MR, Tormos C, Giner V, Chaves J, Iradi A, Saez GT. Antioxidant activities and oxidative stress byproducts in human hypertension. *Hypertension* 2003; 41: 1096–1101.
34. Saez GT, Tormos C, Giner V, Chaves J, Lozano JV, Iradi A, Redon J. Factors related to the impact of antihypertensive treatment in antioxidant activities and oxidative stress by-products in human hypertension. *Am J Hypertens* 2004; 17: 809–816.
35. Simic DV, Mimic-Oka J, Pljesa-Ercegovac M, Savic-Radojevic A, Opacic M, Matic D, Ivanovic B, Simic T. Byproducts of oxidative protein damage and antioxidant enzyme activities in plasma of patients with different degrees of essential hypertension. *J Hum Hypertens* 2006; 20: 149–155.
36. Minuz P, Patrignani P, Gaino S, Seta F, Capone ML, Tacconelli S, Degan M, Faccini G, Fornasiero A, Talamini G, Tommasoli R, Arosio E, Santonastaso CL, Lechi A, Patrono C. Determinants of platelet activation in human essential hypertension. *Hypertension* 2004; 43: 64–70.
37. Yasunari K, Maeda K, Nakamura M, Yoshikawa J. Oxidative stress in leukocytes is a possible link between blood pressure, blood glucose, and C-reacting protein. *Hypertension* 2002; 39: 777–780.
38. Wang D, Strandgaard S, Iversen J. Asymmetric dimethylarginine, oxidative stress, and vascular nitric oxide synthase in essential hypertension. *Am J Physiol Regul Integr Comp Physiol* 2009; 296: R195–R200.
39. Touyz RM, Schiffrin EL. Increased generation of superoxide by angiotensin II in smooth muscle cells from resistance arteries of hypertensive patients: role of phospholipase D-dependent NAD(P)H oxidase-sensitive pathways. *J Hypertens* 2001; 19: 1245–1254.
40. Duffy SJ, Gokce N, Holbrook M, Huang A, Frei B, Keaney JF, Jr., Vita JA. Treatment of hypertension with ascorbic acid. *Lancet* 1999; 354: 2048–2049.
41. Bates CJ, Walmsley CM, Prentice A, Finch S. Does vitamin C reduce blood pressure? Results of a large study of people aged 65 or older. *J Hypertens* 1998; 16: 925–932.
42. Ward NC, Hodgson JM, Croft KD, Burke V, Beilin LJ, Puddey IB. The combination of vitamin C and grape seed polyphenols increases blood pressure: a randomized, double-blind, placebo-controlled trial. *J Hypertens* 2005; 23: 427–434.

43. Stephens NG, Parsons A, Schofield PM, Kelly F, Cheeseman K, Mitchinson MJ. Randomised controlled trial of vitamin E in patients with coronary disease: Cambridge Heart Antioxidant Study (CHAOS). *Lancet* 1996; 347: 781–786.
44. Alpha-Tocopherol, Beta Carotene Cancer Prevention Study Group. The effect of vitamin E and beta carotene on the incidence of lung cancer and other cancers in male smokers. *N Engl J Med* 1994; 330: 1029–1035.
45. GISSI-Prevenzione Investigators. Dietary supplementation with n-3 polyunsaturated fatty acids and vitamin E after myocardial infarction: results of the GISSI-Prevenzione trial. *Lancet* 1999; 354: 447–455.
46. Yusuf S, Dagenais G, Pogue J, Bosch J, Sleight P. Vitamin E supplementation and cardiovascular events in high-risk patients. The Heart Outcomes Prevention Evaluation Study Investigators. *N Engl J Med* 2000; 342: 154–160.
47. Smilde TJ, van Wissen S, Wollersheim H, Trip MD, Kastelein JJ, Stalenhoef AF. Effect of aggressive versus conventional lipid lowering on atherosclerosis progression in familial hypercholesterolaemia (ASAP): a prospective, randomised, double-blind trial. *Lancet* 2001; 357: 577–581.
48. Vaziri ND, Rodriguez-Iturbe B. Mechanisms of disease: oxidative stress and inflammation in the pathogenesis of hypertension. *Nature Clin Prac Nephrol* 2006; 2: 582–593.
49. Paravicini TM, Touyz RM. NADPH oxidases, reactive oxygen species and hypertension: clinical implications and therapeutic possibilities. *Diabetes Care* 2008; 1(Suppl 2): S170–S180.
50. Grossman E. Does increased oxidative stress cause hypertension. *Diabetes Care* 2008; 31 (Suppl 2): S185–S189.
51. Luo Z, Chen Y, Chen S, Welch WJ, Andresen BT, Jose PA, Wilcox CS. Comparison of inhibitors of superoxide generation in vascular smooth muscle cells. *Br J Pharmacol* 2009; 157: 935–943.
52. Quiroz Y, Ferrebuz A, Romero F, Vaziri ND, Rodriguez-Iturbe B. Melatonin ameliorates oxidative stress, inflammation, proteinuria and progression of renal damage in rats with renal mass reduction. *Am J Physiol Renal Physiol* 2008; 294: F336–F344.
53. Sindhu RK, Ehdaie A, Farmand F, Dhaliwal KK, Nguyen T, Zhan CD, Roberts CK, Vaziri ND. Expression of catalase and glutathione peroxidase in renal insufficiency. *Biochim Biophys Acta* 2005; 1743(1–2): 86–92.
54. Quiroz Y, Ferrebuz A, Vaziri ND, Rodriguez-Iturbe B. Effect of chronic antioxidant therapy with superoxide dismutase-mimetic drug, tempol, on progression of renal disease in rats with renal mass reduction. *Nephron Exp Nephrol* 2009; 112: e31–e42.
55. Wilcox CS, Welch WJ. Oxidative stress: cause or consequence of hypertension. *Exp Biol Med* 2001; 226: 619–620.
56. Ungvari Z, Csiszar A, Huang A, Kaminski PM, Wolin MS, Koller A. High pressure induces superoxide production in isolated arteries via protein kinase C-dependent activation of NAD(P)H oxidase. *Circulation* 2003; 108: 1253–1258.
57. Vaziri ND, Ni Z. Expression of NOX-I, gp91phox, p47phox and P67phox in the aorta segments above and below coarctation. *Biochim Biophys Acta* 2005; 1723: 321–327.
58. Boo YC, Jo H. Flow-dependent regulation of endothelial nitric oxide synthase: role of protein kinases. *Am J Physiol Cell Physiol* 2003; 285: C499–C508.
59. Inoue N, Ramasamy S, Fukai T. Shear stress modulates expression of Cu/Zn superoxide dismutase in human aortic endothelial cells. *Circ Res* 1996; 79: 32–37.
60. Harrison DG, Widder J, Grumbach I, Chen W, Weber M, Searles C. Endothelial mechanotransduction, nitric oxide and vascular inflammation. *J Intern Med* 2006; 259: 351–363.
61. De Keulenaer GW, Chappell DC, Ishizaka N, Nerem RM, Alexander RW, Griendling KK. Oscillatory and steady laminar shear stress differentially affect human endothelial redox state: role of a superoxide-producing NADH oxidase. *Circ Res* 1998; 82: 1094–2101.

62. Ali MH, Pearlstein DP, Mathieu CE, Schumacker PT. Mitochondrial requirement for endothelial responses to cyclic strain: implications for mechanotransduction. *Am J Physiol* 2004; 287: L486–L496.
63. Harrison D, Griendling KK, Landmesser U, Hornig B, Drexler H. Role of oxidative stress in atherosclerosis. *Am J Cardiol* 2003; 91: 7A–11A.
64. Wilcox CS. Oxidative stress and nitric oxide deficiency in the kidney: a critical link to hypertension? *Am J Physiol Regul Integr Comp Physiol* 2005; 289: R913–R935.
65. Hamilton CA, Brosnan MJ, McIntyre M, Graham D, and Dominiczak AF. Superoxide excess in hypertension and aging: a common cause of endothelial dysfunction. *Hypertension* 2001; 37: 529–534.
66. Halcox JP, Schenke WH, Zalos G, Mincemoyer R, Prasad A, Waclawiw MA, Nour KR, Quyyumi AA. Prognostic value of coronary vascular endothelial dysfunction. *Circulation* 2002; 106: 653–658.
67. Fichtlscherer S, Breuer S, Zeiher AM. Prognostic value of systemic endothelial dysfunction in patients with acute coronary syndromes: further evidence for the existence of the "vulnerable" patient. *Circulation* 2004; 110: 1926–1932.
68. Chamiot-Clerc P, Renaud JF, Safar ME. Pulse pressure, aortic reactivity and endothelium dysfunction in old hypertensive rats. *Hypertension* 2001; 37: 313–321.
69. Payne JA, Reckeloff JF, Khalil RA. Roleof oxidative stress in age-related reduction of NO-cGMP-mediated vascular relaxation in SHR. *Am J Physiol Integr Comp Physiol* 2003; 285: R542–R551.
70. Singh N, Prasad S, Singer DR, MacAllister RJ. Ageing is associated with impairment of nitric oxide and prostanoid dilator pathways in the human forearm. *Clin Sci (Lond)* 2002; 102: 595–600.
71. Feng MG, Dukacz SA, Kline RL. Selective effect of tempol on renal medullary hemodynamics in spontaneously hypertensive rats. *Am J Physiol Regul Integr Comp Physiol* 2001; 281: R1420–R1425.
72. Shastri S, Gopalakrishnan V, Poduri R, Di Wang H. Tempol selectively attenuates angiotensin II evoked vasoconstrictor responses in spontaneously hypertensive rats. *J Hypertens 2002;* 20: 1381–1391.
73. Schnackenberg CG, Wilcox CS. Two-week administration of tempol attenuates both hypertension and renal excretion of 8-isoprostaglandin F2α. *Hypertension* 1999; 33: 424–428.
74. Pezeshk A, Derick Dalhouse A. Vitamin E, membrane fluidity, and blood pressure in hypertensive and normotensive rats. *Life Sci* 2000; 67: 1881–1889.
75. Vasdev S, Ford CA, Parai S, Longerich L, Gadag V. Dietary vitamin C supplementation lowers blood pressure in spontaneously hypertensive rats. *Mol Cell Biochem* 2001; 218: 97–103.
76. Vasdev S, Gill V, Parai S, Longerich L, Gadag V. Dietary vitamin E supplementation lowers blood pressure in spontaneously hypertensive rats. *Mol Cell Biochem* 2002*;* 238: 111–117.
77. Li JM, Shah AM. Endothelial cell superoxide generation: regulation and relevance for cardiovascular physiology. *Am J Physiol Reg Integ Comp Physiol* 2004; 287: R1014–R1030.
78. Ray R, Shah AM. NADPH oxidase and endothelial cell function. *Clin Sci* 2005; 109: 217–226.
79. Rodriguez-Iturbe B, Vaziri ND, Herrera-Acosta J, Johnson RJ. Oxidative stress, renal infiltration of immune cells and salt-sensitive hypertension: all for one and one for all. *Am J Physiol Renal Physiol* 2004; 286: F606–F616.
80. Schnackenberg CG. Physiological and pathophysiological roles of oxygen radicals in the renal microvasculature. *Am J Physiol Regul Integr Comp Physiol* 2002; 282(2): R335–R342.
81. Iida Y, Katusic ZS. Mechanisms of cerebral arterial relaxations to hydrogen peroxide. Stroke 2000; 31: 2224–2230.

82. Sobey CG, Heistad DD, Faraci FM. Mechanisms of bradykinin-induced cerebral vasodilatation in rats. Evidence that reactive oxygen species activate K channels. *Stroke* 1997; 28: 2290–2294.
83. Barlow RS, White RE. Hydrogen peroxide relaxes porcine coronary arteries by stimulating BKCa channel activity. *Am J Physiol* 1998; 275: H1283–H1289.
84. Matoba T, Shimokawa H, Nakashima M, Hirakawa Y, Mukai Y, Hirano K, Kanaide H, Takeshita A. Hydrogen peroxide is an endothelia-derived hyperpolarizing factor in mice. *J Clin Invest* 2000; 106: 1521–1530.
85. Miura H, Bosnjak JJ, Ning G, Saito T, Miura M, Gutterman DD. Role for hydrogen peroxide in flow-induced dilation of coronary arterioles. *Cir Res* 2003; 92: e31–e40.
86. Wilcox CS, Pearlman A. Chemistry and antihypertensive effects of tempol and other nitroxides. *Pharmacol Rev* 2008; 60: 418–469.
87. Chen Y, Pearlman A, Luo Z, Wilcox CS. Hydrogen peroxide mediates a transient vasorelaxation with tempol during oxidative stress. *Am J Physiol Heart Circ Physiol* 2007; 293: H2085–H2092.
88. Chen YF, Cowley AW, Jr., Zou AP. Increased H2O2 counteracts the vasodilator and natriuretic effects of superoxide dismutation by tempol in renal medulla. *Am J Physiol Renal Physiol* 2003; 285: R827–R833.
89. Makino A, Skelton MM, Zou AP, Cowley AW, Jr. Increased renal medullary H_2O_2 leads to hypertension. *Hypertension* 2003; 42: 25–30.
90. Sorescu D, Somers MJ, Lassegue B, Grant S, Harrison DG, Griendling KK. Electron spin resonance characterization of the NAD(P)H oxidase in vascular smooth muscle cells. *Free Radic Biol Med* 2001; 30: 603–612.
91. Lassegue B, Sorescu D, Szocs K, Yin Q, Akers M, Zhang Y, Grant SL, Lambeth JD, Griendling KK. Novel gp91(phox) homologues in vascular smooth muscle cells: nox1 mediates angiotensin II-induced superoxide formation and redox-sensitive signaling pathways. *Circ Res* 2001; 88: 888–894.
92. Schnackenberg CG, Welch WJ, Wilcox CS. TP receptor mediated vasoconstriction in microperfused afferent arterioles: roles of O_2 and NO. *Am J Physiol Renal Physiol* 2000; 279: F302–F308.
93. Zou AP, Li N, Cowley AW, Jr. Production and actions of superoxide in the renal medulla. *Hypertension* 2001; 37: 547–553.
94. Lounsbury KM, Hu Q, Ziegelstein RC. Calcium signaling and oxidant stress in the vasculature. *Free Radic Biol Med* 2000; 28: 1362–1369.
95. Chen YF, Li PL, Zou AP. Oxidative stress enhances the production and actions of adenosine in the kidney. *Am J Physiol Reg Integr Comp Physiol* 2001; 280: R1808–R1816.
96. Zou MH, Ullrich V. Peroxynitrite formed by simultaneous generation of nitric oxide and superoxide selectively inhibits bovine aortic prostacyclin synthase. *FEBS Lett* 1996; 382: 101–104.
97. Takahashi K, Nammour TM, Fukunaga M, Ebert J, Morrow JD, Roberts LJ, Hoover RL, Badr KF. Glomerular actions of a free radical-generated novel prostaglandin, 8-epiprostaglandin F2α in the rat. Evidence for interaction with thromboxane A2 receptors. *J Clin Invest* 1992; 90: 136–141.
98. Kawazoe T, Kosaka H, Yoneyama H, Hata Y. Acute production of vascular superoxide by angiotensin II but not by catecholamines. *J Hypertens* 2000; 18: 179–185.
99. Kawazoe T, Kosaka H, Yoneyama H, Hata Y. Involvement of superoxide in acute reaction of angiotensin II in mesenteric microcirculation. *Jpn J Physiol* 1999; 49: 437–443.
100. Diikhorst-Oei LT, Stroes ES, Koomans HA, Rabelink TJ. Acute simultaneous stimulation of nitric oxide and oxygen radicals by angiotensin II in humans in vivo. *J Cardiovasc Pharmacol* 1999; 33: 420–424.
101. Ushio-Fukai M, Alexander RW, Aker M, Griendling KK. p38 mitogen-activated protein kinase is a critical component of the redox-sensitive signaling pathways activated by

angiotensin II. Role in vascular smooth muscle cell hypertrophy. *J Biol Chem* 1998; 273: 15022–15029.

102. Touyz RM, He G, El Mabrouk M, Schiffrin EL. P38 MAP kinase regulates vascular smooth muscle cell collagen synthesis by angiotensin II in SHE but not in WKY. *Hypertension* 2001; 37: 574–580.
103. Ushio-Fukai M, Griendling KK, Becker PL, Hilenski L, Halleran S, Alexander RW. Epidermal growth factor receptor transactivation by angiotensin II requires reactive oxygen species in vascular smooth muscle cells. *Arterioscler Thromb Vasc Biol* 2001; 21: 489–495.
104. Tonks NK. Redox redux: revisiting PTPs and the control of cell signaling. *Cell* 2005; 121: 667–670.
105. Rodriguez-Iturbe B, Romero F, Johnson RJ. Pathophysiologic mechanisms of salt-dependent hypertension. *Am J Kidney Dis* 2007; 50: 655–672.
106. Kitiyakara C, Chabrashvili T, Chen Y, Blau J, Karber A, Aslam S, Welch WJ, Wilcox CS. Salt intake, oxidative stress and renal expression of NADPH oxidase and superoxide dismutase. *J Am Soc Nephrol* 2003; 14: 2775–2782.
107. Ni Z, Vaziri ND: Effect of salt loading on nitric oxide synthase expression in normotensive rats. *Am J Hypertens* 2001; 14: 155–163.
108. Chandramohan G, Bai Y, Norris K, Rodriguez-Iturbe B, Vaziri ND. Effects of dietary salt on angiotensin system, NAD(P)oxidase, COX-2, MCP-1, PAI-1 and NFkB in salt sensitive and salt resistant rat kidneys. *Am J Nephro*l 2008; 28: 158–167.
109. Herrera M, Garvin JL. A high salt diet stimulates thick ascending limb eNOS expression by raising medullary osmolality and increasing release of endothelin-1. *Am J Physiol Renal Physiol* 2005; 288: F58–F64.
110. Zewde T, Wu F, Mattson DL. Influence of dietary NaCl on L-arginine transport in the renal medulla. *Am J Physiol Regul Integr Comp Physiol* 2004; 286: R89–R93.
111. Cowley AW. Renal medullary oxidative stress, pressure-natriuresis and hypertension. *Hypertension* 2008; 52: 777–786.
112. Johnson RJ, Gordon KL, Giachelli C, Kurth T, Skelton MM, Cowley AM. Tubulointerstitial injury and loss of nitric oxide synthases parallel the development of hypertension in the Dahl SS rat. *J Hypertens* 2000; 18: 1497–1505.
113. Lu S, Mattson DL, Cowley AW, Jr. Renal medullary captopril delivery lowers blood pressure in spontaneously hypertensive rats. *Hypertension* 1994; 23: 337–345.
114. Miyata N, Cowley AW, Jr. Renal medullary interstitial infusion of L-arginine prevents reduction of medullary blood flow and hypertension in Dahl salt-sensitive rats. *Hypertension* 1999; 33: 446–450.
115. Makino A, Skelton MM, Zou AP, Roman RJ, Cowley AW, Jr. Increased renal medullary oxidative stress produces hypertension. *Hypertension* 2002; 39: 667–672.
116. Chen YF, Cowley AW, Jr, Zou AP. Increased H2O2 counteracts the vasodilator and natriuretic effects of superoxide dismutation by tempol in renal medulla. *Am J Physiol Reg Integr Comp Physiol* 2003; 285; R827–R833.
117. Nakanishi K, Mattson DL, Cowley AW, Jr. Role of medullary blood flow in the development of L-NAME hypertension in rats. *Am J Physiol* 1995; 268: R317–R323.
118. Guarasci G, Kline RL. Pressure-natriuresis following acute and chronic inhibition of nitric oxide synthase in rats. *Am J Physiol* 1996; 270: R469–R478.
119. Laursen JB, Rajagopalan S, Galis Z, Tarpey M, Freeman BA, Harrison DG. Role of superoxide in angiotensin II-induced but not catecholamine-induced hypertension. *Circulation* 1997; 95: 588–593.
120. Nakazono K, Watanabe N, Matsuno K, Sasaki J, Sato T, Inoue M. Does superoxide underlie the pathogenesis of hypertension? *Proc Natl Acad Sci U S A* 1991; 88: 10045–10048.
121. Rajagopalan S, Kurz S, Munzel T, Tarpey M, Freeman BA, Griendling KK, Harrison DG. Angiotensin II-mediated hypertension in the rat increases vascular superoxide production via membrane NADH/NADPH oxidase activation: contribution to alterations of vasomotor tone. *J Clin Invest* 1996; 97: 1916–1923.

122. Gill PS, Wilcox CS. NADPH oxidases in the kidney. *Antioxid Redox Signal* 2006; 8: 1597–1607.
123. Laude K, Cai H, Fink B, Hoch N, Weber DS, McCann L, Kojda G, Fukai T, Schmidt HH, Dikalov S, Ramasamy S, Gamez G, Griendling KK, Harrison DG. Hemodynamic and biochemical adaptations to vascular smooth muscle overexpression of p22phox in mice. *Am J Physiol Heart Circ Physiol* 2005; 288: H7–H12.
124. Cifuentes ME, Rey FE, Carretero OA, Pagano PJ. Upregulation of p67(phox) and gp91 (phox) in aortas from angiotensin II-infused mice. *Am J Physiol Heart Circ Physiol* 2000; 279: H2234–H2240.
125. Kobori H, Nangaku M, Navar LG, Nishiyama A. The intrarenal reninangiotensin system: from physiology to the pathobiology of hypertension and kidney disease. *Pharmacol Rev* 2007; 59: 251–287.
126. Kobori H, Ozawa Y, Suzaki Y, Prieto-Carrasquero MC, Nishiyama A, Shoji T, Cohen EP, Navar LG. Intratubular angiotensinogen in hypertension and kidney diseases. *Am J Hypertens* 2006; 19: 541–550.
127. Nishiyama A, Seth DM, Navar LG. Renal interstitial fluid concentrations of angiotensin I and II in anesthesized rats. *Hypertension* 2002; 39: 129–134.
128. Franco M, Martinez F, Rodriguez-Iturbe B, Johnson RJ, Santamaria J, Montoya A, Nepomuceno T, Bautista R, Tapia E, Herrera-Acosta J. Angiotensin II, interstitial inflammation, and the pathogenesis of salt-sensitive hypertension. *Am J Physiol Renal Physiol* 2006; 291: F1281–F1287.
129. Alvarez V, Quiroz Y, Nava M, Pons H, Rodriguez-Iturbe B. Overload proteinuria is followed by salt-sensitive hypertension caused by renal infiltration of immune cells. *Am J Physiol Renal Physiol* 2002; 283: F1132–F1141.
130. Rodríguez-Iturbe B, Pons H, Quiroz Y, Gordon KL, Rincón J, Chávez M, Parra G, Herrera-Acosta J, Gomez-Garre D, Largo R, Egido J, Johnson RJ. Mycophenolate mofetil prevents salt-sensitive hypertension resulting from angiotensin II exposure. *Kidney Int* 2001; 59: 2222–2232.
131. Rodriguez-Iturbe B, Quiroz Y, Nava M, Bonet L, Chavez M, Herrera-Acosta J, Johnson RJ, Pons HA. Reduction of renal immune cell infiltration results in blood pressure control in genetically hypertensive rats. *Am J Physiol Renal Physiol* 2002; 282: F191–F201.
132. Vanegas V, Ferrebuz A, Quiroz Y, Rodriguez-Iturbe B. Hypertension in Page (cellophane wrapped) kidney is due to interstitial nephritis. *Kidney Int* 2005; 68: 1161–1170.
133. Hoch NE, Guzik TJ, Chen W, Deans T, Maalouf SA, Gratze P, Weyand C, Harrison DG. Regulation of T-cell function by endogenously produced angiotensin II. *Am J Physiol Regul Integr Comp Physiol* 2009; 296: R208–R216.
134. Kobori H, Harrison-Bernard LM, Navar LG. Urinary excretion of angiotensinogen reflects intrarenal angiotensinogen production. *Kidney Int* 2002; 61: 579–585.
135. Kobori H, Nishiyama A, Harrison-Bernard LM, Navar LG. Urinary angiotensinogen as an indicator of intrarenal angiotensin status in hypertension. *Hypertension* 2003; 41: 42–49.
136. Kobori H, Alper B, Jr., Shenava R, Katsurada A, Saito T, Ohashi N, Urushihara M, Miyata K, Satou R, Hamm LL, Navar LG. Urinary angiotensinogen as a novel biomarker of the intrarenal renin-angiotensin system status in hypertensive patients. *Hypertension* 2009; 53 (Pt 2): 344–350.
137. Hehner SP, Breitkreutz R, Shubinsky G, Unsoeld H, Schulze-Osthoff K, Schmitz ML, Dröge W. Enhancement of T cell receptor signaling by mild oxidative shift in the intracellular thiol pool. *J Immunol* 2000; 165: 4319–4328.
138. Los M, Dröge W, Stricker K, Baeuerle PA, Schulze-Osthoff K. Hydrogen peroxide as a potent activator of T lymphocyte functions. *Eur J Immunol* 1995; 25: 159–165.
139. Sen CK, Packer L. Antioxidant and redox regulation of gene transcription. *FASEB J* 1996; 10: 709–720.
140. Polla BS, Bachelet M, Elia G, Santoro MG. Stress proteins in inflammation. *Ann N Y Acad Sci* 1998; 851: 75–85

141. Rodriguez-Iturbe B. The role of immunocompetent cell renal infiltration the pathogenesis of arterial hypertension. *Nefrología* 2008; 28: 483–492.
142. Quiroz Y, Pons H, Gordon KL, Rincón J, Chávez M, Parra G, Herrera-Acosta J, Gómez-Garre D, Largo R, Egido J, Johnson RJ, Rodríguez-Iturbe B. Mycophenolate mofetil prevents salt-sensitive hypertension resulting from short-term nitric oxide synthesis inhibition. *Am J Physiol Renal Physiol* 2001; 281: F38–F47.
143. Rodríguez-Iturbe B, Ferrebuz A, Vanegas V, Quiroz Y, Mezzano S, Vaziri ND. Early and sustained inhibition of nuclear factor kappa B prevents hypertension in spontaneously hypertensive rats. *J Pharmacol Exp Ther* 2005; 315; 51–57.
144. Rodriguez-Iturbe B, Quiroz Y, Ferrebuz A, Parra G, Vaziri ND. Evolution of renal interstitial inflammation and NF-κB activation in spontaneously hypertensive rats. *Am J Nephrol* 2004; 24: 587–594.
145. Bravo Y, Quiroz Y, Ferrebuz A, Vaziri ND, Rodríguez-Iturbe B. Mycophenolate mofetil administration reduces renal inflammation, oxidative stress and arterial pressure in rats with lead-induced hypertension. *Am J Physiol Renal Physiol* 2007; 297: F616–F623.
146. Beswick RA, Zhang H, Marable D, Catravas JD, Hill WD, Webb RC. Long-term antioxidant administration attenuates mineralocorticoid hypertension and renal inflammatory response. *Hypertension* 2001; 37: 781–786.
147. Mattson DL, James L, Berdan EA, Meister CJ. Immune suppression attenuates hypertension and renal disease in Dahl salt-sensitive rat. *Hypertension* 2006; 48: 149–156.
148. Müller DN, Shagdarsuren E, Park JK, Dechend R, Mervaala E, Hampich F, Fiebeler A, Ju X, Finckenberg P, Theuer J, Viedt C, Kreuzer J, Heidecke H, Haller H, Zenke M, Luft FC. Immunosuppressive treatment protects against angiotensin-II induced renal damage. *Am J Pathol* 2002; 161: 1679–1693.
149. Stewart T, Jung FF, Manning J, Vehaskari VM. Kidney immune cell infiltration and oxidative stress contribute to prenatally programmed hypertension. *Kidney Int* 2005; 68: 2180–2188.
150. Tian N, Gu JW, Braddy SJ, Rose RA, Hughson MD, Manning RD, Jr. Immune suppression prevents renal damage and dysfunction and reduces arterial pressure in salt-sensitive hypertension. *Am J Physiol Heart Circ Physiol* 2007; 292: H1018–H1025.
151. Lee DL, Sturgis LC, Labazi H, Osborne JB, Jr, Fleming C, Pollock JS, Manhiani M, Imig JD, Brands MW. Angiotensin II hypertension is attenuated in interleukin-6 knockout mice. *Am J Physiol Heart Circ Physiol* 2006; 290: H935–H940.
152. Khraibi AA, Norman RA, Jr., Dzielak DJ. Chronic immunosuppression attenuates hypertension in Okamoto spontaneously hypertensive rats. *Am J Physiol* 1984; 247: H722–H726.
153. Bataillard P, Freiche J-C, Vincent M, Touraine J-L, Sassard JU. Effects of neonatal thymectomy on blood pressure and immunological characteristics of the genetically hypertensive rats of the Lyon strain. *J Hypertens* 1986; 4(Suppl 3): 5455–5467.
154. Svendsen JG. Spontaneous hypertension and hypertensive vascular disease in the NZB strain of mice. *Acta Pathol Microbiol Scand* (A) 1977; 85(3): 261–268.
155. Norman RA, Jr., Galloway PG, Dzielak DJ, Huang M. Mechanisms of partial renal infarct hypertension. *J Hypertens* 1988; 6: 397–403.
156. Elmarakby AA, Quigley JE, Olearczyk JJ, Sridhar A, Cool AK, Inscho EW, Pollock DM, Imig JD. Chemokine receptor 2b inhibition provides renal protection in angiotensin II salt hypertension. *Hypertension* 2007; 50: 1069–1076.
157. Guzik TJ, Hoch NE, Brown KA, McCann LA, Rahman A, Dikalov S, Goronzy J, Weyand C, Harrison DG. Role of the T cell in the genesis of angiotensin II induced hypertension and vascular dysfunction. *J Exp Med* 2007; 204: 2449–2460.
158. Morales JM. Influence of new immunosuppressive combinations on arterial hypertension after renal transplantation. *Kidney Int Suppl* 2002; 62(Suppl 82): S81–S87.
159. Herrera J, Ferrebuz A, MacGregor EG, Rodriguez-Iturbe B. Mycophenolate mofetil treatment improves hypertension in patients with psoriasis and rheumatoid arthritis. *J Am Soc Nephrol* 2006; 17(Suppl 3): S218–S225.

160. Hughson MD, Gobe GC, Hoy WE, Manning RD, Jr, Douglas-Denton R, Bertram JF. Associations of glomerular number and birth weight with clinicopathological features of African Americans and whites. *Am J Kidney Dis* 2008; 52: 18–28.
161. Ruiz-Ortega M, Esteban V, Ruperez M, Sanchez-Lopez E, Rodriguez-Vita J, Carvajal G, Egido J. Renal and vascular hypertension-induced inflammation: role of angiotensin II. *Curr Opin Nephrol Hypertens* 2006; 15: 159–166.
162. Svendsen UG. Evidence for an initial, thymus independent and a chronic, thymus-dependent phase of DOCA and salt hypertension in mice. *Acta Path Microbiol Scand* 1976; 84: 523–528.
163. Tuttle RS, Boppana DP. Antihypertensive effect of interleukin-2. *Hypertension* 1990; 15: 89–94.
164. Takeichi N, Suzuki K, Kobayashi H. Immunologic depression in spontaneously hypertensive rats. *Clin Exp Immunol* 1980; 40: 120–126.
165. Bendich A, Belisle EH, Strausser HR. Immune system modulation and its effects on blood pressure of the spontaneously hypertensive male and female rat. *Biochem Biophys Res Commun* 1981; 99: 600–607.
166. Rodríguez-Iturbe B, Vaziri ND. Salt sensitive hypertension: update on novel findings. *Nephrol Dial Transplant* 2007; 22: 992–995.
167. Ishizaka N, Aizawa T, Ohno M, Usui Si S, Mori I, Tang SS, Ingelfinger JR, Kimura S, Nagai R. Regulation and localization of HSP70 and HSP25 in the kidney of rats undergoing long-term administration of angiotensin II. *Hypertensio*n 2002; 39: 122–128.
168. Bravo J, Quiroz Y, Pons H, Parra G, Herrera-Acosta J, Johnson RJ, Rodríguez-Iturbe B. Vimentin and heat shock protein expression are induced in the kidney by angiotensin and by nitric oxide inhibition. *Kidney Int* 2003; 64(Suppl 86): S46–S51.
169. Kaufmann SHE. Heat shock proteins and the immune response. *Immunol Today* 1990; 11: 129–136.
170. Young RA. Stress proteins and immunology. *Ann Rev Immunol* 1990; 8: 401–420.
171. Parra G, Quiroz Y, Salazar J, Bravo Y, Pons H, Chavez M, Johnson RJ, Rodriguez-Iturbe B. Experimental induction of salt-sensitive hypertension is associated with lymphocyte proliferative response to HSP70. *Kidney Int Suppl* 2008; 111: S55–S59.

Chapter 8
Uric Acid and Oxidative Stress

Yuri Y. Sautin, Witcha Imaram, Kyung Mee Kim, Alexander Angerhofer, George Henderson, and Richard Johnson

Abstract Biological effects of uric acid, generated in the human body either from food or resulting from the purine degradation pathway via xanthine oxidoreductase, are extremely pleiotropic. They are paradoxically opposing under different experimental conditions. Some of these effects are beneficial and some of them are deleterious. The key feature of uric acid is the ability to be either an antioxidant or pro-oxidant depending on a variety of factors. The complexity of the urate chemistry, including its ability to quench or form various radicals, is a crucial component of the mechanisms underlying the ability of uric acid to induce opposing biological effects. In addition, uric acid is a powerful signaling molecule that can affect intracellular signal transduction, leading to oxidant production via nicotinamide adenine dinucleotide phosphate (NADPH) oxidase and expression of proinflammatory mediators. Beneficial antioxidant effects are manifested in the protection of endothelial cells from external oxidative stress and protection of the central nervous system from oxidative damage in several conditions. Detrimental pro-oxidative effects of uric acid are associated with the metabolic syndrome and cardiovascular and renal diseases.

Keywords Xanthine oxidoreductase · Metabolic syndrome · Endothelium

Uric acid was long thought to be primarily a waste product of purine metabolism that provided a means for efficiently removing nitrogenous wastes [1]. Indeed, animals could be classified based on how they excreted their nitrogenous wastes, with ammonotelic organisms excreting ammonia (NH_4^+), ureotelic organisms excreting urea (which has two nitrogens per molecule), and uricotelic organisms that excreted uric acid (which has four atoms of nitrogen per molecule). Ammonotelic organisms need a lot of water to excrete ammonia (such as freshwater

R. Johnson (✉)
Division of Renal Diseases and Hypertension, University of Colorado, Aurora, CO 80045, USA
e-mail: richard.johnson@ucdenver.edu

T. Miyata et al. (eds.), *Studies on Renal Disorders*, Oxidative Stress in Applied Basic Research and Clinical Practice, DOI 10.1007/978-1-60761-857-7_8,

amphibians and fishes), whereas ureotelic organisms that excrete urea as their main product includes most mammals, and uricotelic organisms includes reptiles and birds in which the uric acid could be excreted via the cloaca with minimal water loss. For humans, who are ureotelic, the excretion of uric acid does not account for much of the nitrogen excretion. However, humans have higher levels of uric acid than most other mammals since they lack the enzyme uricase, which degrades uric acid to allantoin (Fig. 1). As such, uric acid excretion accounts for only a small percentage of overall nitrogen excretion in humans.

The viewpoint of uric acid as primarily a waste product was held for decades. Indeed, the primary concern about uric acid was the well-known observation that high levels can exceed the known solubility (which is 6.8 mg/dL in water) and that it could precipitate into crystals that then could induce a severe inflammatory reaction in joints, causing the disease gout [2]. However, there was then the surprising discovery that, while uric acid is proinflammatory in its crystalline form, it can actually function as an antioxidant [3, 4]. Indeed, Ames et al. documented that it was extremely effective as an antioxidant in multiple settings and in fact may represent the principal antioxidant in the circulation [5]. They also suggested that the genetic mutation that led to a loss of uricase may have provided a survival advantage through its benefit of increased uric acid levels to overall antioxidant potential. Indeed, they proposed that uric acid may help protect animals from the oxidative stress associated with aging and cancer, and that this could explain the increasing longevity observed in species with high uric acid levels. Furthermore, the studies by both Proctor [4] and Ames et al. [5] also suggested that the uricase mutation may have provided an important survival advantage to replace the decrease in vitamin C levels that resulted from the mutation in ascorbate synthesis that had occurred millions of years earlier.

While these studies suggested that uric acid might have a beneficial role in preventing disease, a gnawing and consistent observation was the frequent association of elevated uric acid with cardiovascular disease. Over time, numerous epidemiological studies documented an association of uric acid with increased risk for myocardial infarction, stroke, hypertension, metabolic syndrome, chronic kidney disease, and overall cardiovascular mortality [reviewed in 6, 7]. While some studies found uric acid to be an independent risk factor for cardiovascular disease, other studies could not demonstrate such independence. The inability to resolve the role of uric acid in cardiovascular risk led to great controversies in the literature on the role of uric acid in these conditions, with some authors suggesting that it

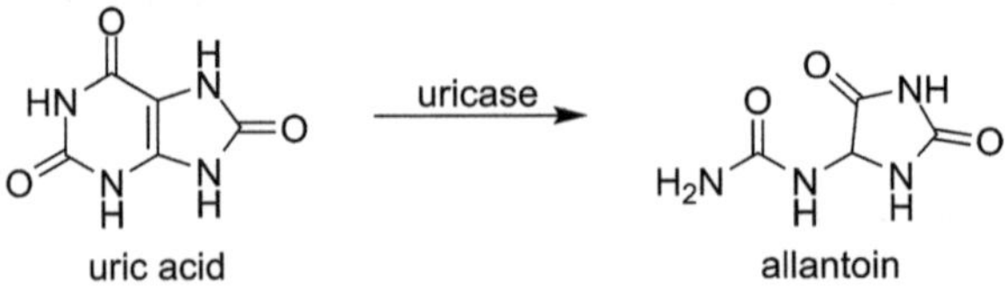

Fig. 1 The enzymatic conversion from uric acid to allantoin by uricase

is simply a marker of risk [8, 9], others stating that it may be a compensatory beneficial response [10–13], and others suggesting that it may increase the risk for cardiovascular complications [14].

The inability to resolve the issues by epidemiology has been followed by a series of experimental studies in cell culture, in animals, and most recently, in humans. Interestingly, some studies suggest that uric acid, while an antioxidant under some conditions, may also function as a pro-oxidant in other situations, especially in the intracellular environment [15, 16]. In this chapter, we review some of these interesting studies that document the paradoxical and opposing effects uric acid may have under different experimental conditions.

1 Uric Acid the Antioxidant

Uric acid is a reactive water soluble antioxidant that will react with a variety of oxidants [17, 18]. Unlike other antioxidants, such as ascorbate, the reaction of uric acid with an oxidant results in its stepwise degradation with the formation of signature end products. Some studies suggest uric acid may be one of the most important antioxidants in the plasma [5, 19].

1.1 Antioxidant Chemistry of Uric Acid

Pioneering experiments by Ames et al. [5] showed that when uric acid is exposed to a variety of different radicals it is degraded, and the authors suggested that allantoin was a product of this degradation. More recently, numerous studies confirmed that antioxidant activity of uric acid is always associated with a certain level of degradation of the urate molecule by a variety of mechanisms depending on the nature of the oxidant [20–22].

1.1.1 Reaction with Superoxide and Hydroxyl Radical

The classical reaction of uric acid is with superoxide ($O_2^{\bullet -}$) and the hydroxyl radical (•OH), resulting in the formation of allantoin (Fig. 2). The observation that allantoin can be detected in the serum and urine of individuals with oxidative stress suggests that this reaction occurs in vivo. Indeed, allantoin levels have been elevated in a variety of conditions, including severe exercise [23, 24], in congestive heart failure [25], in those with arthritis [26], and in subjects with diabetes [27]. We and others [27] have also detected high levels of allantoin in hemodialysis subjects in which oxidative stress is known to be high (unpublished observations).

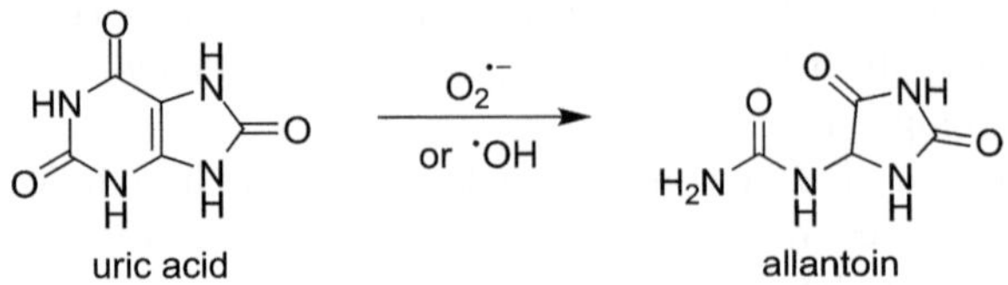

Fig. 2 Reaction of uric acid with superoxide ($O_2^{\bullet-}$) and hydroxyl (•OH)

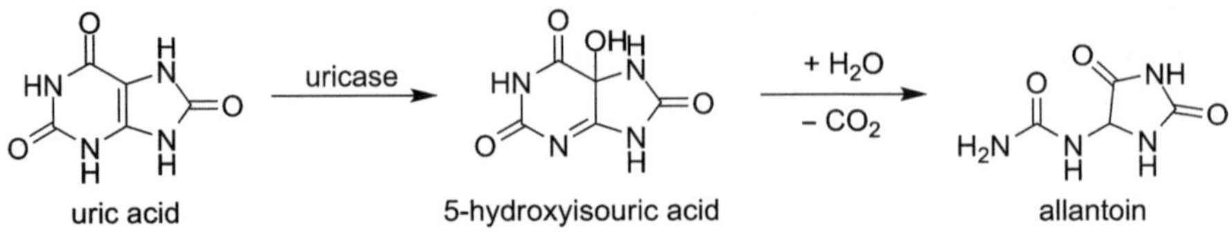

Fig. 3 Non-enzymatic degradation of uric acid to allantoin

Finally, allantoin has also been detected in atherosclerotic plaque [28], consistent with the possibility that uric acid is acting as an antioxidant in inflamed and atherosclerotic tissues.

In mammals that have uricase, uric acid is degraded with the release of oxidants. While most studies have suggested that the uricase product is also allantoin, more recent studies by Kahn et al. suggest the product is 5-hydroxyisouric acid, which is subsequently degraded to allantoin by nonenzymatic reaction (Fig. 3) [29].

1.1.2 Reaction with Peroxynitrite

Peroxynitrite is an important oxidant in biological systems that is formed by the reaction of nitric oxide (NO) with superoxide anion [30]. Peroxynitrite can nitrosylate tissues, which can be detected by the appearance of nitrotyrosine residues as evidenced by mass spectrometry experiments combined with specific antibodies. In this regard, uric acid is known to have very high affinity for peroxynitrite [31, 32], resulting in the formation of triuret (Fig. 4) [31, 33]. Triuret levels have been detected in subjects with oxidative stress, including in preeclampsia [33], smokers [34], and in hemodialysis patients (unpublished observations). In addition to triuret, Skinner et al. reported that uric acid can react with peroxynitrite to form a vasoactive NO donor, likely from a nitrated uric acid derivative [35]. However, triuret appears to be the primary product [31, 36].

1.1.3 Reaction with Nitric Oxide

Uric acid has also been reported to react directly with NO under various conditions. Suzuki reported that uric acid can react with NO under aerobic conditions to form a

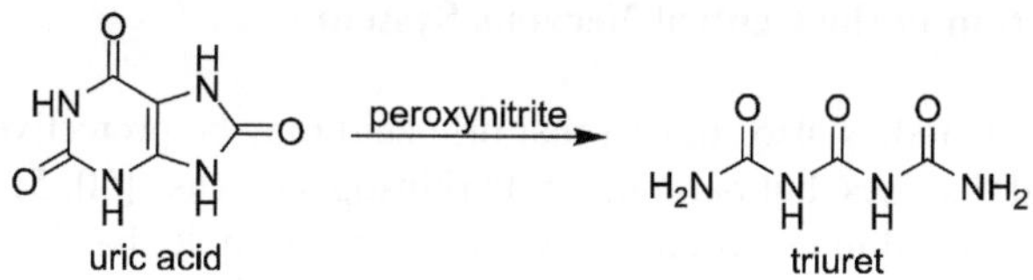

Fig. 4 Peroxynitrite-mediated uric acid oxidation forming triuret

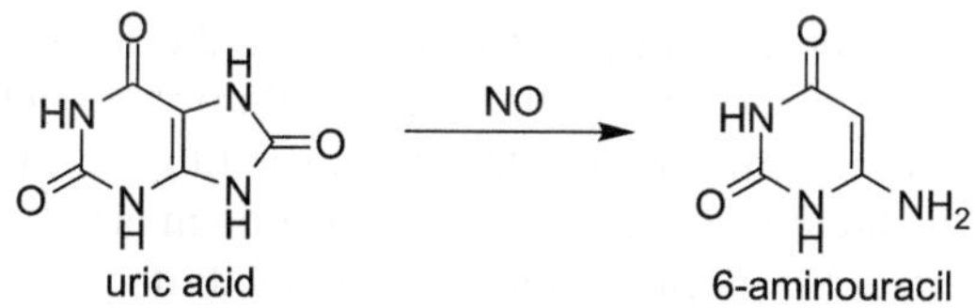

Fig. 5 Formation of 6-aminouracil from uric acid

nitrosated product that is capable of donating NO [37]. In contrast, Gersch et al. reported that uric acid had a reaction with NO that consumed the NO and generated 6-aminouracil [38] (Fig. 5). In preliminary studies, we have also observed elevated levels of 6-aminouracil in preeclampsia patients and subjects on hemodialysis (unpublished data).

1.1.4 Other Antioxidant Reactions

Uric acid may also have other antioxidant actions. For example, uric acid can block ferrous (Fe^{2+})-catalyzed oxidation reactions [39] as well as copper-mediated oxidation of human low density lipoprotein [40, 41]. Uric acid may also increase extracellular superoxide dismutase activity by blocking its degradation by hydrogen peroxide [42]. Finally, an important function of uric acid may be to preserve ascorbate levels by chelating with iron and hence blocking iron-catalyzed ascorbate oxidation [43, 44]. Indeed, ascorbate is preferentially utilized over uric acid in plasma during oxidative stress [45].

1.2 Evidence for Antioxidant Actions of Uric Acid

1.2.1 Preservation of Endothelial Function

Several studies have reported that uric acid may have acute effects to preserve endothelial function, and in particular, endothelial NO production. For example, Kuzkaya et al. reported that uric acid could protect cultured endothelial cells from oxidative stress and that the antioxidant effects were potentiated in the presence of ascorbate [32]. The administration of uric acid to humans has also been associated with an immediate improvement in endothelial dependent (NO-dependent) vasodilation [46–48]. Finally, at least one study has reported that increasing uric acid levels can improve endothelial function in a model of renal insufficiency [49].

1.2.2 Protection of the Central Nervous System

Epidemiological studies also suggest that uric acid may be protective for a variety of neurological diseases, but particularly Parkinson's disease [50, 51] and multiple sclerosis [52]. The administration of uric acid, for example, has been found to be protective in experimental allergic encephalomyelitis (EAE), which is pathogenically similar to multiple sclerosis [52]. Much of the protection in these disorders has been posited to be secondary to the antioxidant properties of uric acid. Uric acid may function in part by preventing the oxidation of dopamine [53–55]. Another mechanism of the protective role of uric acid in EAE is by blocking the blood–brain barrier permeability, thereby preventing the infiltration of inflammatory cells into the brain with a consequent reduction in neuron damage [56]. This latter protective effect has been attributed to the inhibition of peroxynitrite-induced nitrotyrosines [56]. However, ascorbate also blocks nitrotyrosine formation in the CNS in this model but did not block the blood–brain barrier and did not provide neuronal protection. While the authors suggested this may be due to an increased ability for uric acid to block the oxidizing effects of peroxynitrite in the presence of iron, the data would also be consistent with the possibility that the protective effects of uric acid in this model may not be via its antioxidant actions [57]. Uric acid, at least acutely, may also provide some protection against stroke. For example, rats administered uric acid prior to middle artery occlusion show reduced brain injury [54]. In contrast, most studies suggest that chronic hyperuricemia increases the risk for stroke [58].

1.2.3 Other Possible Antioxidant Functions

Several other studies suggest uric acid may function as an antioxidant in disease. For example, uric acid appears to protect the liver from injury during hemorrhagic shock [59]. Uric acid has also been reported to block oxidant mediated cardiac dysfunction in the guinea pig [60].

1.3 Summary

In summary, uric acid can function as a potent antioxidant in plasma, particularly in the setting where ascorbate levels are low [45]. Uric acid can also synergize with ascorbate in some of its antioxidant actions [32]. Most of the antioxidant effects of uric acid can be demonstrated in the test tube by mixing uric acid with oxidants, and in the circulation by the immediate infusion of uric acid. However, an important issue is whether uric acid functions as an intracellular antioxidant. In recent years it has been shown that uric acid can enter cells via various organic anion transporters, most notably URAT1 [61, 62]. As will be discussed below, uric acid acts more commonly as a pro-oxidant inside the cell [16, 41].

2 Uric Acid the Pro-Oxidant

The possibility that uric acid may function as a pro-oxidant has been demonstrated in several in vitro systems. Most notably, under some conditions uric acid can act as a pro-oxidant in the Cu-mediated oxidation of low-lipid density (LDL) cholesterol [63, 64]. Furthermore, a potential deleterious effect of uric acid is that its reaction with oxidants may result in the generation of free radicals. For example, both Santos et al. [22], as well as our group [65] have found that the reaction of uric acid with peroxynitrite results in the formation of at least two carbon-based radicals, most likely the aminocarbonyl radical and the triuretcarbonyl radical (Fig. 6). The observation that radicals can be formed during the reaction of an antioxidant and oxidant has also been observed in other systems, such as with reactions of oxidants with ascorbate or tetrahydrobiopterin. Whether these generated radicals actually induce oxidative injury remains unknown.

2.1 *Intracellular Effects of Uric Acid*

Many of the pro-oxidative effects of uric acid have been shown in cell culture studies. For example, Corry et al. reported that cultured vascular smooth muscle cells incubated with uric acid produce oxidants and angiotensin II [66]. This is also consistent with the studies by Kanellis et al. in which antioxidants were able to prevent uric acid stimulated expression of monocyte chemoattractant protein-1 (MCP-1) [67]. We have also found that uric acid potently stimulates oxidants in cultured endothelial cells, beginning within 5 min of exposure (Min et al., submitted). Sautin et al. also found that uric acid stimulated oxidant production in adipocytes, and this was associated with increased production of oxidized lipids [16]. In the latter study the mechanism was shown to be due to the stimulation of nicotinamide adenine dinucleotide phosphate (NADPH) oxidase. This study also found that the production of oxidants was associated with a reduction in intracellular NO levels with the generation of peroxynitrite-associated nitrotyrosine residues. Thus, in this setting uric acid was inducing peroxynitrite formation as opposed to its classical role to inactivate it.

Evidence that these effects were mediated by intracellular uric acid was shown by experiments in which probenecid was added to the incubation mixture. Probenecid is an organic anion transporter that blocks the uptake of uric acid into the cell [61, 62] and was found to block the effects of uric acid to induce oxidative stress in these cells [16, 67].

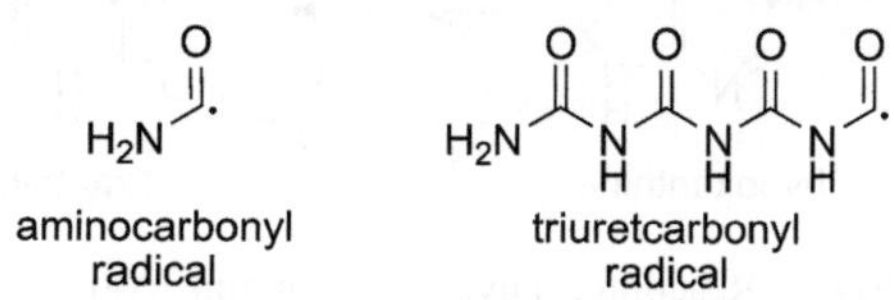

Fig. 6 Aminocarbonyl and triuretcarbonyl radicals

2.2 *Animal Studies*

We have also found that the administration of a uricase inhibitor to rat can result in oxidative stress. The blockade of uricase will actually reduce oxidant generation in the liver, as uricase releases superoxide anion on reaction with uric acid. Nevertheless, we found that the induction of hyperuricemia in rats by uricase inhibition resulted in evidence for intrarenal oxidative stress, as noted by the appearance of 3-nitrotyrosine and 4 hydroxynonnenol adducts in the kidney. An increase in Nox4 expression [68] suggests that the oxidative stress may be secondary to activation of NADPH oxidases, as was demonstrated in the adipocyte cell culture studies [16]. The oxidative stress was associated with the development of systemic and glomerular hypertension and could be reversed by treatment with the superoxide dismutase mimetic, TEMPOL (4-hydroxy-2,2,6,6-tetramethylpiperidine-1-oxyl) [68].

3 Xanthine Oxidoreductase and Uric Acid

While uric acid can have both pro-oxidative and antioxidative effects, many studies examining the role of uric acid in diseased states utilize xanthine oxidoreductase (XOR) inhibitors as a means to evaluate the role of uric acid. XOR is a complex enzyme that may occur either as a xanthine dehydrogenase or xanthine oxidase [69]. The former isoform utilizes NAD as its substrate to form uric acid and NADH, whereas the latter enzyme is used to convert hypoxanthine to xanthine with the production of a superoxide molecule, and a second reaction of xanthine to form uric acid with a second superoxide molecule (Fig. 7) [70, 71]. As such, the two-step reaction will form two oxidant molecules (superoxide) and one uric acid molecule.

Thus, xanthine oxidase inhibitors have the ability to reduce both uric acid formation and xanthine oxidase induced superoxide formation. Interestingly, allopurinol substitutes for xanthine and results in the formation of oxypurinol as opposed to uric acid [72]. As such, allopurinol itself does not block xanthine oxidase-induced oxidants [72]. However, its product, oxypurinol, then acts to irreversibly inhibit xanthine oxidase, thereby reducing superoxide formation [73]. Hence, in this situation it becomes unclear if any benefits observed with xanthine oxidase inhibitors are due to the lowering of uric acid or the reduction in xanthine oxidase-associated oxidants. One approach to resolving these issues is to determining if equivalent benefit can be obtained by lowering uric acid with recombinant

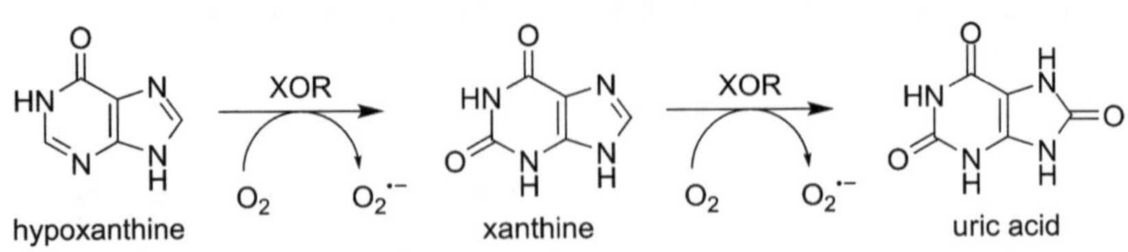

Fig. 7 Reaction catalyzed by xanthine oxidase. Adapted from [71]

uricase or with a uricosuric agent. As will be discussed, most studies in animals suggest it is the uric acid itself that is causing the disease manifestations (to be discussed below), whereas studies in humans have been conflicting. These are briefly summarized below.

4 Uric Acid and Its Role in Hypertension, Metabolic Syndrome, and Renal Disease

4.1 Uric Acid and Hypertension

Hyperuricemia is common in subjects with hypertension and may be present in 25–50% of patients [74]. In adolescents with newly diagnosed primary hypertension, the prevalence is even higher and approaches 90% [75]. Evidence that hyperuricemia may have a causal role in the condition is supported by the finding that elevating uric acid with a uricase inhibitor causes systemic and glomerular hypertension in rats [76]. Importantly, in this model the use of either a uricosuric (benziodarone) or a xanthine oxidase inhibitor (allopurinol) was protective [76]. The mechanism for the hypertension was shown to involve the development of oxidative stress, with a reduction in endothelial NO levels and a stimulation of the local and renal angiotensin system [66, 68, 77, 78]. Evidence that uric acid may also cause hypertension in humans is just emerging. In one small pilot study, the lowering of uric acid in adolescents with newly diagnosed hypertension resulted in a normalization of blood pressure in 86% of children if the uric acid was lowered to less than 5 mg/dL (300 μM); in contrast, only 3% of placebo-treated controls became normotensive [79]. The possibility that the allopurinol was acting by blocking xanthine oxidase-associated oxidants should be considered, as George et al. reported that allopurinol treatment improved endothelial function in subjects with congestive heart failure, whereas the uricosuric, probenecid, did not, despite equivalent lowering of uric acid [80]. However, a recently conducted trial in prehypertensive obese adolescents does appear to confirm a significant blood pressure lowering effect if uric acid is lowered, regardless of whether probenecid or allopurinol is used (Dan Feig, personal communication).

4.2 Uric Acid and Metabolic Syndrome

An elevated uric acid is also common in subjects with metabolic syndrome. Indeed, hyperuricemia and low urinary uric acid excretion are so common in subjects with metabolic syndrome that these features have been considered part of the syndrome [81]. While historically the hyperuricemia was thought to be a consequence of the effects of hyperinsulinemia [82], more recent studies suggest that uric acid may have a role in causing this syndrome. In particular, one of the most effective ways to induce metabolic syndrome in animals is with fructose [83]. Interestingly, one unique feature of fructose is that it generates uric acid. When fructose is metabolized, it undergoes rapid and unregulated phosphorylation with adenosine triphosphate (ATP) consumption and adenosine monophosphate (AMP) accumulation, and the breakdown of the adenine nucleotides coupled with a stimulation of AMP deaminase results in uric acid generation. Serum uric acid rises shortly after a fructose-rich meal [84].

Studies in animals have found that lowering uric acid can ameliorate many features of the metabolic syndrome induced by fructose [85, 86]. Lowering uric acid by either a uricosuric or a xanthine oxidase inhibitor is protective [85]. Uric acid causes metabolic syndrome likely via several mechanisms. First, the ability of uric acid to inhibit endothelial function could contribute to insulin resistance, as insulin requires endothelial NO for maximal effects by increasing blood flow to target organs such as skeletal muscle [85]. Second, uric acid may have direct effects on the adipocyte. We found, for example, that uric acid could induce a proinflammatory state in the adipocyte [16] similar to what has been shown in subjects with metabolic syndrome [87]. Cheung et al. have also reported that XOR is a key enzyme in the adipocyte that is critical for adipogenesis [88]. Furthermore, recent studies suggest uric acid may also regulate the key enzyme in fructose metabolism, fructokinase, as well as augment the effects of fructose to induce ATP depletion (Sanchez-Lozada et al., unpublished).

Few studies in humans have been performed to date. Our group has recently evaluated the effect of lowering uric acid with a xanthine oxidase inhibitor in healthy adult males administered large doses of fructose (200 g/day) [89]. Fructose induced multiple features of metabolic syndrome as well as a significant increase in serum uric acid levels. However, whereas allopurinol was effective at blocking the rise in blood pressure in response to fructose, it had no observable benefits on insulin resistance. In contrast, in a recent controlled study, a Japanese group found that lowering uric acid with a uricosuric could markedly improve insulin sensitivity in subjects with congestive heart failure (Hisatome, personal communication). One potential explanation is that the effects of uric acid to potentiate fructose effects might be less evident under conditions in which very high doses of fructose are used. Obviously, further studies are needed prior to making any conclusion on the role of uric acid in metabolic syndrome in humans.

4.3 Uric Acid and Renal Disease

There is also emerging evidence that lowering uric acid may be beneficial in kidney disease. It has been known for a long time that most patients with chronic kidney disease have high uric acid levels, and it was ascribed to the well-known finding that a reduction in glomerular filtration rate (GFR) will reduce uric acid excretion. Nearly half of dialysis patients have hyperuricemia [90–92]. The possibility that the hyperuricemia might have a role in chronic kidney disease was suggested by experimental studies in animals in which the raising of uric acid was found to induce kidney disease in normal animals [76, 93] and to accelerate chronic kidney disease such as the remnant kidney model [94]. In the latter model both uricosuric agents and xanthine oxidase inhibitors were effective. The mechanism was shown to involve the induction of renal microvascular disease with an alteration in renal autoregulation, the development of renal vascular vasoconstriction, and the stimulation of inflammation within the kidney [95–98]. Further studies showed that lowering uric acid could prevent or ameliorate renal injury in a wide variety of experimental models, including the hyperuricemic remnant kidney model [94], cisplatin associated acute renal failure [99], and diabetic nephropathy [100].

Clinical studies also suggest that uric acid could have a role in causing renal disease. For example, numerous epidemiological studies suggest that an elevated uric acid level can predict the development of renal disease in the general population [101, 102], in subjects with type 1 diabetes [103, 104], and in subjects with glomerular disease and normal renal function [105, 106]. Interestingly, it has been more difficult to show that uric acid is an independent risk factor for renal progression in subjects with preexisting chronic kidney disease [107]. Nevertheless, lowering uric acid with allopurinol has been reported to slow renal disease progression in subjects with chronic kidney disease and asymptomatic hyperuricemia [108]. Talaat et al. also reported that there was a marked exacerbation of hypertension, proteinuria, and renal progression when allopurinol was withdrawn from subjects with chronic kidney disease if they were not on angiotensin-converting enzyme (ACE) inhibitors [109]. This latter study suggests that the use of ACE inhibitors may function similar to a xanthine oxidase inhibitor on renal disease, and this is consistent with experimental studies that suggest that some of the actions of uric acid are via stimulation of the renin angiotensin system [66, 77]. However, there are also some data that uric acid may have effects in addition to stimulating the renin angiotensin system, such as by stimulating endothelin 1 levels [110, 111]. In this regard, in experimental models the addition of ACE inhibitors with allopurinol has been reported to have synergistic effects on blood pressure in rats with fructose-induced metabolic syndrome [112].

5 Conclusion

Uric acid is a molecule with complex chemistry with multiple biological effects. It is not simply a purine waste product that aids in the removal of nitrogenous wastes. Rather it has a mixture of biological effects, of which some are likely beneficial and some are deleterious. The variety of the effects of uric acid and their implications in human disease are summarized in Fig. 8. The antioxidant effects of uric acid are most clearly evident in the extracellular environment immediately after an increase in its concentration. Uric acid acts as a strong secondary reactant with oxidants after ascorbate. It seems possible that the antioxidant effects of uric acid may be most apparent in the circulation and possibly in the central nervous

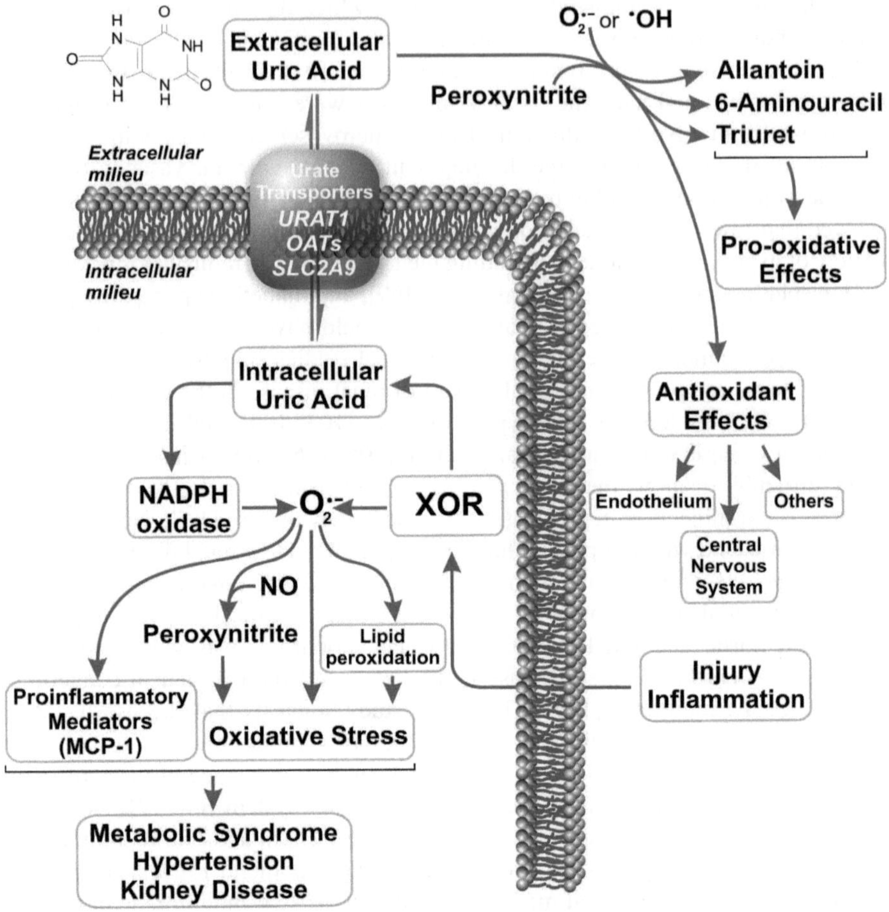

Fig. 8 Pro-oxidant and antioxidant effects of the extracellular and intracellular uric acid. *XOR* xanthine oxidoreductase; *URAT1* major urate reabsorptive transporter; *SLC2A9* sugar transport facilitator family protein, also known as GLUT9; *OAT* organic anion transported family of proteins, which play an important role in the transport of urate; *MCP-1* monocyte chemotactic protein 1

system (CNS). In contrast, most studies suggest uric acid functions primarily as a pro-oxidant in the intracellular environment. In this setting uric acid may induce oxidative stress indirectly via stimulation of NADPH oxidase [16]. However, an equally tenable mechanism could be via the generation of oxidants, radicals, and alkylating agents during the detoxification of peroxynitrite [33] or possibly as a consequence of lowering intracellular NO via direct inactivation [38]. Whatever the mechanism, most data suggest that the intracellular effects of uric acid may be in general deleterious, and this could provide a potential mechanism by which uric acid may contribute to the development of hypertension and microvascular disease, metabolic syndrome and diabetes, and chronic kidney disease.

References

1. Keilin J. The biological significance of uric acid and guanine excretion. Biol Rev Cambridge Phil Soc 1959;34:265–296
2. Terkeltaub RA. Gout. N Engl J Med 2003;349:1647–1655
3. Kellogg EW, III, Fridovich I. Liposome oxidation and erythrocyte lysis by enzymically generated superoxide and hydrogen peroxide. J Biol Chem 1977;252:6721–6728
4. Proctor P. Similar functions of uric acid and ascorbate in man? Nature 1970;228:868
5. Ames BN, Cathcart R, Schwiers E, Hochstein P. Uric acid provides an antioxidant defense in humans against oxidant- and radical-caused aging and cancer: a hypothesis. Proc Natl Acad Sci U S A 1981;78:6858–6862
6. Johnson RJ, Kang DH, Feig D, et al. Is there a pathogenetic role for uric acid in hypertension and cardiovascular and renal disease? Hypertension 2003;41:1183–1190
7. Feig DI, Kang DH, Johnson RJ. Uric acid and cardiovascular risk. N Engl J Med 2008;359:1811–1821
8. Culleton BF. Uric acid and cardiovascular disease: a renal-cardiac relationship? Curr Opin Nephrol Hypertens 2001;10:371–375
9. Culleton BF, Larson MG, Kannel WB, Levy D. Serum uric acid and risk for cardiovascular disease and death: the Framingham Heart Study. Ann Internal Med 1999;131:7–13
10. Reyes AJ, Leary WP. The ALLHAT and the cardioprotection conferred by diuretics in hypertensive patients: a connection with uric acid? Cardiovasc Drugs Ther 2002;16:485–487
11. Reyes AJ, Leary WP. The increase in serum uric acid induced by diuretics could be beneficial to cardiovascular prognosis in hypertension: a hypothesis. J Hypertension 2003;21:1775–1777
12. Nieto FJ, Iribarren C, Gross MD, Comstock GW, Cutler RG. Uric acid and serum antioxidant capacity: a reaction to atherosclerosis? Atherosclerosis 2000;148:131–139
13. Glantzounis GK, Tsimoyiannis EC, Kappas AM, Galaris DA. Uric acid and oxidative stress. Curr Pharm Des 2005;11:4145–4151
14. Johnson RJ, Feig DI, Herrera-Acosta J, Kang DH. Resurrection of uric acid as a causal risk factor in essential hypertension. Hypertension 2005;45:18–20
15. Sautin YY, Johnson RJ. Uric acid: the oxidant-antioxidant paradox. Nucleosides Nucleotides Nucleic Acids 2008;27:608–619
16. Sautin YY, Nakagawa T, Zharikov S, Johnson RJ. Adverse effects of the classic antioxidant uric acid in adipocytes: NADPH oxidase-mediated oxidative/nitrosative stress. Am J Physiol Cell Physiol 2007;293:C584–C596
17. Becker BF. Towards the physiological function of uric acid. Free Radic Biol Med 1993;14:615–631
18. Simie M, Jovanovic S. Antioxidation mechanisms of uric acid. J Am Chem Soc 1989;111:5778–5782
19. Rosell M, Regnstrom J, Kallner A, Hellenius ML. Serum urate determines antioxidant capacity in middle-aged men – a controlled, randomized diet and exercise intervention study. J Intern Med 1999;246:219–226
20. Maples KR, Mason RP. Free radical metabolite of uric acid. J Biol Chem 1988;263:1709–1712
21. Vasquez-Vivar J, Santos AM, Junqueira VB, Augusto O. Peroxynitrite-mediated formation of free radicals in human plasma: EPR detection of ascorbyl, albumin-thiyl and uric acid-derived free radicals. Biochemical J 1996;314(Pt 3):869–876
22. Santos CX, Anjos EI, Augusto O. Uric acid oxidation by peroxynitrite: multiple reactions, free radical formation, and amplification of lipid oxidation. Arch Biochem Biophys 1999;372:285–294
23. Hellsten Y, Tullson PC, Richter EA, Bangsbo J. Oxidation of urate in human skeletal muscle during exercise. Free Radic Biol Med 1997;22:169–174

24. Mikami T, Kita K, Tomita S, et al. Is allantoin in serum and urine a useful indicator of exercise-induced oxidative stress in humans? Free Radic Res 2000;32:235–244
25. Doehner W, Schoene N, Rauchhaus M, et al. Effects of xanthine oxidase inhibition with allopurinol on endothelial function and peripheral blood flow in hyperuricemic patients with chronic heart failure: results from 2 placebo-controlled studies. Circulation 2002;105:2619–2624
26. Grootveld M, Halliwell B. Measurement of allantoin and uric-acid in human-body fluids – a potential index of free-radical reactions in vivo. Biochemical J 1987;243:803–808
27. Kand'ar R, Zakova P, Muzakova V. Monitoring of antioxidant properties of uric acid in humans for a consideration measuring of levels of allantoin in plasma by liquid chromatography. Clin Chim Acta 2006;365:249–256
28. Felici C, Ciari I, Terzuoli L, et al. Purine catabolism in advanced carotid artery plaque. Nucleosides Nucleotides Nucleic Acids 2006;25:1291–1294
29. Kahn K, Serfozo P, Tipton PA. Identification of the true product of the urate oxidase reaction. J Am Chem Soc 1997;119:5435–5442
30. Radi R, Peluffo G, Alvarez MN, Naviliat M, Cayota A. Unraveling peroxynitrite formation in biological systems. Free Radic Biol Med 2001;30:463–488
31. Robinson KM, Morre JT, Beckman JS. Triuret: a novel product of peroxynitrite-mediated oxidation of urate. Arch Biochem Biophys 2004;423:213–217
32. Kuzkaya N, Weissmann N, Harrison DG, Dikalov S. Interactions of peroxynitrite with uric acid in the presence of ascorbate and thiols: implications for uncoupling endothelial nitric oxide synthase. Biochem Pharmacol 2005;70:343–354
33. Gersch C, Palii SP, Imaram W, et al. Reactions of peroxynitrite with uric acid: formation of reactive intermediates, alkylated products and triuret, and in vivo production of triuret under conditions of oxidative stress. Nucleosides Nucleotides Nucleic Acids 2009;28:118–149
34. Kim KM, Henderson GN, Frye RF, et al. Simultaneous determination of uric acid metabolites allantoin, 6-aminouracil, and triuret in human urine using liquid chromatography-mass spectrometry. J Chromatogr B Analyt Technol Biomed Life Sci 2009;877:65–70
35. Skinner KA, White CR, Patel R, et al. Nitrosation of uric acid by peroxynitrite. Formation of a vasoactive nitric oxide donor. J Biol Chem 1998;273:24491–24497
36. Gersch C, Palii SP, Imaram W, Kim KM, Karumanchi SA, Angerhofer A, Johnson RJ, Henderson GN. Reactions of peroxynitrite with uric acid: formation of reactive intermediates, alkylated products and triuret, and in vivo production of triuret under conditions of oxidative stress. Nucleosides Nucleotides Nucleic Acids 2009;28:118–149
37. Suzuki T. Nitrosation of uric acid induced by nitric oxide under aerobic conditions. Nitric Oxide 2007;16:266–273
38. Gersch C, Palii SP, Kim KM, et al. Inactivation of nitric oxide by uric acid. Nucleosides Nucleotides Nucleic Acids 2008;27:967–978
39. Waugh WH. Inhibition of iron-catalyzed oxidations by attainable uric acid and ascorbic acid levels: therapeutic implications for Alzheimer's disease and late cognitive impairment. Gerontology 2008;54:238–243
40. Schlotte V, Sevanian A, Hochstein P, Weithmann KU. Effect of uric acid and chemical analogues on oxidation of human low density lipoprotein in vitro. Free Radic Biol Med 1998;25:839–847
41. Muraoka S, Miura T. Inhibition by uric acid of free radicals that damage biological molecules. Pharmacol Toxicol 2003;93:284–289
42. Hink HU, Santanam N, Dikalov S, et al. Peroxidase properties of extracellular superoxide dismutase: role of uric acid in modulating in vivo activity. Arterioscler Thromb Vasc Biol 2002;22:1402–1408
43. Sevanian A, Davies KJ, Hochstein P. Conservation of vitamin C by uric acid in blood. J Free Radic Biol Med 1985;1:117–124
44. Sevanian A, Davies KJ, Hochstein P. Serum urate as an antioxidant for ascorbic acid. Am J Clin Nutr 1991;54:1129S–1134S

45. Frei B, Stocker R, Ames BN. Antioxidant defenses and lipid peroxidation in human blood plasma. Proc Natl Acad Sci USA 1988;85:9748–9752
46. Waring WS, Convery A, Mishra V, et al. Uric acid reduces exercise-induced oxidative stress in healthy adults. Clin Sci (Lond) 2003;105:425–430
47. Waring WS, McKnight JA, Webb DJ, Maxwell SR. Uric acid restores endothelial function in patients with type 1 diabetes and regular smokers. Diabetes 2006;55:3127–3132
48. Waring WS, Webb DJ, Maxwell SR. Systemic uric acid administration increases serum antioxidant capacity in healthy volunteers. J Cardiovasc Pharmacol 2001;38:365–371
49. Kurra V, Eraranta A, Jolma A, et al. Hyperuricemia, oxidative stress, and carotid artery tone in experimental renal insufficiency. Am J Hypertens 2009;22:694–970
50. Schlesinger I, Schlesinger N. Uric acid in Parkinson's disease. Mov Disord 2008;23:1653–1657
51. Chen H, Mosley TH, Alonso A, Huang X. Plasma urate and Parkinson's disease in the Atherosclerosis Risk in Communities (ARIC) study. Am J Epidemiol 2009;169:1064–1069
52. Hooper DC, Spitsin S, Kean RB, et al. Uric acid, a natural scavenger of peroxynitrite, in experimental allergic encephalomyelitis and multiple sclerosis. Proc Natl Acad Sci USA 1998;95:675–680
53. Guerreiro S, Ponceau A, Toulorge D, et al. Protection of midbrain dopaminergic neurons by the end-product of purine metabolism uric acid: potentiation by low-level depolarization. J Neurochem 2009;109:1118–1128
54. Yu ZF, Bruce-Keller AJ, Goodman Y, Mattson MP. Uric acid protects neurons against excitotoxic and metabolic insults in cell culture, and against focal ischemic brain injury in vivo. J Neurosci Res 1998;53:613–625
55. Church WH, Ward VL. Uric acid is reduced in the substantia nigra in Parkinson's disease: effect on dopamine oxidation. Brain Res Bull 1994;33:419–425
56. Hooper DC, Scott GS, Zborek A, et al. Uric acid, a peroxynitrite scavenger, inhibits CNS inflammation, blood-CNS barrier permeability changes, and tissue damage in a mouse model of multiple sclerosis. FASEB J 2000;14:691–698
57. Spitsin SV, Scott GS, Mikheeva T, et al. Comparison of uric acid and ascorbic acid in protection against EAE. Free Radic Biol Med 2002;33:1363–1371
58. Kim SY, Guevara JP, Kim KM, et al. Hyperuricemia and risk of stroke: a systematic review and meta-analysis. Arthritis Rheum 2009;61:885–892
59. Tsukada K, Hasegawa T, Tsutsumi S, et al. Effect of uric acid on liver injury during hemorrhagic shock. Surgery 2000;127:439–446
60. Becker BF, Reinholz N, Ozcelik T, Leipert B, Gerlach E. Uric acid as radical scavenger and antioxidant in the heart. Pflugers Arch 1989;415:127–135
61. Price KL, Sautin YY, Long DA, et al. Human vascular smooth muscle cells express a urate transporter. J Am Soc Nephrol 2006;17:1791–1795
62. Kang DH, Han L, Ouyang X, et al. Uric acid causes vascular smooth muscle cell proliferation by entering cells via a functional urate transporter. Am J Nephrol 2005;25:425–433
63. Abuja PM. Ascorbate prevents prooxidant effects of urate in oxidation of human low density lipoprotein. FEBS Lett 1999;446:305–308
64. Sanguinetti SM, Batthyany C, Trostchansky A, et al. Nitric oxide inhibits prooxidant actions of uric acid during copper-mediated LDL oxidation. Arch Biochem Biophys 2004;423:302–308
65. Imaram W. PhD Thesis, University of Florida 2008
66. Corry DB, Eslami P, Yamamoto K, et al. Uric acid stimulates vascular smooth muscle cell proliferation and oxidative stress via the vascular renin-angiotensin system. J Hypertens 2008;26:269–275
67. Kanellis J, Watanabe S, Li JH, et al. Uric acid stimulates monocyte chemoattractant protein-1 production in vascular smooth muscle cells via mitogen-activated protein kinase and cyclooxygenase-2. Hypertension 2003;41:1287–1293

68. Sanchez-Lozada LG, Soto V, Tapia E, et al. Role of oxidative stress in the renal abnormalities induced by experimental hyperuricemia. Am J Physiol 2008;295:F1134–F1141
69. Pacher P, Nivorozhkin A, Szabo C. Therapeutic effects of xanthine oxidase inhibitors: renaissance half a century after the discovery of allopurinol. Pharmacol Rev 2006;58:87–114
70. Harrison R. Physiological roles of xanthine oxidoreductase. Drug Metab Rev 2004;36:363–375
71. Berry CE, Hare JM. Xanthine oxidoreductase and cardiovascular disease: molecular mechanisms and pathophysiological implications. J Physiol 2004;555:589–606
72. Galbusera C, Orth P, Fedida D, Spector T. Superoxide radical production by allopurinol and xanthine oxidase. Biochem Pharmacol 2006;71:1747–1752
73. Spector T. Oxypurinol as an inhibitor of xanthine oxidase-catalyzed production of superoxide radical. Biochem Pharmacol 1988;37:349–352
74. Cannon PJ, Stason WB, Demartini FE, Sommers SC, Laragh JH. Hyperuricemia in primary and renal hypertension. N Engl J Med 1966;275:457–464
75. Feig DI, Johnson RJ. Hyperuricemia in childhood primary hypertension. Hypertension 2003;42:247–252
76. Mazzali M, Hughes J, Kim YG, et al. Elevated uric acid increases blood pressure in the rat by a novel crystal-independent mechanism. Hypertension 2001;38:1101–1106
77. Mazzali M, Kanellis J, Han L, et al. Hyperuricemia induces a primary renal arteriolopathy in rats by a blood pressure-independent mechanism. Am J Physiol 2002;282:F991–F997
78. Sanchez-Lozada LG, Tapia E, Lopez-Molina R, et al. Effects of acute and chronic L-arginine treatment in experimental hyperuricemia. Am J Physiol 2007;292:F1238–F1244
79. Feig DI, Soletsky B, Johnson RJ. Effect of allopurinol on the blood pressure of adolescents with newly diagnosed essential hypertension. JAMA 2008;300:922–930
80. George J, Carr E, Davies J, Belch JJ, Struthers A. High-dose allopurinol improves endothelial function by profoundly reducing vascular oxidative stress and not by lowering uric acid. Circulation 2006;114:2508–2516
81. Facchini F, Chen YD, Hollenbeck CB, Reaven GM. Relationship between resistance to insulin-mediated glucose uptake, urinary uric acid clearance, and plasma uric acid concentration. JAMA 1991;266:3008–3011
82. Quinones GA, Natali A, Baldi S, et al. Effect of insulin on uric acid excretion in humans. Am J Physiol 1995;268:E1–E5
83. Johnson RJ, Perez-Pozo SE, Sautin YY, et al. Hypothesis: could excessive fructose intake and uric acid cause type 2 diabetes? Endocr Rev 2009;30:96–116
84. Stirpe F, Della Corte E, Bonetti E, et al. Fructose-induced hyperuricaemia. Lancet 1970;2:1310–1311
85. Nakagawa T, Hu H, Zharikov S, et al. A causal role for uric acid in fructose-induced metabolic syndrome. Am J Physiol 2006;290:F625–F631
86. Reungjui S, Roncal CA, Mu W, et al. Thiazide diuretics exacerbate fructose-induced metabolic syndrome. J Am Soc Nephrol 2007;18:2724–2731
87. Furukawa S, Fujita T, Shimabukuro M, et al. Increased oxidative stress in obesity and its impact on metabolic syndrome. J Clin Invest 2004;114:1752–1761
88. Cheung KJ, Tzameli I, Pissios P, et al. Xanthine oxidoreductase is a regulator of adipogenesis and PPARgamma activity. Cell Metab 2007;5:115–128
89. Perez-Pozo SE, Schold J, Nakagawa T, Sánchez-Lozada LG, Johnson RJ, Lillo JL. Excessive fructose intake induces the features of metabolic syndrome in healthy adult men: role of uric acid in the hypertensive response. Int J Obes (Lond) 2010;34:454–461.
90. Hsu SP, Pai MF, Peng YS, et al. Serum uric acid levels show a "J-shaped" association with all-cause mortality in haemodialysis patients. Nephrol Dial Transplant 2004;19:457–462
91. Suliman ME, Johnson RJ, Garcia-Lopez E, et al. J-shaped mortality relationship for uric acid in CKD. Am J Kidney Dis 2006;48:761–771
92. Lee SM, Lee AL, Winters TJ, et al. Low serum uric acid level is a risk factor for death in incident hemodialysis patients. Am J Nephrol 2009;29:79–85

93. Nakagawa T, Mazzali M, Kang DH, et al. Hyperuricemia causes glomerular hypertrophy in the rat. Am J Nephrol 2003;23:2–7
94. Kang DH, Nakagawa T, Feng L, et al. A role for uric acid in the progression of renal disease. J Am Soc Nephrol 2002;13:2888–2897
95. Sanchez-Lozada LG, Tapia E, Avila-Casado C, et al. Mild hyperuricemia induces glomerular hypertension in normal rats. Am J Physiol 2002;283:F1105–F1110
96. Sanchez-Lozada LG, Tapia E, Bautista-Garcia P, et al. Effects of febuxostat on metabolic and renal alterations in rats with fructose-induced metabolic syndrome. Am J Physiol 2008;294:F710–F718
97. Sanchez-Lozada LG, Tapia E, Santamaria J, et al. Mild hyperuricemia induces vasoconstriction and maintains glomerular hypertension in normal and remnant kidney rats. Kidney Int 2005;67:237–247
98. Sanchez-Lozada LG, Tapia E, Soto V, et al. Effect of febuxostat on the progression of renal disease in 5/6 nephrectomy rats with and without hyperuricemia. Nephron Physiol 2008;108:69–78
99. Roncal CA, Mu W, Croker B, et al. Effect of elevated serum uric acid on cisplatin-induced acute renal failure. Am J Physiol 2007;292:F116–F122
100. Kosugi T, Nakayama T, Heinig M, Zhang L, Yuzawa Y, Sanchez-Lozada LG, Roncal C, Johnson RJ, Nakagawa T. Effect of lowering uric acid on renal disease in the type 2 diabetic db/db mice. Am J Physiol Renal Physiol 2009;297:F481–F488
101. Iseki K, Ikemiya Y, Inoue T, et al. Significance of hyperuricemia as a risk factor for developing ESRD in a screened cohort. Am J Kidney Dis 2004;44:642–650
102. Iseki K, Oshiro S, Tozawa M, et al. Significance of hyperuricemia on the early detection of renal failure in a cohort of screened subjects. Hypertens Res 2001;24:691–697
103. Rosolowsky ET, Ficociello LH, Maselli NJ, et al. High-normal serum uric acid is associated with impaired glomerular filtration rate in nonproteinuric patients with type 1 diabetes. Clin . J Am Soc Nephrol 2008;3:706–713
104. Hovind P, Rossing P, Tarnow L, Johnson RJ, Parving HH. Serum uric acid as a predictor for development of diabetic nephropathy in type 1 diabetes: an inception cohort study. Diabetes 2009;58:1668–1671
105. Syrjanen J, Mustonen J, Pasternack A. Hypertriglyceridaemia and hyperuricaemia are risk factors for progression of IgA nephropathy. Nephrol Dial Transplant 2000;15:34–42
106. Ohno I, Hosoya T, Gomi H, et al. Serum uric acid and renal prognosis in patients with IgA nephropathy. Nephron 2001;87:333–339
107. Madero M, Sarnak MJ, Wang X, et al. Uric acid and long-term outcomes in CKD. Am J Kidney Dis 2009;53:796–803
108. Siu YP, Leung KT, Tong MK, Kwan TH. Use of allopurinol in slowing the progression of renal disease through its ability to lower serum uric acid level. Am J Kidney Dis 2006;47:51–59
109. Talaat KM, El-Sheikh AR. The effect of mild hyperuricemia on urinary transforming growth factor beta and the progression of chronic kidney disease. Am J Nephrol 2007;27:435–440
110. Chao HH, Liu JC, Lin JW, et al. Uric acid stimulates endothelin-1 gene expression associated with NADPH oxidase in human aortic smooth muscle cells. Acta Pharmacol Sin 2008;29:1301–1312
111. Cheng TH, Lin JW, Chao HH, Chen YL, Chen CH, Chan P, Liu JC. Uric acid activates extracellular signal-regulated kinases and thereafter endothelin-1 expression in rat cardiac fibroblasts. Int J Cardiol 2010:139:42–49
112. Roncal CA, Reungjui S, Sánchez-Lozada LG, Mu W, Sautin YY, Nakagawa T, Johnson RJ. Combination of captopril and allopurinol retards fructose-induced metabolic syndrome. Am J Nephrol 2009;30:399–404

Chapter 9
Reactive Oxygen and Nitrogen Species, Oxidative and Nitrosative Stress, and Their Role in the Pathogenesis of Acute Kidney Injury

Eisei Noiri, Francesco Addabbo, and Michael S. Goligorsky

Abstract This chapter briefly summarizes the existing evidence implicating reactive oxygen (ROS) and nitrogen (RNS) species in renal injury and offers insights into the mitochondrial generation of ROS, signaling functions of ROS necessary for survival as well as death decisions, the role of ROS in hormesis, and therapeutic strategies attenuating ROS- and RNS-induced renal injury. Along with this discussion we emphasize the complexity of regulation of ROS and RNS formation and detoxication, physiological functions of these products, as well as molecular mechanisms responsible for caspase-dependent and -independent cell death. Considering the universality of ROS and RNS mechanisms of cell regulation, these data should illuminate basic processes of cell and organ dysfunction.

1 Definitions

The term reactive oxygen species (ROS) encompasses all reactive species derived from the one-electron reduction of molecular oxygen. It includes a short-lived and poorly diffusible across biological membranes superoxide (O_2^-), as well as a long-lived, uncharged, and diffusible hydrogen peroxide (H_2O_2). ROS are produced by plasma membrane-encored enzymes such as nicotinamide adenine dinucleotide phosphate (NADPH) oxidase and uncoupled endothelial nitric oxide synthase (eNOS), cytosolic enzymes like xanthine oxidoreductase (XOR), inducible NOS, cytochrome P450 systems, and mitochondrial enzymes (electron transport chain complexes I and III, uncoupling proteins). Approximately 1–2% of molecular oxygen consumed by mitochondria is converted to O_2^- under physiologic conditions, and this proportion is significantly increased under pathologic conditions such as hypoxia. Protection from cytotoxic effects of ROS is accomplished by *antioxidant systems* of superoxide dismutases (SOD) (extracellular, cytosolic,

E. Noiri (✉)
Department of Medicine, Tokyo University, Tokyo, Japan
e-mail: noiri-tky@umin.ac.jp

T. Miyata et al. (eds.), *Studies on Renal Disorders*, Oxidative Stress in Applied Basic Research and Clinical Practice, DOI 10.1007/978-1-60761-857-7_9,

and mitochondrial isoforms), peroxisomal catalase, glutathione peroxidase, peroxiredoxin, thioredoxin, and glutaredoxin. When net production of ROS outweighs the capacity of antioxidant systems to neutralize them, *oxidative stress* ensues. *Reactive nitrogen species* (RNS) refers to nitric oxide (NO), the product of either of three NOS enzymes: endothelial, neuronal, and inducible. NO modulates the activity of multiple enzymes by binding to their heme moiety, such as soluble guanylate cyclase, or binding to cysteine thiols to form *S*-nitrosothioles (R–S–NO). Biological reduction of oxidized or nitrosated cysteine thiols is accomplished by thioredoxin and glutaredoxin, respectively. The term nitrosative stress implies that NO production has increased to a degree that it overcomes the reductive capacity of antinitrosative systems.

2 Redox Signaling Functions

Generation of ROS is a hallmark of aerobic organisms where they perform multiple signaling functions, as long as processes are reversible. When this precondition is unachievable due to the oxidative stress or depletion of antioxidant defenses, cytotoxic and genotoxic actions of ROS become dominant.

ROS and RNS are critically involved in modulation of functional properties of several diverse families of proteins [1] (Fig. 1). This is accomplished via oxidation of cysteines within a protein (intraprotein disulfide formation), between different proteins (interprotein disulfide formation), and between cysteine and glutathione (S-glutathionylation) and NO-modification of sulfhydryl groups (*S*-nitrosothioles). Trx is capable of reducing protein disulfides, Grx reduces *S*-glutathiones, whereas both enzymes can reduce *S*-nitrosothioles.

These types of protein modification impart on protein phosphatases, protein kinases, transcription factors, other enzymes, and ion channels various signaling functions. Oxidation by H_2O_2 of cysteines in protein phosphatases inhibits their activity. Oxidation also results in the activation of Src kinase, with the subsequent activation of mitogen-activated protein kinase (MAPK) signaling, which induces nuclear export of telomerase reverse transcriptase, thus leading to cell senescence. Protein kinase C activity is dually regulated: its activity is increased when N-terminal regulatory domain cysteines are targeted, but its activity is inhibited upon oxidation of cysteines in the C-terminal catalytic domain. Akt and JNK can undergo S-nitrosylation, which results in inhibition of their activity. Among transcription factors subjected to redox modification, nuclear factor κB (NFκB) is activated by H_2O_2, but inhibited by S-nitrosylation. Oxidation of Keap-1 results in the dissociation of NrF2/Keap-1 complex and increased transcription of *S*-glutathione transferase and heme oxygenase-1. These examples clearly demonstrate an important role of mild-to-moderate oxidative stress in physiological signaling and cell survival.

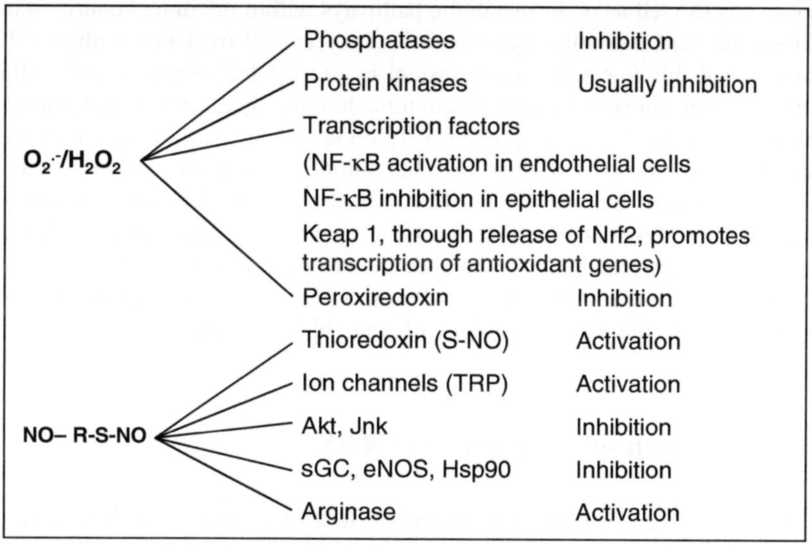

Fig. 1 Redox modifications of protein functions. Modified from [4]

3 Mitochondrial ROS: Generation and Defenses

Mitochondria are intimately involved in energy metabolism of cells by utilizing the products of glycolitic and lipolytic metabolism, play an important role in amino acid metabolism and signaling functions, and regulate cell death pathways. These organelles generate most of the eukaryotic cells energy supply primarily in the form of universal energy currency, electrochemical proton gradient ($\Delta\mu H^+$), which fuels adenosine triphosphate (ATP) production, and a wide range of other processes such as ion transport, cellular metabolism [2], differentiation, growth, and cell death. Generation of the $\Delta\mu H^+$ starts through the series of redox reactions conducted by the four respiratory chain complexes at the ion-impermeable, almost cholesterol-free inner membrane. Reduced nicotinamide adenine dinucleotide (NADH) is the entry point to the complex I (NADH:ubiquinone reductase), whereas the reduced ubiquinol enters the respiratory chain in the second point, complex III (ubiquinol: cytochrome *c* reductase) to reduce cytochrome *c*, the electron carrier to the complex IV, cytochrome *c* oxidase, which reduces molecular oxygen to water. Each of these steps generates $\Delta\mu H^+$ by electrogenic pumping of protons from the mitochondrial matrix to the intermembrane space and is coupled to electrons flowing through the complexes, thus generating the electrical membrane potential of -180 to -220 mV and a pH gradient of 0.4–0.6 units across the inner mitochondrial membrane. Dissipation of the accumulated proton gradient (a net influx of H^+ into the matrix) drives ATP synthesis or protein transport, both prerequisites for maintenance of cell homeostasis as well as other metabolic pathways within the mitochondrial matrix: tricarboxylic acid oxidation cycle and the fatty acid β-oxidation pathway. Each component of this finely regulated system is a target for various harmful stimuli (some of which can be generated within mitochondria themselves). The panoply of noxious stimuli has been comprehensively reviewed by Wallace and Starkov [3]. Reactions in response to diverse noxious stimuli have been compared with the exothermic oxygen combustion and free radical emission. This brief overview on the mitochondrial oxidative stress, extensively reviewed elsewhere [4], is intended to update findings related to the balance between mitochondrial ROS generation and defense systems and mitochondrial targets of excessive ROS, and it discusses their relevance to the pathophysiology of acute kidney injury (AKI).

3.1 Generation of Mitochondrial ROS

Electrons necessary for the generation of ROS are usually derived from the mitochondrial electron transport chain, as well as NADPH oxidases, xantine oxidase, uncoupled eNOS, and cytochrome P450 during univalent reduction of molecular oxygen to superoxide anion (O_2^-), which in turn undergoes spontaneous or enzymatic dismutation to form H_2O_2. It is important to emphasize that all these pathways of ROS formation are interactive and mostly amplify one another. Being charged, O_2^- poorly crosses biological membranes, whereas the uncharged H_2O_2 is readily diffusing, propagating within and outside the organelle of origin. In mitochondria, normal functioning of complex I (NADH dehydrogenase) and complex III (ubiquinone-cytochrome *bc1*) of the electron transport chain results in generation of O_2^- in the amount of 1–2% of the total oxygen consumed and represents the major source of ROS under physiological conditions. Various stimuli, such as integrin ligation, angiotensin II, tumor necrosis factor-alpha (TNF-α), H_2O_2, oxidized low-density lipoprotein (LDL), and electrophilic lipid oxidation products, all induce mitochondrial ROS (reviewed in [1]).

The most notable recent breakthrough in the field is related to the demonstration of the kinetics of superoxide production in mitochondria of living cells or organs [5]. In cultured cells transfected with a novel biosensor of superoxide that is unaffected by H_2O_2, peroxynitrite ($ONOO^-$), hydroxyl radical (OH•), or NO, a circularly permuted yellow fluorescence protein (cpYFP), investigators detected fluorescence flashes arising from individual or a group of functionally coupled mitochondria. Flashes of superoxide production had peak frequency of 3.5 s and dissipated with a half-time of 8.6 s, occurred randomly, and the frequency and amplitude were reduced by a SOD-mimetic or a superoxide scavenger tiron. These superoxide flashes were triggered by the transient openings of mitochondrial permeability transition pores (mPTP) and were unrelated to Ca^{2+} sparks. Transient mPTP-initiated superoxide flashes required functional electron transfer chain (ETC) and were ATP-dependent. Superoxide flashes coincided with a transient decline in mitochondrial membrane potential, matrix acidosis, and mitochondrial swelling.

Anoxia or mild hypoxia decreased the frequency of superoxide flashes in cardiac myocytes, suggesting that elevated ROS levels found under these conditions originate extramitochondrially or are due to the defective antioxidant defense. However, upon reoxygenation – the period of enhanced vulnerability of the cells – flash activity increased and treatment with adenosine preconditioning prevented it. This description of quantal, stochastic, and transient superoxide flashes, driven by the flickers of mPTP in quiescent cells, should profoundly change the current interpretation of such events as being necessarily pathologic, emphasizing the role of graded spatiotemporary ROS production in the normal processes of cell signaling via kinases, phosphatases, channels, and transporters and maintenance of ROS scavenging and detoxifying systems. In addition, these discoveries feed into the deeper appreciation of the complexity of mitochondrial ROS production. Integrating these studies into the existing knowledge will represent the next challenge. Clearly, the bidirectional effects of ROS generated in the cytosol on mitochondria and ROS generated in mitochondria on cytosolic systems will receive additional details. Furthermore, the concept of ROS-induced ROS release involving induction of the mitochondrial permeability transition [6].

Most relevant to the subject of AKI, hypoxia also results in increased ROS formation by the complex III. Importantly, mitochondrial ROS induce stabilization of hypoxia-inducible factor (HIF)-1α in hypoxic endothelial cells [7]. Physiologic concentrations of NO in endothelial cells inhibit cytochrome *c* oxidase and reversibly increase ROS production and signaling, whereas high levels of NO may be responsible for peroxynitrite formation and profound changes in mitochondrial proteins. In this regard, it has been demonstrated that acute ischemic kidney injury is accompanied by uncoupling of eNOS and induction of inducible nitric oxide synthase (iNOS), resulting in a prolonged and supraphysiological generation of NO [8] that, along with development of oxidative stress, is involved in enhanced formation of peroxynitrite [9]. In endothelial cells, a fraction of cellular eNOS is localized to the cytosolic face of the outer mitochondrial membrane [10], and uncoupling of this enzyme would also result in the excessive generation of ROS at the interface of mitochondria and the cytosol. The consequence of such an unopposed oxidative stress would lead to damage of mitochondrial DNA and mitochondrial dysfunction.

3.2 *Defense Against Mitochondrial Oxidative Stress*

Several enzymatic systems guard against accumulation of ROS. These include SOD, catalase, the peroxiredoxin/thioredoxin (Prx/Trx) network, and the glutathione peroxidase/glutathione reductase (Gpx/Grx) network. Mitochondria are endowed with Mn-SOD, Gpx1, Prx3, Trx2, and Grx2. All these systems protect mitochondria from uncontrolled formation of ROS and accumulation of ROS-induced injury to proteins and mitochondrial DNA. Consumption or degradation of these enzymes would inevitably translate into the enhanced oxidative stress.

4 The Concept of Stress-Induced Hormesis and ROS- and RNS-Induced Toxicity

Severity of oxidative or nitrosative stress is a key determinant of the outcome. For instance, mild-to-moderate oxidative stress results in the activation of catalase and Gpx and in the enhanced expression of Trx, whereas severe stress leads to the enhanced degradation of catalase and cathepsin D-induced degradation of Trx. Similarly, elevated levels of peroxynitrite result in the oxidation of BH_4 and Zn-thiolate cluster, uncoupling eNOS; whereas physiological levels of peroxynitrite are important for vasorelaxation by inducing reversible S-glutathionylation of Ca-ATPase in the endoplasmic reticulum [11], and irreversible oxidation of Ca-pump impairs NO-induced vasorelaxation. Hence, the level of ROS and RNS determines cellular and functional outcomes.

4.1 Hormesis

"Stress-response hormesis" refers to the induction of stress-protective mechanisms [12]. Given the toxicological axiom "the dose determines the poison," sublethal exposure to stressors induces a response that results in stress resistance, whereas a lethal level of stressor accelerates cell demise. There are emerging data that a nonlethal oxidative stress may stimulate mitochondrial biogenesis via activation of leucine zipper transcription factors, nuclear factor-E2-related factor (Nrf2) and ATF4, which regulate the expression of antioxidant response element-containing genes such as glutathione-*S*-transferase, glutathione peroxidase, glutathione reductase (all involved in glutathione biosynthesis and cycling), heme oxygenase-1, among others [13]. This process is governed, at least in part, by the mitochondrial ROS stimulating Nrf2 binding to the promoter region (rich in antioxidant response element motifs) of the nuclear respiratory factor-1 (NRF-1) in the Akt-dependent de-repression of Nrf2 nuclear translocation [14]. Activation of NRF-1 is a prerequisite for transcriptional activation of mitochondrial transcription factor A (Tfam) and induction of mitochondrial DNA replication/transcription and mitochondrial biogenesis [15, 16]. Another recently discovered mechanism stimulating mitochondrial biogenesis in response to etoposide-induced ataxia telangiectasia mutated (ATM)-dependent pathway is attributed to adenosine monophosphate activated protein kinase (AMPK) activation [17].

4.2 Oxidative Stress-Induces Programmed Cell Death, Types I and II

High levels of ROS and RNS induce caspase-dependent and -independent cell death via apoptosis and autophagy, respectively. Autophagy, in addition to representing a "self-cannibalism" of cells, is a finely regulated lysosomal pathway involved in the degradation and recycling of oxidized proteins and damaged organelles in the cells, thus providing a defense mechanism protecting cells against oxidative stress [18–21]. There is accumulating evidence that autophagic processing of damaged or excessive organelles, such as peroxisomes, endoplasmic reticulum, and mitochondria, can occur in response to ROS [21–23]. Excessive ROS in quiescent mitochondria poses a risk to this organelle and to the viability of the cell through opening of mitochondrial membrane channels, including the mPTP and the inner membrane anion channel (IMAC) [24], leading to the collapse of mitochondrial membrane potential and a transient increase in ROS generation by the ETC [5]. These events may result in autophagy, apoptosis, or necrosis [24]. In the kidney, autophagy plays an important role. Periyasamy-Thandavan et al. [25] provided in vivo and in vitro evidence that cisplatin treatment induces, in a dose–time dependent fashion, formation of autophagosomes in mouse kidneys and in cultured proximal tubular cells. This effect was partially suppressed by a p53 inhibitor and attenuated in renal cells stably transfected with Bcl-2, suggesting that these proteins play a role in autophagic signaling. These authors suggest that autophagy in AKI plays a role as a protective mechanism for cell survival. A somewhat different picture emerges from another recent report by Isaka et al. [26] who examined the effect of Bcl-2 on renal ischemia/reperfusion (I/R)-induced autophagy. I/R injury, which is associated with marked disruption of normal tubular morphology, induced autophagosome formation. On electron microscopy, the autophagosomes contained mitochondria. Bcl-2 augmentation suppressed the formation of autophagosomes associated with a reduction of tubular damage, confirming the important role for Bcl-2 in I/R injury by suppressing autophagosomal degradation and inhibiting tubular apoptosis.

Apoptosis and autophagy could also be induced by photosensitization of mitochondria or endoplasmic reticulum [27, 28] by employing low and intermediate energy photodynamic treatment (PDT) to stimulate mitochondria ROS production in cultured human epidermoid carcinoma A-431 cells. The approach was based on the visible light excitation of a mitochondria-localized lipophilic cationic photosensitizer (Safranin O), leading to generation of ROS and oxidative damage of cellular components. In this study, low energy PDT of cells failed to show hallmarks of apoptosis, necrosis, or senescence, and the detected loss of viability and reduced proliferation of the cells was ascribed to the enhanced autophagy [28].

5 Mitochondrial Biogenesis

Mechanisms known to regulate mitochondrial biogenesis are multiple. This long-term adaptive process depends on synthesis, mitochondrial protein import, and replication of mtDNA [29]. NRF-1 and -2 control transcription of mitochondrial OxPhos genes in conjunction with a nuclear transcriptional coactivator, PPAR-γ coactivator-1a (PGC-1α), which ultimately integrates various signals to

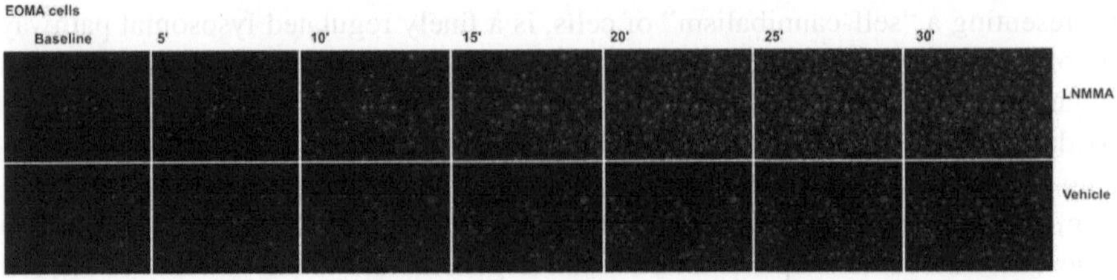

Fig. 2 L-NMMA induced mitochondrial superoxide assessment in vitro in EOMA cells using mito-SOX

enhance mitochondrial biogenesis. Relevant signals include caloric restriction, exposure to cold environment, exercise, employing mediators such as AMPK, sirtuin-1, cAMP, neuregulin, and NO.

In a previous report we have demonstrated that chronic uncoupling of eNOS with the concomitant oxidative stress, in a mouse model of "subclinical" endothelial dysfunction using nonpressor doses of L-NMMA, was associated with the reduction in mitochondrial mass (Fig. 2), as judged by the reduced fluorescence of a fluoroprobe MitoTracker® in the microvascular endothelial cells in vivo (intravital videomicroscopy) and diminished mitochondrial DNA [30].

6 Mitochondrial Oxidative Stress in Acute Kidney Injury

The pathological findings of tubular epithelial cell necrosis and apoptosis are mostly seen in the developing stage of AKI. Many studies, including our own, demonstrated the crucial role of ROS and successful amelioration of AKI using strategies to reduce the level of ROS [9, 31–33].

Under physiological conditions, mitochondria are elongated and filamentous. Upon stress or apoptotic stimulation, mitochondria become fragmented and may contribute to mitochondrial outer membrane permeabilization and the release of factors that induce apoptosis from the mitochondrial intermembrane space. Mitochondrial outer membrane permeabilization is directly related to the change of mitochondrial membrane potential. The proton motive force (Δp; equivalent to $\Delta\mu H^+$) responsible for mitochondrial ATP production consists of $\Delta\Psi$ and ΔpH. Under physiological condition, ΔpH is a small part of Δp and negligible. $\Delta\Psi$ is considered the representative of Δp, and its measurement is sufficient for the evaluation of mitochondrial capability to generate ATP. Under the pathological condition like hypoxia reoxygenation, relative changes of ΔpH and $\Delta\Psi$ will result in the change of Δp. ATP depletion occurs after hypoxia reoxygenation and, therefore, ATP-dependent potassium channels localized at the inner membrane of mitochondria should open to increase potassium permeability. The increase in inner membrane potassium permeability will decrease $\Delta\Psi$ and simultaneously increase ΔpH, which may leave Δp unchanged. If net Δp remained unchanged, it could not account for the energetic deficit after hypoxia reoxygenation. Feldkamp et al. [34] measured Δp and $\Delta\Psi$ and found that $\Delta\Psi$ is still the predominant component of Δp, even after hypoxia reoxygenation, because mitochondrial ΔpH is abrogated after hypoxia reoxygenation, making $\Delta\Psi$ the significant portion of mitochondrial Δp. Thus, renal tubular epithelial cells develop energetic deficiency.

The dynamics of morphological changes in mitochondria may be related to cell apoptosis. Mitochondrial fission involves the constriction and cleavage of mitochondria by fission proteins, such as dynamin-related protein 1 (Drp1) and Fission 1 (Fis1). On the other hand, mitochondrial fusion is the lengthening of mitochondria by tethering and joining together two adjacent mitochondria. Mitofusin-1 and -2 are mainly responsible for outer membrane fusion, while Opa1 is thought to mediate inner membrane fusion. As mentioned above, mitochondria become fragmented upon stress or apoptotic stimulation, as recently reported by Brooks et al. [35]. AKI-associated fragmentation involves the activation of mitochondrial fission via Drp1. These authors provide evidence that the suppression of Drp1 and mitochondrial fragmentation abrogates mitochondrial damage, cytochrome-*c* release, apoptosis, and renal/cellular injury both in vitro and in vivo. In this report, rat renal proximal tubule of wild-type or dominant negative Drp1 was examined under the effect of azide- or cisplatin-induced cell apoptosis. The mitochondrial fragmentation, clearly induced with those stimuli, was reduced with the Bcl-2 but not by pan-caspase inhibition by carbobenzoxy-valyl-alanyl-aspartyl-(*O*-methyl)-fluoromethylketone. A fission protein Drp1 was translocated to mitochondria under these stimuli, but the fragmentation was inhibited in cells with dominant-negative Drp1 in association with decrease in apoptosis. They further demonstrated the efficacy of Drp1 pharmacological inhibitor, mdivi-1, in AKI. As attractive as these studies are, it has been argued that mitochondrial morphology may depend on functional requirements: elongated mitochondria could facilitate signal transduction or reflect the state of active respiration, whereas fragmented mitochondria may be the preferred morphology for their recruitment to distant cellular compartments (reviewed in [36]). Hence, future work should further elucidate the significance of mitochondrial fragmentation in AKI.

Another prominent mechanism for development and perpetuation of mitochondrial dysfunction is related to the induction of an IMM protein, uncoupling protein-2(UCP2) [37]. Oxidative stress results in the accumulation of UCP2 in mitochondria, leading to the inward proton leak that competes with the function of ATP synthase and results in reduction of ATP synthesis from adenosine diphosphate. This situation can be modeled in vitro by application of rotenone, antimycin A, or diethyldithiocarbamate, all increasing mitochondrial O_2^- production, whereas oxidants produced outside mitochondria did not affect UCP2 abundance [38]. This process did not require a parallel increase in ucp2 gene expression or mRNA abundance; rather it is attributed to enhanced translational efficiency and stabilization of UCP2.

In renal ischemia-reperfusion injury, we and others [9, 31, 32, 39] demonstrated that the polyunsaturated fatty acyl group of membrane phospholipids are highly susceptible to O_2^- and a self-propagating chain of reaction produces a wide variety

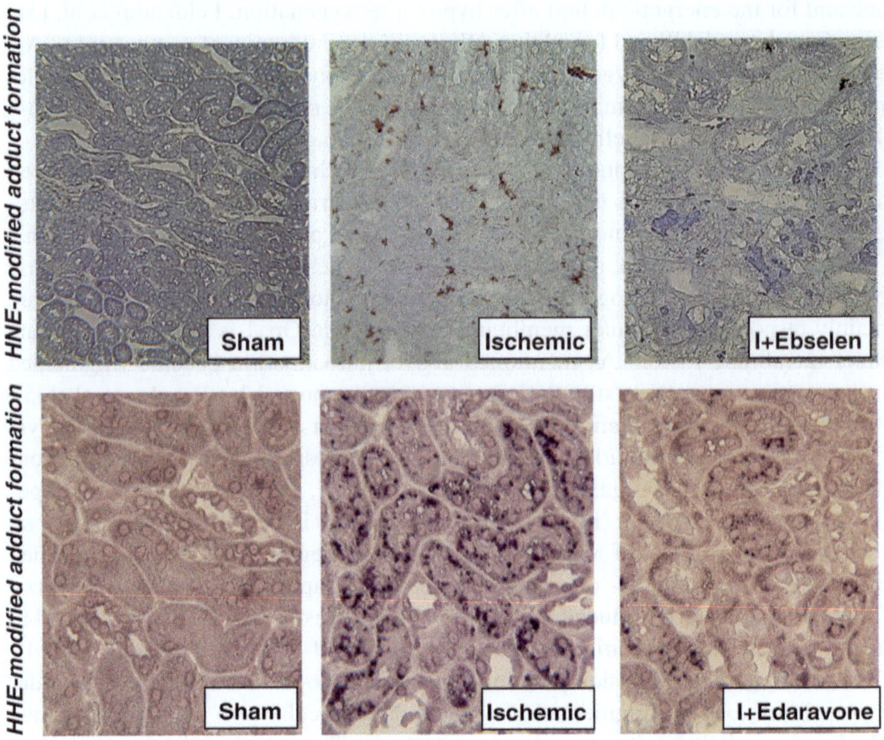

Fig. 3 Antibodies to 4-hydroxy-2-nonenal (HNE)-modified proteins (**a**) and 4-hydroxy hexenol (HHE)-modified proteins (**b**) proved those adduct formations for immunohistochemistry in rat kidney subjected to ischemia-reperfusion. Kidneys were harvested 24 h later. *I* ischemic kidney. Reproduced with permission from the American Physiological Society (**a**) [9] and from Nature Publishing Group (**b**) [39]

of aldehydes, alkenals, and hydroxyalenals, such as malondialdehyde, 4-hydroxy hexenol (HHE), and 4-hydroxy-2-nonenal (HNE) (Fig. 3). Membrane-permeable HNE is noxious and one of the best investigated mediators of free-radical damage. Mitochondrial proteins are targets of HNE, and HNE inactivates the 2-oxoglutarate dehydrogenase, pyruvate dehydrogenase complex, cytochrome-*c* oxidase, and NADH-linked respiration in isolated mitochondria. Echtay et al. [40, 41] proposed a kind of homeostatic role of HNE, inducing mitochondrial uncoupling by specific and inhibitable interactions with UCP1, UCP2, and UCP3, and with the adenine nucleotide translocase. This mild uncoupling decreases mitochondrial ROS generation, which can subsequently promote HNE production. Therefore, it is quite possible that the renal involvement of UCPs might have a dual effect and that UCPs might play a certain role in chronization of AKI.

7 Therapeutic Implications

Attempts to alleviate AKI by controlling renal oxidative and nitrosative stress have been ongoing for more than a decade [31, 42, 43]. The involvement of the blood-derived polymorphonuclear and myelomonocytic cells as well as renal parenchymal cells as the primary sources of renal oxidative stress in AKI poses significant hurdles for targeted therapies. The duality of NO actions further aggravates therapeutic intervention because of the role played by NO in physiological and pathophysiological conditions in diverse cells, such as renal tubular epithelial cells, macrophages, neutrophils, and endothelial cells. We have presented data that demonstrate that an excessive amount of nitric oxide in AKI induced by ischemia-reperfusion can be controlled by the antisense oligonucleotide targeted to iNOS [8]. Mice with renal ischemia showed supra-physiological amount of nitric oxide in ischemic kidneys that was alleviated with the use of antisense oligonucleotide to iNOS (Fig. 4), resulting in functional and structural renoprotection. The histological scoring further confirmed the improvement of ischemic damage when treated by antisense (Fig. 5). The observed induction of iNOS in the face of ongoing oxidative stress represents a powerful mechanism for excessive formation of peroxynitrite. It has been found that hydroxyl radical-like activities are generated from peroxynitrite [44]. We have demonstrated the involvement of oxidative and nitrosative stress in ischemic kidney injury using a cell permeable lecithinized SOD and a specific iNOS inhibitor, L-*N*6-(1-iminoethyl)lysine (L-Nil). While each of these compounds, applied separately, produced a degree of renoprotection, their combination exerted an additive effect [9]. Therapeutic efficacy of controlling oxidonitrosative stress was further confirmed by the use of a potent antioxidant and scavenger of peroxynitrite, ebselen (2-phenyl-1,2-benzinoselenazol-3 (2H)-one). Ebselen is a seleno-organic compound that has no direct effect on NO or superoxide anions, but it acts through activation of glutathione peroxidase and

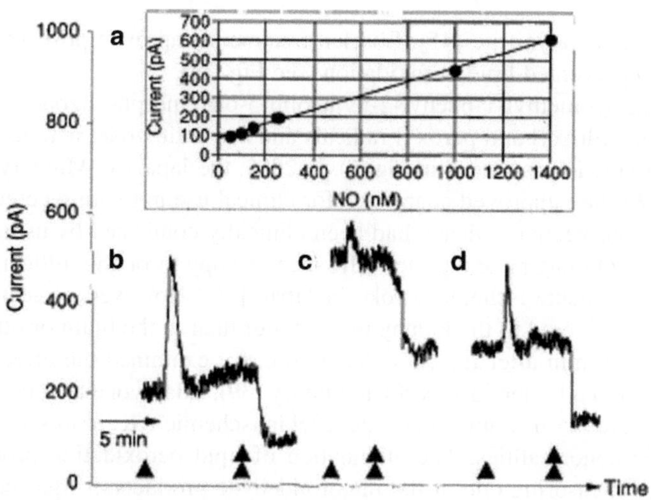

Fig. 4 In vivo nitric oxide monitoring within the renal cortex using an NO-selective electrode. (**a**) A representative calibration curve. The representative recordings obtained from sham (**b**), nontreated ischemic (**c**), and antisense-treated (**d**) ischemic kidneys. *Arrowheads*: *single*, time of bradykinin infusion; *double*, time of L-NAME infusion. Reproduced with permission from the American Society for Clinical Investigation [8]

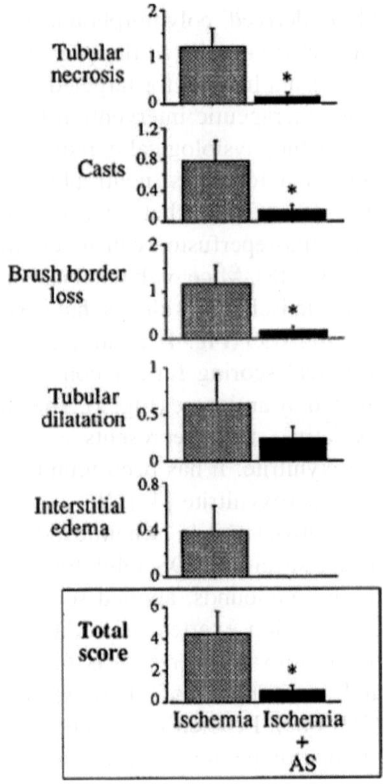

Fig. 5 Pathological scores of acute kidney injury (AKI) in ischemic kidneys and those treated with antisense oligonucleotide for inducible nitric oxide synthase (iNOS). Reproduced with permission from the American Society for Clinical Investigation [8]

scavenging of peroxynitrite [45]. Ebselen treatment improved postischemic renal function and prevented lipid peroxidation (see Fig. 3a).

Edaravone (3-methyl-1-phenyl-2-pyrazolin-5-one; norphenazone) is a potent scavenger of hydroxyl and peroxyl radicals and has antioxidant actions (inhibition of lipid peroxidation as shown in Fig. 6). In 2001, the Japanese Ministry of Health, Labor and Welfare approved edaravone for clinical use in treating cerebral infraction; its neuroprotective efficacy had been clinically confirmed by using magnetic resonance (MR) imaging and proton MR spectroscopy through multicenter, randomized study on acute ischemic stroke in Japan [46]. However, as seen in Fig. 7, edaravone is delivered to the kidney much better than to the brain or other organs, as assessed at 5 min after injection. Therefore, we examined the effectiveness of edaravone to renal ischemia-reperfusion injury [39]. Edaravone at concentrations 5 and 10 mg/kg reduced serum creatinine level in ischemic AKI in rats and prevented morphologic abnormalities. The attenuation of lipid peroxidation products were evaluated by measuring one of the major aldehyde products of lipid peroxidation process, HHE, which modified the protein level. HHE originates exclusively from phospholipid-bound omega-3 unsaturated fatty acids, such as eicosapentaenoic and docosahexaenoic acids, and is one of the most reliable markers of

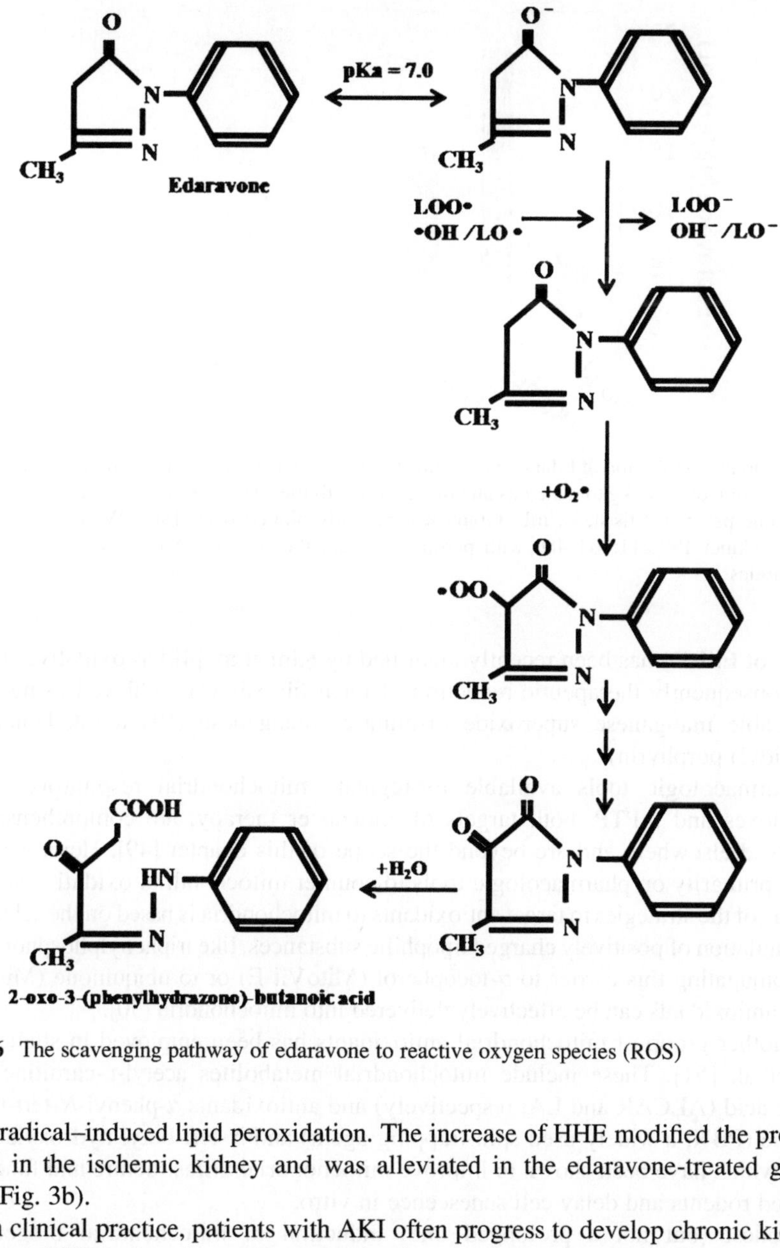

Fig. 6 The scavenging pathway of edaravone to reactive oxygen species (ROS)

free-radical–induced lipid peroxidation. The increase of HHE modified the protein level in the ischemic kidney and was alleviated in the edaravone-treated group (see Fig. 3b).

In clinical practice, patients with AKI often progress to develop chronic kidney disease (CKD), but the epidemiologic data were lacking until recently when Hsu et al. reported the evidence of a carryover of AKI toward CKD and end stage renal disease (ESRD) based on a study population of 39,805. According to their study, CKD stage 3 (but less than 45 mL/min of eGFR) 42% of individuals suffered an episode of dialysis-requiring AKI that developed to ESRD within 30 days of hospital discharge [47]. The mechanistic factor related to the carryover of AKI to

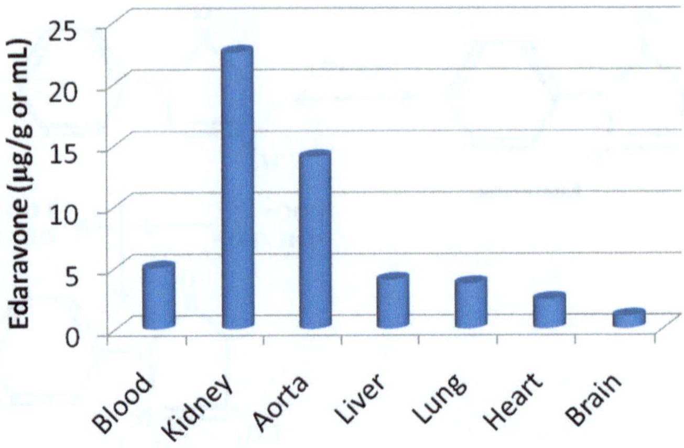

Fig. 7 Organ distribution of Edaravone; 5 min after single injection. Tissues were harvested 5 min after injection of 2 mg/kg to male rats and those radioactivities were converted to µg equivalent to Edaravone per g wet tissue or mL. Graph was partially plotted from Table IV of Drug Metab Pharmacokinet 1996;11:463–480 with permission from the Japanese Society for the Study of Xenobiotics

CKD (or ESRD) has been recently identified by Kim et al. [48] as oxidative stress, and consequently therapeutic reduction of renal fibrosis was achieved using cell-permeable manganese superoxide dismutase (manganese (III) tetrak(1-methyl-4-pyridyl) porphyrin).

Pharmacologic tools available to regulate mitochondrial respiratory chain complexes and mPTP, both targets of anticancer therapy, are comprehensively reviewed elsewhere and are beyond the scope of this chapter [49]. Here, we will focus primarily on pharmacologic tools to counter mitochondrial oxidative stress.

One of the strategies to target antioxidants to mitochondria is based on the selective accumulation of positively charged lipophilic substances, like triphenylphosphonium. By conjugating this carrier to α-tocopherol (MitoVit E) or to ubiquinone (MitoQ), these antioxidants can be effectively delivered into mitochondria [50].

Another group of mitochondrial antioxidants has been employed in studies by Liu et al. [51]. These include mitochondrial metabolites acetyl-L-carnitine and lipoic acid (ALCAR and LA, respectively) and antioxidants α-phenyl-*N*-tert-butyl nitrone (PBN, a widely used spin-trapping agent) and *N*-tert-butyl hydroxylamine [51], which have been shown to improve mitochondrial ultrastructure and function in aged rodents and delay cell senescence in vitro.

Modest (but not a prolonged one) induction of mitochondrial biogenesis has been shown to promote mitochondrial biogenesis in the heart [52] and can be readily achieved using peroxisome proliferator-activated receptor gamma (PPAR-γ) coactivators. Another compound that acts on mitochondrial biogenesis and exhibits antioxidant properties is the plant-derived polyphenol resveratrol (reviewed in [53]). It improves mitochondrial functions in metabolic syndrome by acting on AMPK and activating sirtuin-1 and PGC-1α [54].

Recent studies demonstrated that nitrite anion results in the reduction of ROS produced at the mitochondrial site [55] by inhibiting complex I activity of ETC under conditions of postischemic oxidative stress (but not under normoxic conditions) and by reducing the activation of mPTP. Nitrite anion acts, most probably, via conversion to NO, either through direct reduction or enzymatically by XOR, deoxyhemoglobin, or deoxymyoglobin [56]. This approach may represent a valuable pharmacologic tool to prevent acute postischemic mitochondrial oxidative stress, but may potentially have deleterious effects upon chronic administration.

In ischemia-reperfusion, agents that control potassium at the opening of the inner membrane of mitochondria should prove promising for maintaining mitochondrial membrane potential. Another candidate will be a pharmacologic inhibitor of Drp1, mdivi-1, or an equivalent, which will reduce the possibility of mitochondrial fission during ischemia-reperfusion. This line of investigations awaits confirmation by other laboratories.

References

1. Thomas SR, Witting PK, Drummond GR. Redox control of endothelial function and dysfunction: molecular mechanisms and therapeutic opportunities. Antioxid Redox Signal 2008;10:1713–65.
2. Skulachev VP. Chemiosmotic concept of the membrane bioenergetics: what is already clear and what is still waiting for elucidation? J Bioenerg Biomembr 1994;26:589–98.
3. Wallace KB, Starkov AA. Mitochondrial targets of drug toxicity. Annu Rev Pharmacol Toxicol 2000;40:353–88.
4. Lin MT, Beal MF. Mitochondrial dysfunction and oxidative stress in neurodegenerative diseases. Nature 2006;443:787–95.
5. Wang W, Fang H, Groom L, et al. Superoxide flashes in single mitochondria. Cell 2008;134:279–90.
6. Zorov DB, Filburn CR, Klotz LO, Zweier JL, Sollott SJ. Reactive oxygen species (ROS)-induced ROS release: a new phenomenon accompanying induction of the mitochondrial permeability transition in cardiac myocytes. J Exp Med 2000;192:1001–14.
7. Guzy RD, Hoyos B, Robin E, et al. Mitochondrial complex III is required for hypoxia-induced ROS production and cellular oxygen sensing. Cell Metab 2005;1:401–8.
8. Noiri E, Peresleni T, Miller F, Goligorsky MS. In vivo targeting of inducible NO synthase with oligodeoxynucleotides protects rat kidney against ischemia. J Clin Invest 1996;97:2377–83.
9. Noiri E, Nakao A, Uchida K, et al. Oxidative and nitrosative stress in acute renal ischemia. Am J Physiol Renal Physiol 2001;281:F948–57.
10. Gao S, Chen J, Brodsky SV, et al. Docking of endothelial nitric oxide synthase (eNOS) to the mitochondrial outer membrane: a pentabasic amino acid sequence in the autoinhibitory domain of eNOS targets a proteinase K-cleavable peptide on the cytoplasmic face of mitochondria. J Biol Chem 2004;279:15968–74.
11. Adachi T, Weisbrod RM, Pimentel DR, et al. S-glutathiolation by peroxynitrite activates SERCA during arterial relaxation by nitric oxide. Nat Med 2004;10:1200–7.
12. Gems D, Partridge L. Stress-response hormesis and aging: "that which does not kill us makes us stronger." Cell Metab 2008;7:200–3.
13. Motohashi H, Yamamoto M. Nrf2-Keap1 defines a physiologically important stress response mechanism. Trends Mol Med 2004;10:549–57.

14. Piantadosi CA, Suliman HB. Mitochondrial transcription factor A induction by redox activation of nuclear respiratory factor 1. J Biol Chem 2006;281:324–33.
15. Wu Z, Puigserver P, Andersson U, et al. Mechanisms controlling mitochondrial biogenesis and respiration through the thermogenic coactivator PGC-1. Cell 1999;98:115–24.
16. Gleyzer N, Vercauteren K, Scarpulla RC. Control of mitochondrial transcription specificity factors (TFB1M and TFB2M) by nuclear respiratory factors (NRF-1 and NRF-2) and PGC-1 family coactivators. Mol Cell Biol 2005;25:1354–66.
17. Fu X, Wan S, Lyu YL, Liu LF, Qi H. Etoposide induces ATM-dependent mitochondrial biogenesis through AMPK activation. PLoS One 2008;3:e2009.
18. Donati A, Cavallini G, Paradiso C, et al. Age-related changes in the autophagic proteolysis of rat isolated liver cells: effects of antiaging dietary restrictions. J Gerontol A Biol Sci Med Sci 2001;56:B375–83.
19. Baehrecke EH. Autophagy: dual roles in life and death? Nat Rev Mol Cell Biol 2005;6:505–10.
20. Codogno P, Meijer AJ. Autophagy and signaling: their role in cell survival and cell death. Cell Death Differ 2005;12(Suppl 2):1509–18.
21. Mathew R, Karp CM, Beaudoin B, et al. Autophagy suppresses tumorigenesis through elimination of p62. Cell 2009;137:1062–75.
22. Lemasters JJ. Selective mitochondrial autophagy, or mitophagy, as a targeted defense against oxidative stress, mitochondrial dysfunction, and aging. Rejuvenation Res 2005;8:3–5.
23. Kiffin R, Bandyopadhyay U, Cuervo AM. Oxidative stress and autophagy. Antioxid Redox Signal 2006;8:152–62.
24. Lemasters JJ, Nieminen AL, Qian T, et al. The mitochondrial permeability transition in cell death: a common mechanism in necrosis, apoptosis and autophagy. Biochim Biophys Acta 1998;1366:177–96.
25. Periyasamy-Thandavan S, Jiang M, Wei Q, Smith R, Yin XM, Dong Z. Autophagy is cytoprotective during cisplatin injury of renal proximal tubular cells. Kidney Int 2008;74:631–40.
26. Isaka Y, Suzuki C, Abe T, et al. Bcl-2 protects tubular epithelial cells from ischemia/reperfusion injury by dual mechanisms. Transplant Proc 2009;41:52–4.
27. Kessel D, Reiners JJ, Jr. Apoptosis and autophagy after mitochondrial or endoplasmic reticulum photodamage. Photochem Photobiol 2007;83:1024–8.
28. Sasnauskiene A, Kadziauskas J, Vezelyte N, Jonusiene V, Kirveliene V. Apoptosis, autophagy and cell cycle arrest following photodamage to mitochondrial interior. Apoptosis 2009;14:276–86.
29. Hock MB, Kralli A. Transcriptional control of mitochondrial biogenesis and function. Annu Rev Physiol 2009;71:177–203.
30. Addabbo F, Ratliff B, Park HC, et al. The Krebs cycle and mitochondrial mass are early victims of endothelial dysfunction: proteomic approach. Am J Pathol 2009;174:34–43.
31. Paller MS, Hoidal JR, Ferris TF. Oxygen free radicals in ischemic acute renal failure in the rat. J Clin Invest 1984;74:1156–64.
32. Kato A, Hishida A, Nakajima T. Effects of oxygen free radical scavengers on uranium-induced acute renal failure in rats. Free Radic Biol Med 1994;16:855–9.
33. Portilla D, Dai G, McClure T, et al. Alterations of PPARalpha and its coactivator PGC-1 in cisplatin-induced acute renal failure. Kidney Int 2002;62:1208–18.
34. Feldkamp T, Kribben A, Weinberg JM. Assessment of mitochondrial membrane potential in proximal tubules after hypoxia-reoxygenation. Am J Physiol Renal Physiol 2005;288: F1092–102.
35. Brooks C, Wei Q, Cho SG, Dong Z. Regulation of mitochondrial dynamics in acute kidney injury in cell culture and rodent models. J Clin Invest 2009;119:1275–85.
36. Chan DC. Mitochondria: dynamic organelles in disease, aging, and development. Cell 2006;125:1241–52.

37. Lowell BB, Shulman GI. Mitochondrial dysfunction and type 2 diabetes. Science 2005;307:384–7.
38. Giardina TM, Steer JH, Lo SZ, Joyce DA. Uncoupling protein-2 accumulates rapidly in the inner mitochondrial membrane during mitochondrial reactive oxygen stress in macrophages. Biochim Biophys Acta 2008;1777:118–29.
39. Doi K, Suzuki Y, Nakao A, Fujita T, Noiri E. Radical scavenger edaravone developed for clinical use ameliorates ischemia/reperfusion injury in rat kidney. Kidney Int 2004;65:1714–23.
40. Echtay KS, Esteves TC, Pakay JL, et al. A signalling role for 4-hydroxy-2-nonenal in regulation of mitochondrial uncoupling. EMBO J 2003;22:4103–10.
41. Echtay KS, Pakay JL, Esteves TC, Brand MD. Hydroxynonenal and uncoupling proteins: a model for protection against oxidative damage. Biofactors 2005;24:119–30.
42. Linas SL, Whittenburg D, Parsons PE, Repine JE. Ischemia increases neutrophil retention and worsens acute renal failure: role of oxygen metabolites and ICAM 1. Kidney Int 1995;48:1584–91.
43. Peresleni T, Noiri E, Bahou WF, Goligorsky MS. Antisense oligodeoxynucleotides to inducible NO synthase rescue epithelial cells from oxidative stress injury. Am J Physiol 1996;270: F971–7.
44. Beckman JS, Beckman TW, Chen J, Marshall PA, Freeman BA. Apparent hydroxyl radical production by peroxynitrite: implications for endothelial injury from nitric oxide and superoxide. Proc Natl Acad Sci U S A 1990;87:1620–4.
45. Nakamura Y, Feng Q, Kumagai T, et al. Ebselen, a glutathione peroxidase mimetic selenoorganic compound, as a multifunctional antioxidant. Implication for inflammation-associated carcinogenesis. J Biol Chem 2002;277:2687–94.
46. Houkin K, Nakayama N, Kamada K, Noujou T, Abe H, Kashiwaba T. Neuroprotective effect of the free radical scavenger MCI-186 in patients with cerebral infarction: clinical evaluation using magnetic resonance imaging and spectroscopy. J Stroke Cerebrovasc Dis 1998;7:315–22.
47. Hsu CY, Chertow GM, McCulloch CE, Fan D, Ordonez JD, Go AS. Nonrecovery of kidney function and death after acute on chronic renal failure. Clin J Am Soc Nephrol 2009;4:891–8.
48. Kim J, Seok YM, Jung KJ, Park KM. Reactive oxygen species/oxidative stress contributes to progression of kidney fibrosis following transient ischemic injury in mice. Am J Physiol Renal Physiol 2009;297:F461–70.
49. Toogood PL. Mitochondrial drugs. Curr Opin Chem Biol 2008;12:457–63.
50. Smith RA, Kelso GF, Blaikie FH, et al. Using mitochondria-targeted molecules to study mitochondrial radical production and its consequences. Biochem Soc Trans 2003;31:1295–9.
51. Liu J, Atamna H, Kuratsune H, Ames BN. Delaying brain mitochondrial decay and aging with mitochondrial antioxidants and metabolites. Ann N Y Acad Sci 2002;959:133–66.
52. Lehman JJ, Barger PM, Kovacs A, Saffitz JE, Medeiros DM, Kelly DP. Peroxisome proliferator-activated receptor gamma coactivator-1 promotes cardiac mitochondrial biogenesis. J Clin Invest 2000;106:847–56.
53. Schwartz DR, Sack MN. Targeting the mitochondria to augment myocardial protection. Curr Opin Pharmacol 2008;8:160–5.
54. Lagouge M, Argmann C, Gerhart-Hines Z, et al. Resveratrol improves mitochondrial function and protects against metabolic disease by activating SIRT1 and PGC-1alpha. Cell 2006;127:1109–22.
55. Shiva S, Sack MN, Greer JJ, et al. Nitrite augments tolerance to ischemia/reperfusion injury via the modulation of mitochondrial electron transfer. J Exp Med 2007;204:2089–102.
56. Gladwin MT, Schechter AN, Kim-Shapiro DB, et al. The emerging biology of the nitrite anion. Nat Chem Biol 2005;1:308–14.

Chapter 10
Oxidative Stress in the Kidney: Proximal Tubule Disorders

Sara Terryn and Olivier Devuyst

Abstract Redox changes initiate various cellular signals in the cells, and the redox environment can determine if a cell will proliferate, differentiate, or die. Imbalance of the redox status such as during oxidative stress can trigger a series of events, leading to cellular dysfunction. Reactive oxygen species (ROS) are increasingly considered as being involved in the initiation and progression of chronic renal disease. The proximal tubule is a major site of ROS production, due to its high transport activity supported by an oxygen consuming metabolism. Various congenital and acquired renal disorders induce ROS in proximal tubule cells (PTC), which are characterized by their capacity for receptor-mediated endocytosis of albumin. Recent studies suggest that albumin exerts a dual effect on the proximal tubule. In physiological conditions, endocytosis of albumin is a survival factor, protecting the cells against oxidant injury. In case of glomerular proteinuria, exposure of the PTC to an excess of albumin induces oxidative stress and causes renal injury. Furthermore, defects in the receptor-mediated endocytic uptake of albumin are also related to a state of increased oxidative stress, causing tubulointerstitial fibrosis and renal failure. In this chapter, we review available evidence about the nature of stimuli and initial events involved in ROS generation and antioxidant mechanisms in PTC.

Keywords Diabetic nephropathy · Proteinuria · Endocytosis · Albumin · Carbonic anhydrase type III

O. Devuyst (✉)
Laboratory of Nephrology, Université catholique de Louvain Medical School, 54 Avenue Hippocrate, UCL 5450, 1200 Brussels, Belgium
e-mail: olivier.devuyst@uclouvain.be

T. Miyata et al. (eds.), *Studies on Renal Disorders*, Oxidative Stress in Applied Basic Research and Clinical Practice, DOI 10.1007/978-1-60761-857-7_10,

1 Introduction

1.1 Redox Status Determines Cell Function

Increasing experimental evidence suggests that the redox status (i.e., the balance between oxidants and antioxidants within cells), plays an important role in determining cell function by regulating the gene expression or modification of proteins, lipids, enzymes, and transporters [1, 2]. Normally, cells preserve a reducing environment through enzymes that maintain the reduced state with a constant input of metabolic energy [3]. The cellular redox status is reflected by the ratio between redox couples such as glutathione and oxidized glutathione (GSH/GSSG), thioredoxin (TrxSS/Trx$(SH)_2$), nicotinamide adenine dinucleotide phosphate ($NADP^+$/NADPH) or cysteine/cystine (Cys/CySS). A higher level of cellular peroxide (ROOH) will increase the level of oxidized cysteine residues, leading to protein modification and cellular dysfunction. Therefore, cellular peroxide is kept in a reduced state through the activities of catalase, glutathione peroxidases (GPx), and the thioredoxin-assisted peroxidases (peroxiredoxins, Prx). Another important factor is the reducing capacity of the redox couple, that is, the extent of the redox buffering system and the availability of antioxidant enzymes (Fig. 1) [4]. Oxidants are continuously produced by the oxidative metabolism that provides adenosine triphosphate (ATP) for the cell. Oxidative stimuli are quickly captured by antioxidant enzymes and, therefore, do not always move the cell to a more oxidized state. Mild oxidant stimuli can induce normal cells to divide and seem to be normal regulators of cellular transcription [5, 6], although whether reactive oxygen/nitrogen species (ROS/RNS) can directly serve as signaling molecules is debated [7]. Important physiological functions that involve free radicals or their derivates include regulation of vascular tone, oxygen sensing (like erythropoietin (EPO) production in renal cells, or ventilatory control by carotid bodies), enhancement of signal transduction from various membrane receptors (like epidermal growth factor signaling), and oxidative stress responses that aim to maintain redox homeostasis.

When the capacity to detoxify the produced oxidants is exceeded, the redox status of a cell becomes imbalanced. An oxidized state can then trigger a series of events that alter cell function and promote signals that direct targeted cells toward proliferation, differentiation, transformation, or death by apoptosis (see Fig. 1) [8]. Fully differentiated tissues like the liver, kidney, brain, and intestine are characterized by cells arrested in the quiescent state, and imposition of severe oxidative stress often results in a cytotoxic endpoint. However, for the kidney, cell death is not necessarily the endpoint of oxidative stress. In contrast to the brain and heart, proximal tubule cells (PTC) can recover from an ischemic or toxic insult. After cell death by necrosis and apoptosis, the surviving cells dedifferentiate and proliferate to eventually replace the injured cells and restore tubular integrity [9, 10]. The mechanisms by which renal cells survive and repair tubular damage or undergo cell death after an oxidant injury need to be determined. However, the coordinated activation of a set of signaling pathways, transcription factors, and

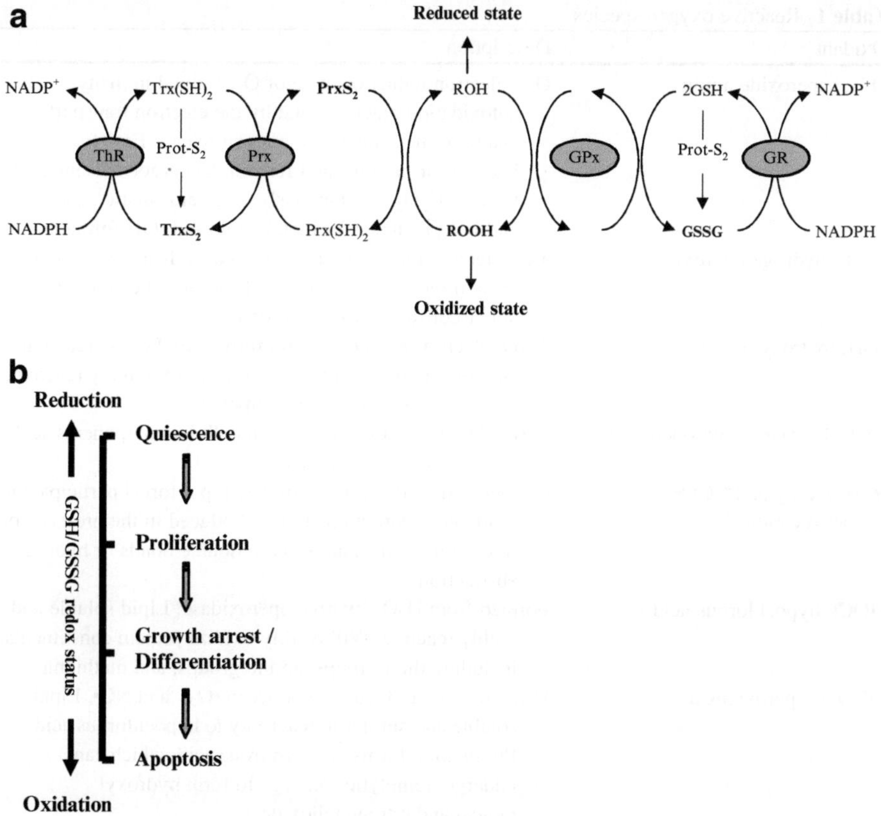

Fig. 1 Cellular responses to GSH/GSSG redox status and steady-state redox potential (Eh) for the cellular GSH/GSSG pool at different cell stages. (**a**) The cell steady-state peroxide (ROOH) levels are largely maintained by the activities of glutathione peroxidases (GPx) and the thioredoxin-assisted peroxidases (Prx). The system also maintains cellular protein cysteines (Prot-S2) in a reduced state. During oxidative stress, where cellular peroxides are elevated, there is an increased level of oxidized protein cysteines eventually leading to cellular dysfunction. *GR* glutathione reductase; *ThR* thioredoxin reductase. (**b**) Fully differentiated cells are typically arrested in the quiescent state (i.e., reduced cellular GSH/GSSG redox status). Cells undergo a progression from a reduced to an oxidized state as they go from proliferation to growth arrest/differentiation to apoptosis. An overwhelming oxidative stimulus results in apoptosis

antioxidant defense mechanisms, which will be discussed in the following sections, might be involved in this capacity to recover.

1.2 Oxidative Stress

Oxidative stress is the result of an imbalance between the production of reactive oxygen derived species (ROS/RNS) and the ability to detoxify the reactive

Table 1 Reactive oxygen species

Oxidant	Description
O_2^-, superoxide anion	One-electron reduction state of O_2, formed in many autoxidation reactions and by the electron transport chain. Rather unreactive but can release Fe^{2+} from iron–sulfur proteins and ferritin. Undergoes dismutation to form H_2O_2 spontaneously or by enzymatic catalysis and is a precursor for metal-catalyzed •OH formation
H_2O_2, hydrogen peroxide	Two-electron reduction state, formed by dismutation of $•O_2^-$ or by direct reduction of O_2. Lipid soluble and thus able to diffuse across membranes
•OH, hydroxyl radical	Three-electron reduction state, formed by Fenton reaction and decomposition of peroxynitrite. Extremely reactive, will attack most cellular components
ROOH, organic hydroperoxide	Formed by radical reactions with cellular components such as lipids and nucleobases
RO•, alkoxy and ROO•, peroxy radicals	Oxygen centered organic radicals. Lipid forms participate in lipid peroxidation reactions. Produced in the presence of oxygen by radical addition to double bonds or hydrogen abstraction
HOCl, hypochlorous acid	Formed from H_2O_2 by myeloperoxidase. Lipid soluble and highly reactive. Will readily oxidize protein constituents, including thiol groups, amino groups, and methionine
$ONOO^-$, peroxynitrite	Formed in a rapid reaction between $•O_2^-$ and NO•. Lipid soluble and similar in reactivity to hypochlorous acid. Protonation forms peroxynitrous acid, which can undergo homolytic cleavage to form hydroxyl radical and nitrogen dioxide

intermediates or repair the resulting damage. The ROS include the free radicals, superoxide ($^•O_2^-$), hydroxyl (OH), and peroxyl ($^•RO_2$), and nonradical species, hydrogen peroxide (H_2O_2) and hydrochlorous acid (HOCl) (Table 1). The superoxide anion (O_2^-) is formed by the univalent reduction of triplet state molecular oxygen ($^3O_2$). This process is mediated by enzymes such as nicotinamide adenine dinucleotide phosphate (NADPH) oxidases and xanthine oxidase or by redox-reactive compounds such as those involved in the mitochondrial electron transport chain (see below). Superoxide can be converted into the nonradical species H_2O_2 and singlet oxygen ($^1O_2$). In the presence of reduced transition metals (e.g., ferrous or cuprous ions), H_2O_2 can be converted into the highly reactive hydroxyl radical ($^•OH$). Antioxidant enzymes can also convert H_2O_2 into water. Most of the oxidative effects are not directly mediated by superoxide but rather by its ROS derivates. Reactive nitrogen species can be produced by the same pathways and include the radicals nitric oxide ($^•NO$) and nitrogen dioxide ($^•NO_2^-$) as well as the nonradical peroxynitrite ($ONOO^-$), nitrous oxide (HNO_2), and alkyl peroxynitrates ($RONOO^-$). The nitric oxide radical ($^•NO$) is produced by oxidation of one of the terminal guanidonitrogen atoms of L-arginine, a process catalyzed by nitric oxide synthase (NOS) isoforms. Depending upon the microenvironment, NO can be converted to various other RNS.

There are different enzymatic and nonenzymatic sources of ROS/RNS (Fig. 2). A major site for ROS production is the mitochondria. Under physiological

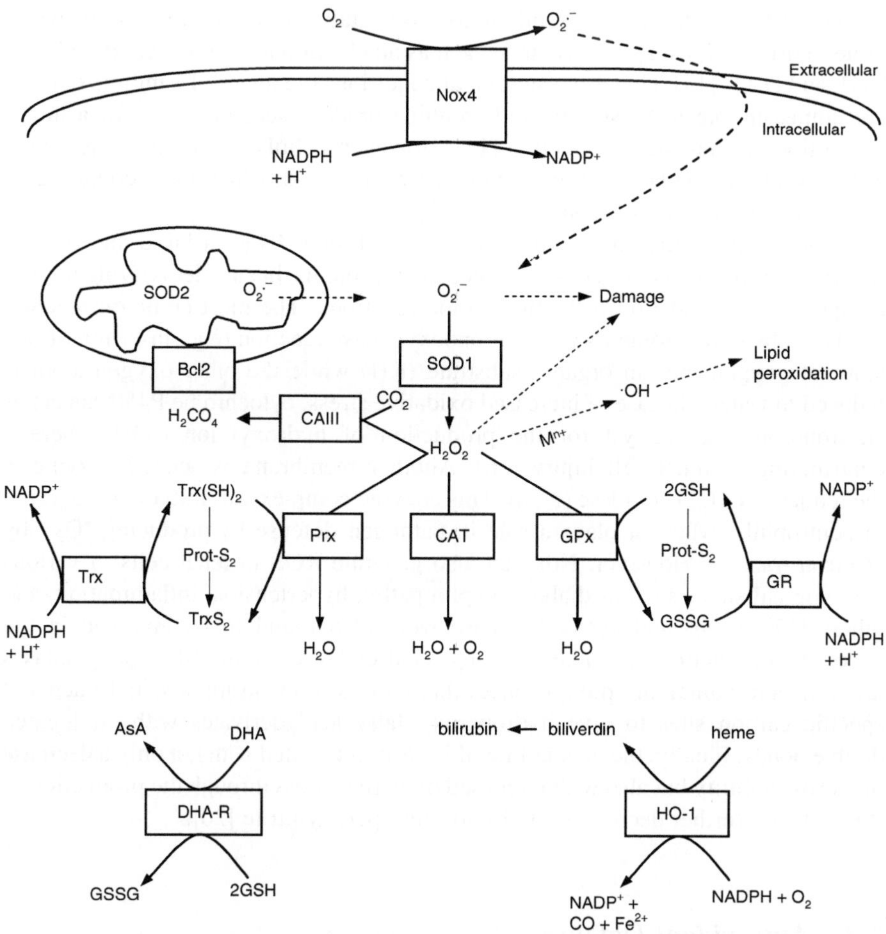

Fig. 2 The formation of ROS and anti-oxidant defense systems in renal epithelial cells. Superoxide ($^{\bullet}O_2^-$) is primarily formed from different reactions, the most important being the mitochondrial oxidative phosphorylation and the membrane-associated Nox4. This radical is highly unstable and can rapidly react to form other ROS or is metabolized by SOD1 to H_2O_2. This reactive species can be detoxified by CAII, CAT, or the glutathione peroxidases and the thioredoxin-assisted peroxidases or induce various signaling pathways, leading to cellular dysfunction. *Nox4* NADPH oxidase; *SOD* superoxide dismutase; *CAIII* carbonic anhydrase type III; *CAT* catalase; *Prx* peroxiredoxin; *Trx(SH)*$_2$ reduced thioredoxin; *TrxS*$_2$ oxidized thioredoxin; *ThR* thioredoxin reductase; *GPx* glutathione peroxidase; *GSH* reduced glutathione; *GSSG* oxidized glutathione; *GR* glutathione reductase; *AsA* ascorbate; *DHA* dehydroascorbate; *DHA-R* dehydroascorbate reductase; *HO-1* heme oxygenase 1; *Mn*$^+$ metal ion

conditions, $^{\bullet}O_2^-$ is predominantly generated in the mitochondria through oxidative phosphorylation of glucose and free fatty acids. Although the respiratory chain is tightly regulated, up to 1–2% of oxygen is partially reduced to $^{\bullet}O_2^-$ instead of water [11, 12]. Excessive ROS in quiescent mitochondria induces mitochondrial dysfunction through the opening of mitochondrial membrane channels, causing a collapse of the mitochondrial membrane potential and further increasing ROS generation by the electron transfer chain [13]. Besides the mitochondria, there are also cytosolic sources of ROS. The enzyme xanthine oxidase catalyzes the oxidation of hypoxanthine to uric acid, using molecular oxygen as an electron acceptor, liberating $^{\bullet}O_2^-$, $^{\bullet}OH$, and H_2O_2 [14]. Under physiological conditions the levels of xanthine oxidase are unmeasurable in most cells and xanthine oxidase accounts only for a minor proportion of total ROS production [11]. However, cellular (oxidative) stress may induce xanthine oxidase and the production of uric acid, which itself could play a role in the progression of renal disease [15].

Another enzymatic source for ROS is cytochrome P450, a large and diverse family of primarily membrane-associated hemoproteins that are detoxifying foreign compounds by oxidation or hydroxylation reactions. The most common reaction catalyzed by cytochrome P450 is a monooxygenase reaction (e.g., insertion of one atom of oxygen into an organic substrate (RH) while the other oxygen atom is reduced to water). In case of increased oxidative stress, cytochrome P450 can act as an iron-donating catalyst for the production of hydroxyl ion ($^{\bullet}OH$), thereby contributing to renal cell injury [16]. Another membrane-associated enzyme is the reduced NADPH oxidase (Nox). This enzyme complex was initially discovered in neutrophils, where it plays a role in pathogen defense by producing $^{\bullet}O_2^-$ by electron transfer. However, Nox can also generate ROS in renal cells in various pathological states such as diabetic nephropathy, hypertension, inflammation, and others [17]. Lipo- and cyclooxygenase were also found to be involved in the excessive production of oxidants during renal cell stress [18, 19]. Lipoxygenases are nonheme containing dioxygenases that oxidize polyunsaturated fatty acids at specific carbon sites to give hydroperoxy fatty acid derivates with conjugated double bonds. Finally, the uncoupling of NOS is associated with not only a decrease in NO availability but also with increased oxidative stress through the production of $^{\bullet}O_2^-$, which further decreases NO by forming peroxynitrite [20].

1.3 Antioxidant Defense

Cells are normally able to defend themselves against ROS damage through antioxidant enzymes such as superoxide dismutase (SOD), catalases, GPx, thioredoxines (Trx), Prx, and heme oxygenases (HO-1) (see Fig. 2). Small antioxidant molecules such as ascorbic acid (vitamin C), tocopherol (vitamin E), and the redox couples of glutathione, thioredoxin, NADPH, and cysteine, also play important roles as cellular antioxidants [2, 4]. The first line of defense against ROS is SODs. Manganese SOD (SOD2) present inside the mitochondrion, the copper/zinc SOD

(SOD1) in the cytosol and the extracellular copper/zinc SOD (SOD3) catalyze the dismutation of $\bullet O_2^-$ into O_2 and H_2O_2 [21, 22]. Detoxification of $\bullet O_2^-$ by SOD produces H_2O_2, which is subsequently dismutated by catalase into O_2 and H_2O. Besides catalase, the selenium-containing peroxidases – Prx, Gpx, and Trx – can scavenge H_2O_2 [4]. As said above, these enzymes work in tandem to maintain the cellular redox state and to keep the cysteine residues in a reduced state (see Fig. 1). Another antioxidant enzyme that participates in the defense against a wide range of cell injuries is heme-oxygenase 1 (HO-1). HO-1 (or Hsp32) belongs to the class of heat-shock proteins and is the rate-limiting enzyme in heme degradation. It participates in the defense against oxidative stress by catabolizing carbon monoxide and free iron [23]. Furthermore, HO-1 is considered the most sensitive and reliable indicator of cellular oxidative stress [24].

The wide range and high specificity of the cellular antioxidant defense mechanisms means that oxidative stress and even toxicity of redox active species may only occur if the level of oxidants exceeds a certain magnitude or overwhelms the mechanisms. Excessive radical formation would then cause maladaptive responses, facilitating aberrant activation of signaling pathways and transcription factors [25], contributing to cellular dysfunction by causing damage to cell components. The most important modifications induced by ROS/RNS are damage of DNA, oxidation of polydesaturated fatty acids in lipids (lipid peroxidation), oxidation of amino acids in proteins (protein nitration), oxidation of carbohydrates (advanced glycated end products), and inactivation of specific enzymes. Furthermore, ROS/RNS can also alter the cytoskeleton, activate leukocytes and induce cytokine production, and cause endothelial cell damage, all factors contributing to tissue dysfunction [26].

1.4 Redox-Sensitive Signaling

Redox signaling is used to describe a regulatory process in which the signal is delivered through redox chemistry. Redox signaling requires that the balance between ROS production and scavenging capacity is (briefly) disturbed. The cellular response to oxidative stress is complex and often involves multiple signaling pathways that act in concert to determine cell fate. Physiological manifestations of redox regulation typically involve a temporary increase or a temporary shift of the intracellular thiol/disulfide redox state toward more oxidative conditions. Most of these physiological redox-responsive regulatory mechanisms serve to protect against oxidative stress and to reestablish redox homeostasis. For instance, redox signaling is involved in the production of NO, the control of ventilation (where ROS acts as a sensor for changes in oxygen concentration and redox regulation of EPO production), and the regulation of cell adhesion, immune responses, or programmed cell death [27]. When sustained ROS production overcomes the antioxidant response, causing a shift to more oxidative conditions, and pathological events may ensue.

Different targets in signaling cascades exhibit redox sensitivity. Oxidants can trigger phosphorylation cascades that lead to activation of mitogen-activated protein kinases (MAPKs) and the nuclear factor κB (NFκB) [28]. However, a direct role for redox activation of the MAPK signaling pathways remains unresolved despite their sensitivity to oxidants and the finding that MAPKs can be activated following GSH oxidation. Numerous transcription factors exhibit redox sensitivity, the most extensively characterized are NFκB, activated protein 1 (AP-1), protein 53 (p53), hypoxia inducible factor 1 (HIF-1), and nuclear factor E2-related factor-2 (Nrf2) [2].

1.4.1 NFκB and AP-1

MAPK signaling cascades are regulated by phosphorylation and dephosphorylation on serine or threonine residues and respond to activation of receptor tyrosine kinases, protein tyrosine kinases, receptors of cytokines and growth factors, and heterotrimeric G-protein coupled receptors. Numerous studies have revealed that MAPK species JNK and p38 are strongly activated by ROS or by a mild shift of the intracellular thiol/disulfide redox state [29]. Not only do the species in the MAPK signaling cascade, but the MAPK-regulated transcription factors also exhibit redox sensitivity.

The redox-sensitive NFκB is among the potential MAPK-regulated transcription factors activated in response to oxidative stress, suggesting that it plays a role in determining cell fate during oxidative stress. NFκB is retained in the cytoplasm by inhibitors of NFκB, or IκB. NFκB is activated through phosphorylation of IκB by IκB kinase (IKK) and subsequent sequestration of IκB. This releases NFκB, followed by translocation to the nucleus and regulation of transcription of different genes by binding to discrete DNA sequences, the κB elements. NFκB can activate different pathways, including the canonical pathway, which is induced by various types of cellular stress [30]. Increased oxidative stress has been shown to activate mitogen-activated-protein kinase kinase-7 (MKK7) and Janus kinase (JNK), which activates the canonical NFκB pathway that leads to apoptosis and induction of antioxidant response. NFκB is thus a prominent factor in the balance between cell death and cell survival, the final effect probably depending on the nature and duration of the stimulus and on the cell type.

The activator protein 1 transcription factor is composed of heterodimers of the Fos/activating transcription factor (ATF) and Jun subfamilies of basic-region leucine-zipper (B-ZIP) proteins (c-Fos/c-Jun complexes). The oxidative activation of AP-1 transcription factor activity is based on oxidative activation of JNK. This MAPK phosphorylates serine residues of the NH_2-terminal transactivation domain of c-Jun, required for functional activation. AP-1 proteins are implicated in the regulation of different cellular processes, including proliferation and survival, differentiation, growth, apoptosis, cell migration, and transformation. Which process is activated or downregulated by AP-1 depends on the abundance of dimerization partners, dimer-composition, posttranslational regulation, and interaction with accessory proteins [31].

The role of redox signaling in the regulation of NFκB and AP-1 activity is still debated. Both transcription factors regulate many cellular processes and genes potentially involved in synthesis and release of cytokines, chemokines, growth factors, and extracellular matrix proteins that may activate signaling cascades that induce and propagate renal injuries. Oxidative stress and activation of NFκB and AP-1 are also believed to be associated with activation of fibrogenic molecules, such as transforming growth factor beta-1 (TGF-β1), leading to renal fibroproliferative diseases [32]. Furthermore, the MAPK-dependent pathway is involved in the amplification of TGF-β1 signaling by ROS in renal epithelial cells [33]. These data suggest that ROS could not only trigger but also sustain disease progression through continuous signal amplification.

1.4.2 Protein 53 (p53)

Much of what is known about p53 comes from studies in cancer cells. In unstressed cells, p53 levels are usually kept low through a continuous degradation. A protein called Mdm2 binds to p53, preventing its action and transporting it from the nucleus to the cytosol. Mdm2 also acts as ubiquitin ligase and covalently attaches ubiquitin to p53, thus marking p53 for degradation by the proteasome [34]. Activation of p53 arises in response to many stress types, including DNA damage, oxidative stress, osmotic shock, and deregulated oncogene expression. This activation leads to a quick accumulation of p53 in the cell through a drastic increase in its half-life, as well as the phosphorylation of its N-terminal domain [35]. The activation of p53 by ROS can be achieved by two mechanisms. First, reactive species can damage DNA via cross-linking and formation of DNA adducts, which can activate ATM (ataxia telangiectasia-mutated) and ATR (ATM-related), two apical DNA damage responsive protein kinases [36, 37]. Upon activation, ATM and ATR may directly phosphorylate and activate p53 [38]. A second mechanism is through the activation of MAPK (JNK1-3, ERK1-2, p38 MAPK), which directly phosphorylates p53. Phosphorylation leads to p53 stabilization and accumulation in cells. Although the mechanism is debated, phosphorylation may actually prevent the mdm2 binding of p53 and its targeting to degradation [39]. Certainly further studies should investigate these DNA damage responsive pathways and their possible regulation by oxidative stress. Activation of p53 triggers, via transcription-dependent and -independent pathways, numerous targets, apoptosis being one of them [40]. In addition, p53 may have antioxidant properties by modulating antioxidant enzymes, as shown by the increased oxidative stress in the p53 knockout mouse model [41, 42].

1.4.3 HIF and Nrf2

Unlike NFκB and AP-1, activation of HIF and Nrf2 seems to protect renal cells from oxidant injury [43, 44]. HIF is a basic helix-loop-helix transcription factor composed of an oxygen-sensing α-subunit and a constitutively expressed β-subunit [45]. HIF is a central coordinator in the response against hypoxia [43], which is important in end stage renal disease [46]. In normoxia, the α-subunit of HIF undergoes prolyl hydroxylation, interaction with the von Hippel–Lindau (VHL) protein, and proteosomal degradation [47, 48]. During hypoxia, the α-subunit binds to the β-subunit, and the heteromeric HIF translocates to the nucleus and binds to the hypoxia-response element in target genes that aim to restore oxygen levels in cells and tissues [49]. HIF not only responds to hypoxic conditions but also regulates gene expression during oxidative stress [50]. HIF promotes the activation of genes responsible for angiogenesis in order to restore blood flow and oxygen supply in hypoxic tissues. One such HIF target gene is connective tissue growth factor (CTGF). Although CTGF is an angiogenic factor, it also has profibrotic activities and has been shown to be an important regulator of epithelial–mesenchymal transition of PTC [51]. So possibly, an inappropriate long-term activation of HIF may lead to tubulointerstitial fibrosis in an overall profibrotic setting. Indeed, HIF is activated in chronic kidney disease [52], and there is evidence that activation of HIF signaling may promote epithelial–mesenchymal transition and tubulointerstitial fibrosis [53, 54]. On the other hand, HIF was shown to have a renoprotective effect as evidenced by the reduction of oxidative stress, inflammation, and fibrosis in the rat remnant kidney [55].

The Nrf2-antioxidant response element (ARE) signaling pathway is also activated by oxidative stress. Under oxidative stress, Nrf2 controls inducible gene expression, while in nonstress conditions, constitutive gene expression is regulated and both effects are mediated by ARE. This dual role of Nrf2 is established through the Nrf2-Keap1 regulatory pathway. Nrf2 activity is regulated by the actin-associated Keap-1 protein in a dynamic process that is regulated by the cullin-3-dependent pathway that enables Nrf2 to control both basal and inducible gene expression [44]. The ARE is found in several genes involved in protecting cells from ROS, including genes involved in glutathione biosynthesis and cycling and heme oxygenase-1 [56, 57]. The importance of Nrf2-ARE signaling pathway in the detoxification of ROS has been demonstrated in the Nrf2-deficient mouse model [58]. Although the development of Nrf2−/− mice is normal, the induction of a wide range of detoxifying enzymes is impaired, which was shown to be responsible for a more severe disease phenotype [59]. For instance when Nrf2−/− mice were subjected to acute kidney injury, they had an increased mortality due to more severe kidney dysfunction as compared to wild-type Nrf2 mice [57].

2 The Proximal Tubule as a ROS Target

The renal PTC are the major determinants of kidney oxygen consumption. They possess a highly oxidative neoglucogenic metabolism to provide ATP for the Na^+/K^+-ATPase that drives their numerous transport processes [60]. The high metabolic activity of the PTCs renders them particularly sensitive to oxidative stress.

2.1 ROS and Renal Disease: A Focus on the Proximal Tubule

Reactive oxygen and nitrogen species are increasingly implicated in the progression of a wide range of renal diseases, including diabetes, toxic-induced nephropathy, ischemia-reperfusion injury, and proteinuria [61–65]. That oxidative stress is involved in chronic renal disease is evidenced by the marked elevation of lipid peroxidation products, AGEs and ALEs, in the plasma of patients with renal failure, pointing to the enhanced interaction of ROS with lipid and protein molecules [66–68]. Furthermore, enhanced ROS-mediated nitric oxide inactivation, protein nitration, and functional NO deficiency, leading to hypertension, support the involvement of RNS in chronic renal disease [69]. As mentioned before, ROS can be generated by the mitochondria or in the cytosol. PTC have numerous mitochondria to provide sufficient ATP for their high transport activity. Normal mitochondrial metabolism produces ROS in the respiratory chain. However, dysfunction of the mitochondrial respiratory chain causes excessive mitochondrial ROS generation, resulting in proximal tubular dysfunction with a more or less complete de Toni–Debre–Fanconi syndrome [70–72].

2.1.1 Diabetic Nephropathy

Diabetic nephropathy is one of the most studied renal diseases with regard to the involvement of ROS in the disease process. The proximal tubule plays a central role in the pathogenesis of the diabetic kidney. Glucose that is filtered by the glomerulus is reabsorbed by PTC through the sodium-dependent glucose cotransporter [73]. Hyperglycemia and consequently excessive filtration and reabsorption of glucose enhance ROS production in PTC. Excessive substrate availability and enhanced glycolysis favor mitochondrial ROS production [74]. Electron transfer through complexes I, III, and IV generates a proton gradient that drives ATP synthase (complex V). The probability that molecular oxygen is reduced to superoxide rather than water is increased when the electrochemical potential difference generated by this proton gradient is high, since the life of superoxide-generating electron transport intermediates such as ubisemiquinone is prolonged. Because the proton gradient is coupled to the conversion of ADP into ATP, mitochondrial ROS generation is particularly strong if the availability of ADP is low. These conditions are present in PTC of the diabetic kidney. Normally, an excess of substrates is prevented by the tight regulation of phosphofructokinase (PFK), one of the early enzymes in this pathway. PFK is inhibited by ATP and citrate, thereby coupling the generation of pyruvate and acetyl CoA to the cellular energy demand. This negative control of the glycolytic pathway is overridden in the diabetic kidney by an excess of fructose-6-phosphate, thereby uncoupling glycolysis from cellular energy demand. The enhanced glucose uptake increases cellular ROS through protein kinase C (PKC)-dependent activation of NADPH oxidase. The renal isoform of the plasma-membrane associated NADPH oxidases, Nox4, is highly expressed in

PTC [75, 76]. Under normal redox state, Nox4 is believed to have a role in oxygen sensing, important for the regulation of EPO production [77, 78].

Superoxide is also generated by the process of glucose auto-oxidation that is associated with the formation of glycated proteins. The interaction of AGEs with cell surface receptors stimulates ROS production and decreases intracellular glutathione levels. Furthermore, high-glucose–induced ROS generation stimulates the production of angiotensin II (ATII), which in turn triggers the production of intracellular ROS. Nox4 is the central enzyme complex involved in ATII- and AGE-mediated ROS production in the diabetic kidney. Therefore, AGE and ATII induce a positive feedback amplifying ROS formation and aggravating disease progression [79]. The increase in intracellular ROS stimulates the MAP kinase and Janus kinase phosphorylation cascade, leading to the activation of NFκB, AP-1, SP-1, and the signal transducer and activator of transcription (STAT) signaling pathways. These transcription factors are implicated in the redox signaling activating TGF-β1 [80]. Since TGF-β1 is an important factor in the development of renal interstitial fibrosis, the activation of TGF-β1 in the diabetic kidney implies that increased oxidative stress can lead to the activation of fibrogenic pathways. TGF-β1 is a major inducer of epithelial–mesenchymal transition in which epithelial cells lose their epithelial phenotype, acquire mesenchymal markers, and start to produce excessive amounts of extracellular matrix proteins, leading to tubulointerstitial fibrosis [81, 82]. The degree of tubulointerstitial fibrosis, being closely related to the progression of renal disease [83], the activation of TGF-β1, and the continuous autocrine amplification of ROS production may eventually progress to end stage renal disease. The role of ROS to induce signaling pathways that lead to the activation of fibrogenic pathways has also been involved in chronic renal diseases such as ischemia/reperfusion injury, toxic-induced renal injury, and proteinuria (see below) [64, 84–86].

2.1.2 Toxic-Induced Nephropathies

The proximal tubule is a prime target for many toxic-induced nephropathies. Nephrotoxic substances that directly affect PTC include aminoglycoside antibiotics (e.g., gentamicin), cisplatin, radio-contrast media, immunoglobulins, aristocholic acid, and heavy metals (e.g., cadmium). Cisplatin-induced renal injury is a well-studied nephrotoxicity model [87]. During cisplatin nephrotoxicity, oxidative stress (ROS and RNS) is induced in PTC [85]. After entering the cell, cisplatin turns into highly reactive platinum species. These species may crosslink with DNA and react with other nucleophiles such as glutathione, lipids, and proteins, resulting in the disruption of cellular homeostasis followed by cell injury and death [37]. Importantly, amelioration of oxidative stress by a variety of ROS scavengers can protect the cells against cisplatin injury [87, 88].

Heavy metals like cadmium (Cd^{2+}) are known to induce ROS in the proximal tubule. Cd^{2+} induces a variety of processes that disrupt normal cellular function, as recently reviewed [65]. In brief, Cd^{2+} interferes with Ca^{2+}-, cAMP-, NO-, ROS-,

MAP kinase-, PKB/Akt-, NFκB-, and developmental signaling processes by acting on cell surface receptors or inducing changes in levels of second messengers. Furthermore, Cd^{2+} may induce mitochondrial damage by inhibiting the electron transfer chain. Cd^{2+} by itself cannot generate redox reactions, since it is not a Fenton metal. However, free Cd^{2+} can change the cellular redox status by interfering with SH groups and Zn^{2+} moieties in enzymes or cofactors, thereby inhibiting antioxidant enzymes. By binding to SH groups, Cd^{2+} can also deplete intracellular radical scavengers such as glutathione and protein sulfhydryls like thioredoxin, leading to enhanced ROS formation. It can also induce hydroxyl radicals in the presence of Fenton metals such as Fe^{2+}. Measurement of reactive species with fluorescent probes indicated that intracellular ROS levels are increased within minutes after Cd^{2+} exposure. In turn, increased oxidative stress can activate downstream events, leading to signaling cascades involved in a wide myriad of cellular responses. The latter include induction of pro-apoptotic and adaptive responses (e.g., induction of antioxidant defense), but also DNA damage, lipid peroxidation, and oxidatively modified proteins, which can lead to cellular dysfunction or cell death.

The implication of ROS in the early phase and the progression of renal diseases associated with proteinuria or defective protein reabsorption will be addressed in the following section.

2.1.3 Genetic Disorders of the Proximal Tubule

Oxidative stress is not only induced in acquired renal disease, it has also been documented in patients with renal Fanconi syndrome due to cystinosis [89] and in mice lacking the chloride transporter, ClC-5 [90]. Cystinosis is a rare, autosomal recessive disease caused by a defect in the transport of cystine across the lysosomal membrane and characterized by early onset of renal proximal tubular dysfunction. The disease is caused by mutations in the *CTNS* gene that encodes cystinosin, the lysosomal cystine transporter [91, 92]. The most severe and frequent form of cystinosis is characterized by a proximal tubule (PT) dysfunction that appears a few months after birth. In absence of treatment, the tubulopathy rapidly progresses (within a few years) toward renal failure and end stage renal disease. Cystinosis is also the most common inherited cause of the renal Fanconi syndrome. Earlier studies have revealed that renal Fanconi syndrome was caused by reduced mitochondrial ATP generation, leading to a decreased activity of tubular Na^+–K^+-ATPase and consequent diminution of Na^+-coupled proximal transport of glucose, amino acids, ions, and water [93, 94]. More recent studies showed that the mitochondrial energy generation capacity appears to be normal in cystinosis, suggesting that other mechanisms might account for reduced transport activity [89, 95]. Furthermore, cystinotic cells seem to be more susceptible to undergo apoptosis even with a normal mitochondrial energy–generating capacity [96]. An alternate hypothesis is that cystine accumulation induces an oxidative stress, which accounts for the cellular damage and apoptosis, as suggested by the effect of injecting cystine dimethyl ester into rat kidneys [97]. The increased ROS hypothesis is also supported by the following observations. First, cystine markedly increased the cytotoxic response of bacteria to H_2O_2, suggesting that cystine might impair the cellular defense machinery against H_2O_2 [98]. Furthermore, an elevated glutathione disulfide to total glutathione (GSSG/total GSH) ratio was observed in cystinotic cells, suggesting that increased oxidative stress might be present in cystinotic cells [89]. GSH depletion in cystinotic PTC may be related to enhanced apoptosis [95]. However, the link between the cellular defect and ROS remains speculative and needs to be further elucidated.

Dent's disease is an X-linked renal Fanconi syndrome associated with low-molecular-weight (LMW) proteinuria, nephrocalcinosis, nephrolithiasis, rickets, and renal failure in some patients [99]. The disease is caused by inactivating mutations in the *CLCN5* gene, which encodes the endosomal Cl^-/H^+ exchanger ClC-5 [100]. ClC-5 is primarily localized in PTC endosomes, where it codistributes with the vacuolar H^+-ATPase (V-ATPase) [101]. Genetic inactivation of *Clcn5* in mouse causes renal tubular defects that mimic human Dent's disease, including severe PT dysfunction with impaired endocytosis and trafficking defects of megalin and cubilin, leading to LMW proteinuria [102–104]. We recently demonstrated higher cell proliferation (cyclin E, PCNA, KI67) and oxidative stress (type 1 SOD, thioredoxin) in the PTC of ClC-5-deficient (*Clcn5*$^{Y/-}$) mice, as well as the specific induction of type III carbonic anhydrase (CAIII), which may participate in oxidative scavenging in PT cells [90]. The upregulation of CAIII in parallel with proliferative and oxidative stress markers was confirmed in human kidney samples harboring an inactivating mutation of *CLCN5* [90]. These studies suggest that generalized PT dysfunction is associated with increased cell proliferation, dedifferentiation, and oxidative stress. The potential role of the induction of CAIII, which is an early mesodermal marker, is discussed below.

2.2 *CAIII: A Novel Player in the Antioxidant Defense in the Proximal Tubule*

PTC possess a wide range of antioxidant enzymes to counteract the production of ROS (see Fig. 2). Besides the "classical" antioxidant enzymes (described in Sect. 1.2), renal CAIII has recently been implicated as a ROS scavenger and is in the kidney exclusively distributed in scattered PTC (Fig. 3) [90]. Unlike other CA isoforms, CAIII is characterized by a very low carbon dioxide hydration ability, reflecting nonconservation of the catalytic site [105]. However, Kim et al. [106] suggested that CAIII might function as percarbonic acid anhydrase detoxifying H_2O_2 into percarbonic acid, a powerful oxidizing compound. Indeed, CAIII expression decreased the level of ROS when expressed in NIH-3T3 cells, lacking endogenous CAIII, and protected them against cytotoxic concentrations of H_2O_2 [107]. These observations were further supported by our results showing that CAIII is specifically induced in HK-2 and OK-cells exposed to H_2O_2 [90]. The possible link between CAIII and oxidative stress was further demonstrated in kidneys from

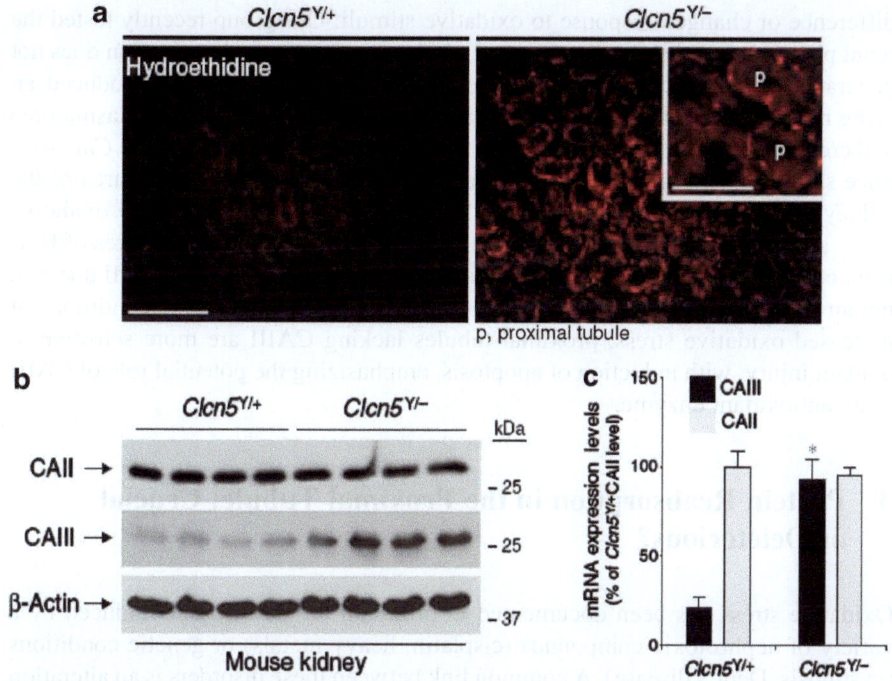

Fig. 3 Proximal tubule dysfunction: oxidative stress and CAIII induction. (**a**) Proximal tubule cells (PTC) from the ClC-5 knockout mouse ($Clcn5^{Y/-}$) show increased levels of oxidative stress as detected by a stronger signal for the red fluorescent ethidium bromide in $Clcn5^{Y/-}$ proximal tubule as compared to $Clcn5^{Y/+}$. *Bars* = 100 μm (*inset*, 50 μm). (**b–c**) The kidneys lacking ClC-5 ($Clcn5^{Y/-}$) are characterized by a specific induction of the carbonic anhydrase type III (CAIII) at the protein (**b**) and mRNA (**c**) level. Modified from Gailly et al. [90]

ClC-5 knockout mice. PTC from these mice has an increased state of oxidative stress, as shown by the increased production of $^{\bullet}O_2^-$ and the induction of antioxidant enzymes SOD1 and Trx [90]. Other studies also suggested a protective role of CAIII against oxidative stress. CAIII can undergo S-glutathionylation, a hallmark of proteins involved in oxidative stress [108]. Oxidative modification of CAIII involves two reactive SH groups, Cys183 and Cys188, that interact with glutathione [109]. A second possible action of CAIII in PTC exposed to stress could involve cellular proliferation and dedifferentiation. The PTC with an increased CAIII expression also shows proliferation markers and a more rapid growth [90, 107]. Considering that CAIII is an early mesodermal marker [110], increased expression of CAIII may reflect cell dedifferentiation and, along with other genes encoding growth factors and transcription factors, recapitulate the expression pattern observed during nephrogenesis [111]. To study the mechanisms of CAIII induction in PTC and its involvement as an antioxidant or inducer of cell dedifferentiation/proliferation, a CAIII knockout mouse model can be used. Kim et al. [106] generated a *Car3* knockout (*Car3−/−*) mouse, which show no apparent phenotypic difference or changed response to oxidative stimuli. Our group recently tested the renal phenotype of the *Car3−/−* mouse. Although *Car3* gene inactivation does not generate any apparent renal phenotype, administration of cisplatin produced an acute renal injury in *Car3−/−*, which was more severe (albuminuria, plasma urea and creatinine, apoptosis index) than in *Car3+/+* mice. Furthermore, the *Car3−/−* mice showed a significantly higher level of oxidative stress in cortical area of the kidneys (unpublished observations) [112]. The increased sensitivity to oxidative stress and ROS-induced apoptosis were also confirmed in primary cultures of PTC isolated from *Car3−/−* mice. These data suggest that inactivation of CAIII does not impair renal proximal tubules under normal conditions. However, in conditions of increased oxidative stress, proximal tubules lacking CAIII are more sensitive to oxidant injury, with induction of apoptosis, emphasizing the potential role of CAIII as an antioxidant enzyme.

3 Protein Reabsorption in the Proximal Tubule: Crucial or Deleterious?

Oxidative stress has been documented in proximal tubule disorders induced by a variety of nephrotoxic compounds (cisplatin, heavy metals) or genetic conditions (cystinosis, Dent's disease). A common link between these disorders is an alteration of the receptor-mediated endocytic uptake of albumin and LMW proteins [113]. Therefore, the possibility of a link between oxidative stress and defective endocytosis in PTC could be raised. On the other hand, excessive filtration of albumin, leading to increased endocytic uptake in PTC, has been show to induce oxidative stress in PTC [61]. Thus, defective or excessive albumin endocytosis may result in oxidative stress, raising the issue of the exact role of albumin in the redox state of PTC.

3.1 Antioxidant Properties of Albumin

The reabsorption of albumin through the megalin-cubilin-amnionless receptor pathway has a dual effect on PTC. At physiological concentrations, albumin is a survival factor for renal PTC, by activation of the membrane-bound serine/threonine kinase PKB and phosphorylation of the Bad protein, known to inhibit apoptosis [114]. Furthermore, albumin is able to act as an antioxidant, most likely through scavenging of ROS by the free hydroxyl groups of albumin and its free sulfhydryl group at Cys^{34}, a unique feature of albumin (Fig. 4) [115]. Other plasma proteins such as transferrin and ceruloplasmin also have antioxidant properties. Like albumin, they can bind to transition metals Fe^{2+} and Cu^{2+}, thereby preventing the generation of $^{\cdot}OH$ via the Fenton reaction [116]. Albumin and other proteins can also act as sacrificial sinks for ROS attack, either directly through oxidation of amino acid side chains or indirectly through reaction with lipid species and radicals arising from the peroxidation of cell

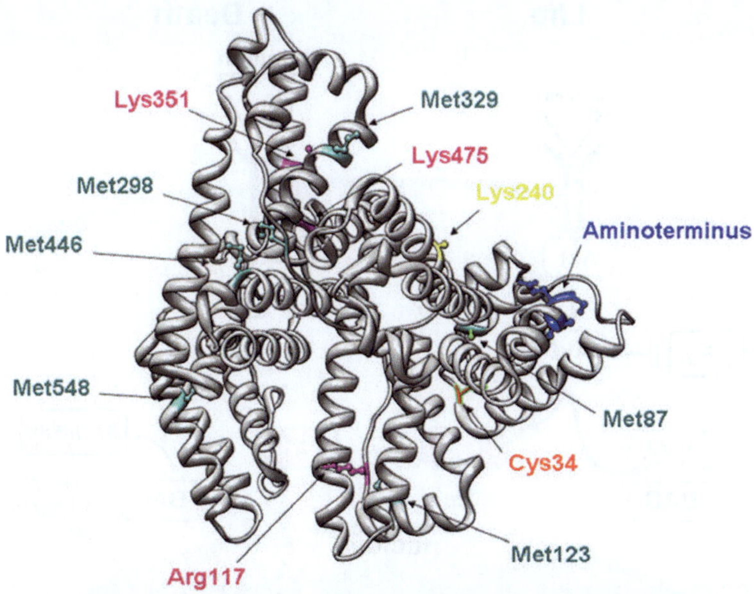

Fig. 4 Main sites in albumin involved in its antioxidant activity. The four amino acids (*in blue*) of the N-terminus of albumin are involved in the metal binding of the protein. The sole free cysteine in the protein (Cys34) is shown *in red*. Lateral carbon chain of residues involved in albumin antioxidant properties are developed and colored. From Roche et al. [119]

membrane lipids. The amino acids cysteine, histidine, methionine, tyrosine, and tryptophan are particularly susceptible to direct oxidative attack, whereas lysine is most susceptible to attack by malondialdehyde, one of the principal products of lipid peroxidation [117]. The role of albumin as a ROS scavenger has been confirmed in cell-free systems with the use of a variety of oxidative species, including HOCl, H_2O_2, $^{\bullet}OH$, carbon radicals, and peroxynitrite [118].

Recent studies in the ClC-5 and megalin knockout mouse models, both having a major defect in the megalin-cubilin-mediated endocytosis, show that the lack of protective response elicited by the uptake of albumin is associated with increased susceptibility to disease progression or tubulointerstitial fibrosis, potentially mediated by oxidative stress [90, 119, 120]. In particular, the protective effects of albumin in PTC seem to be mediated by megalin-mediated signaling. As said above, megalin binds to and activates PKB (Akt), thereby promoting cell survival and inhibiting apoptosis [114]. However, the protective effects of albumin seem to be limited to low, physiological concentrations. Indeed, exposure to high concentrations of albumin decreases the expression of megalin at the plasma membrane, decreasing Akt activity and Bad phosphorylation, thereby inducing albumin-induced apoptosis (Fig. 5) [114]. Similarly, exposure of LLC-PK1 cells, a model of the proximal tubular epithelium, to high concentrations of albumin induces apoptosis [121].

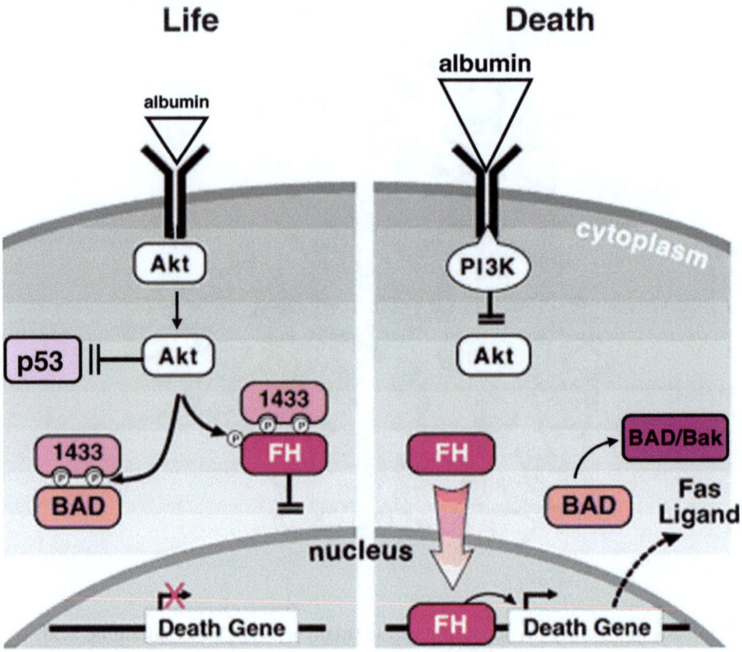

Fig. 5 Model of the dual effect of albumin on cell survival/cell death in proximal tubule. *Left*: At physiological concentrations, albumin uptake by megalin activates Akt, leading to phosphorylation and sequestration by 1,433 of BAD and the forkhead transcription factor (FH). *Right*: In case of albumin overload, Akt remains inactive through a PI3 kinase-dependent mechanism (PI3K), leading to an activation of FH and BAD. Active FH and BAD induce gene expression related to apoptosis and cell death. Adapted from an image by Steven Moskowitz, Advanced Medical Graphics

3.2 Oxidative Capacity of Albumin

Albumin is the major component of proteinuria, a factor that has a strong predictive value in chronic nephropathies [122]. Protein overload caused by a primary glomerular defect induces the expression of a wide array of proinflammatory and profibrotic mediators in renal tubular cells mainly through the activation of the transcription factors NFκB and AP-1 and the STAT signaling pathway [123–125]. The activation of both transcription factors seems to be dependent on an increase in oxidative stress [84, 124, 126]. Excessive endocytic uptake of albumin activates Nox4 through PKC, resulting in the formation of free oxygen radicals (O_2^-). Detoxification of O_2^- increases cytosolic levels of H_2O_2, which induces STAT, NFκB, and AP-1 intracellular signaling cascades, mainly leading to inflammation and fibrosis. H_2O_2 also induces apoptosis signal-regulating kinase 1 (Ask1), which activates the p38 MAPK pathway. MAPK regulates apoptosis and survival. Another regulator of the MAPK pathway in response to chronic protein overload is glia maturation factor-B (GMF-B), although little information regarding this molecule is available [127]. GMF-B induction increases the susceptibility to oxidative stress due to an upregulation of H_2O_2-producing enzymes (CuZn-SOD) and a downregulation of H_2O_2-reducing enzymes (catalase, GPx, GSH). Although albumin, as a molecule, is capable of inducing oxidative stress, it is also thought that albumin-bound fatty acids (FAs) exert extra deleterious effects on PTC [128]. Albumin-bound FAs increase the production of mitochondrial ROS due to an impaired mitochondrial ROS-scavenging process (SOD2). Taken together, these data suggest that, depending upon its concentration in the tubular lumen and its ligation to beneficial vs. potentially harmful factors, albumin can act as a potent survival factor in PTC or promote tubulointerstitial damage.

3.3 *Future Directions*

New mouse models, and in particular the podocyte-specific Nphs2 knockout mouse [129], may be of interest to characterize the effects of albumin under normal vs. pathologic conditions. Mutations in the *NPHS2* gene, which encodes the slit diaphragm protein podocin, account for approximately 40% of familial and 10% of sporadic forms of nephritic syndrome [130]. Using the established Cre-loxP technology, Mollet et al. [129] inactivated podocin in the adult mouse kidney in a podocyte-specific manner. This inactivation is strictly time-controlled, as it is induced by injection of tamoxifen, which activates the tamoxifen-responsive Cre recombinase in podocytes, leading to progressive decrease in podocin expression within 7 days, with ensuing proteinuria. Therefore, the effects of overload proteinuria can be investigated at any given time after the induction of podocin inactivation. A detailed investigation of these mice will allow us to study the early effects of albumin on PTC, including the oxidative stress that is thought to play a role in the early phase of kidney injury.

In addition to the above model, further studies will address the effect of defective endocytosis – and thus defective albumin uptake – on the oxidative stress and differentiation of PTC. We have observed that PTC of ClC-5 knockout mouse are in a state of oxidative stress and increased proliferation [90]. Furthermore, these PTC are also more sensitive to induction of pro-fibrotic factors such as TGF-β1 (unpublished observations) [131], which could indicate that the lack of albumin uptake is associated with an increased sensitivity to disease progression.

4 Conclusion

Emerging evidence shows that ROS, oxidative stress, and redox state are key players in the initiation and progression of various renal diseases, including diabetic nephropathy, toxic-induced renal injury, and inherited disorders associated with proximal tubule dysfunction. Oxidative stress triggers myriad cellular responses.

Various signaling pathways and transcription factors are involved, which can regulate the expression of genes involved in apoptosis, fibrosis, and inflammation or, alternatively, induce antioxidant mechanisms in an attempt to rescue the cell. Whether pro- or antioxidant pathways are activated most likely depends on the nature and duration of the stimulus as well as the cell or tissue that is targeted. For instance, a series of evidence, obtained both in vivo and in vitro, pointed to the high sensibility of PTC to oxidative damage and the potential role of a tightly regulated albumin endocytosis in the redox state of these cells. Investigation of the nature of the stimuli and initial events involved in ROS generation and the antioxidant mechanisms in PTC will probably help to develop novel diagnostic and therapeutic strategies for renal disease.

References

1. Sen CK, Packer L (1996) Antioxidant and redox regulation of gene transcription. FASEB J 10(7): 709–720
2. Jackson MJ, Papa S, Bolanos J et al. (2003) Antioxidants, reactive oxygen and nitrogen species, gene induction and mitochrondrial function. Mol Aspects Med 23: 209–285
3. Schafer FQ, Buettner GR (2001) Redox environment of the cell as viewed through the redox state of the glutathione disulfide/glutathione couple. Free Radic Biol Med 30(11): 1191–1212
4. Day BJ (2009) Catalase and glutathione peroxidase mimics. Biochem Pharmacol 77: 285–296
5. Jones DP (2006) Redefining oxidative stress. Antioxid Redox Signal 8(9–10): 1865–1879
6. Groeger G, Quiney C, Cotter TG (2009) Hydrogen peroxide as a cell survival signaling molecule. Antioxid Redox Signal 11(11): 2655–2671
7. D'Autreaux B, Toledano MB (2007) ROS as signalling molecules: mechanisms that generate specificity in ROS homeostasis. Nat Rev Mol Cell Biol 8: 813–824
8. Aw TY (2003) Cellular redox: a modulator of intestinal epithelial cell proliferation. News Physiol Sci 18: 201–204
9. Lin F, Moran A, Igarashi P (2005) Intrarenal cells, not bone marrow-derived cells, are the major source for regeneration in postischemic kidney. J Clin Invest 115: 1756–1764
10. Hallman MA, Zhuang S, Schnellmann RG (2008) Regulation of dedifferentiation and redifferentiation in renal proximal tubular cells by the epidermal growth factor receptor. J Pharmacol Exp Ther 325(2): 520–528
11. Chance B, Sies H, Boveris A (1979) Hydroperoxide metabolism in mammalian organs. Physiol Rev 59(3): 527–605
12. Addabbo F, Montagnani M, Goligorsky MS (2009) Mitochondria and reactive oxygen species. Hypertension 53(6): 885–892
13. Wang W, Fang H, Groom L et al. (2008) Superoxide flashes in single mitochondria. Cell 134: 279–290
14. Valko M, Rhodes CJ, Moncol J et al. (2006) Free radicals, metals and antioxidants in oxidative-stress induced cancer. Chem Biol Interact 160(1): 1–40
15. Kang DH, Nakagawa T, Feng L et al. (2002) A role for uric acid in the progression of renal disease. J Am Soc Nephrol 13: 2888–2897
16. Al-Ghamdi SS, Chatterjee PK, Raftery MJ et al. (2004) Role of cytochrome P4502E1 activation in proximal tubular cell injury induced by hydrogen peroxide. Ren Fail 26(2): 103–110

17. Vaziri ND, Dicus M, Ho ND et al. (2003) Oxidative stress and dysregulation of superoxide dismutase and NADPH oxidase in renal insufficiency. Kidney Int 63(1): 179–185
18. Taccone-Gallucci M, Manca-di-Villahermosa S, Battistini L et al. (2006) N-3 PUFAs reduce oxidative stress in ESRD patients on maintenance HD by inhibiting 5-lipoxygenase activity. Kidney Int 69(8): 1450–1454
19. Villeneuva S, Céspedes C, González AA (2007) Effect of ischemic acute renal damage on the expression of COX-2 and oxidative stress-related elements in rat kidney. Kidney Int 292: F1364–F1371
20. Passauer J, Pistrosch F, Büssemaker E (2005) Nitric oxide in chronic renal failure. Kidney Int 67(5): 1665–1667
21. Mates JM, Sanchez-Jimenez F (1999) Anitoxidant enzymes and their implications in pathological processes. Front Biosci 4: D339–D345
22. Mates JM (2000) Effects of antioxidant enzymes in the molecular control of reactive oxygen species toxicology. Toxicology 153: 83–104
23. Akagi R, Takahashi T, Sassa S (2005) Cytoprotective effects of heme-oxygenase in acute renal failure. Contrib Nephrol 148: 70–85
24. Poss KD, Tonegawa S (1997) Reduced stress defence in heme oxygenase 1-deficient cells. Proc Natl Acad Sci USA 94(20): 10925–10930
25. Abid MR, Razzaque MS, Taguchi T (2005) Oxidant stress in renal pathophysiology. Contrib Nephrol 148: 135–153
26. Bonventre JV (1993) Mechanisms of ischemic acute renal failure. Kidney Int 43: 1160–1178
27. Droge W (2002) Free radicals in the regulation of physiological functions. Physiol Rev 82: 47–95
28. Tian W, Zhang Z, Cohen DM (2002) MAPK signaling and the kidney. Am J Physiol 279: F593–F604
29. Guyton KZ, Liu Y, Gorospe M et al. (1996) Activation of mitogen-activated protein kinase by H_2O_2. Role in cell survival following oxidant injury. J Biol Chem 271: 4138–4142
30. Hayden MS, Ghosh S (2004) Signaling to NF-kappaB. Genes Dev 18: 2195–2224
31. Vesely PW, Staber PB, Hoefler G et al. (2009) Translational regulation mechanisms of AP-1 proteins. Mutat Res 682(1): 7–12
32. Cochrane AL, Ricardo SD (2003) Oxidant stress and regulation of chemokines in the development of renal interstitial fibrosis. Contrib Nephrol 139: 102–119
33. Rhyu DY, Yang Y, Ha H et al. (2005) Role of reactive oxygen species in TGF-β1-induced mitogen-activated protein kinase activation and epithelial-mesenchymal transition in renal tubular epithelial cells. J Am Soc Nephrol 16: 667–675
34. Lee JT, Gu W (2010) The multiple levels of regulation by p53 ubiquitination. Cell Death Differ 17(1):86–92
35. Kruse JP, Gu W (2009) Modes of p53 regulation. Cell 137(4): 609–622
36. Shiloh Y (2003) ATM and related protein kinases: safeguarding genome integrity. Nat Rev Cancer 3(3): 155–168
37. Siddik ZH (2003) Cisplatin: mode of cytotoxic action and molecular basis of resistance. Oncogene 22(47): 7265–7279
38. Norbury CJ, Zhivotovsky B (2004) DNA damage-induced apoptosis. Oncogene 23(16): 2797–2808
39. Bode AM, Dong Z (2004) Post-translational modification of p53 in tumorigenesis. Nat Rev Cancer 4(10): 793–805
40. Menendez D, Inga A, Resnick MA (2009) The expanding universe of p53 targets. Nat Rev Cancer 9(10): 724–737
41. Cano CE, Gommeaux J, Pietri S et al. (2009) Tumor protein 53-induced nuclear protein 1 is a major mediator of p53 antioxidant function. Cancer Res 69(1): 219–226
42. Pouyet L, Carrier A (2010) Mutant mouse models of oxidative stress. Transgenic Res 19(2):155–164

43. Nangaku M, Inagi R, Miyata T et al. (2008) Hypoxia and hypoxia-inducible factor in renal disease. Nephron Exp Nephrol 110: e1–e7
44. Nguyen T, Nioi P, Pickett CB (2009) The Nrf2-Antioxidant response element signaling pathway and its activation by oxidative stress. J Biol Chem 284(20): 13291–13295
45. Wang GL, Jiang BH, Rue EA et al. (1995) Hypoxia-inducible factor 1 is a basic-helix-loop-helix-PAS heterodimer regulated by cellular O_2 tension. Proc Natl Acad Sci USA 92(12): 5510–5514
46. Fine LG, Norman JT (2008) Chronic hypoxia as a mechanism of progression of chronic kidney diseases: from hypothesis to novel therapeutics. Kidney Int 74(7): 867–872
47. Ivan M, Kondo K, Yang H (2001) HIFalpha targeted for VHL-mediated destruction by proline hydroxylation: implications for O_2 sensing. Science 292(5516): 464–468
48. Jaakkola P, Mole DR, Tian YM et al. (2001) Targeting of HIF-alpha to the von Hippel-Lindau ubiquitylation complex by O_2-regulated prolyl hydroxylation. Science 292(5516): 468–472
49. Wenger RH, Stiehl DP, Camenisch G et al. (2005) Integration of oxygen signaling at the consensus HRE. Sci STKE 2005(306): re12
50. Kietzmann T, Görlach A (2005) Reactive oxygen species in the control of hypoxia-inducible factor-mediated gene expression. Semin Cell Dev Biol 16(4–5): 474–486
51. Zhang C, Meng X, Zhu Z, Liu J et al. (2004) Connective tissue growth factor regulates the key events in tubular epithelial to myofibroblast transition in vitro. Cell Biol Int 28(12): 863–873
52. Hill P, Shukla D, Tran MG et al. (2008) Inhibition of hypoxia inducible factor hydroxylases protects against renal ischemia-reperfusion injury. J Am Soc Nephrol 19(1): 39–46
53. Higgins DF, Kimura K, Bernhardt WM et al. (2007) Hypoxia promotes fibrogenesis in vivo via HIF-1 stimulation of epithelial-to-mesenchymal transition. J Clin Invest 117(12): 3810–3820
54. Nangaku M, Eckhardt KU (2007) Hypoxia and the HIF system in kidney disease. J Mol Med 85(12): 1325–1330
55. Song YR, You SJ, Lee YM et al. (2010) Activation of hypoxia-inducible factor attenuates renal injury in rat remnant kidney. Nephrol Dial Transplant 25(1):77–85
56. Chen J, Shaikh ZA (2009) Activation of Nrf2 by cadmium and its role in protection against cadmium-induced apoptosis in rat kidney cells. Toxicol Appl Pharmacol 241(1): 81–89
57. Liu M, Grigoryev DN, Crow MT et al. (2009) Transcription factor Nrf2 is protective during ischemic and nephrotoxic acute kidney injury in mice. Kidney Int 76(3): 277–285
58. Itoh K, Chiba T, Takahashi S et al. (1997) An Nrf2/small Maf heterodimer mediates the induction of phase II detoxifying enzyme genes through antioxidant response elements. Biochem Biophys Res Commun 236(2): 313–322
59. Chan K, Kan YW (1999) Nrf2 is essential for protection against acute pulmonary injury in mice. Proc Natl Acad Sci USA 96: 12731–12736
60. Mandel LJ (1985) Metabolic substrates, cellular energy production, and the regulation of proximal tubular transport. Annu Rev Physiol 47: 85–101
61. Shalamanova L, McArdle F, Amara AB et al. (2007) Albumin overload induces adaptive responses in human proximal tubular cells through oxidative stress but not via angiotensin II type 1 receptor. Am J Physiol 292: F1846–F1857
62. Forbes JM, Coughlan MT, Cooper ME (2008) Oxidative stress as a major culprit in kidney disease in diabetes. Diabetes 57: 1446–1454
63. Cachofeiro V, Goicochea M, Garcia de Vinuesa S et al. (2008) Oxidative stress and inflammation, a link between chronic kidney disease and cardiovascular disease. Kidney Int 74(Suppl 111): S4–S9
64. Kim J, Seok YM, Jung K-J et al. (2009) Reactive oxygen species/oxidative stress contributes to progression of kidney fibrosis following transient ischemic injury in mice. Am J Physiol 297: F461–F470

65. Thevenod F (2009) Cadmium and cellular signaling cascades: to be or not to be? Toxicol Appl Pharmacol 238(3): 221–239
66. Paul JL, Man NK, Moatti N et al. (1991) Membrane phospholipid peroxidation in renal insufficiency and chronic hemodialysis. Nephrologie 12: 4–7
67. Vaziri ND, Oveisi F, Ding Y (1998) Role of increased oxygen free radical activity in the pathogenesis of uremic hypertension. Kidney Int 53: 1748–1754
68. Miyata T, Kurokawa K, Van Ypersele de Strihou C (2000) Advanced glycation and lipoxidation end products: role of reactive carbonyl compounds generated during carbohydrate and lipid metabolism. J Am Soc Nephrol 11: 1744–1752
69. Vaziri ND, Ni Z, Oveisi F, Liang K (2002) Enhanced nitric oxide inactivation and protein nitration by reactive oxygen species in chronic renal insufficiency. Hypertension 39: 135–141
70. Rotig A (2003) Renal disease and mitochondrial genetics. J Nephrol 16(2): 286–292
71. Martin-Hernandez E, Garcia-Silva MT, Vara J et al. (2005) Renal pathology in children with mitochondrial diseases. Pediatr Nephrol 20: 1299–1305
72. Verkaart S, Koopman WJ, van Ernst-de Vries SE et al. (2007) Superoxide production is inversely related to complex I activity in inherited complex I deficiency. Biochim Biophys Acta 1772: 373–381
73. Bakris GL, Fonseca VA, Sharma K et al. (2009) Renal sodium-glucose transport: role in diabetes mellitus and potential clinical implications. Kidney Int 75(12): 1272–1277
74. Nishikawa T, Edelstein D, Du XL et al. (2000) Normalizing mitochondrial superoxide production blocks three pathways of hyperglycaemic damage. Nature 404: 787–790
75. Sachse A, Wolf G (2007) Angiotensin II-induced reactive oxygen species and the kidney. J Am Soc Nephrol 18: 2439–2446
76. Gorin Y, Block K, Abboud HE (2009) Subcellular localization of Nox4 and regulation in diabetes. Proc Natl Acad Sci USA 106(34): 14385–14390
77. Geiszt M, Kopp JB, Varnai P et al. (2000) Identification of renox, an NADPH oxidase in kidney. Proc Natl Acad Sci USA 97(14): 8010–8014
78. Orient A, Donko A, Szabo A et al. (2007) Novel sources of reactive oxygen species in the human body. Nephrol Dial Transplant 22: 1281–1288
79. Lee HB, Yu M-R, Yang Y et al. (2003) Reactive oxygen species-regulated signaling pathways in diabetic nephropathy. J Am Soc Nephrol 14: S241–S245
80. Hu T, Ramachandrarao SP, Siva S et al. (2005) ROS production via NADPH oxidase mediates TGF-beta-induced cytoskeletal alterations in endothelial cells. Am J Physiol 289: F816–F825
81. Zhang A, Jia Z, Guo X et al. (2007) Aldosterone induces epithelial-mesenchymal transition via ROS of mitochondrial origin. Am J Physiol 293(3): F723–F731
82. Strutz FM (2009) EMT and proteinuria as progression factors. Kidney Int 75: 475–481
83. Eddy AA, Neilson EG (2006) Chronic kidney disease progression. J Am Soc Nephrol 17: 2964–2966
84. Morigi M, Macconi D, Zoja C et al. (2002) Protein overload-induced NF-κB activation in proximal tubular cells requires H_2O_2 through a PKC-dependent pathway. J Am Soc Nephrol 13: 1179–1189
85. Chirino YI, Pedraza-Chaverri J (2009) Role of oxidative and nitrosative stress in cisplatin-induced nephrotoxicity. Exp Toxicol Pathol 61(3): 223–242
86. Okamura DM, Himmelfarb J (2009) Tipping the redox balance of oxidative stress in fibrogenic pathways in chronic kidney disease. Pediatr Nephrol 24(12): 2309–2319
87. Pabla N, Dong Z (2008) Cisplatin nephrotoxicity: mechanisms and renoprotective strategies. Kidney Int 73: 994–1007
88. Ramesh G, Reeves WB (2005) p38 MAP kinase inhibition ameliorates cisplatin nephrotoxicity in mice. Am J Physiol 289: F166–F174
89. Wilmer MJ, de Graaf-Hess A, Blom HJ et al. (2005) Elevated oxidized glutathione in cystinotic proximal tubular epithelial cells. Biochem Biophys Res Commun 337: 610–614

90. Gailly P, Jouret F, Martin D et al. (2008) A novel renal carbonic anhydrase type III plays a role in proximal tubule dysfunction. Kidney Int 74: 52–61
91. Town M, Jean G, Cherqui S et al. (1998) A novel gene encoding an integral membrane protein is mutated in nephropathic cystinosis. Nat Genet 18: 319–324
92. Kalatzis V, Antignac C (2003) New aspects of the pathogenesis of cystinosis. Pediatr Nephrol 18: 207–215
93. Ben-Nun A, Bashan N, Potashnik R et al. (1993) Cystine loading induces Fanconi's syndrome in rats: in vivo and vesicle studies. Am J Physiol 265: 839–844
94. Foreman JW, Benson LL, Wellons M et al. (1995) Metabolic studies of rat renal tubule cells loaded with cystine: the cystine dimethyl ester model of cystinosis. J Am Soc Nephrol 6: 269–272
95. Laube GF, Shah V, Stewart VC et al. (2006) Glutathione depletion and increased apoptosis rate in human cystinotic proximal tubular cells. Pediatr Nephrol 21: 503–509
96. Park MA, Helip-Wooley A, Thoene J (2002) Lysosomal cystine storage increases apoptosis in cultured human fibroblasts and renal proximal tubule epithelial cells. J Am Soc Nephrol 13: 2878–2887
97. Rech VC, Feksa LR, Arevalo do Amaral MF et al. (2007) Promotion of oxidative stress in kidney of rats loaded with cystine dimethyl ester. Pediatr Nephrol 22: 1121–1128
98. Cantoni O, Brandi G, Albano A et al. (1995) Action of cystine in the cytotoxic response of *Escherichia coli* cells exposed to hydrogen peroxide. Free Radic Res 22: 275–283
99. Scheinman SJ (1998) X-linked hypercalciuric nephrolithiasis: clinical syndromes and chloride channel mutations. Kidney Int 53: 3–17
100. Lloyd SE, Pearce SH, Fisher SE et al. (1996) A common molecular basis for three inherited kidney stone diseases. Nature 379: 445–449
101. Devuyst O, Christie PT, Courtoy PJ et al. (1999) Intra-renal and subcellular distribution of the human chloride channel, CLC-5, reveals a pathophysiological basis for Dent's disease. Hum Mol Genet 8: 247–257
102. Piwon N, Gunther W, Schwake M et al. (2000) ClC-5 Cl – channel disruption impairs endocytosis in a mouse model for Dent's disease. Nature 408: 369–373
103. Wang SS, Devuyst O, Courtoy PJ et al. (2000) Mice lacking renal chloride channel, CLC-5, are a model for Dent's disease, a nephrolithiasis disorder associated with defective receptor-mediated endocytosis. Hum Mol Genet 9: 2937–2945
104. Christensen EI, Devuyst O, Dom G et al. (2003) Loss of chloride channel ClC-5 impairs endocytosis by defective trafficking of megalin and cubilin in kidney proximal tubules. Proc Natl Acad Sci USA 100: 8472–8477
105. Lehtonen J, Shen B, Vihinen M et al. (2004) Characterization of CA XIII, a novel member of the carbonic anhydrase isozyme family. J Biol Chem 279: 2719–2727
106. Kim G, Lee TH, Wetzel P et al. (2004) Carbonic anhydrase III is not required in the mouse for normal growth, development and life span. Mol Cell Biol 24: 9942–9947
107. Raisanen SR, Lehenkari P, Tasanen M et al. (1999) Carbonic anhydrase III protects cells from hydrogen peroxide-induced apoptosis. FASEB J 13(3): 513–522
108. Cabiscol E, Levine RL (1995) Carbonic anhydrase III. Oxidative modification in vivo and loss of phosphatase activity during aging. J Biol Chem 270: 14742–14747
109. Mallis RJ, Poland BW, Chatterjee TK et al. (2000) Crystal structure of *S*-glutathiolated carbonic anhydrase III. FEBS Lett 482(3): 237–241
110. Lyons GE, Buckingham ME, Tweedie S et al. (1991) Carbonic anhydrase III, an early mesodermal marker, is expressed in embryonic mouse skeletal muscle and notochord. Development 111: 233–244
111. Bonventre JV (2003) Dedifferentiation and proliferation of surviving epithelial cells in acute renal failure. J Am Sol Nephrol 14: S55–S61
112. Parreira KS, Terryn S, Lima WR et al. (2008) Induction of type III carbonic anhydrase in proximal tubule cells attenuates oxidative stress and apoptosis in cisplatin nephrotoxicity. J Am Soc Nephrol 19: 631A

113. Birn H, Christensen EI (2006) Renal albumin absorption in physiology and pathophysiology. Kidney Int 69(3): 440–449
114. Caruso-Neves C, Pinheiro AAS, Cai H (2006) PKB and megalin determine the survival or death of renal proximal tubule cells. Proc Natl Acad Sci USA 103(49): 18810–18815
115. Kabosha-Iwatsuki M, Miyamoto M, Inoue M (1997) Effect of nitric oxide on the ligand-binding activity of albumin. Arch Biochem Biophys 345: 237–242
116. Halliwell B, Gutteridge JMC (1990) Role of free radical and catalytic metal ions in human disease: an overview. Methods Enzymol 186: 1–85
117. Halliwell B, Gutteridge JMC (1989) Free radicals in biology and medicine. Oxford: Clarendon
118. Theilig F, Kriz W, Jerichow T et al. (2007) Abrogation of protein uptake through megalin-deficient proximal tubules does not safeguard against tubulointerstitial injury. J Am Soc Nephrol 18: 1824–1834
119. Roche M, Rondeau P, Singh NR et al. (2008) The antioxidant properties of serum albumin. FEBS Lett 582: 1783–1787
120. Ludwig M, Utsch B, Monnens LAH (2006) Recent advances in understanding the clinical and genetic heterogeneity of Dent's disease. Nephrol Dial Transplant 21: 2708–2717
121. Erkan E, De Leon M, Devarajan P (2001) Albumin overload induces apoptosis in LLC-PK (1) cells. Am J Physiol 280(6): F1107–F1114
122. Remuzzi G, Bertani T (1998) Pathophysiology of progressive nephropathies. N Engl J Med 339: 1448–1456
123. D'Amico G, Bazzi C (2003) Pathophysiology of proteinuria. Kidney Int 63(3): 809–825
124. Nakajima H, Takenaka M, Kaimori JY et al. (2004) Activation of the signal transducer and activator of transcription signalling pathway in renal proximal tubular cells by albumin. J Am Soc Nephrol 15(2): 276–285
125. Abbate M, Zoja C, Remuzzi G (2006) How does proteinuria cause progressive renal damage? J Am Soc Nephrol 17(11): 2974–2984
126. Imai E, Nakajima H, Kaimori J-Y (2004) Albumin turns on a vicious spiral of oxidative stress in renal proximal tubules. Kidney Int 66: 2085–2087
127. Kaimori J, Takenaka M, Nakajima H et al. (2003) Induction of glia maturation factor in proximal tubular cells leads to vulnerability to oxidative injury through the p38 pathway and changes in antioxidant enzyme activities. J Biol Chem 278(35): 33519–33527
128. Ishola DA, Post JA, van Timmeren MM et al. (2006) Albumin-bound fatty acids induce mitochondrial oxidant stress and impair antioxidant responses in proximal tubular cells. Kidney Int 70: 724–731
129. Mollet G, Ratelade J, Boyer O (2009) Podocin inactivation in mature kidneys causes focal segmental glomerulosclerosis and nephrotic syndrome. J Am Soc Nephrol 20(10): 2181–2189
130. Machuca E, Hummel A, Nevo F et al. (2009) Clinical and epidemiological assessment of steroid-resistant nephrotic syndrome associated with the NPHS2 R229Q variant. Kidney Int 75(7): 727–735
131. Terryn S, Parreira KP, Devuyst O (2009) Oxidative stress and increased propensity to develop fibrosis in proximal tubule cells lacking ClC-5. J Am Soc Nephrol 20(10): 381A

Chapter 11
Iron Metabolism and Oxidative Stress

Abolfazl Zarjou, József Balla, György Balla, and Anupam Agarwal

Abstract Iron is essential for almost all living organisms. It participates in a wide variety of fundamental metabolic processes, including oxygen transport, DNA synthesis, and electron transport. However, when present in excess, iron poses a threat to cells and tissues. The toxicity of iron is largely based on its ability to catalyze the generation of free radicals, which attack and damage cellular membranes, protein, and DNA. Under physiological circumstances this threat is reduced because meticulously regulated mechanisms have evolved to move iron across biological membranes and to ensure that its distribution in multicellular organisms is carefully orchestrated. These mechanisms are responsible to keep free iron concentrations at the lowest sufficient level under healthy conditions. High tissue iron concentrations have been associated with the development and progression of several pathological conditions, including certain cancers, liver and cardiovascular diseases, diabetes, hormonal derangements, skeletal abnormalities, several neurodegenerative disorders, and immune system dysfunctions. In this chapter, we discuss oxidative stress related to iron metabolism and provide an overview of several diseases that are linked to iron overload and toxicity.

Keywords Diabetic nephropathy · Proteinuria · Endocytosis · Albumin · Carbonic anhydrase type III

1 Iron Absorption and Distribution

Iron is able to readily accept and donate electrons, interconverting between ferric Fe^{3+} and ferrous Fe^{2+} states (+772 mV at neutral pH). The ability of this transition metal to exist in two redox states makes it an important candidate for numerous

A. Agarwal (✉)
Division of Nephrology, THT 647, University of Alabama at Birmingham,
1900 University Boulevard, Birmingham, Alabama 35294, USA
e-mail: agarwal@uab.edu

T. Miyata et al. (eds.), *Studies on Renal Disorders*, Oxidative Stress in Applied Basic Research and Clinical Practice, DOI 10.1007/978-1-60761-857-7_11,

biological activities as the catalytic center of fundamental biochemical reactions [1, 2]. Iron has an essential role in the formation of hemoglobin and myoglobin, cytochromes, and many other metalloproteins involved in neurotransmitter production, collagen formation, and immune system function. The bioavailability of iron is generally limited, because under aerobic conditions, Fe^{2+} is readily oxidized in solution to Fe^{3+}, which is virtually insoluble at physiological pH. The average adult human's body contains 3–4 g of iron, which makes it the most abundant transition metal. Women have lower levels of iron because of their recurrent blood loss through menstruation.

Since humans have no excretory pathway for iron, it is essential for the intestine to tightly regulate the absorption of iron to replace the loss through menstruation and desquamation of enterocytes. Absorptive enterocytes close to the gastroduodenal junction are responsible for almost all iron absorption (Fig. 1). Because there is little, if any, paracellular iron transport under normal circumstances, iron must traverse both the apical and basolateral membranes to gain access to the circulation. The low pH of gastric secretion provides a proton-rich environment and helps to "dissolve" ingested iron. This facilitates enzymatic reduction of ferric iron to its

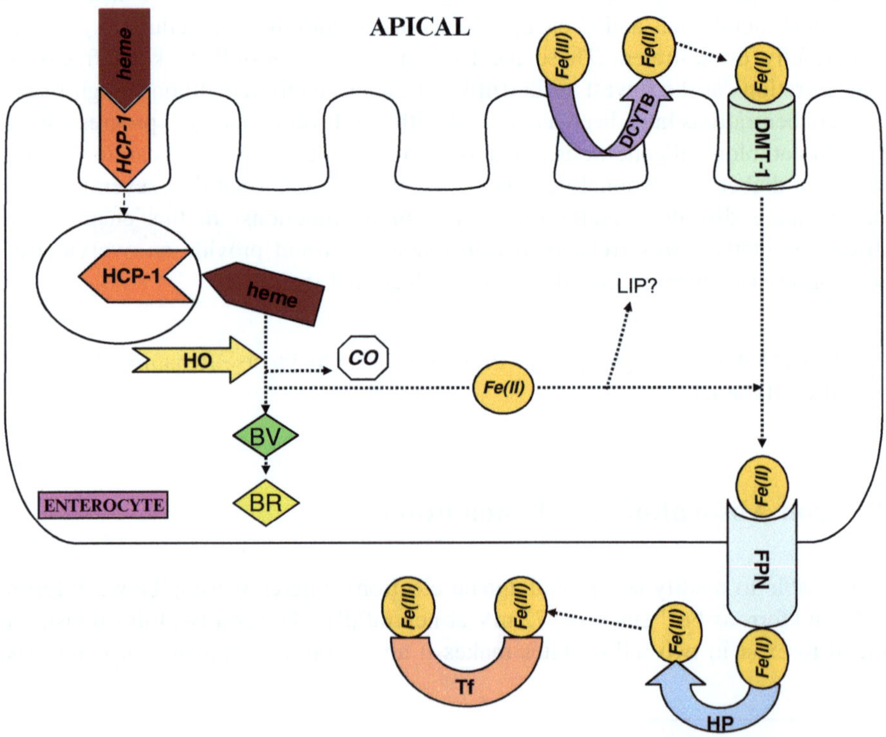

Fig. 1 Mechanism of intestinal iron absorption. *HCP-1* heme carrier protein-1; *TF* transferrin; *DCYTB* duodenal cytochrome b; *DMT-1* divalent metal transporter; *HP* hephaestin; *HO* heme-oxygenase; *BV* biliverdin; *BR* bilirubin; *CO* carbon monoxide; *FPN* ferroportin

ferrous form by duodenal cytochrome b (DCYTB), the first identified intestinal ferrireductase present on the apical surface of the enterocytes. DCYTB levels are increased in iron deficiency, which signifies its importance in iron absorption [3]. A divalent metal transporter that is nonspecific to iron (DMT-1) transfers iron across the apical membrane and into the cell through a proton-coupled process.

Heme is a biologically important iron-containing compound and is a key source of dietary iron. Heme results from the breakdown of hemoglobin and myoglobin found in meat products. Heme iron appears to be transported intact from the gut lumen into enterocytes, through the activity of heme carrier protein-1 (HCP-1). HCP-1 is expressed at high levels in the duodenum [4] and, upon binding of heme to HCP-1 on the cell surface, the complex is internalized by receptor-mediated endocytosis. The resultant endosomal vesicles are thought to migrate to the endoplasmic reticulum, where heme induces the synthesis of heme oxygenase-1 (HO-1), the rate-limiting enzyme in the catabolism of heme [5]. HO-1 cleaves the porphyrin ring at the alpha-methene bridge to form biliverdin (subsequently converted to bilirubin by biliverdin reductase) and carbon monoxide and also releases free redox active iron. Although the exact mechanism is as yet unclear, studies with HCP-1 demonstrated that heme transport is posttranscriptionally regulated by iron stores [4].

Independent of the source of iron, the Fe^{2+} is then exported through the basolateral membrane of the enterocyte to the interstitial space by the exporter ferroportin [6]. Ferroportin is critically important for maternoembryonic iron transfer, for basolateral transport of iron out of enterocytes, and for the export of iron from tissue macrophages. It appears to play a lesser role in the export of iron from hepatocytes. Iron is subsequently oxidized by the ferroxidase activity of a ceruloplasmin homologue called hephaestin [7, 8]. Once the iron moves out of the cell it binds with high affinity to transferrin, which is synthesized by hepatocytes and secreted by the liver to the plasma [9]. Transferrin is an 80-kDa protein with two iron binding sites [10]. Under physiological circumstances transferrin is in excess in relation to iron, and therefore, the concentration of "free" iron in the plasma is virtually zero.

Approximately 80% of the body's iron content is incorporated into hemoglobin in developing erythroid precursors and mature red blood cells. Hence the erythroid bone marrow is the largest consumer of transferrin-associated iron and approximately a billion iron atoms are used each day to form hemoglobin in new red blood cells. Sequestration of senescent red blood cells by specialized macrophages in the spleen, liver, and bone marrow (reticuloendothelial system) liberates similar amounts of iron from hemoglobin, generating a turnover of 1 billion iron atoms (35 mg) between erythropoiesis and reticuloendothelial systems. The heme moiety of hemoglobin is catabolized by heme oxygenase, and the released iron is sequestered by ferritin. The expression of ferroportin-1 in reticuloendothelial macrophages supports the notion that this iron-export protein is essential in iron recycling from senescent erythrocytes. Ceruloplasmin with ferroxidase activity is required for oxidation and incorporation of ferrous iron into transferrin to form ferric transferrin [11].

Iron transport by transferrin serves two main functions: one is to render the very water-insoluble Fe^{3+} into the soluble diferric–transferrin complex, and the other is to allow Fe^{3+}–transferrin complex to be recognized by the TfR located on the cell surface. There are two forms of this receptor, namely TfR-1, which is carried by almost all cells, with the highest density on erythroid precursors, on normal continuously dividing cells, and on neoplastic cells. Only mature red blood cells lack this receptor. The other form, TfR-2, has ~66% sequence similarity to TfR-1. It is constitutively expressed predominantly in liver and localized on the basolateral membrane surface of hepatocytes. However, it is important to note that TfR-2 cannot compensate for TfR-1 for iron metabolism [2]. The Fe^{3+}–transferrin–TfR complex is internalized in clathrin-coated pits that form endocytic vesicles. Once inside the cells, the complex in the endosome is acidified by a vacuolar H^+ adenosine triphosphatase (ATPase) that lowers the luminal pH to about 5.5. This acidification process induces conformational changes in transferrin–TfR complex with consequent release of Fe^{3+}. The transport of released iron across the endosomal membrane into the cytosol requires the activity of a ferrireductase, and the resultant divalent metal is translocated to the cytosol by the action of DMT-1, which is then utilized or stored in ferritin. Apotransferrin (transferrin without iron) still bound to the transferrin receptor is recycled back to the cell surface, where it is released into the serum.

One imperative regulator of intracellular iron levels is ferritin. The ferritins are a family of proteins characterized by highly conserved three-dimensional structures similar to spherical shells, designed to sequester and store large amounts of iron in a safe, soluble, and bioavailable form. It is comprised of 24 subunits of two types (H [heavy] and L [light] chain), whose proportion depends on the iron status of the cell, the tissue, and the organ [12]. The two ferritin polypeptides are related, but H-ferritin carries a ferroxidase activity to oxidize Fe^{2+} to Fe^{3+}, allowing incorporation of iron into the shell. Ferritin shells can store up to 4,500 iron atoms. Ferritin acts as a depot, sequestering excess iron and allowing for the mobilization of iron when needed [13]. The expression of both ferritin and TfR is delicately regulated by a labile iron pool (LIP) at the posttranscriptional level [14]. The observation of early embryonic lethality of H-ferritin knockout mice indicates its indispensable nature for life forms [15].

There is substantial evidence that suggests the mechanisms by which expression of ferritins and TfR are regulated are via iron regulatory proteins (IRPs) [16]. IRP-1 and IRP-2 are cytoplasmic proteins that bind with high affinity and specificity to iron responsive elements (IREs), which are conserved hairpin structures found in the untranslated regions of mRNA. They control expression of several important proteins in iron uptake, storage, utilization, and export, for instance, TfR, DMT-1, ferritin H- and L-chains, ferroportin, erythroid 5-aminolevulinic acid synthase (the first enzyme of heme biosynthesis), and several others. Depending on the position of IRE (5′ or 3′ untranslated region), IRP binding has an opposite effect on the synthesis of iron-regulated proteins. Binding of IRPs to 3′ IREs stabilizes mRNA, resulting in increased protein synthesis, whereas binding to 5′ IREs prevents the translation of mRNA, resulting in decreased protein synthesis [16]. Once the cell requires iron, IRP binds to IRE and stabilizes the mRNA for TfR while also decreasing translation of mRNA for ferritin. This leads to increased expression of TfR and decreased ferritin, which eventually translates to increased iron uptake and availability within the cell. On the contrary, high intracellular iron levels decrease IRE-binding activity, leading to efficient translation of ferritin and decreased stability of TfR mRNA, ultimately enhancing iron sequestration over uptake. Regulation of DMT-1 and ferroportin is similar to that of TfR and ferritin, respectively.

Iron–sulfur cluster (3Fe–4S↔4Fe–4S) is assembled in IRP-1, which acts as an iron sensor [17]. The ratio of (3Fe–4S) and (4Fe–4S) clusters depends upon the intracellular iron level. A high intracellular iron level shifts the ratio toward the (4Fe–4S) cluster. IRP-1 containing (4Fe–4S) cluster cannot bind mRNA but assumes an enzymatic function as a cytoplasmic aconitase similar to that of mitochondrial aconitase, the enzyme that reversibly isomerizes citrate to isocitrate through a *cis*-aconitate intermediate in the citric acid cycle [18]. At low intracellular iron levels the ratio shifts toward the (3Fe–4S) cluster. IRP-1 containing (3Fe–4S) cluster binds mRNA but does not exhibit an aconitase activity. IRP-2 does not contain an iron sulfur cluster and is rapidly ubiquitinated and degraded in the presence of excess iron [19]. Apart from iron there are several candidate signals of iron repletion for IRP-2, leading to its proteosomal degradation; these include heme, nitric oxide, as well as cytosolic iron bound to cysteines near the N-terminus [20]. It has been suggested that all of the identified IRP-2 regulatory mechanisms may be relevant, but to different degrees in different cells or situations [20, 21].

2 Hepcidin, Ferroportin, and Systemic Iron Homeostasis

Systemic iron homeostasis requires tightly regulated processes to provide iron as needed yet avoid the toxicity associated with excess iron. This is achieved by regulating the iron flow into the plasma, which includes release of iron from macrophages recycling senescent red blood cells, release of stored iron from hepatocytes, and absorption of dietary iron by duodenal enterocytes, and during fetal development, transfer of iron from mother to fetus across the placenta. Evidence indicates that iron is released from all these tissues into plasma through the membrane iron exporter ferroportin, which is posttranslationally regulated by hepcidin [22]. This means that functional hepcidin (a 25-amino-acid protein), which is primarily produced by hepatocytes and excreted by the kidneys, is the fundamental regulator of systemic iron metabolism. Hepcidin is derived from an 84-amino-acid precursor, has eight cysteine residues, is similar to the antimicrobial defensin proteins, and has measurable antimicrobial activity [23]. Although it is mainly produced by hepatocytes, other tissues such as hematopoietic cells, heart, and pancreas can also produce hepcidin [24]. Hepcidin acts by directly binding to ferroportin, causing ferroportin to be internalized and degraded in lysosomes [25, 26]. The loss of ferroportin from cell membrane consequently ablates cellular iron export [27]. Evidence suggests that increased hepcidin levels can lead to functional iron deficiency and subsequently anemia by interrupting intestinal iron absorption and inducing the retention of iron within recycling reticuloendothelial macrophages [28]. On the other hand, it was observed that targeted disruption of the hepcidin gene (Hamp1$^{-/-}$ mice) causes severe iron overload [29]. Synthesis of hepcidin is homeostatically increased by iron loading and decreased by anemia and hypoxia. Hepcidin is also elevated during infections and inflammation, causing a decrease in serum iron levels and contributing to the development of anemia or inflammation, probably as a host defense mechanism to limit the availability of iron to invading microorganisms [30, 31], thus completing the homeostatic loop controlling systemic iron levels. Renal injury that commonly leads to chronic kidney disease (CKD) causes production of several inflammatory cytokines, in particular interleukin (IL)-6 and IL-8. Such inflammatory conditions lead to elevated levels of hepatic production of hepcidin, which in turn restricts the iron absorption and release from macrophages. These mechanisms and others (discussed in more detail below) predispose patients with CKD to require intravenous (IV) iron administration to correct their decreased iron bioavailability and subsequent anemia. In fact combined measurements of serum and urine hepcidin have proven to correlate well with the level of anemia, and for these reasons hepcidin levels can become not only an essential tool in the diagnosis of inflammatory anemia but also might become a target of therapy, since lowering hepcidin may aid in improving the gastrointestinal uptake of iron and its release from macrophages, thus limiting the need for IV iron.

3 Iron: Friend and Foe

Excess iron can saturate the threshold for its sequestration, increasing the free labile form and ultimately posing a threat for the living organism. Such injurious effects have been confirmed, for example, by showing that the amount of hydrogen peroxide (H_2O_2) required to kill *Staphylococcus aureus* decreases 1,000-fold if the bacteria are raised in iron-rich media [32], and on the contrary, depletion of cellular iron protects both prokaryotic and eukaryotic cells against oxidative stress [33]. In fact the very same chemical properties of iron that make it indispensable for life are also responsible for the hazardous effects of iron. This property of being able to accept and donate electrons readily can lead to the formation of reactive oxygen species (ROS) [34]. These ROS can be free radicals, which means they have an unpaired electron in their outer shell (e.g., $O_2^{\bullet-}$ [superoxide radical] and $OH^{\bullet}$ [hydroxyl radical]) or nonradicals (e.g., H_2O_2); they can be anions (e.g., O_2^{-} [superoxide] and $ONOO^{-}$ [peroxynitrite]) or nonions (e.g., H_2O_2, $NO^{\bullet}$ or $OH^{\bullet}$). Among them $OH\bullet$ is the most reactive oxygen radical known, reacting instantly with molecules in its immediate vicinity, which explains its great destructive power. $O_2^{\bullet-}$ is less reactive than $OH^{\bullet}$ and does not readily react with most biological molecules. Since $O_2^{\bullet-}$ does not easily cross membranes, a prompt reaction with another radical requires that the particular radical for such a reaction is formed in the same subcellular compartment or travels to this compartment. In quantitative terms, the mitochondrial respiratory chain is the most important site of $O_2^{\bullet-}$ generation [35]. H_2O_2 is produced continuously in all cells through the enzymatic action of superoxide dismutase. It diffuses within and in between cells. Since it is only a weak oxidizing or reducing agent, it can play a role in signal transduction, but it can also be cytotoxic at micromolar concentrations.

Anytime iron exceeds the metabolic needs of the cell it may form a low molecular weight pool, otherwise referred to as the LIP (labile iron pool). This unbound and free iron can catalyze production of the highly toxic $OH^{\bullet}$ via entering the Fenton chemistry converting normal by-products of cell respiration, like $O_2^{\bullet-}$ and H_2O_2, into highly damaging $OH^{\bullet}$ or equally aggressive ferryl ions or even oxygen-bridged Fe^{2+}/Fe^{3+} complexes where iron itself is a reactant rather than a catalyst. These free radicals are highly reactive species and may induce oxidation of proteins, lipids and lipoproteins, nucleic acids, carbohydrates, and other cellular components [36, 37]. Oxidative damage to the vital cellular components might have in turn a deleterious effect at cellular and tissue levels, leading to cell death, tissue necrosis, and degenerative diseases or cell phenotype changes and even malignant transformation.

Protection from tissue-damaging effects that results from the reactions of iron with oxygen is accomplished in large part through vigilant iron sequestration; but whenever there is free iron it can participate in the Fenton's reaction, catalyzing the formation of the $OH^{\bullet}$:

$$O_2^{\bullet-} + Fe^{3+} \rightarrow O_2 + Fe^{2+}$$

$$Fe^{2+} + H_2O_2 \rightarrow Fe^{3+} + OH^{\bullet} + HO^{-} \quad \text{(Fenton reaction)}$$

One important target of free radicals is the polyunsaturated fatty acids (PUFA). The $OH^{\bullet}$ can abstract a hydrogen atom from PUFA (LH) to initiate lipid peroxidation.

$$OH^{\bullet} + LH \rightarrow H_2O + L^{\bullet}$$

$$L^{\bullet} + O_2 \rightarrow LOO^{\bullet}$$

$$LOO^{\bullet} + LH \rightarrow LOOH + L^{\bullet}$$

Once lipid hydroperoxides (LOOH) accumulate, free iron may directly initiate additional lipid peroxidation.

$$Fe^{2+} + LOOH \rightarrow Fe^{3+} + LO^{\bullet} + OH^{-}$$

The produced alkoxyl radical $LO^{\bullet}$ can undergo fragmentation to produce lipid-aldehyde or can react with O_2 to produce epoxy-allylic peroxyl radicals ($OLOO^{\bullet}$), which can further propagate the reaction. The accumulation of such lipid hydroperoxides can damage the integrity and function of the cell membrane [37]. Such events have been shown in experimental models of iron overload in vivo to greatly increase PUFA oxidation of hepatic mitochondria, as well as cause lysosomal fragility [38]. In another in vitro study involving iron loading of primary cultures of newborn rat cardiomyocytes, similar iron-mediated PUFA oxidation and mitochondrial dysfunction were observed [39]. More importantly, iron overload in experimental animals results in oxidative damage to lipids [40], whereas chelation of iron by pyridoxal isonicotinoyl hydrazone prevents the Fe-citrate-dependent formation of lipid peroxides [41]. One critical feature required for iron-mediated damage to cells is intrusion of the metal into lipid compartments. Cells exposed to highly lipophilic heme (iron protoporphyrin IX) or 8-hydroxyquinoline-iron chelate accumulate catalytically active iron and are exquisitely sensitive to both endogenous and exogenous oxidant stresses [42, 43]. Another possible target of iron-induced oxidative damage is low density lipoprotein (LDL). Heme iron is taken up by LDL particles and promotes the oxidative modification of lipid moieties as well as the apoprotein. This is of enormous importance in the context of endothelial dysfunction, CKD, and atherosclerosis and will be discussed in more detail below.

There is mounting evidence that proteins might be early targets of ROS, and that the altered proteins can in turn damage other biomolecules. Although the role of LIP in induction of protein peroxides in vivo is not so unequivocal (which may at least in part be due to the inherent in vivo antioxidant mechanisms), there are several lines of in vitro evidence that underscore the essential role of iron in ROS-mediated protein damage and modification. Formation of protein hydroperoxides prior to the formation of lipid peroxides has been shown in U937 cells exposed to peroxyl radicals [44], and other reports also indicate that the Fenton chemistry-mediated formation of protein hydroperoxides clearly takes place in vitro [45]. Mechanism of iron-catalyzed protein damage includes oxidative scission, bityrosine crosslinks, the introduction of carbonyl groups, loss of histidine residues, and the formation of protein-centered alkyl, alkylperoxyl, and alkoxyl radicals [46]. With a few exceptions, protein damage is likely to be a reparable and nonlethal event for a cell. However, there is convincing evidence that two mitochondrial proteins – aconitase and adenine nucleotide translocase – may be important targets of long-term oxidative damage [47].

Because mutagenesis caused by oxidative damage of DNA has been widely recognized as one of the primary steps in carcinogenesis, the correlation between iron, ROS, and cancer has received great attention in the past few decades. In the absence of transition metals such as iron and copper, DNA is quite unreactive with oxidants such as H_2O_2. However, in the presence of added iron, DNA scission occurs [48], preferentially in internucleosomal linker regions [49], which in turn produces "ladders" resembling those typical of apoptosis [50]. The products of iron-mediated DNA damage are not fully characterized but include strand breaks, oxidatively modified bases, DNA–protein crosslinks [51], covalent reactions with lipid peroxidation products [52], and other structurally uncharacterized bulky DNA adducts [53]. After the demonstration of the existence of a LIP pool in the cell nucleus [54], it became clear that the nuclear redox-active iron may indeed be involved in DNA damage induction by hydrogen peroxide and other oxidizing compounds.

In support of this hypothesis significant correlation between cellular LIP level and the yield of 8-oxo-7,8-dihydro-2′-deoxyguanosine, a typical marker of ROS-induced DNA damage, was found in human lymphocytes [55]. Strong correlation has also been found between the LIP level and DNA breaks estimated by the comet assay in cells treated with H_2O_2 [56]. Conversely, chelation of intracellular iron with deferoxamine prevents the formation of such DNA breaks and cytotoxicity induced by H_2O_2 to both the nuclear [57] and mitochondrial [58] DNA.

Another important concept for free radical-induced biological damage is the site-specificity for inciting injury. Fe^{3+} ions that are bound to biological molecules such as DNA and proteins can undergo cyclic reduction and reoxidation. This concept is distinguished from "free" iron in that bound iron is not diffusible, but it explains the funneling of free radical damage to specific sites and the possible "multihit" effect on the molecule, which can have deleterious effects on DNA. In support, ferritin sequestering iron and exhibiting ferroxidase activity was demonstrated to prevent bleomycin-provoked cytotoxicity and cellular DNA strand scission in cancer cell lines [59, 60]. Moreover, it was recently observed that H-ferritin (traditionally known as a cytoplasmic protein) is found in the nucleus of corneal epithelial cells, whereby sequestering iron and converting it to a more stable form (Fe^{3+}) protects the DNA against UV light-induced oxidative damage [61]. On the other hand, another tenable hypothesis concerning the etiology of cell and organ damage arising from iron overload is that excess iron selectively targets mitochondria and, perhaps, the mitochondrial genome. The mitochondrial electron transport chain "leaks" 1–2% of its electrons into O_2^-/H_2O_2. This leak of electrons is responsible for ~90% of the activated oxygen generated by most cell types. Therefore, in iron overload states, it is likely that mitochondria and in particular its DNA (mtDNA) would be most extensively damaged. There may be several reasons for the apparent fragility of mtDNA, which include: (a) mitochrondria is a major source of ROS; (b) mitochondria are intrinsically rich in iron (since DNA is unreactive with H_2O_2 in the absence of iron or another transition metal, the preferential damage to mtDNA probably reflects the presence of larger amounts of reactive iron within even the normal mitochondrion); (c) mtDNA is deficient in histones, which normally provide partial protection against oxidant damage; and (d) repair of damage to mtDNA is less effective and requires more time.

4 Iron Overload and Pathological Conditions

Iron overload states present a clear revelation of hazardous effects of excess iron. Excessive body iron can have primary and secondary causes. Primary iron overload results from an inherent defect in iron regulation and continuous overabsorption of iron from the gastrointestinal tract. Hereditary hemochromatosis is the most common cause of primary iron overload [62], and due to deposition of excess iron in the skin and pancreas it can damage the insulin producing β-cell sometimes referred to as "bronze diabetes." Secondary iron overload results from excessive intake of iron from multiple blood transfusions, parenteral iron administration, dietary intake, or iron supplements. Secondary iron overload can also occur in patients who have chronic disorders of erythropoiesis or hemolytic anemias (such as sideroblastic and sickle-cell anemias and β-thalassemia) [63]. Iron accumulates in the parenchyma of various organs, particularly in the liver, pancreas, heart, gonads, and skeletal system, and eventually causes end-organ damage. We will herein briefly discuss some of these clinical conditions and provide evidence of the central role that iron may play in the pathogenesis of such conditions.

4.1 Cardiovascular System

Iron-derived ROS are implicated in the pathogenesis of numerous vascular disorders. Under physiological conditions, circulating free iron is almost nonexistent, but in contrast one abundant source of redox-active iron is heme [64]. Of all sites in the body, the vasculature – and in particular the endothelial lining – may be at greatest risk of exposure to free heme. This is because of high concentration of heme in the form of hemoglobin in erythrocytes, which are vulnerable to unexpected lysis. The extracellular hemoglobin is easily oxidized to methemoglobin (Fe^{3+}) [65], which is the form of hemoglobin that can readily release its heme [66]. Heme itself is potentially toxic to the endothelial cells, and this toxicity is enhanced by the hydrophobic nature of heme, which allows it to cross and enter cell membranes. Heme uptake by endothelial cells can exacerbate their damage by activated polymorphonuclear leukocytes (PMN). Once within the cell, heme can promote oxidative damage either directly or, more importantly, through the liberation of iron, which may initially lodge within the hydrophobic interstices of the phospholipid bilayer [43]. Within this highly oxidizable matrix, iron acts as an especially active catalyst of oxidation of cell membrane constituents. Under physiological conditions the heme iron in the hydrophobic pocket of the hemoglobin is in a ferrous state and quite stable. Nevertheless, the ferrous heme of normal functional hemoglobin continuously undergoes autoxidation, producing Fe^{3+} hemoglobin and a superoxide anion radical:

$$\mathrm{Hb(Fe^{2+})O_2} \rightarrow \mathrm{Hb(Fe^{3+})} + \mathrm{O_2}^{\bullet -}$$

This reaction can be amplified by the presence of ROS and produces the very unstable ferrihemoglobin, which readily releases its heme moieties. One important source of such ROS are activated PMN. For instance when activated with phorbol ester, PMN rapidly oxidize ferrohemoglobin to ferrihemoglobin [65]. Accordingly, ferrohemoglobin oxidized to ferrihemoglobin by activated PMN can shower the endothelium with heme, which greatly enhances cellular susceptibility to oxidant-mediated cell injury [65]. Another candidate for generating ferrihemoglobin and subsequent release of heme is nitric oxide (NO). Reaction of NO with free hemoglobin produces ferrihemoglobin and leads to decreased NO bioavailability, which in turn may cause damage to the vasculature, pulmonary hypertension, and end-organ injury [67]. Facing such potential deleterious outcomes, endothelial cells instantly upregulate their antioxidant defense mechanisms, namely HO-1/ferritin system after exposure to heme. In fact different products of heme catabolism (i.e., carbon monoxide, biliverdin, and bilirubin) have been shown to be essential in maintaining vascular homeostasis [68], whereas ferritin is a crucial antioxidant that protects endothelium from oxidant-mediated injury and does so in a dose-responsive manner [69]. Such defense mechanisms are extremely fundamental since endothelial injury and functional derangements are considered the primary events in the development of atherosclerosis. Furthermore, heme can also threaten vascular endothelial cell integrity indirectly by its ability to mediate the oxidative modification of LDL [70].

Oxidized LDL (and not the native form) contributes to atherogenesis in several ways, including being readily ingested by macrophages through the scavenger receptor that is distinct from the LDL receptor, being chemotactic for the circulating monocytes and increasing monocyte adhesion, inhibiting the motility of macrophages already present in the lesion, stimulating release of cytokines and growth factors, and being cytotoxic to endothelial and smooth muscle cells [71]. It is well known that in vitro, transition metals such as iron and copper will facilitate LDL oxidation. Whereas these metals are unlikely to exist in free form in normal body fluids, heme in the form of hemoglobin is abundant within the vasculature. The process of heme-mediated LDL oxidation involves coupled interactions between LDL, heme, oxidants, and antioxidants [64]. The initial step in these complex reactions is the spontaneous insertion of heme into LDL particles. The inserted heme directly promotes extensive oxidative modification of LDL; such modification can be amplified by trace amounts of H_2O_2, PMN-derived oxidants, or preformed lipid hydroperoxides within the LDL. Heme can oxidatively modify both the lipid moiety of LDL and the apoprotein. During these oxidative reactions between heme, LDL, and peroxides, the heme ring (protoporphyrin IX) is degraded, with resultant release of free iron. The subsequent release of free iron results in iron catalysis of oxidation of further heme, fatty acids, cholesterol, and apolipoprotein B-100 in LDL particles [70]. Free iron is also a significant risk factor in ischemia/reperfusion. Although ischemia ultimately evokes cell death, a sudden reconstitution of blood flow also induces severe cell and tissue injury. This is mainly due to high levels of ROS that are generated during reperfusion, and the importance of iron and Fenton reaction catalysts was confirmed by using systemic iron chelators that significantly reduce the level of tissue necrosis after reperfusion [72].

Another important aspect of iron-induced cardiovascular damage is restrictive (infiltrative) cardiomyopathy. This is an irreversible and devastating condition in which accumulation of iron can cause the walls of the ventricles to become stiff and noncompliant, but not necessarily thickened, and resist normal filling with blood during diastole. In fact patients with *HFE* gene mutations (which eventually leads to systemic iron overload) when compared with the age- and sex-matched population without *HFE* mutations, have significantly higher risk of cardiomyopathy [73].

4.2 Iron and Chronic Kidney Disease

Iron-restricted erythropoiesis is a common clinical condition in patients with CKD. There are several factors that contribute to the pathogenesis of CKD-induced anemia which include: (a) decreased levels of erythropoietin production; (b) blood loss during hemodialysis, from gastrointestinal tract and cannula puncture sites following hemodialysis, and too frequent diagnostic blood tests (hemodialysis patients lose an average of 2 g of iron per year) [74]; and (c) inadequate intestinal iron absorption and inhibition of iron release from macrophages, commonly seen in chronic inflammatory diseases such as CKD, wherein cytokines released by activated leukocytes and other cells exert multiple effects [75]. These contribute to the increased hepatic synthesis of hepcidin, which in turn binds to ferroportin and prevents egress of iron from reticuloendothelial macrophages and from intestinal epithelial cells, leading to "functional iron deficiency." Based on these facts both erythropoietin and iron replacement are essential to treat anemia in CKD patients. However, since oral iron therapy is often not sufficient for CKD patients, parenteral administration of iron is necessary to optimally care for these patients (Fig. 2).

Intravenous iron can be given to CKD patients as long as the therapy is performed according to international recommendations and guidelines. There are

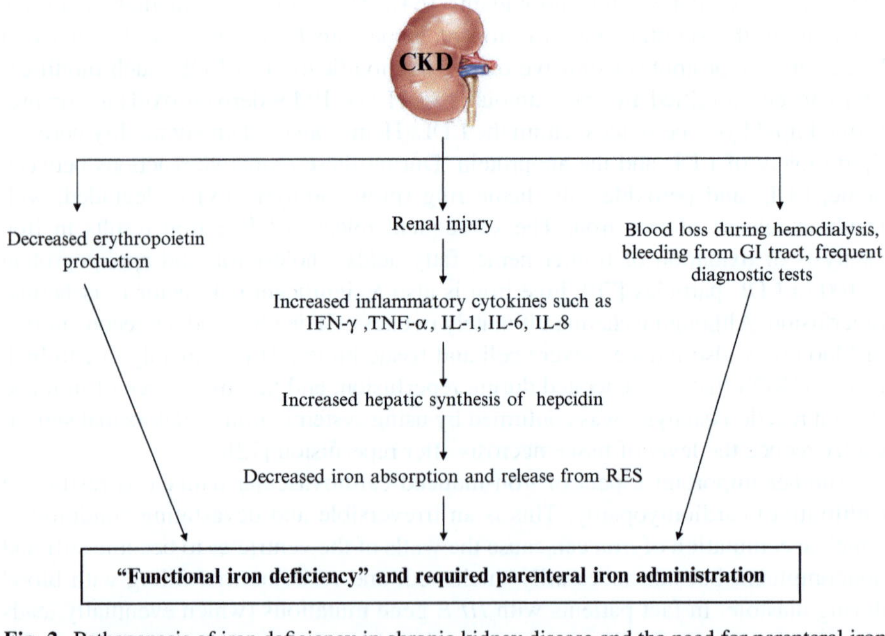

Fig. 2 Pathogenesis of iron deficiency in chronic kidney disease and the need for parenteral iron RES; reticulo-endothelial system, GI; gastrointestinal tract

three different forms of iron that have been widely used worldwide. These include Venofer (iron sucrose), Ferrlecit (sodium ferric gluconate), and InFeD (iron dextran). Several studies have investigated the safety issues with IV iron administration. Potential risk factors associated with IV iron therapy can be divided into acute, which includes allergic reactions such as rash, dyspnea, wheezing, or even anaphylaxis, as well as long-term complications, which are mainly due to the generation of ROS, initiation and propagation of lipid peroxidation, nephrotoxicity, endothelial dysfunction, propagation of vascular smooth muscle cell proliferation, and inhibition of cellular host defense [76]. Iron dextran has been implicated in the development of type I hypersensitivity reactions, and the dextran moiety is believed to be the underlying cause. On the contrary, iron sucrose carries the lowest risk for hypersensitivity reactions [77]. Acute iron administration, whether IV [78] or oral [79], evidently causes oxidative stress. An exacerbation of oxidative stress occurs after IV infusion of iron sucrose in dialysis patients, as demonstrated by an increase in plasma concentrations of malondialdehyde, which is a marker of lipid peroxidation [80]. In addition, Ferric gluconate modifies β2-microglobulin as a marker for oxidative stress [81].

In a randomized controlled trial it was concluded that IV iron produces oxidative stress that can cause renal tubular damage and is associated with transient proteinuria [82]. In another comparative study of different iron formulations, it was demonstrated that iron sucrose is a highly potent pro-oxidant capable of inducing tubular and endothelial cell death with lesser toxicity by ferric gluconate and the lowest toxicity by iron dextran [83]. Another study found similar results both in vivo and in vitro showing that iron sucrose is more toxic than iron dextran [84]. Such safety concerns with the use of IV iron have led to the development of a new form of iron known as ferumoxytol that was recently FDA approved (July 2009). Ferumoxytol (Feraheme) is a superparamagnetic iron oxide nanoparticle with a polyglucose sorbitol carboxymethylether coating [85]. It is isotonic, and preliminary data suggest that it contains less free iron than other IV iron preparations [85]. These physicochemical properties may explain why ferumoxytol can be given rapidly at relatively high dosages [86]. In addition, ferumoxytol appears quickly in circulating red blood cells, suggesting ready bioavailability for erythropoiesis. Ferumoxytol, administered as an IV in two doses of 510 mg within 5 ± 3 days, was well tolerated, with no reported adverse events of hypotension or hypersensitivity. In fact, the proportion of patients with related adverse events was lower in the ferumoxytol group (10.6%) compared with oral iron (24.0%). There was no increase in the incidence of adverse events among patients who received a second course of IV ferumoxytol. In contrast, other IV preparations administered rapidly or at higher dosages (e.g., >200 mg/dose) have been associated with a higher rate of adverse events [87, 88]. These properties and the fact that less free iron is available to induce oxidative damage with the use of ferumoxytol may unravel a promising new regimen for the treatment of anemia in CKD and may lead to the replacement of other currently used iron regimens with ferumoxytol in the near future.

4.3 Iron and the Nervous System

Iron accumulation in the brain is commonly associated with several neurodegenerative disorders and also plays a role in cellular damage following hemorrhagic stroke and traumatic brain injury. In vivo studies demonstrate induction of lipid peroxidation and free radical formation after intracerebral hemorrhage [89]. Considering the substantial amount of iron released during a hemorrhagic event, it is likely that many of the homeostatic mechanisms responsible for hemoglobin and iron metabolism and binding become saturated; it is the free and excess iron that is recognized as the major cause of such toxicity [90]. In support of these findings it was found that injection of 1 mM ferric iron into rat cerebral cortex induces a large amount of neuronal loss at the site of the injection [91]. In addition, iron chelation treatment with deferoxamine was shown to be effective in attenuating the neuronal damage following a hemorrhagic stroke [92].

Neurodegenerative disorders include a variety of pathological conditions, which share similar critical metabolic processes such as protein aggregation and oxidative stress. There are several forms of such disorders but in this section we will mainly discuss the two more common forms of neurodegenerative disorders: Alzheimer's disease (AD) and Parkinson's disease (PD). Although it is still unclear whether iron accumulation is a primary cause or secondary event in the former group, there is no doubt that iron-induced oxidative stress contributes to neurodegeneration. It must be noted that one important characteristic is the fact that brain iron metabolism is evidently separated from systemic iron metabolism by the blood–brain barrier (BBB), accounting for only little iron accumulation in diseases with systemic iron overload [93]. Transport of iron across the BBB is complex due to the restricted diffusion of transferrin from blood to brain. One study demonstrated that iron is transported across the BBB at a faster rate than transferrin [94], supporting the idea that nontransferrin-bound iron enters the brain and that much of the serum transferrin is effluxed or transcytosed out of the brain [95]. To allow for immobilization of this potentially free iron species, oligodendrocytes and choroid plexus cells produce transferrin within the brain [96].

Approximately 50% of total brain iron is stored within glial cells [97] and is bound to ferritin in its soluble form [98]. However, iron accumulation in the brain may induce neuronal damage even after it is bound to ferritin because iron can be released in its ferrous form under the acidic conditions present in extracellular fluid [91] and through interaction with components such as excess superoxide radicals [99], ascorbate [100], and intracerebral hemorrhage accompanied with decreased pH [101]. Increased iron levels in the substantia nigra have been described in numerous postmortem studies of PD patients [102], and studies based on different magnetic resonance imaging methods [103] and on transcranial ultrasound [104] have verified this observation.

Evidence to support the role of iron-caused oxidative stress in neurodegeneration is overwhelming. For instance, as discussed earlier, the homozygous H-ferritin mutant is lethal to mice, while heterozygotes have less than half the levels of

H-ferritin compared to wild-type mice, and an increase in transferrin, transferrin receptor, L-ferritin, DMT-1, and ceruloplasmin. This is similar to that found in the brains of patients with PD and AD. The neurons in these mice had decreased superoxide dismutase activity, absence of H-ferritin and L-ferritin staining in the neurons, and increased caspase-3 and Bax, indicating that the loss of these iron homeostatic mechanisms contributes to neurodegeneration. This model may serve as the animal model for oxidative stress relevant to disorders such as PD [105]. In PD, accumulation of high iron concentrations directly correlates with dopaminergic cell loss and disease progression. Injection of iron into the substantia nigra mimics increased levels of the LIP and induces a disruption in dopamine metabolism exacerbated by dopamine autoxidation [106] and significant lipid peroxidation, followed by neurodegeneration [107, 108]. Overall, high iron accumulation that results in oxidative stress and decreased glutathione levels within nigral neuromelanin have been suggested to directly contribute to dopaminergic neuronal toxicity, leading to PD [109, 110].

Genetic, biochemical, and immunological evidence supports a mechanistic role for amyloid-β peptide in the pathophysiology of AD. Furthermore, one of the earliest events in AD is the generation of oxidative stress, which may be related to the generation of free radicals by the excess iron that is observed in the disease. In fact brains from patients with AD show accumulation of iron in senile plaques (mainly composed of amyloid-β) and an altered distribution of iron transport and storage proteins. In addition, evidence suggests a central role for amyloid-β in AD, but also for the ability of iron to alter its properties. AD brains show accumulation of iron within senile plaques (~1 mM) [111] and neurofibrillary tangles [112], concurrent with decreased levels of transferrin receptor expression [113]. Therefore, these brains are subject to high levels of oxidative stress [114], as indicated by significantly elevated activities measured for nicotinamide adenine dinucleotide phosphate (NADPH) oxidase and HO-1 [115].

In an intriguing investigation it was demonstrated that amyloid-β is only toxic to cultured cells when iron is present within the culture media, indicating that the iron influences the toxicity of amyloid-β and indeed may produce an artifactual cytotoxicity that is related to the iron rather than the amyloid-β itself [116]. This finding received support from a subsequent study where it was shown that pretreatment of amyloid-β with deferoxamine significantly reduced the neuronal loss after amyloid-β was added to cultured neurons, showing that synthetic amyloid-β preparations also appear to be contaminated with iron [94]. Under these conditions, amyloid-β toxicity may be caused by the promotion of aggregation by iron; however, there may also be a direct production of free radicals by the iron. For instance, the binding of iron to amyloid-β causes the generation of H_2O_2 [117]. It has also been shown that when iron and amyloid-β are mixed and then added to brain homogenates, ROS are produced [118]. In addition, recent studies have demonstrated that the previously reported spontaneous production of radicals by amyloid-β may be attributed to low-level contamination with iron [119]. In spite of the fact that brain iron content and accumulation often do not mimic systemic iron levels, it has nonetheless been found that patients with hemochromatosis develop AD at an earlier age than nonhemochromatosis patients. These findings lend support to the concept that general iron accumulation may accelerate the AD process [115].

5 Iron: A Potential Risk Factor for Diabetes

The pancreatic β-cell is our one and only source of insulin production, but at the same time the β-cell is extremely susceptible to oxidative stress. This is because it uses glucose catabolism to meet its energy needs, which also provides the energy required for insulin biosynthesis and exocytosis. Simultaneously, the β-cell generates the signal for glucose-induced insulin secretion. The integration of these two tasks requires meticulous harmony in which the needs for adequate energy supply of the cell do not impede the signal-generating function of glucose metabolism for insulin secretion. There is strong evidence that a decrease in the ROS-inactivating capacity in the β-cells, owing to genetic variations, even in the absence of autoimmune background, results in defective insulin secretion and deterioration in glucose tolerance [120, 121].

There are two major findings that underscore the role of iron in the pathogenesis of diabetes: (a) an increased incidence of type 2 diabetes in diverse causes of iron overload and (b) reversal or improvement in diabetes (glycemic control) with a reduction in iron load achieved using either phlebotomy or iron chelation therapy. Moreover, frequent blood donation and decreased iron stores are associated with the improvement in insulin sensitivity and insulin secretion [122]. On the contrary, high meat intake (due to its high heme content) is reported to be associated with diabetes [123]. In addition, in a mouse model of hemochromatosis, iron excess and oxidative stress mediate apoptosis of pancreatic islets with a resultant decrease in insulin secretory capacity [124]. Also, β-cells have a relatively high expression of DMT, which additionally predisposes them for more accumulation of iron than other cells [125] and potentiates the danger from iron catalyzed oxidative stress. As discussed earlier, transfusional iron overload is the most common cause of acquired iron overload and is typically seen in transfusion-dependent chronic hemolytic anemia such as β-thalassemia. Higher insulin resistance has been confirmed in these patients [126] as well as other causes of high body iron stores [127].

6 Iron and Carcinogenesis

Increased formation of ROS in the cells can contribute to carcinogenesis either directly through genotoxic effects or indirectly via modification of signaling pathways that lead to altered expression of numerous genes. There is strong experimental evidence to support the involvement of oxidative stress in carcinogenesis [128, 129]. In particular, ROS-induced modulation of cell signaling pathways can activate transcriptional factors, such as HIF-1, AP-1, and nuclear factor κB (NFκB) among others [130, 131]. Early studies showed that the clinically used iron chelator, deferoxamine, had some activity at inhibiting the growth of neuroblastoma and leukemia in cell culture and clinical trials [132, 133]. Other studies have investigated the mechanism by which iron depletion leads to G1/S stage arrest and apoptosis. These studies reveal that a multitude of cell cycle control molecules are regulated by iron [134, 135]. Compared to normal cells, neoplastic cells require greater amounts of iron because generally they proliferate at a greater rate than their normal counterparts. This is reflected by the higher expression of TfR [136] and the higher rate of iron uptake from transferrin in cancer cells [137]. Furthermore, neoplastic cells express high levels of ribonucleotide reductase, making them more susceptible to the action of iron chelators than normal cells [138]. Although epidemiologic studies regarding iron and human cancer remain largely inconclusive until now, carcinogenicity of iron compounds in animal models has been clearly demonstrated. Studies reveal that mice exposed to iron oxide dust develop pulmonary tumors [139]. Other reports include soft tissue sarcoma induction by injection of iron dextran [140], renal cell carcinoma models produced by intraperitoneal injection of iron chelates [141], and malignant mesothelioma by repeated intraperitoneal injection of ferric saccharate [142].

The role of available iron in cell transformation and apoptosis was also stressed by the finding that the protein encoded by c-Myc proto-oncogene suppressed the expression of H-ferritin and stimulated the expression of IRP-2, which in turn elevates the intracellular LIP [143]. Furthermore, this coordinated regulation of genes controlling the intracellular iron availability was shown to be required for c-Myc-induced cell transformation, indicating the essential role of iron in this process. It is worth noting that iron availability at the systemic level is also restricted in cancer patients, probably through the concerted action of the iron-regulating hormone hepcidin, which is overexpressed in these patients [144]. It is plausible that in this way nature takes advantage of the elevated iron requirements of cancer cells in order to slow down tumor growth. This consideration raises questions about the broadly used IV iron administration in cancer patients. Although several lines of evidence suggest a central role of iron in carcinogenesis and cancer progression, further investigation is needed in order to elucidate the exact role of the overall iron homeostasis and its relation to cancer.

7 Concluding Remarks

Iron is essential for life, but even trace amounts of free iron can catalyze production of a highly toxic hydroxyl radical via Fenton reaction. Here we provided brief insight into molecular mechanisms of iron-mediated oxidative damage and have discussed some of the common pathophysiological conditions related to iron overload. Although several important questions regarding iron metabolism and oxidative stress have been answered, this still remains a field of many unknowns and speculations that require further elucidation. This will help us better understand different aspects of iron homeostasis and provide new avenues for planning better strategies in combating conditions related to iron overload.

Acknowledgments The work was supported by NIH grants DK059600, DK075332, DK079337, and DK071875 (to AA).

References

1. Lieu PT, Heiskala M, Peterson PA, Yang Y. The roles of iron in health and disease. Mol Aspects Med 2001; 22:1–87.
2. Hentze MW, Muckenthaler MU, Andrews NC. Balancing acts: molecular control of mammalian iron metabolism. Cell 2004; 117:285–97.
3. McKie AT, Barrow D, Latunde-Dada GO, Rolfs A, Sager G, Mudaly E, Mudaly M, Richardson C, Barlow D, Bomford A, Peters TJ, Raja KB, Shirali S, Hediger MA, Farzaneh F, Simpson RJ. An iron-regulated ferric reductase associated with the absorption of dietary iron. Science 2001; 291:1755–9.
4. Shayeghi M, Latunde-Dada GO, Oakhill JS, Laftah AH, Takeuchi K, Halliday N, Khan Y, Warley A, McCann FE, Hider RC, Frazer DM, Anderson GJ, Vulpe CD, Simpson RJ, McKie AT. Identification of an intestinal heme transporter. Cell 2005; 122:789–801.
5. Maines MD. The heme oxygenase system: past, present, and future. Antioxid Redox Signal 2004; 6:797–801.
6. Donovan A, Lima CA, Pinkus JL, Pinkus GS, Zon LI, Robine S, Andrews NC. The iron exporter ferroportin/Slc40a1 is essential for iron homeostasis. Cell Metab 2005; 1:191–200.
7. Chen H, Attieh ZK, Su T, Syed BA, Gao H, Alaeddine RM, Fox TC, Usta J, Naylor CE, Evans RW, McKie AT, Anderson GJ, Vulpe CD. Hephaestin is a ferroxidase that maintains partial activity in sex-linked anemia mice. Blood 2004; 103:3933–9.
8. Vulpe CD, Kuo YM, Murphy TL, Cowley L, Askwith C, Libina N, Gitschier J, Anderson GJ. Hephaestin, a ceruloplasmin homologue implicated in intestinal iron transport, is defective in the Sla mouse. Nat Genet 1999; 21:195–9.
9. Aisen P, Leibman A, Zweier J. Stoichiometric and site characteristics of the binding of iron to human transferrin. J Biol Chem 1978; 253:1930–7.
10. Bali PK, Zak O, Aisen P. A new role for the transferring receptor in the release of iron from transferrin. Biochemistry 1991; 30:324–8.
11. Mareschal JC, Rama R, Crichton RR. The role of ceruloplasmin in Fe(III)-transferrin formation in vitro. FEBS Lett 1980; 110:268–70.
12. Harrison PM, Arosio P. The ferritins: molecular properties, iron storage function and cellular regulation. Biochim Biophys Acta 1996; 1275:161–203.
13. Theil EC. Ferritin: structure, gene regulation, and cellular function in animals, plants, and microorganisms. Annu Rev Biochem 1987; 56:289–315.
14. Casey JL, Hentze MW, Koeller DM, Caughman SW, Rouault TA, Klausner RD, Harford JB. Iron-responsive elements: regulatory RNA sequences that control mRNA levels and translation. Science 1988; 240:924–8.
15. Ferreira C, Bucchini D, Martin ME, Levi S, Arosio P, Grandchamp B, Beaumont C. Early embryonic lethality of H ferritin gene deletion in mice. J Biol Chem 2000; 275:3021–4.

16. Pantopoulos K. Iron metabolism and the IRE/IRP regulatory system: an update. Ann N Y Acad Sci 2004; 1012:1–13.
17. Rouault TA, Stout CD, Kaptain S, Harford JB, Klausner RD. Structural relationship between an iron-regulated RNA-binding protein (IRE-BP) and aconitase: functional implications. Cell 1991; 64:881–3.
18. Kaptain S, Downey WE, Tang C, Philpott C, Haile D, Orloff DG, Harford JB, Rouault TA, Klausner RD. A regulated RNA binding protein also possesses aconitase activity. Proc Natl Acad Sci U S A 1991; 88:10109–13.
19. Muckenthaler MU, Galy B, Hentze MW. Systemic iron homeostasis and the iron-responsive element/iron-regulatory protein (IRE/IRP) regulatory network. Annu Rev Nutr 2008; 28:197–213.
20. Wallander ML, Leibold EA, Eisenstein RS. Molecular control of vertebrate iron homeostasis by iron regulatory proteins. Biochim Biophys Acta 2006; 1763:668–89.
21. Rouault TA. The role of iron regulatory proteins in mammalian iron homeostasis and disease. Nat Chem Biol 2006; 2:406–14.
22. Oates PS. The role of hepcidin and ferroportin in iron absorption. Histol Histopathol 2007; 22:791–804.
23. Nemeth E, Ganz T. Regulation of iron metabolism by hepcidin. Annu Rev Nutr 2006; 26:323–42.
24. Ilyin G, Courselaud B, Troadec MB, Pigeon C, Alizadeh M, Leroyer P, Brissot P, Loréal O. Comparative analysis of mouse hepcidin 1 and 2 genes: evidence for different patterns of expression and co-inducibility during iron overload. FEBS Lett 2003; 542:22–6.
25. Delaby C, Pilard N, Gonçalves AS, Beaumont C, Canonne-Hergaux F. Presence of the iron exporter ferroportin at the plasma membrane of macrophages is enhanced by iron loading and down-regulated by hepcidin. Blood 2005; 106:3979–84.
26. Nemeth E, Tuttle MS, Powelson J, Vaughn MB, Donovan A, Ward DM, Ganz T, Kaplan J. Hepcidin regulates cellular iron efflux by binding to ferroportin and inducing its internalization. Science 2004; 306:2090–3.
27. Knutson MD, Oukka M, Koss LM, Aydemir F, Wessling-Resnick M. Iron release from macrophages after erythrophagocytosis is up-regulated by ferroportin 1 overexpression and down-regulated by hepcidin. Proc Natl Acad Sci USA 2005; 102:1324–8.
28. Nicolas G, Bennoun M, Porteu A, Mativet S, Beaumont C, Grandchamp B, Sirito M, Sawadogo M, Kahn A, Vaulont S. Severe iron deficiency anemia in transgenic mice expressing liver hepcidin. Proc Natl Acad Sci U S A 2002; 99:4596–601.
29. Viatte L, Lesbordes-Brion JC, Lou DQ, Bennoun M, Nicolas G, Kahn A, Canonne-Hergaux F, Vaulont S. Deregulation of proteins involved in iron metabolism in hepcidin-deficient mice. Blood 2005; 105:4861–4.
30. Nicolas G, Chauvet C, Viatte L, Danan JL, Bigard X, Devaux I, Beaumont C, Kahn A, Vaulont S. The gene encoding the iron regulatory peptide hepcidin is regulated by anemia, hypoxia, and inflammation. J Clin Invest 2002; 110:1037–44.
31. Weinstein DA, Roy CN, Fleming MD, Loda MF, Wolfsdorf JI, Andrews NC. Inappropriate expression of hepcidin is associated with iron refractory anemia: implications for the anemia of chronic disease. Blood 2002; 100:3776–81.
32. Repine JE, Fox RB, Berger EM. Hydrogen peroxide kills *Staphylococcus aureus* by reacting with staphylococcal iron to form hydroxyl radical. J Biol Chem 1981; 256:7094–6.
33. Gannon DE, Varani J, Phan SH, Ward JH, Kaplan J, Till GO, Simon RH, Ryan US, Ward PA. Source of iron in neutrophil-mediated killing of endothelial cells. Lab Invest 1987; 57:37–44.
34. Papanikolaou G, Pantopoulos K. Iron metabolism and toxicity. Toxicol Appl Pharmacol 2005; 202:199–211.
35. Turrens JF. Mitochondrial formation of reactive oxygen species. J Physiol 2003; 552:335–44.

36. Halliwell B, Gutteridge JMC. Oxygen toxicity, oxygen radicals, transition metals and disease. Biochem J 1984; 219:1–14.
37. Nelson SK, McCord JM. Iron, oxygen radicals, and disease. Adv Mol Cell Biol 1998; 25:157–83.
38. Britton RS, Ramm GA, Olynyk J, Singh R, O'Neill R, Bacon BR. Pathophysiology of iron toxicity. Adv Exp Med Biol 1994; 356:239–253.
39. Link G, Konijn AM, Hershko C. Cardioprotective effect of α-tocopherol, ascorbate, deferoxamine, and deferiprone: mitochondrial function in cultured, iron-loaded heart cells. J Lab Clin Med 1999; 133:179–188.
40. Sochaski MA, Bartfay WJ, Thorpe SR, Baynes JW, Bartfay E, Lehotay DC, Liu PP. Lipid peroxidation and protein modification in a mouse model of chronic iron overload. Metabolism 2002; 51:645–51.
41. Santos NC, Castilho RF, Meinicke AR, Hermes-Lima M. The iron chelator pyridoxal isonicotinoyl hydrazone inhibits mitochondrial lipid peroxidation induced by Fe(II)-citrate. Eur J Pharmacol 2001; 428:37–44.
42. Balla G, Vercellotti GM, Eaton JW, Jacob HS. Iron loading of endothelial cells augments oxidant damage. J Lab Clin Med 1990; 116:546–54.
43. Balla G, Vercellotti GM, Muller-Eberhard U, Eaton J, Jacob HS. Exposure of endothelial cells to free heme potentiates damage mediated by granulocytes and toxic oxygen species. Lab Invest 1991; 64:648–55.
44. Gieseg S, Duggan S, Gebicki JM. Peroxidation of proteins before lipids in U937 cells exposed to peroxyl radicals. Biochem J 2000; 350:215–8.
45. Gebicki S, Gebicki JM. Formation of peroxides in amino acids and proteins exposed to oxygen free radicals. Biochem J 1993; 289:743–9.
46. Stadtman ER. Protein oxidation and aging. Free Radic Res 2006; 40:1250–8.
47. Yan LJ, Sohal RS. Mitochondrial adenine nucleotide translocase is modified oxidatively during aging. Proc Natl Acad Sci USA 1998; 95:12896–901.
48. Jornot L, Petersen H, Junod AF. Hydrogen peroxide-induced DNA damage is independent of nuclear calcium but dependent on redox-active ions. Biochem J 1998; 335:85–94.
49. Enright HU, Miller WJ, Hebbel RP. Nucleosomal histone protein protects DNA from iron-mediated damage. Nucleic Acids Res 1992; 20:3341–6.
50. Enright H, Hebbel RP, Nath KA. Internucleosomal cleavage of DNA as the sole criterion for apoptosis may be artifactual. J Lab Clin Med 1994; 124:63–8.
51. Zastawny TH, Altman SA, Randers-Eichhorn L, Madurawe R, Lumpkin JA, Dizdaroglu M, Rao G. DNA base modifications and membrane damage in cultured mammalian cells treated with iron ions. Free Radic Biol Med 1995; 18:1013–22.
52. Vaca CE, Harms-Ringdahl M. Interaction of lipid peroxidation products with nuclear macromolecules. Biochim Biophys Acta 1989; 1001:35–43.
53. Carmichael PL, Hewer A, Osborne MR, Strain AJ, Phillips DH. Detection of bulky DNA lesions in the liver of patients with Wilson's disease and primary haemochromatosis. Mutat Res 1995; 326:235–43.
54. Petrat F, de Groot H, Rauen U. Subcellular distribution of chelatable iron: a laser scanning microscopic study in isolated hepatocytes and liver endothelial cells. Biochem J 2001; 356:61–9.
55. Gackowski D, Kruszewski M, Bartlomiejczyk T, Jawien A, Ciecierski M, Olinski R. The level of 8-oxo-7,8-dihydro-2′-deoxyguanosine is positively correlated with the size of the labile iron pool in human lymphocytes. J Biol Inorg Chem 2002; 7:548–50.
56. Lipiński P, Drapier JC, Oliveira L, Retmańska H, Sochanowicz B, Kruszewski M. Intracellular iron status as a hallmark of mammalian cell susceptibility to oxidative stress: a study of L5178Y mouse lymphoma cell lines differentially sensitive to H(2)O(2). Blood 2000; 95:2960–6.
57. Barbouti A, Doulias PT, Zhu BZ, Frei B, Galaris D. Intracellular iron, but not copper, plays a critical role in hydrogen peroxide-induced DNA damage. Free Radic Biol Med 2001; 31:490–8.
58. Itoh H, Shioda T, Matsura T, Koyama S, Nakanishi T, Kajiyama G, Kawasaki T. Iron ion induces mitochondrial DNA damage in HTC rat hepatoma cell culture – role of antioxidants

in mitochondrial DNA protection from oxidative stresses. Arch Biochem Biophys 1994; 313:120–5.
59. Cermak J, Balla J, Jacob HS, Balla G, Enright H, Nath K, Vercellotti GM. Tumor cell heme uptake induces ferritin synthesis resulting in altered oxidant sensitivity: possible role in chemotherapy efficacy. Cancer Res 1993; 53:5308–13.
60. Byrnes RW, Petering DH. Inhibition of bleomycin-induced cellular DNA strand scission by 1,10-phenanthroline. Biochem Pharmacol 1991; 41:1241–8.
61. Cai C, Ching A, Lagace C, Linsenmayer T. Nuclear ferritin-mediated protection of corneal epithelial cells from oxidative damage to DNA. Dev Dyn 2008; 237:2676–83.
62. Powell LW, George DK, McDonnell SM, Kowdley KV. Diagnosis of hemochromatosis. Ann Intern Med 1998; 129:925–31.
63. Whitlock EP, Garlitz BA, Harris EL, Beil TL, Smith PR. Screening for hereditary hemochromatosis: a systematic review for the U.S. Preventive Services Task Force. Ann Intern Med 2006; 145:209–23.
64. Balla J, Vercellotti GM, Jeney V, Yachie A, Varga Z, Jacob HS, Eaton JW, Balla G. Heme, heme oxygenase, and ferritin: how the vascular endothelium survives (and dies) in an iron-rich environment. Antioxid Redox Signal 2007; 9:2119–37.
65. Balla J, Jacob HS, Balla G, Nath K, Eaton JW, Vercellotti GM. Endothelial-cell heme uptake from heme proteins: induction of sensitization and desensitization to oxidant damage. Proc Natl Acad Sci USA 1993; 90:9285–9.
66. Bunn HF, Jandl JH. Exchange of heme among hemoglobins and between hemoglobin and albumin. J Biol Chem 1968; 243:465–75.
67. Gladwin MT, Crawford JH, Patel RP. The biochemistry of nitric oxide, nitrite, and hemoglobin: role in blood flow regulation. Free Radic Biol Med 2004; 36:707–17.
68. Loboda A, Jazwa A, Grochot-Przeczek A, Rutkowski AJ, Cisowski J, Agarwal A, Jozkowicz A, Dulak J. Heme oxygenase-1 and the vascular bed: from molecular mechanisms to therapeutic opportunities. Antioxid Redox Signal 2008; 10:1767–812.
69. Balla G, Jacob HS, Balla J, Rosenberg M, Nath K, Apple F, Eaton JW, Vercellotti GM. Ferritin: a cytoprotective antioxidant stratagem of endothelium. J Biol Chem 1992; 267: 18148–53.
70. Balla G, Jacob HS, Eaton JW, Belcher JD, Vercellotti GM. Hemin: a possible physiological mediator of low density lipoprotein oxidation and endothelial injury. Arterioscler Thromb 1991; 11:1700–11.
71. Nilsson J, Regnström J, Frostegård J, Stiko A. Lipid oxidation and atherosclerosis. Herz 1992; 17:263–9.
72. Hedlund BE, Hallaway PE. High-dose systemic iron chelation attenuates reperfusion injury. Biochem Soc Trans 1993; 21:340–3.
73. Niederau C, Fischer R, Sonnenberg A, Stremmel W, Trampisch HJ, Strohmeyer G. Survival and causes of death in cirrhotic and in noncirrhotic patients with primary hemochromatosis N Engl J Med 1985; 313:1256–62.
74. National Kidney Foundation. K/DOQI clinical practice guidelines for chronic kidney disease: evaluation, classification, and stratification. Am J Kidney Dis 2002; 39:S1–266.
75. Zarychanski R, Houston DS. Anemia of chronic disease: a harmful disorder or an adaptive, beneficial response? CMAJ 2008; 179:333–7.
76. KDOQI; National Kidney Foundation. KDOQI clinical practice guidelines and clinical practice recommendations for anemia in chronic kidney disease. Am J Kidney Dis 2006; 47:S11–145.
77. Bailie GR, Clark JA, Lane CE, Lane PL. Hypersensitivity reactions and deaths associated with intravenous iron preparations. Nephrol Dial Transplant 2005; 20:1443–9.
78. Salahudeen AK, Oliver B, Bower JD, Roberts LJ, II. Increase in plasma esterified F2-isoprostanes following intravenous iron infusion in patients on hemodialysis. Kidney Int 2001; 60:1525–31.

79. Breuer W, Ronson A, Slotki IN, Abramov A, Hershko C, Cabantchik ZI. The assessment of serum nontransferrin-bound iron in chelation therapy and iron supplementation. Blood 2000; 95:2975–82.
80. Lim PS, Wei YH, Yu YL, Kho B. Enhanced oxidative stress in haemodialysis patients receiving intravenous iron therapy. Nephrol Dial Transplant 1999; 14:2680–7.
81. Michelis R, Sela S, Kristal B. Intravenous iron-gluconate during haemodialysis modifies plasma beta2-microglobulin properties and levels. Nephrol Dial Transplant 2005; 20:1963–9.
82. Agarwal R, Vasavada N, Sachs NG, Chase S. Oxidative stress and renal injury with intravenous iron in patients with chronic kidney disease. Kidney Int 2004; 65:2279–89.
83. Zager RA, Johnson AC, Hanson SY, Wasse H. Parenteral iron formulations: a comparative toxicologic analysis and mechanisms of cell injury. Am J Kidney Dis 2002; 40:90–103.
84. Zager RA, Johnson AC, Hanson SY. Parenteral iron nephrotoxicity: potential mechanisms and consequences. Kidney Int 2004; 66:144–56.
85. Spinowitz BS, Kausz AT, Baptista J, Noble SD, Sothinathan R, Bernardo MV, Brenner L, Pereira BJ. Ferumoxytol for treating iron deficiency anemia in CKD. J Am Soc Nephrol 2008; 19:1599–605.
86. Landry R, Jacobs PM, Davis R, Shenouda M, Bolton WK. Pharmacokinetic study of ferumoxytol: a new iron replacement therapy in normal subjects and hemodialysis patients. Am J Nephrol 2005; 25:400–10.
87. Agarwal R, Rizkala AR, Bastani B, Kaskas MO, Leehey DJ, Besarab A. A randomized controlled trial of oral versus intravenous iron in chronic kidney disease. Am J Nephrol 2006; 26:445–54.
88. Macdougall IC, Roche A. Administration of intravenous iron sucrose as a 2-minute push to CKD patients: a prospective evaluation of 2,297 injections. Am J Kidney Dis 2005; 46:283–9.
89. Wagner KR, Sharp FR, Ardizzone TD, Lu A, Clark JF. Heme and iron metabolism: role in cerebral hemorrhage. J Cereb Blood Flow Metab 2003; 23:629–52.
90. Levy YS, Streifler JY, Panet H, Melamed E, Offen D. Hemin-induced apoptosis in PC12 and neuroblastoma cells: implications for local neuronal death associated with intracerebral hemorrhage. Neurotox Res 2002; 4:609–16.
91. Bishop GM, Robinson SR. Quantitative analysis of cell death and ferritin expression in response to cortical iron: implications for hypoxia-ischemia and stroke. Brain Res 2001; 907:175–87.
92. Hua Y, Nakamura T, Keep RF, Wu J, Schallert T, Hoff JT, Xi G. Long-term effects of experimental intracerebral hemorrhage: the role of iron. J Neurosurg 2006; 104:305–12.
93. Burdo JR, Connor JR. Brain iron uptake and homeostatic mechanisms: an overview. Biometals 2003; 16:63–75.
94. Banks WA, Kastin AJ, Fasold MB, Barrera CM, Augereau G. Studies of the slow bidirectional transport of iron and transferrin across the blood-brain barrier. Brain Res Bull 1988; 21:881–5.
95. Moroo I, Ujiie M, Walker BL, Tiong JW, Vitalis TZ, Karkan D, Gabathuler R, Moise AR, Jefferies WA. Identification of a novel route of iron transcytosis across the mammalian blood-brain barrier. Microcirculation 2003; 10:457–62.
96. Bartlett WP, Li XS, Connor JR. Expression of transferrin mRNA in the CNS of normal and jimpy mice. J Neurochem 1991; 57:318–22.
97. Morris CM, Candy JM, Keith AB, Oakley AE, Taylor GA, Pullen RG, Bloxham CA, Gocht A, Edwardson JA. Brain iron homeostasis. J Inorg Biochem 1992; 47:257–65.
98. Koeppen AH. The history of iron in the brain. J Neurol Sci 1995; 134:1–9.
99. Siesjö BK, Agardh CD, Bengtsson F. Free radicals and brain damage. Cerebrovasc Brain Metab Rev 1989; 1:165–211.

100. Baader SL, Bruchelt G, Carmine TC, Lode HN, Rieth AG, Niethammer D. Ascorbic-acid-mediated iron release from cellular ferritin and its relation to the formation of DNA strand breaks in neuroblastoma cells. J Cancer Res Clin Oncol 1994; 120:415–21.
101. Ying W, Han SK, Miller JW, Swanson RA. Acidosis potentiates oxidative neuronal death by multiple mechanisms. J Neurochem 1999; 73:1549–56.
102. Götz ME, Double K, Gerlach M, Youdim MB, Riederer P. The relevance of iron in the pathogenesis of Parkinson's disease. Ann N Y Acad Sci 2004; 1012:193–208.
103. Atasoy HT, Nuyan O, Tunc T, Yorubulut M, Unal AE, Inan LE. T2-weighted MRI in Parkinson's disease; substantia nigra pars compacta hypointensity correlates with the clinical scores. Neurol India 2004; 52:332–7.
104. Ward RJ, Kühn LC, Kaldy P, Florence A, Peters TJ, Crichton RR. Control of cellular iron homeostasis by iron-responsive elements in vivo. Eur J Biochem 1994; 220:927–31.
105. Thompson K, Menzies S, Muckenthaler M, Torti FM, Wood T, Torti SV, Hentze MW, Beard J, Connor J. Mouse brains deficient in H-ferritin have normal iron concentration but a protein profile of iron deficiency and increased evidence of oxidative stress. J Neurosci Res 2003; 71:46–63.
106. Mohanakumar KP, de Bartolomeis A, Wu RM, Yeh KJ, Sternberger LM, Peng SY, Murphy DL, Chiueh CC. Ferrous-citrate complex and nigral degeneration: evidence for free-radical formation and lipid peroxidation. Ann N Y Acad Sci 1994; 738:392–9.
107. Sengstock GJ, Olanow CW, Menzies RA, Dunn AJ, Arendash GW. Infusion of iron into the rat substantia nigra: nigral pathology and dose-dependent loss of striatal dopaminergic markers. J Neurosci Res 1993; 35:67–82.
108. Sziráki I, Mohanakumar KP, Rauhala P, Kim HG, Yeh KJ, Chiueh CC. Manganese: a transition metal protects nigrostriatal neurons from oxidative stress in the iron-induced animal model of parkinsonism. Neuroscience 1998; 85:1101–11.
109. Faucheux BA, Martin ME, Beaumont C, Hauw JJ, Agid Y, Hirsch EC. Neuromelanin associated redox-active iron is increased in the substantia nigra of patients with Parkinson's disease. J Neurochem 2003; 86:1142–8.
110. Jurma OP, Hom DG, Andersen JK. Decreased glutathione results in calcium-mediated cell death in PC12. Free Radic Biol Med 1997; 23:1055–66.
111. Lovell MA, Robertson JD, Teesdale WJ, Campbell JL, Markesbery WR. Copper, iron and zinc in Alzheimer's disease senile plaques. J Neurol Sci 1998; 158:47–52.
112. Castellani RJ, Smith MA, Nunomura A, Harris PL, Perry G. Is increased redox-active iron in Alzheimer disease a failure of the copper-binding protein ceruloplasmin? Free Radic Biol Med 1999; 26:1508–12.
113. Thompson KJ, Shoham S, Connor JR. Iron and neurodegenerative disorders. Brain Res Bull 2001; 55:155–64.
114. Smith MA, Hirai K, Hsiao K, Pappolla MA, Harris PL, Siedlak SL, Tabaton M, Perry G. Amyloid-beta deposition in Alzheimer transgenic mice is associated with oxidative stress. J Neurochem 1998; 70:2212–5.
115. Núñez-Millacura C, Tapia V, Muñoz P, Maccioni RB, Núñez MT. An oxidative stress-mediated positive-feedback iron uptake loop in neuronal cells. J Neurochem 2002; 82:240–8.
116. Goswami T, Rolfs A, Hediger MA. Iron transport: emerging roles in health and disease. Biochem Cell Biol 2002; 80:679–89.
117. Huang X, Atwood CS, Hartshorn MA, Multhaup G, Goldstein LE, Scarpa RC, Cuajungco MP, Gray DN, Lim J, Moir RD, Tanzi RE, Bush AI. The A beta peptide of Alzheimer's disease directly produces hydrogen peroxide through metal ion reduction. Biochemistry 1999; 38:7609–16.
118. Bondy SC, Guo-Ross SX, Truong AT. Promotion of transition metal-induced reactive oxygen species formation by beta-amyloid. Brain Res 1998; 799:91–6.
119. Turnbull S, Tabner BJ, El-Agnaf OM, Twyman LJ, Allsop D. New evidence that the Alzheimer beta-amyloid peptide does not spontaneously form free radicals: an ESR study using a series of spin-traps. Free Radic Biol Med 2001; 30:1154–62.

120. Zraika S, Aston-Mourney K, Laybutt DR, Kebede M, Dunlop ME, Proietto J, Andrikopoulos S. The influence of genetic background on the induction of oxidative stress and impaired insulin secretion in mouse islets. Diabetologia 2006; 49:1254–63.
121. Freeman H, Shimomura K, Horner E, Cox RD, Ashcroft FM. Nicotinamide nucleotide transhydrogenase: a key role in insulin secretion. Cell Metab 2006; 3:35–45.
122. Fernández-Real JM, López-Bermejo A, Ricart W. Iron stores, blood donation, and insulin sensitivity and secretion. Clin Chem 2005; 51:1201–5.
123. Schulze MB, Manson JE, Willett WC, Hu FB. Processed meat intake and incidence of type 2 diabetes in younger and middle-aged women. Diabetologia 2003; 46:1465–73.
124. Cooksey RC, Jouihan HA, Ajioka RS, Hazel MW, Jones DL, Kushner JP, McClain DA. Oxidative stress, beta-cell apoptosis, and decreased insulin secretory capacity in mouse models of hemochromatosis. Endocrinology 2004; 145:5305–12.
125. Andrews NC. The iron transporter DMT1. Int J Biochem Cell Biol 1999; 31:991–4.
126. Merkel PA, Simonson DC, Amiel SA, Plewe G, Sherwin RS, Pearson HA, Tamborlane WV. Insulin resistance and hyperinsulinemia in patients with thalassemia major treated by hypertransfusion. N Engl J Med 1988; 318:809–14.
127. Tuomainen TP, Nyyssönen K, Salonen R, Tervahauta A, Korpela H, Lakka T, Kaplan GA, Salonen JT. Body iron stores are associated with serum insulin and blood glucose concentrations. Population study in 1,013 eastern Finnish men. Diabetes Care 1997; 20:426–8.
128. Halliwell B. Oxidative stress and cancer: have we moved forward? Biochem J 2007; 401:1–11.
129. Gius D, Spitz DR. Redox signaling in cancer biology. Antioxid Redox Signal 2006; 8:1249–52.
130. Arrigo AP. Gene expression and the thiol redox state. Free Radic Biol Med 1999; 27:936–44.
131. Bensaad K, Vousden KH. Savior and slayer: the two faces of p53. Nat Med 2005; 11:1278–9.
132. Donfrancesco A, Deb G, Dominici C, Pileggi D, Castello MA, Helson L. Effects of a single course of deferoxamine in neuroblastoma patients. Cancer Res 1990; 50:4929–30.
133. Kaplinsky C, Estrov Z, Freedman MH, Gelfand EW, Cohen A. Effect of deferoxamine on DNA synthesis, DNA repair, cell proliferation, and differentiation of HL-60 cells. Leukemia 1987; 1:437–41.
134. Gao J, Richardson DR. The potential of iron chelators of the pyridoxal isonicotinoyl hydrazone class as effective antiproliferative agents, IV: The mechanisms involved in inhibiting cell-cycle progression. Blood 2001; 98:842–50.
135. Le NT, Richardson DR. Iron chelators with high antiproliferative activity up-regulate the expression of a growth inhibitory and metastasis suppressor gene: a link between iron metabolism and proliferation. Blood 2004; 104:2967–75.
136. Larrick JW, Cresswell P. Modulation of cell surface iron transferrin receptors by cellular density and state of activation. J Supramol Struct 1979; 11:579–86.
137. Richardson DR, Baker E. The uptake of iron and transferrin by the human malignant melanoma cell. Biochim Biophys Acta 1990; 1053:1–12.
138. Takeda E, Weber G. Role of ribonucleotide reductase in expression in the neoplastic program. Life Sci 1981; 28:1007–14.
139. Campbell JA. Effects of precipitated silica and of iron oxide on the incidence of primary lung tumours in mice. Br Med J 1940; 2:275–80.
140. Richmond HG. Induction of sarcoma in the rat by iron–dextran complex. Br Med J 1959; 1:947–9.
141. Liu M, Okada S. Induction of free radicals and tumors in the kidneys of Wistar rats by ferric ethylenediamine-N,N′-diacetate. Carcinogenesis 1994;15:2817–21.
142. Okada S, Hamazaki S, Toyokuni S, Midorikawa O. Induction of mesothelioma by intraperitoneal injections of ferric saccharate in male Wistar rats. Br J Cancer 1989; 60:708–11.
143. Wu KJ, Polack A, Dalla-Favera R. Coordinated regulation of iron-controlling genes, H-ferritin and IRP2, by c-myc. Science 1999; 283:676–9.
144. Dunn LL, Rahmanto YS, Richardson DR. Iron uptake and metabolism in the new millennium. Trends Cell Biol 2007; 17:93–100.

Chapter 12
Hypoxia, Oxidative Stress, and the Pathophysiology of Contrast-Media-Induced Nephropathy

Samuel N. Heyman, Seymour Rosen, Mogher Khamaisi, Jean-Marc Odee, and Christian Rosenberger

Abstract Radiocontrast-induced nephropathy (CIN) is a leading cause of acute kidney injury among hospitalized patients. Two principal mechanisms conceivably participating in the pathogenesis of CIN are the induction of renal parenchymal hypoxic injury and a direct endothelial/vascular and tubular toxicity. Reactive oxygen species (ROS) are thought to take part in both injurious pathways and may serve as a link between these processes. Risk factors for the development of CIN are characterized by a predisposition to develop renal parenchymal hypoxia and by the propensity for enhanced ROS formation. In this chapter the evidence for the development of renal parenchymal hypoxia and of the formation of ROS during radiocontrast studies is detailed, and the role they possibly play in the development of CIN is discussed. In this perspective, strategies currently adopted in the clinical practice for the prevention of CIN, namely hydration, alkalization, and the administration of *N*-acetyl-cysteine, are critically reviewed.

Keywords Acute kidney injury · Reactive oxygen species · Medulla · HIF · Nitric oxide

1 Introduction

Iodinated contrast media (CM) used for intravascular injection are derivatives of the 1,3,5 triiodobenzoic acid. The first generation of CM, introduced over 50 years ago, comprises high-osmolar (~1,600 mOsm/L) ionic compounds and consisted of a single benzene ring. New generations include low-osmolar (600–700 mOsm/L) and iso-osmolar (~300 mOsm/L) CM, generated by the condensation of two

S.N. Heyman (✉)
Department of Medicine, Hadassah University Hospital, Mount Scopus, 24035, Jerusalem 91240, Israel
e-mail: heyman@cc.huji.ac.il

T. Miyata et al. (eds.), *Studies on Renal Disorders*, Oxidative Stress in Applied Basic Research and Clinical Practice, DOI 10.1007/978-1-60761-857-7_12,

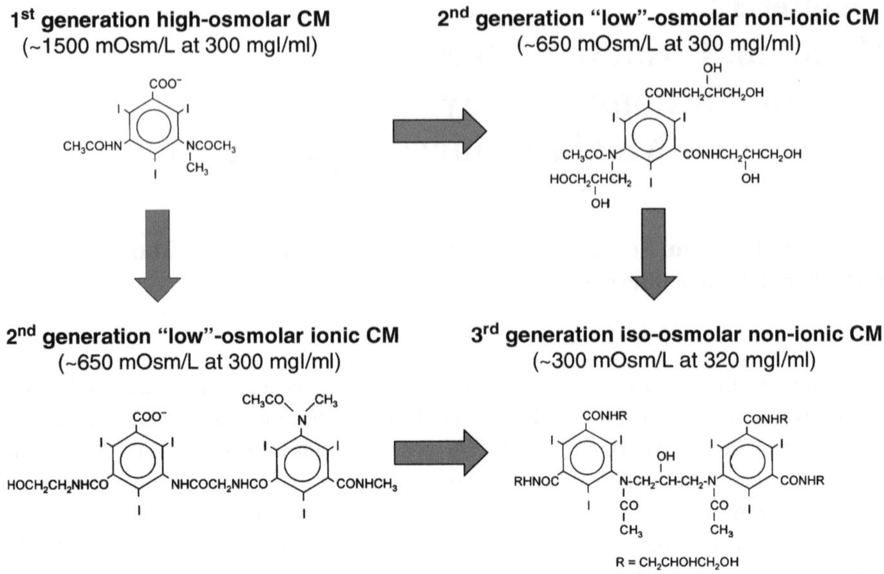

Fig. 1 The evolution of contrast media (CM) classes. The first-generation *high-osmolar ionic agents* (such as iothalamate or diatrizoate) consist of a single triiodobenzene ring with a carboxylic acid residue. Second-generation "low" osmolar agents were developed by the condensation of two triiodobenzene rings, with one carboxylic acid residue (*ionic "low" osmolar* dimers, such as ioxaglate) or by the substitution of the carboxylic acid with polyhydroxylated alkyl side chains (*nonionic "low" osmolar* monomers, such as iohexol, iopromide, or ioversol). The development of third generation agents is based on the combination of these two approaches, with the generation of *iso-osmolar nonionic* dimmers (such as iodixanol). Displayed are samples of molecular structures for each class of CM

iodinated benzene rings or with the substitution of the ionic carboxylic group with polyhydroxylated alkyl side chains (Fig. 1).

The use of CM in diagnostic and interventional medicine is rapidly growing. It is estimated that about 9.2 million liters of CM were consumed worldwide in 2008, corresponding to some 106.6 million contrast injection studies (JM Idee, Guerbet, personal communication). Not surprisingly, CM-induced nephropathy (CIN) remains a leading cause of iatrogenic acute kidney injury (AKI) [1, 2], accounting for some 10% of in-hospital AKI, despite adherence to protocols of risk assessment and prevention strategies.

CIN is usually nonoliguric, manifested by an asymptomatic transient decline in glomerular filtration rate (GFR, detected as rising plasma creatinine), that develops 24–72 h after the CM study. However, a small fraction of patients with CIN may develop severe, oliguric renal failure, requiring temporary, or even chronic, renal replacement therapy, particularly among those with marked baseline renal impairment.

The pathogenesis of CIN is not fully understood. The predisposition to this disorder to patients with well-defined risk factors underscores potential intrinsic

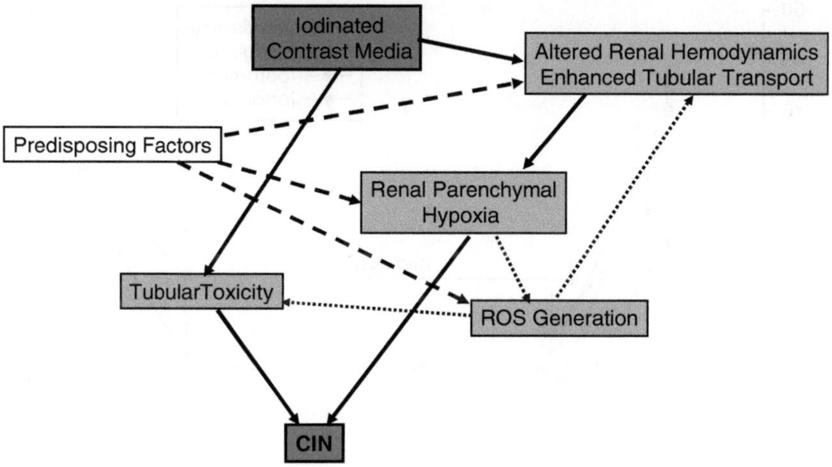

Fig. 2 Reactive oxygen species (ROS) and the pathogenesis of contrast-induced nephropathy (CIN). Iodinated contrast media may exert direct tubular injury and compromise renal oxygenation, through altered systemic and renal hemodynamics, and by transiently enhanced oxygen consumption for tubular transport. CIN probably reflects a synergic effect of both hypoxic injury and direct tissue damage (*full arrows*). As illustrated by the fine-dashed arrows, ROS likely links these two processes together. Formed during hypoxic stress, ROS probably injure all renal parenchymal elements and may cause microcirculatory dysfunction and enhance tubular transport. This can lead to a vicious circle of intensification of hypoxia, ROS formation, and damage. As illustrated by the coarsely dashed arrows, various morbidities predisposing to CIN are characterized by altered renal oxygenation, enhanced ROS generation, or defective ROS deactivation

defects in renal physiology, challenged and unmasked by CM. It is currently believed that the two leading mechanisms are the induction of renal parenchymal hypoxic injury and a direct endothelial/vascular and tubular toxicity. Reactive oxygen species (ROS) are thought to take part in both injurious pathways and may serve as a link between these processes (Fig. 2). In this chapter we will appraise these two deleterious trails, assess the data regarding the role for ROS in the pathogenesis of CIN, and evaluate the place of ROS scavenging in its prevention.

2 Radiocontrast Agents Hamper Renal Oxygenation

The special anatomy and function of the renal medulla, providing urine concentration, leads to physiologic medullary pO_2 as low as 25 mmHg, reflecting limited regional oxygen supply, hardly matching high local oxygen consumption for tubular reabsorption [3]. Protective mechanisms, including prostaglandins, nitric oxide (NO), and adenosine, adjust medullary tubular transport activity to the limited available oxygen supply, acting by both the enhancement of regional blood flow and downregulation of distal tubular transport, particularly in medullary thick ascending limbs (mTAL). Decreased GFR also reduces downstream transport activity and oxygen consumption.

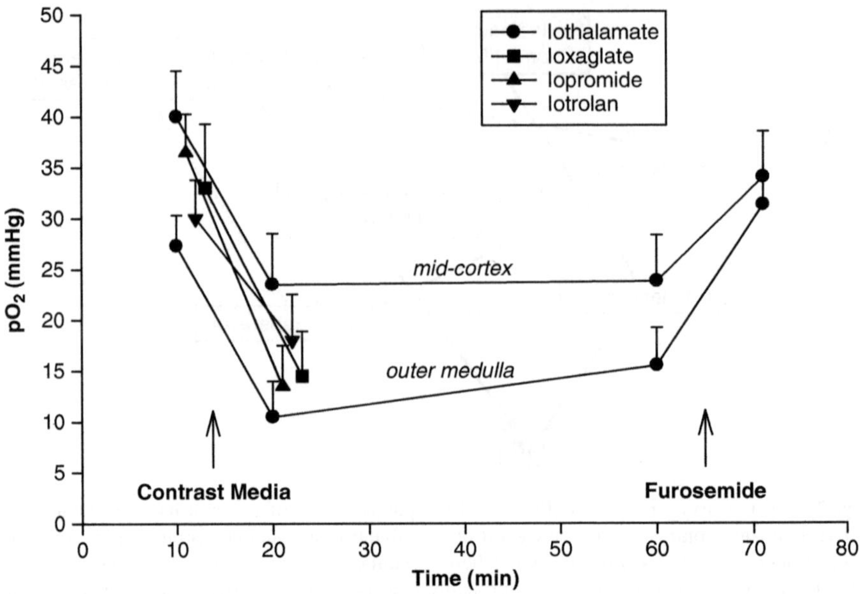

Fig. 3 Radiocontrast-induced outer medullary hypoxia. Renal outer medullary oxygenation, determined with oxygen microelectrodes before and after the administration of radiocontrast agents in rats. Outer medullary PO_2 markedly declines after the administration of the high-osmolar contrast medium iothalamate, the low-osmolar ionic (ioxaglate), and nonionic (iopromide) agents, as well as the iso-osmolar radiocontrast iotrolan. Radiocontrast-induced medullary hypoxia is protracted and relived only after transport inhibition with the loop diuretic furosemide. Cortical measurements are shown as well for iothalamate only. Data obtained from Heyman et al. [5] and Liss et al. [6], adapted with permission from Heyman et al. [2]

The overall effect of these mechanisms in maintenance of medullary oxygen sufficiency [3] and their failure may evolve into AKI [2, 4].

Renal parenchymal oxygenation is adversely affected by CM (Fig. 3). Using oxygen microelectrodes inserted into the renal parenchyma, Brezis et al. found that outer medullary pO_2, already low under normal physiologic conditions [3], markedly declined in rats following the administration of the high osmolar CM sodium iothalamate to levels as low as 9 mmHg [5]. Comparable observations were reported by Liss et al. [6], looking at the impact of high- as well as low- and iso-osmolar CM. Renal cortical oxygenation declined as well, but rather moderately, in part due to higher baseline measurements [5]. Shorter periods of outer medullary pO_2 decline, of about 10 min, were reported in dogs using smaller volumes of CM injected into the renal artery [7]. CM-induced decline of renal medullary oxygenation has also been detected noninvasively in experimental models [8] as well as in humans [9], using blood oxygen level dependent (BOLD) functional magnetic resonance imaging (MRI). Experimental CIN immunohistochemistry revealed renal medullary accumulation of pimonidazole adducts [10], which occurs at regions with pO_2 below 10 mmHg. Furthermore, hypoxia-inducible transcription factors(HIFs) were upregulated, suggesting adaptational responses to medullary hypoxia [11].

Tissue oxygenation reflects the balance between regional oxygen supply and demand. Tubular transport and oxygen consumption are most intense in the outer medulla (principally by mTALs), where renal sodium gradient is formed for the generation of urine concentrating capacity. Paradoxically, this region receives only 10% of total renal blood flow, delivered through vasa recta emerging from juxta-medullary nephrons. Thus, low medullary pO_2 under normal physiologic conditions underscores limited regional oxygen supply hardly sufficient for intense tubular transport and oxygen consumption [3]. Importantly, medullary oxygen sufficiency is maintained by systems that match regional oxygen supply and demand, including prostaglandin and NO synthesis and the generation of adenosine during the breakdown of adenosine triphosphate (ATP) [3].

CM-associated renal medullary hypoxia principally reflects imbalance in regional oxygen supply and demand, in part through the deactivation of systems that maintain renal oxygen sufficiency, in particular in patients predisposed to CIN with a priori alterations in these systems [2]. In the following sections we will outline the mechanisms by which renal tissue oxygenation is hampered by contrast media and evaluate the potential role of ROS in these processes.

3 Radiocontrast-Mediated Changes in Renal Blood Supply

Renal blood flow and intrarenal microcirculation are substantially altered by CM [2]. The extent and distribution of renal hemodynamical changes seem to depend on the species investigated, as well as on the type, volume, and rate of CM administration. Furthermore, renal hemodynamic effects of CM conceivably depend on the hydration state and on additional inherent predisposing factors and comorbidities that may have an effect on the renal microcirculation and vascular tone regulation, such as chronic kidney disease, diabetes, aging, or inflammation.

In one of the earliest studies in dogs subjected to a high-osmolar CM, total renal blood flow transiently increased for 5–15 min, followed by a protracted decline by 25% below baseline, extending beyond a couple of hours [12]. Rats also displayed reduced renal blood flow in response to a high- and low-osmolar CM [13, 14].

In healthy humans renal blood flow fell 8% over 30 min after the intravenous administration of conventional doses of CM [15]. By contrast, in patients with chronic kidney disease undergoing coronary angiography, a transient brief enhancement of renal blood flow was replaced by a substantial 40% decline, lasting over 3 h [16]. Notably, in a comparable group of patients with renal impairment undergoing coronary intervention, the decline noted in renal blood flow showed marked variability, ranging between 4 and 40%. The authors suggested that these differences might reflect individual predisposition, as well as the varied doses of injected CM [17]. Underlining this variability is an additional report showing stable renal blood flow in patients with chronic renal impairment given high-osmolar ionic CM [18].

Since only a small fraction of total renal blood flow is delivered to the medulla, further studies were designed to separately determine the effect of CM upon cortical and medullary microcirculation. Using laser flow probes and video microscopy of transilluminated papillary vasa recta, Nygren and Liss and their colleagues found a marked decline in papillary blood flow after CM [19, 20]. By contrast, data regarding changes in outer medullary flow is inconsistent. Agmon and Heyman and their colleagues [21, 22] reported an increase following injection of large volume of the high-osmolar CM iothalamate, whereas Myers and Efrati and their colleagues found declining medullary flow [23, 24]. Increased outer medullary flow has also been reported by Palm et al. [25], using the low-osmolar CM iopromide, while Liss et al. found a dose-related response, with a decline in regional flow at low and intermediate volumes of CM, but enhancement at high volumes [26, 27].

Altogether, these findings indicate that CM-induced accentuation of inner medullary hypoxia is caused by a decline in regional blood flow and oxygen supply. By contrast, CM-induced decline in outer medullary pO_2 might represent both altered regional blood flow (at low doses of CM) as well as enhanced oxygen consumption, not fully compensated by increased regional oxygen delivery (with very large volumes of CM). Conceivably, the former mechanism is the clinically relevant one [27].

4 Mechanisms Involved in Radiocontrast-Induced Altered Renal Microcirculation

CM injure vascular endothelial cells in vitro [28–31] and affect endothelial interactions with circulating blood constituents [2]. Indeed, red blood cell aggregation was noted in vasa recta following CM injection in rats [20]. However, isolated vasa recta, perfused with contrast media, respond with vasoconstriction [32], suggesting that altered renal microcirculation is, to a large extent, mediated by local alterations in vascular tone. Systemic neurohumoral responses are also activated [2]. Altogether, CM can turn on renal vasoconstrictors (vasopressin, histamine), vasodilators (nitric oxide, natriuretic peptides, prostaglandin E_2 (PGE_2), prostacyclin (PGI_2)), and factors producing varying responses at different renal vascular beds (endothelin, adenosine). The most studied mediators affected by contrast media are adenosine, prostaglandins, endothelin, and nitric oxide.

Renal adenosine rises following CM administration [33, 34], conceivably reflecting ATP breakdown during evolving hypoxia and perhaps as a result of enhanced tubular transport. Adenosine exerts renal cortical vasoconstriction and outer medullary vasodilation, mediated by adenosine A1 and A2 receptors, respectively [35], with an overall decline in total renal blood flow and GFR [35, 36].

PGE_2 is the most important medullary cyclooxygenase product, maintaining medullary oxygen sufficiency [3]. Although a sevenfold increase was noted in urinary PGE_2 in rats following contrast administration [13], renal parenchymal PGE_2, detected with microdialysis probes or by enzyme-linked immunosorbent assay (ELISA) in dissected tissues remained unchanged [23, 24]. These conflicting findings probably reflect rapid washout of PGE_2 because of enhanced urine flow. The effect of CM on PGI_2 production by renal endothelial cells is debated, with one report showing declining levels of 6-keto-PGF1α in renal veins in dogs [12], another study demonstrating rising plasma levels in humans subjected to high-osmolar ionic and low-osmolar nonionic CM [37], and a third report of stable renal parenchymal 6-keto-PGF1α levels [23].

CM also enhance plasma and renal endothelin-1 [38, 39], possibly through the induction of endothelin-converting enzyme (ECE)-1 [40]. Like adenosine, endothelin-1 also exerts renal cortical vasoconstriction and medullary vasodilation, mediated by endothelin ETA and ETB receptors, respectively [41].

Activation of the above-mentioned three systems illustrates the physiologic response to maintain medullary oxygenation [2, 3], by both the preservation of regional blood flow and the reduction of local transport activity (by decreasing cortical blood flow, GFR, and solute delivery to the distal nephron, as well as by the direct inhibition of tubular transport). Some of these effects might be concealed by the complex interacting physiologic responses. For instance, attenuation of CM-induced decline in renal blood flow with an endothelin ETA receptor antagonist is revealed only following the inhibition of prostaglandin synthesis [13].

Renal NO, a most important regulator of medullary blood flow oxygenation [21, 42], is also altered by CM. Direct measurements in vivo using specific NO electrodes illustrate that cortical NO declines, in parallel with a fall in renal blood flow, probably related to reduced nitric oxide synthase (NOS) activity [22]. Myers et al. reported a drop in both cortical and medullary NO, using microdialysis probes [23]. Furthermore, vasa recta NO declines in vitro when flushed by contrast medium [32]. Indeed, Efrati and colleagues report that in humans urinary NO metabolites are reduced following CM [43]. They propose that reduced NO might reflect its conversion to peroxynitrite, coupled with reduced NOS activity. Moreover, Goodman and colleagues report an increase in iNOS but a drop in eNOS in a rat model of CIN [44]. Finally, the NO precursor, L-arginine, prevented the decline in renal blood flow and renal impairment in hypercholesterolemic rats subjected to CM [45] and attenuated hypoxic medullary damage in a CIN rat model [46]. Taken together, these data imply that observed renal hemodynamic abnormalities is related, to a large extent, to a CM-induced endothelial/vascular dysfunction with a decline in NO generation. Interestingly, outer medullary NO recorded with an NO electrode paradoxically increases following radiocontrast, while medullary NOS activity (determined at normoxic conditions) remains unchanged [22]. Suggested increased NO bioavailability within the hypoxic medullary milieu [47] might be too simplistic, given the complex control of NO synthesis, transport, and consumption at varying regional oxygen, ROS, and hemoglobin concentrations [48], all altered by CM.

Physical-mechanical factors also play a role in CM-induced altered renal microcirculation. Enhanced diuresis and increasing contrast-related urine viscosity, as urine concentration grows along the distal nephron [49], particularly with nonionic iso-osmolar agents [50, 51], may lead to an expansion of tubular lumens and kidney volume and increase renal interstitial pressure [52]. This might further be affected by intratubular obstruction by precipitation of uric acid and oxalate crystals or by coprecipitating with Tamm-Horsfall [53, 54] and Bence Jones proteins [55] in patients with multiple myeloma. Blood flow within long, small-diameter vasa recta is conceivably reduced due to the subsequent rise in renal interstitial pressure and increased blood viscosity. Interestingly, the decline in medullary blood flow [7] and pO_2 [56], as well as renal retention of CM [57] was found to be most pronounced with iso-osmolar agents, characterized by the highest inherent viscosity.

In the presence of acute urine outflow obstruction renal blood flow may be further reduced with the administration of CM [58], with an especially compromised medullary blood flow [59]. Again, compression of vasa recta by rising renal interstitial pressure might play a role in this phenomenon.

5 Changes in Renal Oxygen Consumption

The injection of contrast media leads to abrupt water diuresis and natriuresis [60], caused by plasma volume expansion and a brief rise in atrial natriuretic peptide (ANP) [38], by a transiently enhanced GFR [5], and perhaps by an osmotic effect of CM. Induction of endothelin probably also plays an important role in contrast-induced natriuresis, inhibiting proximal tubular transport [61]. It is tempting to assume that such enhanced diuresis and sodium delivery to the distal nephron increases medullary tubular transport and oxygen consumption. Two observations support this possibility. The first is the pronounced decline in outer medullary oxygenation [5], noted despite an enhancement of regional microcirculation [21]. The second is the marked attenuation of hypoxic tubular injury with a loop diuretic in a rat model of CIN, detailed below [62]. Nevertheless, in the absence of data regarding differential effects of CM on sodium transport in proximal versus distal nephron segments, the role of enhanced tubular transport in enhancing medullary hypoxia remains so far speculative.

6 Radiocontrast Agents and Direct Tubular Cell Toxicity

Following injection into the bloodstream, CM are filtered in the glomerulus and do not undergo tubular or paracellular transport. Their clearance via glomerular filtration is fast, with an elimination half-life of about 2 h in healthy subjects. Since most filtered fluids are reabsorbed, urine concentration of CM markedly increases downstream along the nephron. Low or declining GFR, combined with enhanced urine viscosity, related to increasing concentration of CM, conceivably prolongs intraluminal transit time and exposure of tubular cells to high concentration of CM. In fact, persistent nephrogram, a radiological hallmark of CIN [63], reflects protractedly retained CM at high concentration within distal nephrons.

Prolonged exposure of tubular cells to high concentrations of CM brings about the possibility of direct tubular toxicity. Indeed, proximal tubular vacuolization has been encountered in human biopsies of CIN [64–66] and in animal studies [67–69] following all types of CM. However, such changes are a consistent finding following CM administration, irrespective to the development of renal dysfunction. Furthermore, these vacuoles were found to represent out-pouching of membranes of the lateral cellular interdigitations, perhaps caused by radiocontrast in the paracellular space [5]. Noteworthy, however, gradual accumulation of acid phosphatase in these vacuoles [68, 69] suggests a cellular response to retained small amounts of CM [70] and merging with lysosomes.

To assess possible CM-induced direct tubular cell toxicity, studies were conducted in vitro [71] in isolated tubular segments or in cultured tubular cell lines. When incubated for 45 min with CM, isolated proximal tubular cells disclosed no evidence of injury or lipid peroxidation [72]. By contrast, prolonged (20-h) incubation of kidney-proximal tubule cell line LLC-PK1 and Madin–Darby canine kidney (MDCK) tubular cells with CM at concentrations of 10–100 mg iodine/mL disclosed a concentration-dependent cellular injury, which was more prominent with high- and low-osmolar ionic CM, as compared with a nonionic CM [73]. In an additional study in MDCK cells, high-osmolar CM was more toxic than low-osmolar CM, irrespective to ionic strength, implying that hypertonicity is a major determinant [74]. However, hyperosmolar control solutions were only marginally toxic or noncytotoxic [72, 75]. Injury was also noted in LLC-PK1 cells incubated with nonionic low- and iso-osmolar CM for 6 h only, followed by 24 h free of CM [76]. Higher concentrations (100–200 mg iodine/mL) of low- and iso-osmolar CM induced injury already within 15 min in LLC-PK1, MDCK, and HEK293 cells [77]. Iodine alone and hyperosmolar solution had no such effect. Apoptosis, associated with activated caspases 3 and 9, and poly (adenosine diphosphate (ADP) ribose) fragmentation were noted [76, 77], indicating a role for intrinsic apoptotic pathways. Indeed, pro-apoptotic members of the Bcl2 family were upregulated [77]. Cell viability was reduced in MDCK cell lines, incubated for 12 h with ionic CM. Redistribution of the E-cadherin, ZO-1, and occludin was reported, associated with impaired monolayer integrity [78].

In summary, findings in vitro support the likelihood that CM at high concentrations, as conceivably present in the distal nephron, are tubulotoxic, in part via activation of apoptotic pathways. It should be emphasized, however, that these isolated tubular preparations are very remote from the complex tubular-interstitial-vascular relationship that characterizes the intact kidney.

7 Risk Factors Predisposing to CIN: A Role for Renal Oxygenation Imbalance

Unlike most nephrotoxic injuries, where AKI invariably develops dose-dependently, healthy subjects very rarely develop CIN, even when subjected to large doses of CM. By contrast, as outlined in Table 1, there are well-defined risk

Table 1 Risk factors for contrast nephropathy (reviewed in [29]): proposed mechanisms predisposing to renal hypoxic injury and evidence for enhanced ROS formation

Intrinsic patient-related predisposing factors		Mechanisms causing medullary hypoxia	Evidence for enhanced ROS formation (references)
	Preexisting renal failure	EW, MV	[113–115, 117, 126]
	Diabetes	NO, MV	[121]
	Effective blood volume depletion	VS	
	Dehydration, hypotension		[107]
	Heart failure, cirrhosis, nephrosis		[118–120]
	Aging	PG	[181]
	Hypertension	NO	[123–125]
	Hypelipidemia	NO	[116, 128, 182]
	Atherosclerosis	NO	[129]
	Anemia	OD	[126]
	Transplanted kidney		[133, 134]
	Myeloma		–
	Male gender		[127]
	Other nephrotoxins		
	Exogenous: drugs	NO, PG	[135]
	Endogenous: heme pigments	NO	[136]
	Systemic inflammation	NO	[130–132]
Procedure-related predisposing factors	Dye type (HOA $\gg$ LOA $\approx$ IOA)		–
	Dye volume	TW	[112]
	Injection site (artery>vein)		–
	Repeated exposure within 72 h		–
	Primary coronary intervention/ emergency procedure[a]	VS	[183, 184]

Possible predisposing mechanisms: altered defense systems – nitric oxide (NO) and prostaglandins (PG); systemic vasoconstrictive stimuli (VS); enhanced tubular workload (TW), structural changes of renal microvasculature (MV), and reduced systemic oxygen delivery (OD). Dye type refers to high-osmolar (HOA)-, low-osmolar (LOA)-, and iso-osmolar agents (IOA)

[a]A putative risk factor among patients with preexisting renal failure, as indirectly extrapolated from clinical studies [168, 185–187]. This might be related to hemodynamic instability and the insufficient time for the implementation of preventive measures

factors that predispose to the development of renal impairment following CM studies [2], categorized as intrinsic and procedure related. The most significant patient-related risk factors are preexisting renal disease, diabetes, dehydration, or compromised systemic hemodynamics, with effective blood volume depletion. Principal procedure-related risk factors are the CM class (high-osmolar ionic CM being the most nephrotoxic) and the dosage of CM. The risk to develop CIN is thus quite predictable, directly proportional to the number of existing predisposing factors [79–81]. In a patient with several combined risk factors, for instance diabetes, hypertension, congestive heart failure, and advanced chronic kidney disease, the risk for CIN may exceed 80%, with a high probability of requiring renal replacement therapy.

Dehydration is a risk factor shared by most nephrotoxins, leading to higher tubular intraluminal concentrations and longer transit periods of the noxious agent, increasing its uptake and the potential for direct tubular cell damage. Enhanced vasoconstrictive stimuli in dehydrated patients may also enhance toxin-related renal vasoconstriction. CM are somewhat unique, with their high viscosity and lack of uptake by tubular cells. It has been suggested that dehydration intensifies CM-induced urine viscosity [7, 56]. This might be the reason for the nonsuperiority [82–84] or even inferiority [85] of highly viscous iso-osmolar CM, as compared with low-osmolar CM, regarding the risk for CIN. Although never studied in a controlled prospective way, volume expansion seemingly is a most efficacious preventive measure in patients at risk [86], conceivably by reducing intraluminal concentration of the CM and subsequent urine viscosity.

Looking closely at most other intrinsic risk factors (see Table 1), there is a plausible association with a predisposition for enhanced medullary hypoxia and hypoxic damage [87, 88]. Regulatory systems that govern medullary oxygen balance are disturbed in diabetic, hypertensive, atherosclerotic, and hyperlipemic patients (NO synthesis) or in elderly patients or those taking nonsteroidal anti-inflammatory drugs NSAIDs (prostaglandins). Anemia leads to reduced renal oxygen delivery. Chronic kidney disease is characterized by a loss in peritubulary capillary meshwork and by interstitial fibrosis, which reduces oxygen diffusion. Furthermore, transport activity in remnant functioning hypertrophic tubules may increase. Enhanced medullary tubular transport-related oxygen consumption may also be noted in uncontrolled diabetes or in hypertension. Indeed, reduced medullary oxygenation has been encountered in subjects with chronic kidney disease, in the diabetic and hypertensive kidney, in anemic patients, and among elderly but otherwise healthy individuals [88].

Other risk factors for CIN are characterized by enhanced systemic and renal vasoconstrictive stimuli. Increased sympathetic activity, vasopressin release, and the upregulation of the renin-angiotensin system occur in patients with effective volume depletion. Endothelin-1 synthesis is enhanced in diabetes, presumably due to the upregulation of ECE-1. Interestingly, ECE-1, enhanced five times in the experimentally diabetic renal medulla, is further markedly enhanced three times following CM, culminating in 15 times higher levels as compared with control animals [40]. These findings are in line with rising plasma endothelin specifically in diabetic patients undergoing CM studies [39]. Altered renal microcirculation in the diabetic kidney may be related to enhanced endothelin synthesis as well as to altered nitrovasodilation, leading to intensified vasoconstriction in response to various stimuli, including adenosine, generated following CM administration [89]. The impact of enhanced systemic and renal vasoconstriction upon renal oxygenation profiles varies, depending on the vascular bed affected [4]. Renal oxygenation may deteriorate in the medulla and cortex, but oxygenation-gradient might invert and medullary hypoxia may even paradoxically improve if medullary perfusion is maintained while cortical vasoconstriction predominates, with consequent reduction in GFR and tubular transport load [4].

Thus, dyregulation of medullary oxygen balance is noted in conditions predisposing to CIN, related to altered renal parenchymal structure, to endothelial/vascular dysfunction, or to enhanced tubular transport and oxygen consumption.

Renal hypoxia caused by radiocontrast agents might be especially pronounced under such circumstances, particularly when systems designed to maintain medullary oxygen balance are hampered. In these perspectives, it is not surprising that the decline in renal medullary blood flow caused by CM markedly varies among patients, even under standardized administration protocols [17].

8 Risk Factors, Renal Hypoxia, and CIN: Lessons from Animal Models

Like humans, intact animals do not develop CIN. They either die with excessive volumes of CM or maintain kidney function and integrity. As detailed elsewhere [2, 90], it was found that the development of CIN models requires the induction of predisposing factors, based on clinically relevant scenarios, such as the reduction of functional renal mass (uninephrectomy) with compensatory hypertrophy of the contralateral kidney, subtotal (five sixths) nephrectomy, chronic salt depletion or dehydration, heart failure, angiotensin II infusion, transiently enhanced GFR, diabetes, hypercholesterolemia, acute urinary outflow obstruction, or short transient global renal ischemia. Another adopted strategy has been the inactivation of NO or prostaglandin synthesis, thus preventing adaptive renal responses designed to maintain medullary oxygenation [2, 90]. Many rat model studies in our laboratories were complemented by pO_2 determination, by hemodynamic and functional studies, and by detailed renal morphology and immunohistochemistry for hypoxia response in kidneys perfusion fixed in vivo. We found that CM-related reduction in medullary oxygenation is intensified by prior inactivation of NO or prostaglandin synthesis [21]. In these models medullary blood flow has been severely compromised, above and beyond the impact of radiocontrast alone. Hypoxia has also been documented by the accumulation of pimonidazole adducts and by the induction of HIF-mediated medullary hypoxia responses, both in tubular segments and in vascular endothelial and interstitial cells [11]. These findings underscore the significance of medullary protection by prostaglandins and NO in healthy individuals subjected to CM. In all these AKI models, renal functional impairment, manifested by declining GFR, develops over 24 h, with a gradual subsequent recovery over a few days [11, 67]. Sodium MRI detects tubular dysfunction within a few hours after the induction of CIN, with a loss of the corticomedullary sodium gradient [91], associated with a subsequent decline in fractional tubular sodium reabsorption [67].

Morphologic evaluation reveals hypoxic medullary injury that can be traced as early as 15 min after the administration of CM as condensed “dark” cells, (apoptotic changes) principally involving mTALs in the inner stripe of the outer medulla [5]. Injury gradient pattern is evident, principally affecting tubules at the midinterbundle zone, most remote from vasa recta and oxygen supply. mTAL damage progresses into a reversible phase of mitochondrial swelling and nuclear pyknosis, culminating in frank cell necrosis with cell membrane disruption [5, 21, 67]. Extensive apoptotic cell death has been documented as well [92–94]. Damage often spread to S3 segments of the proximal tubules in the adjacent outer stripe and medullary rays and to the renal papilla [21, 59, 67]. Collecting ducts are usually preserved, and inflammation is very limited. Proximal tubular vacuolar changes, detailed above, are unequivocally noted as well, but bear no relevance to the development of renal failure.

Importantly, this pattern of distal tubular injury is not specific to CM but also occurs in rats with advanced heart failure and advanced renal failure, following the inactivation of NO and prostaglandin synthesis [95], as well as in hypoxic isolated perfused kidneys [4, 96]. Interestingly, such injury is fully prevented by the inhibition of tubular transport activity [97]. Hypoxic injury parallels the distribution of pimonidazole adducts [11, 98]. This type of medullary hypoxic damage, predominantly affecting distal nephron segments, is quite distinct from the commonly utilized ischemia-reperfusion AKI models, produced by protracted total cessation of renal blood flow and characterized principally by ischemic proximal S3 tubular injury. Conceivably, in this latter type of AKI model, mTALs are paradoxically protected, since they can endure protracted severe hypoxia as long as transport activity is abolished [96].

As in humans, the degree of renal dysfunction is proportional to the number of applied predisposing perturbations [2]. In severe hypoxic distal nephron models with extensive tubular necrosis, the degree of renal dysfunction is directly proportional to the extent of medullary hypoxic damage (Fig. 4). By contrast, in conceivably more clinically relevant models with limited focal tubular injury, the decline in kidney function may be unproportionately high, suggesting altered renal hemodynamics in response to medullary hypoxia, perhaps through the activation of tubuloglomerular feedback mechanisms. This fits well with the observation of rather high tubular sodium reabsorption, often encountered in patients with CIN.

Over the past 5 years the same concepts led us to develop two more highly clinically relevant models of CIN in rats with experimental diabetes [99] or chronic tubulointerstitial disease [100]. Both conditions lead to chronic renal parenchymal hypoxia, the former through enhanced tubular transport and altered renal hemodynamics, and the latter via rarefaction of peritubular capillaries and putative oxygen diffusion defect caused by interstitial fibrosis [88]. Indeed, in both models we found a somewhat more pronounced renal dysfunction following exposure to radiocontrast. However, unexpectedly, the extent of hypoxic medullary injury was comparable to that developed in control animals. One explanation for this paradox might be that chronic hypoxia upregulates HIF-driven cellular protective mechanisms [100, 101], providing tissue tolerance to the acute CM-associated hypoxic insult.

Finally, the relevance of experimental models with distal nephron injury to clinical CIN, and to human AKI in general, is strongly supported by enhanced urinary excretion by mTALs of Tamm Horsfall protein following CM [54] and by increased urinary neutrophil gelatinase-associated lipocalin (NGAL), a urinary AKI biomarker predominantly derived from distal tubular segments, within 2 h after contrast administration, specifically in patients that subsequently develop CIN [102].

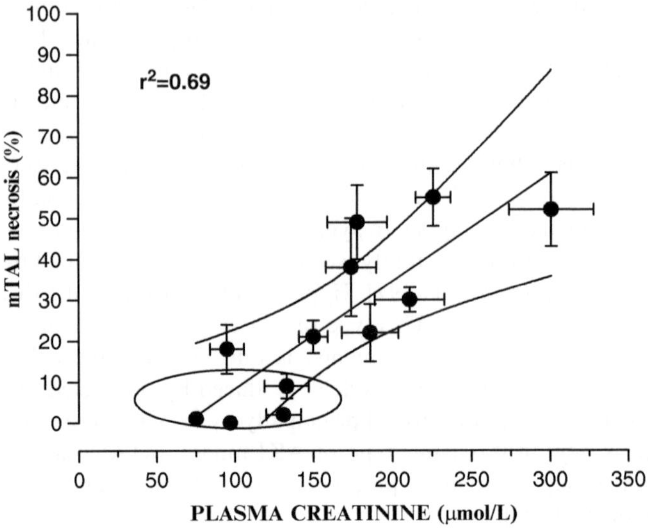

Fig. 4 Function-morphology correlation in animal models of contrast-induced nephropathy (CIN). Overall, the decline in creatinine clearance is proportional to the extent of outer medullary hypoxic injury, expressed as the percentage of necrotic mTALs. However, in models with limited overt tissue injury (*within the ellipse*), the decline in kidney function might yet be substantial, possibly reflecting altered renal hemodynamics

9 Evidence for Enhanced ROS Production in CIN

ROS, principally generated in the outer medulla [103], may increase following contrast administration, given the decline in medullary blood flow and oxygenation and the increase in tubular transport activity. Bakris et al., using anesthetized dogs, reported that renal venous malondialdehyde (MDA), a marker of lipid peroxidation, raised almost fourfold following contrast administration [104]. The administration of the xanthine-oxidase inhibitor allopurinol or superoxide dismutase (SOD) effectively prevented the drop in renal blood flow and GFR and reduced MDA in the renal vein. Toprak et al. also reported increased serum MDA and reduced thiol groups following CM administration in dehydrated rats [105], and Cetin et al. found increments in renal parenchymal MDA, which was prevented by ascorbic acid [106]. Similarly, Yoshioka and colleagues found that phosphatidylcholine and phosphatidylethanolamine hydroperoxide, additional specific products of membrane lipid peroxidation, more than doubled in salt-depleted rats that developed renal impairment following radiocontrast injection [107]. Efrati and colleagues describe a fivefold increase in cortical and medullary isoprostan throughout 30 min after the injection of high-osmolar agent, which has been modestly attenuated by *N*-acetyl-L-cysteine (NAC) at the 15-min time point [24]. Goodman and colleagues reported a substantial rise in renal parenchymal heme and an 80% increase in superoxide ion content in uninephrectomized, salt-depleted rats subjected to indomethacin and contrast [44]. This group and others [57] also reported increased renal expression of heme oxygenase (HO)-1, which suppresses ROS formation by multiple pathways [108, 109]. Indeed HO-1 inhibition in these animals subjected to radiocontrast further enhanced renal parenchymal superoxide formation [44].

Clinical data regarding CM-induced ROS formation, though limited, are in line with the above findings. In patients undergoing coronary angiography Efrati and colleagues showed that urinary F2 isoprostane excretion, a marker of lipid peroxidation, increased 28% [43]. Drager et al. also reported a threefold increase in this metabolite in the urine following coronary angiography in patients with stable chronic renal failure, which was prevented by NAC [110]. Urinary xanthine was also found to increase in patients given high-osmolar contrast agents, suggesting enhanced generation of ROS during adenosine degradation. Furthermore, the administration of the xanthine oxidase inhibitor allopurinol in a small cohort of such patients with low magnesium levels attenuated the decline in kidney function [111]. Finally, Fiaccadori et al. reported a brief twofold rise in urinary 3-nitrotyrosine, a stable metabolite of peroxinitrite, immediately following coronary angiography, which was proportional to the volume of injected CM [112]. This implies that CM administration results in the generation of superoxide, with subsequent formation of peroxinitrite by its chemical interaction with NO.

As illustrated in Table 1 and in many sections in this book, ROS formation is enhanced in most conditions known to predispose to CIN. Patients with chronic renal failure have increased oxidative stress and defective antioxidant systems [113–115], associated with inflammation and altered nitrovasodilation [116, 117]. Dehydration leads to a substantial downregulation of renal cortical SOD and catalase in rats, with GSH levels somewhat declining as well [107], rendering these animals susceptible to a deleterious effect of acutely formed ROS during contrast administration. Enhanced ROS formation has also been reported in other scenarios of effective volume depletion, such as heart failure [118] and liver disease [119, 120]. Diabetes is also characterized by enhanced ROS generation, presumably responsible for the development of renal parenchymal hypoxia [121]. Hypertension may also be associated with reduced medullary oxygenation [122] and enhanced transport-dependent [123, 124] or transport-independent [125] superoxide production by mTALs. Enhanced ROS has also been associated with anemia [126] and in the male gender [127], in hypercholesterolemia and atherosclerosis [116, 128, 129], in critically ill patients with systemic inflammation [130–132], in transplanted kidneys [133, 134], and in nephrotoxic AKI [135] and rhabdomyolysis [136]. As outlined below, it is conceivable that in all these circumstances, a priori ROS generation alters medullary oxygen balance and inactivates the homeostatic mechanisms that would have restored medullary oxygenation following CM administration in healthy subjects.

10 ROS and the Pathophysiology of CIN

CM-induced ROS generation could be directly toxic to tubules or alter renal hemodynamics, leading to secondary tubular hypoxic damage. The former mechanism is not supported by in vitro studies: LLC-PK1 necrosis and MDCK apoptosis were noted in cell cultures subjected to ionic high- and low-osmolar agents for 24 h and were intensified by hypoxia. However, hydrogen peroxide, superoxide anion, or MDA levels were not increased in both cell lines [75]. Additionally, NAC, ascorbic acid, α-tocopherol, glutathione, β-carotene, allopurinol, cimetidine, and citric acid did not attenuate the tubular cell damage. In another cell culture study, CM did not cause lipid peroxidation, and cell injury was not affected by antioxidants or pro-oxidant interventions [72]. Renal glutathione was also unaffected in a model of renal cortical slices incubated with CM [137]. Nonetheless, there were indications that contrast-induced mitochondrial impairment is a key factor in direct tubular cell toxicity. Interestingly, radiocontrast-induced apoptosis has been attenuated by increasing endogenous cAMP synthesis, and activation of A kinase/PI 3-kinase/Akt pathway, followed by phosphorylation of cyclic AMP response element binding protein (CREB) and enhanced expression of Bcl-2 [138]. The recent report by Romano et al. [77] provides the only hint for a direct CM-induced ROS-mediated tubular injury, showing attenuation of damage with NAC and ascorbic acid in LLC-PK1 and MDCK cells, subjected to high concentrations of CM (100–200 mg iodine/mL) for 3 h.

Thus, with the exception of Romano's report, most in vitro studies suggest that CM do not induce ROS generation directly in tubular cells. However, locally formed ROS in vivo during radiocontrast-induced medullary hypoxia and reoxygenation might intensify cellular hypoxic damage by membranal oxidative attack. Furthermore, ROS-induced DNA damage may activate high-energy–consuming reparative processes such poly-(ADP-ribose) polymerase (PARP), which may, in turn, initiate a vicious circle of additional intracellular energy store depletion and tubular damage. Indeed, PARP inhibition has been found to attenuate oxidative damage in vitro [139] and to improve renal dysfunction in a rat model of CIN [140]. ROS may also enhance medullary oxygen expenditure by disinhibition of transport in mTALs [141, 142]. In that way, enhanced mitochondrial generation of ROS, most prominent in mTALs, may, again, initiate a self-perpetuating system of ROS formation and cellular damage.

Perhaps most important in the pathogenesis of CIN is the adverse impact of ROS, formed during hypoxia on nitric oxide bioavailability and renal hemodynamics [143, 144]. Intense vasoconstriction, associated with reduced NO synthesis, has been noted in isolated vasa recta perfused with contrast agents and was prevented by the ROS scavenger tempol (4-hydroxy-2,2,6,6-tetramethylpiperidine-1-oxyl) [32]. Furthermore, these blood vessels displayed hyperresponsiveness to angiotensin II following contrast administration, which has been attenuated by tempol. Treatment with SOD also prevented contrast-induced altered NO generation in rats [23] and attenuated the transient decline in glomerular filtration in dogs infused with

CM [104]. Activation of other antioxidant mechanisms was also found to be beneficial. Allopurinol was as renoprotective as SOD in the dog model [104], while the induction of HO-1 in rats subjected to CIN prevented the rise in superoxide and the subsequent development of renal dysfunction and tubular cell apoptosis [44].

Importantly, a reciprocal effect exist of NO upon ROS. NO prevents ROS-mediated endothelial cell injury [145]. It also blunts flow- and transport-dependent ROS formation in mTALs [146]. Thus, increased urine tubular flow after CM injection, particularly in patients with altered NO synthesis, might enhance ROS formation and initiate the vicious circle, illustrated in Fig. 2, of ROS formation, intensified hypoxia, and medullary injury.

Finally, ROS might directly affect the activation of adaptive cellular mechanisms during hypoxia. A few reports indicate ROS-mediated attenuation of HIF response under certain conditions [147, 148], and we found that in diabetic rats, tempol intensified medullary HIF expression without affecting regional hypoxia [101]. However, ROS-mediated inhibition of HIF-associated tissue response remains a highly controversial issue. The other way around, upregulation of adaptive cellular mechanisms, such as HIF or its downstream products HO-1 and erythropoietin, were found to prevent experimental hypoxic renal injury, and erythropoietin was shown to attenuate CIN in vivo [94, 149] and to reduce contrast-related tubular cell injury in vitro [76]. It is possible that induction of antioxidants plays a role in these protective effects [44, 150, 151].

In summary, ROS, present in many predisposing clinical conditions and generated during the evolution of CIN, conceivably play a role in its pathophysiology. This has led to the development of preventive strategies, aimed at the attenuation of ROS formation and action.

11 Clinical Trials: The ROS Perspective

The complexity of animal CIN models, with limited consistency and large variability (as in humans), led to the initiation of clinical trials of antioxidant regimens without preceding animal studies, most often conducted in large cohorts of high-risk patients with preexisting renal failure. The outcome of such clinical trials, later supplemented by laboratory studies, provides important insight regarding the potential role of ROS in the pathophysiology of CIN. Interfering with ROS generation and action has been the principal approach in these clinical trials. As an example, a recent report of 204 patients with chronic renal failure suggests that the antioxidant probucol may prevent CIN [152]. The incidence of CIN in probucol-treated and control subjects was 8 and 15%, respectively, but even with this selected high-risk population much larger numbers of observations may be needed to statistically confirm probucol efficacy in preventing CIN. Few additional studies using ascorbic acid as an antioxidant provided conflicting results [153–156]. NAC and bicarbonate hydration are, however, the two most intensively explored ROS-relevant putative protective interventions in clinical CIN studies.

Since the first report by Tepel et al. in 2000 [157], a great dispute has been generated regarding the potential protective properties of NAC in high-risk patients undergoing radiocontrast studies, with some studies illustrating clear beneficial effects while others show none. Multiple meta-analyses gave conflicting messages, some claim that NAC is effective [158–160], while others underscore the inability to clearly prove NAC efficacy in such heterogeneous studies [161–165]. Indeed, great variability exists in these trials regarding patients' characteristics, undertaken radiological or interventional procedures, radiocontrast type and dose, hydration protocols, NAC dosage, timing, and mode of administration. Furthermore, even the definition of CIN differs between these studies, and a publication bias clearly affects the conclusions [166]. Nevertheless, we feel that based on large, well-controlled randomized prospective trials in a well-characterized population with chronic renal impairment [155, 167, 168], NAC, given in repeated large doses of 1,200 mg each, started a sufficient time before contrast administration, is likely to be renoprotective in patients with chronic renal failure, particularly when large volumes of contrast material are used. The incidence of CIN can roughly fall 50% with NAC prophylaxis under such circumstances.

NAC, a thiol-containing antioxidant, could serve as an ROS scavenger either directly or by replenishing intracellular antioxidants, such as glutathione. Indeed, NAC and ascorbic acid (but not sodium bicarbonate) dose dependently protected tubular cell cultures after 3 h' incubation with very high concentrations (200 mg iodine/mL) of both low- and iso-osmolar contrast agents [77]. NAC can also directly induce renal vasodilation following radiocontrast [169] and may attenuate the decline in medullary blood flow if given before the radiocontrast agent [24]. A substantial increase in renal PGE_2 in NAC-treated rats paralleled this effect [24], but NAC-induced vasodilation could also be mediated by nitric oxide. Indeed, NAC administration to patients undergoing coronary angiography provided renoprotection while preventing the decline in urinary nitric oxide end products [43]. It has been suggested that NAC enhances eNOS and NO generation and potentiates the effect of NO by binding to it and forming a more stable active substance. Lipid peroxidation (reflected by urinary isoprostanes) was unaffected by NAC in these series in humans [43] and was only mildly and transiently attenuated in rats (renal parenchymal isoprostanes) [24]. Thus some studies indicate that the protective impact of NAC is not entirely an antioxidant effect. This conclusion is further supported by clinical trials that show the lack of protection provided by ascorbic acid given alone or in addition to NAC [155, 156]. Finally, it has been suggested that NAC modify tubular handling of creatinine and might lead to the false impression of renal protection [170, 171], but the magnitude of this impact is perhaps negligible [172].

Bicarbonate infusion has also been proposed to be renoprotective in patients undergoing contrast studies. The rationale was that urine alkalization might reduce the formation of hydroxyl radicals, which usually takes place in acidic milieu. Furthermore, it has been suggested that bicarbonate might directly trap ROS, such as peroxynitrite.

In a pioneer report, Merten and colleagues studied the effect of sodium bicarbonate infusion, started 1 h before radiocontrast studies, and found that in patients with chronic renal failure the incidence of CIN, defined as a $\geq$25% rise in plasma creatinine, fell from 13.6% in control saline-treated high-risk patients to 1.7% in subjects given bicarbonate [173]. In a large prospective study, Briguori et al. found that in patients with chronic renal failure undergoing coronary angiography and interventions, the combination of NAC and bicarbonate hydration protocol was superior to NAC with saline hydration, with or without ascorbic acid [155]. The incidence of CIN fell from about 10% in the two latter groups to the incredibly low figure of 1.9% in patients given NAC and bicarbonate.

As with NAC clinical trials, subsequent studies revealed conflicting outcomes, with some reporting no effect at all [174], and one large retrospective analysis even suggesting a triple increase in the risk to develop CIN in patients treated with bicarbonate [175]. However, over the past few years repeated meta-analyses consistently showed a better outcome with bicarbonate, as compared with saline hydration, with about a 50% reduction in the incidence of CIN. Yet, there was no impact on the need for dialysis and in-hospital mortality [176–178]. Noteworthy, as with NAC, these meta-analyses suffer from pooling heterogeneous population with diverse patient- and procedure-related variables, and there might also be a publication bias. Thus, very large, multicenter, randomized prospective studies are needed to assess beyond a doubt the individual and combined protective potential of NAC and bicarbonate infusion, controlled for all these parameters.

If high urinary pH is instrumental in preventing CIN via attenuation of ROS formation in the tubular lumen, inhibition of renal carbonic anhydrase might be more effective than bicarbonate infusion. Indeed, urine alkalization with the administration of acetazolamide was found to ameliorate renal dysfunction in a CIN rat model [179]. Furthermore, in children with stable renal impairment undergoing radiocontrast studies, acetazolamide was found to better increase urine pH and to completely prevent CIN, as compared with an 8% incidence of CIN in a control bicarbonate infusion group of patients [180].

Two points of caution should be mentioned regarding bicarbonate and acetazolamide treatments and CIN. First, the impact of these strategies on vascular and tubular intracellular formation and action of ROS remains controversial, as pointed out by From et al. [175]. Second, bicarbonate treatment may hamper renal oxygenation since it alters O_2 dissociation from hemoglobin. Indeed, aggressive bicarbonate treatment for diabetic ketoacidosis may increase lactate, an indicator for poor oxygen delivery to tissues. Therefore, systemic bicarbonate could have a narrow renal therapeutic range.

Thus, clinical trials, using NAC, bicarbonate infusion, acetazolamide, and probucol provide circumstantial evidence that ROS might be involved in the pathogenesis of CIN. Noteworthy, however, the efficacy of these interventions, as well as their mode of putative beneficial action, remain controversial.

12 Conclusion

As illustrated in Fig. 2, the complex pathogenesis of CIN is a paradigm of hypoxic-toxic injury, involving altered renal microcirculation and hypoxia, as well as ROS-mediated cellular injury. Medullary hypoxic damage probably develops, especially in high-risk patients, in whom renal protective mechanisms that maintain medullary oxygen balance and prevent ROS generation and action are hampered. Formation of ROS results from the evolving hypoxia and reoxygenation and perpetuates renal hypoxia and damage through the initiation of endothelial/vascular dysfunction, the upregulation of tubular transport, the induction of oxygen-consuming reparative mechanisms, and by interfering with hypoxia-adaptive cell responses. ROS-mediated direct tubular toxicity seemingly is a secondary phenomenon in the pathogenesis of CIN.

Improvement of medullary oxygenation and inhibition of ROS formation and ROS scavenging are, therefore, reasonable therapeutic interventions. However, the protective properties attributed to NAC and to bicarbonate infusion and their putative action through disarming oxidative stress are yet to be proven.

Acknowledgements This work was supported by a grant from the Israeli Science Foundation (1473/08) and by the Harvard Medical Faculty Physicians at Beth Israel Deaconess Medical Center, Boston, MA

References

1. Tepel M, Aspelin P, Lameire N. Contrast-induced nephropathy: a clinical and evidence-based approach. Circulation 2006; 113:1799–1806
2. Heyman SN, Rosen S, Rosenberger C. Renal parenchymal hypoxia, hypoxia adaptation and the pathogenesis of radiocontrast nephropathy. Clinical J Am Soc Nephrol 2008; 3:288–296
3. Brezis M, Rosen S. Hypoxia of the renal medulla: implications for diseases. N Engl J Med 1995; 332:647–655
4. Rosenberger C, Rosen S, Heyman SN. Renal parenchymal oxygenation and hypoxia adaptation in acute kidney injury. Clin Exp Pharmacol Physiol 2006; 33:980–988
5. Heyman SN, Brezis M, Epstein FH, et al. Early renal medullary hypoxic injury from radiocontrast and indomethacin. Kidney Int 1991; 40:632–642
6. Liss P, Nygren A, Erikson U, et al. Injection of low and iso-osmolar contrast medium decreases oxygen tension in the renal medulla. Kidney Int 1998; 53:698–702
7. Lancelot E, Idee JM, Lacledere C, et al. Effects of two dimeric iodinated contrast media on renal medullary blood perfusion and oxygenation in dogs. Invest Radiol 2002; 37:368-375
8. Prasad PV, Priatna A, Spokes K, et al. Changes in intrarenal oxygenation as evaluated by BOLD MRI in a rat kidney model for radiocontrast nephropathy. J Magn Reson Imaging 2001; 13:744–747
9. Hofmann L, Simon-Zoula S, Nowak A, et al. BOLD-MRI for the assessment of renal oxygenation in humans: acute effect of nephrotoxic xenobiotics. Kidney Int 2006; 70:144–150
10. Rosenberger C, Rosen S, Paliege A, et al. Pimonidazole adduct immunohistochemistry in the rat kidney: detection of tissue hypoxia. Methods Mol Biol 2009; 466:161-174

11. Rosenberger C, Heyman SN, Rosen S, et al. Upregulation of HIF in acute renal failure–evidence for a protective transcriptional response to hypoxia. Kidney Int 2005; 67:531–542
12. Workman RJ, Shaff MI, Jackson RV, et al. Relationship of renal hemodynamic and functional changes following intravascular contrast to the renin-angiotensin system and renal prostacyclin in the dog. Invest Radiol 1983; 18:160–166
13. Cantley LG, Spokes K, Clark B, et al. Role of endothelin and prostaglandins in radiocontrast-induced renal artery constriction. Kidney Int 1993; 44:1217–1223
14. Idée JM, Lancelot E, Berthommier C, et al. Effects of non-ionic monomeric and dimeric iodinated contrast media on renal and systemic haemodynamics in rats. Fundam Clin Pharmacol 2000; 14:11–18
15. Hackstein N, Schneider C, Eichner G, et al. Effect of IV injection of radiographic contrast media on human renal blood flow. Am J Radiol 2007; 188:1367–1372
16. Tumlin JA, Wang A, Murray PT, et al. Fenoldopam mesylate blocks reductions in renal plasma flow after radiocontrast dye infusion: a pilot trial in the prevention of contrast nephropathy. Am Heart J 2002; 143:894–903
17. Möckel M, Radovic M, Kühnle Y, et al. Acute renal haemodynamic effects of radiocontrast media in patients undergoing left ventricular and coronary angiography. Nephrol Dial Transpl 2008; 23:1588–1594
18. Weisberg LS, Kurnik PB, Kurnik BR. Radiocontrast-induced nephropathy in humans: role of renal vasoconstriction. Kidney Int 1992; 41:1408–1415
19. Nygren A, Ulfendahl HR, Hansell P, et al. Effects of intravenous contrast media on cortical and medullary blood flow in the rat kidney. Invest Radiol 1988; 23:753–761
20. Liss P, Nygren A, Olsson U, et al. Effects of contrast media and mannitol on renal medullary blood flow and red cell aggregation in the rat kidney. Kidney Int 1996; 49:1268–1275
21. Agmon Y, Peleg H, Greenfeld Z, et al. Nitric oxide and prostanoids protect the renal outer medulla from radiocontrast toxicity in the rat. J Clin Invest 1994; 94:1069–1075
22. Heyman SN, Goldfarb M, Karmeli F, et al. The effect of radiocontrast on intra-renal nitric oxide (NO) and NO synthase activity. Exp Nephrol 1998; 6:557–562
23. Myers SI, Wang L, Liu F, et al. Iodinated contrast induced renal vasoconstriction is due in part to the downregulation of renal cortical and medullary nitric oxide synthesis. J Vasc Surg 2006; 44:383–391
24. Efrati S, Berman S, Ilgiyeav I, et al. Differential effects of *N*-acetylcysteine, theophylline or bicarbonate on contrast-induced rat renal vasoconstriction. Am J Nephrol 2009; 29:181–191
25. Palm F, Carlsson PO, Fasching A, et al. Effects of the contrast medium iopromide on renal hemodynamics and oxygen tension in the diabetic rat kidney. Adv Exp Med Biol 2003; 530:653–659
26. Liss P, Nygren A, Hansell P. Hypoperfusion in the renal outer medulla after injection of contrast media in rats. Acta Radiol 1999; 40:521–527
27. Liss P, Aukland K, Carlsson PO, et al. Influence of iothalamate on renal medullary perfusion and oxygenation in the rat. Acta Radiol 2005; 46:823–829
28. Zhang H, Holt CM, Malik N, et al. Effects of radiographic contrast media on proliferation and apoptosis of human vascular endothelial cells. Br J Radiol 2000; 73:1034–1041
29. Franke RP, Fuhrmann R, Park JW, et al. The effect of radiographic contrast media on the morphology of human venous endothelial cells. Clin Hemorheol Microcirc 2007; 37:329–338
30. Franke RP, Fuhrmann R, Hiebl B, et al. Influence of various radiographic contrast media on the buckling of endothelial cells. Microvasc Res 2008; 76:110–113
31. Zhao J, Huang Y, Song Y, et al. Low osmolar contrast medium induces cellular injury and disruption of calcium homeostasis in rat glomerular endothelial cells in vitro. Toxicol Lett 2009; 185:124–131

32. Sendeski M, Patzak A, Pallone TL, et al. Iodixanol, constriction of medullary descending vasa recta, and risk for contrast medium-induced nephropathy. Radiology 2009; 251 (3):697–704
33. Arakawa K, Suzuki H, Naitoh M, et al. Role of adenosine in the renal responses to contrast medium. Kidney Int 1996; 49:1199–1206
34. Pflueger A, Larson TS, Nath KA, et al. Role of adenosine in contrast media-induced acute renal failure in diabetes mellitus. Mayo Clin Proc 2000; 75:1275–1283
35. Agmon Y, Dinour D, Brezis M. Disparate effects of adenosine A1 and A2-receptor agonists on intrarenal blood flow. Am J Physiol 1993; 265:F802–F806
36. Arend LJ, Bakris GL, Burnett JC, et al. Role for intrarenal adenosine in the renal hemodynamic response to contrast media. J Lab Clin Med 1987; 110:406–411
37. Parvez Z, Marsan RE, Moncada R, et al. Effect of contrast media on prostaglandin synthesis in vivo. Invest Radiol 1988; 23 (Suppl 1): S178–S181
38. Heyman SN, Clark BA, Kaiser N, et al. Radiocontrast agents induce endothelin release in vivo and in vitro. J Am Soc Nephrol 1992; 3:58–65
39. Clark BA, Kim D, Epstein FH. Endothelin and atrial natriuretic peptide levels following radiocontrast exposure in humans. Am J Kidney Dis 1997; 30:82–86
40. Khamaisi M, Raz I, Shilo V, et al. Diabetes and radiocontrast media increase endothelin converting enzyme-1 in the kidney. Kidney Int 2008; 74:91–100
41. Gurbanov K, Rubinstein I, Hoffman A, et al. Differential regulation of renal regional blood flow by endothelin-1. Am J Physiol 1996; 271: F1166–1172
42. Brezis M, Heyman SN, Dinour D, et al. Role of nitric oxide in renal medullary oxygen balance. Studies in isolated and intact rat kidneys. J Clin Invest 1991; 88:390–395
43. Efrati S, Dishy V, Averbukh M, et al. The effect of *N*-acetylcysteine on renal function, nitric oxide, and oxidative stress after angiography. Kidney Int 2003; 64:2182–2188
44. Goodman A, Olszanecki R, Yang LM, et al. Heme oxygenase-1 protects against radiocontrast induced acute kidney injury by regulating anti-apoptotic proteins. J Am Soc nephrol 2007; 72:945–953
45. Andrade L, Campos SB, Seguro AC. Hypercholesterolemia aggravates radiocontrast nephrotoxicity: protective role of L-arginine. Kidney Int 1998; 53:1736–1742
46. Touati C, Idee JM, Deray G, et al. Modulation of the renal effects of contrast media by endothelium- derived nitric oxide in the rat. Invest Radiol 1993; 28:814–820
47. Heyman SN, Goldfarb M, Darmon D, et al. Tissue oxygenation modifies nitric oxide bioavailability. Microcirculation 1999; 6:199–203
48. Zhang W, Pibulsonggram T, Edwards A. Determinants of basal nitric oxide concentration in the renal medullary microcirculation. Am J Physiol 2004; 287:F1189–F1203
49. Ueda J, Nygren A, Sjoquist M, et al. Iodine concentrations in the rat kidney measured by X-ray microanalysis. Comparison of concentrations and viscosities in the proximal tubules and renal pelvis after intravenous injections of contrast media. Acta Radiol 1998; 39:90–95
50. Seeliger E, Flemming B, et al. Viscosity of contrast media perturbs renal hemodynamics. J Am Soc Nephrol 2007; 18:2912–2920
51. Persson PB, Hansell P, Liss P. Pathophysiology of contrast medium-induced nephropathy. Kidney Int 2005; 68:14–22
52. Ueda J, Nygren A, Hansell P, et al. Effect of intravenous contrast media on proximal and distal tubular hydrostatic pressure in the rat kidney. Acta Radiol 1993; 34:83–87
53. Dawnay AB, Thornley C, Nockler I, et al. Tamm-Horsfall glycoprotein excretion and aggregation during intravenous urography. Relevance to acute renal failure. Invest Radiol 1985; 20:53–57
54. Bakris GL, Gaber AO, Jones JD. Oxygen free radical involvement in urinary Tamm-Horsfall protein excretion after intrarenal injection of contrast medium. Radiology 1990; 175:57–60
55. Holland MD, Galla JH, Sanders PW, et al. Effect of urinary pH and diatrizoate on Bence Jones protein nephrotoxicity in the rat. Kidney Int 1985; 27:46–50

56. Seeliger E, Flemming B, Wronski T, et al. Viscosity of contrast media perturbs renal hemodynamics. J Am Soc Nephrol 2007; 18:2912–2920
57. Jost G, Pietsch H, Sommer J, et al. Retention of iodine and expression of biomarkers for renal damage in the kidney after application of iodinated contrast media in rats. Invest Radiol 2009; 44:114–123
58. Unal B, Kara S, Bilgili Y. Contrast media induces hypoperfusion in kidneys with ureteral stone: Doppler US study. Ultrasound Med Biol 2005; 31:31–37
59. Heyman SN, Fuchs S. Yafe R, et al. Renal microcirculation and tissue damage during acute ureteral obstruction: the effect of saline infusion, indomethacin and radiocontrast. Kidney Int 1997; 51:653–663
60. Russo D, Minutolo R, Cianciaruso B, et al. Early effects of contrast media on renal hemodynamics and tubular function in chronic renal failure. J Am Soc Nephrol 1995; 6:1451–1458
61. Perico N, Cornejo RP, Benigni A, et al. Endothelin induces diuresis and natriuresis in the rat by acting on proximal tubular cells through a mechanism mediated by lipoxygenase products. J Am Soc Nephrol 1991; 2:57–69
62. Heyman SN, Brezis M, Greenfeld Z, et al. Protective role of furosemide and saline in radiocontrast-induced acute renal failure in the rat. Am J Kidney Dis 1989; 14:377–385
63. Love L, Lind JA Jr, Olson MC. Persistent CT nephrogram: significance in the diagnosis of contrast nephropathy. Radiology 1989; 172:125–129
64. Moreau JF, Droz D, Sabto J, et al. Osmotic nephrosis induced by water-soluble tri-iodinated contrast media in man. Radiology 1975; 115:329–336
65. Moreau JF, Droz D, Noel LH. Nephrotoxicity of metrizamide in man. Lancet 1978; 1:1201
66. Moreau JF, Noel LH, Droz D, et al. Nephrotoxicity of ioxaglic acid (AG 6227 or P286) in humans. Invest Radiol 1978; 13:554–555
67. Heyman SN, Brezis M, Reubinoff CA, et al. Acute renal failure with selective medullary injury in the rat. J Clin Invest 1988; 82:401–412
68. Tervahartiala P. Contrast media-induced renal tubular vacuolization after dehydration. A light and electron microscopic study in rats. Invest Radiol 1992; 27:114–118
69. Tervahartiala P, Kivisaari L, Kivisaari R, et al. Structural changes in the renal proximal tubular cells induced by iodinated contrast media. Nephron 1997; 76:96–102
70. Dobrota M, Powell CJ, Holtz E, et al. Biochemical and morphological effects of contrast media on the kidney. Acta Radiologica Suppl 1995; 399:196–203
71. Haller C, Hizoh I. The cytotoxicity of iodinated radiocontrast agents on renal cells in vitro. Invest Radiol 2004; 39:149–154
72. Zager RA, Johnson ACM, Hanson SY. Radiographic contrast media–induced tubular injury: evaluation of oxidant stress and plasma membrane integrity. Kidney Int 2003; 64:128–139
73. Andersen KJ, Christensen EI, Vik H. Effects of iodinated X-ray contrast media on renal epithelial cells. Invest Radiol 1994; 29:955–962
74. Schick CS, Haller C. Comparative cytotoxicity of ionic and non-ionic radiocontrast agents on MDCK cell monolayers in vitro. Nephrol Dial Transplant 1999; 14:342–347
75. Garofalo AS, Borges FT, Dalboni MA, et al. Reactive oxygen species independent cytotoxicity induced by radiocontrast agents in tubular cells (LLC-PK1 and MDCK). Ren Fail 2007; 29:121–131
76. Kolyada AY, Liangos O, Madias NE, et al. Protective effect of erythropoietin against radiocontrast-induced renal tubular epithelial cell injury. Am J Nephrol 2008; 28:203–209
77. Romano G, Briguori C, Quintavalle C, et al. Contrast agents and renal cell apoptosis. Eur Heart J 2008; 29(20):2569–2576
78. Schick CS, Bangert R, Kübler W, et al. Ionic Radiocontrast Media Disrupt intercellular contacts via an extracellular calcium-independent mechanism. Exp Nephrol 2002; 10:209–215

79. Rich MW, Crecelius CA, Incidence, risk factors, and clinical course of acute renal insufficiency after cardiac catheterization in patients 70 years of age or older. A prospective study. Arch Int Med 1990; 150:1237–1242
80. McCullough PA, Wolyn R, Rocher LL, et al. Acute renal failure after coronary intervention: incidence, risk factors, and relationship to mortality. Am J Med 1997; 103:368–375
81. Marenzi G, Lauri G, Assanelli E, et al. Contrast-induced nephropathy in patients undergoing primary angioplasty for acute myocardial infarction. J Am Coll Cardiol 2004; 44:1780–1785
82. Solomon R, Briguori C, Biguori C, et al. Selection of contrast media. Kidney Int Suppl 2006; 100:S39–S45
83. Solomon RJ, Natarajan MK, Doucet S, et al. Cardiac Angiography in Renally Impaired Patients (CARE) study: a randomized double-blind trial of contrast-induced nephropathy in patients with chronic kidney disease. Circulation 2007; 115:3189–3196
84. Heinrich MC, Haberle L, Muller V, et al. Nephrotoxicity of iso-osmolar iodixanol compared with nonionic low-osmolar contrast media: meta-analysis of randomized controlled trials. Radiology 2009; 250:68–86
85. Liss P, Persson PB, Hansell P, et al. Renal failure in 57925 patients undergoing coronary procedures using iso-osmolar or low-osmolar contrast media. Kidney Int 2006; 70:1811–1817
86. Weisbord SD, Palevsky PM. Prevention of contrast-induced nephropathy with volume expansion. Clin J Am Soc Nephrol 2008; 3:273–280
87. Nangaku M. Chronic hypoxia and tubulointerstitial injury: a final common pathway to end-stage renal failure. J Am Soc Nephrol 2006; 17:17–25
88. Heyman SN, Khamaisi M, Rosen S, et al. Renal parenchymal hypoxia, hypoxia response and the progression of chronic kidney disease. Am J Nephrol 2008; 28:998–1006
89. Pflueger A, Abramowitz D, Calvin AD. Role of oxidative stress in contrast-induced acute kidney injury in diabetes mellitus. Med Sci Monit 2009; 15:RA125–RA136
90. Heyman SN, Reichman J, Brezis M. The pathophysiology of contrast nephropathy: a role for medullary hypoxia. Invest Radiol 1999; 34:685–691
91. Maril N, Margalit R, Rosen S, et al. Detection of evolving acute tubular necrosis with renal ^{23}Sodium MRI: studies in Rats. Kidney Int 2006; 69:765–768
92. Beeri R, Symon Z, Brezis M, et al. Rapid DNA fragmentation from hypoxia along the thick ascending limb of rat kidneys. Kidney Int 1995; 47:1806–1810
93. Damianovich M, Ziv I, Heyman SN, et al. ApoSense: a novel technology for imaging of cell death in acute renal tubular necrosis. Eur J Nucl Med Mol Imaging 2006; 33:281–291
94. Yokomaku Y, Sugimoto T, Kume S, et al. Asialoerythropoietin prevents contrast-induced nephropathy. J Am Soc Nephrol 2008; 19:321–328
95. Goldfarb M, Abassi Z, Rosen S, et al. Chronic heart failure predisposes to hypoxic outer medullary tubular injury: studies in rats. Kidney Int 2001; 60:607–613
96. Heyman SN, Rosenberger C, Rosen S. Experimental ischemia-reperfusion – biases and myths: the proximal vs. distal hypoxic tubular injury debate revisited. Kidney Int 2010; 77(1):9–16
97. Brezis M, Rosen S, Silva P, et al. Transport activity modifies thick ascending limb damage in the isolated perfused kidney. Kidney Int 1984; 25:65–72
98. Rosenberger C, Shina A, Rosen S, et al. Hypoxia inducible factors and tubular cell survival in isolated perfused kidneys. Kidney Int 2006; 70:60–70
99. Rosenberg C, Goldfarb M, Khamaisi M, et al. Acute kidney injury in the diabetic rat: studies in the isolated perfused and intact kidney. Am J Nephrol 2008; 28:831–839
100. Goldfarb M, Rosenberger C, Abassi Z, et al. Acute-on-chronic renal failure in the rat: functional compensation and hypoxia tolerance. Am J Nephrol 2006; 26:22–33
101. Rosenberger C, Khamaisi M, Abassi Z, et al. Adaptation to hypoxia in the diabetic rat kidney. Kidney Int 2008; 73:34–42
102. Hirsch R, Dent C, Pfriem H, et al. NGAL is an early predictive biomarker of contrast-induced nephropathy in children. Pediatr Nephrol 2007; 22:2089–2095

103. Zou AP, Li N, Cowley AW Jr. Production and actions of superoxide in the renal medulla. Hypertension 2001; 37:547–553
104. Bakris GL, Lass N, Gaber AO, et al. Radiocontrast medium induced declines in renal function: a role for oxygen free radicals. Am J Physiol 1990; 258:F115–F120
105. Toprak O, Cirit M, Tanrisev M, et al. Preventive effect of nebivolol on contrast-induced nephropathy in rats. Nephrol Dial Transplant 2008; 23:853–859
106. Cetin M, Devrim E, Serin Kilicoglu S, et al. Ionic High-osmolar contrast medium causes oxidant stress in kidney tissue: partial Protective Role of Ascorbic Acid. Renal Failure 2008; 30:567–572
107. Yoshioka T, Fogo A, Beckman FK. Reduced activity of antioxidant enzymes underlies contrast media-induced renal injury in volume depletion. Kidney Int 1992; 41:1008–1015
108. Abraham NG, Kappas A. Heme oxygenase and the cardiovascular–renal system. Free Radic Biol Med 2005; 39:1–25
109. Vera T, Kelsen S, Yanes LL, et al. HO-1 induction lowers blood pressure and superoxide production in the renal medulla of angiotensin II hypertensive mice. Am J Physiol 2007; 292: R1472–R1478
110. Drager L, Andrade L, Barros de Toledo JF, et al. Renal effects of *N*-acetylcysteine in patients at risk for contrast nephropathy: decrease in oxidant stress-mediated renal tubular injury. Nephrol Dial Transplant 2004; 19:1803–1807
111. Katholi RE, Woods WT Jr, Taylor GJ, et al. Oxygen free radicals and contrast nephropathy. Am J Kidney Dis 1998; 32:64–71
112. Fiaccadori E, Maggiore U, Rotelli C, et al. Plasma and urinary free 3-nitrotyrosine following cardiac angiography procedures with non-ionic radiocontrast media. Nephrol Dial Transplant 2004; 19:865–869
113. Martin-Mateo MC. Oxidative stress in chronic renal failure. Ren Fail 1999; 21:155–167
114. Yoshimura S. Plasma glutathione peroxidase deficiency caused by renal dysfunction. Nephron 1996; 73:207–211
115. Mimic Oka J. Alteration in plasma antioxidant capacity in various degrees of renal failure. Clin Nephrol 1999; 51:233–241
116. Okamura DM, Pennathur S, Pasichnyk K, et al. CD36 regulates oxidative stress and inflammation in. hypercholesterolemic CKD. J Am Soc Nephrol 2009; 20:495-505
117. Cachofeiro V, Goicochea M, de Vinuesa SG, et al. Oxidative stress and inflammation, a link between chronic kidney disease and cardiovascular disease. Kidney Int Suppl 2008; 111: S4–S9
118. Tojo A, Onozato ML, Kobayashi N, et al. Angiotensin II and oxidative stress in Dahl Salt-sensitive rat with heart failure. Hypertension 2002; 40:834–839
119. Gracia-Sancho J, Lavina B, Rodriguez-Vilarrupla A, et al. Increased oxidative stress in cirrhotic rat livers: a potential mechanism contributing to reduced nitric oxide bioavailability. Hepatology 2008; 47:1248–1256
120. Vidali M, Occhino G, Ivaldi A, et al. Combination of oxidative stress and steatosis is a risk factor for fibrosis in alcohol-drinking patients with chronic hepatitis C. Am J Gastroenterol 2008; 103:147–153
121. Palm F, Cederberg J, Hansell P, et al. Reactive oxygen species cause diabetes-induced decrease in renal oxygen tension. Diabetologia 2003; 46:1153–1160
122. Welch WJ, Baumgartl H, Lubbers D, et al. Nephron PO_2 and renal oxygen usage in the hypertensive rat kidney. Kidney Int 2001; 59:230–237
123. Silva GB, Garvin JL. Angiotensin II-dependent hypertension increases Na transport-related oxygen consumption by the thick ascending limb. Hypertension 2008; 52:1091–1098
124. Abe M, O'Connor P, Kaldunski M, et al. Effect of sodium delivery on superoxide and nitric oxide in the medullary thick ascending limb. Am J Physiol 2006; 291:F350–F357
125. O'Connor PM, Lu L, Schreck C, et al. Enhanced amiloride-sensitive superoxide production in renal medullary thick ascending limb of Dahl salt-sensitive rats. Am J Physiol 2008; 295: F726–F733

126. Rossert J, Froissart M. Role of anemia in progression of chronic kidney disease. Semin Nephrol 2006; 26:283–289
127. Wang X, Desai K, Juurlink BH, et al. Gender-related differences in advanced glycation endproducts, oxidative stress markers and nitric oxide synthases in rats. Kidney Int 2006; 69:281–287
128. Chade AR, Mushin OP, Zhu X, et al. Pathways of renal fibrosis and modulation of matrix turnover in experimental hypercholesterolemia. Hypertension 2005; 46:772–779
129. Chade AR, Lerman A, Lerman LO. Kidney in early atherosclerosis. Hypertension 2005; 45:1042–1049
130. Biesalski HK, McGregor GP. Antioxidant therapy in critical care – is the microcirculation the primary target? Crit Care Med 2007; 35 (9 Suppl):S577–S583
131. Motoyama T, Okamoto K, Kukita I, et al. Possible role of increased oxidant stress in multiple organ failure after systemic inflammatory response syndrome. Crit Care Med 2003; 31:1048–1052
132. Goode HF, Cowley HC, Walker BE, et al. Decreased antioxidant status and increased lipid peroxidation in patients with septic shock and secondary organ dysfunction. Crit Care Med 1995; 23:646–651
133. Raj DS, Lim G, Levi M, et al. Advanced glycation end products and oxidative stress are increased in chronic allograft nephropathy. Am J Kidney Dis 2004; 43:154–160
134. Djamali A. Oxidative stress as a common pathway to chronic tubulointerstitial injury in kidney allografts. Am J Physiol 2007; 293:F445–F455
135. Baliga R, Ueda N, Walker PD, et al. Oxidant mechanisms in toxic acute renal failure. Am J Kidney Dis 1997; 29:465–477
136. Nath KA, Balla G, Vercellotti GM, et al. Induction of heme oxygenase is a rapid, protective response in rhabdomyolysis in the rat. J Clin Invest 1992; 90:267–270
137. Harmon RC, Duffy SP, Terneus MV, et al. Characterization of a novel model for investigation of radiocontrast nephrotoxicity. Nephrol Dial Transplant 2009; 24:763–268
138. Yano T, Itoh Y, Kubota T, et al. Prostacyclin analog prevents radiocontrast nephropathy via phosphorylation of cyclic AMP response element binding protein. Am J Pathol 2005; 166:1333–1342
139. Filipovic DM, Meng X, Reeves WB. Inhibition of PARP prevents oxidant-induced necrosis but not apoptosis in LLC-PK1 cells. Am J Physiol 1999; 277:F428–F436
140. Darmon D, Goldfarb M, Shina A, et al. The effect of poly-(ADP-ribose) polymerase (PARP) inhibition on outer medullary hypoxic damage. Nephron Physiol 2003; 95:1–9
141. Juncos R, Garvin JL. Superoxide enhances Na-K-2Cl cotransporter activity in the thick ascending limb. Am J Physiol 2005; 288:F982–F987
142. Evans RG, Fitzgerald SM. Nitric oxide and superoxide in the renal medulla: a delicate balancing act. Curr Opin Nephrol Hypertens 2005; 14:9–15
143. Schnackenberg CG. Physiological and pathophysiological roles of oxygen radicals in the renal microvasculature. Am J Physiol 2002; 282: R335–R342
144. Mori T, O'Connor PM, Abe M, et al. Enhanced superoxide production in renal outer medulla of Dahl salt-sensitive rats reduces nitric oxide tubular-vascular cross-talk. Hypertension 2007; 49:1336–1341
145. Cortese-Krott MM, Suschek CV, Wetzel W, et al. Nitric oxide-mediated protection of endothelial cells from hydrogen peroxide is mediated by intracellular zinc and glutathione. Am J Physiol 2009; 296: C811–C820
146. Hong NJ, Garvin JL. Nitric oxide reduces flow-induced superoxide production via cGMP-dependent protein kinase in thick ascending limbs. Am J Physiol 2009; 296: F1061–F1066
147. Yang ZZ, Zhang AY, Yi FX, et al. Redox regulation of HIF-1alpha levels and HO-1 expression in renal medullary interstitial cells. Am J Physiol 2003; 284:F1207–F1215
148. Zou AP Cowley AW. Reactive oxygen species and molecular regulation of renal oxygenation. Acta Physiol Scand 2003; 179:233–241

149. Goldfarb M, Rosenberger C, Shina A, et al. A role for erythropoietin in the attenuation of contrast medium-induced acute renal failure in rats Renal Failure 2006; 28:345–350
150. Chakraborty M, Ghosal J, Biswas T, et al. Effect of erythropoietin on membrane lipid peroxidation, superoxide dismutase, catalase, and glutathione peroxidase of rat RBC. Biochem Med Metab Biol 1988; 40:8–18
151. Cetin H, Olgar S, Oktem F, et al. Novel evidence suggesting an anti-oxidant property for erythropoietin on vancomycin-induced nephrotoxicity in a rat model. Clin Exp Pharmacol Physiol 2007; 34:1181–1185
152. Li G, Yin L, Liu T, et al. Role of probucol in preventing contrast-induced acute kidney injury after coronary interventional procedure. Am J Cardiol 2009; 103:512–514
153. Spargias K, Alexopoulos E, Kyrzopoulos S, et al. Ascorbic acid prevents contrast-mediated nephropathy in patients with renal dysfunction undergoing coronary angiography or intervention. Circulation 2004; 110:2837–2842
154. Boscheri A, Weinbrenner C, Botzek B, et al. Failure of ascorbic acid to prevent contrast-media induced nephropathy in patients with renal dysfunction. Clin Nephrol 2007; 68:279–286
155. Briguori C, Airoldi F, D'Andrea D, et al. Renal insufficiency following contrast media administration trial (REMEDIAL). A randomized comparison of 3 preventive strategies. Circulation 2007; 115:1211–1217
156. Jo SH, Koo BK, Park JS, et al. N-acetylcysteine versus ascorbic acid for preventing contrast-induced nephropathy in patients with renal insufficiency undergoing coronary angiography. NASPI study – a prospective randomized controlled trial. Am Heart J 2009; 157:576–583
157. Tepel M, van der Giet M, Schwarzfeld C, et al. Prevention of radiographic-contrast-agent-induced reductions in renal function by acetylcysteine. N Engl J Med 2000; 343:180–184
158. Birck R, Krzossok S, Markowetz F, et al. Acetylcysteine for prevention of contrast nephropathy: meta-analysis. Lancet 2003; 362:598–603
159. Alonso A, Lau J, Jaber BL, et al. Prevention of radiocontrast nephropathy with *N*-acetylcysteine in patients with chronic kidney disease: a meta-analysis of randomized, controlled trials. Am J Kid Dis 2004; 43:1–9
160. Kelly AM, Dwamena B, Cronin P, et al. Meta-analysis: effectiveness of drugs for preventing contrast-induced nephropathy. Ann Intern Med 2008; 148:284–294
161. Zagler A, Azadpour M, Mercado C, et al. *N*-acetylcysteine and contrast-induced nephropathy: a meta-analysis of 13 randomized trials. Am Heart J 2006; 151:140–145
162. Nallamothu BK, Shojania KG, Saint S, et al. Is acetylcysteine effective in preventing contrast-related nephropathy? A meta-analysis. Am J Med 2004; 117:938–947
163. Pannu N, Manns B, Lee H, et al. Systematic review of the impact of *N*-acetylcysteine on contrast nephropathy. Kidney Int 2004; 65:1366–1374
164. Gonzales DA, Norsworthy KJ, Kern SJ, et al. A meta-analysis of *N*-acetylcysteine in contrast-induced nephrotoxicity: unsupervised clustering to resolve heterogeneity. BMC Med 2007; 5:32
165. Kshirsagar AV, Poole C, Mottl A, et al. *N*-acetylcysteine for the prevention of radiocontrast induced nephropathy: a meta-analysis of prospective controlled trials. J Am Soc Nephrol 2004; 15:761–769
166. Vaitkus PT, Brar C. *N*-acetylcysteine in the prevention of contrast-induced nephropathy: publication bias perpetuated by meta-analyses. Am Heart J 2007; 153:275–280
167. Briguori C, Colombo A, Violante A, et al. Standard vs. double dose of *N*-acetylcysteine to prevent contrast agent associated nephrotoxicity. Eur Heart J 2004; 20:206–211
168. Marenzi G, Assanelli E, Marana I, et al. *N*-acetylcysteine and contrast-induced nephropathy in primary angioplasty. N Engl J Med 2006; 354:2773–2782
169. Heyman SN, Goldfarb M, Shina A, et al. *N*-acetylcysteine (NAC) ameliorates renal microcirculation: studies in rats. Kidney Int 2003; 63:634–641
170. Hoffmann U, Fischereder M, Kruger B, et al. The value of *N*-acetylcysteine in the prevention of radiocontrast agent-induced nephropathy seems questionable. JASN 2004; 15:407–410

171. Poletti PA, Saudan P, Platon A, et al. I.v. *N*-acetylcysteine and emergency CT: use of serum creatinine and cystatin C as markers of radiocontrast nephrotoxicity. AJR 2007; 189:687–692
172. Rehman T, Fought J, Solomon R. *N*-acetylcysteine effect on serum creatinine and cystatin C levels in CKD patients. Clin J Am Soc Nephrol 2008; 3:1610–1614
173. Merten GJ, Burgess WP, Gray LV, et al. Prevention of contrast-induced nephropathy with sodium bicarbonate: a randomized controlled trial. JAMA 2004; 291:2328–2334
174. Brar SS, Shen AY, Jorgensen MB, et al. Sodium bicarbonate vs. sodium chloride for the prevention of contrast medium-induced nephropathy in patients undergoing coronary angiography: a randomized trial. JAMA 2008; 300:1038–1046
175. From AM, Bartholmai BJ, Williams AW, et al. Sodium bicarbonate is associated with an increased incidence of contrast nephropathy: a retrospective cohort study of 7977 patients at Mayo Clinic. Clin J Am Soc Nephrol 2008; 3:10–18
176. Hogan SE, L'Allier P, Chetcuti S, et al. Current role of sodium bicarbonate-based preprocedural hydration for the prevention of contrast-induced acute kidney injury: a meta-analysis. Am Heart J 2008; 156:414–421
177. Navaneethan SD, Singh S, Appasamy S, et al. Sodium bicarbonate therapy for prevention of contrast-induced nephropathy: a systematic review and meta-analysis. Am J Kidney Dis 2009; 53:617–627
178. Meier P, Ko DT, Tamura A, et al. Sodium bicarbonate-based hydration prevents contrast-induced nephropathy: a meta-analysis. BMC Med 2009; 7:23
179. Hellman RN, Novikov M, Campos S, et al. Acetazolamide is renoprotective in a rat model of high dose contrast media induced renal failure. J Am Soc Nephrol 2004; 15:717A
180. Assadi F. Acetazolamide for prevention of contrast-induced nephropathy: a new use for an old drug. Pediatr Cardiol 2006; 27:238–242
181. Zhou XJ, Rakheja D, Yu X, et al. The aging kidney. Kidney Int 2008; 74:710–720
182. Deji N, Kume S, Araki S, et al. Structural and functional changes in the kidneys of high-fat diet-induced obese mice. Am J Physiol 2009; 296: F118–F126
183. Reilly MP, Delanty N, Roy L, et al. Increased formation of the isoprostanes IPF2alpha-I and 8-epi-prostaglandin F2alpha in acute coronary angioplasty: evidence for oxidant stress during coronary reperfusion in humans. Circulation 1997; 96:3314–3320
184. Shimomura H, Ogawa H, Takazoe K, et al. Comparison of urinary biopyrrin levels in acute myocardial infarction (after reperfusion therapy) versus stable angina pectoris and their usefulness in predicting subsequent cardiac events. Am J Cardiol 2002; 90:108–111
185. Dragu R, Behar S, Sandach A, et al. Should primary percutaneous coronary intervention be the preferred method of reperfusion therapy for patients with renal failure and ST-elevation acute myocardial infarction? Am J Cardiol 2006; 97:1142–1145
186. Cooper HA, Monge C, Panza JA. Patients with end-stage renal disease and acute myocardial infarction have poor short-term outcomes despite modern cardiac intensive care. Coron Artery Dis 2008; 19:231–235
187. Kowalczyk J, Lenarczyk R, Kowalski O, et al. Different types of renal dysfunction in patients with acute myocardial infarction treated with percutaneous coronary intervention. J Interv Cardiol 2007; 20:143–152

Chapter 13
Cardiovascular Complications in Renal Failure: Implications of Advanced Glycation End Products and Their Receptor RAGE

Hidenori Koyama and Yoshiki Nishizawa

Abstract Cardiovascular disease is the major cause of death in patients with renal insufficiency, accounting for 50% of all deaths in renal replacement therapy patients and in recipients of renal transplants. Mortality from cardiovascular diseases in patients with renal failure is approximately 9% per year, which is about 30 times the risk in the general population. Evidence has emerged pointing to a potential role of oxidative stress in generation of advanced glycation end products (AGEs) in patients with end stage renal disease (ESRD). Moreover, interaction of the receptor for AGEs (RAGE) with AGEs leads to crucial biomedical pathway, generating intracellular oxidative stress and inflammatory mediators, which could result in further amplification of the pathway involved in AGE generation. AGEs and RAGE can profoundly be involved in cardiovascular diseases through regulation of (a) atherogenesis, (b) angiogenic response, (c) vascular injury, and (d) inflammatory response. Recently, numerous truncated forms of RAGE have been described, and the C-terminally truncated soluble form of RAGE has received much attention. Soluble RAGE consists of several forms including endogenous secretory RAGE (esRAGE), which is a spliced variant of RAGE, and a shedded form derived from cell surface RAGE. These heterogeneous forms of soluble RAGE, carrying all of the extracellular domains but devoid of the transmembrane and intracytoplasmic domains, bind ligands including AGEs and are capable of antagonizing RAGE signaling in vitro and in vivo. Enzyme-linked immunosorbent assay (ELISA) systems to measure plasma esRAGE and total soluble RAGE have been developed, and the pathophysiological roles of soluble RAGE have begun to be unveiled clinically. In this chapter, we will summarize the recent findings of AGEs/RAGE/soluble RAGE axis as a crucial mediator of oxidative stress and cardiovascular disease and discuss their potential usefulness as therapeutic targets and biomarkers for the disease.

H. Koyama (✉)
Department of Metabolism, Endocrinology and Molecular Medicine
Osaka City University School of Medicine, 1-4-3, Asahi-machi,
Abeno, Osaka 545-8585, Japan
e-mail: hidekoyama@med.osaka-cu.ac.jp

T. Miyata et al. (eds.), *Studies on Renal Disorders*, Oxidative Stress in Applied Basic Research and Clinical Practice, DOI 10.1007/978-1-60761-857-7_13,

Keywords AGE · Chronic kidney disease · End stage renal disease · Atherosclerosis

1 Introduction

The risk of cardiovascular disease in patients with chronic renal disease appears to be far greater than in the general population [1]. In 1974 Lindner et al. first described the higher risk for cardiovascular diseases in patients with maintenance hemodialysis [2]. They prospectively followed the cohort of hemodialysis patients, and as many as 50% of total death were due to cardiovascular complications, among which 80% (40% of total deaths) were from cardiovascular deaths and 75% of cardiovascular deaths (30% of total) were from coronary artery disease. This result was supported by several epidemiological findings. Foley et al. showed that among patients treated by hemodialysis or peritoneal dialysis, the prevalence of coronary artery disease is approximately 40% and the prevalence of left ventricular hypertrophy is approximately 75% [1]. Cardiovascular mortality has been estimated to be approximately 9% per year. Even after stratification by age, gender, race, and the presence or absence of diabetes, cardiovascular mortality in dialysis patients is 10–20 times higher than in the general population [1]. A national random sample of new end stage renal disease (ESRD) in the United States in 1996/1997 (n = 4,025) revealed the prevalence of coronary artery disease was 38% [3]. The United States Renal Data System reported that prevalence of coronary artery diseases in hemodialysis patients is 42%, which is much higher that healthy population (5–12%) [4]. In Japanese population, Nishimura et al. reported that 71.5% of the hemodialysis patients had significant coronary stenosis [5].

It appears that despite the absence of cardiac symptoms, patients with chronic renal failure are already in a very high risk group for coronary artery stenosis at the initiation of renal replacement therapy. Joki et al. [6] performed coronary angiography within 1 month of initiation of maintenance hemodialysis in 24 patients, and found that 62.5% of the patients showed significant coronary artery disease ($\geq$75% stenosis). Ohtake et al. [7] showed that significant coronary artery stenosis (>50% stenosis) was seen in 53.3% of 30 asymptomatic uremic patients on coronary angiography at the start of renal replacement therapy. Hase et al. [8] also showed that in 112 predialysis uremic patients with no cardiac history, asymptomatic coronary artery disease was present in 47 patients (42%). Moreover, in subsequent follow-up for a median of 24 months, the patients with asymptomatic coronary artery disease had a significantly higher cumulative rate for major adverse cardiac events than those without coronary artery disease (49 vs. 3%) [8]. In addition, recent evidence also suggests that this process of cardiovascular damage starts very early during progression in well-defined chronic kidney disease (CKD), long before ESRD is developed (i.e., stages 1 and 2 according to glomerular filtration rate [GFR]) [9].

The links between CKD and cardiovascular disease could be numerous, since both share a number of common etiological factors [10]. Indeed, these patients exhibit an excess of traditional cardiovascular risk factors (i.e., age, gender, hypertension, diabetes, and dyslipidemia). However, even after adjusting for these factors, the CKD and prevalence of cardiovascular disease are still significantly associated [11]. Therefore, it is likely that nontraditional risk factors, identified and as yet unidentified, could be involved as well. Advanced glycation end products (AGEs) are an essential nontraditional risk factor associated with the development of long-term complications of ESRD. Evidence has emerged pointing to a potential role of oxidative stress in generation of AGEs in patients with ESRD. Moreover, interaction of the receptor for AGEs (RAGE) with AGEs leads to crucial biomedical pathway, generating intracellular oxidative stress and inflammatory mediators, which could result in further amplification of the pathway involved in AGE generation. In this chapter we will focus on the current understanding of the role of AGEs–RAGE axis in the pathogenesis of cardiovascular complications in patients with CKD.

2 AGEs in Patients with CKD

Advanced glycation end products are proteins generated by a series of reactions termed the Maillard reaction (Fig. 1). Classically, AGE formation has been described by a nonenzymatic reaction between proteins and glucose [12, 13]. AGEs derive from the spontaneous reaction of carbohydrates with amino group of proteins, which undergo the formation of reversible products (Schiff base adducts) to the generation of more stable products (Amadori products). Subsequently, complex reactions occur, including intermolecular crosslink formation and cleavage through oxidation, dehydration, condensation, cyclization. Other reactions follow, with generation of AGEs through a late reaction characterized by fluorescent and brown coloration and molecular crosslinkage. Other than diabetes mellitus patients, high plasma and tissue levels of AGEs are observed in patients with ESRD. It has been reported that no difference was noted in blood AGEs levels between those with and without diabetes mellitus among chronic renal failure patients on hemodialysis, which is believed to enhance production and accumulation of AGEs in conditions other than hyperglycemia. Local accumulation of AGEs is also observed in patients with Alzheimer's disease, rheumatoid arthritis, arteriosclerosis, cancer, and other diseases, suggesting the involvement of inflammation and oxidative stress in the formation of AGEs.

Early observations suggest that circulating AGEs in patients with renal failure are low molecular weight, which are removed by dialysis therapy [14]. Serum levels of low molecular weight AGEs, AGE free adducts, generated from proteolysis of glycated proteins, increase with loss of kidney function, and both serum and tissue AGE levels are markedly increased in patients with ESRD [15, 16]. AGE-free adducts are filtered through glomerulus, reabsorbed, and degraded by renal proximal tubular cells [17]. In patients with diabetes, AGE levels are related to decreased

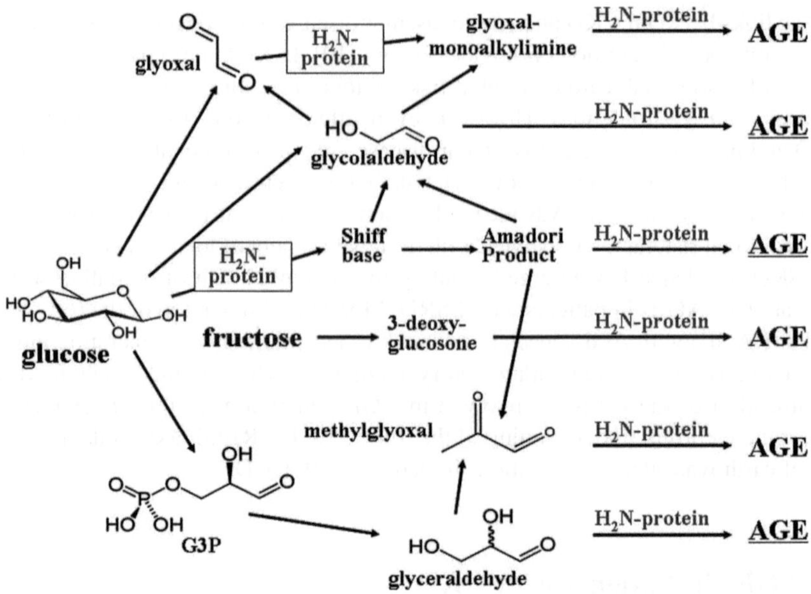

Fig. 1 Alternative routes for the formation of immunochemically distinct AGEs. Advanced glycation end products (AGEs) arise from the decomposition of Amadori products, the glycolysis intermediate product glyceraldehyde, the Schiff base fragmentation product glycolaldehyde, the triose phosphate and the Amadori product fragmentation product methylglyoxal, the autoxidation product of glucose glyoxal, and decomposition product of Amadori products and fructose-3-phosphate to 3-deoxyglucosone. Modified from ([207], Fig. 6)

kidney function and the presence of diabetic nephropathy [14]. On one hand (Fig. 2), serum pentosidine levels are significantly correlated with loss of renal function, with the level markedly high in patients with CKD stage 5. Serum levels of carboxy-methyl-lysine (CML), on the other hand, are marginally affected by renal dysfunction in patients with diabetes. Patients without diabetes with mild chronic kidney failure were also shown to have AGE levels dependent on kidney impairment. AGE levels are shown to predict the progression of kidney impairment in patients with diabetes independent of other risk factors [18, 19]. Of particular importance, it is now well established that most of the AGEs in uremia are protein bound and are only partly removed by dialysis therapy [15, 20–22].

2.1 Pathways of AGE Generation in Patients with ESRD

In addition to the formation of carbohydrate intermediates, there is increasing evidence that AGEs are also formed through lipid-derived intermediates, resulting in advanced lipoxidation products [23]. AGEs might be formed directly by autoxidation of free glucose [24, 25]. In this pathway, known as autoxidative

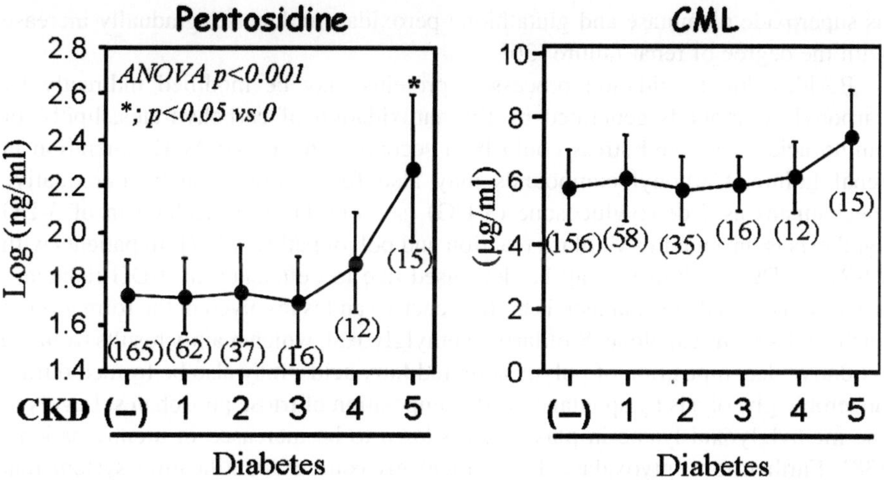

Fig. 2 Serum pentosidine and carboxymethyl-lysine (CML) in diabetic patients with different degrees of renal dysfunction. Chronic kidney disease (CKD) stage was based on glomerular filtration rate (GFR) estimated by MDRD equation: I; GFR $\geq$ 90, II; GFR 60–89, III; GFR 30–59, IV; 15–29, V; GFR < 15 mL/min)

glycosylation, such reactive oxygen species as hydrogen peroxide were identified as both products and catalysts of autoxidation of sugars.

In renal failure, plasma and tissue AGE levels are increased in normoglycemic patients and are completely independent of serum glucose levels. Thus, the increase in AGE levels in patients with ESRD cannot be attributed to only increased glycation of proteins. In this context, evidence suggests a potential role of oxidative stress in generation of AGEs in patients with ESRD. The formation of glycoxidation products, such as CML and pentosidine, is considered to be the result of a chemical reaction dependent on the concentration of carbohydrate precursors and of reactive oxygen species (oxidative stress). In contrast to the diabetic patients, in whom the accumulation of AGEs in protein may be attributable to both glycative and oxidative stress, increased AGEs in ESRD could be primarily regulated by the second mechanism, oxidative stress [26]. ESRD is a condition of increased (intracellular) oxidative stress, indicated by increased lipid peroxidation and a decrease in the ratio of oxidized glutathione to reduced glutathione [27–29]. The concentration of lipid peroxidation marker, malondialdehyde lysine is also increased in uremic patients [22]. Oxidative stress in these patients has been attributed to the processes of loss of renal function or renal replacement therapy [29, 30]. Suboptimal biocompatibility in dialysis membranes may acutely further aggravate oxidative stress and related endothelial dysfunction [31]. However, even before the start of renal replacement therapy, renal impairment is associated with a state of increased oxidative stress [32]. Moreover, accelerated oxidative stress, together with a decrease in superoxide excavenger capacity, is already present in early stages of CKD as well [33–35], and alterations in antioxidant system components (such as superoxide dismutase and glutathione peroxidase/reductase) gradually increase with the degree of renal failure [36].

Besides direct oxidation processes, proteins may be modified indirectly by carbonyl compounds generated by the autoxidation of carbohydrates, lipids, or amino acids. Carbonyl stress could be a second pathway for AGE generation in renal failure. Carbonyl compounds may also form through such nonoxidative mechanisms as 3-deoxyglucosone (3-DG) (see Fig. 1). The mechanism of 3-DG synthesis is through the Maillard reaction and polyol pathway [37]. In patients with ESRD, 3-DG catabolism may be decreased due to deficiency of 3-DG reductase activity. The 3-DG mechanism is highly reactive and is involved in the formation of such AGEs as imidazolone. Similarly, methylglyoxal, which mainly results from the oxidative decomposition of polyunsaturated fatty acids, may also be formed during anaerobic glycolysis by spontaneous decomposition of triose phosphates. Levels of the methylglyoxal lysine in plasma are shown to be increased in uremic patients [38]. Furthermore, glyoxalase 1, a critical glyoxalase detoxification system that prevents glycation reactions mediated by methylglyoxal/glyoxal [39], was found to be deficient in dialysis patients [40]. Thus, there would not only be an increase in oxidative stress, but also an increase in carbonyl stress in uremia, which could be attributable to the accelerated accumulation of AGEs in patients with ESRD.

Decreased renal clearance of serum AGEs could also contribute to increased accumulation of AGEs in patients with uremia [15]. Both a decrease in glomerular filtration rate, leading to a corresponding decrease in glomerular filtration of AGEs, and a decreased tubular catabolism of AGEs can play a role in decreased clearance of AGEs. The role of renal tubular catabolism of AGEs is shown by the findings of tubular accumulation of AGEs in rat experimental proteinuria despite relatively well-preserved kidney function [41]. Absorption of toxic AGEs from food [42] and the impact of smoking [43] on aggravate AGE accumulation suggest environmental risk factors also contribute to the accumulation of AGEs. Metabolic abnormalities, such as insulin resistance in uremia [44], may increase reactive oxygen species formation [45] and thus may further increase AGE formation in patients without diabetes with ESRD.

2.2 AGE Accumulation and Cardiovascular Complications in Patients with ESRD

AGE accumulation occurs in tissues in patients with ESRD, such as in atherosclerotic plaques [46] and in cardiac tissues [47]. Alternatively, AGE modifications of lipoproteins may increase vascular deposition of low-density lipoprotein as a consequence of impaired low-density lipoprotein receptor–mediated clearance [48–51], which could be attributable to the augmented atherogenesis. Plasma pentosidine levels are shown to be associated with accelerated carotid atherosclerosis as determined by intima-media thickness measured with ultrasound in ESRD patients at the initiation of dialysis therapy [52]. Increased AGE levels are also associated with extensive coronary artery calcification in uremic patients [53], which is a predictor for cardiovascular events.

AGEs are also involved in arterial stiffness that results from nonenzymatic protein glycation to form irreversible crosslinks between long-lived proteins such as collagen and elastin [54, 55]. AGE-linked extracellular matrix is stiffer and less susceptible to hydrolytic turnover, resulting in accumulation of structurally inadequate matrix molecules. Aortic stiffening, which is increased in ESRD patients, is an important pathophysiological aspect of large artery damage and is a predictor of all-cause and cardiovascular mortality [56–58]. We have recently measured skin accumulation of AGEs using the autofluorescence reader (AFR), developed a noninvasive device to estimate accumulation of AGEs in humans [59], and examined its association with arterial stiffness as measured by pulse wave velocity in 120 nondiabetic ESRD patients and 110 age- and gender-matched controls with neither renal disease nor diabetes [60]. As shown in Fig. 3a, skin autofluorescence was significantly associated with age in the group of all patients (Rs = 0.255, Spearman's rank correlation test) and that of controls (Rs = 0.493), but not in the group of ESRD patients (Rs = 0.046), suggesting that the effect of age on AGE accumulation is overwhelmed by the effect of uremia in ESRD patients. Pulse wave velocity (PWV) was significantly and positively associated with skin autofluorescence both in the group of all patients (Rs = 0.335), controls (Rs = 0.246), and that of ESRD patients (Rs = 0.205) (see Fig. 3b). Multiple regression analyses showed that in the group of all patients, association of skin autofluorescence with PWV was significant even after adjustment for other covariates, including the presence of ESRD and age. Moreover, for ESRD patients, a significant association between skin autofluorescence and PWV was found, independent of age, suggesting that AGE accumulation could be an important predictor for arterial stiffening in ESRD patients. Experimental studies also showed that inhibition or breaking of AGEs prevents cardiac hypertrophy and arterial stiffness and may restore cardiac function [61]. AGE crosslink breaker is also successfully shown to improve arterial compliance in humans as well [62].

Clinical data assessing AGEs as a cardiovascular risk factor in patients with ESRD are limited. Recent observations show that the tissue accumulation of AGEs as determined by skin autofluorescence has been implicated as a risk predictor for cardiovascular mortality in patients with ESRD as well as diabetes [63, 64]. As opposed to these observations for tissue AGEs accumulation, the relationship between serum AGEs and mortality in hemodialysis patients is rather controversial: some observed a strong relationship between serum AGE levels and survival [65, 66], whereas others did not [67, 68]. It is not clear at present how to explain these discrepant observations: different methods to determine the AGE levels (immunoassay vs. serum fluorescence), influence of dialysis modality and timing, or effects of food and smoking may partly contribute to these results. Altered explanation could be that high serum AGE levels may reflect better nutritional support (e.g., high serum albumin) and low inflammation (e.g., high C-reactive protein), both of which are known predictors for mortality in ESRD patients [69, 70].

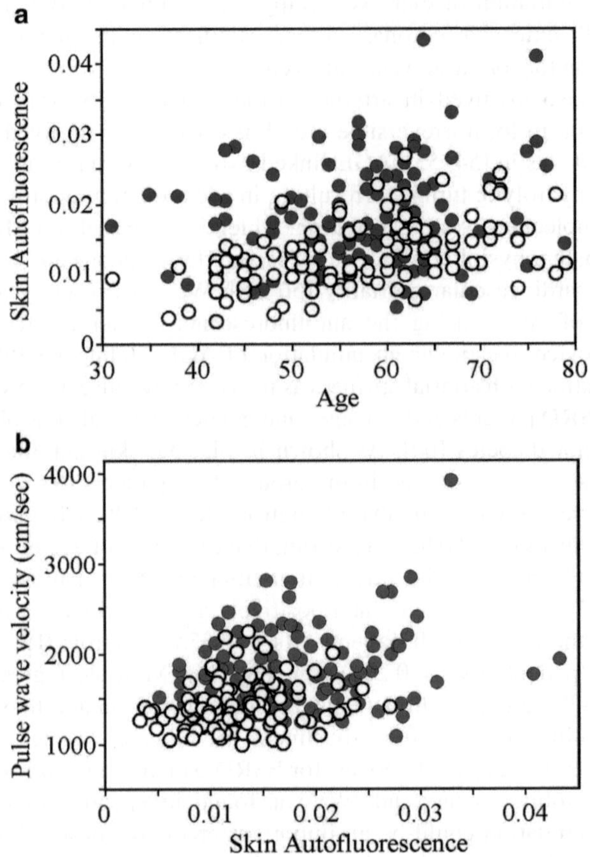

Fig. 3 Aging, arterial stiffness and AGEs accumulation. Skin accumulation of advanced glycation end products (AGEs) was measured by using the autofluorescence reader, recently developed noninvasive device [59]. Arterial stiffness was measured as pulse wave velocity (PWV). (**a**) Skin autofluorescence was significantly associated with age in the group of all subjects (Rs = 0.255, Spearman's rank correlation test) and that of control subjects (Rs = 0.493), but not in the group of end stage renal disease (ESRD) subjects (Rs = 0.046), suggesting that the effect of age on AGE accumulation is overwhelmed by the effect of uremia in ESRD subjects. (**b**) PWV was significantly and positively associated with skin autofluorescence both in the group of all subjects (Rs = 0.335), controls (Rs = 0.246), and that of ESRD subjects (Rs = 0.205). Open circles; control subjects, closed circles; ESRD subjects. Modified from [60]

3 Receptor for Advanced Glycation End Products

Receptor for advanced glycation end products is a multiligand cell-surface protein that was isolated from bovine lung in 1992 by Schmidt et al. [71, 72]. RAGE belongs to the immunoglobulin superfamily of cell surface molecules and has an extracellular region containing one V-type immunoglobulin domain and two C-type immunoglobulin domains [71, 72] (Fig. 4). The extracellular portion of the receptor is followed by a hydrophobic transmembrane spanning and then by a highly charged, short cytoplasmic domain, which is essential for intracellular RAGE signaling. RAGE is initially identified as a receptor for CML modified proteins [73], a major AGE in vivo [74]. The three-dimensional structure of the recombinant AGE-binding domain analyzed by using multidimensional heteronuclear magnetic resonance spectroscopy revealed that the domain assumes a structure similar to those of other immunoglobulin V-type domains [75, 76]. Three distinct surfaces of the V domain were identified to mediate AGE–V domain interactions [75]. The site-directed mutagenesis studies identified the basic amino acids that play a key role in the AGE-binding activities [76]. RAGE also interacts with other nonglycated peptide ligands, including S100/calgranulin [77], amphoterin (also known as high mobility group box 1 protein [HMGB1]) [78, 79], amyloid fibrils [80], transthyretin [81], and a leukocyte integrin, Mac-1 [82]. The common characteristics of these ligands are the presence of multiple β-sheets [82–84]. RAGE is thought to interact with these ligands through their shared three-dimensional structure.

Activation of RAGE is associated with diabetic microvascular complications, including nephropathy, retinopathy, and neuropathy. By use of RAGE-

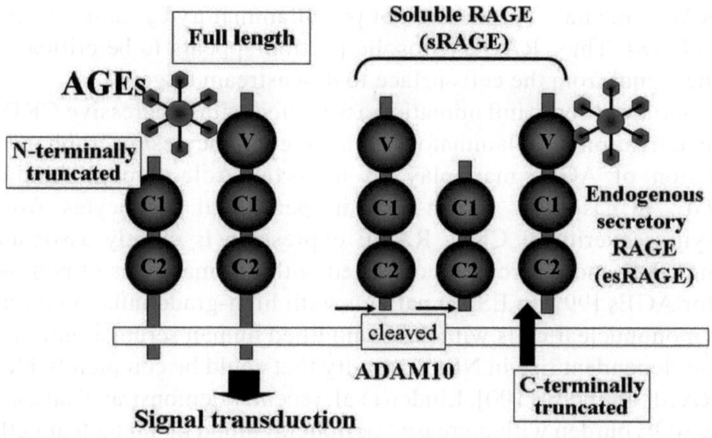

Fig. 4 Numerous truncated forms of receptor for advanced glycation end products (RAGE). There are three major spliced variants of RAGE: full length, N-terminally truncated, and C-terminally truncated. The C-terminally truncated form of RAGE is secreted from the cell and is named endogenously secreted RAGE (esRAGE). The esRAGE has a V-domain, which is essential for binding with ligands and is capable of competing with RAGE signaling as a decoy receptor. There are other forms of soluble RAGE (sRAGE) that are cleaved from cell-surface RAGE by a shaddase ADAM10. The enzyme-linked immunosorbent assay (ELISA) assay for sRAGE measures all soluble forms including esRAGE in human plasma, while the ELISA for esRAGE measures only esRAGE, using polyclonal antibody raised against the unique C-terminus of the esRAGE sequence

overexpressing and deficient mice, RAGE is shown to have a pathological role in both early and advanced phase diabetic nephropathy [85, 86]. RAGE expression in the peripheral nervous system rises cumulatively and relates to progressive pathological changes, and mice lacking RAGE have attenuated features of neuropathy and limited activation of potentially detrimental signaling pathways [87, 88]. In mesangial [89] and endothelial [90] cells, RAGE activation results in a burst of reactive oxygen species (ROS). The exact mechanism for this is unknown, but it is thought to involve nicotinamide adenine dinucleotide phosphate (reduced) (NADPH) oxidase [91], which could alone contribute to cellular oxidative stress and dysfunction. In addition, RAGE signals via phosphatidylinositol-3 kinase (PI-3K), Ki-Ras, and the mitogen-activated protein kinases (MAPKs), Erk1 and Erk2 [89, 92]. These signaling pathways initiate and sustain the translocation of nuclear factor κB (NFκB) from the cytoplasm to the nucleus in a number of cell types, including circulating monocytes and endothelial cells [93–95], leading to prolonged inflammation and resulting in a RAGE-dependent expression of proinflammatory mediators such as monocyte chemoattractant protein-1 (MCP-1) and vascular cell adhesion molecule-1 (VCAM-1) and organ damage [77, 96–98]. RAGE has a short cytosolic portion that contains 43 amino acids [72]. So far, adaptors or scaffold proteins that interact with the cytosolic tail of RAGE have barely been identified. The RAGE mutant lacking the 43-residue C-terminal tail fails to activate NFκB, and expression of the mutant receptor results in a dominant negative effect against RAGE-mediated production of proinflammatory cytokines from macrophages [77, 78]. Thus, RAGE cytosolic portion appears to be critical in transducing the signal from the cell surface to downstream targets.

Inappropriate chronic inflammation associated with progressive CKD reflects sustained activation of inflammatory cells, like monocytes/macrophages, where accumulation of AGEs may play an important role through binding with the RAGE. It has been shown that in peripheral monocytes from those with varying severity of CKD, RAGE expression is closely associated with worsening CKD and is strongly correlated with plasma levels of pentosidine, a marker for AGEs [99]. In ESRD patients with high-grade inflammation, stimulation of mononuclear cells with AGE-modified human serum albumin causes a rapid dose-dependent rise in NFκB activity that could be completely blocked by an anti-RAGE antibody [100]. Linden et al. recently demonstrated an association of excess AGE burden with increased peripheral blood mononuclear cell mRNA RAGE and in vivo endothelial dysfunction in patients with CKD [101]. They hypothesized that endothelial dysfunction in CKD may be partly mediated by AGE-induced inhibition of endothelial nitric oxide synthase through RAGE activation. At the site of endothelium, in contrast to normal endothelial cells that do not constitutively express RAGE, arterial and capillary endothelial cells of uremic patients do express RAGE [102]. Thus, enhanced RAGE expression in ESRD may amplify AGE-induced perturbation and contribute to systemic and local vascular inflammatory disease, causing atherosclerotic and nonatherosclerotic vascular lesions.

3.1 RAGE and Atherosclerosis

Markedly progressed atherosclerosis in patients with ESRD is thought to contribute to the high incidence of cardiovascular diseases. Ultrasound-determined intimal-medial thickness of carotid and femoral artery, a surrogate marker for atherosclerosis, is found to be significantly increased in patients with ESRD than age-matched healthy patients [103]. Diabetic ESRD patients are shown to have higher intimal-medial thickness than nondiabetic ESRD patients.

Possible involvement of RAGE in atherosclerosis has been extensively examined in diabetic humans and animals. It has been demonstrated that RAGE is upregulated in human diabetic atherosclerotic plaques, the expression of which colocalized with inflammatory markers such as cyclooxygenase-2 (COX-2) and matrix metalloproteinases (MMPs), particularly in macrophages at the vulnerable regions of the atherosclerotic plaques [104, 105]. In a murine model of accelerated atherosclerosis, apolipoprotein (apo) E null mice, induction of diabetes by streptozotocin for 6 weeks was found to be associated with a significant increase in atherosclerotic lesion area at the aortic sinus compared with nondiabetic apoE null mice [106]. This diabetes-associated atherosclerotic lesion has been found to exhibit increased accumulation of AGEs and S100/calgranulins and enhanced expression of RAGE [107]. Of importance, daily treatment of the mice with genetically engineered murine soluble RAGE (sRAGE) suppressed diabetes-associated acceleration of lesion area and complexity with the effect independent of glycemic and lipid profiles [106]. Similarly, sRAGE has been found to prevent progression of atherosclerosis in apoE null mice with insulin-resistant type 2 diabetes (db/db background). Of note, vascular inflammatory phenotype such as overexpression of VCAM-1 tissue factor and matrix metalloproteinase in mice was also prevented by administration of sRAGE [108]. Bucciarelli et al. [109] also found that administration of sRAGE stabilized atherosclerotic lesion area and complexity at advanced phase in diabetic apoE null mice, and suppressed inflammatory markers such as expression of COX-2, VCAM-1, JE-MCP-1, MMP-9 activity, tissue factor, and phosphorylation of p38 MAPK.

Involvement of RAGE in accelerated atherosclerosis in diabetes has recently been confirmed by using RAGE-deficient mice [110]. It is becoming apparent that RAGE is also involved in atherogenesis even in nondiabetic conditions in a mouse model of atherosclerosis [110, 111]. Harja et al. demonstrated critical roles for RAGE and its ligands in vascular inflammation, endothelial dysfunction, and atherosclerotic plaque development and highlighted that endothelial RAGE and its ligands mediate vascular and inflammatory stresses that culminate in atherosclerosis in the vulnerable vessel wall [111]. Similarly by using apoE-deficient mice, Soro-Paavonen et al. [110] recently showed that in diabetic conditions, RAGE deficiency significantly decreased the atherosclerotic lesion formation associated with attenuation of leukocyte recruitment, decreased expression of proinflammatory mediators, and reduced oxidative stress in aorta. In their report, RAGE-deficiency significantly suppressed the atherosclerotic lesion area even in a nondiabetic condition, which was associated with aortic inflammatory gene expression. Administration of sRAGE was also found to significantly stabilize atherosclerotic lesion area and complexity in nondiabetic apoE null mice [109]. Thus, these reports implicated the functional role of RAGE in endothelial cells in atherogenesis, possibly through regulation of inflammatory signaling.

It is still not clear whether RAGE accumulation is increased in atherosclerotic lesions in uremia. However, chronic renal failure markedly accelerates atherogenesis in apoE-deficient mice [112, 113], while blockade of RAGE reduces the proatherogenic effects of uremia, possibly through a systemic decrease in oxidative stress [114].

3.2 RAGE and Impaired Angiogenic Response

In diabetes, progressive vasodegeneration in microvascular beds is the major underlying factor in initiation and progression of vascular complications [115–117]. Recent observations in diabetic patients and diabetic animals indicate that an increase in cardiovascular events and severity in diabetes can be associated with impairment in angiogenic response or development of new collateral vessels in response to local ischemia or inflammation [118–121]. The AGE–RAGE axis has been found to be involved in impaired angiogenic response in diabetes. Goova et al. [122] demonstrated that blockade of RAGE by sRAGE restores effective wound healing in diabetic mice. Tamarat et al. [123] demonstrated that blockade of AGEs formation by aminoguanidine improves ischemia-induced angiogenesis in diabetic mice. Recently, ablation of galectin-3, a receptor for AGEs, has been shown to abolish the AGE-mediated increase in retinal ischemia and restore the neovascular response to that seen in controls [121]. To directly examine the role of RAGE in impairment of angiogenic response in diabetes, we compared angiogenic response in RAGE null or WT mice in diabetic condition [124]. We found that angiogenic response determined by Matrigel-plague assay containing vascular endothelial growth factor (VEGF) in streptozotocin-treated diabetic mice was significantly decreased as compared with that of the nondiabetic mice. In sharp contrast in RAGE-null mice, angiogenic response of diabetic mice was not found to be significantly different from that of nondiabetic mice (Fig. 5). The impairment of angiogenic response in diabetes was associated with RAGE-mediated decrease in proliferation and increase in cell death. The changes in cellular function were associated with activation of NFκB in vascular cells. In this model, expression of VEGF was found to be decreased in diabetic mice. The decrease in expression of VEGF in diabetes, however, was also observed in RAGE-null mice. This result was in agreement with an in vitro finding that a decrease in VEGF secretion in smooth muscle cells cultured on glycated type 1 collagen was also independent of the presence of RAGE [125]. These results suggest that RAGE-mediated decrease in angiogenic response in diabetes is not necessarily explained by regulation of proangiogenic factors such as VEGF. Although angiogenesis in renal failure has not been well elucidated, augmented AGE interaction with RAGE may similarly impair angiogenic response and contribute to an increase in cardiovascular events and severity in patients with renal failure.

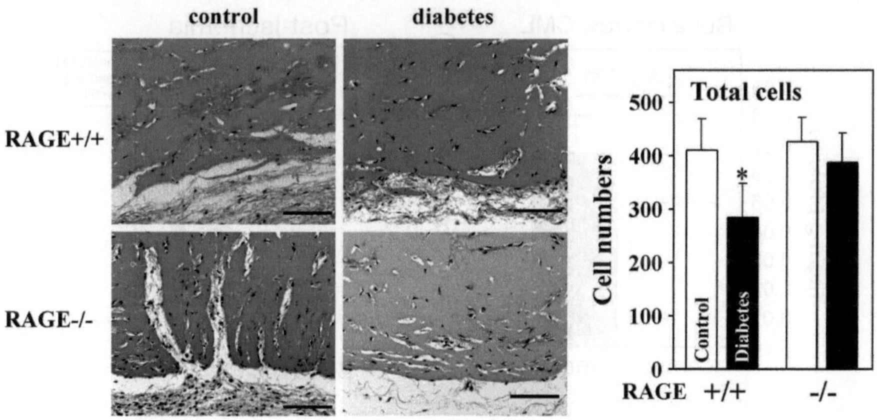

Fig. 5 Decrease in angiogenic response in diabetes is dependent on the presence of receptor for advanced glycation end products (RAGE). Angiogenic response in control and streptozotocin-induced diabetic RAGE+/+ or RAGE−/− mice was evaluated through the Matrigel assay. VEGF (300 ng/mL) was suspended in Matrigel, and the Matrigel was implanted subcutaneously into the abdominal midline of mice at age 13 weeks. The plug was removed 2 weeks later and processed for histochemical analysis. Bars represent 50 μm. Summary of the quantitative analyses is shown in the right panel. Data are shown as mean ± standard deviation. * $p < 0.05$ vs. control, ANOVA with multiple comparison (Scheffe's type). Modified from ([124], Fig. 2)

In addition to the problems of local ischemic tissue (so called *soil*), *seeds* (*angiogenic progenitor cells*) for neovascularization could be impaired in patients with ESRD. Endothelial progenitor cells (EPCs) attract much attention, since they play a key role in the maintenance and repair of vascular integrity in response to endothelial injury [126–128]. The majority, but not all, of the studies have shown that the numbers of circulating EPCs are decreased in patients with ESRD [129–133] and diabetes [134, 135], the underlying mechanisms of which remain poorly characterized. Age-dependent depression in circulating endothelial progenitor cells is also reported and is implicated in the risk for cardiovascular diseases [136, 137]. Common features for aging, diabetes and ESRD, are an accumulation of AGEs, which may be an important predictor for decreased EPCs in these patients. A potential support for this is that continuous ambulatory peritoneal dialysis treatment, which we recently identified as a positive regulator of EPCs and progenitor cells [133], is shown to decrease accumulation of AGEs than hemodialysis [15]. Indeed, we have recently found that the circulating EPC numbers were significantly and inversely associated with the levels of tissue accumulation of AGEs [138]. In the 5/6 nephrectomized uremic rat model, progenitor cell mobilization induced by hind limb ischemia was inversely associated with carboxymethyl-lysine levels in the bone marrow (Fig. 6), suggesting that mobilization of EPCs in response to systemic stimuli may be attenuated by AGE accumulation in bone marrow. If that is the case, RAGE expressed in progenitor cells might play a fundamental role in regulation of circulating EPCs in response to accumulated AGEs in bone marrow [139].

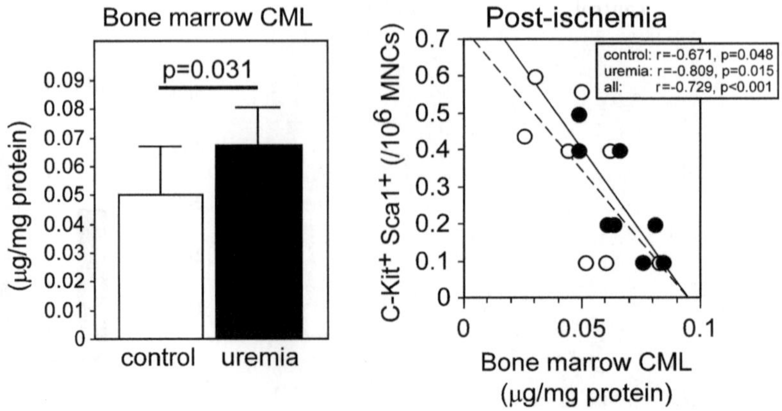

Fig. 6 Progenitor cell (PC) mobilization is inversely associated with bone marrow carboxymethyllysine (CML) levels in uremic rats. The 5/6 nephrectomized uremic ($n = 9$) and sham-operated control rats ($n = 9$) were subjected to unilateral hind limb ischemia by complete resection of the entire left femoral artery and vein. Peripheral blood PCs as determined by double expression of stem cell antigen-1 (Sca1) and c-kit antigens with direct two-color flow cytometry and bone marrow CML levels were measured 24 h after the operation. *Left* panel shows CML levels in bone marrows in uremic and sham-operated rats. Columns represent mean ± SD. *p* value was calculated by student's *t*-test. *Right* panel shows association between circulating PCs following hind limb ischemia and bone marrow CML levels. Open circles represents sham-operated, and closed circles, uremic rats. Solid line: linear regression line for uremic rats, dotted line: sham-operated rats

3.3 RAGE and Neointimal Expansion Following Arterial Injury

As has been discussed, both ESRD and lesser degrees of renal dysfunction are associated with accelerated rates of atherosclerosis and a high incidence of cardiovascular morbidity and mortality. Percutaneous angioplasty is an effective therapy for symptomatic coronary atherosclerosis, but in patients with ESRD it is associated with high complication rates and poor long-term outcomes [140, 141]. Results of percutaneous intervention in patients with severe renal disease remain suboptimal despite the increased use of coronary stents. In patients with ESRD rates of target vessel as well as target lesion revascularization after coronary stenting are high [142]. Similarly, diabetic patients demonstrate accelerated responses to arterial injury (i.e., therapeutic angioplasty) [143, 144], which is associated with an increase in neointimal formation, enhanced vascular smooth muscle cell proliferation and migration, and production of extracellular matrix. The augmented response to arterial injury in diabetes has been shown to be associated with RAGE, since administration of sRAGE caused decreased neointimal expansion in hyperglycemic fatty Zucker rats [145]. Of interest, Sakaguchi et al. [146] showed by using RAGE-null mice that RAGE also plays an important role in arterial injury-induced neointimal expansion in nondiabetic mice. Even in nondiabetic mice, RAGE transcripts in smooth muscle cells were found to be increased by arterial endothelial denudation, which was associated with induction of S100 transcripts and CML-modified AGE adducts [146].

3.4 RAGE Modifies Functions of Inflammatory Cells

It is now well recognized that mononuclear cells, such as lymphocytes and monocyte/macrophage, play fundamental roles in the progression of atherosclerosis [147]. AGEs, S100A12. and HMGB1 can bind to mononuclear phagocytes through RAGE and modify their functions, including cell activation, chemotaxisis, transendothelial migration, and expression of key inflammatory mediators [77, 95, 148, 149]. Inhibition of RAGE has been found to markedly decrease infiltration of the immune and inflammatory cells, such as $CD4^{+}$ T-cells, leading to suppression of autoimmune encephalitis [150], autoimmune diabetes [151], and allograft rejection [152]. HMGB1, an essential component of DNA-containing immune complexes, stimulated cytokine production in plasmacytoid dendritic cells, and B-cells are also found to involve RAGE and Toll-like receptor 9, which is implicated in the pathogenesis of systemic autoimmune diseases [153].

RAGE is also found to function as an endothelial adhesion receptor, promoting leukocyte recruitment [82]. In an animal model of thioglycollate-induced acute peritonitis, leukocyte recruitment was significantly impaired in RAGE-deficient mice as opposed to wild-type mice, both in nondiabetic and diabetic mice. They also found that RAGE-dependent leukocyte adhesion to endothelial cells is mediated by a direct interaction of RAGE with the β2-integrin Mac-1, and the RAGE-Mac-1 interaction is augmented by the proinflammatory S100 protein. Orlova et al. [154] define a novel function of HMGB1 in promoting Mac-1-dependent neutrophil recruitment, which require the presence of RAGE on neutrophils but not on endothelial cells. HMGB1 is also found to enhance the interaction between Mac-1 and RAGE. Consistently, HMGB1 stimulates activation of Mac-1, Mac-1-mediated adhesive and migratory functions, and activation of NFκB of neutrophils in a RAGE-dependent manner. Taken together, RAGE plays a pivotal role as a pro-inflammatory molecule acting both at the sides of inflammatory and endothelial cells, which may underlie part of the mechanism in progression of atherosclerosis.

4 C-Terminally Truncated form of RAGE (Soluble RAGE)

Numerous truncated forms of RAGE have been described [155–161] (see Fig 4). Two major spliced variants of RAGE mRNA, N-terminal and C-terminal truncated forms, have been most extensively characterized [156]. The N-truncated isoform of RAGE mRNA codes for a 303-amino-acid protein lacking the

N-terminal signal sequence and the first V-like extracellular domain. The N-truncated form is incapable of binding with AGEs, since the V-domain is critical for binding of the ligand [71]. The N-truncated form of RAGE appears to be expressed on the cell surface similar to the full-length RAGE, although its biological roles remain to be elucidated [162]. It has been suggested that this form of RAGE could be involved in angiogenic regulation in a fashion independent of the classical RAGE signaling pathway [162]. The C-terminal truncated form of RAGE lacks the exon 10 sequences that encode the transmembrane and intracytoplasmic domains [156]. This spliced variant mRNA of RAGE encodes a protein consisting of 347 amino acids with a 22-amino-acid signal sequence and is released from cells. This C-truncated form is now known to be present in human circulation and is named endogenous secretory RAGE (esRAGE), which was found to be capable of neutralizing the effects of AGEs on endothelial cells in culture [156]. Adenoviral overexpression of esRAGE in vivo in mice reverses diabetic impairment of vascular dysfunction [124]. Thus, the decoy function of esRAGE may exhibit a feedback mechanism by which esRAGE prevents the activation of RAGE signaling. It has also been suggested that some soluble RAGE (sRAGE) isoforms that could act as decoy receptors may be cleaved proteolytically from the native RAGE expressed on the cell surface [163], suggesting heterogeneity of the origin and nature of sRAGE. This proteolytic generation of sRAGE was initially described as occurring in mice [164]. Recent findings by screening chemical inhibitors and genetically modified mouse embryonic fibroblasts suggest that a disintegrin and metalloprotease 10 (ADAM10) and MMP-9 as membrane proteases responsible for RAGE cleavage [165, 166]. ADAM is known to shed several inflammatory receptors and can be involved in regulation of RAGE/sRAGE balance. Thus, the molecular heterogeneity of the diverse types of sRAGE in human plasma could exert significant protective effects against RAGE-mediated toxicity. However, the endogenous action of sRAGE may not be confined to a decoy function against RAGE signaling. In an HMGB1-induced arthritis model, for example, sRAGE is found to interact with Mac-1 and act as an important proinflammatory and chemotactic molecule [167]. Further analyses are warranted to understand more about the endogenous activity of sRAGE.

4.1 sRAGE and esRAGE as Biomarkers for Cardiovascular and Metabolic Diseases

Since sRAGE and esRAGE may be involved in feedback regulation of the toxic effects of RAGE-mediated signaling, recent clinical studies have focused on the potential significance of circulating sRAGE and esRAGE in a variety of pathophysiological conditions. Table 1 summarizes the currently available findings. First, Falcone et al. [168] reported that total sRAGE levels are significantly lower in patients with angiographically proven coronary artery disease than in age-matched healthy controls. The association between circulating sRAGE and

Table 1 Levels of circulating soluble RAGE in cardiovascular and metabolic diseases

sRAGE		References
CAD (non-DM)	Decreased	[168, 200]
	Increased	[187]
CAD, atherosclerosis (DM)	Decreased	[208]
	Increased	[172, 173]
Heart failure (poor prognosis)		[209]
Diabetes (type 1)	Increased	[185]
Diabetes (type 2)	Increased	[186, 187]
	Decreased	[183, 184]
Hypertension	Decreased	[182]
Alzheimer's disease	Decreased	[169]
Chronic kidney disease	Increased	[177, 186, 189, 190, 210]
Oxidative stress and inflammatory markers	Positive association	[173, 211–213]
	Inverse association	[184]
esRAGE		
Metabolic syndrome	Decreased	[171]
Diabetes (type 1)	Decreased	[170, 174]
Diabetes (type 2)	Decreased	[171, 178]
Hypertension	Decreased	[171]
Atherosclerosis (IMT)	Inverse association	[171, 174–176]
	No association	[177]
CAD (DM)	Decreased	[178]
Chronic kidney disease	Increased	[181, 210]
Inflammatory markers	Inverse association	[214]

CAD coronary artery disease; *DM* diabetes mellitus; *IMT* intimal-medial thickness

angiographic observations was shown to be dose dependent, with individuals in the lowest quartile of sRAGE exhibiting the highest risk for coronary artery disease. Importantly, this cohort consisted of a nondiabetic population, suggesting that the potential significance of sRAGE is not confined to diabetes. Falcone et al. also showed that the association between sRAGE and the risk of coronary artery disease was independent of other classical risk factors. The same research group also showed that patients with Alzheimer's disease have lower levels of sRAGE in plasma than patients with vascular dementia and controls, suggesting a role for the RAGE axis in this clinical entity as well [169]. Following development of an enzyme-linked immunosorbent assay (ELISA) system to specifically measure human esRAGE [170], we measured plasma esRAGE level and cross-sectionally examined its association with atherosclerosis in 203 type 2 diabetic and 134 nondiabetic age- and gender-matched patients [171]. The esRAGE levels were inversely correlated with carotid and femoral atherosclerosis, as measured as intimal-medial thickness (IMT) by arterial ultrasound. Stepwise regression analyses revealed that plasma esRAGE was the third strongest and an independent factor inversely associated with carotid IMT, following age and systolic blood pressure [171].

In contrast to nondiabetic population, evidence so far with regard to the association between soluble forms of RAGE and vascular disease in diabetes is contradictory. Several studies shows that serum sRAGE is positively associated with coronary artery disease in types 1 and 2 diabetic patients [172, 173]. In contrast, Katakami et al. found that plasma esRAGE, but not sRAGE, is inversely associated with carotid atherosclerosis in both type 1 [174, 175] and type 2 [176] diabetics. On the other hand, when nondiabetic and diabetic groups were separately analyzed, inverse correlation between plasma esRAGE level and IMT was significant in nondiabetic population only, but not observed in type 2 diabetics (Fig. 7). No association of plasma esRAGE with IMT in diabetes was reported in another study with 110 Caucasian type 2 diabetics [177]. Recently, Katakami's group also longitudinally examined the predictive significance of plasma esRAGE and sRAGE on progression of carotid atherosclerosis and found that low circulating esRAGE level, as well as sRAGE level, was an independent risk factor for the progression of carotid IMT in type 1 diabetics. In Chinese type 2 diabetic patients, plasma esRAGE is recently shown to be decreased in angiographically proved patients with coronary artery disease as opposed to those without it [178]. The reasons for these contradictory findings are not clear but may be explained by

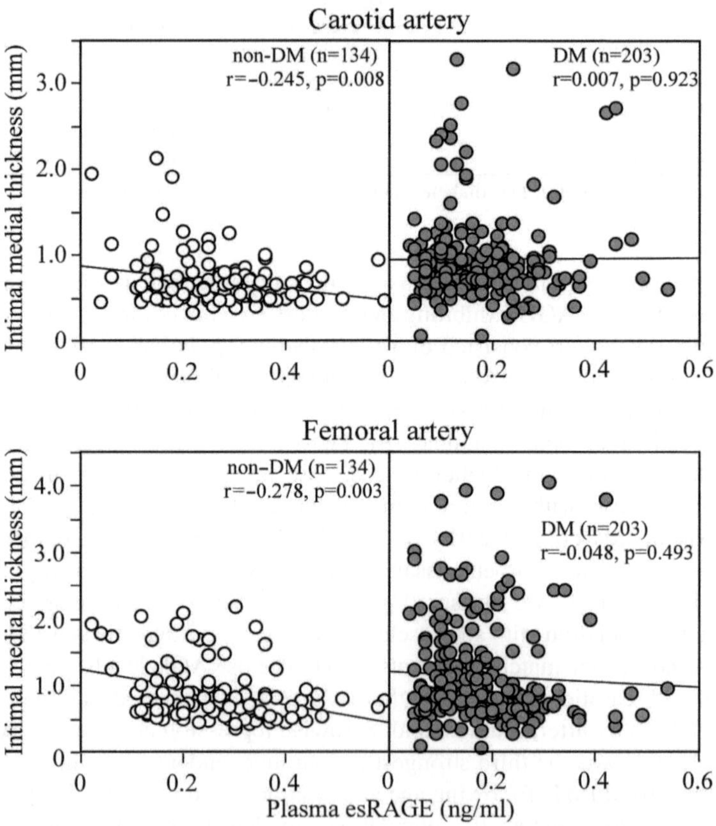

Fig. 7 Plasma endogenous secretory receptor for advanced glycation end products (esRAGE) level is inversely correlated with carotid and femoral atherosclerosis in a nondiabetic population, not in a diabetic population. Atherosclerosis was determined as intimal-medial thickness measured by arterial ultrasound. $n = 337$ including 203 type 2 diabetic patients. Modified from ([125], Fig. 5)

the fact that different variants of sRAGE are detected by different assays used in different studies, or by different type of subjects enrolled in the study (i.e., renal complications). Nevertheless, the association between plasma sRAGE and atherosclerosis needs to be confirmed in larger studies, particularly with diabetic patients. Association of circulating sRAGE with vascular diseases in CKD or ESRD patients is also an important topic to be elucidated.

Several metabolic components have been well established as risk factors for cardiovascular diseases and have been shown to be associated with altered plasma sRAGE or esRAGE levels. We have shown that plasma esRAGE levels are decreased in patients with metabolic syndrome and are inversely correlated with several components of metabolic syndrome, including body mass index, blood pressure, insulin resistance index, fasting plasma glucose, serum triglycerides, and lower high-density lipoprotein cholesterol levels [171]. The majorities of these correlations remained significant even when the nondiabetic or type 2 diabetic subpopulation was extracted for analyses. An inverse correlation between esRAGE (or sRAGE) and body mass index was also found for controls [179], those with type 1 diabetes [180], and those with ESRD [181]. Patients with hypertension have been found to have lower plasma sRAGE or esRAGE levels [171, 182].

The findings regarding plasma levels of the soluble form of RAGE in diabetes are quite confusing. We and other groups have found that plasma esRAGE level is significantly lower in types 1 and 2 diabetic patients than in nondiabetic controls [171, 174]. A recent study with large numbers of type 2 diabetic and nondiabetic patients in a Chinese population ($n = 1{,}320$) also confirmed these findings [178]. Plasma sRAGE levels have been shown to be decreased in diabetics [183, 184], although conflicting findings have also been reported for type 1 [185] and type 2 diabetes [186, 187]. We examined plasma sRAGE levels by different ELISA system using esRAGE as a standard protein and different sets of antibodies vs. a whole RAGE molecule [188]. The type 2 diabetic patients without overt nephropathy (0.60 ± 0.28 ng/mL) exhibited significantly ($p < 0.001$, Student's t-test) lower plasma sRAGE level than nondiabetic controls (0.77 ± 0.34 ng/mL). Of note, when diabetic patients alone were extracted for analyses, a direct association was not observed between plasma soluble RAGE (both sRAGE and esRAGE) levels and the status of glycemic control (i.e., glycated hemoglobin A1c) [171, 177, 180, 183, 189]. Thus, these contradictory findings in diabetic patients suggest that levels of plasma in soluble forms of RAGE are not determined simply by status of glycemic control, and that even plasma esRAGE and sRAGE levels may be under the control of distinct mechanisms.

4.2 Soluble RAGE, CKD, and Cardiovascular Disease

Another important component that can affect plasma sRAGE is the presence of CKD, which may explain controversial findings of plasma sRAGE in diabetes. It has been shown that, in peripheral monocytes from those with varying severities of CKD, RAGE expression is closely associated with worsening of CKD and is

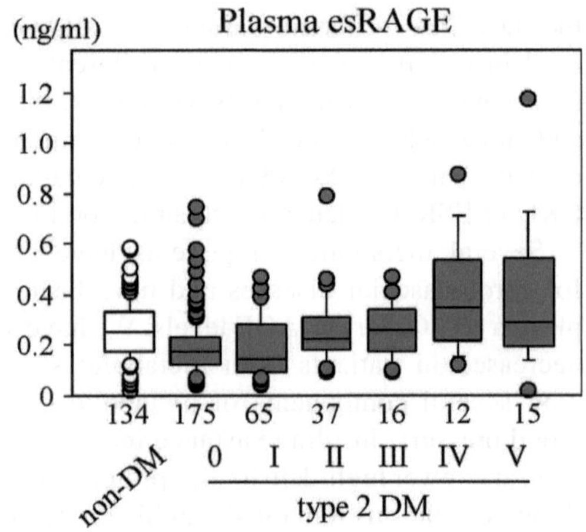

Fig. 8 Plasma endogenous secretory receptor for advanced glycation end products (esRAGE) is influenced by the presence of chronic kidney disease (CKD) in patients with type 2 diabetes. CKD categories are based on glomerular filtration rate (GFR) estimated by MDRD equation – I: GFR $\geq$ 90, II: GFR 60–89, III: GFR 30–59, IV: 15–29, V: GFR < 15 mL/min). Numbers of subjects were described in X-axis. DM, diabetes mellitus. Modified from ([125], Fig. 6)

strongly correlated with plasma levels of pentosidine, a marker for AGEs [99]. Circulating sRAGE levels have been shown to be increased in patients with decreased renal function, particularly those with ESRD [177, 186, 190]. As shown in Fig. 8, plasma esRAGE levels in type 2 diabetic patients without CKD are lower than for nondiabetic controls. However, the level gradually elevated in accordance with progression of CKD. Thus, plasma sRAGE and esRAGE are markedly affected by the presence of CKD, which might make the previous findings regarding comparison between nondiabetics and diabetics quite controversial. It remains to be determined whether the increase in plasma esRAGE in CKD is caused by decreased renal function alone or whether esRAGE levels are upregulated to protect against toxic effects of the RAGE ligands. Successful kidney transplantation resulted in significant decrease in plasma sRAGE [191], implying that the kidneys play a role in sRAGE removal.

We recently reported an observational cohort study in patients with ESRD and longitudinally evaluated the effect of plasma esRAGE on cardiovascular mortality [181]. Patients with ESRD have been reported to have a substantially increased rate of cardiovascular mortality. The cohort in that study included 206 ESRD patients, who had been treated by regular hemodialysis for more than 3 months. The median follow-up period was 111 months. At the end of follow-up, 132 patients were confirmed to be alive on hemodialysis and 74 had died, 34 of which were due to fatal cardiovascular events. Even though the plasma esRAGE levels at baseline were higher in those in ESRD than in those without kidney disease, those in the lowest tertile of plasma esRAGE levels exhibited significantly higher cardiovascular mortality, but not noncardiovascular mortality (Fig. 9). Importantly, even in the subpopulation of nondiabetics alone, low circulating esRAGE level was a predictor of cardiovascular mortality, independent of the other classical risk factors. Our findings thus suggest that low circulating esRAGE level is a predictor for atherosclerosis and

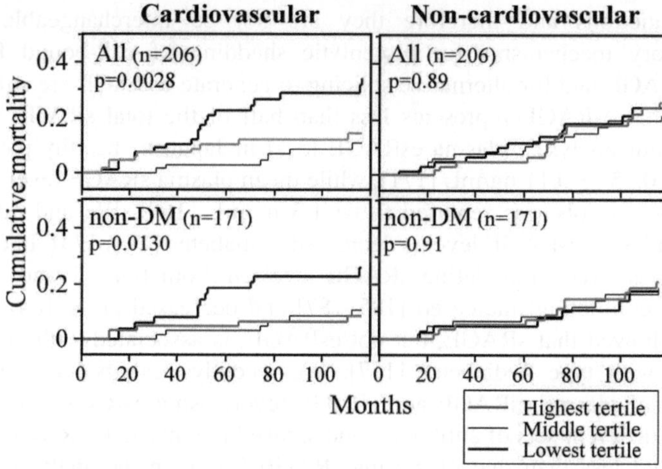

Fig. 9 Low plasma endogenous secretory receptor for advanced glycation end products (esRAGE) level is a predictor of cardiovascular mortality in patients with end stage renal disease ESRD. Cumulative mortalities in subjects with the lowest, middle, and highest tertiles of plasma esRAGE levels were estimated by Kaplan Meier analysis and the log-rank test. Modified from [181]

cardiovascular events in patients with ESRD. In a cohort of 591 patients after transplantation, low sRAGE levels are also shown to be associated with a 2–3-times higher risk for mortality, especially after correction for creatinine clearance, again implying lack of sRAGE as a risk factor for mortality in renal transplant recipients [192]. A larger prospective cohort study will be warranted to examine whether decreased sRAGE is also a risk predictor for general population and in different disease entities.

It is not known at present how esRAGE is involved in cardiovascular mortality. In our ESRD cohort, neither plasma pentosidine nor carboxymethyl-lysine level predicted cardiovascular mortality. Moreover, the inverse correlation between low circulating esRAGE level and cardiovascular mortality was not dependent on plasma AGEs levels. Thus, the protective effect of esRAGE against cardiovascular mortality may not be entirely dependent on neutralization of toxic AGEs. Other endogenous ligands for RAGE, such as S100A12, may also be involved in the function of esRAGE. The plasma level of S100A12 has been shown to be increased in diabetes and inversely correlated with serum sRAGE level [183, 193].

4.3 *sRAGE vs. esRAGE: Any Differences?*

It is unclear at present whether the pathophysiological significances of circulating esRAGE and sRAGE are distinct in different clinical settings. These two assays measure different pools of sRAGE, and different variants of sRAGE may have different functions and therefore they are not be interchangeable. Indeed, the regulatory mechanisms for proteolytic shedding of cell-bound RAGE to generate sRAGE and for alternative splicing to generate esRAGE are still unclear. It appears that esRAGE represents less than half of the total sRAGE in human plasma. In our analyses, plasma esRAGE level in Japanese healthy patients was found to be 0.25 ± 0.11 ng/mL [171], while mean plasma sRAGE level in Caucasian healthy controls was reported to be 1.3 ng/mL [168]. We and others have shown that plasma esRAGE level is decreased in diabetes [171, 174]. In contrast to the case of esRAGE, circulating sRAGE levels in both types 1 and 2 diabetic patients are conflicting: increased [185–187] and decreased [183, 184]. Humpert et al. also showed that sRAGE, but not esRAGE, is associated with albuminuria in patients with type 2 diabetes [177]. We recently described a head-to-head comparison of plasma esRAGE and sRAGE levels using esRAGE as a standard protein and different sets of antibodies and showed that plasma esRAGE level was about twofold less than that of plasma sRAGE [188]. In the analysis, esRAGE and sRAGE levels were positively correlated, with a stronger correlation in healthy patients than in type 1 diabetic patients. Since regulatory mechanisms are not understood for alternative splicing to generate esRAGE and for proteolytic shedding of cell surface RAGE to generate sRAGE, the possibility of distinct roles for them in certain disease conditions requires further examination. Furthermore, new splice variants have been recently discovered [160, 161], and further studies are necessary to establish which variants of sRAGE are measured by these assays and whether different soluble forms of RAGE have different pathophysiological functions.

4.4 Soluble RAGE as a Therapeutic Target?

As described earlier, the potential usefulness of soluble RAGE for prevention and treatment of inflammatory diseases has been demonstrated in many animal models. Blockade of RAGE by administration of genetically engineered sRAGE successfully prevented the development of micro- [87, 194] and macrovascular complications in diabetes [106, 107, 109]. We have also shown that adenoviral overexpression of esRAGE successfully restored the impaired angiogenic response in diabetic mice [124]. Sakaguchi et al. found that administration of sRAGE markedly suppressed neointimal formation following arterial injury in nondiabetic mice [146]. Soluble RAGE has also been shown to effectively prevent the development of diabetes [151], protect against tumor growth and metastasis [79], improve the outcome of colitis [77], restore impaired wound healing [122], and suppress Alzheimer's disease-like conditions [195]. These effects of soluble RAGE in animal models could be explained by its decoy function, inhibiting RAGE interaction with its proinflammatory ligands, which might be applicable to human diseases as well.

Further application of soluble RAGE to the treatment of human diseases will require answers to several questions. Most importantly, limited findings are available regarding the mechanisms of regulation of circulating esRAGE or sRAGE in humans. A tissue microarray technique using a wide variety of adult normal human preparations obtained from surgical and autopsy specimens revealed that esRAGE was widely distributed in tissues, including vascular endothelium, monocyte/macrophage, pneumocytes, and several endocrine organs [196]. However, it is unclear at present from which organ or tissue plasma sRAGE or esRAGE originates. Other splice variants recently discovered may also play significant parts [160, 161]. Circulating AGEs may be involved in regulation of the secretion or production of soluble RAGE, since AGEs are known to upregulate RAGE expression in vitro [197]. Levels of esRAGE could be simultaneously upregulated by AGEs and act as a negative feedback loop to compensate for the damaging effects of AGEs. We and others have found positive correlations between plasma sRAGE or esRAGE and AGEs [179–181, 186]. As shown in Fig. 10, the correlation between plasma esRAGE and pentosidine was modest in all patients ($r = 0.616$; $p < 0.001$), and was also statistically significant in hemodialysis and nonhemodialysis patients alone. Plasma esRAGE was also significantly correlated with plasma CML ($r = 0.492$; $p < 0.001$), with the significant correlation lost in hemodialysis and nonhemodialysis patients alone. AGEs-mediated regulation of

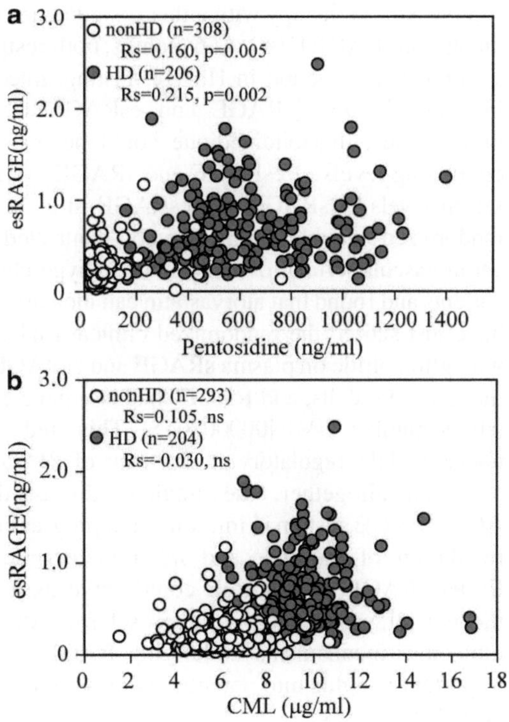

Fig. 10 Correlation between plasma endogenous secretory receptor for advanced glycation end products (esRAGE) and advanced glycation end products (AGEs), pentosidine (**a**) and carboxymethyllysine (CML) (**b**). In all subjects, correlation between plasma esRAGE and both pentosidine ($r = 0.616$) and CML ($r = 0.492$) were significant ($p < 0.001$). Modified from ([125], Fig. 8)

soluble RAGE is further supported by the findings that the suppression of sRAGE expression in diabetic rat kidney is reversed by blockade of AGEs accumulation with alagebrium [198]. Other inflammatory mediators, such as S100, tumor necrosis factor-α, and C-reactive protein, could also be potential candidates for regulation of the plasma level of soluble RAGE in humans [183, 197, 199]. Moreover, Geroldi et al. [200] showed that high serum sRAGE is associated with extreme longevity, suggesting that understanding the intrinsic regulation of RAGE and sRAGE is important for longevity and antiaging strategies. Without doubt, further understanding of the regulation of soluble RAGE will be most helpful in delineating potential targets for therapeutic application of sRAGE.

Second, it would be important to determine whether currently available pharmacologic agents can regulate plasma sRAGE or esRAGE. Forbes et al. [201] showed that inhibition of angiotensin-converting enzyme (ACE) in rats increased renal expression of sRAGE, and that this was associated with decreases in expression of renal full-length RAGE protein. They also showed that plasma sRAGE levels were significantly increased by inhibition of ACE in both diabetic rats and in humans with type 1 diabetes. Thus, one attractive scenario is that the protective effect of ACE inhibition against progression of renal dysfunction is mediated through regulation of RAGE vs. sRAGE production. Other potential agents that may affect circulating soluble RAGE include the thiazolidinediones [202] and statins [203, 204], both of which are known to modulate the AGEs–RAGE system in culture. A randomized, open-label, parallel group study was performed with 64 participants randomized to receive add-on therapy with either rosiglitazone or sulfonylurea to examine the effect on plasma sRAGE [205]. At 6 months, both rosiglitazone and sulfonylurea resulted in a significant reduction in HbA1c, fasting glucose, and AGE. However, significant increases in total sRAGE and esRAGE were only seen in the rosiglitazone group. Thus, thiazolidinedione could be one promising candidate that increases circulating levels of esRAGE and sRAGE. Tam et al. recently reported changes in serum levels of sRAGE and esRAGE in archived serum samples from a previous randomized double-blind, placebo-controlled clinical trial that explored the cardiovascular effects of atorvastatin in hypercholesterolemic Chinese type 2 diabetic patients and found that atorvastatin can increase circulating esRAGE levels [206]. We have just started the randomized clinical trial comparing the effect of pioglitazone with glimepiride on plasma sRAGE and esRAGE, expression of RAGE on peripheral mononuclear cells, and RAGE shaddase gene expression in type 2 diabetic patients (study number UMIN000002055). This study will be of particular importance to understand the regulatory mechanisms of sRAGE and esRAGE in a clinical setting.

Taken altogether, the findings discussed here implicate pivotal role of AGEs–RAGE system in initiation and progression of cardiovascular disease through regulation of vascular and inflammatory cells in patients with renal failure. Either sRAGE or esRAGE could serve as a novel biomarker for estimation of the risk of progression of atherosclerotic disorders. Further examination of the molecular mechanisms underlying RAGE and esRAGE regulation will provide important insights into potential targets for the prevention and treatment of cardiovascular diseases.

Acknowledgments We thank our colleagues in the Osaka City University Graduate School of Medicine and Kanazawa University Graduate School of Medical Science for their unflagging support of our projects. We apologize to all colleagues whose work we could not cite other than indirectly through other publications, due to limitation of space. This study was supported in part by a Grant-in-Aid for Scientific Research (20591067 to HK) from the Ministry of Education, Culture, Sports, Science and Technology, Japan.

References

1. Foley RN, Parfrey PS, Sarnak MJ. Epidemiology of cardiovascular disease in chronic renal disease. *J Am Soc Nephrol* 1998; **9**: S16–S23.
2. Lindner A, Charra B, Sherrard DJ, Scribner BH. Accelerated atherosclerosis in prolonged maintenance hemodialysis. *N Engl J Med* 1974; **290**: 697–701.
3. Stack AG, Bloembergen WE. Prevalence and clinical correlates of coronary artery disease among new dialysis patients in the United States: a cross-sectional study. *J Am Soc Nephrol* 2001; **12**: 1516–1523.
4. USRDS: the United States Renal Data System. *Am J Kidney Dis* 2003; **42**: 1–230.
5. Nishimura M, Hashimoto T, Kobayashi H, Fukuda T, et al. Myocardial scintigraphy using a fatty acid analogue detects coronary artery disease in hemodialysis patients. *Kidney Int* 2004; **66**: 811–819.
6. Joki N, Hase H, Nakamura R, Yamaguchi T. Onset of coronary artery disease prior to initiation of haemodialysis in patients with end-stage renal disease. *Nephrol Dial Transplant* 1997; **12**: 718–723.
7. Ohtake T, Kobayashi S, Moriya H, Negishi K, et al. High prevalence of occult coronary artery stenosis in patients with chronic kidney disease at the initiation of renal replacement therapy: an angiographic examination. *J Am Soc Nephrol* 2005; **16**: 1141–1148.
8. Hase H, Tsunoda T, Tanaka Y, Takahashi Y, et al. Risk factors for de novo acute cardiac events in patients initiating hemodialysis with no previous cardiac symptom. *Kidney Int* 2006; **70**: 1142–1148.
9. Vanholder R, Massy Z, Argiles A, Spasovski G, et al. Chronic kidney disease as cause of cardiovascular morbidity and mortality. *Nephrol Dial Transplant* 2005; **20**: 1048–1056.
10. Hostetter TH. Chronic kidney disease predicts cardiovascular disease. *N Engl J Med* 2004; **351**: 1344–1346.
11. Redon J, Cea-Calvo L, Lozano JV, Fernandez-Perez C, et al. Kidney function and cardiovascular disease in the hypertensive population: the ERIC-HTA study. *J Hypertens* 2006; **24**: 663–669.
12. Monnier VM, Cerami A. Nonenzymatic browning in vivo: possible process for aging of long-lived proteins. *Science* 1981; **211**: 491–493.
13. Dyer DG, Blackledge JA, Thorpe SR, Baynes JW. Formation of pentosidine during nonenzymatic browning of proteins by glucose. Identification of glucose and other carbohydrates as possible precursors of pentosidine in vivo. *J Biol Chem* 1991; **266**: 11654–11660.
14. Makita Z, Radoff S, Rayfield EJ, Yang Z, et al. Advanced glycosylation end products in patients with diabetic nephropathy. *N Engl J Med* 1991; **325**: 836–842.
15. Miyata T, Ueda Y, Yoshida A, Sugiyama S, et al. Clearance of pentosidine, an advanced glycation end product, by different modalities of renal replacement therapy. *Kidney Int* 1997; **51**: 880–887.
16. Shaw S, Akyol M, Bell J, Briggs JD, et al. Effects of continuous ambulatory peritoneal dialysis and kidney transplantation on advanced glycation endproducts in the skin and peritoneum. *Cell Mol Biol (Noisy-le-grand)* 1998; **44**: 1061–1068.

17. Gugliucci A, Bendayan M. Renal fate of circulating advanced glycated end products (AGE): evidence for reabsorption and catabolism of AGE-peptides by renal proximal tubular cells. *Diabetologia* 1996; **39**: 149–160.
18. Genuth S, Sun W, Cleary P, Sell DR, et al. Glycation and carboxymethyllysine levels in skin collagen predict the risk of future 10-year progression of diabetic retinopathy and nephropathy in the diabetes control and complications trial and epidemiology of diabetes interventions and complications participants with type 1 diabetes. *Diabetes* 2005; **54**: 3103–3111.
19. Gerrits EG, Lutgers HL, Kleefstra N, Graaff R, et al. Skin autofluorescence: a tool to identify type 2 diabetic patients at risk for developing microvascular complications. *Diabetes Care* 2008; **31**: 517–521.
20. Sell DR, Lapolla A, Odetti P, Fogarty J, et al. Pentosidine formation in skin correlates with severity of complications in individuals with long-standing IDDM. *Diabetes* 1992; **41**: 1286–1292.
21. Friedlander MA, Wu YC, Elgawish A, Monnier VM. Early and advanced glycosylation end products. Kinetics of formation and clearance in peritoneal dialysis. *J Clin Invest* 1996; **97**: 728–735.
22. Miyata T, Fu MX, Kurokawa K, van Ypersele de Strihou C, et al. Autoxidation products of both carbohydrates and lipids are increased in uremic plasma: is there oxidative stress in uremia? *Kidney Int* 1998; **54**: 1290–1295.
23. Thorpe SR, Baynes JW. Maillard reaction products in tissue proteins: new products and new perspectives. *Amino Acids* 2003; **25**: 275–281.
24. Wolff SP, Dean RT. Glucose autoxidation and protein modification. The potential role of “autoxidative glycosylation” in diabetes. *Biochem J* 1987; **245**: 243–250.
25. Harding JJ, Beswick HT. The possible contribution of glucose autoxidation to protein modification of diabetes. *Biochem J* 1988; **249**: 617–618.
26. Miyata T, Maeda K, Kurokawa K, van Ypersele de Strihou C. Oxidation conspires with glycation to generate noxious advanced glycation end products in renal failure. *Nephrol Dial Transplant* 1997; **12**: 255–258.
27. Canestrari F, Galli F, Giorgini A, Albertini MC, et al. Erythrocyte redox state in uremic anemia: effects of hemodialysis and relevance of glutathione metabolism. *Acta Haematol* 1994; **91**: 187–193.
28. Miyata T, Wada Y, Cai Z, Iida Y, et al. Implication of an increased oxidative stress in the formation of advanced glycation end products in patients with end-stage renal failure. *Kidney Int* 1997; **51**: 1170–1181.
29. Vaziri ND. Oxidative stress in uremia: nature, mechanisms, and potential consequences. *Semin Nephrol* 2004; **24**: 469–473.
30. Galli F, Varga Z, Balla J, Ferraro B, et al. Vitamin E, lipid profile, and peroxidation in hemodialysis patients. *Kidney Int Suppl* 2001; **78**: S148–S154.
31. Kosch M, Levers A, Fobker M, Barenbrock M, et al. Dialysis filter type determines the acute effect of haemodialysis on endothelial function and oxidative stress. *Nephrol Dial Transplant* 2003; **18**: 1370–1375.
32. Diepeveen SH, Verhoeven GH, van der Palen J, Dikkeschei BL, et al. Oxidative stress in patients with end-stage renal disease prior to the start of renal replacement therapy. *Nephron Clin Pract* 2004; **98**: c3–c7.
33. Witko-Sarsat V, Friedlander M, Nguyen Khoa T, Capeillere-Blandin C, et al. Advanced oxidation protein products as novel mediators of inflammation and monocyte activation in chronic renal failure. *J Immunol* 1998; **161**: 2524–2532.
34. Fortuno A, Beloqui O, San Jose G, Moreno MU, et al. Increased phagocytic nicotinamide adenine dinucleotide phosphate oxidase-dependent superoxide production in patients with early chronic kidney disease. *Kidney Int Suppl* 2005; (**99**): S71–S75.
35. Yilmaz MI, Saglam M, Caglar K, Cakir E, et al. The determinants of endothelial dysfunction in CKD: oxidative stress and asymmetric dimethylarginine. *Am J Kidney Dis* 2006; **47**: 42–50.

36. Ceballos-Picot I, Witko-Sarsat V, Merad-Boudia M, Nguyen AT, et al. Glutathione antioxidant system as a marker of oxidative stress in chronic renal failure. *Free Radic Biol Med* 1996; **21**: 845–853.
37. Niwa T. 3-Deoxyglucosone: metabolism, analysis, biological activity, and clinical implication. *J Chromatogr B Biomed Sci Appl* 1999; **731**: 23–36.
38. Williamson JR, Chang K, Frangos M, Hasan KS, et al. Hyperglycemic pseudohypoxia and diabetic complications. *Diabetes* 1993; **42**: 801–813.
39. Thornalley PJ. Glyoxalase I – structure, function and a critical role in the enzymatic defence against glycation. *Biochem Soc Trans* 2003; **31**: 1343–1348.
40. Miyata T, van Ypersele de Strihou C, Imasawa T, Yoshino A, et al. Glyoxalase I deficiency is associated with an unusual level of advanced glycation end products in a hemodialysis patient. *Kidney Int* 2001; **60**: 2351–2359.
41. Waanders F, Greven WL, Baynes JW, Thorpe SR, et al. Renal accumulation of pentosidine in non-diabetic proteinuria-induced renal damage in rats. *Nephrol Dial Transplant* 2005; **20**: 2060–2070.
42. Koschinsky T, He CJ, Mitsuhashi T, Bucala R, et al. Orally absorbed reactive glycation products (glycotoxins): an environmental risk factor in diabetic nephropathy. *Proc Natl Acad Sci USA* 1997; **94**: 6474–6479.
43. Cerami C, Founds H, Nicholl I, Mitsuhashi T, et al. Tobacco smoke is a source of toxic reactive glycation products. *Proc Natl Acad Sci U S A* 1997; **94**: 13915–13920.
44. DeFronzo RA, Alvestrand A, Smith D, Hendler R, et al. Insulin resistance in uremia. *J Clin Invest* 1981; **67**: 563–568.
45. Park K, Gross M, Lee DH, Holvoet P, et al. Oxidative stress and insulin resistance: the coronary artery risk development in young adults study. *Diabetes Care* 2009; **32**: 1302–1307.
46. Sakata N, Imanaga Y, Meng J, Tachikawa Y, et al. Increased advanced glycation end products in atherosclerotic lesions of patients with end-stage renal disease. *Atherosclerosis* 1999; **142**: 67–77.
47. Yoshida S, Yamada K, Hamaguchi K, Nishimura M, et al. Immunohistochemical study of human advanced glycation end-products (AGE) and growth factors in cardiac tissues of patients on maintenance dialysis and with kidney transplantation. *Clin Nephrol* 1998; **49**: 273–280.
48. Bucala R, Makita Z, Vega G, Grundy S, et al. Modification of low density lipoprotein by advanced glycation end products contributes to the dyslipidemia of diabetes and renal insufficiency. *Proc Natl Acad Sci U S A* 1994; **91**: 9441–9445.
49. Oranje WA, Wolffenbuttel BH. Lipid peroxidation and atherosclerosis in type II diabetes. *J Lab Clin Med* 1999; **134**: 19–32.
50. Drueke TB, Nguyen Khoa T, Massy ZA, Witko-Sarsat V, et al. Role of oxidized low-density lipoprotein in the atherosclerosis of uremia. *Kidney Int Suppl* 2001; **78**: S114–S119.
51. Stocker R, Keaney JF, Jr. Role of oxidative modifications in atherosclerosis. *Physiol Rev* 2004; **84**: 1381–1478.
52. Suliman ME, Stenvinkel P, Jogestrand T, Maruyama Y, et al. Plasma pentosidine and total homocysteine levels in relation to change in common carotid intima-media area in the first year of dialysis therapy. *Clin Nephrol* 2006; **66**: 418–425.
53. Taki K, Takayama F, Tsuruta Y, Niwa T. Oxidative stress, advanced glycation end product, and coronary artery calcification in hemodialysis patients. *Kidney Int* 2006; **70**: 218–224.
54. Tarsio JF, Reger LA, Furcht LT. Decreased interaction of fibronectin, type IV collagen, and heparin due to nonenzymatic glycation. Implications for diabetes mellitus. *Biochemistry* 1987; **26**: 1014–1020.
55. Klein DJ, Oegema TR, Jr., Brown DM. Release of glomerular heparan-35SO4 proteoglycan by heparin from glomeruli of streptozocin-induced diabetic rats. *Diabetes* 1989; **38**: 130–139.

56. Blacher J, Guerin AP, Pannier B, Marchais SJ, et al. Impact of aortic stiffness on survival in end-stage renal disease. *Circulation* 1999; **99**: 2434–2439.
57. Shoji T, Emoto M, Shinohara K, Kakiya R, et al. Diabetes mellitus, aortic stiffness, and cardiovascular mortality in end-stage renal disease. *J Am Soc Nephrol* 2001; **12**: 2117–2124.
58. Kitahara T, Ono K, Tsuchida A, Kawai H, et al. Impact of brachial-ankle pulse wave velocity and ankle-brachial blood pressure index on mortality in hemodialysis patients. *Am J Kidney Dis* 2005; **46**: 688–696.
59. Meerwaldt R, Graaff R, Oomen PH, Links TP, et al. Simple non-invasive assessment of advanced glycation endproduct accumulation. *Diabetologia* 2004; **47**: 1324–1330.
60. Ueno H, Koyama H, Tanaka S, Fukumoto S, et al. Skin autofluorescence, a marker for advanced glycation end product accumulation, is associated with arterial stiffness in patients with end-stage renal disease. *Metabolism* 2008; **57**: 1452–1457.
61. Wolffenbuttel BH, Boulanger CM, Crijns FR, Huijberts MS, et al. Breakers of advanced glycation end products restore large artery properties in experimental diabetes. *Proc Natl Acad Sci USA* 1998; **95**: 4630–4634.
62. Kass DA, Shapiro EP, Kawaguchi M, Capriotti AR, et al. Improved arterial compliance by a novel advanced glycation end-product crosslink breaker. *Circulation* 2001; **104**: 1464–1470.
63. Meerwaldt R, Hartog JW, Graaff R, Huisman RJ, et al. Skin autofluorescence, a measure of cumulative metabolic stress and advanced glycation end products, predicts mortality in hemodialysis patients. *J Am Soc Nephrol* 2005; **16**: 3687–3693.
64. Meerwaldt R, Lutgers HL, Links TP, Graaff R, et al. Skin autofluorescence is a strong predictor of cardiac mortality in diabetes. *Diabetes Care* 2007; **30**: 107–112.
65. Roberts MA, Thomas MC, Fernando D, Macmillan N, et al. Low molecular weight advanced glycation end products predict mortality in asymptomatic patients receiving chronic haemodialysis. *Nephrol Dial Transplant* 2006; **21**: 1611–1617.
66. Wagner Z, Molnar M, Molnar GA, Tamasko M, et al. Serum carboxymethyllysine predicts mortality in hemodialysis patients. *Am J Kidney Dis* 2006; **47**: 294–300.
67. Schwedler SB, Metzger T, Schinzel R, Wanner C. Advanced glycation end products and mortality in hemodialysis patients. *Kidney Int* 2002; **62**: 301–310.
68. Busch M, Franke S, Muller A, Wolf M, et al. Potential cardiovascular risk factors in chronic kidney disease: AGEs, total homocysteine and metabolites, and the C-reactive protein. *Kidney Int* 2004; **66**: 338–347.
69. Pecoits-Filho R, Lindholm B, Stenvinkel P. The malnutrition, inflammation, and atherosclerosis (MIA) syndrome – the heart of the matter. *Nephrol Dial Transplant* 2002; **17:(Suppl 11)**: 28–31.
70. Rambod M, Bross R, Zitterkoph J, Benner D, et al. Association of Malnutrition-Inflammation Score with quality of life and mortality in hemodialysis patients: a 5-year prospective cohort study. *Am J Kidney Dis* 2009; **53**: 298–309.
71. Schmidt AM, Vianna M, Gerlach M, Brett J, et al. Isolation and characterization of two binding proteins for advanced glycosylation end products from bovine lung which are present on the endothelial cell surface. *J Biol Chem* 1992; **267**: 14987–14997.
72. Neeper M, Schmidt AM, Brett J, Yan SD, et al. Cloning and expression of a cell surface receptor for advanced glycosylation end products of proteins. *J Biol Chem* 1992; **267**: 14998–15004.
73. Kislinger T, Fu C, Huber B, Qu W, et al. N(epsilon)-(carboxymethyl)lysine adducts of proteins are ligands for receptor for advanced glycation end products that activate cell signaling pathways and modulate gene expression. *J Biol Chem* 1999; **274**: 31740–31749.
74. Reddy S, Bichler J, Wells-Knecht KJ, Thorpe SR, et al. N epsilon-(carboxymethyl)lysine is a dominant advanced glycation end product (AGE) antigen in tissue proteins. *Biochemistry* 1995; **34**: 10872–10878.

75. Xie J, Reverdatto S, Frolov A, Hoffmann R, et al. Structural basis for pattern recognition by the receptor for advanced glycation end products (RAGE). *J Biol Chem* 2008; **283**: 27255–27269.
76. Matsumoto S, Yoshida T, Murata H, Harada S, et al. Solution structure of the variable-type domain of the receptor for advanced glycation end products: new insight into AGE-RAGE interaction. *Biochemistry* 2008; **47**: 12299–12311.
77. Hofmann MA, Drury S, Fu C, Qu W, et al. RAGE mediates a novel proinflammatory axis: a central cell surface receptor for S100/calgranulin polypeptides. *Cell* 1999; **97**: 889–901.
78. Hori O, Brett J, Slattery T, Cao R, et al. The receptor for advanced glycation end products (RAGE) is a cellular binding site for amphoterin. Mediation of neurite outgrowth and co-expression of rage and amphoterin in the developing nervous system. *J Biol Chem* 1995; **270**: 25752–25761.
79. Taguchi A, Blood DC, del Toro G, Canet A, et al. Blockade of RAGE-amphoterin signalling suppresses tumour growth and metastases. *Nature* 2000; **405**: 354–360.
80. Yan SD, Chen X, Fu J, Chen M, et al. RAGE and amyloid-beta peptide neurotoxicity in Alzheimer's disease. *Nature* 1996; **382**: 685–691.
81. Sousa MM, Yan SD, Stern D, Saraiva MJ. Interaction of the receptor for advanced glycation end products (RAGE) with transthyretin triggers nuclear transcription factor κB (NF-κB) activation. *Lab Invest* 2000; **80**: 1101–1110.
82. Chavakis T, Bierhaus A, Al-Fakhri N, Schneider D, et al. The pattern recognition receptor (RAGE) is a counterreceptor for leukocyte integrins: a novel pathway for inflammatory cell recruitment. *J Exp Med* 2003; **198**: 1507–1515.
83. Krieger M, Stern DM. Series introduction: multiligand receptors and human disease. *J Clin Invest* 2001; **108**: 645–647.
84. Schmidt AM, Yan SD, Yan SF, Stern DM. The multiligand receptor RAGE as a progression factor amplifying immune and inflammatory responses. *J Clin Invest* 2001; **108**: 949–955.
85. Yamamoto Y, Kato I, Doi T, Yonekura H, et al. Development and prevention of advanced diabetic nephropathy in RAGE-overexpressing mice. *J Clin Invest* 2001; **108**: 261–268.
86. Myint KM, Yamamoto Y, Doi T, Kato I, et al. RAGE control of diabetic nephropathy in a mouse model: effects of RAGE gene disruption and administration of low-molecular weight heparin. *Diabetes* 2006; **55**: 2510–2522.
87. Bierhaus A, Haslbeck KM, Humpert PM, Liliensiek B, et al. Loss of pain perception in diabetes is dependent on a receptor of the immunoglobulin superfamily. *J Clin Invest* 2004; **114**: 1741–1751.
88. Toth C, Rong LL, Yang C, Martinez J, et al. Receptor for advanced glycation end products (RAGEs) and experimental diabetic neuropathy. *Diabetes* 2008; **57**: 1002–1017.
89. Xu D, Kyriakis JM. Phosphatidylinositol 3'-kinase-dependent activation of renal mesangial cell Ki-Ras and ERK by advanced glycation end products. *J Biol Chem* 2003; **278**: 39349–39355.
90. Yan SD, Schmidt AM, Anderson GM, Zhang J, et al. Enhanced cellular oxidant stress by the interaction of advanced glycation end products with their receptors/binding proteins. *J Biol Chem* 1994; **269**: 9889–9897.
91. Wautier MP, Chappey O, Corda S, Stern DM, et al. Activation of NADPH oxidase by AGE links oxidant stress to altered gene expression via RAGE. *Am J Physiol Endocrinol Metab* 2001; **280**: E685–E694.
92. Dukic-Stefanovic S, Gasic-Milenkovic J, Deuther-Conrad W, Munch G. Signal transduction pathways in mouse microglia N-11 cells activated by advanced glycation endproducts (AGEs). *J Neurochem* 2003; **87**: 44–55.
93. Li J, Schmidt AM. Characterization and functional analysis of the promoter of RAGE, the receptor for advanced glycation end products. *J Biol Chem* 1997; **272**: 16498–16506.
94. Bierhaus A, Schiekofer S, Schwaninger M, Andrassy M, et al. Diabetes-associated sustained activation of the transcription factor nuclear factor-kappaB. *Diabetes* 2001; **50**: 2792–2808.

95. Shanmugam N, Kim YS, Lanting L, Natarajan R. Regulation of cyclooxygenase-2 expression in monocytes by ligation of the receptor for advanced glycation end products. *J Biol Chem* 2003; **278**: 34834–34844.
96. Schmidt AM, Hori O, Chen JX, Li JF, et al. Advanced glycation endproducts interacting with their endothelial receptor induce expression of vascular cell adhesion molecule-1 (VCAM-1) in cultured human endothelial cells and in mice. A potential mechanism for the accelerated vasculopathy of diabetes. *J Clin Invest* 1995; **96**: 1395–1403.
97. Vlassara H. The AGE-receptor in the pathogenesis of diabetic complications. *Diabetes Metab Res Rev* 2001; **17**: 436–443.
98. Basta G, Lazzerini G, Massaro M, Simoncini T, et al. Advanced glycation end products activate endothelium through signal-transduction receptor RAGE: a mechanism for amplification of inflammatory responses. *Circulation* 2002; **105**: 816–822.
99. Hou FF, Ren H, Owen WF, Jr, Guo ZJ, et al. Enhanced expression of receptor for advanced glycation end products in chronic kidney disease. *J Am Soc Nephrol* 2004; **15**: 1889–1896.
100. Rodriguez-Ayala E, Anderstam B, Suliman ME, Seeberger A, et al. Enhanced RAGE-mediated NFkappaB stimulation in inflamed hemodialysis patients. *Atherosclerosis* 2005; **180**: 333–340.
101. Linden E, Cai W, He JC, Xue C, et al. Endothelial dysfunction in patients with chronic kidney disease results from advanced glycation end products (AGE)-mediated inhibition of endothelial nitric oxide synthase through RAGE activation. *Clin J Am Soc Nephrol* 2008; **3**: 691–698.
102. Greten J, Kreis I, Wiesel K, Stier E, et al. Receptors for advance glycation end-products (AGE) – expression by endothelial cells in non-diabetic uraemic patients. *Nephrol Dial Transplant* 1996; **11**: 786–790.
103. Kawagishi T, Nishizawa Y, Konishi T, Kawasaki K, et al. High-resolution B-mode ultrasonography in evaluation of atherosclerosis in uremia. *Kidney Int* 1995; **48**: 820–826.
104. Cipollone F, Iezzi A, Fazia M, Zucchelli M, et al. The receptor RAGE as a progression factor amplifying arachidonate-dependent inflammatory and proteolytic response in human atherosclerotic plaques: role of glycemic control. *Circulation* 2003; **108**: 1070–1077.
105. Burke AP, Kolodgie FD, Zieske A, Fowler DR, et al. Morphologic findings of coronary atherosclerotic plaques in diabetics: a postmortem study. *Arterioscler Thromb Vasc Biol* 2004; **24**: 1266–1271.
106. Park L, Raman KG, Lee KJ, Lu Y, et al. Suppression of accelerated diabetic atherosclerosis by the soluble receptor for advanced glycation endproducts. *Nat Med* 1998; **4**: 1025–1031.
107. Kislinger T, Tanji N, Wendt T, Qu W, et al. Receptor for advanced glycation end products mediates inflammation and enhanced expression of tissue factor in vasculature of diabetic apolipoprotein E-null mice. *Arterioscler Thromb Vasc Biol* 2001; **21**: 905–910.
108. Wendt T, Harja E, Bucciarelli L, Qu W, et al. RAGE modulates vascular inflammation and atherosclerosis in a murine model of type 2 diabetes. *Atherosclerosis* 2006; **185**: 70–77.
109. Bucciarelli LG, Wendt T, Qu W, Lu Y, et al. RAGE blockade stabilizes established atherosclerosis in diabetic apolipoprotein E-null mice. *Circulation* 2002; **106**: 2827–2835.
110. Soro-Paavonen A, Watson AM, Li J, Paavonen K, et al. Receptor for advanced glycation end products (RAGE) deficiency attenuates the development of atherosclerosis in diabetes. *Diabetes* 2008; **57**: 2461–2469.
111. Harja E, Bu DX, Hudson BI, Chang JS, et al. Vascular and inflammatory stresses mediate atherosclerosis via RAGE and its ligands in apoE-/- mice. *J Clin Invest* 2008; **118**: 183–194.
112. Massy ZA, Ivanovski O, Nguyen-Khoa T, Angulo J, et al. Uremia accelerates both atherosclerosis and arterial calcification in apolipoprotein E knockout mice. *J Am Soc Nephrol* 2005; **16**: 109–116.
113. Phan O, Ivanovski O, Nguyen-Khoa T, Mothu N, et al. Sevelamer prevents uremia-enhanced atherosclerosis progression in apolipoprotein E-deficient mice. *Circulation* 2005; **112**: 2875–2882.

114. Bro S, Flyvbjerg A, Binder CJ, Bang CA, et al. A neutralizing antibody against receptor for advanced glycation end products (RAGE) reduces atherosclerosis in uremic mice. *Atherosclerosis* 2008; **201**: 274–280.
115. Mizutani M, Kern TS, Lorenzi M. Accelerated death of retinal microvascular cells in human and experimental diabetic retinopathy. *J Clin Invest* 1996; **97**: 2883–2890.
116. Stitt A, Gardiner TA, Anderson NL, Canning P, et al. The AGE inhibitor pyridoxamine inhibits development of retinopathy in experimental diabetes. *Diabetes* 2002; **51**: 2826–2832.
117. Hammes HP, Du X, Edelstein D, Taguchi T, et al. Benfotiamine blocks three major pathways of hyperglycemic damage and prevents experimental diabetic retinopathy. *Nat Med* 2003; **9**: 294–299.
118. Abaci A, Oguzhan A, Kahraman S, Eryol NK, et al. Effect of diabetes mellitus on formation of coronary collateral vessels. *Circulation* 1999; **99**: 2239–2242.
119. Rivard A, Silver M, Chen D, Kearney M, et al. Rescue of diabetes-related impairment of angiogenesis by intramuscular gene therapy with adeno-VEGF. *Am J Pathol* 1999; **154**: 355–363.
120. Waltenberger J. Impaired collateral vessel development in diabetes: potential cellular mechanisms and therapeutic implications. *Cardiovasc Res* 2001; **49**: 554–560.
121. Stitt AW, McGoldrick C, Rice-McCaldin A, McCance DR, et al. Impaired retinal angiogenesis in diabetes: role of advanced glycation end products and galectin-3. *Diabetes* 2005; **54**: 785–794.
122. Goova MT, Li J, Kislinger T, Qu W, et al. Blockade of receptor for advanced glycation end-products restores effective wound healing in diabetic mice. *Am J Pathol* 2001; **159**: 513–525.
123. Tamarat R, Silvestre JS, Huijberts M, Benessiano J, et al. Blockade of advanced glycation end-product formation restores ischemia-induced angiogenesis in diabetic mice. *Proc Natl Acad Sci USA* 2003; **100**: 8555–8560.
124. Shoji T, Koyama H, Morioka T, Tanaka S, et al. Receptor for advanced glycation end products is involved in impaired angiogenic response in diabetes. *Diabetes* 2006; **55**: 2245–2255.
125. Koyama H, Yamamoto H, Nishizawa Y. RAGE and soluble RAGE: potential therapeutic targets for cardiovascular diseases. *Mol Med* 2007; **13**: 625–635.
126. Asahara T, Murohara T, Sullivan A, Silver M, et al. Isolation of putative progenitor endothelial cells for angiogenesis. *Science* 1997; **275**: 964–967.
127. Rafii S, Lyden D. Therapeutic stem and progenitor cell transplantation for organ vascularization and regeneration. *Nat Med* 2003; **9**: 702–712.
128. Rauscher FM, Goldschmidt-Clermont PJ, Davis BH, Wang T, et al. Aging, progenitor cell exhaustion, and atherosclerosis. *Circulation* 2003; **108**: 457–463.
129. Choi JH, Kim KL, Huh W, Kim B, et al. Decreased number and impaired angiogenic function of endothelial progenitor cells in patients with chronic renal failure. *Arterioscler Thromb Vasc Biol* 2004; **24**: 1246–1252.
130. de Groot K, Bahlmann FH, Sowa J, Koenig J, et al. Uremia causes endothelial progenitor cell deficiency. *Kidney Int* 2004; **66**: 641–646.
131. Herbrig K, Pistrosch F, Oelschlaegel U, Wichmann G, et al. Increased total number but impaired migratory activity and adhesion of endothelial progenitor cells in patients on long-term hemodialysis. *Am J Kidney Dis* 2004; **44**: 840–849.
132. Schlieper G, Hristov M, Brandenburg V, Kruger T, et al. Predictors of low circulating endothelial progenitor cell numbers in haemodialysis patients. *Nephrol Dial Transplant* 2008; **23**: 2611–2618.
133. Ueno H, Koyama H, Fukumoto S, Tanaka S, et al. Dialysis modality is independently associated with circulating endothelial progenitor cells in end-stage renal diseases patients. *Nephrol Dial Transplant* 2010; 25: 581–586.

134. Tepper OM, Galiano RD, Capla JM, Kalka C, et al. Human endothelial progenitor cells from type II diabetics exhibit impaired proliferation, adhesion, and incorporation into vascular structures. *Circulation* 2002; **106**: 2781–2786.
135. Fadini GP, Sartore S, Agostini C, Avogaro A. Significance of endothelial progenitor cells in subjects with diabetes. *Diabetes Care* 2007; **30**: 1305–1313.
136. Scheubel RJ, Zorn H, Silber RE, Kuss O, et al. Age-dependent depression in circulating endothelial progenitor cells in patients undergoing coronary artery bypass grafting. *J Am Coll Cardiol* 2003; **42**: 2073–2080.
137. Heiss C, Keymel S, Niesler U, Ziemann J, et al. Impaired progenitor cell activity in age-related endothelial dysfunction. *J Am Coll Cardiol* 2005; **45**: 1441–1448.
138. Ueno H, Koyama H, Fukumoto S, Tanaka S, et al. Advanced glycation end-products, carotid atherosclerosis and circulating endothelial progenitor cells in patients with end-stage renal disease. *Metabolism* 2010 in press
139. Chavakis E, Hain A, Vinci M, Carmona G, et al. High-mobility group box 1 activates integrin-dependent homing of endothelial progenitor cells. *Circ Res* 2007; **100**: 204–212.
140. Kahn JK, Rutherford BD, McConahay DR, Johnson WL, et al. Short- and long-term outcome of percutaneous transluminal coronary angioplasty in chronic dialysis patients. *Am Heart J* 1990; **119**: 484–489.
141. Ahmed WH, Shubrooks SJ, Gibson CM, Baim DS, et al. Complications and long-term outcome after percutaneous coronary angioplasty in chronic hemodialysis patients. *Am Heart J* 1994; **128**: 252–255.
142. Azar RR, Prpic R, Ho KK, Kiernan FJ, et al. Impact of end-stage renal disease on clinical and angiographic outcomes after coronary stenting. *Am J Cardiol* 2000; **86**: 485–489.
143. Rozenman Y, Sapoznikov D, Mosseri M, Gilon D, et al. Long-term angiographic follow-up of coronary balloon angioplasty in patients with diabetes mellitus: a clue to the explanation of the results of the BARI study. Balloon Angioplasty Revascularization Investigation. *J Am Coll Cardiol* 1997; **30**: 1420–1425.
144. Abizaid A, Kornowski R, Mintz GS, Hong MK, et al. The influence of diabetes mellitus on acute and late clinical outcomes following coronary stent implantation. *J Am Coll Cardiol* 1998; **32**: 584–589.
145. Zhou Z, Wang K, Penn MS, Marso SP, et al. Receptor for AGE (RAGE) mediates neointimal formation in response to arterial injury. *Circulation* 2003; **107**: 2238–2243.
146. Sakaguchi T, Yan SF, Yan SD, Belov D, et al. Central role of RAGE-dependent neointimal expansion in arterial restenosis. *J Clin Invest* 2003; **111**: 959–972.
147. Ross R. Atherosclerosis – an inflammatory disease. *N Engl J Med* 1999; **340**: 115–126.
148. Schmidt AM, Yan SD, Brett J, Mora R, et al. Regulation of human mononuclear phagocyte migration by cell surface-binding proteins for advanced glycation end products. *J Clin Invest* 1993; **91**: 2155–2168.
149. Rouhiainen A, Kuja-Panula J, Wilkman E, Pakkanen J, et al. Regulation of monocyte migration by amphoterin (HMGB1). *Blood* 2004; **104**: 1174–1182.
150. Yan SS, Wu ZY, Zhang HP, Furtado G, et al. Suppression of experimental autoimmune encephalomyelitis by selective blockade of encephalitogenic T-cell infiltration of the central nervous system. *Nat Med* 2003; **9**: 287–293.
151. Chen Y, Yan SS, Colgan J, Zhang HP, et al. Blockade of late stages of autoimmune diabetes by inhibition of the receptor for advanced glycation end products. *J Immunol* 2004; **173**: 1399–1405.
152. Moser B, Szabolcs MJ, Ankersmit HJ, Lu Y, et al. Blockade of RAGE suppresses alloimmune reactions in vitro and delays allograft rejection in murine heart transplantation. *Am J Transplant* 2007; **7**: 293–302.
153. Tian J, Avalos AM, Mao SY, Chen B, et al. Toll-like receptor 9-dependent activation by DNA-containing immune complexes is mediated by HMGB1 and RAGE. *Nat Immunol* 2007; **8**: 487–496.

154. Orlova VV, Choi EY, Xie C, Chavakis E, et al. A novel pathway of HMGB1-mediated inflammatory cell recruitment that requires Mac-1-integrin. *Embo J* 2007; **26**: 1129–1139.
155. Malherbe P, Richards JG, Gaillard H, Thompson A, et al. cDNA cloning of a novel secreted isoform of the human receptor for advanced glycation end products and characterization of cells co-expressing cell-surface scavenger receptors and Swedish mutant amyloid precursor protein. *Brain Res Mol Brain Res* 1999; **71**: 159–170.
156. Yonekura H, Yamamoto Y, Sakurai S, Petrova RG, et al. Novel splice variants of the receptor for advanced glycation end-products expressed in human vascular endothelial cells and pericytes, and their putative roles in diabetes-induced vascular injury. *Biochem J* 2003; **370**: 1097–1109.
157. Schlueter C, Hauke S, Flohr AM, Rogalla P, et al. Tissue-specific expression patterns of the RAGE receptor and its soluble forms – a result of regulated alternative splicing? *Biochim Biophys Acta* 2003; **1630**: 1–6.
158. Park IH, Yeon SI, Youn JH, Choi JE, et al. Expression of a novel secreted splice variant of the receptor for advanced glycation end products (RAGE) in human brain astrocytes and peripheral blood mononuclear cells. *Mol Immunol* 2004; **40**: 1203–1211.
159. Ding Q, Keller JN. Splice variants of the receptor for advanced glycosylation end products (RAGE) in human brain. *Neurosci Lett* 2005; **373**: 67–72.
160. Hudson BI, Carter AM, Harja E, Kalea AZ, et al. Identification, classification, and expression of RAGE gene splice variants. *FASEB J* 2008; **22**: 1572–1580.
161. Kalea AZ, Reiniger N, Yang H, Arriero M, et al. Alternative splicing of the murine receptor for advanced glycation end-products (RAGE) gene. *FASEB J* 2009; **23**: 1766–1774.
162. Bierhaus A, Humpert PM, Morcos M, Wendt T, et al. Understanding RAGE, the receptor for advanced glycation end products. *J Mol Med* 2005; **83**: 876–886.
163. Hudson BI, Harja E, Moser B, Schmidt AM. Soluble levels of receptor for advanced glycation endproducts (sRAGE) and coronary artery disease: the next C-reactive protein? *Arterioscler Thromb Vasc Biol* 2005; **25**: 879–882.
164. Hanford LE, Enghild JJ, Valnickova Z, Petersen SV, et al. Purification and characterization of mouse soluble receptor for advanced glycation end products (sRAGE). *J Biol Chem* 2004; **279**: 50019–50024.
165. Raucci A, Cugusi S, Antonelli A, Barabino SM, et al. A soluble form of the receptor for advanced glycation endproducts (RAGE) is produced by proteolytic cleavage of the membrane-bound form by the sheddase a disintegrin and metalloprotease 10 (ADAM10). *FASEB J* 2008; **22**: 3716–3727.
166. Zhang L, Bukulin M, Kojro E, Roth A, et al. Receptor for advanced glycation end products is subjected to protein ectodomain shedding by metalloproteinases. *J Biol Chem* 2008; **283**: 35507–35516.
167. Pullerits R, Brisslert M, Jonsson IM, Tarkowski A. Soluble receptor for advanced glycation end products triggers a proinflammatory cytokine cascade via beta2 integrin Mac-1. *Arthritis Rheum* 2006; **54**: 3898–3907.
168. Falcone C, Emanuele E, D'Angelo A, Buzzi MP, et al. Plasma levels of soluble receptor for advanced glycation end products and coronary artery disease in nondiabetic men. *Arterioscler Thromb Vasc Biol* 2005; **25**: 1032–1037.
169. Emanuele E, D'Angelo A, Tomaino C, Binetti G, et al. Circulating levels of soluble receptor for advanced glycation end products in Alzheimer disease and vascular dementia. *Arch Neurol* 2005; **62**: 1734–1736.
170. Sakurai S, Yamamoto Y, Tamei H, Matsuki H, et al. Development of an ELISA for esRAGE and its application to type 1 diabetic patients. *Diabetes Res Clin Pract* 2006; **73**: 158–165.
171. Koyama H, Shoji T, Yokoyama H, Motoyama K, et al. Plasma level of endogenous secretory RAGE is associated with components of the metabolic syndrome and atherosclerosis. *Arterioscler Thromb Vasc Biol* 2005; **25**: 2587–2593.

172. Nakamura K, Yamagishi S, Adachi H, Kurita-Nakamura Y, et al. Elevation of soluble form of receptor for advanced glycation end products (sRAGE) in diabetic subjects with coronary artery disease. *Diabetes Metab Res Rev* 2007; **23**: 368–371.
173. Nin JW, Ferreira I, Schalkwijk CG, Prins MH, et al. Levels of soluble receptor for AGE are cross-sectionally associated with cardiovascular disease in type 1 diabetes, and this association is partially mediated by endothelial and renal dysfunction and by low-grade inflammation: the EURODIAB prospective complications study. *Diabetologia* 2009; **52**: 705–714.
174. Katakami N, Matsuhisa M, Kaneto H, Matsuoka TA, et al. Decreased endogenous secretory advanced glycation end product receptor in type 1 diabetic patients: its possible association with diabetic vascular complications. *Diabetes Care* 2005; **28**: 2716–2721.
175. Katakami N, Matsuhisa M, Kaneto H, Matsuoka TA, et al. Endogenous secretory RAGE but not soluble RAGE is associated with carotid atherosclerosis in type 1 diabetes patients. *Diab Vasc Dis Res* 2008; **5**: 190–197.
176. Katakami N, Matsuhisa M, Kaneto H, Yamasaki Y. Serum endogenous secretory RAGE levels are inversely associated with carotid IMT in type 2 diabetic patients. *Atherosclerosis* 2007; **190**: 22–23.
177. Humpert PM, Djuric Z, Kopf S, Rudofsky G, et al. Soluble RAGE but not endogenous secretory RAGE is associated with albuminuria in patients with type 2 diabetes. *Cardiovasc Diabetol* 2007; **6**: 9.
178. Lu L, Pu LJ, Zhang Q, Wang LJ, et al. Increased glycated albumin and decreased esRAGE levels are related to angiographic severity and extent of coronary artery disease in patients with type 2 diabetes. *Atherosclerosis* 2009; **206**: 540–545.
179. Yamagishi S, Adachi H, Nakamura K, Matsui T, et al. Positive association between serum levels of advanced glycation end products and the soluble form of receptor for advanced glycation end products in nondiabetic subjects. *Metabolism* 2006; **55**: 1227–1231.
180. Miura J, Yamamoto Y, Osawa M, Watanabe T, et al. Endogenous secretory receptor for advanced glycation endproducts levels are correlated with serum pentosidine and CML in patients with type 1 diabetes. *Arterioscler Thromb Vasc Biol* 2007; **27**: 253–254.
181. Koyama H, Shoji T, Fukumoto S, Shinohara K, et al. Low circulating endogenous secretory receptor for AGEs predicts cardiovascular mortality in patients with end-stage renal disease. *Arterioscler Thromb Vasc Biol* 2007; **27**: 147–153.
182. Geroldi D, Falcone C, Emanuele E, D'Angelo A, et al. Decreased plasma levels of soluble receptor for advanced glycation end-products in patients with essential hypertension. *J Hypertens* 2005; **23**: 1725–1729.
183. Basta G, Sironi AM, Lazzerini G, Del Turco S, et al. Circulating soluble receptor for advanced glycation end products is inversely associated with glycemic control and S100A12 protein. *J Clin Endocrinol Metab* 2006; **91**: 4628–4634.
184. Devangelio E, Santilli F, Formoso G, Ferroni P, et al. Soluble RAGE in type 2 diabetes: association with oxidative stress. *Free Radic Biol Med* 2007; **43**: 511–518.
185. Challier M, Jacqueminet S, Benabdesselam O, Grimaldi A, et al. Increased serum concentrations of soluble receptor for advanced glycation endproducts in patients with type 1 diabetes. *Clin Chem* 2005; **51**: 1749–1750.
186. Tan KC, Shiu SW, Chow WS, Leng L, et al. Association between serum levels of soluble receptor for advanced glycation end products and circulating advanced glycation end products in type 2 diabetes. *Diabetologia* 2006; **49**: 2756–2762.
187. Nakamura K, Yamagishi SI, Adachi H, Kurita-Nakamura Y, et al. Elevation of soluble form of receptor for advanced glycation end products (sRAGE) in diabetic subjects with coronary artery disease. *Diabetes Metab Res Rev* 2007; **23**: 368–71.
188. Yamamoto Y, Miura J, Sakurai S, Watanabe T, et al. Assaying soluble forms of receptor for advanced glycation end products. *Arterioscler Thromb Vasc Biol* 2007; **27**: e33–e34.
189. Humpert PM, Kopf S, Djuric Z, Wendt T, et al. Plasma sRAGE is independently associated with urinary albumin excretion in type 2 diabetes. *Diabetes Care* 2006; **29**: 1111–1113.

190. Kalousova M, Hodkova M, Kazderova M, Fialova J, et al. Soluble receptor for advanced glycation end products in patients with decreased renal function. *Am J Kidney Dis* 2006; **47**: 406–411.
191. Kalousova M, Bartosova K, Zima T, Skibova J, et al. Pregnancy-associated plasma protein a and soluble receptor for advanced glycation end products after kidney transplantation. *Kidney Blood Press Res* 2007; **30**: 31–37.
192. Gross S, van Ree RM, Oterdoom LH, de Vries AP, et al. Low levels of sRAGE are associated with increased risk for mortality in renal transplant recipients. *Transplantation* 2007; **84**: 659–663.
193. Kosaki A, Hasegawa T, Kimura T, Iida K, et al. Increased plasma S100A12 (EN-RAGE) levels in patients with type 2 diabetes. *J Clin Endocrinol Metab* 2004; **89**: 5423–5428.
194. Wendt TM, Tanji N, Guo J, Kislinger TR, et al. RAGE drives the development of glomerulosclerosis and implicates podocyte activation in the pathogenesis of diabetic nephropathy. *Am J Pathol* 2003; **162**: 1123–1137.
195. Arancio O, Zhang HP, Chen X, Lin C, et al. RAGE potentiates Abeta-induced perturbation of neuronal function in transgenic mice. *Embo J* 2004; **23**: 4096–4105.
196. Cheng C, Tsuneyama K, Kominami R, Shinohara H, et al. Expression profiling of endogenous secretory receptor for advanced glycation end products in human organs. *Mod Pathol* 2005; **18**: 1385–1396.
197. Tanaka N, Yonekura H, Yamagishi S, Fujimori H, et al. The receptor for advanced glycation end products is induced by the glycation products themselves and tumor necrosis factor-alpha through nuclear factor-kappa B, and by 17beta-estradiol through Sp-1 in human vascular endothelial cells. *J Biol Chem* 2000; **275**: 25781–25790.
198. Coughlan MT, Thallas-Bonke V, Pete J, Long DM, et al. Combination therapy with the advanced glycation end product cross-link breaker, alagebrium, and angiotensin converting enzyme inhibitors in diabetes: synergy or redundancy? *Endocrinology* 2007; **148**: 886–895.
199. Zhong Y, Li SH, Liu SM, Szmitko PE, et al. C-Reactive protein upregulates receptor for advanced glycation end products expression in human endothelial cells. *Hypertension* 2006; **48**: 504–511.
200. Geroldi D, Falcone C, Minoretti P, Emanuele E, et al. High levels of soluble receptor for advanced glycation end products may be a marker of extreme longevity in humans. *J Am Geriatr Soc* 2006; **54**: 1149–1150.
201. Forbes JM, Thorpe SR, Thallas-Bonke V, Pete J, et al. Modulation of soluble receptor for advanced glycation end products by angiotensin-converting enzyme-1 inhibition in diabetic nephropathy. *J Am Soc Nephrol* 2005; **16**: 2363–2372.
202. Marx N, Walcher D, Ivanova N, Rautzenberg K, et al. Thiazolidinediones reduce endothelial expression of receptors for advanced glycation end products. *Diabetes* 2004; **53**: 2662–2668.
203. Okamoto T, Yamagishi S, Inagaki Y, Amano S, et al. Angiogenesis induced by advanced glycation end products and its prevention by cerivastatin. *Faseb J* 2002; **16**: 1928–1930.
204. Cuccurullo C, Iezzi A, Fazia ML, De Cesare D, et al. Suppression of RAGE as a basis of simvastatin-dependent plaque stabilization in type 2 diabetes. *Arterioscler Thromb Vasc Biol* 2006; **26**: 2716–2723.
205. Tan KC, Chow WS, Tso AW, Xu A, et al. Thiazolidinedione increases serum soluble receptor for advanced glycation end-products in type 2 diabetes. *Diabetologia* 2007; **50**: 1819–1825.
206. Tam HL, Shiu SW, Wong Y, Chow WS, et al. Effects of atorvastatin on serum soluble receptors for advanced glycation end-products in type 2 diabetes. *Atherosclerosis* 2010; **209**: 173–177.
207. Takeuchi M, Yanase Y, Matsuura N, Yamagishi Si S, et al. Immunological detection of a novel advanced glycation end-product. *Mol Med* 2001; **7**: 783–791.
208. Katakami N, Matsuhisa M, Kaneto H, Matsuoka TA, et al. Serum endogenous secretory RAGE level is an independent risk factor for the progression of carotid atherosclerosis in type 1 diabetes. *Atherosclerosis* 2009; **204**: 288–292.

209. Koyama Y, Takeishi Y, Niizeki T, Suzuki S, et al. Soluble Receptor for advanced glycation end products (RAGE) is a prognostic factor for heart failure. *J Card Fail* 2008; **14**: 133–139.
210. Kalousova M, Jachymova M, Mestek O, Hodkova M, et al. Receptor for advanced glycation end products – soluble form and gene polymorphisms in chronic haemodialysis patients. *Nephrol Dial Transplant* 2007; **22**: 2020–2026.
211. Nakamura K, Yamagishi S, Adachi H, Kurita-Nakamura Y, et al. Serum levels of sRAGE, the soluble form of receptor for advanced glycation end products, are associated with inflammatory markers in patients with type 2 diabetes. *Mol Med* 2007; **13**: 185–189.
212. Nakamura K, Yamagishi SI, Adachi H, Matsui T, et al. Circulating advanced glycation end products (AGEs) and soluble form of receptor for AGEs (sRAGE) are independent determinants of serum monocyte chemoattractant protein-1 (MCP-1) levels in patients with type 2 diabetes. *Diabetes Metab Res Rev* 2008; **24**: 109–114.
213. Nakamura K, Yamagishi S, Adachi H, Matsui T, et al. Serum levels of soluble form of receptor for advanced glycation end products (sRAGE) are positively associated with circulating AGEs and soluble form of VCAM-1 in patients with type 2 diabetes. *Microvasc Res* 2008; **76**: 52–56.
214. Choi KM, Yoo HJ, Kim HY, Lee KW, et al. Association between endogenous secretory RAGE, inflammatory markers and arterial stiffness. *Int J Cardiol* 2009; **132**: 96–101.

Chapter 14
Infection and the Kidney

Chih-Wei Yang

Abstract Reactive oxygen species (ROS) are formed as a result of physiologic cellular metabolism, and a homeostatic balance exists between the formation of ROS and their removal by antioxidant scavenging compounds. Host–pathogen interaction results in a variety of responses, which include phagocytosis of the pathogen, release of cytokines, secretion of toxins, as well as production of ROS, leading to oxidative stress in the kidney. Oxidative stress is widely recognized as an important feature of urinary tract infection, pyelonephritis, and in sepsis kidney. Innate immunity toll-like receptors (TLR) are involved in the pathogen recognition and ROS may mediate its initiation and function. Efforts must be made to identify the precise contribution of these factors in infection process in order to clarify the mechanisms associated with kidney disease. This will certainly lead to discovery of therapeutic strategies in the future.

Keywords Reactive oxygen species · Urinary tract infection · Pyelonephritis · Sepsis · Toll-like receptor · Antioxidant

1 Introduction

Reactive oxygen species (ROS) are formed as a result of physiologic cellular metabolism, and a homeostatic balance exists between the formation of ROS and their removal by antioxidant scavenging compounds [1]. Oxidative stress occurs when this balance is disrupted by excessive production of ROS, including superoxide, hydrogen peroxide and hydroxyl radicals, and by inadequate antioxidative defenses,

C.-W. Yang (✉)
Kidney Research Center, Department of Nephrology,
Chang Gung Memorial Hospital,
199, Tun-Hua North Road, Taipei, Taiwan
e-mail: cwyang@ms1.hinet.net

T. Miyata et al. (eds.), *Studies on Renal Disorders*, Oxidative Stress in Applied Basic Research and Clinical Practice, DOI 10.1007/978-1-60761-857-7_14,

including superoxide dismutase (SOD), catalase, vitamins C and E, and reduced glutathione (GSH). These conditions occur in infection and in sepsis [2].

Plasma membrane–associated oxidases have been implicated as the abundant sources of growth factor or cytokine stimulated oxidant production. The best characterized plasma membrane oxidases is the phagocytic nicotinamide adenine dinucleotide phosphate (NADPH) oxidase, which serves a specialized function in phagocytic cells for host defense against invading microorganisms. A variety of cytokines and growth factors that bind receptors of different classes have been reported to generate ROS in nonphagocytic cells, including kidney cells.

This chapter reviews the current knowledge of how oxidative stress may be involved in infection of the kidney.

2 Host–Pathogen Interaction and Oxidative Stress

Host–pathogen interaction results in a variety of responses, which include phagocytosis of the pathogen, release of cytokines, secretion of toxins, as well as production of ROS. Although ROS are essential for signal transduction, when produced in excess amount, they may contribute to the pathogenesis of many diseases. It is widely recognized that oxidative stress is an important feature of many diseases, including host–pathogen interaction.

It is now well established that ROS generated by leukocytes have bactericidal activity [3]. Additional reactive oxygen metabolites are formed as a result of the metabolism of hydrogen peroxide by neutrophil-derived myeloperoxidase (MPO) to produce highly reactive toxic products, including hypochlorous acid. MPO reacting with hydrogen peroxide forms an enzyme substrate complex that can oxidize various halides to produce highly reactive toxic products. Because of the wide distribution of chloride ion in biologic systems, the formation of hypochlorous acid is probably one of the most significant products during infection [4–6].

Recent studies have indicated that many pathogens exert control and regulate apoptosis in the host. Apoptosis may be induced upon infection, which results from a complex interaction of pathogen proteins with cellular host proteins. Inhibition of host cell to apoptosis is often beneficial for the pathogen and results in a successful host invasion to ensure the survival of the pathogen and may establish latent infection. However, in some cases, induction of apoptosis in the infected cells significantly protects the host from the pathogen. There is a strong correlation between apoptosis and the host protein translation machinery, including oxidative stress, and this requires future understanding [7].

3 Infection and ER Stress

The endoplasmic reticulum (ER) is a membrane-bound intracellular organelle involved in lipid and protein biosynthesis. The ER is a well-orchestrated protein-folding machinery composed of protein chaperones, catalyzes for protein folding, and sensors that detect misfolded or unfolded proteins. The accumulation of unfolded proteins in the ER represents a cellular stress induced by multiple stimuli and pathological conditions. These include hypoxia, oxidative injury, high-fat diet, hypoglycemia, protein inclusion bodies, and infection. ER stress triggers unfolded protein response (UPR) in which a conserved series of signal-transduction events occur to ameliorate the accumulation of unfolded proteins in the ER. However, when these events are severe or protracted, they can induce cell death [8].

Accumulating evidence suggests that protein folding and generation of ROS as a by-product of protein oxidation in the ER are closely linked events. Activation of the UPR on exposure to oxidative stress is an adaptive mechanism to preserve cell function and survival. These include changes in intraluminal calcium, altered glycosylation, nutrient deprivation, pathogen infection, expression of folding-defective proteins, and changes in redox status. Persistent oxidative stress and protein misfolding initiate apoptotic cascades and are now known to play predominant roles in the pathogenesis of multiple human diseases, including diabetes, atherosclerosis, neurodegenerative diseases, and infection [9].

As the recognition of an association between ER stress and human diseases increases and the understanding of the diverse underlying molecular mechanisms improves, novel targets for drug discovery and new strategies for therapeutic intervention are ready to emerge.

4 Oxidative Stress in Urinary Tract Infection

4.1 Urinary Tract Infection

Oxidative stress is widely recognized as an important feature of urinary tract infection. Urinary tract infections may appear from asymptomatic bacteriuria to urosepsis. The infection may induce severe inflammation, transient impairment in renal function, and scar formation, ranging in severity from acute symptomatic pyelonephritis to chronic pyelonephritis. A variety of oxidation products are found in urine and thought to mirror local and systemic oxidative stress [10, 11].

Following infection, phagocytes (macrophages and polymorphonuclear neutrophils) may accumulate at the site of infection; these cells are known to produce ROS. The inflammatory response during urinary tract infection consists of three principal steps: (a) uroepithelial cell activation with transmembrane signaling, resulting in the production of inflammatory mediators; (b) attraction of the innate immune cells to the infectious focus; and (c) local destruction and elimination of the invading bacteria, mediated by the generation of ROS or the release of preformed antimicrobial peptides [12].

4.1.1 Lipid Peroxidation and Urinary Tract Infection

Lipid peroxidation is one of the most important expressions of oxidative stress induced by ROS. It was observed that malondialdehyde (MDA), an indicator of lipid peroxidation levels, was increased while catalase and SOD activities were decreased in positive urine cultures for infecting microorganisms includes *Escherichia coli*, *Proteus mirabilis*, *Klebsiella pneumoniae*, *Candida* spp, *Staphylococcus saprophyticus*, and *Pseudomonas aeruginosa* compared to negative cultures [13]. This possible damage by ROS can be prevented by the endogenous antioxidant enzymes such as SOD and catalase [14, 15].

In diabetic patients with urinary tract infection, decreased antioxidant capacity and the increased levels of lipid peroxidation were profoundly high. It is suggested that urinary tract infection aggravates the oxidative stress in diabetic patients, therefore, these patients may need antioxidant treatment [16].

4.1.2 Inducible NOS in Urinary Tract Infection

Expression of inducible nitric oxide synthase (iNOS) following inflammation or infection acts as a pivotal component of the host immune response against virulent pathogens [17]. In rodent macrophages and neutrophils, iNOS has been reported to produce large quantities of nitric oxide (NO) that can modulate immune, inflammatory, and cardiovascular responses [18, 19]. However, the production of NO may lead to reaction with the superoxide anion to form peroxynitrite anion, which is a highly toxic molecule that leads to DNA damage and protein modification [20].

4.1.3 Urinary Tract Infection, Oxidative Stress, and Carcinogenesis

Urinary tract infection, when associated with increased oxidative stress in the bladder, may increase the incidence of bladder cancer. Treatment with killed *E. coli* or lipopolysaccharide (LPS), a major cell wall component of *E. coli*, had a tumor-enhancing effect when treated together with *N*-methyl-*N*-nitrosourea. The enhancing effects were associated with a marked increase in the numbers of polymorphonuclear leukocytes and an increase in the hydrogen peroxide concentration in the bladder lumen. This result indicated that oxidative stress in uroepithelium exposed to carcinogen and a proliferative response to the inflammatory stimulation by LPS appeared to play a significant role in tumor enhancement [21].

4.1.4 Antioxidant for Urinary Tract Infection

Among various sources of antioxidants, cranberries are a powerful source of high-quality antioxidants [22] and cranberries are effective not only in the prevention of urinary tract infection but also for the prevention of oxidative stress [23].

4.2 Pyelonephritis

Pyelonephritis is the most common urinary tract infection in females and is the result of an inflammatory process in the renal parenchyma, which may occur as a result of excessive ROS or impaired antioxidant capacity produced by infecting micro-organisms.

A clear correlation between extent of ROS generation and subsequent lipid peroxidation and DNA damage in kidneys was observed during pyelonephritis. Administration of antioxidants, SOD, catalase, and dimethylsulfoxide significantly reversed the histopathologic changes, reduced the extent of lipid peroxidation in renal brush border membrane, and also reversed the altered enzyme activities to near normal [24].

In a model of acute pyelonephritis induced by *P. aeruginosa*, increase production of ROS in urine, bladder, and renal tissue following infection was observed, which correlated with bacterial load, neutrophil recruitment, and MDA. Treatment of mice with *N*-acetylcysteine (NAC), a potent antioxidant, lead to significant amelioration of oxidative stress and subsequent decrease in bacterial titer, neutrophil influx, MDA, as well as tissue pathology, highlighting the important role of free radicals in *P. aeruginosa*–induced pyelonephritis [25].

In pyelonephritic rats with elevated serum tumor necrosis factor-alpha (TNF-α), lactate dehydrogenase (LDH), serum urea nitrogen (BUN), and creatinine levels, melatonin treatment reversed all these biochemical indices, as well as histopathologic alterations induced by acute pyelonephritis. The protective effects of melatonin can be ascribed to its ability to inhibit neutrophil infiltration, to balance the oxidant–antioxidant status, and to regulate the generation of inflammatory mediators, suggesting a future role for melatonin in the treatment of acute pyelonephritis and the rationale in treating pyelonephritis with antioxidant and in reducing the cellular infiltration [26].

These results clearly suggest that interaction of ROS with various cellular organelles in kidneys has a significant deleterious effect, and this could be the underlying mechanism for renal dysfunction in pyelonephritis.

5 Oxidative Stress in Sepsis Kidney

Dysregulation of the immunoinflammatory response, as seen in sepsis, may culminate in host cell and organ damage. LPS from Gram-negative bacterial cell walls induces gene activation and subsequent inflammatory mediator expression. Infection-induced activation of phagocytes is associated with oxidative stress, not only because ROS are released but also because activated phagocytes may also release pro-oxidant cytokines, such as TNF, and interleukin-1 (IL-1). Gene activation is regulated by a number of transcription factors at the nuclear level, of which nuclear factor κB (NFκB) appears to have a central role. In sepsis, a state of severe oxidative stress is encountered, with host endogenous antioxidant defenses overwhelmed.

In sepsis, there are several potential sources of ROS, including the mitochondrial respiratory electron transport chain, xanthine oxidase activation as a result of ischemia and reperfusion, the respiratory burst associated with neutrophil activation, and arachidonic acid metabolism. Activated neutrophils produce superoxide as a cytotoxic agent as part of the respiratory burst via the action of membrane-bound NADPH oxidase. Neutrophils also produce the free radical nitric oxide, which can react with superoxide to produce peroxynitrite, a powerful oxidant, which may decompose to form the hydroxyl radicals. During oxidative stress, damage mediated by ROS can occur. Oxidation of DNA and proteins may take place, along with membrane damage, because of lipid peroxidation, leading to alterations in membrane permeability, modification of protein structure, and functional changes. Oxidative damage to the mitochondrial membrane can also occur, resulting in membrane depolarization and the uncoupling of oxidative phosphorylation, with altered cellular respiration [27].

Endotoxemia-induced sepsis has been frequently studied in the rat model by LPS administration [28]. LPS causes enhanced formation of ROS, which are produced by the increased inflammatory infiltrates and damaged resident cells [29, 30], and finally contributes to multiple organ dysfunction syndrome. Endotoxemia results in a rapid, but transient, decline in the expression of both mRNA and protein of Cu/Zn-SOD as well as an increase in the expression of the mRNA of Mn-SOD in the kidney. Endotoxemia for 6 h also caused hypotension, acute kidney injury, hepatocellular injury, pancreatic injury, and an increase in the plasma levels of nitrite/nitrate [31].

Septic shock, systemic inflammatory response syndrome (SIRS), and acute respiratory distress syndrome (ARDS) are mainly due to endotoxemia. The vascular pathology of sepsis/SIRS and ARDS is initiated through the uncontrolled production of ROS, characterized by disordered vascular control, systemic hypotension, and peripheral vasodilation refractory to intravascular volume resuscitation and vasopressor therapy [32].

In animal models of critical diseases, antioxidant therapy was shown to reduce mortality. Treatment with SOD mimetics, which selectively mimic the catalytic activity of the human SOD enzymes, have been shown to prevent in vivo shock and the cellular energetic failure associated with shock [33].

On the other hand, ischemic proximal tubular injury primes mice to endotoxin-induced TNF-α generation and systemic release. Under modest ischemia reperfusion injury and in vitro, HK-2 cell mitochondrial inhibition dramatically sensitized the kidney proximal tubule cells to LPS-mediated TNF-α generation and increases in TNF-α mRNA. It is of interest that ischemia itself can "prime" tubules to LPS response(s) and thus could have potential important implications for sepsis syndrome, concomitant renal ischemia, and for induction of acute kidney injury [34].

Despite intense research into potential treatments by antioxidants, the associated development of acute kidney injury and the incidence of sepsis mortality remain high. Targeting the inflammatory response and sepsis-induced alterations in the microcirculation are therapeutic strategies. Modulating the downstream mechanisms responsible for organ system dysfunction provides another approach. Activation of iNOS during sepsis leads to increased NO levels that influence renal hemodynamics and cause tubular injury through the local generation of reactive nitrogen species and peroxynitrite. In many organs iNOS is not constitutively expressed; however, it is constitutively expressed in the kidney. In humans, a relationship between the upregulation of renal iNOS and proximal tubular injury during systemic inflammation has been demonstrated [35]. Various animal studies have demonstrated that selective iNOS inhibition attenuates sepsis-induced renal dysfunction and improves survival. Therefore, the selective inhibition of renal iNOS might have important implications for the treatment of sepsis-induced acute kidney injury that warrants investigation in human clinical trials [36]. It can be anticipated that the measurement of a cluster of assays representative of the quantification of reactive species or of antioxidants may improve the usefulness of therapeutic intervention and increase knowledge of pathophysiological alterations [37].

6 Toll-Like Receptor and Oxidative Stress

Oxidants play an important role in inflammation and pathology; however, little is known how oxidants are sensed and propagate inflammation. Several lines of evidence show that exogenous oxidants are sensed by the Toll-like receptor (TLR), thereby initiating inflammatory processes [38–40].

6.1 Recognition of Exogenous and Endogenous Ligands by Toll-Like Receptors

Members of the TLR family play key roles in innate immunity and connect to adaptive immune responses [41]. These receptors are capable of directing innate immune response to diverse pathogen-associated molecular patterns such as bacterial lipopeptide, LPS, viral RNA, CpG-containing DNA, and flagellin [42]. TLRs are also able to mediate responses to endogenous molecules. To date, evidence has shown that TLRs are involved in the development of many pathological conditions, including infectious diseases, tissue damage, autoimmune, neurodegenerative diseases, and cancer. At least ten members of the TLR family have been identified in humans [42]. Currently the best understood are TLR2, which recognizes bacterial lipoteichoic acid and peptidoglycan [43]; TLR4, which binds LPS [44]; TLR5, which binds bacterial flagellin [45]; and TLR9, which recognizes unmethylated bacterial CpG DNA [46].

There is strong evidence of a role for TLR2 and TLR4 in renal ischemia-reperfusion injury, with the effects probably mediated by endogenous ligands. In systemic lupus erythematosus (SLE), stimulation of TLR7 and TLR9 by host-derived nucleic acids is important. TLR7 stimulation exacerbates disease, but the role of TLR9 is complex. In autoimmune glomerular disease, endogenous ligands for TLR7 and TLR9 are crucial in the loss of tolerance to autoantigens in SLE [47].

6.2 ROS and Toll-Like Receptor Activation

Activation of TLR triggers downstream pathways for activation of inflammation. ROS has been detected as a ligand of TLR or the component of TLR downstream pathways [39] (Fig. 1). TLR2 is essential for the sensing of oxidants during inflammation. The cells "sense" soluble oxidants, which is translated into an inflammatory reaction. Oxidants evoked the release of CXCL8 from monocytes/macrophages, and this was abrogated by pretreatment with NAC or binding antibodies to TLR2 [39]. Oxidants added to HEK293 cells transfected with TLR2, TLR1/2, or TLR2/6 but not TLR4/MD2-CD14 or control HEK nulls resulted in the release of CXCL8. In animals, TLR2(−/−) mice did not respond to oxidant challenge. Therefore, there is evidence that oxidant-TLR2 interactions provide a

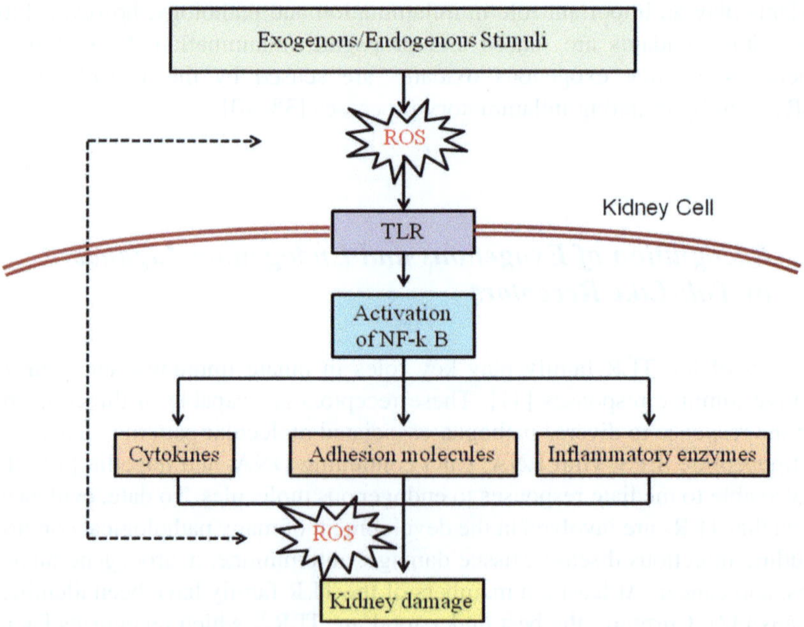

Fig. 1 Participation of oxidative stress in kidney cell injury related to Toll-like receptor (TLR), in response to exogenous and endogenous stimuli. *ROS* reactive oxygen species; *TLR* Toll-like receptors; *NFκB* nuclear factor κB

signal that initiates the inflammatory response in vitro and in vivo. TLR2 alone or in association with TLR1 or TLR6 is required for the induction of CXCL8 release by oxidants. Furthermore, TLR2 was required for oxidant-induced inflammation in vivo [39].

ROS production after vascular injury was attenuated in TLR2 knockout mice compared with control mice. These findings suggested the regulation of vascular inflammation in injured vessels and the vascular injury–mediated cytokine expression, ROS production, as well as subsequent neointimal formation, which requires TLR2-mediated signaling pathway in vivo [48].

Emerging evidence suggests that ROS contribute to LPS-induced cascade and give diverse signaling pathways. TLR4 was shown to induce or upregulate a variety of gene products. Purified *E. coli* LPS, a highly specific TLR4 agonist, elicited an oxidative burst in the monocyte-like cell line in a time- and dose-dependent manner (Fig. 2). This oxidative burst was shown to be dependent on the presence of TLR4 through transfection studies in HEK cells, which do not normally express this protein, and with bone marrow–derived macrophages from C3H/HeJ mice, which express a mutated TLR4 protein. LPS-stimulated IL-8 expression and could be blocked by the antioxidants *N*-acetylcysteine and dimethyl sulfoxide at both the

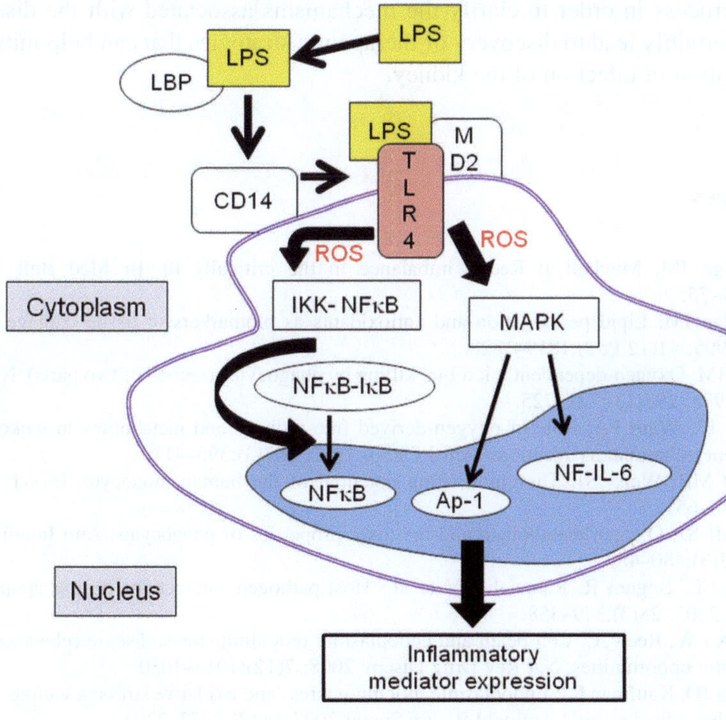

Fig. 2 Reactive oxygen species (ROS) participate in the lipopolysaccharide (LPS) activation of inflammatory mediator expression. *NFκB* nuclear factor κB; *IκB* inhibitor of NFκB; *MAPK* mitogen-activated protein kinases; *IL* interleukin

protein and mRNA levels. These antioxidants also blocked LPS-induced IL-8 promoter transactivation as well as the nuclear translocation of NFκB. These data provide evidence that ROS regulate immune signaling through TLR4 via their effects on NFκB activation [49]. In summary, ROS are actively involved in the regulation of TLR-related pathways in response to exogenous stimulation produced from micro-organisms.

7 Conclusion

In the past, reactive oxygen and reactive nitrogen species were mainly known as harmful agents, but recent investigations have given new insight into the pathophysiologic importance of these substances. In fact, ROS may also serve as powerful messenger molecules involved in gene regulation, thereby enabling the synthesis of cytokines or adhesion molecules necessary for defending inflammatory processes as occurs in innate immunity.

In conclusion, ROS are involved in cell signaling and kidney injury in infection. Efforts must be made to identify the precise contribution of these factors in infection process in order to clarify the mechanisms associated with the disease. This will certainly lead to discovery of therapeutic strategies that can help mitigate ROS activation in infection of the kidney.

References

1. Gutteridge JM, Mitchell J: Redox imbalance in the critically ill. Br Med Bull 1999; 55(1):49–75.
2. Gutteridge JM: Lipid peroxidation and antioxidants as biomarkers of tissue damage. Clin Chem 1995; 41(12 Pt 2):1819–1828.
3. Babior BM: Oxygen-dependent microbial killing by phagocytes (second of two parts). N Engl J Med 1978; 298(13):721–725.
4. Fantone JC, Ward PA: Role of oxygen-derived free radicals and metabolites in leukocyte-dependent inflammatory reactions. Am J Pathol 1982; 107(3):395–418.
5. Lampert MB, Weiss SJ: The chlorinating potential of the human monocyte. Blood 1983; 62(3):645–651.
6. Klebanoff SJ: Oxygen metabolism and the toxic properties of phagocytes. Ann Intern Med 1980; 93(3):480–489.
7. Hasnain SE, Begum R, Ramaiah KV et al.: Host-pathogen interactions during apoptosis. J Biosci 2003; 28(3):349–358.
8. Kim I, Xu W, Reed JC. Cell death and endoplasmic reticulum stress: disease relevance and therapeutic opportunities. Nat Rev Drug Discov 2008; 7(12):1013–1030.
9. Malhotra JD, Kaufman RJ: Endoplasmic reticulum stress and oxidative stress: a vicious cycle or a double-edged sword? Antioxid Redox Signal 2007; 9(12):2277–2293.
10. Kirschbaum B: Total urine antioxidant capacity. Clin Chim Acta 2001; 305(1–2):167–173.
11. Nathan C, Xie QW: Nitric oxide synthases: roles, tolls, and controls. Cell 1994; 78(6):915–918.

12. Mulvey MA, Schilling JD, Martinez JJ, Hultgren SJ: Bad bugs and beleaguered bladders: interplay between uropathogenic *Escherichia coli* and innate host defenses. Proc Natl Acad Sci USA 2000; 97(16):8829–8835.
13. Kurutas EB, Ciragil P, Gul M, Kilinc M: The effects of oxidative stress in urinary tract infection. Mediators Inflamm 2005; 2005(4):242–244.
14. Mates JM: Effects of antioxidant enzymes in the molecular control of reactive oxygen species toxicology. Toxicology 2000; 153(1–3):83–104.
15. McGirr LG, Hadley M, Draper HH: Identification of *N* alpha-acetyl-epsilon-(2-propenal) lysine as a urinary metabolite of malondialdehyde. J Biol Chem 1985; 260(29):15427–15431.
16. Gul M, Kurutas E, Ciragil P et al.: Urinary tract infection aggravates oxidative stress in diabetic patients. Tohoku J Exp Med 2005; 206(1):1–6.
17. Smith SD, Wheeler MA, Weiss RM: Nitric oxide synthase: an endogenous source of elevated nitrite in infected urine. Kidney Int 1994; 45(2):586–591.
18. Kolodziejska KE, Burns AR, Moore RH, Stenoien DL, Eissa NT: Regulation of inducible nitric oxide synthase by aggresome formation. Proc Natl Acad Sci USA 2005; 102(13):4854–4859.
19. Rumbo M, Courjault-Gautier F, Sierro F, Sirard JC, Felley-Bosco E: Polarized distribution of inducible nitric oxide synthase regulates activity in intestinal epithelial cells. FEBS J 2005; 272(2):444–453.
20. Jablonska E, Puzewska W, Marcinczyk M, Grabowska Z, Jablonski J: iNOS expression and NO production by neutrophils in cancer patients. Arch Immunol Ther Exp (Warsz) 2005; 53(2):175–179.
21. Kawai K, Yamamoto M, Kameyama S et al.: Enhancement of rat urinary bladder tumorigenesis by lipopolysaccharide-induced inflammation. Cancer Res 1993; 53(21):5172–5175.
22. Vinson JA, Bose P, Proch J, Al Kharrat H, Samman N: Cranberries and cranberry products: powerful in vitro, ex vivo, and in vivo sources of antioxidants. J Agric Food Chem 2008; 56(14):5884–5891.
23. Valentova K, Stejskal D, Bednar P et al.: Biosafety, antioxidant status, and metabolites in urine after consumption of dried cranberry juice in healthy women: a pilot double-blind placebo-controlled trial. J Agric Food Chem 2007; 55(8):3217–3224.
24. Gupta A, Sharma S, Nain CK, Sharma BK, Ganguly NK: Reactive oxygen species-mediated tissue injury in experimental ascending pyelonephritis. Kidney Int 1996; 49(1):26–33.
25. Mittal R, Sharma S, Chhibber S, Harjai K: Contribution of free radicals to *Pseudomonas aeruginosa* induced acute pyelonephritis. Microb Pathog 2008; 45(5–6):323–330.
26. Sener G, Tugtepe H, Velioglu-Ogunc A et al.: Melatonin prevents neutrophil-mediated oxidative injury in *Escherichia coli*-induced pyelonephritis in rats. J Pineal Res 2006; 41(3):220–227.
27. Macdonald J, Galley HF, Webster NR: Oxidative stress and gene expression in sepsis. Br J Anaesth 2003; 90(2):221–232.
28. Khan RZ, Badr KF: Endotoxin and renal function: perspectives to the understanding of septic acute renal failure and toxic shock. Nephrol Dial Transplant 1999; 14(4):814–818.
29. Gullo A: Sepsis and organ dysfunction/failure. An overview. Minerva Anestesiol 1999; 65(7–8):529–540.
30. Wiesel P, Patel AP, DiFonzo N et al.: Endotoxin-induced mortality is related to increased oxidative stress and end-organ dysfunction, not refractory hypotension, in heme oxygenase-1-deficient mice. Circulation 2000; 102(24):3015–3022.
31. Lee JG, Lee SH, Park DW et al.: Toll-like receptor 9-stimulated monocyte chemoattractant protein-1 is mediated via JNK-cytosolic phospholipase A2-ROS signaling. Cell Signal 2008; 20(1):105–111.
32. Azevedo LC, Janiszewski M, Soriano FG, Laurindo FR: Redox mechanisms of vascular cell dysfunction in sepsis. Endocr Metab Immune Disord Drug Targets 2006; 6(2):159–164.
33. Salvemini D, Cuzzocrea S: Oxidative stress in septic shock and disseminated intravascular coagulation. Free Radic Biol Med 2002; 33(9):1173–1185.

34. Zager RA, Johnson AC, Hanson SY, Lund S: Ischemic proximal tubular injury primes mice to endotoxin-induced TNF-alpha generation and systemic release. Am J Physiol Renal Physiol 2005; 289(2):F289–F297.
35. Heemskerk S, Pickkers P, Bouw MP et al.: Upregulation of renal inducible nitric oxide synthase during human endotoxemia and sepsis is associated with proximal tubule injury. Clin J Am Soc Nephrol 2006; 1(4):853–862.
36. Heemskerk S, Masereeuw R, Russel FG, Pickkers P: Selective iNOS inhibition for the treatment of sepsis-induced acute kidney injury. Nat Rev Nephrol 2009; 5(11):629–640.
37. Roth E, Manhart N, Wessner B: Assessing the antioxidative status in critically ill patients. Curr Opin Clin Nutr Metab Care 2004; 7(2):161–168.
38. Frantz S, Kelly RA, Bourcier T: Role of TLR-2 in the activation of nuclear factor kappaB by oxidative stress in cardiac myocytes. J Biol Chem 2001; 276(7):5197–5203.
39. Paul-Clark MJ, McMaster SK, Sorrentino R et al.: Toll-like receptor 2 is essential for the sensing of oxidants during inflammation. Am J Respir Crit Care Med 2009; 179(4):299–306.
40. Remer KA, Brcic M, Jungi TW: Toll-like receptor-4 is involved in eliciting an LPS-induced oxidative burst in neutrophils. Immunol Lett 2003; 85(1):75–80.
41. Lemaitre B, Nicolas E, Michaut L, Reichhart JM, Hoffmann JA: The dorsoventral regulatory gene cassette spatzle/Toll/cactus controls the potent antifungal response in *Drosophila* adults. Cell 1996; 86(6):973–983.
42. Medzhitov R: Toll-like receptors and innate immunity. Nat Rev Immunol 2001; 1 (2):135–145.
43. Schwandner R, Dziarski R, Wesche H, Rothe M, Kirschning CJ: Peptidoglycan- and lipoteichoic acid-induced cell activation is mediated by toll-like receptor 2. J Biol Chem 1999; 274 (25):17406–17409.
44. Poltorak A, He X, Smirnova I et al.: Defective LPS signaling in C3H/HeJ and C57BL/10ScCr mice: mutations in TLR4 gene. Science 1998; 282(5396):2085–2088.
45. Hayashi F, Smith KD, Ozinsky A et al.: The innate immune response to bacterial flagellin is mediated by toll-like receptor 5. Nature 2001; 410(6832):1099–1103.
46. Hemmi H, Takeuchi O, Kawai T et al.: A toll-like receptor recognizes bacterial DNA. Nature 2000; 408(6813):740–745.
47. Robson MG: Toll-like receptors and renal disease. Nephron Exp Nephrol 2009; 113(1):e1–e7.
48. Shishido T, Nozaki N, Takahashi H et al.: Central role of endogenous toll-like receptor-2 activation in regulating inflammation, reactive oxygen species production, and subsequent neointimal formation after vascular injury. Biochem Biophys Res Commun 2006; 345 (4):1446–1453.
49. Ryan KA, Smith MF, Jr., Sanders MK, Ernst PB: Reactive oxygen and nitrogen species differentially regulate toll-like receptor 4-mediated activation of NF-kappa B and interleukin-8 expression. Infect Immun 2004; 72(4):2123–2130.

Chapter 15
Oxidative/Carbonyl Stress in the Renal Circulation and Cardiovascular Renal Injury

Takefumi Mori, Toshio Miyata, and Sadayoshi Ito

Abstract Both clinical and animal studies have demonstrated that oxidative and carbonyl stress are involved in the pathogenesis of hypertension, diabetes, and chronic kidney disease (CKD). Not only does direct biochemical action cause renal injury, but oxidative/carbonyl stress also participates in the physiological role on sodium homeostasis and regulation of blood pressure. The physiological role of oxidative stress relies on the balance between nitric oxide (NO) and reactive oxygen species (ROS). The balance between NO and ROS regulates renal medullary circulation and regulates blood pressure. In physiological conditions, NO is largely produced in the tubules and diffuses into the surrounding capillaries such as the vasa recta. These NO can be counter-balanced by the ROS produced in the tubules of renal medulla such as the medullary thick ascending limb. This tubulovascular NO crosstalk in the medulla regulates medullary blood flow (MBF). Reduction of MBF with reduction of NO and increased ROS increases sodium reabsorption and thereby develops hypertension. This is one of the mechanisms of salt-sensitive hypertension seen in Dahl-salt sensitive rats. Since oxygen tension of renal medulla is flow limited, reduction of renal MBF induces hypoxia as well as ischemia. Therefore, the outer medulla is an early target for enhanced oxidative stress, such as in hypertension and diabetes. Enhanced oxidative/carbonyl stress is also involved in insulin sensitivity, which is a key pathogenesis of CKD and metabolic syndrome. An increase in renal perfusion pressure results in the enhanced expression of molecules related to oxidative stress, inflammation, and wound healing. These mechanisms may had an advantage for lifestyle in ancient days but have become a silent killer in modern culture. This paradigm shift of antiaging could explain why oxidative/carbonyl stress is enhanced in modern lifestyle diseases.

T. Mori (✉)
Division of Nephrology, Endocrinology and Vascular Medicine,
Tohoku University Graduate School of Medicine, 1-1 Seiryocho,
Aoba-ku, Sendai 980-8574, Japan
e-mail: tmori@med.tohoku.ac.jp

T. Miyata et al. (eds.), *Studies on Renal Disorders*, Oxidative Stress in Applied Basic Research and Clinical Practice, DOI 10.1007/978-1-60761-857-7_15,

Keywords Angiotensin II · Oxidative stress · Carbonyl compoureactive oxygen species · Hypertension · Renal injury · Superoxide · Hydrogen peroxide · Advanced glycation end products · Carbonyl substances

1 Introduction

Chronic kidney disease (CKD) is a risk factor of cardiovascular disease (CVD), as suggested by clinical studies, independent of classical risk factors such as hypertension, hyperlipidemia, and diabetes [1, 2]. Thus, diagnosis and treatment of CKD are important not only for renal injury but also other organs such as heart, brain, and vasculature. Since glomerular filtration rate (GFR) can be gradually reduced over time, any person could reach a level of CKD without any specific diagnosis of other renal disease. Reduction rate of GFR could be enhanced by hypertension, diabetes, and other glomerular diseases. Thus, the major purpose of the treatment of CKD could be an anti-aging of renal function and protection from additional diseases. Interestingly, even in those with normal renal function or no history of hypertension or diabetes, microalbuminturia is an independent risk factor of both cardiovascular and noncardiovascular events and death, similar to a level of those with hypertension and diabetes [3, 4]. However, the mechanism that could completely explain the connection between albuminuria and CVD has not yet been explored.

Hypertension and diabetes are the major diseases that could cause both CKD and CVD. Among many pathways involved in hypertension and diabetes, oxidative/carbonyl stress plays an important role in the pathogenesis of these diseases [5–7]. Oxidative/carbonyl stress has been demonstrated to be involved in the pathogenesis of CKD [7, 8]. However, in many studies, oxidative stress has been demonstrated to induce many molecules and pathways that are related to cellular injury and apoptosis [5–8]. Besides these actions, oxidative/carbonyl stress could participate in the physiologic role of renal and vascular function. In this chapter, the focus will be on the connection between the pathophysiological role of oxidative/carbonyl stress in CVD and CKD.

2 Physiological Role of Oxidative/Carbonyl Stress in the Regulation of Renal Function and Blood Pressure

Oxidative/carbonyl stress is involved in the pathogenesis of renal injury in many animal models such as remnant kidney model, ischemia reperfusion model, Thy-1 nephritis model, OLETF models, angiotensin II infused models, spontaneously hypertensive rats (SHR) and Dahl salt sensitive (Dahl S) rats. For example,

Nishiyama et al. [9] gave superoxide dismutase (SOD) mimetic TEMPOL (hydroxy-2,2,6,6-tetramethylpiperidine-1-oxyl) orally to Dahl S rats with high salt diet and compared that to those of rats treated with hydralazine. Both TEMPOL and hydralazine reduced blood pressure to a similar level and reduced glomerular injury and proteinuria. However, TEMPOL reduced further proteinuria compared to hydralazine, indicating that oxidative stress participates in the renal injury independent of blood pressure [9]. Moreover, renal injury of ischemia reperfusion model was reduced in rats with overexpressed glyoxalase I, where carbonyl stress is reduced by the reduction of methylglyoxal (MGO) [10].

Superoxide, a molecule that initiates oxidative stress, is produced in the kidney. Zou et al. [11] determined the source of superoxide in each segments of the kidney in normal Sprague-Dawaley rats. Nicotinamide adenine dinucleotide (NADH) oxidase and mitochondrial enzyme activity for superoxide production was the major source of superoxide production. In contrast, nicotinamide adenine dinucleotide phosphate (NADPH) and xanthine oxidase were minor. NADH oxidase mediated superoxide production was higher in both cortex and outer medulla compared to those of inner medulla. Mitochondrial enzyme activity for superoxide production was highest in the outer medulla. Thus, outer medulla is the region that enzymatic production of superoxide is high.

In contrast to the direct action to renal injury, pathophysiological role of oxidative stress to the renal function and regulation of blood pressure has also been determined [12–14]. In rat model of hypertension such as SHR and Dahl S rats, oxidative stress played a role in the regulation of blood pressure.

Hypertension of both angiotensin II–infused rats or SHR was attenuated in rats treated with intravenous infusion of SOD mimetic TEMPOL, suggesting that superoxide was responsible for development of hypertension [13, 14]. Welch et al. [15] have demonstrated that tubuloglomerular feedback (TGF) is blunted when TEMPOL was microperfused into efferent arteriole of SHR rats, indicating that superoxide is responsible for TGF.

The physiological role of oxidative stress to the development of hypertension and renal circulation was well determined in Dahl S rats [12, 16, 17]. Meng et al. [18] demonstrated that oxidative stress indicated by urinary excretion of prostaglandin F2α was increased in Dahl S rats compared to those of control Dahl salt resistant (Dahl R) rats. This was associated with the decease of SOD in the kidney of Dahl S rats than Dahl R rats. Production of superoxide was determined in homogenized renal tissue using lucigenin chemiluminescence and was enhanced when Dahl S rats were treated with a high salt diet. Treatment with intravenous infusion of TEMPOL reduced the blood pressure of high salt–fed Dahl S rats without alteration of total renal blood flow and GFR and decreased glomerulosclerosis and proteinuria [12]. These results indicate that superoxide is responsible for pathogenesis of CKD and hypertension in Dahl S rats. Role of oxidative stress in the renal medulla in Dahl S rat was also determined. Higher expression of NAD(P)H oxidase mRNA was observed in the renal outer medulla of Dahl S rats as compared to those of SS13BN rats, a rat model in which chromosome 13 of normotensive Brown Norway rats were introgressed into Dahl S background and salt sensitivity of blood pressure was

Table 1 Role of nitric oxide (NO) and superoxide on renal circulation, sodium reabsorption and blood pressure

Infusion	Microdialysis	MBF	CBF	UNa	BP
L-NAME r.i. (SD)	NO ↓	↓	–	↓	↑
DETC r.i.(SD)	O_2^- ↑	↓	–	↓	↑
H_2O_2 r.i. (SD)	H_2O_2 ↑	↓	–	↓	↑
Catalase r.i. (DahlS + HS)	H_2O_2	–	–	–	–
Tiron r.i. (Dahl + HS)	O_2^-	–	–	–	–
AngII i.v.(SD)	NO ↑	–	–		–
AngII i.v.(SD) + L-NAME(SD)	NO^-	↓	–	↓	↑
AngII i.v. (Dahl S)	NO	↓	–		↑

Responses to renal and intravenous infusion of drugs were summarized. *MBF* medullary blood flow; *CBF* cortical blood flow; *UNa* sodium excretion; *BP* blood pressure; *r.i.*, renal interstitial infusion; *AngII* angiotensin II; *HS* high salt diet

reduced [19]. Taken together with the result that enzymatic production of superoxide is highest in the outer medulla, superoxide seems to participate in the regulation of renal medullary circulation.

To determine the role of superoxide to the medullary circulation, SOD inhibitor diethyldithiocarbamic acid (DETC) or TEMPOL was directly infused into the renal medullary interstitial space from a small implanted catheter (Table 1). Local blood flow was determined through an implanted optical fiber attached to a laser Doppler flowmeter. When DETC was infused and superoxide increased in the renal medulla of anesthetized Sprague Dawley (SD) rats, reduction of medullary blood flow(MBF) and sodium excretion was observed independent of cortical blood flow (CBF) [11]. In contrast, renal interstitial infusion (r.i.) of TEMPOL and decreased renal medullary superoxide resulted in the increase of MBF and sodium excretion. These results suggest that superoxide is responsible for medullary circulation and sodium excretion [11]. Makino et al. have infused DETC in SD rats from a chronically implanted catheter in the medulla. An increase in outer medullary superoxide was confirmed by fluorescent indicator dihydroethidium applied to an interstitial fluid corrected by microdialysis technique. Chronic RI of DETC, which increased renal superoxide in SD rats, reduced MBF and showed sustained hypertension [20]. These results indicate that superoxide in the renal interstitial fluid could play a role in the regulation of MBF.

The physiological role of superoxide in the development of hypertension was also determined and compared between Dahl S rats and SS13BN rats. As described earlier, SS13BN rat is a consomic rat in which chromosome 13 of Brown Norway rats were introgressed into Dahl S rat background. SS13BN rats had a lower response of blood pressure to a high-salt diet and less renal interstitial fibrosis compared to those of Dahl S rats [19]. The allele difference between Dahl S rats vs. SS13BN rats with a genome-wide scan using 265 polymorphic simple sequence length polymorphism markers was less than 2%, while that between Dahl S and Dahl R rats was approximately 30% [21]. Thus, SS13BN rats were used for specific control for Dahl S rats in regard to salt-sensitive hypertension and renal injury research. In spite of comparable genetic backgrounds, Dahl S rats exhibit higher renal medullary oxidative stress, which is involved in the salt sensitivity of blood pressure and renal injury. Renal medullary superoxide level determined with microdialysis method and tissue homogenate was higher in Dahl S rats compared to SS13BN rats. Increased superoxide level in Dahl S rats was reduced by NAD(P)H oxidase inhibitor diphenylene iodonium. A mitochondrial uncoupler, dinitrophenol, eliminated the remaining superoxide production, but not with xanthine oxidase, nitric oxide (NO) synthase, and cycloxygenase inhibitors [17]. As is similar to SD rats, these results suggests that NADPH oxidase and mitochondrial enzyme were responsible for production of superoxide in the outer medulla. Interestingly, the level of L-arginine and nitric oxide synthase (NOS), SOD, catalase, and glutathione peroxidase activities was not different between Dahl S and SS13BN rats. NAD(P)H oxidase inhibitor apocynin chronically infused into the renal medullary interstitial space inhibited the blood pressure response to high salt diet [17]. Therefore, NAD(P)H oxidase plays a major role in production of superoxide in the outer medulla and salt-sensitive hypertension.

Role of NO to the development of hypertension was also determined. In SD rats, NOS activity was higher in the renal inner medulla (26 times higher than cortex) and outer medulla (4 times higher) compared to that of cortex [22]. In addition, NOS activity was high in the inner medullary collecting duct (IMCD), vasa recta, and medullary thick ascending limbs. There are differences in NOS isoforms expressed in each segments: NOS I (nNOS) is expressed in thick ascending limb, collecting ducts, glomeruli, vasa recta, and particularly heavily in the macula densa; NOS II (iNOS) is expressed in the thick ascending limb, collecting ducts, and S3 region of proximal tubules; NOS III (eNOS) is expressed in thick ascending limb, collecting ducts, glomeruli, and vasa recta [22, 23]. Nitric oxide has been shown to be involved in the myogenic response and TGF as a glomerular function [24]. It has also been shown to participate in the regulation of medullary circulation and development of hypertension. Acute renal medullary interstitial infusion of NOS inhibitor L-NAME (N^{G}-nitro-L-arginine methyl ester) reduced MBF without altering CBF and decreased sodium excretion [25]. Chronic infusion of L-NAME into renal medulla reduced MBF independent of CBF and resulted in sustained hypertension [26]. Thus, renal NO no doubt plays a role in medullary circulation. A subpressor dose of intravenous infusion of angiotensin II does not alter MBF or CBF. However, when a small subpressor dose of L-NAME was preinfused with the same dose of angiotensin II, reduction of MBF was observed in both acute and chronic studies [27, 28]. As summarized in Table 1, an increase in blood pressure was observed when subpressor dose of angiotensin II was intravenously infused with renal medullary interstitial infusion of a subpressor dose of L-NAME, indicating that NO was protecting against reduction of MBF with physiological dose of angiotensin II [28]. An increase of NO in the microdialysate of the renal medulla was observed after acute intravenous infusion of angiotensin II [27]. These results suggest that NO is produced in the interstitium of renal medulla and is responsible for MBF and development of hypertension [29].

How does the NO and superoxide get produced and where does it act? The role of superoxide and NO in the renal medulla was determined using microdissected isolated thin tissue strips and florescent indicators. Angiotensin II stimulates NO production in medullary thick ascending limb of SD rats. It has been shown that NO can diffuse into pericytes of vasa recta. Nitric oxide participates in the dilation of pericyte and thereby may increase MBF [30]. TEMPOL increased NO in the vasa recta and attenuated the vasoconstriction with angiotensin II [30]. As shown in Table 1, angiotensin II stimulates both superoxide and NO production in medullary thick ascending limb [31–33]. Interestingly, NO and superoxide can diffuse into the pericyte of vasa recta. Since angiotensin II failed to increase NO and superoxide directly in the vasa recta, tubular NO and superoxide were the main free radical sources for the MBF response to angiotensin II [31, 32]. Consistent to the in vivo studies, superoxide response to angiotensin II in the medullary thick ascending limb was inhibited by diphenilene iodonium (DPI) and apocynin and confirmed that NADPH oxidase was responsible for superoxide production [32]. Diffusion of NO was enhanced when superoxide was reduced by TEMPOL. In contrast, superoxide cannot diffuse normally in the outer medulla of SD rats; however, when a NO scavenger carboxy-PTIO was applied, superoxide produced in the medullary thick ascending limb was able to diffuse into vasa recta [32]. These results suggest that NO and superoxide counteract each other for diffusion in the outer medulla. This tubulovascular cross-talk of the free radicals plays an important role in the regulation of MBF and development of hypertension (see Table 1 and Figure 1) [8, 29]. Diffusion of NO in response to angiotensin II is dominant in normotensive SD rats and SS13BN rats, while superoxide is dominant in Dahl S rats [33]. Not only angiotensin II but solutes such as sodium and glucose are responsible

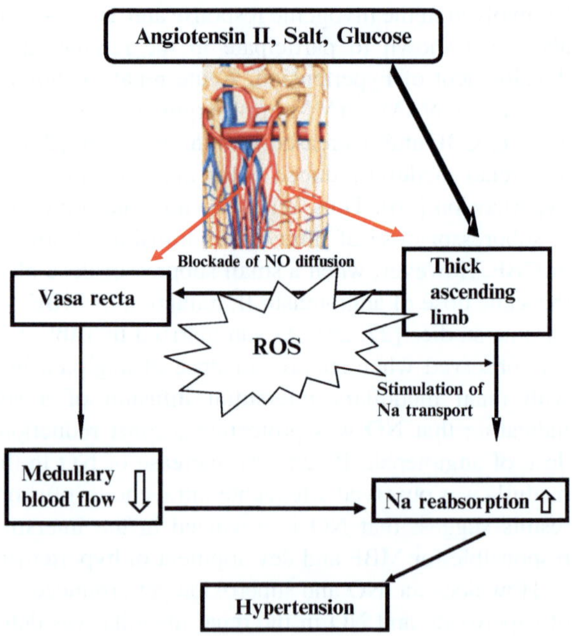

Fig. 1 Physiological role of oxidative stress in the renal medulla. Both nitric oxide (NO) and superoxide diffuse from tubules to the vasa recta and regulate medullary blood flow (MBF). Diffusion of NO was attenuated by enhanced oxidative stress, which is seen in Dahl salt sensitive rats

for superoxide production in the medullary thick ascending limb [34, 35]. Since blockade of sodium exchangers or pumps could inhibit the superoxide production by these solutes, metabolic rate and transport seem to be responsible for the induction of oxidative stress in this region [35].

In addition to superoxide, hydrogen peroxide has been shown to participate in the regulation of MBF. Chen et al. [36] infused hydrogen peroxide directly into the interstitial space of renal medulla and observed a reduction of MBF and sodium excretion in anesthetized rats. Chronic infusion of hydrogen peroxide into the renal medulla through a small chronically implanted catheter in SD rats also reduced MBF and resulted in sustained hypertension [37]. Renal medullary hydrogen peroxide is increased in Dahl S rats compared to those of control SS13BN rats [38].Reduction of renal medullary hydrogen peroxide with renal medullary infusion of catalase into Dahl S rats attenuated salt sensitivity of hypertension [38].

Other than regulation of MBF, oxidative stress in the renal medulla contributes to the direct action of tubular transport. This was determined using isolated tubules. NO reduced NaCl reabsorption in both proxymal tubules and the thick ascending limb [39, 40]. When L-NAME was added to the chamber of the isolated perfused thick ascending limb, NaCl absorption was increased [40]. In contrast, both superoxide and hydrogen peroxide increased NaCl reabsorption in the thick ascending limb [41]. Ortiz and Garvin perfused TEMPOL into the luminal side of the thick ascending limb and demonstrated that NaCl reabsorption was reduced [41]. These results suggest that oxidative stress is responsible for direct tubular NaCl transport.

The plasma level of MGO, a dicarbonyl compound derived from glycolysis pathway that initiates carbonyl stress, is increased in patients with CKD, regardless of diabetes [42]. Carbonyl stress is involved in the pathogenesis of CKD and uremia [5, 42]. As described above, renal injury of ischemia reperfusion model was attenuated in rats with overexpressed glyoxalase I, which metabolizes MGO [10]. Increased plasma level of MGO was observed in remnant kidney rats and chronic Thy-1 nephritis rats (unpublished observation). Nakayama demonstrated in an in vitro study, using electron resonance spin-trapping method, that large amounts of reactive radicals are formed by hydrogen peroxide and MGO, suggesting the interaction between carbonyl and oxidative stress [43]. Thus, the role of the interaction between oxidative and carbonyl stress in the pathogenesis of CKD has been determined, as demonstrated in Table 2. Salt sensitivity of hypertension

Table 2 Carbonyl stress induces salt sensitivity and insulin resistance

	Renal CEL	Renal 3′ nitrotirosine	Urinary TBARS	Salt sensitive hypertension	Insulin resistance
MGO	+	+	+		+
Salt		+	+		
MGO + salt	+	++	++	+	

+ indicates increase in each parameters

Immunohistochemical analysis of renal carbonyl/oxidative stress was determined. Salt sensitivity was determined by the blood pressure difference between normal salt diet and high salt diet. Insulin sensitivity was determined with glucose clamp technique and insulin resistance is commonly seen in CKD. Although an 8% salt diet does not significantly influence blood pressure in SD rats, when 1% MGO, a dose that does not influence blood pressure, was added to drinking water, it increased blood pressure [44]. This result suggests that MGO enhanced salt sensitive hypertension. When SD rats were treated with high-salt diet, it was demonstrated that renal oxidative stress was increased [45]. A further increase in renal oxidative stress, especially in the outer medulla, was observed when 1% MGO was given in drinking water with high-salt diet. One percent MGO in drinking water was used because this was the dose that increased plasma level of MGO (approximately ten times) and was equivalent to those of CKD stage 5. Since an increase in oxidative stress in the renal medulla causes a reduction of MBF and salt sensitivity of hypertension, as described above, carbonyl stress could interact with oxidative stress and participate in the salt sensitivity of hypertension [44].

It has been demonstrated that carbonyl stress was also involved in insulin resistance, which is also commonly observed in CKD. Sprague Dawley rats were given 1% MGO in drinking water and insulin sensitivity was determined by glucose clamp technique [44]. MGO reduced salt sensitivity, which was completely recovered by treatment of either *N*-acetyl cysteine (NAC) or TM2002, which can inhibit both carbonyl stress and oxidative stress. These results indicate that increased carbonyl stress in the CKD also plays a role in metabolic disturbances such as insulin resistance.

Increased renal hydrogen peroxide in Dahl S rats may also interact with carbonyl stress. In Dahl S rats, 1% MGO in drinking water increased blood pressure and cardiovascular–renal injury (unpublished observation). Enhanced renal oxidative and carbonyl stress and increase in inflammatory cells were observed strongly in the outer medulla. The same dose of MGO did not increase blood pressure in SD rats, which has a relatively lower level of hydrogen peroxide. These results may suggest that carbonyl and oxidative stress has a cardiovascular–renal connection in CKD.

3 Role of Oxidative/Carbonyl Stress in CKD: Clinical Point of View

Enhanced oxidative stress is observed in hypertension and diabetes and is a major cause of CKD and CVD. Ogawa et al. [46] determined oxidative stress and inflammatory markers in hypertensive diabetes patients. Albuminuria and these markers were compared in patients treated with either thiazide diuretics or angiotensin II receptor blockers for 8 weeks. No differences in blood pressure and HbA1c was observed between groups. The urinary albumin to creatinine ratio was still increased in thiazide diuretics group, but those of the angiotensin II receptor blockers group was significantly reduced. Urinary oxidative stress markers such as 8epi-prostaglandin F2α and 8-OHdG were also significantly reduced in the angiotensin II receptor blockers group but not in the diuretics group. This was followed by the plasma inflammatory markers, monocyte chemoattractant protein-1 (MCP-1) and interleukin (IL)-6, which were only reduced in the angiotensin II receptor blockers group. Significant correlations were found between the changes in urinary albumin to creatinine ratio and urinary oxidative stress markers. Interestingly, patients with a higher baseline level of urinary oxidative stress markers had further reduction in their urinary albumin to creatinine ratios. Furthermore, patients who were treated with angiotensin II converting enzyme inhibitors prior to the study were able to reduce albuminuria, oxidative stress, and inflammatory markers even with the thiazide diuretics treatment. Inflammatory markers were also reduced by treatment of angiotensin II receptor blockers in hypertensive patients, which was demonstrated in the European Trial on Olmesartan and Pravastatin in Inflammation and Atherosclerosis (EUTOPIA) study [47]. Taken together, these results suggest that oxidative stress and inflammation derived by renin-angiotensin system play a major role in the renal injury of hypertensive and diabetic patients.

Carbonyl/oxidative stress is also involved in the pathogenesis of CKD, especially diabetic nephropathy. Increased carboxymethyllysine (CML) and pentosidine accumulation was observed in the expanded mesangial matrix and nodular lesions of the biopsy samples of diabetic nephropathy patients [6]. These accumulations were colocalized with malondialdehyde-lysine (MDA-lysine), a lipoxidation product, indicating that carbonyl stress could be associated with oxidative stress. This is also confirmed by the absence of pyrraline accumulation, which is another advanced glycated end product (AGE) marker, independent of oxidative stress, in the glomeruli of diabetic nephropathy [6].

4 Role of Renal Perfusion Pressure and Oxidative Stress on the Progression of Hypertensive Renal Injury

Results from clinical studies have indicated that hypertension is a strong independent risk factor of renal injury. However, as described above, carbonyl and oxidative stress and humoral factors such as renin-angiotensin system play an important role in the pathogenesis of hypertensive renal injury. Thus, there is some difficulty in determining the specific role of blood pressure to renal injury in humans. Mori et al. have established a system that can determine a specific role of renal perfusion pressure to the renal injury in rats. An inflatable occluder cuff was implanted in the aorta between two renal arterial blanches. By using the servo-control of the cuff pressure with a downstream blood pressure, renal perfusion pressure of left kidney was controlled to a normal pressure for 3 weeks [16, 48]. Pressure-induced renal injury was determined by the differences of renal injury between the right and left kidneys. When this system was applied to angiotensin II–infused rats, a comparison of renal injury between the right and left kidneys attributed to the renal perfusion pressure, while comparison between the left kidney and a kidney of angiotensin II noninfused sham rats was attributed to angiotensin II [48]. Interestingly, the role of renal perfusion pressure to hypertensive renal injury was heterogeneous: 70% of renal medullary tubule-interstitial injury was attributed to renal perfusion pressure; 90% of renal injury in the outer cortical glomeruli was attributed to angiotensin II; and 80% of juxtamedullary glomerular injury was attributed to renal perfusion pressure. This can be explained by a lesser amount of autoregulation and sensitivity to angiotensin II in the preglomerular vessels of juxtamedullary nephron. In contrast, in Dahl S rats, renal injury attributed to renal perfusion pressure was more than 90% in most of the regions [16], including outer cortical glomeruli. Since Dahl S rats have blunted autoregulation in all glomeruli, this could be why blood pressure contributes strongly to renal injury in Dahl S rats. In these studies, molecules differentially expressed between high and normal perfusion pressure were determined with cDNA microarray, real-time polymerase chain reaction and immunohistochemistry. Among the various molecules determined, molecules related to oxidative stress (tumor growth factor-β (TGF-β), nuclear factor κB (NFκB), osteopontin), inflammation (ED-1), and wound healing (MMP-2, TIMP-1, αSMA) were observed, as shown in Fig. 2 [16, 48]. Jin et al. confirmed the production of reactive oxygen species (ROS) with increased renal perfusion pressure in anesthetized rats. Hydrogen peroxide concentration of renal interstitial fluid, determined with microdialysis technique, was enhanced when renal perfusion pressure was increased by the occluder below and above renal artery [49].

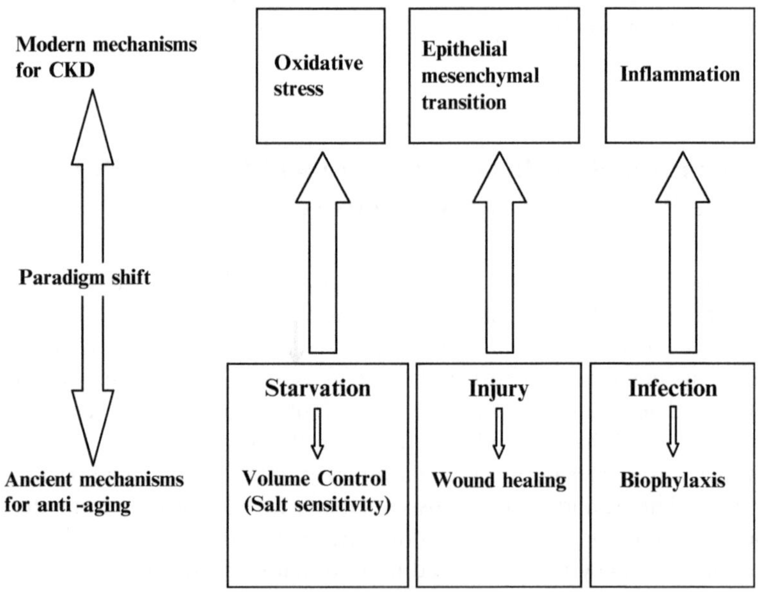

Fig. 2 Mechanism of chronic kidney disease (CKD) could be associated with the food and lifestyle of the ancients. Oxidative stress, wound healing, and inflammation are major mechanisms for antiaging that protect us from starvation, injury, and infection, respectively. Because of a paradigm shift of antiaging, these mechanisms have become a threat for the pathogenesis of CKD and hypertension

Thus, in addition to the renin-angiotensin system, as shown above, renal perfusion pressure can also induce oxidative stress.

5 Strain Vessel Hypothesis Could Explain the Pathophysiological Connections Between CKD and CVD

The risk of CVD is increased not only through a reduction of renal function, but also from even a small increase in albuminuria without a reduction of renal function. The results of the Prevention of Renal and Vascular End-stage Disease (PREVEND) study have demonstrated that microalbuminuria is a risk of CVD, even in subjects without any history of hypertension or diabetes [3]. Although there are proposed mechanisms related to endothelial dysfunction or oxidative stress, even without adjustment of these parameters albuminuria is still a risk factor of CVD [50]. In spite of significant epidemiological evidence, there is no clear mechanism that explains why albuminuria could be associated with CVD. Ito et al. [51] have proposed a hypothesis that anatomical and functional relationships between the specific vessels could explain the connection between the kidney, heart, and brain. Hypertensive renal vascular damages occur first and more severely in juxtamedullary glomeruli located deep in the cortex, because their afferent arterioles arise directly from large arcuate arteries, and, therefore, are exposed to very high pressure as compared with same-size arterioles in peripheral circulation (Fig. 3). Since the glomerular capillary pressure is maintained to 50 mmHg, these afferent arterioles and preglomerular vessels have to create huge pressure gradients in short distances. These hemodynamic conditions are analogous to those of perforating arteries in the central nervous system or retinal arteries in the eye [51]. Therefore, albuminuria may be an early sign of vascular damage imposed on small vessels that are exposed to high pressure and are destined to create huge pressure gradients; we have named these vasculatures strain vessels. Coronary circulation is also under unique hemodynamic conditions, where entire epicardial segments of coronary arteries are exposed to a very high pressure during systolic phase. As described above, high perfusion pressure in the kidney could induce oxidative stress. Thus it is possible to speculate that oxidative stress in the strain vessel could also be involved in the pathogenesis of cerebrovascular–renal connection, and this requires further investigation.

6 Paradigm Shift of Antiaging Between Ancient and Modern Times

From the evolutional point of view, there could be a reason why oxidative/carbonyl stress had played a role in CKD and metabolic syndrome. When we consider lifestyles and antiaging regimens in ancients time, we may find an answer as to

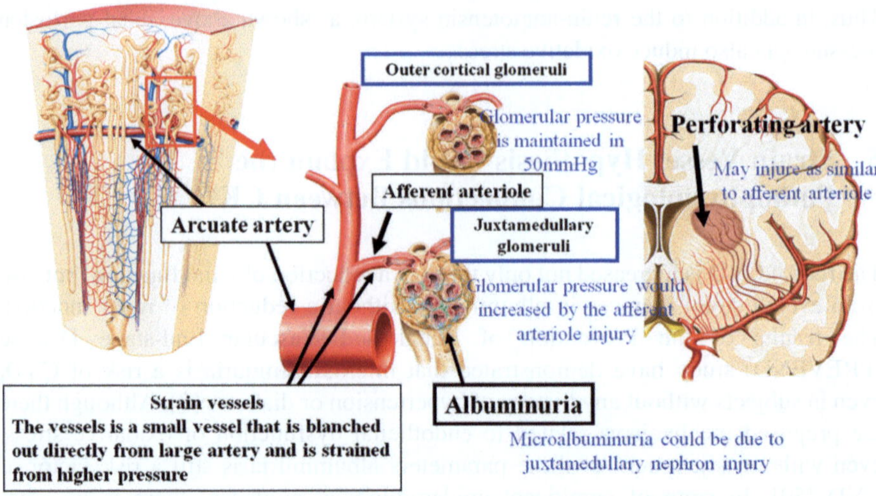

Fig. 3 Strain vessel hypothesis, which would explain why albuminuria is a marker of early vascular injury. To control glomeruli pressure to 50 mmHg, the preglomerular artery of the juxtamedullary nephron has to create huge pressure gradients in short distances. Albuminuria is a marker of juxtamedullary injury. These hemodynamic conditions are analogous to those of perforating arteries in the central nervous system, which could explain why albuminuria is an independent risk of cardiovascular disease. Figure adapted and refigured [51]

why oxidative/carbonyl stress is enhanced in modern diseases. As summarized in Fig. 3, we can speculate that animals in ancient times seem to have been threatened with starvation, injury, and infection, and those who had a greater ability to protect themselves from these had an advantage to live longer. Three major mechanism involved in the pathogenesis of hypertensive renal injury can be attribute to the above three threatening aspects (see Fig. 2). To protect ourselves from dehydration, starvation, and infection, enhanced sodium reabsorption by renal oxidative stress was an advantage. For infection protection, an enhanced inflammatory mechanism was required for phylaxis and cure. Since, animals in ancient times had extensive chance of getting injured, enhanced mechanism for wound healing were also an advantage to live longer in ancient times. Our ancestors adapted to the tough lifestyle of the ancients and may have had an enhanced mechanism of oxidative stress, inflammation, and wound healing compared to those who did not survive enough to leave offspring. Although they had enhanced oxidative stress, inflammation, and wound healing abilities, they did not develop CKD or CVD. They had a different lifestyle than we now have, and salt and pepper for food was not an option. They didn't even cook meat but ate it raw for many years. To eat raw meat, they had to hunt every day, which resulted in a good exercise. However, our ancestors found salt as a spice and started to cook with fire. These were good tools to enable them to live longer, since salted meat preserved well, tasted better, and protected them from dehydration. Since carbonyl compounds and advanced glycation end products increase taste, cooked meats tasted better and could be preserved more easily. In addition to less-frequent hunting, the invention of farming also prevented starvation but resulted in les exercise. Environmental pollution by modern industries may also have altered the body's abilities for oxidative/carbonyl stress, inflammation, and wound healing. Moreover, in ancient times strain vessels were mandatory for survival under the conditions of difficult access to salt and water and the danger of hypoperfusion. However, in modern times we are threatened by hypertension and diabetes, and stain vessels have become the target of hypertensive organ damage.

All of this taken together, we propose that this paradigm shift of antiaging could explain why oxidative/carbonyl stress, inflammation, and wound healing are involved in the pathogenesis of CKD, CVD, and metabolic syndrome. Since our genetic backgrounds and anatomy are not much different from those of our ancestors, we do not seem to be adapting well enough to our modern environment. It is also possible to speculate that CKD and CVD may be targeting those who had an advantage in the ancient lifestyle, which could be a disadvantage in the modern lifestyle. This paradigm shift of antiaging was seen in the past in which dinosaurs became extinct because of a failure to adapt to changes of their environments. This should be a warning for us about our future lifestyle and environments.

7 Conclusion

Oxidative/carbonyl stress, inflammation, and wound healing play a key role in the pathogenesis of hypertension and CKD. This mechanism, which was advantageous in the life-threatening environments of the ancients, has now become an evil subject for those with a modern lifestyle. Thus, enhanced markers of oxidative stress, inflammation, and urinary albumin excretion could be a warning. In spite of advanced medication available today, we are still not able to inhibit ancient-derived mechanisms. To completely protect against CKD and CVD, we may have to take into a consideration the ancient lifestyle and counteract against modern practices to stop the antiaging paradigm shift.

Acknowledgments We thank Yoshimi Yoneki, Hiroko Ito, and Kiyomi Kisu for their expert technical assistance.

References

1. Sarnak MJ, Levey AS, Schoolwerth AC, Coresh J, Culleton B, Hamm LL, McCullough PA, Kasiske BL, Kelepouris E, Klag MJ, Parfrey P, Pfeffer M, Raij L, Spinosa DJ, Wilson PW. Kidney disease as a risk factor for development of cardiovascular disease: a statement from the American Heart Association Councils on Kidney in Cardiovascular Disease, High Blood Pressure Research, Clinical Cardiology, and Epidemiology and Prevention. Hypertension 2003 Nov;42(5):1050–65.

2. Ibsen H, Olsen MH, Wachtell K, Borch-Johnsen K, Lindholm LH, Mogensen CE, Dahlof B, Snapinn SM, Wan Y, Lyle PA. Does albuminuria predict cardiovascular outcomes on treatment with losartan versus atenolol in patients with diabetes, hypertension, and left ventricular hypertrophy? The LIFE study. Diabetes Care 2006 Mar;29(3):595–600.
3. Hillege HL, Fidler V, Diercks GF, van Gilst WH, de Zeeuw D, van Veldhuisen DJ, Gans RO, Janssen WM, Grobbee DE, de Jong PE. Urinary albumin excretion predicts cardiovascular and noncardiovascular mortality in general population. Circulation 2002 Oct;106 (14):1777–82.
4. Cubeddu LX, Hoffmann IS, Aponte LM, Nunez-Bogesits R, Medina-Suniaga H, Roa M, Garcia RS. Role of salt sensitivity, blood pressure, and hyperinsulinemia in determining high upper normal levels of urinary albumin excretion in a healthy adult population. Am J Hypertens 2003 May;16(5 Pt 1):343–9.
5. Miyata T, Ueda Y, Yamada Y, Izuhara Y, Wada T, Jadoul M, Saito A, Kurokawa K, van Ypersele de Strihou C. Accumulation of carbonyls accelerates the formation of pentosidine, an advanced glycation end product: carbonyl stress in uremia. J Am Soc Nephrol 1998 Dec;9 (12):2349–56.
6. Suzuki D, Miyata T, Saotome N, Horie K, Inagi R, Yasuda Y, Uchida K, Izuhara Y, Yagame M, Sakai H, Kurokawa K. Immunohistochemical evidence for an increased oxidative stress and carbonyl modification of proteins in diabetic glomerular lesions. J Am Soc Nephrol 1999 Apr;10(4):822–32.
7. Miyata T, Kurokawa K, van Ypersele de Strihou C. Relevance of oxidative and carbonyl stress to long-term uremic complications. Kidney Int Suppl 2000 Aug;76:S120–5.
8. Mori T, Cowley AW, Jr., Ito S. Molecular mechanisms and therapeutic strategies of chronic renal injury: physiological role of angiotensin II-induced oxidative stress in renal medulla. J Pharmacol Sci 2006 Jan;100(1):2–8.
9. Nishiyama A, Yoshizumi M, Hitomi H, Kagami S, Kondo S, Miyatake A, Fukunaga M, Tamaki T, Kiyomoto H, Kohno M, Shokoji T, Kimura S, Abe Y. The SOD mimetic tempol ameliorates glomerular injury and reduces mitogen-activated protein kinase activity in Dahl salt-sensitive rats. J Am Soc Nephrol 2004 Feb;15(2):306–15.
10. Kumagai T, Nangaku M, Kojima I, Nagai R, Ingelfinger JR, Miyata T, Fujita T, Inagi R. Glyoxalase I overexpression ameliorates renal ischemia-reperfusion injury in rats. Am J Physiol Renal Physiol 2009 Apr;296(4):F912–21.
11. Zou AP, Li N, Cowley AW, Jr. Production and actions of superoxide in the renal medulla. Hypertension 2001 Feb;37(2 Pt 2):547–53.
12. Meng S, Cason GW, Gannon AW, Racusen LC, Manning RD, Jr. Oxidative stress in Dahl salt-sensitive hypertension. Hypertension 2003 Jun;41(6):1346–52.
13. Nishiyama A, Fukui T, Fujisawa Y, Rahman M, Tian RX, Kimura S, Abe Y. Systemic and regional hemodynamic responses to tempol in angiotensin II-infused hypertensive rats. Hypertension 2001 Jan;37(1):77–83.
14. Schnackenberg CG, Wilcox CS. Two-week administration of tempol attenuates both hypertension and renal excretion of 8-Iso prostaglandin f2alpha. Hypertension 1999 Jan;33(1 Pt 2):424–8.
15. Welch WJ, Tojo A, Wilcox CS. Roles of NO and oxygen radicals in tubuloglomerular feedback in SHR. Am J Physiol Renal Physiol 2000 May;278(5):F769–76.
16. Mori T, Polichnowski A, Glocka P, Kaldunski M, Ohsaki Y, Liang M, Cowley AW, Jr. High perfusion pressure accelerates renal injury in salt-sensitive hypertension. J Am Soc Nephrol 2008 Aug;19(8):1472–82.
17. Taylor NE, Glocka P, Liang M, Cowley AW, Jr. NADPH oxidase in the renal medulla causes oxidative stress and contributes to salt-sensitive hypertension in Dahl S rats. Hypertension 2006 Apr;47(4):692–8.
18. Meng S, Roberts LJ, II, Cason GW, Curry TS, Manning RD, Jr. Superoxide dismutase and oxidative stress in Dahl salt-sensitive and -resistant rats. Am J Physiol Regul Integr Comp Physiol 2002 Sep;283(3):R732–8.

19. Cowley AW, Jr., Roman RJ, Kaldunski ML, Dumas P, Dickhout JG, Greene AS, Jacob HJ. Brown Norway chromosome 13 confers protection from high salt to consomic Dahl S rat. Hypertension 2001 Feb;37(2 Pt 2):456–61.
20. Makino A, Skelton MM, Zou AP, Roman RJ, Cowley AW, Jr. Increased renal medullary oxidative stress produces hypertension. Hypertension 2002 Feb;39(2 Pt 2):667–72.
21. Cowley AW, Jr., Stoll M, Greene AS, Kaldunski ML, Roman RJ, Tonellato PJ, Schork NJ, Dumas P, Jacob HJ. Genetically defined risk of salt sensitivity in an intercross of Brown Norway and Dahl S rats. Physiol Genomics 2000 Apr;2(3):107–15.
22. Wu F, Park F, Cowley AW, Jr., Mattson DL. Quantification of nitric oxide synthase activity in microdissected segments of the rat kidney. Am J Physiol 1999 Jun;276(6 Pt 2):F874–81.
23. Tojo A, Kimoto M, Wilcox CS. Renal expression of constitutive NOS and DDAH: separate effects of salt intake and angiotensin. Kidney Int 2000 Nov;58(5):2075–83.
24. Ito S, Carretero OA, Abe K. Role of nitric oxide in the control of glomerular microcirculation. Clin Exp Pharmacol Physiol 1997 Aug;24(8):578–81.
25. Mattson DL, Lu S, Nakanishi K, Papanek PE, Cowley AW, Jr. Effect of chronic renal medullary nitric oxide inhibition on blood pressure. Am J Physiol 1994 May;266(5 Pt 2): H1918–26.
26. Nakanishi K, Mattson DL, Cowley AW, Jr. Role of renal medullary blood flow in the development of L-NAME hypertension in rats. Am J Physiol 1995 Feb;268(2 Pt 2):R317–23.
27. Zou AP, Wu F, Cowley AW, Jr. Protective effect of angiotensin II-induced increase in nitric oxide in the renal medullary circulation. Hypertension 1998 Jan;31(1 Pt 2):271–6.
28. Szentivanyi M, Jr., Maeda CY, Cowley AW, Jr. Local renal medullary L-NAME infusion enhances the effect of long-term angiotensin II treatment. Hypertension 1999 Jan;33(1 Pt 2):440–5.
29. Cowley AW, Jr., Mori T, Mattson D, Zou AP. Role of renal NO production in the regulation of medullary blood flow. Am J Physiol Regul Integr Comp Physiol 2003 Jun;284(6):R1355–69.
30. Rhinehart KL, Pallone TL. Nitric oxide generation by isolated descending vasa recta. Am J Physiol Heart Circ Physiol 2001 Jul;281(1):H316–24.
31. Dickhout JG, Mori T, Cowley AW, Jr. Tubulovascular nitric oxide crosstalk: buffering of angiotensin II-induced medullary vasoconstriction. Circ Res 2002 Sep;91(6):487–93.
32. Mori T, Cowley AW, Jr. Angiotensin II-NAD(P)H oxidase-stimulated superoxide modifies tubulovascular nitric oxide cross-talk in renal outer medulla. Hypertension 2003 Oct;42 (4):588–93.
33. Mori T, O'Connor PM, Abe M, Cowley AW, Jr. Enhanced superoxide production in renal outer medulla of Dahl salt-sensitive rats reduces nitric oxide tubular-vascular cross-talk. Hypertension 2007 Jun;49(6):1336–41.
34. Abe M, O'Connor P, Kaldunski M, Liang M, Roman RJ, Cowley AW, Jr. Effect of sodium delivery on superoxide and nitric oxide in the medullary thick ascending limb. Am J Physiol Renal Physiol 2006 Aug;291(2):F350–7.
35. Mori T, Cowley AW, Jr. Renal oxidative stress in medullary thick ascending limbs produced by elevated NaCl and glucose. Hypertension 2004 Feb;43(2):341–6.
36. Chen YF, Cowley AW, Jr., Zou AP. Increased H(2)O(2) counteracts the vasodilator and natriuretic effects of superoxide dismutation by tempol in renal medulla. Am J Physiol Regul Integr Comp Physiol 2003 Oct;285(4):R827–33.
37. Makino A, Skelton MM, Zou AP, Cowley AW, Jr. Increased renal medullary H_2O_2 leads to hypertension. Hypertension 2003 Jul;42(1):25–30.
38. Taylor NE, Cowley AW, Jr. Effect of renal medullary H_2O_2 on salt-induced hypertension and renal injury. Am J Physiol Regul Integr Comp Physiol 2005 Dec;289(6):R1573–9.
39. Liang M, Berndt TJ, Knox FG. Mechanism underlying diuretic effect of L-NAME at a subpressor dose. Am J Physiol Renal Physiol 2001 Sep;281(3):F414–9.
40. Ortiz PA, Garvin JL. Interaction of O(2)(−) and NO in the thick ascending limb. Hypertension 2002 Feb;39(2 Pt 2):591–6.

41. Ortiz PA, Garvin JL. Superoxide stimulates NaCl absorption by the thick ascending limb. Am J Physiol Renal Physiol 2002 Nov;283(5):F957–62.
42. Nakayama K, Nakayama M, Iwabuchi M, Terawaki H, Sato T, Kohno M, Ito S. Plasma alpha-oxoaldehyde levels in diabetic and nondiabetic chronic kidney disease patients. Am J Nephrol 2008;28(6):871–8.
43. Nakayama M, Saito K, Sato E, Nakayama K, Terawaki H, Ito S, Kohno M. Radical generation by the non-enzymatic reaction of methylglyoxal and hydrogen peroxide. Redox Rep 2007;12 (3):125–33.
44. Guo Q, Mori T, Jiang Y, Hu C, Osaki Y, Yoneki Y, Sun Y, Hosoya T, Kawamata A, Ogawa S, Nakayama M, Miyata T, Ito S. Methylglyoxal contributes to the development of insulin resistance and salt sensitivity in Sprague-Dawley rats. J Hypertens 2009 Aug;27(8):1664–71.
45. Endo S, Mori T, Yoneki Y, Nakamichi T, Hosoya T, Ogawa S, Tokudome G, Miyata T, Ito S. Blockade of angiotensin II type-1 receptor increases salt sensitivity in Sprague-Dawley rats. Hypertens Res 2009 Jun;32(6):513–9.
46. Ogawa S, Mori T, Nako K, Kato T, Takeuchi K, Ito S. Angiotensin II type 1 receptor blockers reduce urinary oxidative stress markers in hypertensive diabetic nephropathy. Hypertension 2006 Apr;47(4):699–705.
47. Fliser D, Buchholz K, Haller H. Antiinflammatory effects of angiotensin II subtype 1 receptor blockade in hypertensive patients with microinflammation. Circulation 2004 Aug;110 (9):1103–7.
48. Mori T, Cowley AW, Jr. Role of pressure in angiotensin II-induced renal injury: chronic servo-control of renal perfusion pressure in rats. Hypertension 2004 Apr;43(4):752–9.
49. Jin C, Hu C, Polichnowski A, Mori T, Skelton M, Ito S, Cowley AW, Jr. Effects of renal perfusion pressure on renal medullary hydrogen peroxide and nitric oxide production. Hypertension 2009 Jun;53(6):1048–53.
50. Gerstein HC, Mann JF, Yi Q, Zinman B, Dinneen SF, Hoogwerf B, Halle JP, Young J, Rashkow A, Joyce C, Nawaz S, Yusuf S. Albuminuria and risk of cardiovascular events, death, and heart failure in diabetic and nondiabetic individuals. JAMA 2001 Jul;286(4):421–6.
51. Ito S, Nagasawa T, Abe M, Mori T. Strain vessel hypothesis: a viewpoint for linkage of albuminuria and cerebro-cardiovascular risk. Hypertens Res 2009 Feb;32(2):115–21.

Part III
Current Therapy Targeting Oxidative tress

Chapter 16
The Renin Angiotensin System

Josephine M. Forbes and Mark E. Cooper

Abstract The renin angiotensin system (RAS) is a hormonal cascade that is thought to act as a master controller of blood pressure and fluid balance within the body. In addition to the systemic RAS, there is a fully functional intrarenal RAS, which is postulated to play a central role in the development of chronic kidney disease. Indeed there is unprecedented diversity in the pathways affected by the RAS, which include generation of a number of reactive oxygen species (ROS) and effects on antioxidant enzymes. There are many therapeutics, which interrupt different steps in the RAS to primarily lower blood pressure, although these compounds also have numerous beneficial nonhemodynamic effects on both structural and functional parameters in progressive kidney disease. In this chapter we will discuss the effects of the RAS and therapies that interrupt various components of the RAS and on renal redox imbalances that may affect the development and progression of chronic kidney disease, particularly in the context of diabetes.

1 Interruption of the Renin Angiotensin System

The renin angiotensin system (RAS) is increasingly appreciated to be a more complex pathway than originally postulated, with the major arm involving the generation of the powerful vasoconstrictor, angiotensin II (AngII). However, over the past decade it has become apparent that the vasodilator arm, which includes the generation of other angiotensins such as angiotensin 1-7, may also be important [1]. Nevertheless, treatment targets have as yet only focused on interrupting key components of the constrictor arm (Fig. 1, red crosses). The most widely recognized are those that interrupt the conversion of angiotensin I to its active metabolite AngII, namely angiotensin-converting enzyme-1 inhibitors (ACEI) or agents that compete with AngII for binding to the angiotensin type 1 (AT1) receptor subtype, AT1

J.M. Forbes (✉)
Diabetes Complications Division, Baker IDI Heart and Diabetes Institute, Glycation and Diabetes, Baker IDI Heart and Diabetes Institute, 75 Commercial Road, Melbourne, VIC 3004, Australia
e-mail: Josephine.Forbes@bakeridi.edu.au

T. Miyata et al. (eds.), *Studies on Renal Disorders*, Oxidative Stress in Applied Basic Research and Clinical Practice, DOI 10.1007/978-1-60761-857-7_16,

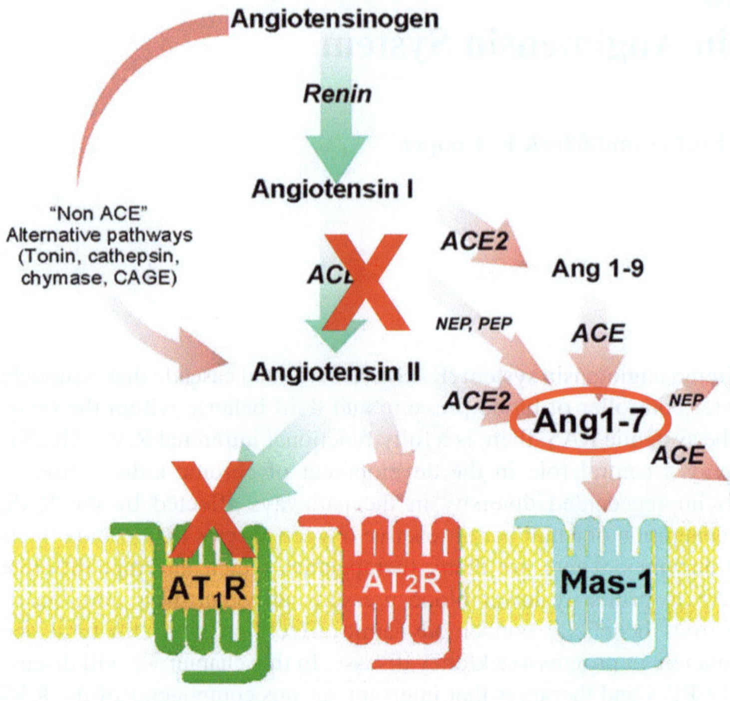

Fig. 1 Schematic representation of the renin-angiotensin cascade. Red crosses represent sites for current therapeutic intervention. MAS-1 - receptor for Ang1-7

antagonists (ARB). The large-scale clinical trials performed with these agents ascribe the profound beneficial effects of blockade of RAS to renoprotection that goes beyond mere blood pressure reduction [2]. For example, in one study, candesartan showed dose-related reductions in proteinuria independent of further lowering of systolic blood pressure with plasma concentrations in patients with chronic kidney disease (CKD), reaching a plateau when doses were increased up to 96 mg/d [2]. While the exact mechanism of action responsible for the renoprotective effects of these agents in previous studies was thought to be via amelioration of intraglomerular hypertension and lowering of systolic hypertension, it is evident that blockade of RAS, using ACEIs and ARBs, also conferred a number of nonhemodynamic benefits, including amelioration of chronic hypoxia and attenuation of renal oxidative or redox imbalances.

RAS blockers have also shown renoprotective effects against mitochondrial dysfunction as the result of aging, which is postulated to be via increases in mitochondrial hydrogen peroxide generation [3]. However, in other chronic diseases, one cannot exclude that RAS blockade may not fully suppress reactive oxygen species (ROS) generation, particularly from other sources such as mitochondria [4], and indeed this could explain the persistent progression, albeit at a slower rate, seen in those with diabetic nephropathy concomitantly treated with agents that interrupt the RAS. Indeed, we have previously shown that in diabetic nephropathy, there is no increase in mitochondrial hydrogen peroxide [5], but rather in superoxide since magnesium superoxide dismutase (MnSOD) activity is very low and therefore there is little conversion of superoxide to hydrogen peroxide. Therefore, it is worth considering as an important strategy for identification of therapeutic targets that could lead to new treatments that confer synergistic effects with those seen with RAS blockade.

More recently, direct renin inhibitors, such as aliskiren, have been used clinically, in concert with currently available RAS blockers for more complete blood pressure lowering [6]. Furthermore, as seen with other strategies that interrupt the RAS, it appears as if renin inhibitors also have a number of beneficial nonhemodynamic effects in renal disease, a feature characteristic of this class of agents that ultimately decreases the generation or action of AngII.

Not surprisingly, inhibitors of the RAS are recognized by many clinical practice guidelines as first line agents for the treatment of CKD. These agents maximize the protection of residual renal function, and a large number of prospective, randomized, controlled studies have demonstrated the renoprotective effects of ACEIs or ARBs. This is yet to be fully elucidated for the direct renin inhibitor aliskiren, although this is likely to be addressed by the ASPIRE HIGHER clinical program [7], which is designed to assess whether the promising pharmacological properties of aliskiren translate into long-term clinical renal benefits. Within that program, the main subtrial to address the renal benefits of these agents in the diabetic context is known as the ALTITUDE trial [8], which is currently in progress.

2 Angiotensinogen

Mostly hepatic in origin, circulating concentrations of angiotensinogen are the source of angiotensin I in humans and other animals (see Fig.1). A posttranslationally glycosylated protein of ranging molecular weight (52–60 kDa), angiotensinogen is transcribed from the human angiotensinogen gene [9]. In addition, the angiotensinogen gene is expressed within the kidney and is most abundantly seen within the renal cortex, predominantly within proximal tubules [10, 11]. Indeed, urinary angiotensinogen excretion may have utility as a biomarker of the severity of CKD [12] since it appears to reflect renal RAS activation [13], although this remains controversial. Nevertheless, in type 1 diabetic subjects, elevations in the urinary excretion of angiotensinogen have been shown to precede the development of albuminuria [14]. Furthermore, patients with hypertension also have elevations in urinary angiotensinogen excretion [15]. Angiotensinogen is most commonly secreted in response to stresses such as tissue injury and bacterial infection, which are also conditions associated with redox imbalances. Angiotensinogen secretion can also occur as part of a feedback loop, via stimulation by the final active metabolite of the RAS, AngII. Indeed, AngII can also stimulate intrarenal expression of the angiotensinogen gene. Furthermore, the angiotensinogen gene is stimulated by nuflear factor κB (NFκB) activation, which is sensitive to the redox ratio, providing a positive feedback loop that can upregulate AngII production. Therefore, it is not difficult to envisage AngII driven by effects of ROS (discussed later in this chapter), on the renal and hepatic expression of angiotensinogen, which might contribute to a pathological feedback loop in CKD. In support of this, mice studies have suggested that superoxide is a mediator of vascular dysfunction in mice that express human renin and angiotensinogen [16].

3 Renin

Renin is synthesized and stored by granular cells within the juxtaglomerular (JG) apparatus of the kidney as an inactive precursor, preprorenin [17, 18]. Following cleavage, prorenin from the renal JG is secreted into the bloodstream via fenestrated capillaries, which is postulated as the major source of circulating renin. A series of events, involving cleavage and glycosylation, result in biologically active renin (see Fig. 1). Like angiotensinogen, AngII also inhibits renin secretion and renin gene transcription in a negative feedback loop [19], which is a likely point of interplay between oxidative stress and renin in the kidney.

Several mechanisms are evoked as controllers of renin secretion, including renal pressure sensors (baroreceptors), mechanisms involving the macula densa, neural mechanisms, endocrine/paracrine pathways, and intracellular mechanisms. For example, renin secretion is inhibited by increased blood pressure or stretch within the afferent arteriole of the normal kidney. Not surprisingly, under these circumstances, renin secretion increases in response to decreased stretch, including alterations in renal perfusion pressure. In patients with CKD, the baroreceptors in the kidney are exposed to myriad perfusion pressures, primarily as the result of luminal narrowing of some afferent preglomerular vessels [2]. As a consequence, a proportion of glomeruli sense inappropriately low perfusion pressures and consequently activate renin secretion, independent of the systemic blood pressure.

Interestingly, a number of previous renal studies in rodents and in vitro experiments have shown that mechanical stretch, where intrarenal renin concentrations are likely to be elevated, is associated with increases in inflammatory markers [20, 21] and other modulators of oxidative stress such as the IκB–NFκB pathway [20]. These data would suggest that there may be concerns surrounding marked elevations in tissue renin concentrations following use of the direct renin inhibitor aliskiren, which could paradoxically increase blood pressure or intraglomerular pressure and lead to increased oxidative stress. To date this concern appears unfounded, based first upon rodent studies showing that aliskiren provides equivalent renoprotection via lowering of oxidative stress (3-nitrotyrosine and the nicotinamide adenine dinucleotide phosphate (NADPH) oxidase subunit, nox-2) to that seen with the ARB irbesartan [22] and second from analyses of hypertensive patients participating in clinical trials to date. Indeed, studies in animals and humans indicate that aliskiren accumulates in renal tissue and interferes with the deleterious cellular effects of AngII by mechanisms that may include enzymatic blockade of renin and prorenin at the site of the prorenin receptor [23]. These data suggest that lowering of intrarenal renin concentrations may not be necessary to acquire renal benefits, such as a reduction in oxidative stress, and that blockade of the prorenin receptor may have more specific effects that those seen with the renin inhibitor aliskiren. Indeed, this whole area of research has been recently transformed first by the identification of a putative (pro)renin receptor that is widely expressed including in the kidney [24] and second through the positive renal benefits described, including in the setting of diabetic nephropathy using a peptide that is postulated to inhibit renin binding to its receptor [25].

This link between renin and oxidative stress has been further examined in recent studies, albeit in a nonrenal context. For example, in a rodent model of atherosclerosis, administration of aliskiren improved impaired nitric oxide bioavailability and protected against atherosclerotic changes [26]. Vascular superoxide and peroxynitrite levels were significantly reduced by combination therapy using aliskiren and the ARB valsartan in that study [26], supporting the postulate that inhibition of the actions of renin is likely to have beneficial effects on renal oxidative parameters in chronic renal disease.

Renal renin secretion is also affected by the composition of tubular fluid seen by the macula densa [27]. However, the macula densa is not in direct contact with the renin-secreting granular cells within the afferent arteriole wall, suggesting that ion transport, in particular sodium and chloride ions, and second messengers are necessary for this interaction. Interestingly, plasma renin activity is suppressed by exogenous administration of molecules containing chloride ions. This is the paradox of salt! Indeed, in support of this, a recent study has shown that low salt diets activate the renin angiotensin system, which increases oxidative stress via NADPH oxidase and decreases nitric oxide (NO) bioavailability in the heart [28], consistent with the higher mortality that has been reported in some patients consuming low salt diets. Certainly, oxidative stress has been implicated in many forms of AngII-related hypertension, which one could postulate would be associated with lower renal renin concentrations, given the previously discussed findings. In support of this, there is also substantial oxidative stress in the tissues or blood vessels of rats with deoxycorticosterone acetate–salt hypertension in which the circulating renin-angiotensin-aldosterone system is suppressed profoundly. Furthermore, the use of the adenosine receptor antagonist, 1,3-dipropyl-8-sulfophenylxanthine in Wistar rats, which would also lower renal renin concentrations, induces profound hypertension which can be prevented with blockade of NADPH oxidase or glutathione peroxidase, but not TEMPOL (hydroxy-2,2,6,6-tetramethylpiperidine-1-oxyl). These findings have been interpreted to indicate the involvement of cytosolic superoxide and hydrogen peroxide as pathogenic mediators of hypertension in this model [29].

There is also neurological regulation of renin secretion from the kidney. Nebivolol is a β1 receptor blocker with nitric oxide-potentiating vasodilatory effects, which is used in the treatment of hypertension. Nebivolol lowers blood pressure by reducing peripheral vascular resistance and significantly increasing stroke volume with preservation of cardiac output [30]. Recently, studies in the Ren-2 transgenic rat, which overexpresses the RAS, have shown that nebivolol reduces proteinuria and lowers renal NADPH oxidase dependent generation of ROS [31]. In addition, β1 receptor blockade with nebivolol also improved renal endothelial nitric oxide synthase activity, which would likely translate into increased NO bioavailability.

Activation of the sympathetic nervous system can also modulate renal renin secretion. Indeed, administration of norepinephrine, which is known to be released as a consequence of sympathetic nervous system activation, increases afferent arteriolar sensitivity to AngII by means of alpha receptor activation, which would affect renin secretion [32]. In addition, oxidative stress, which is evident during CKD, stimulates renal sympathetic nerve activation during the development of hypertension, which also would alter renal renin concentrations [33, 34].

4 Angiotensin-Converting Enzyme

Angiotensin-converting enzyme-1 (ACE-1) is a large protein (200 kDa), which is responsible for the cleavage of AngI into the pressor substance AngII, although this enzyme has renowned promiscuity and also cleaves other small peptides such as substance P, luteinizing hormone, and bradykinin. Indeed it is possible that the success of ACE inhibitors as agents to treat CKD in addition to hypertension and cardiovascular disease, derives from the diversity of substrates cleaved by ACE. However, this remains controversial since there is minimal evidence in humans to indicated that ACE inhibitors confer superior renoprotective effects to other more specific strategies such as ARBs [35]. ACE is localized primarily but not exclusively to the glomerular endothelium and proximal tubule cell brush border in the kidney where it is postulated to facilitate cleavage of peptides for subsequent resorption. It is also likely that ACE can alter tubular resorption via production of AngII in the proximal tubular urinary filtrate.

It has recently been identified that a novel regulator of ACE expression exists in the kidney and lung, termed midkine [36]. Of interest is that oxidative stress can modulate renal and lung midkine expression following renal ablation (5/6 nephrectomy), which subsequently alters circulating AngII concentrations via the expression of NADPH oxidase. Indeed, in this model of CKD, there is a perpetuation of ROS- and RAS-mediated pathways mediated by midkine. Therefore, in the future, agents that regulate the expression of midkine may have utility in CKD, where they may have dual actions on both AngII concentrations and renal oxidative stress.

Angiotensin-converting enzyme-2 (ACE-2) is a homologue of ACE and is thought to have originated from a common ancestral gene. Although both ACE and ACE-2 cleave AngI, their actions are unique, where cleavage of AngI by ACE-2 to remove one amino acid results in the peptide Ang 1-9. ACE then cleaves Ang 1-9 into Ang 1-7 by removing two amino acids, resulting in a vasodilator. In addition, ACE-2 can cleave AngII to form Ang1-7, which may be a biological brake for AngII-mediated vasoconstriction. Recent studies in rodents have shown that human recombinant ACE-2 slows the progression of diabetic nephropathy [37]. The administration of

ACE-2 not only lowered renal AngII concentrations, but also attenuated AngII induced NADPH oxidase activation. ACE-2 also attenuates nuclear ROS formation, resulting from AngII within kidneys from aged sheep as compared to kidneys from young sheep [38]. In cell nuclei, it is postulated that ACE-2 provides a unique protective mechanism against oxidative stress and cell damage as a result of decreases in AngII-mediated ROS generation.

5 Angiotensins

AngII has a very short half-life within the circulation. It is rapidly cleaved by aminopeptidases, resulting in the heptapeptide AngIII, which retains some vasoconstrictor actions and is postulated as the principal effector of the RAS within the brain. As is well documented for AngII, AngIII also activates the redox sensitive transcription factor NFκB, which appears to be modulated via the production of ROS following ligation to AT2 receptors [39, 40]. AngIII has an even shorter half-life than AngII and is rapidly converted to AngIV by other aminopeptidases [41]. AngIV was originally thought to be an inactive peptide; however, studies have shown that AngIV can also bind to AT4 receptors to regulate inflammation via NFκB [42], in addition to stimulating the endothelial release of plasminogen activator inhibitor-1 (PAI-1) [41]. Perhaps, not surprisingly, ROS also mediate PAI-1 expression in diabetic nephropathy [43].

Angiotensins can also be hydrolysed by endopeptidases, including trypsin and neutral endopeptidase (NEP). In particular, NEP can directly cleave AngI to Ang 1-7 (see Fig. 1, circled), which is thought to have opposing effects to the vasoconstricting and growth promoting actions of AngII. However, although it is likely that Ang 1-7 could also act to limit oxidative stress in chronic renal disease [44], there is recent evidence that infusion of Ang 1-7 causes renal inflammation and oxidative stress [45]. In support of this, chronic injection of Ang 1-7 also accelerates renal disease in streptozotocin diabetic rats via activation of oxidative pathways [46].

6 Angiotensin II Receptors

AngII mediates its biological effects via at least two specific receptor subtypes, the AT1 and angiotensin type 2 (AT2) receptors [47], which are each expressed in the kidney. Most of the pressor actions attributed to AngII, including vasoconstriction and activation of the sympathetic nervous system, appear to be mediated via ligation to the AT1 receptor. In addition, activation of the AT1 receptor has most often been associated with CKD and is known to have specific effects on oxidative parameters. In particular this is via ligation with AngII, which stimulates the expression of NADPH oxidase and subsequently produces superoxide. Indeed, even short-term AngII infusion induces oxidative stress via AT1 receptors, producing superoxide and decreasing the expression of extracellular superoxide dismutase in rats [48]. In hypertensive rodent models such as renal ablation (5/6 nephrectomy), AngII/AT1 mediated effects, which induce oxidative stress, are also pathological [49].

Oxidative stress is also seen in other CKDs including obstructive nephropathy. In rodents, unilateral ureteral obstruction (UUO) led to decreased SOD activity and elevations in the concentration of hydroxyl radicals (OH^-) and superoxide (O_2^-) when compared to sham control rats. Losartan decreased ROS generation, facilitated the recovery of renal SOD activity, and increased the expression of heat shock proteins [50]. Other studies in UUO have also demonstrated AngII/AT1 dependent effects on renal oxidative parameters [51].

Diabetes is the most common cause of CKD in developed nations. Diabetic nephropathy can often progress rather slowly, particularly if aggressively treated with blood pressure and glucose lowering agents, and is significantly influenced by all components of the RAS and by oxidative stress [52]. Indeed, diabetic nephropathy appears to occur as a result of complex interplay between hemodynamic and metabolic pathways, which we have previously reviewed extensively [53].

Oxidative stress is known to develop early in patients with focal segmental glomerulosclerosis [54]. In particular, increases in glomerular and extracellular malondialdehyde levels in such patients are attenuated with AT1 receptor antagonism. However, in other studies in patients with glomerulosclerosis, despite oxidative damage to urinary protein and lipids being reduced with additional AngII receptor blockade, no changes were seen in malondialdehyde and other plasma oxidative markers. Indeed, these studies have led to the hypothesis that urinary measurements of markers of oxidative damage to lipids and proteins appear to be more sensitive biomarkers of disease than plasma measurements in patients with CKD [55].

Renal concentrations of AngII also have effects on sodium and water resorption from the proximal tubules, which directly inhibit renin secretion from granular cells of the JG via the AT1 receptor. The downstream effects of modulation of renin secretion are discussed earlier in this chapter. Indeed, blockade of the AT1 receptor increases salt sensitivity in AngII-infused Sprague-Dawley rats, which is reversed by diuretics. However, although AT1 blockade lowered the activity of NADPH oxidase, it did not affect urinary markers of oxidative stress [56]. Interestingly this phenomenon has also been shown in Zucker obese hyperglycemic rats (ZDFn Gm-fa/fa), where the vasopeptidase inhibitor AVE7688 was much more effective than the AT1 antagonist losartan in lowering urinary excretion of oxidative and carbonyl stress–derived protein modifications [57]. The spontaneously hypertensive rat (SHR) also has increased oxidative stress and enhanced O_2 usage relative to tubular sodium transport, resulting in CKD. In this model, AngII acting via AT1-receptors (AT1-R) causes renal oxidative stress and functional NO deficiency that enhances O_2 usage [58]. Inefficient utilization of O_2 for Na^+ transport in the SHR kidney can also be ascribed to the effects of AT1 receptor and appear to be mostly independent of blood pressure [58].

The AT2 receptor subtype is characterized by its high affinity to certain compounds such as PD123319, PD123177, and CGP42112. The AT2 receptor gene is located on the X chromosome and AT2 receptor shares only 34% of sequence homology with the AT1 receptor, despite having equal binding affinity for AngII [47]. It adult kidneys, the AT2 receptor is localized primarily to glomeruli but is also found at low levels in cortical tubules and interstitial cells [59]. Since the AT2 receptor is highly abundant in fetal tissues, it is believed to play an important role in nephrogenesis. Indeed, mouse strains deficient in AT2 receptors show defective apoptosis of undifferentiated mesenchymal cells in the area surrounding the developing kidney and urinary tract. This abnormal apoptosis may well interfere with the normal interaction between the ureteral bud and metanephric blastema and is hypothesized to play a key role in the development of congenital obstructive nephropathies and uropathies [60]. Additional studies in humans indicate that many infants with congenital anomalies of the kidney and urinary tract have a significant mutation of the *AT2* gene [61].

Although AT2 deficient mice develop normally, their drinking response is impaired and they exhibit increased vasopressor response to AngII and elevated blood pressure [62, 63]. They also have impaired production of NO [64]. AT2 deficient mice are also, in general, more susceptible to atherosclerosis [65] and renal disease [66] and have enhanced production of ROS, which interestingly appears to be via activation of NADPH oxidase [65]. However, this situation may be more complex, with a recent study by our group suggesting that in the diabetic setting, the AT2 receptor may be proatherosclerotic, based on studies involving not only pharmacological blockade but also studies in AT2 receptor knockout mice [67].

7 Conclusion

Over time, it has become increasingly apparent that there is not just one effector of the RAS, namely AngII, as was first postulated. Indeed, the complexity of the RAS, in particular the local renal RAS and its contribution to CKD, is clearly evident. It is also well documented that both the activation of the RAS via oxidative stress, in addition to the production of ROS and dampening of antioxidant defenses by RAS components each have important pathological roles in the development of CKD. Until these are fully elucidated, it will remain difficult to ascertain exactly which targets are likely to be most useful as synergistic additions to RAS blockade for the treatment of CKDs such as diabetic nephropathy.

References

1. Burrell LM, Johnston CI, Tikellis C, Cooper ME: ACE2, a new regulator of the renin-angiotensin system. *Trends Endocrinol Metab* 15:166–169, 2004
2. Nangaku M, Fujita T: Activation of the renin-angiotensin system and chronic hypoxia of the kidney. *Hypertens Res* 31:175–184, 2008
3. de Cavanagh EM, Ferder L, Toblli JE, Piotrkowski B, Stella I, Fraga CG, Inserra F: Renal mitochondrial impairment is attenuated by AT1 blockade in experimental type I diabetes. *Am J Physiol Heart Circ Physiol* 294:H456–H465, 2008

4. Coughlan MT, Thallas-Bonke V, Pete J, Long DM, Gasser A, Tong DC, Arnstein M, Thorpe SR, Cooper ME, Forbes JM: Combination therapy with the advanced glycation end product cross-link breaker, alagebrium, and angiotensin converting enzyme inhibitors in diabetes: synergy or redundancy? *Endocrinology* 148:886–895, 2007
5. Coughlan MT, Thorburn DR, Penfold SA, Laskowski A, Harcourt BE, Sourris KC, Tan AL, Fukami K, Thallas-Bonke V, Nawroth PP, Brownlee M, Bierhaus A, Cooper ME, Forbes JM: RAGE-induced cytosolic ROS promote mitochondrial superoxide generation in diabetes. *J Am Soc Nephrol* 20(4):742–752, 2009
6. Parving HH, Persson F, Lewis JB, Lewis EJ, Hollenberg NK: Aliskiren combined with losartan in type 2 diabetes and nephropathy. *N Engl J Med* 358:2433–2446, 2008
7. Lee HY, Oh BH: Cardio-renal protection with aliskiren, a direct renin inhibitor, in the ASPIRE HIGHER program. *Expert Rev Cardiovasc Ther* 7:251–257, 2009
8. Parving HH, Brenner BM, McMurray JJ, de Zeeuw D, Haffner SM, Solomon SD, Chaturvedi N, Ghadanfar M, Weissbach N, Xiang Z, Armbrecht J, Pfeffer MA: Aliskiren trial in type 2 diabetes using cardio-renal endpoints (ALTITUDE): rationale and study design. *Nephrol Dial Transplant* 24:1663–1671, 2009
9. Fukamizu A, Takahashi S, Seo MS, Tada M, Tanimoto K, Uehara S, Murakami K: Structure and expression of the human angiotensinogen gene. Identification of a unique and highly active promoter. *J Biol Chem* 265:7576–7582, 1990
10. Ingelfinger JR, Zuo WM, Fon EA, Ellison KE, Dzau VJ: In situ hybridization evidence for angiotensinogen messenger RNA in the rat proximal tubule. An hypothesis for the intrarenal renin angiotensin system. *J Clin Invest* 85:417–423, 1990
11. Terada Y, Tomita K, Nonoguchi H, Marumo F: PCR localization of angiotensin II receptor and angiotensinogen mRNAs in rat kidney. *Kidney Int* 43:1251–1259, 1993
12. Kobori H, Ohashi N, Katsurada A, Miyata K, Satou R, Saito T, Yamamoto T: Urinary angiotensinogen as a potential biomarker of severity of chronic kidney diseases. *J Am Soc Hypertens* 2:349–354, 2008
13. Kobori H, Harrison-Bernard LM, Navar LG: Urinary excretion of angiotensinogen reflects intrarenal angiotensinogen production. *Kidney Int* 61:579–585, 2002
14. Saito T, Urushihara M, Kotani Y, Kagami S, Kobori H: Increased urinary angiotensinogen is precedent to increased urinary albumin in patients with type 1 diabetes. *Am J Med Sci* 338:478–480, 2009
15. Kobori H, Alper AB, Jr., Shenava R, Katsurada A, Saito T, Ohashi N, Urushihara M, Miyata K, Satou R, Hamm LL, Navar LG: Urinary angiotensinogen as a novel biomarker of the intrarenal renin-angiotensin system status in hypertensive patients. *Hypertension* 53:344–350, 2009
16. Didion SP, Ryan MJ, Baumbach GL, Sigmund CD, Faraci FM: Superoxide contributes to vascular dysfunction in mice that express human renin and angiotensinogen. *Am J Physiol Heart Circ Physiol* 283:H1569–H1576, 2002
17. Griendling KK, Murphy TJ, Alexander RW: Molecular biology of the renin-angiotensin system. *Circulation* 87:1816–1828, 1993
18. Hackenthal E, Paul M, Ganten D, Taugner R: Morphology, physiology, and molecular biology of renin secretion. *Physiol Rev* 70:1067–1116, 1990
19. Burns KD, Homma T, Harris RC: The intrarenal renin-angiotensin system. *Semin Nephrol* 13:13–30, 1993
20. Gruden G, Setti G, Hayward A, Sugden D, Duggan S, Burt D, Buckingham RE, Gnudi L, Viberti G: Mechanical stretch induces monocyte chemoattractant activity via an NF-kappaB-dependent monocyte chemoattractant protein-1-mediated pathway in human mesangial cells: inhibition by rosiglitazone. *J Am Soc Nephrol* 16:688–696, 2005
21. Gruden G, Thomas S, Burt D, Lane S, Chusney G, Sacks S, Viberti G: Mechanical stretch induces vascular permeability factor in human mesangial cells: mechanisms of signal transduction. *Proc Natl Acad Sci U S A* 94:12112–12116, 1997

22. Whaley-Connell AT, Nistala R, Habibi J, Hayden MR, Schneider RI, Johnson MS, Tilmon R, Rehmer N, Ferrario CM, Sowers JR: Comparative effect of direct renin inhibition and AT1R blockade on glomerular filtration barrier injury in the transgenic Ren2 rat. *Am J Physiol Renal Physiol* 298(3):F655–F661, 2009
23. Pinto R, Gradman AH: Direct renin inhibition: an update. *Curr Hypertens Rep* 11:456–462, 2009
24. Nguyen G, Delarue F, Burckle C, Bouzhir L, Giller T, Sraer JD: Pivotal role of the renin/prorenin receptor in angiotensin II production and cellular responses to renin. *J Clin Invest* 109:1417–1427, 2002
25. Ichihara A, Hayashi M, Kaneshiro Y, Suzuki F, Nakagawa T, Tada Y, Koura Y, Nishiyama A, Okada H, Uddin MN, Nabi AH, Ishida Y, Inagami T, Saruta T: Inhibition of diabetic nephropathy by a decoy peptide corresponding to the "handle" region for nonproteolytic activation of prorenin. *J Clin Invest* 114:1128–1135, 2004
26. Imanishi T, Tsujioka H, Ikejima H, Kuroi A, Takarada S, Kitabata H, Tanimoto T, Muragaki Y, Mochizuki S, Goto M, Yoshida K, Akasaka T: Renin inhibitor aliskiren improves impaired nitric oxide bioavailability and protects against atherosclerotic changes. *Hypertension* 52:563–572, 2008
27. Lorenz JN, Greenberg SG, Briggs JP: The macula densa mechanism for control of renin secretion. *Semin Nephrol* 13:531–542, 1993
28. Suematsu N, Ojaimi C, Recchia FA, Wang Z, Skayian Y, Xu X, Zhang S, Kaminski P, Sun D, Wolin MS, Kaley G, Hintze TH: Potential mechanisms of low sodium diet-induced cardiac disease. superoxide-NO in the heart. *Circ Res* 106(3):593–600, 2009
29. Sousa T, Pinho D, Morato M, Marques-Lopes J, Fernandes E, Afonso J, Oliveira S, Carvalho F, Albino-Teixeira A: Role of superoxide and hydrogen peroxide in hypertension induced by an antagonist of adenosine receptors. *Eur J Pharmacol* 588:267–276, 2008
30. Kamp O, Sieswerda GT, Visser CA: Comparison of effects on systolic and diastolic left ventricular function of nebivolol versus atenolol in patients with uncomplicated essential hypertension. *Am J Cardiol* 92:344–348, 2003
31. Whaley-Connell A, Habibi J, Johnson M, Tilmon R, Rehmer N, Rehmer J, Wiedmeyer C, Ferrario CM, Sowers JR: Nebivolol reduces proteinuria and renal NADPH oxidase-generated reactive oxygen species in the transgenic Ren2 rat. *Am J Nephrol* 30:354–360, 2009
32. Lai EY, Fahling M, Ma Z, Kallskog O, Persson PB, Patzak A, Persson AE, Hultstrom M: Norepinephrine increases calcium sensitivity of mouse afferent arteriole, thereby enhancing angiotensin II-mediated vasoconstriction. *Kidney Int* 76:953–959, 2009
33. Campese VM, Shaohua Y, Huiquin Z: Oxidative stress mediates angiotensin II-dependent stimulation of sympathetic nerve activity. *Hypertension* 46:533–539, 2005
34. Ye S, Zhong H, Yanamadala S, Campese VM: Oxidative stress mediates the stimulation of sympathetic nerve activity in the phenol renal injury model of hypertension. *Hypertension* 48:309–315, 2006
35. Yusuf S, Teo KK, Pogue J, Dyal L, Copland I, Schumacher H, Dagenais G, Sleight P, Anderson C: Telmisartan, ramipril, or both in patients at high risk for vascular events. *N Engl J Med* 358:1547–1559, 2008
36. Hobo A, Yuzawa Y, Kosugi T, Kato N, Asai N, Sato W, Maruyama S, Ito Y, Kobori H, Ikematsu S, Nishiyama A, Matsuo S, Kadomatsu K: The growth factor midkine regulates the renin-angiotensin system in mice. *J Clin Invest* 119:1616–1625, 2009
37. Oudit GY, Liu GC, Zhong J, Basu R, Chow FL, Zhou J, Loibner H, Janzek E, Schuster M, Penninger JM, Herzenberg AM, Kassiri Z, Scholey JW: Human recombinant angiotensin converting enzyme 2 reduces the progression of diabetic nephropathy. *Diabetes* 59 (2):529–538, 2009
38. Gwathmey TM, Pendergrass KD, Reid SD, Rose JC, Diz DI, Chappell MC: Angiotensin-(1-7)-angiotensin-converting enzyme 2 attenuates reactive oxygen species formation to angiotensin II within the cell nucleus. *Hypertension* 55:166–171

39. Lorenzo O, Ruiz-Ortega M, Suzuki Y, Ruperez M, Esteban V, Sugaya T, Egido J: Angiotensin III activates nuclear transcription factor-kappaB in cultured mesangial cells mainly via AT(2) receptors: studies with AT(1) receptor-knockout mice. *J Am Soc Nephrol* 13:1162–1171, 2002
40. Ruiz-Ortega M, Lorenzo O, Egido J: Angiotensin III increases MCP-1 and activates NF-kappaB and AP-1 in cultured mesangial and mononuclear cells. *Kidney Int* 57:2285–2298, 2000
41. Kerins DM, Hao Q, Vaughan DE: Angiotensin induction of PAI-1 expression in endothelial cells is mediated by the hexapeptide angiotensin IV. *J Clin Invest* 96:2515–2520, 1995
42. Ruiz-Ortega M, Esteban V, Egido J: The regulation of the inflammatory response through nuclear factor-kappaB pathway by angiotensin IV extends the role of the renin angiotensin system in cardiovascular diseases. *Trends Cardiovasc Med* 17:19–25, 2007
43. Lee EA, Seo JY, Jiang Z, Yu MR, Kwon MK, Ha H, Lee HB: Reactive oxygen species mediate high glucose-induced plasminogen activator inhibitor-1 up-regulation in mesangial cells and in diabetic kidney. *Kidney Int* 67:1762–1771, 2005
44. Benter IF, Yousif MH, Dhaunsi GS, Kaur J, Chappell MC, Diz DI: Angiotensin-(1-7) prevents activation of NADPH oxidase and renal vascular dysfunction in diabetic hypertensive rats. *Am J Nephrol* 28:25–33, 2008
45. Esteban V, Heringer-Walther S, Sterner-Kock A, de Bruin R, van den Engel S, Wang Y, Mezzano S, Egido J, Schultheiss HP, Ruiz-Ortega M, Walther T: Angiotensin-(1-7) and the g protein-coupled receptor MAS are key players in renal inflammation. *PLoS One* 4: e5406, 2009
46. Shao Y, He M, Zhou L, Yao T, Huang Y, Lu LM: Chronic angiotensin (1-7) injection accelerates STZ-induced diabetic renal injury. *Acta Pharmacol Sin* 29:829–837, 2008
47. de Gasparo M, Husain A, Alexander W, Catt KJ, Chiu AT, Drew M, Goodfriend T, Harding JW, Inagami T, Timmermans PB: Proposed update of angiotensin receptor nomenclature. *Hypertension* 25:924–927, 1995
48. Chabrashvili T, Kitiyakara C, Blau J, Karber A, Aslam S, Welch WJ, Wilcox CS: Effects of Ang II type 1 and 2 receptors on oxidative stress, renal NADPH oxidase, and SOD expression. *Am J Physiol Regul Integr Comp Physiol* 285:R117–R124, 2003
49. Vaziri ND, Bai Y, Ni Z, Quiroz Y, Pandian R, Rodriguez-Iturbe B: Intra-renal angiotensin II/AT1 receptor, oxidative stress, inflammation, and progressive injury in renal mass reduction. *J Pharmacol Exp Ther* 323:85–93, 2007
50. Manucha W, Carrizo L, Ruete C, Molina H, Valles P: Angiotensin II type I antagonist on oxidative stress and heat shock protein 70 (HSP 70) expression in obstructive nephropathy. *Cell Mol Biol (Noisy-le-grand)* 51:547–555, 2005
51. Sugiyama H, Kobayashi M, Wang DH, Sunami R, Maeshima Y, Yamasaki Y, Masuoka N, Kira S, Makino H: Telmisartan inhibits both oxidative stress and renal fibrosis after unilateral ureteral obstruction in acatalasemic mice. *Nephrol Dial Transplant* 20:2670–2680, 2005
52. Forbes JM, Coughlan MT, Cooper ME: Oxidative stress as a major culprit in kidney disease in diabetes. *Diabetes* 57:1446–1454, 2008
53. Cooper ME: Interaction of metabolic and haemodynamic factors in mediating experimental diabetic nephropathy. *Diabetologia* 44:1957–1972, 2001
54. Kuo HT, Kuo MC, Chiu YW, Chang JM, Guh JY, Chen HC: Increased glomerular and extracellular malondialdehyde levels in patients and rats with focal segmental glomerulosclerosis. *Eur J Clin Invest* 35:245–250, 2005
55. Agarwal R: Proinflammatory effects of oxidative stress in chronic kidney disease: role of additional angiotensin II blockade. *Am J Physiol Renal Physiol* 284:F863–F869, 2003
56. Endo S, Mori T, Yoneki Y, Nakamichi T, Hosoya T, Ogawa S, Tokudome G, Hosoya T, Miyata T, Ito S: Blockade of angiotensin II type-1 receptor increases salt sensitivity in Sprague-Dawley rats. *Hypertens Res* 32:513–519, 2009
57. Portero-Otin M, Pamplona R, Boada J, Jove M, Gonzalo H, Buleon M, Linz W, Schafer S, Tack I, Girolami JP: Inhibition of renin angiotensin system decreases renal protein oxidative damage in diabetic rats. *Biochem Biophys Res Commun* 368:528–535, 2008

58. Welch WJ, Baumgartl H, Lubbers D, Wilcox CS: Renal oxygenation defects in the spontaneously hypertensive rat: role of AT1 receptors. *Kidney Int* 63:202–208, 2003
59. Ozono R, Wang ZQ, Moore AF, Inagami T, Siragy HM, Carey RM: Expression of the subtype 2 angiotensin (AT2) receptor protein in rat kidney. *Hypertension* 30:1238–1246, 1997
60. Pope JCt, Brock JW, 3rd, Adams MC, Stephens FD, Ichikawa I: How they begin and how they end: classic and new theories for the development and deterioration of congenital anomalies of the kidney and urinary tract, CAKUT. *J Am Soc Nephrol* 10:2018–2028, 1999
61. Pope JCt, Nishimura H, Ichikawa I: Role of angiotensin in the development of the kidney and urinary tract. *Nephrologie* 19:433–436, 1998
62. Ichiki T, Labosky PA, Shiota C, Okuyama S, Imagawa Y, Fogo A, Niimura F, Ichikawa I, Hogan BL, Inagami T: Effects on blood pressure and exploratory behaviour of mice lacking angiotensin II type-2 receptor. *Nature* 377:748–750, 1995
63. Siragy HM, Carey RM: The subtype 2 (AT2) angiotensin receptor mediates renal production of nitric oxide in conscious rats. *J Clin Invest* 100:264–269, 1997
64. Siragy HM, Inagami T, Ichiki T, Carey RM: Sustained hypersensitivity to angiotensin II and its mechanism in mice lacking the subtype-2 (AT2) angiotensin receptor. *Proc Natl Acad Sci U S A* 96:6506–6510, 1999
65. Iwai M, Chen R, Li Z, Shiuchi T, Suzuki J, Ide A, Tsuda M, Okumura M, Min LJ, Mogi M, Horiuchi M: Deletion of angiotensin II type 2 receptor exaggerated atherosclerosis in apolipoprotein E-null mice. *Circulation* 112:1636–1643, 2005
66. Sourris KC, Morley AL, Koitka A, Samuel P, Coughlan MT, Penfold SA, Thomas MC, Bierhaus A, Nawroth PP, Yamamoto H, Allen TJ, Walther T, Hussain T, Cooper ME, Forbes JM. Receptor for AGEs (RAGE) blockade may exert its renoprotective effects in patients with diabetic nephropathy via induction of the angiotensin II type 2 (AT2) receptor. Diabetologia. 2010 Jul 15. [Epub ahead of print]
67. Koitka A, Cao Z, Koh P, Watson AM, Sourris KC, Loufrani L, Soro-Paavonen A, Walther T, Woollard KJ, Jandeleit-Dahm KA, Cooper ME, Allen TJ: Angiotensin II subtype 2 receptor blockade and deficiency attenuate the development of atherosclerosis in an apolipoprotein E-deficient mouse model of diabetes. *Diabetologia* 53(3):584–592, 2009

Chapter 17
Oxidative Stress in Kidney Injury: Peroxisome Proliferator-Activated Receptor-γ Agonists Are in Control

Li-Jun Ma and Agnes B. Fogo

Abstract The thiazolidinediones (TZDs) are agonists for peroxisome proliferator-activated receptor-γ (PPARγ), and they promote insulin sensitization and improve dyslipidemia in patients with type 2 diabetes mellitus. PPARγ is widely expressed, both in circulating and renal parenchymal cells. The TZDs are widely used clinically to improve metabolic syndrome, but may also have beneficial effects on progressive renal damage beyond diabetes. PPARγ agonists also have direct benefits on progressive renal injury independent of altering the metabolic profile, including effects on inflammation and oxidative stress. We will review selected evidence of such actions of PPARγ agonists beyond metabolism and examine interactions with other key pro-fibrotic mediators, including transforming growth factor-β (TGF-β) and the renin-angiotensin system (RAS).

Keywords Thiazolidinedione · Chronic kidney disease · Diabetes · Renin-angiotensin system · Aging · Inflammation

1 Experimental Studies

1.1 Peroxisome Proliferator-Activated Receptor-γ Agonists in Chronic Kidney Disease Animal Models

We investigated the potential benefits of peroxisome proliferator-activated receptor-γ (PPARγ) agonists in the 5/6 nephrectomy (5/6 Nx) rat model of chronic kidney disease (CKD) [1]. Hypertension, progressive proteinuria, segmental

A.B. Fogo (✉)
Division of Renal Pathology, Vanderbilt University School of Medicine,
C-3310 Medical Center North Department of Pathology,
Vanderbilt University Medical Center, Nashville, TN 37232-2561, USA
e-mail: agnes.fogo@vanderbilt.edu

T. Miyata et al. (eds.), *Studies on Renal Disorders*, Oxidative Stress in Applied Basic Research and Clinical Practice, DOI 10.1007/978-1-60761-857-7_17,

glomerulosclerosis, and tubulointerstitial fibrosis develop in this model without dysmetabolic features. We treated the 5/6 Nx rats with the PPARγ agonist troglitazone from the onset of 5/6 nephrectomy. Troglitazone ameliorated the progression of glomerulosclerosis, despite only a modest antihypertensive effect. Nonspecific antihypertensive treatment alone (reserpine, hydrochlorothiazide, and hydralazine) did not affect sclerosis in this model, but provided better blood pressure control and augmented the protective effects of troglitazone on sclerosis. These beneficial effects were linked to regulation of glomerular cell proliferation, hypertrophy, and decreased expression of the pro-fibrotic mediators plasminogen activator inhibitor-1 (PAI-1) and transforming growth factor-β (TGF-β). Troglitazone also attenuated both glomerulosclerosis and aortic medial thickening in spontaneously hypertensive rats with added kidney injury induced by the 5/6 nephrectomy model, despite only modest effects on hypertension [2].

Progression of chronic kidney disease not only correlates with increased pro-fibrotic mediators and accumulation of extracellular matrix, but also tubular atrophy and rarefaction of peritubular capillaries. Loss of the microvasculature implies an important mechanism for hypoxia in CKD [3]. Hypoxia has direct effects on the expression of genes involved in fibrogenesis. In human renal fibroblasts, hypoxia simultaneously stimulated production of collagen alpha1, increased tissue inhibitor of metalloproteinase-1 (TIMP-1), but decreased expression of collagenase [4]. Hypoxia aggravates kidney injury by both activating oxidative stress and by suppressing antioxidative pathways [5]. Hypoxia-inducible factor (HIF) is a transcription factor that regulates cellular hypoxic responses. Treatment for 4 weeks with dimethyloxalylglycine (DMOG) to activate HIF attenuated renal injury in the rat remnant kidney. This protection was associated with decreased plasma malondialdehyde (MDA), a marker of oxidative stress, and increased renal expression of catalase, an antioxidant [6].

Hypoxia-induced inflammatory responses in the diseased kidney is also mediated through its inhibition of PPARγ. Hypoxia reduced PPARγ expression in human proximal renal tubular epithelial cells. Further, hypoxia inhibited reduction of monocyte chemoattractant protein-1 (MCP-1) production by PPARγ's activation [7]. On the other hand, rosiglitazone attenuated chronic hypoxia-induced pulmonary hypertension in a mouse model [8].

Thiazolidinediones (TZDs) also had beneficial effects in a nonhypertensive CKD model with primary podocyte injury, the puromycin aminonucleoside nephropathy (PAN) model [9]. We had previously observed increased PPARγ expression in podocytes in sclerotic conditions, both in human disease and animal models, including diabetic nephropathy and chronic allograft nephropathy [10, 11]. However, although increased PPARγ was associated with sclerosis, we postulated that this increase could be counter-regulatory and a response to podocyte injury. These hypotheses were investigated further in parallel studies in vitro and in vivo. We studied the PAN model of primary podocyte injury, which is characterized by an acute phase of nephrotic syndrome from about day 5 to day 10 after administration of PAN, followed by progressive sclerosis and low-grade proteinuria [9]. We started treatment with the PPARγ agonist pioglitazone after the acute nephrotic syndrome. Untreated PAN rats showed increased PPARγ expression in podocytes, mesangial cells, and in glomeruli with segmental sclerosis, whereas WT-1 staining, a podocyte differentiation marker, was decreased. Pioglitazone treatment decreased sclerosis and restored podocyte differentiation. The latter was associated with increased vascular endothelial growth factor (VEGF) expression in glomeruli. The PPARγ angiopoietin-related gene, also called angiopoietin-like protein 4 (Aglp4), is a target of PPARγ and PPARα and hypoxia-inducible factor-1α, and inhibits actions of VEGF to promote angiogenesis in vivo. Aglp4 was increased in the kidneys of PAN rats, perhaps due to ischemia related to sclerosis. Treatment with PPARγ decreased Aglp4, which, along with restored VEGF, would promote angiogenesis and capillary growth, and thus decrease sclerosis. This hypothesis is supported by the increase in number of glomerular endothelial cells in PAN rats treated with pioglitazone compared to untreated PAN rats. PPARγ agonist also had effects on inflammation and extracellular matrix (ECM) modulators, with reduced infiltrating macrophages and PAI-1. Of note, TGF-β or TIMP-1 were not altered by pioglitazone in this model [9].

In our parallel in vitro studies, immortalized mouse podocytes expressed PPARγ at baseline. When injured with PAN [12], PPARγ expression was decreased. Pioglitazone treatment in these injured podocytes resulted in significantly increased PPARγ mRNA and activity, assessed by a PPRE3 – thymidine kinase (TK) – luciferase reporter assay, with amelioration of PAN-induced podocyte injury. The increased podocyte PPARγ induced by pioglitazone treatment not only restored podocyte differentiation but also suppressed apoptosis and necrosis induced by PAN, leading to restoration of balance of anti-apoptotic versus pro-apoptotic molecules. Thus, pioglitazone increased Bcl-x_L and p27 expression, both of which were decreased after injury. Activated caspase-3, an effector of apoptosis, was significantly increased by PAN and was decreased by pioglitazone. These studies show that PPARγ is normally expressed in podocytes and is activated after injury, an event that is protective against PAN-induced apoptosis and necrosis.

In addition to beneficial effects on insulin sensitization, both rosiglitazone and pioglitazone ameliorated development of albuminuria and renal injury in rodent models of type 2 diabetes [13]. Interestingly, although metformin resulted in similar glycemic control, there was no decrease in microalbuminuria, supporting that additional nonmetabolic actions contribute to this effect. TZDs clearly have direct effects on the kidney parenchyma, including beneficial effects on podocytes, as detailed above. TZDs may also affect endothelial cells indirectly by activating VEGF or decreasing Aglp4. Mesangial cell, fibroblast, and infiltrating macrophage functions are also modulated by PPARγ agonists (see below). These results point to cell and injury-specific interactions of PPARγ agonist with key modulators of fibrosis.

1.2 Peroxisome Proliferator-Activated Receptor-γ and Transforming Growth Factor-β

CKD has largely been targeted by aiming for blockade of components of the renin-angiotensin system (RAS). Transforming growth factor-β is also widely increased in progressive renal disease, plays an important role in mediating epithelial-to-mesenchymal transdifferentiation (EMT), and is thus an additional attractive target. Importantly, these molecules interact with PAI-1, which inhibits both fibrinolysis and proteolysis. Thus, angiotensin results in increased TGF-β and PAI-1, and TGF-β may also induce PAI-1. However, attempts to inhibit TGF-β have shown the complexity of its actions. Transforming growth factor-β is not only pro-fibrotic, it also modulates immune function. Thus, a high dose of pan-anti-TGF-β antibody failed to protect against injury in the PAN nephropathy model, whereas a low dose was protective, perhaps related to suppression of TGF-β immune-modulatory effects and increased macrophage infiltration consequent to high-dose inhibition [14]. However, combination antagonism of TGF-β and the RAS with an anti-TGF-β antibody and an angiotensin-converting enzyme (ACE) inhibitor did provide added benefit by diminishing renal damage and normalizing proteinuria in diabetic rats [15]. The possibility that TZDs might directly or indirectly down-regulate TGF-β is therefore of great interest.

Studies in experimental diabetic models show that TZDs could decrease TGF-β and inflammation, likely contributed to by inhibition of nuclear factor-κB (NFκB) [16]. Additional studies in an obese type 2 diabetic rat model showed that pioglitazone had superior renoprotective effects to insulin, resulting in less proteinuria, improved renal function, and less histologic injury [17]. In vitro studies indicated that insulin itself could enhance TGF-β mRNA and protein from cultured rat proximal tubular cells. These data identify an important potential mechanism whereby hyperinsulinemia could drive progressive renal injury and, conversely, how TZDs may be renoprotective [17].

Of note, we found that TZDs did not uniformly decrease TGF-β in all CKD rat models, with a decrease in the 5/6 nephrectomy model, but with no significant decrease in vivo in the puromycin model. Contrasting the lack of in vivo effects of pioglitazone on overall cortical expression of TGF-β, in cultured podocytes, PPARγ agonist decreased the augmented TGF-β induced by PAN. The results further indicate that the protective effect of PPARγ agonist in podocytes may be mediated in part by effects of PPARγ on p27 and TGF-β expression. In vitro studies of human mesangial cells showed that TGF-β increased ECM synthesis and down-regulated PPARγ, effects that were reversed by PPARγ agonist treatment [18]. Mesangial cell proliferation in vitro in response to platelet-derived growth factor was also inhibited by the TZD rosiglitazone [19]. Similarly, PPARγ agonist inhibited proliferation of renal fibroblasts in vitro and decreased ECM accumulation, without affecting TGF-β or TIMP activity [20]. PPARγ agonists blocked the Smad downstream signaling pathway induced by TGF-β in cultured renal fibroblasts, likely contributed to by the induction of antifibrotic hepatocyte growth factor [21].

New evidence suggests that PPARγ blocks Smad-mediated transcriptional responses by preventing p300 recruitment and histone H4 hyperacetylation, resulting in the inhibition of TGF-β-induced collagen gene expression [22].

1.3 Peroxisome Proliferator-Activated Receptor-γ and the Renin–Angiotensin System

PPARγ agonist may also interact with angiotensin type 1 receptor blockers (ARBs). Some ARBs, particularly telmisartan, function as PPARγ ligands, although their modulation of PPARγ is weak [23]. Renin-angiotensin system and PPARγ interactions were further explored in the low-density lipoprotein (LDL) receptor-knockout mouse model of atherosclerosis [24]. Chimeric LDL receptor−/− mice were created by bone marrow transplantation, with wild-type bone marrow or macrophage-specific PPARγ-knockout bone marrow. The LDL receptor−/− mice that also had PPARγ-deficient bone marrow had worsened atherosclerotic lesions. Adding increased angiotensin exposure by exogenous infusion further enhanced the atherosclerotic lesions in these mice. Mice with PPARγ-replete bone marrow did not show enhanced lesions when challenged with exogenous angiotensin. Treatment with ARBs in this model significantly decreased atherosclerotic lesions [24]. These PPARγ-deficient macrophages showed increased expression of the angiotensin AT1 receptor. Further, the PPARγ-deficient macrophages functionally had a greater migratory response to angiotensin exposure, about 50% greater than in PPARγ-replete macrophages. Whether pharmacological inhibition, rather than genetic deletion, of PPARγ would also modulate the AT1 receptor expression remains to be determined.

Interestingly, the interactions between ARB and PPARγ occur not only in macrophages, but also in endothelial cells and adipocytes. In human cultured microvascular endothelial cells (ECs), telmisartan treatment suppressed expression of the receptor for advanced glycation end products (RAGE). The protection was prevented by a PPARγ inhibitor [25]. Endothelial progenitor cells promote the repair of damaged vascular endothelial cells and are intriguing targets in the treatment of not only ischemic diseases but also chronic kidney disease. Telmisartan induced proliferation of cultured human endothelial progenitor cells and might contribute to endothelial integrity in disease states. The actions of telmisartan on endothelial progenitor cells were mediated via the PPARγ-dependent PI3K/Akt pathway [26]. Using cultured preadipocytes, we and others have also shown that ARB induces in vitro adipogenesis through PPARγ activation [27, 28].

1.4 Peroxisome Proliferator-Activated Receptor-γ and Aging

Aging in most species is associated with impaired adaptive and homeostatic mechanisms, leading to susceptibility to environmental or internal stresses with increasing rates of disease and death. Oxidative stress, caused by an imbalance between oxidant production and antioxidants, is widely believed to be a central player in aging. Substantial evidence indicates PPARγ agonists also protect against age-related glomerulosclerosis [29].

We tested the impact of the PPARγ agonist, pioglitazone, on renal injury in an aging model in rats [30]. Our data show pioglitazone is renoprotective in aging by reducing proteinuria, improving glomerular filtration rate, decreasing sclerosis, and alleviating cell senescence, paralleled by increased local expression of PPARγ. The protective effects of pioglitazone are mediated by three important mechanisms, including increased Klotho, decreased mitochondrial injury and oxidative stress, and regulated $p66^{SHC}$ phosphorylation [30].

Klotho is a newly identified antiaging factor that functions through inhibition of the insulin/IGF-1 signaling pathway and reactive oxygen species (ROS). Klotho-deficient mice had a drastically shortened lifespan and developed aging-associated injury in many organs [31]. In contrast, overexpression of Klotho in mice increased lifespan [32]. In the kidney, Klotho is expressed predominantly in renal distal convoluted tubules. A recent study reported that PPARγ increases renal tubular Klotho mRNA and protein expression through two noncanonical PPARγ binding sites upstream of the Klotho gene [33]. We found that pioglitazone treatment in aging rats increased renal Klotho expression by more than 60%, another possible contributor to its effects on senescence and ROS [30].

1.5 Peroxisome Proliferator-Activated Receptor-γ Agonist and Oxidative Stress

Mitochondria are a major source of ROS in cells. Elevated levels of both oxidant-damaged DNA and protein are found in aged organisms. Since the mtDNA codes for some electron carriers, mtDNA damage may indirectly inhibit respiration and stimulate ROS formation. We found increased urinary and renal lipid oxidation in aging rats, and pioglitazone attenuated those aging-related changes. In our study, pioglitazone had a trend to decrease mtDNA deletion, and significantly increased mitochondrial inner membrane integrity [30].

Several mechanisms have been proposed for TZD's effect on mitochondria. First, pioglitazone increases mtDNA [34]. Second, PPARγ and nuclear respiratory factor 1 transcription factors share a common coactivator, PPAR coactivator 1α, and nuclear respiratory factor 1 can promote mitochondrial biogenesis [35]. Third, TZDs may affect coupling-uncoupling dynamics, which could in turn increase glucose utilization and influence free radical production/oxidative stress [36]. Lastly, several findings suggest that activation of PPARγ may exert an antioxidant activity by favorably altering the expression of specific enzymes participating in the production or elimination of reactive oxygen species in mitochondria, including nicotinamide adenine dinucleotide phosphate (NADPH) and catalase [37].

The gene $p66^{Shc}$ is newly recognized and associated with longevity in mice, and it also serves as an integration point for many signaling pathways that affect mitochondrial function [38, 39]. Once activated, $p66^{Shc}$ oxidizes cytochrome C and induces opening of the mitochondrial permeability transition pore, generates more ROS, and releases more pro-apoptotic factors into the cytosol. Therefore, suppression of $p66^{Shc}$ activation confers protection against ROS-induced injury. Mice lacking $p66^{Shc}$ had extended lifespan that correlated with a decrease in mitochondrial metabolism and reactive oxygen species production [40]. Inhibiting or silencing protein kinase C-β protects cells against hydrogen peroxide (H_2O_2) challenge. In contrast, overexpression of protein kinase C-β reproduces the mitochondrial fragmentation and Ca^{2+} signaling defect in cells expressing $p66^{Shc}$, but not in cells lacking $p66^{Shc}$ [41]. Cells expressing a mutant form of $p66^{Shc}$ ($p66^{Shc}$S36A) that cannot be phosphorylated also lack the early mitochondrial response to protein kinase C-β activity, indicating a requirement for $p66^{Shc}$ phosphorylation. We showed that pioglitazone also decreased total cytochrome C oxidase activity, although not its expression [30]. More importantly, we found that phosphorylation of $p66^{SHC}$, a key modulator of mitochondrial function, was decreased in pioglitazone-treated rats.

1.6 Peroxisome Proliferator-Activated Receptor-γ Agonists and Macrophages

The protective mechanisms of PPARγ agonists are also mediated through their anti-inflammatory effects. Activation of PPARγ suppresses pro-inflammatory signaling pathways (including NFκB, AP-1, and STAT), and decreases production of proinflammatory cytokines and chemokines. In experimental models of diabetic nephropathy, PPARγ agonist ameliorates renal injury associated with anti-inflammatory effects, presumably mostly by altering macrophages [16, 42].

Macrophage infiltration into the kidney is a common feature of many kidney diseases [43, 44]. However, macrophages show significant heterogeneity in functions depending on the local microenvironment and stage of tissue injury [45, 46]. Macrophages are generally classified into two polarization states: M1 and M2. M1 or "classically activated" macrophages are induced by classical immune pathways, such as lipopolysaccharide (LPS) and interferon-γ (IFN-γ), and have important roles in inflammation by enhancing pro-inflammatory cytokine production, such as tumor necrosis factor-α (TNF-α), interleukin-1β (IL-1β), or IL-6. M2 or "alternatively activated" macrophages are generated by exposure to IL-4 and IL-13. M2 macrophages are important in resolution of inflammation and tissue repair through synthesis of high levels of anti-inflammatory cytokines IL-10 and IL-1 decoy receptor and have high endocytic clearance capacities [47–49].

Macrophage phenotypes and function are critical determinants of the balance of promoting tissue injury versus resolution of injury [48, 49]. Experimental approaches to inhibit proinflammatory macrophages have been successful in reducing kidney injury [50–53]. Strategies that modulate M2 macrophage function and phenotype have produced beneficial effects in models of experimental chronic inflammation and kidney injury [54–56].

Adipose tissue is a complex endocrine organ that synthesizes and releases proteins and adipokines, including PPARγ [57–65]. Macrophages recruited into adipose tissue of obese individuals participate in inflammation and contribute to insulin resistance [57, 66–75]. Obesity induces a macrophage phenotypic switch [76], leading to more M1 activation in adipose tissue of obese animals [47, 73, 76, 77]. An elegant study using macrophage-specific PPARγ deficient mice showed recently that PPARγ is a key transcriptional factor in controlling maturation of alternatively activated macrophages [78]. Disruption of PPARγ in macrophages impairs alternative macrophage activation and predisposes animals to development of diet-induced obesity and insulin resistance [78]. By contrast, activation of PPARγ with rosiglitazone promotes infiltration of alternatively activated macrophages into adipose tissue [79]. Conditioned media from rosiglitazone-treated alternatively activated macrophages neutralized the inhibitory effect of macrophages on 3T3-L1 adipocyte differentiation, suggesting that alternatively activated macrophages may be involved in mediating the effects of rosiglitazone on adipogenesis [79]. Although there is increasing interest in the possibility that phenotypic alteration of macrophages modulates renal response to injury, there is no information on whether this is relevant to obesity-associated CKD. The impact of PPARγ in controlling and modulation of macrophage phenotypic switch in obesity-related CKD merits further investigation.

2 Human Studies

2.1 Peroxisome Proliferator-Activated Receptor-γ and Human Chronic Kidney Disease

Specific studies of effects of PPARγ agonists, either alone or in synergy with other interventions, on progression of human nondiabetic CKD are lacking. Additional complexity in considering such combination therapies arises from the recent controversy regarding possible increased, albeit nonsignificant, risk of death from cardiovascular cause in patients with type 2 diabetes treated with rosiglitazone [80–82]. Studies of TZDs in humans with CKD have been largely performed in diabetic nephropathy. Benefits in these patients on kidney disease progression could well be secondary to improvement of metabolic syndrome. Beneficial effects of TZDs observed include reduction of blood pressure and microalbuminuria, as reviewed by Sarafidis and Bakris [83]. Their analysis suggests that improvement in microalbuminuria likely does not merely reflect improved glycemic control, based on the observations that other active treatments that achieved equivalent blood glucose levels did not similarly reduce microalbuminuria. Other studies showed that improved glycemic control in patients with type 2 diabetes with ARB and PPARγ agonist combination resulted in better renal function, with decreased proteinuria and serum creatinine [84]. Some data suggest that ARBs may even influence development of diabetes or its severity, perhaps in part related to effects to decrease PAI-1 [85]. Effects of ARBs and TZDs are complex and could alter many potentially injurious factors that impact on oxidative stress, inflammation, and other pro-fibrotic stimuli. Short-term treatment for 3 months with rosiglitazone resulted in decreased free fatty acids and TNF-α and increased adiponectin in plasma of type 2 diabetic patients. These changes were each linked independently to decreased albuminuria [86]. Experimental studies and some studies in humans with diabetes, as reviewed above, suggest additional actions of PPARγ agonists on podocytes, mesangial cells, tubular cells, growth factors, inflammation, endothelial growth and function, and cell proliferation. Our recent data add beneficial effects on mitochondrial injury and oxidative stress to the actions of PPARγ agonists.

3 Conclusion

Peroxisome proliferator-activated receptor-γ has manifold effects beyond control of glucose and lipid metabolism. These include direct effects on key pathways of ECM turnover, inflammation, cell senescence, and ROS. Thus, PPARγ agonists may provide a novel therapeutic tool for treatment not only of progression of CKD, but also of underlying contributors to development of CKD.

References

1. Ma LJ, Marcantoni C, Linton MF, Fazio S, Fogo AB. Peroxisome proliferator-activated receptor-gamma agonist troglitazone protects against nondiabetic glomerulosclerosis in rats. Kidney Int 2001; 59: 1899–1910.
2. Yoshida K, Kohzuki M, Xu HL, Wu XM, Kamimoto M, Sato T. Effects of troglitazone and temocapril in spontaneously hypertensive rats with chronic renal failure. J Hypertens 2001; 19: 503–510.
3. Fine LG, Norman JT. Chronic hypoxia as a mechanism of progression of chronic kidney diseases: from hypothesis to novel therapeutics. Kidney Int 2008; 74: 867–872.
4. Norman JT, Clark IM, Garcia PL. Hypoxia promotes fibrogenesis in human renal fibroblasts. Kidney Int 2000; 58: 2351–2366.
5. Son D, Kojima I, Inagi R, Matsumoto M, Fujita T, Nangaku M. Chronic hypoxia aggravates renal injury via suppression of Cu/Zn-SOD: a proteomic analysis. Am J Physiol Renal Physiol 2008; 294: F62–F72.
6. Song YR, You SJ, Lee YM, Chin HJ, Chae DW, Oh YK, Joo KW, Han JS, Na KY. Activation of hypoxia-inducible factor attenuates renal injury in rat remnant kidney. Nephrol Dial Transplant 2009; 25(1): 77–85.
7. Li X, Kimura H, Hirota K, Sugimoto H, Kimura N, Takahashi N, Fujii H, Yoshida H. Hypoxia reduces the expression and anti-inflammatory effects of peroxisome proliferator-activated

receptor-gamma in human proximal renal tubular cells. Nephrol Dial Transplant 2007; 22: 1041–1051.
8. Nisbet RE, Bland JM, Kleinhenz DJ, Mitchell PO, Walp ER, Sutliff RL, Hart CM. Rosiglitazone attenuates chronic hypoxia-induced pulmonary hypertension in a mouse model. Am J Respir Cell Mol Biol 2009; 42(4):482–490.
9. Yang HC, Ma LJ, Ma J, Fogo AB. Peroxisome proliferator-activated receptor-gamma agonist is protective in podocyte injury-associated sclerosis. Kidney Int 2006; 69: 1756–1764.
10. Paueksakon P, Revelo MP, Ma LJ, Marcantoni C, Fogo AB. Microangiopathic injury and augmented PAI-1 in human diabetic nephropathy. Kidney Int 2002; 61: 2142–2148.
11. Revelo MP, Federspiel C, Helderman H, Fogo AB. Chronic allograft nephropathy: expression and localization of PAI-1 and PPAR-gamma. Nephrol Dial Transplant 2005; 20: 2812–2819.
12. Kanjanabuch T, Ma LJ, Chen J, Pozzi A, Guan Y, Mundel P, Fogo AB. PPAR-gamma agonist protects podocytes from injury. Kidney Int 2007; 71: 1232–1239.
13. Okada T, Wada J, Hida K, Eguchi J, Hashimoto I, Baba M, Yasuhara A, Shikata K, Makino H. Thiazolidinediones ameliorate diabetic nephropathy via cell cycle-dependent mechanisms. Diabetes 2006; 55: 1666–1677.
14. Ma LJ, Jha S, Ling H, Pozzi A, Ledbetter S, Fogo AB. Divergent effects of low versus high dose anti-TGF-beta antibody in puromycin aminonucleoside nephropathy in rats. Kidney Int 2004; 65: 106–115.
15. Benigni A, Zoja C, Corna D, Zatelli C, Conti S, Campana M, Gagliardini E, Rottoli D, Zanchi C, Abbate M, Ledbetter S, Remuzzi G. Add-on anti-TGF-beta antibody to ACE inhibitor arrests progressive diabetic nephropathy in the rat. J Am Soc Nephrol 2003; 14: 1816–1824.
16. Ohga S, Shikata K, Yozai K, Okada S, Ogawa D, Usui H, Wada J, Shikata Y, Makino H. Thiazolidinedione ameliorates renal injury in experimental diabetic rats through anti-inflammatory effects mediated by inhibition of NF-kappaB activation. Am J Physiol Renal Physiol 2007; 292: F1141–F1150.
17. Ohtomo S, Izuhara Y, Takizawa S, Yamada N, Kakuta T, van Ypersele de Strihou C, Miyata T. Thiazolidinediones provide better renoprotection than insulin in an obese, hypertensive type II diabetic rat model. Kidney Int 2007; 72: 1512–1519.
18. Maeda A, Horikoshi S, Gohda T, Tsuge T, Maeda K, Tomino Y. Pioglitazone attenuates TGF-beta(1)-induction of fibronectin synthesis and its splicing variant in human mesangial cells via activation of peroxisome proliferator-activated receptor (PPAR)gamma. Cell Biol Int 2005; 29: 422–428.
19. Nicholas SB, Kawano Y, Wakino S, Collins AR, Hsueh WA. Expression and function of peroxisome proliferator-activated receptor-gamma in mesangial cells. Hypertension 2001; 37: 722–727.
20. Zafiriou S, Stanners SR, Saad S, Polhill TS, Poronnik P, Pollock CA. Pioglitazone inhibits cell growth and reduces matrix production in human kidney fibroblasts. J Am Soc Nephrol 2005; 16: 638–645.
21. Li Y, Wen X, Spataro BC, Hu K, Dai C, Liu Y. Hepatocyte growth factor is a downstream effector that mediates the antifibrotic action of peroxisome proliferator-activated receptor-gamma agonists. J Am Soc Nephrol 2006; 17: 54–65.
22. Ghosh AK, Bhattacharyya S, Wei J, Kim S, Barak Y, Mori Y, Varga J. Peroxisome proliferator-activated receptor-gamma abrogates Smad-dependent collagen stimulation by targeting the p300 transcriptional coactivator. FASEB J 2009; 23: 2968–2977.
23. Erbe DV, Gartrell K, Zhang YL, Suri V, Kirincich SJ, Will S, Perreault M, Wang S, Tobin JF. Molecular activation of PPARgamma by angiotensin II type 1-receptor antagonists. Vascul Pharmacol 2006; 45: 154–162.
24. Suganuma E, Babaev VR, Fogo AB, Linton MF, Fazio S, Ichikawa I, Kon V. Macrophage deficient in peroxisome proliferator-activated receptor gamma (PPARgamma) accelarate angiotenin II (Ang II)-induced atherosclerosis. J Am Soc Nephrol 2004; 15: 5A.
25. Yamagishi S, Matsui T, Nakamura K, Takeuchi M, Inoue H. Telmisartan inhibits advanced glycation end products (AGEs)-elicited endothelial cell injury by suppressing AGE receptor

(RAGE) expression via peroxisome proliferator-activated receptor-gamma activation. Protein Pept Lett 2008; 15: 850–853.
26. Honda A, Matsuura K, Fukushima N, Tsurumi Y, Kasanuki H, Hagiwara N. Telmisartan induces proliferation of human endothelial progenitor cells via PPARgamma-dependent PI3K/Akt pathway. Atherosclerosis 2009; 205: 376–384.
27. Corsa B, Fogo AB, Ma LJ. Promotion of recruitment and differentiation of adipocytes are affected by both angiotensin type 1 and type 2 receptors. Diabetes 2009; 58: A450.
28. Janke J, Schupp M, Engeli S, Gorzelniak K, Boschmann M, Sauma L, Nystrom FH, Jordan J, Luft FC, Sharma AM. Angiotensin type 1 receptor antagonists induce human in-vitro adipogenesis through peroxisome proliferator-activated receptor-gamma activation. J Hypertens 2006; 24: 1809–1816
29. Zhang R, Zheng F. PPAR-gamma and aging: one link through klotho? Kidney Int 2008; 74: 702–704.
30. Yang HC, Deleuze S, Zuo Y, Potthoff SA, Ma LJ, Fogo AB. The PPAR{gamma} Agonist pioglitazone ameliorates aging-related progressive renal injury. J Am Soc Nephrol 2009; 20 (11): 2380–2388.
31. Kuro-o M. Klotho as a regulator of oxidative stress and senescence. Biol Chem 2008; 389: 233–241.
32. Kurosu H, Yamamoto M, Clark JD, Pastor JV, Nandi A, Gurnani P, McGuinness OP, Chikuda H, Yamaguchi M, Kawaguchi H, Shimomura I, Takayama Y, Herz J, Kahn CR, Rosenblatt KP, Kuro-o M. Suppression of aging in mice by the hormone Klotho. Science 2005; 309: 1829–1833.
33. Zhang H, Li Y, Fan Y, Wu J, Zhao B, Guan Y, Chien S, Wang N. Klotho is a target gene of PPAR-gamma. Kidney Int 2008; 74: 732–739.
34. Ghosh S, Patel N, Rahn D, McAllister J, Sadeghi S, Horwitz G, Berry D, Wang KX, Swerdlow RH. The thiazolidinedione pioglitazone alters mitochondrial function in human neuron-like cells. Mol Pharmacol 2007; 71: 1695–1702.
35. Scarpulla RC. Transcriptional activators and coactivators in the nuclear control of mitochondrial function in mammalian cells. Gene 2002; 286: 81–89.
36. Kelly LJ, Vicario PP, Thompson GM, Candelore MR, Doebber TW, Ventre J, Wu MS, Meurer R, Forrest MJ, Conner MW, Cascieri MA, Moller DE. Peroxisome proliferator-activated receptors gamma and alpha mediate in vivo regulation of uncoupling protein (UCP-1, UCP-2, UCP-3) gene expression. Endocrinology 1998; 139: 4920–4927.
37. Inoue I, Goto S, Matsunaga T, Nakajima T, Awata T, Hokari S, Komoda T, Katayama S. The ligands/activators for peroxisome proliferator-activated receptor alpha (PPARalpha) and PPARgamma increase Cu2+,Zn2+-superoxide dismutase and decrease p22phox message expressions in primary endothelial cells. Metabolism 2001; 50: 3–11.
38. Hajnoczky G, Hoek JB. Cell signaling. Mitochondrial longevity pathways. Science 2007; 315: 607–609.
39. Camougrand N, Rigoulet M. Aging and oxidative stress: studies of some genes involved both in aging and in response to oxidative stress. Respir Physiol 2001; 128: 393–401.
40. Giorgio M, Migliaccio E, Orsini F, Paolucci D, Moroni M, Contursi C, Pelliccia G, Luzi L, Minucci S, Marcaccio M, Pinton P, Rizzuto R, Bernardi P, Paolucci F, Pelicci PG. Electron transfer between cytochrome c and p66Shc generates reactive oxygen species that trigger mitochondrial apoptosis. Cell 2005; 122: 221–233.
41. Pinton P, Rimessi A, Marchi S, Orsini F, Migliaccio E, Giorgio M, Contursi C, Minucci S, Mantovani F, Wieckowski MR, Del Sal G, Pelicci PG, Rizzuto R. Protein kinase C beta and prolyl isomerase 1 regulate mitochondrial effects of the life-span determinant p66Shc. Science 2007; 315: 659–663.
42. Ko GJ, Kang YS, Han SY, Lee MH, Song HK, Han KH, Kim HK, Han JY, Cha DR. Pioglitazone attenuates diabetic nephropathy through an anti-inflammatory mechanism in type 2 diabetic rats. Nephrol Dial Transplant 2008; 23: 2750–2760.
43. Eddy AA. Interstitial macrophages as mediators of renal fibrosis. Exp Nephrol 1995; 3: 76–79.

44. Kluth DC, Erwig LP, Rees AJ. Multiple facets of macrophages in renal injury. Kidney Int 2004; 66: 542–557.
45. Gordon S, Taylor PR. Monocyte and macrophage heterogeneity. Nat Rev Immunol 2005; 5: 953–964.
46. Mantovani A, Sica A, Sozzani S, Allavena P, Vecchi A, Locati M. The chemokine system in diverse forms of macrophage activation and polarization. Trends Immunol 2004; 25: 677–686.
47. Gordon S. Alternative activation of macrophages. Nat Rev Immunol 2003; 3: 23–35.
48. Ricardo SD, van Goor H, Eddy AA. Macrophage diversity in renal injury and repair. J Clin Invest 2008; 118: 3522–3530.
49. Kluth DC. Pro-resolution properties of macrophages in renal injury. Kidney Int 2007; 72: 234–236.
50. Wang Y, Mahajan D, Tay YC, Bao S, Spicer T, Kairaitis L, Rangan GK, Harris DC. Partial depletion of macrophages by ED7 reduces renal injury in adriamycin nephropathy. Nephrology (Carlton) 2005; 10: 470–477.
51. Duffield JS, Tipping PG, Kipari T, Cailhier JF, Clay S, Lang R, Bonventre JV, Hughes J. Conditional ablation of macrophages halts progression of crescentic glomerulonephritis. Am J Pathol 2005; 167: 1207–1219.
52. Lim AK, Ma FY, Nikolic-Paterson DJ, Thomas MC, Hurst LA, Tesch GH. Antibody blockade of c-fms suppresses the progression of inflammation and injury in early diabetic nephropathy in obese db/db mice. Diabetologia 2009; 52: 1669–1679.
53. Usui HK, Shikata K, Sasaki M, Okada S, Matsuda M, Shikata Y, Ogawa D, Kido Y, Nagase R, Yozai K, Ohga S, Tone A, Wada J, Takeya M, Horiuchi S, Kodama T, Makino H. Macrophage scavenger receptor-a-deficient mice are resistant against diabetic nephropathy through amelioration of microinflammation. Diabetes 2007; 56: 363–372.
54. Kluth DC, Ainslie CV, Pearce WP, Finlay S, Clarke D, Anegon I, Rees AJ. Macrophages transfected with adenovirus to express IL-4 reduce inflammation in experimental glomerulonephritis. J Immunol 2001; 166: 4728–4736.
55. Wilson HM, Stewart KN, Brown PA, Anegon I, Chettibi S, Rees AJ, Kluth DC. Bone-marrow-derived macrophages genetically modified to produce IL-10 reduce injury in experimental glomerulonephritis. Mol Ther 2002; 6: 710–717.
56. Wang Y, Wang YP, Zheng G, Lee VW, Ouyang L, Chang DH, Mahajan D, Coombs J, Wang YM, Alexander SI, Harris DC. Ex vivo programmed macrophages ameliorate experimental chronic inflammatory renal disease. Kidney Int 2007; 72: 290–299.
57. Hotamisligil GS, Shargill NS, Spiegelman BM. Adipose expression of tumor necrosis factor-alpha: direct role in obesity-linked insulin resistance. Science 1993; 259: 87–91.
58. Zhang Y, Proenca R, Maffei M, Barone M, Leopold L, Friedman JM. Positional cloning of the mouse obese gene and its human homologue. Nature 1994; 372: 425–432.
59. Ahima RS, Flier JS. Adipose tissue as an endocrine organ. Trends Endocrinol Metab 2000; 11: 327–332.
60. Scherer PE, Williams S, Fogliano M, Baldini G, Lodish HF. A novel serum protein similar to C1q, produced exclusively in adipocytes. J Biol Chem 1995; 270: 26746–26749.
61. Evans RM, Barish GD, Wang YX. PPARs and the complex journey to obesity. Nat Med 2004; 10: 355–361.
62. Saltiel AR. You are what you secrete. Nat Med 2001; 7: 887–888.
63. Yamauchi T, Kamon J, Waki H, Terauchi Y, Kubota N, Hara K, Mori Y, Ide T, Murakami K, Tsuboyama-Kasaoka N, Ezaki O, Akanuma Y, Gavrilova O, Vinson C, Reitman ML, Kagechika H, Shudo K, Yoda M, Nakano Y, Tobe K, Nagai R, Kimura S, Tomita M, Froguel P, Kadowaki T. The fat-derived hormone adiponectin reverses insulin resistance associated with both lipoatrophy and obesity. Nat Med 2001; 7: 941–946.
64. Loskutoff DJ. A slice of PAI. J Clin Invest 1993; 92: 2563.

65. Liang X, Kanjanabuch T, Mao SL, Hao CM, Tang YW, Declerck PJ, Hasty AH, Wasserman DH, Fogo AB, Ma LJ. Plasminogen activator inhibitor-1 modulates adipocyte differentiation. Am J Physiol Endocrinol Metab 2006; 290: E103-E113.
66. Wellen KE, Hotamisligil GS. Obesity-induced inflammatory changes in adipose tissue. J Clin Invest 2003; 112: 1785–1788.
67. Neels JG, Olefsky JM. Inflamed fat: what starts the fire? J Clin Invest 2006; 116: 33–35.
68. Weisberg SP, McCann D, Desai M, Rosenbaum M, Leibel RL, Ferrante AW, Jr.. Obesity is associated with macrophage accumulation in adipose tissue. J Clin Invest 2003; 112: 1796–1808.
69. Xu H, Barnes GT, Yang Q, Tan G, Yang D, Chou CJ, Sole J, Nichols A, Ross JS, Tartaglia LA, Chen H. Chronic inflammation in fat plays a crucial role in the development of obesity-related insulin resistance. J Clin Invest 2003; 112: 1821–1830.
70. Cinti S, Mitchell G, Barbatelli G, Murano I, Ceresi E, Faloia E, Wang S, Fortier M, Greenberg AS, Obin MS. Adipocyte death defines macrophage localization and function in adipose tissue of obese mice and humans. J Lipid Res 2005; 46: 2347–2355.
71. Kanda H, Tateya S, Tamori Y, Kotani K, Hiasa K, Kitazawa R, Kitazawa S, Miyachi H, Maeda S, Egashira K, Kasuga M. MCP-1 contributes to macrophage infiltration into adipose tissue, insulin resistance, and hepatic steatosis in obesity. J Clin Invest 2006; 116: 1494–1505.
72. Weisberg SP, Hunter D, Huber R, Lemieux J, Slaymaker S, Vaddi K, Charo I, Leibel RL, Ferrante AW, Jr. CCR2 modulates inflammatory and metabolic effects of high-fat feeding. J Clin Invest 2006; 116: 115–124.
73. Lumeng CN, Deyoung SM, Bodzin JL, Saltiel AR. Increased inflammatory properties of adipose tissue macrophages recruited during diet-induced obesity. Diabetes 2007; 56: 16–23.
74. Inouye KE, Shi H, Howard JK, Daly CH, Lord GM, Rollins BJ, Flier JS. Absence of CC chemokine ligand 2 does not limit obesity-associated infiltration of macrophages into adipose tissue. Diabetes 2007; 56: 2242–2250.
75. Surmi BK, Hasty AH. Macrophage infiltration into adipose tissue: initiation, propagation and remodeling. Future Lipidol 2008; 3: 545–556.
76. Lumeng CN, Bodzin JL, Saltiel AR. Obesity induces a phenotypic switch in adipose tissue macrophage polarization. J Clin Invest 2007; 117: 175–184.
77. Lumeng CN, DelProposto JB, Westcott DJ, Saltiel AR. Phenotypic switching of adipose tissue macrophages with obesity is generated by spatiotemporal differences in macrophage subtypes. Diabetes 2008; 57: 3239–3246.
78. Odegaard JI, Ricardo-Gonzalez RR, Goforth MH, Morel CR, Subramanian V, Mukundan L, Eagle AR, Vats D, Brombacher F, Ferrante AW, Chawla A. Macrophage-specific PPAR-gamma controls alternative activation and improves insulin resistance. Nature 2007; 447: 1116–1120.
79. Stienstra R, Duval C, Keshtkar S, van der Laak J, Kersten S, Muller M. Peroxisome proliferator-activated receptor gamma activation promotes infiltration of alternatively activated macrophages into adipose tissue. J Biol Chem 2008; 283: 22620–22627.
80. Nissen SE, Wolski K. Effect of rosiglitazone on the risk of myocardial infarction and death from cardiovascular causes. N Engl J Med 2007; 356: 2457–2471.
81. Doggrell SA. Does rosiglitazone increase cardiovascular outcomes? Expert Opin Pharmacother 2007; 8: 2665–2669.
82. Hanefeld M. Is rosiglitazone associated with increased risk for cardiovascular events? Nat Clin Pract Cardiovasc Med 2007; 4: 648–649.
83. Sarafidis PA, Bakris GL. Protection of the kidney by thiazolidinediones: an assessment from bench to bedside. Kidney Int 2006; 70: 1223–1233.
84. Jin HM, Pan Y. Renoprotection provided by losartan in combination with pioglitazone is superior to renoprotection provided by losartan alone in patients with type 2 diabetic nephropathy. Kidney Blood Press Res 2007; 30: 203–211

85. Ma LJ, Mao SL, Taylor KL, Kanjanabuch T, Guan Y, Zhang Y, Brown NJ, Swift LL, McGuinness OP, Wasserman DH, Vaughan DE, Fogo AB. Prevention of obesity and insulin resistance in mice lacking plasminogen activator inhibitor 1. Diabetes 2004; 53: 336–346
86. Miyazaki Y, Cersosimo E, Triplitt C, DeFronzo RA. Rosiglitazone decreases albuminuria in type 2 diabetic patients. Kidney Int 2007; 72: 1367–1373.

Chapter 18
Current Therapy Targeting Oxidative Stress: Statin

Ravi Nistala and Adam Whaley-Connell

Abstract Oxidative stress is central to many pathological kidney processes. Redox states in the kidney determine the levels of oxidative stress as well as functional status of the various cell types in the kidney. Excessive generation of free radicals under the influence of certain cellular, extracellular, and environmental cues tilts the balance of redox toward damaging processes. For example, the renin-angiotensin-aldosterone system can be turned on excessively, leading to generation of free radicals in the form of superoxide anion, which is one of the potent oxidizing agents. Oxidation of lipids, proteins and even nitric oxide can lead to further generation of cell damaging products. Dyslipidemia has been shown to be intricately involved in the generation of oxidative stress. Statins improve dyslipidemia through the blockade of 3-hydroxy-3-methylglutaryl coenzyme A reductase and through decreased participation of downstream moieties in the generation of oxidative stress. In addition, statins affect the activation of transcription factors such as nuclear factor κB to mediate their anti-inflammatory and antifibrotic effects. This "pleiotropism" is important in contributing to statins overall salutary effects in chronic kidney disease.

Keywords Dyslipidemia · Chronic kidney disease · Oxidative stress

1 Introduction

Chronic kidney disease (CKD) is a significant global public health dilemma. The predicted future disease and economic burden from progression to end stage renal disease (ESRD) warrants interventions to effectively prevent or slow down this

R. Nistala (✉)
Division of Nephrology, Department of Internal Medicine, University of Missouri-Columbia School of Medicine, CE416, 5 Hospital Drive, Columbia, MO 65212, USA
e-mail: nistalar@health.missouri.edu

T. Miyata et al. (eds.), *Studies on Renal Disorders*, Oxidative Stress in Applied Basic Research and Clinical Practice, DOI 10.1007/978-1-60761-857-7_18,

course. Chronic kidney disease is associated with both traditional risk factors such as dyslipidemia, hypertension, diabetes, cardiovascular disease (CVD), and with nontraditional risk factors such as oxidative stress, inflammation, endothelial dysfunction, and proteinuria [1]. Tight control of traditional risk factors such as diabetes and hypertension has been shown to be associated with maximal risk reduction for CKD progression. Reduction of dyslipidemia is associated with improved renal endpoints in select patient populations [1]. However, the choice of patient population, intervention chosen, and coexistence of multiple risk factors have minimized the beneficial effects of 3-hydroxy-3-methylglutaryl coenzyme A reductase inhibitors (HMG-CoA inhibitors, statins) [1]. Statins not only block the formation of the end product cholesterol but also enhance the uptake of formed cholesterol by upregulating low-density lipoprotein cholesterol (LDL-C) receptors (LDLR). In addition, the formation of intermediates including geranyl and farnesyl phosphates is blocked, which are important moieties mediating oxidative stress (Fig. 1). These extra-lipid lowering effects have been termed "pleiotropic effects," and statins have been proposed to have other such effects including modulation of nuclear factor κB (NFκB), a transcription factor with a central role in inflammation and fibrosis [2]. Statins have been recently shown to have inhibitory role on nicotinamide adenine dinucleotide phosphate dehydrogenase (NADPH) oxidase and the generation of oxidative stress [3, 4]. Ultimately, reduction of inflammation, fibrosis and even apoptosis results, leading to improvements in glomerular filtration barrier integrity, interstitial fibrosis, and tubular toxicity. In this chapter, we will review the current evidence linking dyslipidemia, oxidative stress, and renal damage as well as the potential benefit that statins may have in slowing the progression of CKD.

2 Dyslipidemia of CKD

The lipid profile of CKD patients is markedly different from the general population without kidney disease. Most of the studies involving statins have used measurements of total cholesterol (TC), low density lipoprotein (LDL), high density lipoprotein (HDL), and triglycerides (TG) as markers of dyslipidemia. Total cholesterol and LDL levels may be normal or even low when compared to those without kidney disease. Low LDL levels are attributed to decreased production [5]. Moreover, triglyceride rich as opposed to cholesterol rich (delipidated) small dense LDL predominates in CKD, which has been shown to be highly atherogenic and associated with significant cardiovascular risk [5, 6]. Low HDL levels and moderately elevated triglycerides characterize the lipid profile in CKD patients. Moreover, there is accumulation of triglyceride rich HDL, which is also called nonprotective HDL. Impaired maturation of HDL is attributed to decreased hepatic lipase and increased plasma cholesteryl ester transfer protein (CETP) [5]. Other less often measured derangements include increased very low density lipoprotein (VLDL) and chylomicron remnants, posttranslationally modified lipoproteins, and abnormal concentrations of Lp(a). Importantly, some of these lipoprotein derangements lead to increased superoxide formation,

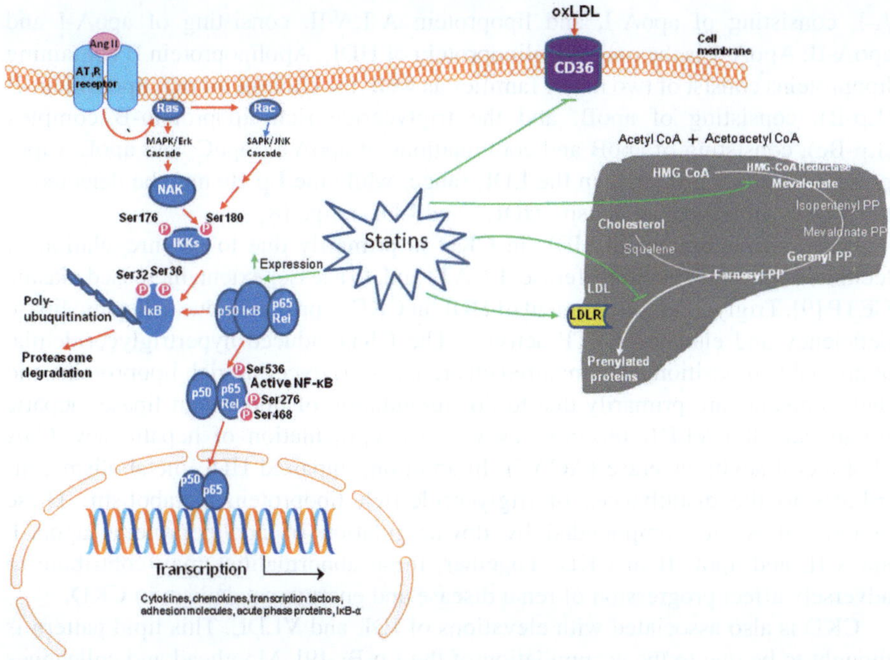

Fig. 1 Schematic showing the role of statins in modulating major pathways. Statins block the conversion of 3-hydroxy-3-methylglutaryl coenzyme A reductase inhibitors (HMG-CoA) to mevalonate by inhibition of HMG-CoA reductase. This reaction effectively blocks the downstream production of geranyl and FPPs. Geranylgeranlylation and farnesylation of small G-proteins is also inhibited, thereby blocking several pathways that use these isoprenylated proteins such as activation of nicotinamide adenine dinucleotide phosphate (NADPH) oxidase complex, trafficking of endothelin-1. In addition, statins block CD36 mediated oxidative low-density lipoprotein (oxLDL) uptake, increase uptake of LDL via LDL receptor (LDLR) into the hepatocytes and increase expression of IκB, leading to stabilization of the complex with p50/p65 subunits of NFκB

inflammation, and endothelial dysfunction. For example, TG-enriched VLDL and ApoCIII as well as elevated LDL levels, directly stimulate NADPH oxidase mediated superoxide production likely via NFκB, which then feeds back to increased reactive oxygen species (ROS) generation and so on [6]. Moreover, the ratio of TC/HDL-C and ApoB were predictors of greater risk for cardiovascular events than either TC or LDL alone in a Women's Health Study in which multiple biomarkers were simultaneously evaluated [7].

Lipoproteins, consisting of lipids and proteins called apolipoproteins, function to transport water insoluble lipids from the gut to areas of utilization. The apolipoproteins not only serve to provide structure and stability to the lipoprotein molecule, but also serve as ligands for hepatic uptake receptors and activate lipase enzymes. In CKD, there is a reduction in most apolipoprotein A (apoA) containing lipoproteins and an increase in the apolipoprotein B (apoB) containing lipoproteins [8]. Apolipoprotein A containing lipoproteins consist of two major families: lipoprotein

A-I, consisting of apoA-I, and lipoprotein A-I:A-II, consisting of apoA-I and apoA-II. ApoA-1 is the major apolipoprotein of HDL. Apolipoprotein B containing lipoproteins consist of two major families as well: the cholesterol rich lipoprotein-B (Lp-B), consisting of apoB, and the triglyceride rich lipoprotein-B complex (Lp-Bc), consisting of apoB and combinations of apoA-II, apoC, and apoE. Lipoprotein B occurs primarily in the LDL range, while the Lp-Bc may be detected in the VLDL, intermediate density (IDL), and LDL range [8].

Impaired maturation of HDL in CKD is primarily due to downregulation of lecithin-cholesterol acyltransferase (LCAT) and, to a lesser extent, increased plasma CETP [9]. Triglyceride enrichment of HDL in CKD is primarily due to hepatic lipase deficiency and elevated CETP activity. The CKD-induced hypertriglyceridemia, abnormal composition, and impaired clearance of triglyceride-rich lipoproteins and their remnants are primarily due to downregulation of lipoprotein lipase, hepatic lipase, and the VLDL receptor, as well as, upregulation of hepatic acyl-CoA cholesterol acyltransferase (ACAT). In addition, impaired HDL metabolism contributes to the disturbances of triglyceride-rich lipoprotein metabolism. These abnormalities are compounded by downregulation of apolipoproteins apoA-I, apoA-II, and apoC-II in CKD. Together, these abnormalities may contribute to adversely affect progression of renal disease and energy metabolism in CKD.

CKD is also associated with elevations of IDL and VLDL. This lipid pattern is thought to be due to the accumulation of the Lp-Bc [9]. Moorhead and colleagues speculated that this pattern was due to urinary losses of lipoprotein lipase activators, leading to impaired lipolysis of triglycerides [10]. However, recent evidence suggests that resistance to lipolysis by these lipid particles may be more important [11]. The impaired lipolysis of these compounds may be due to the increased apoC-III, an inhibitor of lipoprotein lipase, relative to apoC-II, a promoter of lipoprotein lipase [9, 11].

Lipoprotein (a) [Lp(a)] is an LDL-like particle that consists of LDL covalently bound to apo(a). The apo(a) molecule is highly polymorphic, with over 30 different isoforms grouped into two categories: high and low molecular weight [12]. In healthy patients, the subgroup with the high molecular weight isoforms tends to have low levels of Lp(a), while those with the low molecular weight isoforms have elevated levels. Among patients with nonnephrotic CKD, only those with the high molecular weight isoform tend to have Lp(a) levels increase as GFR declines [13]. Among nephrotic patients, all Lp(a) isoforms are elevated [14].

The apoA-IV levels are elevated in chronic kidney disease as opposed to apoA-I [15]. The apoA-IV also participates in reverse cholesterol transport and activates LCAT and lipases. Despite this function, investigators found that elevated apoA-IV predicts progression of kidney disease in patients with mild to moderate renal failure [16].

To summarize, the dyslipidemia of nonnephrotic CKD is characterized by an accumulation of triglyceride-rich lipoproteins, particularly IDL and VLDL, small dense (delipidated) LDL (despite normal LDL levels), Lp(a), and apoA-IV. HDL levels are reduced. The apoCIII is enriched in apoB containing triglyceride-rich lipoproteins in CKD stages 3 and 4 [17]. This particular lipid pattern likely plays a role in both the progression of CKD and the cardiovascular risk associated with CKD.

3 Oxidative Stress in the Kidney

Oxidative stress is a major pathway for kidney damage and CKD. Loss of redox balance leads to generation of oxidative stress. Free radicals such as superoxide anion and peroxynitrite, when generated in excessive amounts, contribute to oxidative stress. This may happen in common precursor conditions for CKD, including diabetes, hypertension, and cardiometabolic syndrome. Several mechanisms are at play in these disease states, including excessive activation of renin-angiotensin-aldosterone system (RAAS), hyperglycemia, insulin resistance, dyslipidemia, and inflammation of obesity and proteinuria. The NADPH oxidase enzyme is thought to be the major source for superoxide anion and plays a central role in the generation of ROS and kidney oxidative stress damage. Excessive ROS can in turn activate NADPH oxidase enzyme at the expression level, creating a feed-forward loop that leads to oxidative stress. All the subunits for NADPH oxidase enzyme are expressed in the kidney [18]. However, the isoforms differ based on the cell type and function. For example, the macula densa cells express both Nox2 and Nox4 but not Nox1, and Nox2 has been shown to be the primary isoform responsible for NaCl-induced superoxide generation [19]. In the glomeruli ROS can downregulate nephrin, and possibly other podocyte specific markers including podocin and desmin [3], and this can lead to glomerular filtration barrier injury and proteinuria. Both angiotensin II (AngII) and aldosterone have been implicated in this type of injury as blockade with specific blockers ameliorating the pathology [20, 21]. In addition, blockade of ROS generation with TEMPOL (hydroxy-2,2,6,6-tetramethylpiperidine-1-oxyl) and statins, improved glomerular filtration barrier (GFB) injury as well without any hemodynamic effects [3]. Ang II activates mitogen activated protein kinase (MAPK) and extracellular regulated kinase ERK1/ERK2 via ROS and this in turn activates transforming growth factor-β (TGF-β1) and fibronectin, leading to fibrotic changes in the glomeruli. Angiotensin II can also activate other growth and proliferation pathways via ROS including vascular endothelial growth factor (VEGF), Akt/PKB, prostaglandin receptor (EP1), and epithelial growth factor receptor (EGFR). Angiotensin II mediates LDL-induced superoxide generation and hypertrophy of mesangial cells [22]. Reactive oxygen species may induce tissue damage by initiating lipid peroxidation (LPO) [23]. In turn, this leads to adduct formation between breakdown products of LPO with structural proteins, such as formation of malondialdehyde (MDA) or 4-hydroxynonenal-lysine adducts. Type IV collagen was specifically identified as being modified by MDA adducts. When LPO was inhibited by pretreatment of passive Heymann nephritis (PHN) rats with the antioxidant probucol, proteinuria was reduced by approximately 85%, and glomerular immunostaining for dialdehyde adducts was markedly reduced [23].

Reactive oxygen species can also directly damage the tubules and lead to both acute and chronic kidney disease. Similar to mesangial cells and glomeruli, proximal tubule cells can generate ROS in response to AngII, and that can lead to activation of p42/p44 MAPK, cell cycle regulator p27(Kip1), and cellular hypertrophy [24]. Angiotensin II increases the expression of VEGF and monocyte chemoattractant protein-1 (MCP-1) via ROS, and inhibitors of NADPH oxidase blunted this pathway [18]. Reactive oxygen species can be generated via activation of other oxidases in the kidney, including mitochondrial respiratory cycle enzymes, xanthine oxidase, cyclooxygenase, and lipoxygenase [18]. To further illustrate this point, xanthine oxidase can be stimulated to generate ROS in a hypercholesterolemic pig model used to study renal hemodynamics, and inhibition of xanthine oxidase led to improved renal plasma flows [25].

Activation of NFκB, a transcription factor central to the mediation of inflammation along with other transcription factors such as activator protein-1 (AP-1), is a major mechanism by which ROS can promote cell growth and hypertrophy, migration, fibrosis, and apoptosis. Multiple cues can activate NFκB such as active RAAS, hyperglycemia, hyperinsulinemia, tumor necrosis factor-α (TNF-α), interleukin-1β (IL-1β), and important for this discussion, LDL. The p65/p50/Rel subunit of NFκB exists in an inactive state in the cytoplasm when bound to IκB [26]. Phosphorylation of IκB by inflammatory cues sets this moiety toward degradation and facilitates translocation of p65/p50/Rel to the nucleus, which then activates the expression of pro-inflammatory genes such as TNF-α and pro-fibrotic genes such as TGF-β1. Increased TGF-β1 then mediates deposition of fibronectin and collagen IV in the interstitium, leading to fibrotic changes in the long run. Inflammation in the interstitium also promotes the migration of inflammatory cells such as monocytes and macrophages that in turn secrete inflammatory cytokines and damage the tubules, glomeruli.

4 Statins Modulate Oxidative Stress

Diabetes is associated with increased cholesterol levels, elevated triglycerides, LDL and oxLDL levels, and reduction in HDL levels. Statins can reverse this pattern of dyslipidemia in diabetes, thereby protecting the kidneys and slowing progression of CKD [27]. Furthermore, renoprotection by statins may be a summation of modulatory effects on multiple pathways, including dyslipidemia, inflammation, and oxidant stress. First, statins act by blocking HMG-CoA reductase, thereby inhibiting synthesis of mevalonic acid, a precursor of many nonsteroidal isoprenoid compounds such as farnesyl pyrophosphate (FPP) and geranylgeranyl pyrophosphate (GGPP) involved in subcellular localization and intracellular trafficking of several membrane-bound proteins involved in oxidative stress injury (Rho, Ras, Rac, Rab, Ral, and Rap) (see Fig. 1) [7, 28]. Reduction in GGPP and geranylation of Rac1 inhibits translocation of Rac1 to the plasma membrane, which mediates the activation and generation of ROS by NADPH oxidase enzyme [5]. The statin blocked geranylgeranylation of Rho GTPase and decreased levels of surface protein endothelin-1, a potent vasoconstrictor and mitogen, and may play a role in retarding glomerulosclerosis [29]. Rho-GTPases are involved in several cellular processes, including cytoskeletal remodeling, membrane trafficking, transcriptional activation, and cell growth control. For example, reduced geranylgeranylation of Rho can lead to decreased leukocyte adhesion and fibrinolytic activity [7]. Second, statins suppress the expression of protein subunits of NADPH oxidase (p22phox, Nox2, Nox4, and p47phox), thereby reducing the overall activity of the enzyme, and this was evidenced from decrease NADPH oxidase enzyme activity (Fig. 2) [3]. Third, statins prolong eNOS mRNA half-life and eNOS expression and upregulate eNOS activity, thereby reducing hypertension-induced glomerular injury by inhibiting the isoprenylation of Rho [30]. Fourth, statins interfere with LDL oxidation by several mechanisms [31]: (1) blocking isoprenylation of Rac1, an

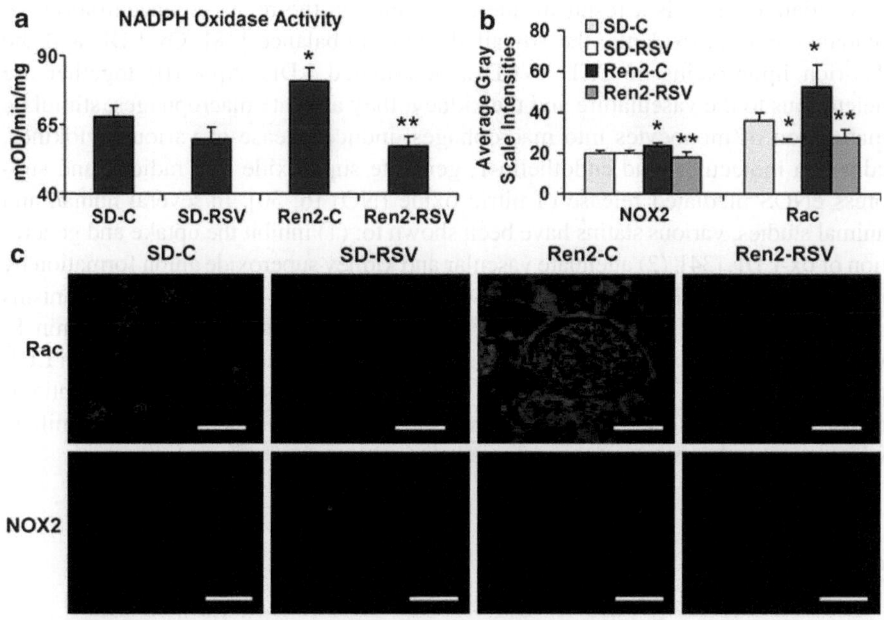

Fig. 2 Rosuvastatin attenuation of nicotinamide adenine dinucleotide phosphate (NADPH) oxidase in the transgenic Ren2 rat. Rosuvastatin was given 20 mg/kg intra-peritoneal (IP). (**a**) Ren2 untreated (Ren2-C) rats had increased NADPH oxidase activity when compared to Sprague Dawley controls (SD-C) and this increase was abrogated by rosuvastatin (Ren2-RSV). *Asterisk* indicates $p < 0.05$ for Ren2-C versus SD-C; *double asterisk* indicates $p < 0.05$ for Ren2-C versus Ren2-RSV treated with rosuvastatin. (**b**) Immunohistochemistry with Nox2 and Rac1 antibodies demonstrates increased expression of these subunits in the Ren-C when compared to SD-C and abrogation with rosuvastatin. (**c**) Average gray-scale intensities for *panel B*. *Asterisk* indicates $p < 0.05$ for Ren2-C versus SD-C, *double asterisk* indicates $p < 0.05$ for Ren2-C versus Ren2-RSV treated with rosuvastatin

important component of NADPH oxidase complex; (2) reducing expression of NADPH oxidase subunits; and (3) reducing serum levels of LDL available for oxidation. The active o-hydroxy metabolite of atorvastatin found in plasma was shown to enhance the resistance of LDL to oxidative modification at pharmacological levels. This effect is specific to the o-hydroxy metabolite and is not demonstrated by the parent atorvastatin compound nor other statins such as lovastatin, simvastatin, pravastatin, or rosuvastatin. Fifth, by suppressing the expression of scavenger receptor CD36 on monocytes, statins inhibit their uptake of oxidized LDL and subsequent conversion to macrophages [32]. Last but not least, statins suppress the nuclear translocation of NFκB subunit p65/p50, thereby downregulating the activation of several pro-inflammatory and pro-fibrotic genes [33]. This may occur via increasing the expression of IκB [33]. Interestingly, Rho and Rac can phosphorylate IκB and activate NFκB [7]. Regardless of the mechanisms, there is convincing evidence to support the theory that statin therapy in humans ameliorates oxidative stress.

Oxidative stress is a result of altered balance in the relative concentrations of oxidants and antioxidants, the so-called redox imbalance [18]. Ox-LDL and the TG-rich lipoproteins of CKD (VLDL, delipidated LDL, ApoCIII) together are deleterious to the vasculature and the kidney; they activate macrophages, stimulate maturation of monocytes into macrophages, induce release of various cytokines, adhesion molecules, and endothelin-1, generate superoxide free radical, and suppress eNOS mediated release of nitric oxide (NO) [6, 30]. In several human and animal studies, various statins have been shown to: (1) inhibit the uptake and generation of ox-LDL [34], (2) attenuate vascular and kidney superoxide anion formation by inhibition of NADPH oxidases via Rho-dependent and Rho-independent mechanisms as described above [7, 30, 33], and (3) preserve the relative levels of vitamin E, vitamin C, and endogenous antioxidants such as ubiquinone and glutathione in LDL particles [35–37]. Thus, statins not only decrease oxidants but also restore antioxidants, thereby possibly reducing the level of oxidative stress in the vascular milieu, which may explain some of the observed clinical beneficial effects [38].

5 Statins and the Kidneys

Statins have been shown to attenuate renal injury in both in vivo and in vitro studies. Renal injury from dyslipidemia and oxidative stress initiates inflammatory cascades that involve similar cellular events as seen in vascular tissue. Statins inhibit key events in this process that alter the progression of renal injury. Pravastatin has been shown to ameliorate the structural and functional changes of diabetic nephropathy in hyperglycemic insulin-deficient diabetic rats [39]. Furthermore, statins have been demonstrated to decrease TGF-ß production and suppress the enhanced Ras-dependent activation of MAPK cascade in these diabetic rats. In glomerular mesangial cells, pravastatin suppressed oxidative stress and fibronectin expression that were induced by oxLDL [40]. Lovastatin has similar effects on glomerular disease in obese insulin-resistant rats [41]. Lovastatin decreased chemotactic activity of human fetal mesangial cells by inhibition of MCP-1, and these effects were likely via inhibition of isoprenylation as mevalonate reversed the inhibition [42]. In another model of renal injury due to overexpression of AngII, cerivastatin decreased systolic blood pressure, albuminuria, and cortical necrosis [43]. These changes were associated with reduced infiltration of inflammatory cells, diminished expression of adhesion molecules, and lower levels of transcription factor (NFκB) activity. We have shown that rosuvastatin ameliorated glomerular filtration barrier injury in a hypertensive, insulin resistant rat model (TgmRen2) by decreasing podocyte effacement, restoring slit diaphragms, and improving the pore numbers (Fig. 3) [3]. In rats with glomerulonephritis, simvastatin decreased mesangial cell proliferation and monocyte/macrophage infiltration [43]. Statins have been shown to inhibit the proliferative actions of platelet-derived growth factor [44] and TGF-ß [39]. Cytokines released during renal injury activate NFκB and growth-regulating pathways in mesangial and tubular cells. Statins both decrease the levels of cytokines and inhibit the NFκB-dependent gene activation, such as MCP-1 and IL-6 [45]. Fluvastatin protects rat kidneys by decreasing oxidative stress, TGF-β, fibronectin, MCP-1, and LOX-1 expression and increasing NO bioavailability [46]. Streptozotocin-induced diabetic rats have increased Nox4, TGF-β, and connective tissue growth factor (CTGF) expression in their renal cortices followed by increased NADPH oxidase activity and ROS generation [47]. Fasudil, a Rho-kinase inhibitor, as well as statins, ameliorated diabetic nephropathy in these rats by suppressing NADPH oxidase activity and fibrosis. As described before, statins inhibit CD36, a scavenger receptor that facilitates the uptake of oxLDL and promotes foam cell formation and oxidative stress in dialysis patients [32]. In humans, statins also decrease urinary albumin excretion in patients with nephrotic syndrome and in patients with type 2 diabetes [48]. Thus, statins modulate glomerular mesangial and interstitial inflammatory processes independent of lipid reduction. Clinical relevance of these observations is yet to be determined by the ongoing interventional studies.

6 Clinical Trials with Statins in CKD Patients

Over the past 10–15 years, several clinical trials have attempted to elucidate a role for statins in slowing the progression of CKD. Conflicting reports have emerged over the usefulness of statins in slowing down hard endpoints, including doubling of serum creatinine, initiation of dialysis, and death from CKD. Surprisingly, most trials showed a reduction in cholesterol and proteinuria, but only some showed improvement in the hard renal endpoints. The data on CVD endpoints have been more encouraging and consistent. The Cholesterol Treatment Trialists (CTT) Collaborators study was among the first to show that lowering of LDL-cholesterol by 1 mmol/L with statins resulted in a decrease in CVD risk by 20% [49]. Post-hoc analyses of subgroups in the CTT trial with and without CKD and with diabetic kidney disease (DKD) revealed that the benefits of statin therapy are sustained in

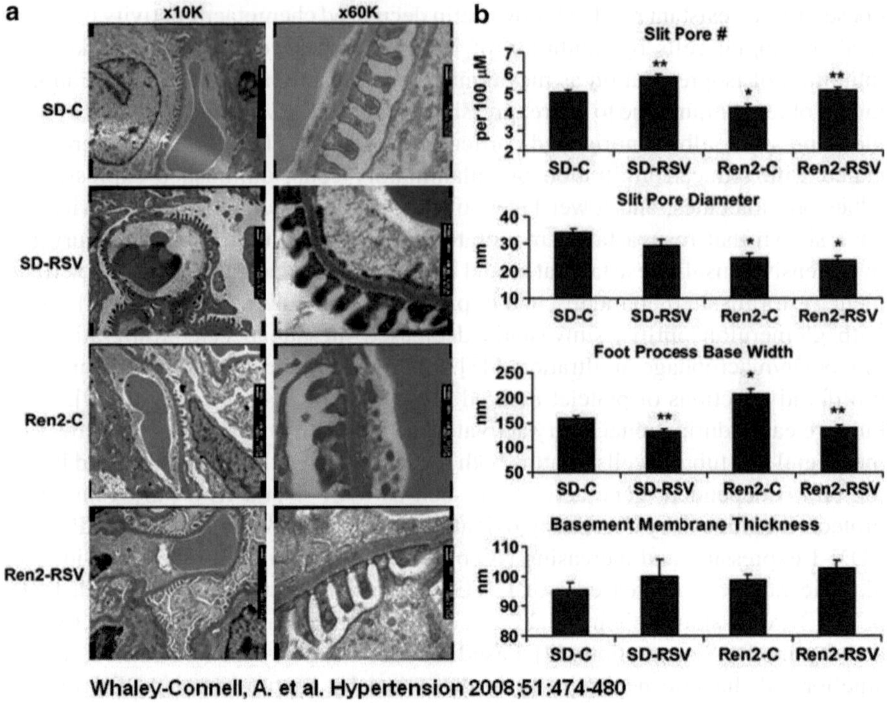

Fig. 3 Rosuvastatin improves indices of podocyte foot process effacement on TEM. (**a**) Ultrastructural measurements by transmission electron microscopy (TEM) performed on the glomerular filtration barrier at 10 K (*left panel*) and 60 K (*right panel*) magnification. The Ren2-C animals have podocyte effacement and highly disordered structure when compared to SD-C. In addition, there is a decrease in the number of slit pores and some obliteration of the slit pore membrane. Rosuvastatin, restores the integrity of the glomerular filtration barrier. (**b**) Quantification of the defects seen in (**a**) reveals that the podocyte base width was significantly increased, while slit pore diameter and slit pore numbers were decreased in the Ren2-C. This derangement was abrogated by rosuvastatin in the Ren2-RSV. *Asterisk* indicates $p < 0.05$ for Ren2-C versus SD-C, *double asterisk* indicates $p < 0.05$ for Ren2-C versus Ren2-RSV treated with rosuvastatin

patients with decreased GFR [50]. Further validation of CTT data was obtained from the Anglo-Scandinavian Cardiac Outcomes Trial-Lipid Lowering Arm (ASCOT-LLA) and the Cholesterol and Recurrent Events (CARE) study groups, which showed similar improvements in CVD endpoints in patients with CKD stages 2 and 3 [51–53]. In addition, subgroup analysis of the Greek Atorvastatin and Coronary Artery Disease Evaluation (GREACE) study revealed that after adjustment for 25 CVD risk factors, there was significant improvement in eGFR in the statin-treated group [54]. Furthermore, the Assessment of Lescol in Renal Transplantation (ALERT) study showed sustained CV benefit of statins when patients were followed for a prolonged period [55]. However, CV benefits as measured by intima-media thickness reduction was only evident in nonrenal patients and not in renal patients, even when the LDL cholesterol was drastically lowered [56]. This was followed by the Die Deutsche Diabetes Dialyze Studie (4D study) that did not show significant CV benefits from lowering cholesterol by 42% in dialysis patients [57] as did a small Scandinavian study [58]. Several reasons have been put forth for this sort of unexpected outcome. Alternative more overriding pathomechanisms such as inflammation or calcification could play a role in the uremic state. The cardiovascular death was from noncoronary causes such as heart failure and sudden death, making the 4D study underpowered for detecting appropriate CV risk. In addition, it is possible that a substantial number of CKD patients may die from coronary events even before they reach ESRD, likely removing the cohort that is most likely to respond to statins.

The initial data on the renal endpoints were highly encouraging and statins were being touted as the next big drug. In the Heart Protection Study (HPS), simvastatin lowered cholesterol and showed a decline in GFR rates when compared to placebo in a 5-year follow-up of approximately 15,000 patients [59]. In a subgroup of patients with GFR <60 mL/min from the CARE study, pravastatin decreased the rate of decline in GFR, and the protection was even greater in persons with GFR <40 mL/min/m^2/year [60]. Importantly, the benefits from simvastatin were greater in the diabetic group when compared to the nondiabetic group. In a meta-analysis of 13 trials, Fried et al. showed that lipid reduction appeared to slow the decline of GFR and proteinuria and there was a correlation with years of follow-up [61]. In another small, randomized, placebo-controlled trial involving CKD patients with idiopathic glomerulopathies, atorvastatin decreased proteinuria by 55% and improved creatinine clearance by 57% [62]. This was then followed by the meta-analysis by Douglas et al. who showed that in the microalbuminuric and macro-albuminuric range, statins decreased proteinuria by close to 50% along with improvements in renal outcomes [63]. However, the benefits of treatment with statins could not be translated to hemodialysis patients. The 4D study showed that there was no benefit to using statins in ESRD patients on hemodialysis, prompting analyses and conclusions that statins may not have any effect "beyond a point" [57]. Surprisingly, there was an increased risk of cerebrovascular disease in statin-treated patients when compared to placebo. This was in contrast to the AURORA study (a study to evaluate the use of rosuvastatin in subjects on regular hemodialysis: an assessment of survival and cardiovascular events), which did not reveal such a risk [64]. However, in keeping with the conclusions of the 4D study, AURORA confirmed that although statins lowered LDL-cholesterol concentrations by 43%, there was no CV or CKD benefit to treating ESRD on hemodialysis patients. The AURORA study was limited by other caveats such as low event rate, high dropout rate (50%), and exclusion of patients who were most likely to benefit from statin treatment [65]. The Study of Heart and Renal Protection (SHARP) is another large study (9,000 patients) focused on investigating the effects of statins on renal end-points. Interestingly, the SHARP cohort includes patients who are not yet on dialysis, patients on peritoneal dialysis, and patients on hemodialysis [65]. The importance of cohort selection will be clear over the next few sentences. The conclusions of the 4D and the AURORA studies have been questioned by experts who believe that CKD patients have a CV burden that is vastly different from the general population. CKD patients have left ventricular hypertrophy and aortic calcifications as opposed to the majority of nondialysis patients who have atheromatous lesions (some in the coronary arteries). While the mortality in the general population is from myocardial infarctions, CKD populations have higher sudden cardiac death and arrhythmias, events for which statins are not even indicated [65]. Many CKD patients have heart failure and statins have not been shown to be beneficial in lowering mortality in heart failure patients. Together, the ineffectiveness of statins in lowering the primary or secondary endpoints in ESRD patients on hemodialysis may be a function of not targeting either the causal pathway or the disease spectrum. Interestingly, an analysis of the U.S. Renal Data System Morbidity and Mortality Wave-2 study, a cohort of about 3,700 new start dialysis patients randomly selected between 1996 and 1998, revealed a significant decrease in all-cause mortality as well as CV-specific mortality [66].

Newer trials such as the Aggressive Lipid-Lowering Initiation Abates New Cardiac Events (ALLIANCE) study have been designed to answer the question whether statins have a role in CKD 1–5 [67]. When atorvastatin was used with an intention to treat (LDL goal <80 mg/dL or up to 80 mg/d) versus standard therapy by primary care physicians, the focused treatment group had better outcomes compared to usual therapy both in patients with (<60 mL/min/1.73 m^2 GFR) and without advanced CKD (>60 mL/min/1.73 m^2 GFR). A review of the Cochrane Database System Reviews in dialysis patients showed that statins are safe in this population group, and that longer treatment duration is necessary to elicit beneficial effects on CVD [64, 68]. In the subgroup of patients not requiring dialysis, statins significantly reduced CV events when compared to the placebo [69].

7 Conclusion

Dysplipidemia of CKD is a major risk factor, and modulation of this dysplipidemia by statins has consistently demonstrated to decrease CVD risk in CKD populations. Evidence is emerging for the use of statins in early stages of CKD 1 through 4. Even in ESRD patients, the use of statins is not discouraged if the patient has a high CVD risk. Measurement of endpoints of these failed trials does not include VLDLs and other lipids intrinsic to CKD, which is the real culprit in diabetic patients, not LDL. In addition, the pleiotropy that statins exhibit when compared to other lipid-lowering agents clearly makes them superior in altering both CVD and CKD risk. Targeting oxidative stress, which may be central to the pathogenesis in diabetes, hypertension, obesity, and the cardiometabolic syndrome and a major mechanism for dyslipidemia in CKD, is very effectively altered by statins. Future trials such as the Study of Heart and Renal Protection (SHARP) hold promise. Moreover, clinical trials should be designed to measure oxidative stress and CKD specific dysplipidemia via the use of biomarkers both to detect incident disease and provide therapy to modulate the process.

References

1. KDOQI Clinical Practice Guidelines and Clinical Practice Recommendations for Diabetes and Chronic Kidney Disease. Am J Kidney Dis, 2007. **49**(2 Suppl 2): S12–154.
2. Epstein, M. and V.M. Campese. Pleiotropic effects of 3-hydroxy-3-methylglutaryl coenzyme a reductase inhibitors on renal function. Am J Kidney Dis, 2005. **45**(1): 2–14.
3. Whaley-Connell, A., J. Habibi, R. Nistala, et al. Attenuation of NADPH oxidase activation and glomerular filtration barrier remodeling with statin treatment. Hypertension, 2008. **51**(2): 474–80.
4. Wassmann, S., U. Laufs, K. Muller, et al. Cellular antioxidant effects of atorvastatin in vitro and in vivo. Arterioscler Thromb Vasc Biol, 2002. **22**(2): 300–5.
5. Romayne Kurukulasuriya, L., G. Athappan, G. Saab, et al. HMG CoA reductase inhibitors and renoprotection: the weight of the evidence. Ther Adv Cardiovasc Dis, 2007. **1**(1): 49–59.
6. Wanner, C. and E. Ritz. Reducing lipids for CV protection in CKD patients-current evidence. Kidney Int Suppl, 2008(111): S24–8.
7. Jain, M.K. and P.M. Ridker. Anti-inflammatory effects of statins: clinical evidence and basic mechanisms. Nat Rev Drug Discov, 2005. **4**(12): 977–87.
8. Attman, P.O., O.G. Samuelsson, J. Moberly, et al. Apolipoprotein B-containing lipoproteins in renal failure: the relation to mode of dialysis. Kidney Int, 1999. **55**(4): 1536–42.
9. Vaziri, N.D. Dyslipidemia of chronic renal failure: the nature, mechanisms, and potential consequences. Am J Physiol Renal Physiol, 2006. **290**(2): F262–72.
10. Moorhead, JF., M.K. Chan, M. El-Nahas, et al. Lipid nephrotoxicity in chronic progressive glomerular and tubulo-interstitial disease. Lancet, 1982. **2**(8311): 1309–11.
11. Ikewaki, K., J.R. Schaefer, M.E. Frischmann, et al. Delayed in vivo catabolism of intermediate-density lipoprotein and low-density lipoprotein in hemodialysis patients as potential cause of premature atherosclerosis. Arterioscler Thromb Vasc Biol, 2005. **25**(12): 2615–22.
12. Kronenberg, F., A. Lingenhel, K. Lhotta, et al. The apolipoprotein(a) size polymorphism is associated with nephrotic syndrome. Kidney Int, 2004. **65**(2): 606–12.
13. Kronenberg, F., E. Kuen, E. Ritz, et al. Lipoprotein(a) serum concentrations and apolipoprotein(a) phenotypes in mild and moderate renal failure. J Am Soc Nephrol, 2000. **11**(1): 105–15.
14. Kronenberg, F., E. Kuen, E. Ritz, et al. Apolipoprotein A-IV serum concentrations are elevated in patients with mild and moderate renal failure. J Am Soc Nephrol, 2002. **13**(2): 461–9.
15. Boes, E., D. Fliser, E. Ritz, et al. Apolipoprotein A-IV predicts progression of chronic kidney disease: the mild to moderate kidney disease study. J Am Soc Nephrol, 2006. **17**(2): 528–36.
16. Schaeffner, E.S., T. Kurth, G.C. Curhan, et al. Cholesterol and the risk of renal dysfunction in apparently healthy men. J Am Soc Nephrol, 2003. **14**(8): 2084–91.
17. Attman, P.O., O. Samuelsson, and P. Alaupovic. Lipoprotein metabolism and renal failure. Am J Kidney Dis, 1993. **21**(6): 573–92.
18. Nistala, R., A. Whaley-Connell, and J.R. Sowers. Redox control of renal function and hypertension. Antioxid Redox Signal, 2008. **10**(12): 2047–89.
19. Zhang, R., P. Harding, J.L. Garvin, et al. Isoforms and functions of NAD(P)H oxidase at the macula densa. Hypertension, 2009. **53**(3): 556–63.
20. Shibata, S., M. Nagase, S. Yoshida, et al. Podocyte as the target for aldosterone: roles of oxidative stress and Sgk1. Hypertension, 2007. **49**(2): 355–64.
21. Whaley-Connell, AT., N.A. Chowdhury, M.R. Hayden, et al. Oxidative stress and glomerular filtration barrier injury: role of the renin-angiotensin system in the Ren2 transgenic rat. Am J Physiol Renal Physiol, 2006. **291**(6): F1308–14.
22. Park, S.Y., C.Y. Song, B.C. Kim, et al. Angiotensin II mediates LDL-induced superoxide generation in mesangial cells. Am J Physiol Renal Physiol, 2003. **285**(5): F909–15.

23. Neale, T.J., P.P. Ojha, M. Exner, et al. Proteinuria in passive Heymann nephritis is associated with lipid peroxidation and formation of adducts on type IV collagen. J Clin Invest, 1994. **94**(4): 1577–84.
24. Hannken, T., R. Schroeder, G. Zahner, et al. Reactive oxygen species stimulate p44/42 mitogen-activated protein kinase and induce p27(Kip1): role in angiotensin II-mediated hypertrophy of proximal tubular cells. J Am Soc Nephrol, 2000. **11**(8): 1387–97.
25. Daghini, E., A.R. Chade, J.D. Krier, et al. Acute inhibition of the endogenous xanthine oxidase improves renal hemodynamics in hypercholesterolemic pigs. Am J Physiol Regul Integr Comp Physiol, 2006. **290**(3): R609–15.
26. Hayden, M.S. and S. Ghosh. Signaling to NF-kappaB. Genes Dev, 2004. **18**(18): 2195–224.
27. Nesto, R.W. Beyond low-density lipoprotein: addressing the atherogenic lipid triad in type 2 diabetes mellitus and the metabolic syndrome. Am J Cardiovasc Drugs, 2005. **5**(6): 379–87.
28. Ceriello, A. Controlling oxidative stress as a novel molecular approach to protecting the vascular wall in diabetes. Curr Opin Lipidol, 2006. **17**(5): 510–8.
29. Campese, V.M., M.K. Nadim, and M. Epstein. Are 3-hydroxy-3-methylglutaryl-CoA reductase inhibitors renoprotective? J Am Soc Nephrol, 2005. **16** (Suppl 1): S11–7.
30. Wolfrum, S., K.S. Jensen, and J.K. Liao. Endothelium-dependent effects of statins. Arterioscler Thromb Vasc Biol, 2003. **23**(5): 729–36.
31. Mason, R.P., M.F. Walter, and R.F. Jacob. Effects of HMG-CoA reductase inhibitors on endothelial function: role of microdomains and oxidative stress. Circulation, 2004. **109** (21 Suppl 1): II34–41.
32. Chmielewski, M., E. Bryl, L. Marzec, et al. Expression of scavenger receptor CD36 in chronic renal failure patients. Artif Organs, 2005. **29**(8): 608–14.
33. Dichtl, W., J. Dulak, M. Frick, et al. HMG-CoA reductase inhibitors regulate inflammatory transcription factors in human endothelial and vascular smooth muscle cells. Arterioscler Thromb Vasc Biol, 2003. **23**(1): 58–63.
34. Aviram, M., G. Dankner, U. Cogan, et al. Lovastatin inhibits low-density lipoprotein oxidation and alters its fluidity and uptake by macrophages: in vitro and in vivo studies. Metabolism, 1992. **41**(3): 229–35.
35. Blaha, V., Z. Zadak, D. Solichova, et al. Hypocholesterolemic effect of pravastatin is associated with increased content of antioxidant vitamin-E in cholesterol fractions. Acta Medica (Hradec Kralove), 1998. **41**(2): 87–90.
36. Laaksonen, R., K. Jokelainen, J. Laakso, et al. The effect of simvastatin treatment on natural antioxidants in low-density lipoproteins and high-energy phosphates and ubiquinone in skeletal muscle. Am J Cardiol, 1996. **77**(10): 851–4.
37. Suzumura, K., E. Kasahara, Y. Ohnishi, et al. Fluvastatin normalizes the decreased turnovers of glutathione and ascorbic acid in Watanabe heritable hyperlipidaemic rabbits. Clin Exp Pharmacol Physiol, 2000. **27**(9): 709–14.
38. McFarlane, S.I., R. Muniyappa, R. Francisco, et al. Clinical review 145: Pleiotropic effects of statins: lipid reduction and beyond. J Clin Endocrinol Metab, 2002. **87**(4): 1451–8.
39. Yokota, T., K. Utsunomiya, Y. Murakawa, et al. Mechanism of preventive effect of HMG-CoA reductase inhibitor on diabetic nephropathy. Kidney Int Suppl, 1999. **71**: S178–81.
40. Chen, H.C., J.Y. Guh, S.J. Shin, et al. Pravastatin suppress superoxide and fibronectin production of glomerular mesangial cells induced by oxidized-LDL and high glucose. Atherosclerosis, 2002. **160**(1): 141–6.
41. O'Donnell, M.P., B.L. Kasiske, Y. Kim, et al. Lovastatin retards the progression of established glomerular disease in obese Zucker rats. Am J Kidney Dis, 1993. **22**(1): 83–9.
42. Kim, S.Y., C. Guijarro, M.P. O'Donnell, et al. Human mesangial cell production of monocyte chemoattractant protein-1: modulation by lovastatin. Kidney Int, 1995. **48**(2): 363–71.
43. Yoshimura, A., K. Inui, T. Nemoto, et al. Simvastatin suppresses glomerular cell proliferation and macrophage infiltration in rats with mesangial proliferative nephritis. J Am Soc Nephrol, 1998. **9**(11): 2027–39.

44. Grandaliano, G., P. Biswas, G.G. Choudhury, et al Simvastatin inhibits PDGF-induced DNA synthesis in human glomerular mesangial cells. Kidney Int, 1993. **44**(3): 503–8.
45. Massy, Z.A., Y. Kim, C. Guijarro, et al. Low-density lipoprotein-induced expression of interleukin-6, a marker of human mesangial cell inflammation: effects of oxidation and modulation by lovastatin. Biochem Biophys Res Commun, 2000. **267**(2): 536–40.
46. Zhou, M.S., I.H. Schuman, E.A. Jaimes, et al. Renoprotection by statins is linked to a decrease in renal oxidative stress, TGF-beta, and fibronectin with concomitant increase in nitric oxide bioavailability. Am J Physiol Renal Physiol, 2008. **295**(1): F53–9.
47. Gojo, A., K. Utsunomiya, K. Taniguchi, et al. The Rho-kinase inhibitor, fasudil, attenuates diabetic nephropathy in streptozotocin-induced diabetic rats. Eur J Pharmacol, 2007. **568**(1–3): 242–7.
48. Tonolo, G., M.G. Melis, M. Formato, et al. Additive effects of Simvastatin beyond its effects on LDL cholesterol in hypertensive type 2 diabetic patients. Eur J Clin Invest, 2000. **30**(11): 980–7.
49. Baigent, C., A. Keech, P.M. Kearney, et al. Efficacy and safety of cholesterol-lowering treatment: prospective meta-analysis of data from 90,056 participants in 14 randomised trials of statins. Lancet, 2005. **366**(9493): 1267–78.
50. Wanner, C. Statin effects in CKD: is there a "point of no return"? Am J Kidney Dis, 2009. **53** (5): 723–5.
51. Tonelli, M., A. Keech, J. Shepherd, et al. Effect of pravastatin in people with diabetes and chronic kidney disease. J Am Soc Nephrol, 2005. **16**(12): 3748–54.
52. Sever, P.S., B. Dahlof, N.R. Poulter, et al. Prevention of coronary and stroke events with atorvastatin in hypertensive patients who have average or lower-than-average cholesterol concentrations, in the Anglo-Scandinavian Cardiac Outcomes Trial–Lipid Lowering Arm (ASCOT-LLA): a multicentre randomised controlled trial. Lancet, 2003. **361**(9364): 1149–58.
53. Tonelli, M., L. Moye, F.M. Sacks, et al. Pravastatin for secondary prevention of cardiovascular events in persons with mild chronic renal insufficiency. Ann Intern Med, 2003. **138**(2): 98–104.
54. Athyros, V.G., D.P. Mikhailidis, A.A. Papageorgiou, et al. The effect of statins versus untreated dyslipidaemia on renal function in patients with coronary heart disease. A subgroup analysis of the Greek atorvastatin and coronary heart disease evaluation (GREACE) study. J Clin Pathol, 2004. **57**(7): 728–34.
55. Holdaas, H., B. Fellstrom, E. Cole, et al. Long-term cardiac outcomes in renal transplant recipients receiving fluvastatin: the ALERT extension study. Am J Transplant, 2005. **5**(12): 2929–36.
56. Fathi, R., N. Isbel, L. Short, et al. The effect of long-term aggressive lipid lowering on ischemic and atherosclerotic burden in patients with chronic kidney disease. Am J Kidney Dis, 2004. **43**(1): 45–52.
57. Wanner, C., V. Krane, W. Marz, et al. Atorvastatin in patients with type 2 diabetes mellitus undergoing hemodialysis. N Engl J Med, 2005. **353**(3): 238–48.
58. Holmberg, B., M. Brannstrom, B. Bucht, et al. Safety and efficacy of atorvastatin in patients with severe renal dysfunction. Scand J Urol Nephrol, 2005. **39**(6): 503–10.
59. MRC/BHF Heart Protection Study of cholesterol lowering with simvastatin in 20,536 high-risk individuals: a randomised placebo-controlled trial. Lancet, 2002. **360**(9326): 7–22.
60. Tonelli, M., L. Moye, F.M. Sacks, et al. Effect of pravastatin on loss of renal function in people with moderate chronic renal insufficiency and cardiovascular disease. J Am Soc Nephrol, 2003. **14**(6): 1605–13.
61. Fried, L.F., T.J. Orchard, and B.L. Kasiske. Effect of lipid reduction on the progression of renal disease: a meta-analysis. Kidney Int, 2001. **59**(1): 260–9.
62. Bianchi, S., R. Bigazzi, A. Caiazza, et al. A controlled, prospective study of the effects of atorvastatin on proteinuria and progression of kidney disease. Am J Kidney Dis, 2003. **41**(3): 565–70.

63. Douglas, K., P.G. O'Malley, and J.L. Jackson. Meta-analysis: the effect of statins on albuminuria. Ann Intern Med, 2006. **145**(2): 117–24.
64. Fellstrom, B.C., A.G. Jardine, R.E. Schmieder, et al. Rosuvastatin and cardiovascular events in patients undergoing hemodialysis. N Engl J Med, 2009. **360**(14): 1395–407.
65. Strippoli, G.F. and J.C. Craig. Sunset for statins after AURORA? N Engl J Med, 2009. **360** (14): 1455–7.
66. Seliger, S.L., N.S. Weiss, D.L. Gillen, et al. HMG-CoA reductase inhibitors are associated with reduced mortality in ESRD patients. Kidney Int, 2002. **61**(1): 297–304.
67. Koren, M.J., M.H. Davidson, D.J. Wilson, et al. Focused atorvastatin therapy in managed-care patients with coronary heart disease and CKD. Am J Kidney Dis, 2009. **53**(5): 741–50.
68. Navaneethan, S.D., S.U. Nigwekar, V. Perkovic, et al. HMG CoA reductase inhibitors (statins) for dialysis patients. Cochrane Database Syst Rev, 2009(3):CD004289.
69. Navaneethan, S.D., F. Pansini, V. Perkovic, et al. HMG CoA reductase inhibitors (statins) for people with chronic kidney disease not requiring dialysis. Cochrane Database Syst Rev, 2009(2):CD007784.

Chapter 19
N-Acetylcysteine in Kidney Disease

Giancarlo Marenzi, Erminio Sisillo, and Antonio L. Bartorelli

Abstract Acute kidney injury (AKI) is a frequent complication of contrast administration (contrast-induced nephropathy) and cardiac surgery and is associated with increased short-term and long-term morbidity and mortality. Acute kidney injury is becoming a growing problem in modern cardiovascular medicine, in parallel with the increase in application of radiological procedures utilizing iodinated contrast media for both diagnostic and therapeutic purposes, and with the increasing number of elderly patients with comorbidities undergoing cardiac surgery. Due to the lack of any effective treatment, prevention of this complication is the key strategy. As oxidative stress seems to play an important role in both contrast-induced nephropathy and postcardiac surgery AKI etiologies, the potential prophylactic effect of the antioxidant agent *N*-acetylcysteine (NAC) has been largely investigated in recent years in patients at risk.

In this chapter, we review the clinical impact of contrast-induced nephropathy and postcardiac surgery AKI and the clinical and investigational evidence for the potential role of NAC in the prevention of these two conditions.

Keywords Acute kidney injury · Contrast-induced nephropathy · Cardiac surgery · Oxidative stress · *N*-acetylcysteine

1 Introduction

There is increasing awareness that acute kidney injury (AKI) represents an important problem in cardiovascular medicine from an epidemiological, clinical, and prognostic perspective.

Diagnosis and treatment of patients with cardiovascular diseases relies heavily on cardiovascular imaging, percutaneous interventions, and cardiac surgery.

G. Marenzi (✉)
Centro Cardiologico Monzino, I.R.C.C.S. Department of Cardiovascular Sciences, University of Milan, Milan, Italy
e-mail: Giancarlo.marenzi@ccfm.it

T. Miyata et al. (eds.), *Studies on Renal Disorders*, Oxidative Stress in Applied Basic Research and Clinical Practice, DOI 10.1007/978-1-60761-857-7_19,

Contrast-mediated imaging studies and procedures are a relevant portion of modern medical practice. An increasing number of patients, estimated at 30 million per year in the United States, receive contrast agents during diagnostic or interventional procedures [1]. However, iodinated contrast media, which play a key role for cardiovascular imaging and diagnostic accuracy, have toxic effects to the kidney. Indeed, one of the most troublesome complications of contrast agent administration is contrast-induced nephropathy (CIN), which represents one of the leading causes of renal impairment and the third cause of in-hospital acquired renal failure [2].

Acute kidney injury is also a serious complication after cardiac surgery and is associated with increased short-term and long-term mortality [3, 4]. This complication occurs in 5–30% of patients who undergo cardiac surgery, depending on the definition used for AKI [5, 6]. Most of the previous studies focused on severe AKI, defined either as need for dialysis or substantial increase in serum creatinine [3]. However, studies reporting an association between small changes in serum creatinine and adverse short-term outcome are emerging in the literature [4–7].

In both CIN and postcardiac surgery AKI, there is evidence that an increase of free-radical production may play a role in the multifactorial etiology of renal injury [8–10]. In a variety of clinical conditions, oxidative stress and reactive oxygen metabolites have been shown to be significantly involved in the pathogenesis of AKI. This led to the evaluation of the potential therapeutic role of antioxidant agents in animal models and in patients with AKI due to cardiac and aortic surgery, sepsis, drug nephrotoxicity (contrast agents, cisplatin, gentamicin, and cyclosporine), as well as rhabdomyolysis [11–18]. Specifically, the potential prophylactic effect of *N*-acetylcysteine (NAC) against AKI has been the rationale for its use in patients at risk of CIN and in those undergoing cardiac surgery, with the aim of pharmacological modulation of oxidative stress and outcome improvement.

In this chapter, the potential role of NAC in the prevention of CIN and postcardiac surgery AKI will be discussed on the basis of the existing clinical and investigational evidence. A brief and updated overview of the clinical impact of these two conditions will be provided in order to better define the clinical scenario in which NAC therapy may exert its positive effects.

2 Contrast-Induced Nephropathy

Contrast-induced nephropathy implies an acute and transient impairment in renal function after intravascular administration of contrast medium. It is usually defined as an absolute (>0.5 mg/dL) or a relative (>25%) increase in serum creatinine within 48–72 h following contrast exposure [19]. In patients with normal renal function, the frequency of CIN is low (<3%), but it may rise up to 50% in high-risk patients.

Development of CIN is associated with a higher risk of cardiovascular complications, prolonged hospitalization, and increased in-hospital and long-term mortality [20, 21]. Spontaneous recovery of renal function ensues within 1–2 weeks in most patients who develop CIN. However, severe nephropathy requiring in-hospital dialysis may also occur, a condition associated with very poor clinical outcome [20–25]. Several studies have demonstrated a relatively high rate of in-hospital morbidity and an increased risk for subsequent adverse events, including death, in patients with CIN. The mechanisms underlying the significantly higher long-term mortality in patients who develop reversible CIN are unclear. It is possible that, despite clinically self-limited changes in renal function following contrast exposure, subclinical renal parenchymal changes may persist even after laboratory renal function returns to normal. These alterations may lead, over time, to progressive kidney function deterioration that can increase the risk of mortality. This hypothesis is supported by a recent study showing that, despite similar baseline and discharge creatinine levels in patients with and without CIN, those who developed this complication had a significantly greater reduction of renal function during postdischarge follow-up [26]. It is noteworthy that several risk factors, including chronic kidney disease (CKD), diabetes mellitus, intravascular volume depletion, and high contrast volume, are associated with an increased risk for CIN [20, 22, 27–29] (Table 1). Thus, CIN is predictable in most cases, and, therefore, the only effective therapeutic approach is the use of preventive strategies. Accordingly, several studies focused on CIN prevention, most often using pharmacologic agents. However, none of the drugs tested showed an unquestionably positive effect, particularly when patients with severe CKD were considered [20, 30–32] (Table 2).

The exact mechanism of CIN has not been completely elucidated and is still under investigation. However, there is increasing evidence that the combination of direct toxic effects on tubular epithelial cells and renal ischemia plays the major role [33].

Table 1 Risk factors for contrast-induced nephropathy

Patient related	Chronic kidney disease (stage 3 or greater)
	Diabetes mellitus (type 1 or type 2)
	Volume depletion
	Older age
	Congestive heart failure (or left ventricular ejection fraction <40%)
	Hypertension
	Anemia and blood loss
	Hypoalbuminemia (<35 g/L)
	Nephrotoxic drug use (NSAIDs, cyclosporine, aminoglycosides)
	Diuretics
	ACE inhibitors
	Hypotension or pre-procedural hemodynamic instability
	Urgent procedure (acute myocardial infarction)
	Intra-aortic balloon pump use
	Renal transplant
Not patient related	Contrast properties
	High osmolar contrast
	Ionic contrast
	Contrast viscosity
	Contrast volume
	Intra-arterial administration of contrast

NSAIDs nonsteroidal anti-inflammatory drugs; *ACE* angiotensin-converting enzyme

Table 2 Strategies evaluated for contrast-induced nephropathy risk reduction

Positive results (potentially beneficial)
Hydration (isotonic saline, sodium bicarbonate)
Theophylline/aminophylline
N-acetylcysteine
Ascorbic acid
Statins
Hemofiltration
Prostaglandin E_1
Trimetazidine
Captopril
Neutral results (no consistent effect)
Fenoldopam
Dopamine
Calcium channel blockers
Amlodipine
Felodipine
Nifedipine
Nitrendipine
Atrial natriuretic peptide
L-arginine
Negative results (potentially detrimental)
Furosemide
Mannitol
Endothelin receptor antagonist
Hemodialysis

Direct toxic effects on the proximal convoluted tubular cells and the inner cortex of the kidneys, presenting with epithelial cell vacuolization, interstitial inflammation, and cellular necrosis, were demonstrated following exposure to a variety of iodinated contrast agents. Concerning ischemic injury several studies demonstrated vasoconstriction and reduction in renal blood flow occurring immediately after contrast medium administration. Other studies showed that the changes in renal plasma flow are not uniform. Indeed, contrast media appear to exert regional effects within the kidney, with increase in blood flow to the renal cortex and simultaneous flow reduction to the outer medulla. The deeper portion of the outer medulla of the kidney is particularly susceptible to ischemic injury, since this area is maintained at the verge of hypoxia, with pO_2 levels often as low as 20 mmHg. In addition to a low oxygen tension, the outer medulla has a high metabolic activity and a relatively high oxygen requirement due to salt reabsorption that account for its vulnerability [34].

Animal studies have suggested that oxidant-mediated injury, due to enhanced production of oxygen-free radicals and lipid peroxidation of biological membranes, may be implicated. The ability to accommodate oxidant injury decreases with age, a factor that may contribute to the increased risk of CIN among older patients. Moreover, increased oxidative stress is present in CKD and in diabetes [35, 36].

2.1 N-Acetylcysteine for Prevention of Contrast-Induced Nephropathy

In recent years, several clinical studies have been performed using antioxidant compounds in an attempt to prevent CIN. The rationale for the use of antioxidant agents is based on animal experiments suggesting a pathogenetic role of reactive oxygen species in the occurrence of CIN [37–40]. In 2004, Drager et al. [41] demonstrated that after radiocontrast exposure, urinary levels of 15-isoprostane F2, a specific marker of oxidative stress, increased significantly over baseline values in patients receiving hydration only. Conversely, they remained essentially unchanged in patients treated with hydration and NAC. Furthermore, NAC treatment led to lower levels of alpha-glutathione S-transferase, a specific proximal tubular injury marker, as compared to hydration only.

The *N*-acetyl derivative of cysteine, known as NAC, is the most widely studied agent of all prophylaxis strategies. It has direct vasodilating effects on kidneys vessels, contributing to improved renal hemodynamics, and may also attenuate endothelial dysfunction. More notably, it is also able to scavenge oxygen-free radicals, thus preventing the direct oxidative tissue damage occurring after contrast administration [41–46].

After the publication of the seminal study by Tepel et al. [17] showing that NAC offers some protection, a large number of studies, in most cases with a relatively small sample sizes, were published. Indeed, Tepel et al. [17] demonstrated that NAC (600 mg orally twice daily) plus hydration before and after contrast administration has a CIN preventive effect in patients with renal insufficiency undergoing computed tomography with a fixed dose (75 mL) of contrast. The results were impressive: the rate of CIN was 21% in the placebo group and only 2% in the NAC group ($P = 0.01$). This finding, however, was supported by some, but not all, subsequent clinical trials investigating the efficacy of NAC as a CIN preventive agent, both in patients with preexisting CKD and in those with normal renal function [48–50].

Several meta-analyses were published on this topic [51–61] (Table 3). By combining the data from available prospective controlled clinical trials that used NAC, they reported an overall significant relative risk reduction in patients with CKD receiving NAC [62]. Nine meta-analyses presented pooled risk estimates suggesting benefit. However, as the available literature is greatly heterogeneous, the benefit of oral NAC among all individuals with renal insufficiency cannot be definitely confirmed [54]. Differences in contrast media type and volumes, definitions of CIN, patient selection, type of intervention, applied hydration regimens, NAC dose (cumulative dosage varied between 1,500 and >10,000 mg in the

Table 3 Meta-analyses of studies on the prophylactic use of *N*-acetylcysteine to prevent contrast-induced nephropathy

Source	Procedure	Type of study	Number of trials	Number of patients	Heterogeneity	Pooled estimate (95% CI)	Author conclusions
Birck et al. [51]	CT or angiography	A	7	805	Present ($P = 0.02$)	RR 0.44 (0.22–0.88)	Beneficial
Isenbarger et al. [52]	CT or angiography	A	7	805	Present ($P = 0.01$)	OR 0.37 (0.16–0.84)	Beneficial
Alonso et al. [53]	CT or angiography	A, B	8	805	Not reported	RR 0.41 (0.22–0.79)	Beneficial
Kshirsager et al. [54]	CT or angiography	A, B	16	1,538	Present ($P < 0.001$)	Not reported	Inconclusive
Pannu et al. [55]	CT or angiography	A, B, D	15	1,776	Present ($P = 0.02$)	RR 0.65 (0.43–1.00)	Inconclusive
Guru and Fremes [56]	CT or angiography	A, C	11	1,213	Present ($P = 0.01$)	OR 0.46 (0.32–0.66)	Beneficial
Bagshaw and Ghail [57]	Angiography	A	14	1,261	Present ($P = 0.03$)	OR 0.54 (0.32–0.91)	Inconclusive
Misra et al. [58]	Angiography	A	5	643	Present ($P = 0.05$)	RR 0.30 (0.11–0.82)	Beneficial
Nallamothu et al. [59]	CT or angiography	A, D	20	2,195	Present ($P = 0.01$)	RR 0.73 (0.52–1.0)	Inconclusive
Liu et al. [60]	CT or angiography	A, B	9	1,028	Present ($P = 0.03$)	RR 0.43 (0.24–0.75)	Beneficial
Duong et al. [61]	CT or angiography	A, C	14	1,584	Present ($P = 0.01$)	RR 0.57 (0.37–0.84)	Beneficial

A randomized controlled trials (articles); *B*, randomized controlled trials (abstracts); *C*, not randomized trials (articles); *D*, unpublished; *CT*, computed tomography; *RR*, relative risk; *OR*, odd ratio; *CI*, confidence intervals. Modified from Bagshaw et al. [137]

different studies), and route of administration (intravenous vs. oral), as well as the timing of the cardiovascular procedure (urgent vs. elective) may have contributed to the heterogeneity (i.e., variation of effect across studies greater than can be expected by chance) observed in the pooled analysis [63].

An interesting point was raised by Briguori et al. [50] who observed that NAC-associated renal protection was restricted to patients receiving small (<140 mL) volumes of dye. They postulated that the discordance in results among different studies may be due to the amount of contrast administered. Indeed, patients undergoing percutaneous coronary interventions (PCI), endovascular aortic aneurysm repair, and other peripheral vascular interventions often require larger contrast doses as compared to that used by Eggebrecht et al. (75 mL) [64].

Additional studies seem to support this hypothesis. In the RAPPID study [65], patients with mild-to-moderate CKD undergoing PCI were randomized to receive NAC and intravenous hydration or intravenous hydration alone. NAC was given intravenously at a dose of 150 mg/kg before contrast exposure, followed by 50 mg/kg over the subsequent 4 hours. Therefore, for a 70-kg patient, the cumulative NAC dose was 14,000 mg, a value much greater than that used by most authors (2,400 mg). In the two groups, the contrast volume was 238 and 222 mL and CIN occurred in 5 and 21% of cases, respectively ($P = 0.045$). In another study by Briguori et al. [66], two different NAC dosages were compared (600 vs. 1,200 mg orally twice daily) before and after contrast administration in patients with mild CKD undergoing coronary or peripheral procedures. The incidence of CIN was lower in patients receiving a double dose of NAC (3.5 vs. 11%; $P = 0.038$). The benefit of double-dose NAC was greater for patients receiving a contrast dose ≥140 mL (5.4 vs. 18.9%; $P = 0.039$) than for those who received a contrast dose <140 mL (1.7 vs. 3.6%; $P = 0.61$). Thus, the emerging concept from these studies is that a greater dose of NAC is probably needed in CKD patients undergoing PCI, suggesting a dose-dependent protective effect of NAC (Fig. 1). Further evidence of a possible dose-dependent effect of NAC derives from a study evaluating its use for the prevention of CIN in patients with ST-segment elevation myocardial infarction (STEMI) undergoing primary PCI [67]. Patients treated with primary PCI represent a population at higher risk for CIN than those undergoing elective PCI because several conditions may contribute to renal injury in this setting. Among them, STEMI-associated hemodynamic impairment, use of a large volume of contrast media, and the impossibility of starting a renal prophylactic therapy are the factors most likely involved [68]. In this study, a total of 352 STEMI patients were randomly assigned to receive placebo (control group, $n = 119$), an intravenous bolus of 600 mg of NAC before PCI, followed by an oral administration (600 mg twice daily) for the following 48 h (NAC total dose = 3,000 mg) (NAC group, $n = 116$), or an intravenous bolus of 1,200 mg of NAC before intervention, followed by an oral administration (1,200 twice daily) for the following 48 h (NAC total dose = 6,000 mg) (high-dose NAC group, $n = 118$). The observed rate of CIN (increase in creatinine ≥25%) was 37% in the control group, 15% in the NAC group, and 8% in the high-dose NAC group ($P < 0.001$). When an absolute rise in creatinine (≥0.5 mg/dL) was considered, the frequency of

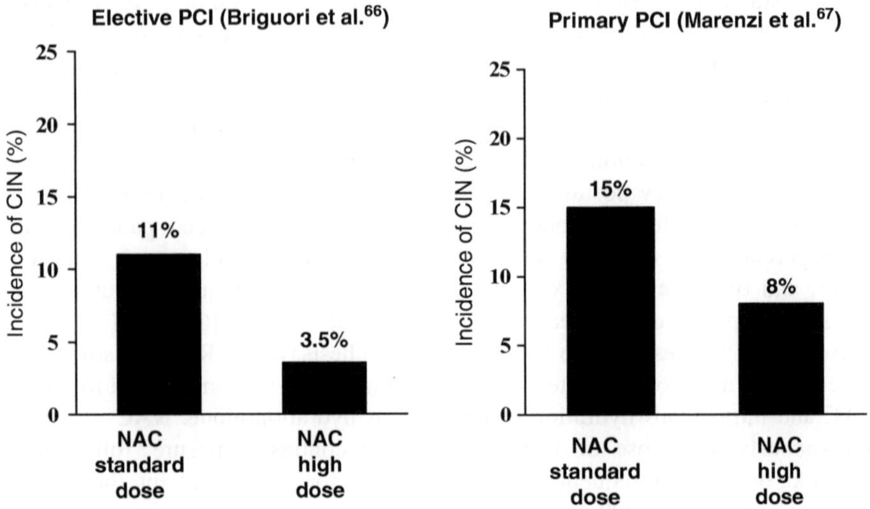

Fig. 1 Dose-dependent protective effect of *N*-acetylcysteine (NAC) against contrast-induced nephropathy (CIN). *PCI* percutaneous coronary intervention

CIN was 18, 6, and 3%, respectively ($P < 0.001$). A significant trend toward a reduction of in-hospital death and other clinical complications in patients receiving NAC was also observed. The mechanisms through which NAC reduces CIN and improves clinical outcomes in this clinical setting, however, remain unclear, and additional studies should investigate whether the extrarenal effects of NAC play some beneficial role. Indeed, in both clinical and experimental acute myocardial infarction studies, intravenous infusion of NAC was associated with decreased infarct size and left ventricular function improvement, possibly due to the antioxidant and free radical scavenger properties of this drug [69, 70]. These cardiac effects may be enhanced in patients treated with primary PCI. This is a clinical setting in which oxidative stress and reperfusion injury were demonstrated to occur and in which these deleterious phenomena are particularly pronounced due to higher coronary patency rates, with more rapid and complete flow restoration. Moreover, it was demonstrated that NAC inhibits platelet aggregation, and this effect too could be relevant during acute coronary thrombosis and mechanical thrombus fragmentation [71].

Positive results were also observed in the RENO study [72] in which hydration with sodium bicarbonate plus NAC, started just before contrast injection and continued for the following 12 h in patients undergoing emergency PCI (primary PCI in 43% of cases), reduced the incidence of CIN (1.8 vs. 21.8%; $P < 0.001$) and anuric AKI (1.8 vs. 12.7%; $P = 0.032$) in comparison to the standard hydration protocol consisting of intravenous isotonic saline for 12 h after PCI [15]. In both groups, two doses of oral NAC were administered the next day.

A recently published meta-analysis evaluating the efficacy of high-dose NAC for the prevention of CIN seems to support this concept [73]. High-dose NAC was a priori defined as a daily dose greater than 1,200 mg or a single periprocedural dose greater than 600 mg, periprocedural being defined as immediately or within 4 h of planned contrast exposure. Sixteen prospective studies of patients (total sample size of 1,677 subjects) randomized to NAC administered either orally or intravenously vs. a control group (842 assigned to high-dose NAC, 835 to the control arm) were included in this meta-analysis. The overall effect size revealed an odds ratio of 0.46 (95% confidence interval [CI]: 0.33–0.63; $P < 0.0001$) for the occurrence of CIN with the use of high-dose NAC, suggesting a significant protective effect of high-dose NAC against CIN (Fig. 2). Notably, in this meta-analysis, the definition of high-dose NAC was arbitrary. It must be noted, however, that no dose finding studies with NAC used as a CIN preventive agent were performed. Indeed, the so-called standard dosing is purely based on the original Tepel et al. [17] study that employed a 600-mg dose 2 times a day. Therefore, the appropriate dose remains uncertain except that a dose greater than that used in this study decreases the risk of CIN. A similar lack of clear guideline limits the choice of the optimal route, orally vs. intravenously, since studies have employed varying routes. Further, and more importantly, the existing data do not address the issue of whether high-dose NAC has any impact on clinical outcome other than that of the incidence of CIN.

The relevant pathogenetic role of oxidative stress in CIN development seems also to be confirmed by the preliminary evidence of the effectiveness of ascorbic acid, statins, and trimetazidine, all agents that were shown to reduce oxidative stress [74–76].

		Incidence of CIN	
Study	**Total N**	**Control^**	**NAC^**
Durham (2002)	79	22.0%	26.3%
Baker (2003)	80	20.5%	4.9%
Kefer (2003)	104	5.9%	3.8%
Oldemeyer (2003)	96	6.4%	8.2%
Efrati (2003)*	49	8.0%	0.0%
Goldenberg (2004)	80	7.7%	9.8%
Miner (2004)	180	22.4%	9.5%
Ochoa (2004)	80	25.0%	8.3%
Rashid (2004)	94	6.2%	6.5%
Balderramo (2004)	61	10.7%	3.0%
Marenzi (2006)	237	32.8%	8.5%
Khalili (2006)	70	17.1%	5.7%
Poletti (2007)	87	16.3%	4.5%
Shaikh (2007)	161	13.8%	9.9%
Shaikh (2007)	159	7.6%	10.0%
Heng (2007)	60	6.2%	3.6%
Summary			

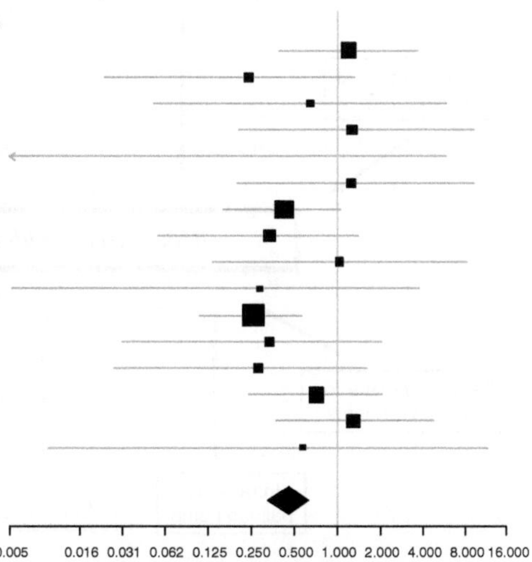

OR 0.46 (95% CI:0.330.63; P<0.0001)

Fig. 2 Forest plot of odds ratios (OR). Horizontal lines represent 95% confidence intervals (CI). *CIN*, contrast-induced nephropathy; *NAC*, *N*-acetylcysteine. Modified from [73]

Based on the evidence of the most recent literature, and given its potential benefit, low-cost and excellent safety profile, the use of high-dose NAC should be recommended in all high-risk patients for the prevention of CIN, particularly when its grim prognosis is considered.

3 Acute Kidney Injury After Cardiac Surgery

Acute kidney injury is a common complication following cardiac surgery that is associated with significant morbidity and mortality and prolonged intensive care unit and hospital stay. This is true for AKI necessitating renal replacement therapies, as well as for AKI not requiring dialysis [77–79]. The frequency of AKI for all patients undergoing cardiac surgery is about 1–8%. In 6–20% of these patients, dialysis is required [77–81]. It was reported that 12–64% of these patients die, compared with 1–5% of those without AKI [77–82]. The mortality risk after cardiac surgery increases with even minor elevations of creatinine value from baseline and may exceed 50% in the most severe cases that require hemodialysis [6]. Several risk factors for postoperative AKI have been entertained. The most consistent are preexisting CKD, advanced age, history of heart failure, diabetes mellitus, prolonged cardiopulmonary bypass (CPB) time, and recent exposure to nephrotoxic agents, such as contrast dye [3, 77, 83–85] (Fig. 3). In patients with moderate and severe CKD, as defined by an estimated creatinine clearance lower than 60 mL/min,

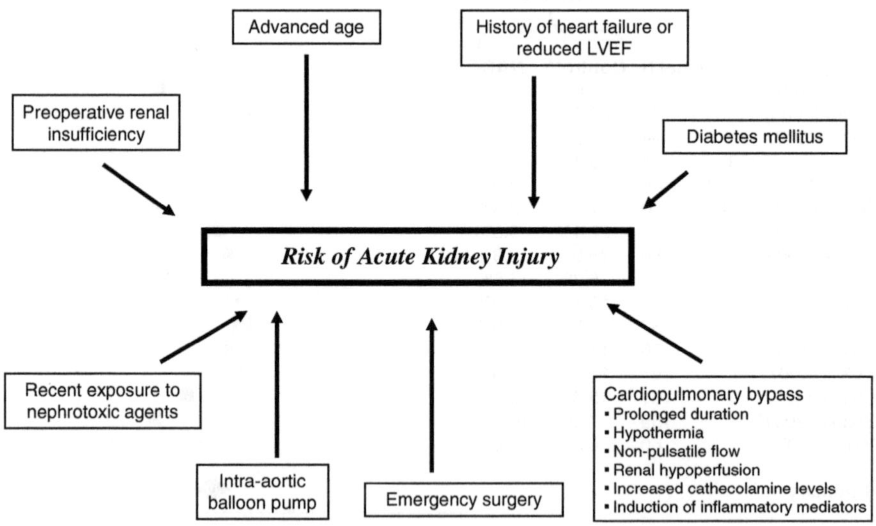

Fig. 3 Risk factors for acute kidney injury after cardiac surgery. *LVEF* left ventricular ejection fraction

postoperative AKI occurs in almost 50% of patients and is associated with a 10-times higher perioperative mortality, increasing to up to more than 30 times for patients requiring a renal replacement therapy [86]. It has been demonstrated that the incidence of both short- and long-term mortality increases in parallel with the increasing severity of the preoperative renal impairment [87–89]. Acute renal dysfunction also has an important impact on nonrenal morbidity. Indeed, patients who develop AKI after cardiac surgery have higher incidence of gastrointestinal bleeding, respiratory failure, infections, and sepsis [92, 93].

Many factors contribute to postcardiac surgery AKI, including exogenous toxins, hemodynamic and metabolic factors, inflammation, neurohormonal activation, vasoconstrictor compounds release induced by CPB, and the interactions between blood components and artificial membranes [4]. All these mechanisms may contribute to vasoconstriction and renal ischemia and result in formation of oxygen-free radicals. In particular, use of CPB during cardiac surgery and systemic exposure to these nonbiologic surfaces liberate oxygen-free radicals from activated neutrophils [92, 93], provoking an oxidative stress response [97–99] that can activate systemic inflammatory processes [100]. Moreover, CPB has been shown to increase renal vascular resistance [77, 101–103] and to release emboli [104]. The use of CPB can also increase serum substances including total peroxide, reactive oxidative metabolites, C reactive protein, and interleukin-6 [97, 105, 106]. This may contribute to the development of postcardiac surgery complications including AKI [107], myocardial injury [70, 44], and atrial fibrillation [108, 109]. Thus, agents with antioxidant properties may attenuate the oxidative stress [92, 107–109] and resultant inflammation [95, 96, 110–112] observed in patients undergoing cardiac surgery and may potentially reduce postoperative complications. In these patients, NAC was demonstrated to reduce oxygen-free radical production, pump-related ischemia–reperfusion injury, and pro-inflammatory cytokine levels. In rats undergoing CPB, NAC ameliorates kidney injury [113].

Temporal trends in cardiac surgery indicate that patients with comorbidities and complex surgical procedures are increasing [114]. Thus, AKI after cardiac surgery is an important health problem whose incidence is expected to increase. Therefore, the prevention of renal deterioration, particularly in patients with preexisting CKD, should improve morbidity and mortality and decrease hospital stay and costs.

As high-risk patients can be easily identified before cardiac surgery, implementation of prophylactic measures represents the best opportunity to prevent AKI, and many pharmacologic strategies have been proposed. Although there are isolated reports suggesting that perioperative administration of fenoldopam, clonidine, natriuretic peptides, sodium nitroprusside, or elective preoperative hemodialysis may prevent AKI, none of these interventions have demonstrated clear efficacy. Thus, exploration of new prophylactic strategies continues.

3.1 N-Acetylcysteine for Prevention of Acute Kidney Injury After Cardiac Surgery

Whether perioperative NAC administration reduces the risk of AKI after cardiac surgery remains unclear. Only a few single-center studies with relatively small sample sizes have addressed this issue. Differences in drug dosing and delivery, patient selection, and evaluation of renal function may explain some disparity. In the first large study by Burns et al. [18], in which 295 high-risk cardiac surgery patients (with one or more of the following criteria: creatinine >1.4 mg/dL, age ≥70 years, diabetes mellitus, ejection fraction <35%, complex surgery or redo) were considered, no difference in the proportion of patients with postoperative AKI was found (30% in NAC-treated patients and 29% in control group). Similarly, no difference in the proportion of patients requiring a renal replacement therapy (0.7 vs. 2.1%, respectively) was observed. Furthermore, in-hospital mortality was not affected by NAC (3.4 vs. 2.7%, respectively). However, the NAC dose used was similar to that initially proposed by Tepel et al. [17] for the prevention of CIN in patients undergoing a low-dose contrast exposure during computed tomography. Given the complex interplay of factors that may impact renal function during cardiac surgery, it is possible that a greater dose of NAC is needed for kidney protection in this setting. In a recent study, Haase et al. [115] used higher doses of NAC (300 mg/kg) in 61 high-risk cardiac surgery patients with similar negative results. In both these studies, however, only a minority of patients (23 and 17%, respectively) had preexisting CKD. Indeed, the renoprotective effect of NAC in patients with risk factors other than renal insufficiency has never been demonstrated. Therefore, due to either the nonhomogeneous population considered [18, 115] or the small number of patients included in these studies, the prophylactic effects of NAC in patients with preexisting CKD undergoing cardiac surgery remains unproved, particularly when significant clinical endpoints are considered.

Sisillo et al. [86] investigated the effects of high-dose intravenous NAC (1,200 mg for 4 times) in 256 patients with baseline CKD (estimated glomerular filtration rate <60 mL/min). A nonstatistically significant reduction in the incidence of AKI in NAC-treated patients was observed in this study as compared to controls (40 vs. 52%; $P = 0.06$). The need for renal replacement therapy and in-hospital mortality rate were not influenced by the drug (8 vs. 5% and 4 vs. 3%, respectively). Notably, when only patients undergoing cardiac surgery with CPB support were considered (90% of the whole study population), NAC-treated patients showed a significantly lower incidence of AKI than controls (40 vs. 54%; $P = 0.03$). It is possible that the difference in AKI incidence reflects the counteracting action of NAC on oxidative stress, reperfusion injury, and systemic inflammatory response associated with CPB. Indeed, severe oxidative stress was demonstrated to occur from the extracorporeal circulation of blood [116], blood cardioplegia, and reperfusion injury [117]. Moreover, off-pump procedures are associated with lower degrees of oxidative stress than on-pump cardiac surgery [118]. Pretreatment with NAC was previously found to reduce the CPB-induced oxidative stress and inflammatory response and to preserve myocardial function after cardioplegic arrest [109]. Attenuation of myocardial dysfunction due to ischemia-reperfusion during CPB may in turn preserve renal perfusion and result in lower incidence of AKI. The deleterious phenomena observed in this setting may be amplified in patients with CKD, a clinical condition associated with increased oxidative stress [35]. Moreover, in the Sisillo et al. study, patients treated with NAC were less likely to require mechanical ventilation for a prolonged period and had a shorter intensive care unit stay, suggesting a positive effect of NAC on CPB-induced ischemia-reperfusion pulmonary injury and, as a result, on the overall clinical course of the patients (Fig. 4).

Fischer et al. [119] found a nonsignificant improvement in creatinine clearance with NAC in 40 patients with normal preoperative renal function. No protection was offered by NAC in a study of 80 patients with preoperative creatinine ranging between 1 and 4 mg/dL [120]. Barr and Kolodner [121] compared fenoldopam, a selective stimulator of the DA-1 dopamine receptor that increases renal blood flow, and NAC (600 mg orally twice a day for 2 days), alone or in combination, with placebo in a prospective, randomized, double-blinded trial that enrolled 79 patients with CKD (creatinine clearance <40 mL/min) who underwent cardiac surgery.

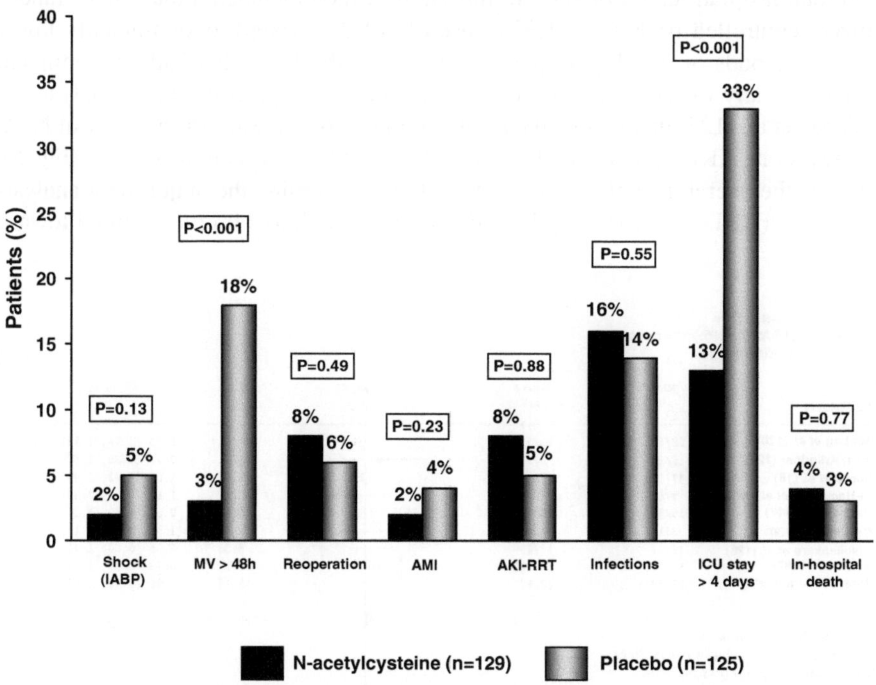

Fig. 4 Perioperative clinical complications in patients with chronic renal insufficiency, undergoing cardiac surgery randomized to *N*-acetylcysteine vs. placebo. *AMI*, acute myocardial infarction; *AKI-RRT* acute kidney injury requiring renal replacement therapy; *IABP*, intra-aortic balloon pump; *ICU*, intensive care unit; *MV*, mechanical ventilation. Modified from [86]

They found that the change in creatinine clearance from preoperative to postoperative day 3 was statistically less for patients treated with fenoldopam or NAC. However, only a trend in favor of patients with the combination of the two drugs was found. Moreover, there was no decrease in length of critical care and hospital stay or hospital costs.

Despite potential effectiveness and increasing interest, only few systematic reviews comprehensively assessed the potential efficacy and adverse effects of perioperative NAC administration in adults undergoing cardiac surgery. In the meta-analysis by Ho and Morgan [122], ten studies involving a total of 1,193 patients undergoing cardiac and noncardiac surgery (elective abdominal aneurysm repair surgery, esophagectomy, total gastrectomy, pancreatectomy, and liver resection surgery) were analyzed. Use of NAC was not associated with a decrease in mortality (odds ratio [OR], 1.05; 95% CI, 0.52–1.92), acute renal failure requiring dialysis (OR, 1.04; 95% CI, 0.45–2.37), AKI defined as a greater than 25% increase in serum creatinine above baseline (OR, 0.84; 95% CI, 0.64–1.11), or length of intensive care unit stay (OR, 0.46; 95% CI, −0.43–1.36). Another meta-analysis by Baker et al. [123] evaluated the potential beneficial effects of perioperative NAC administration in patients undergoing cardiac surgery, in terms of atrial fibrillation occurrence, myocardial infarction, stroke, AKI, need for renal replacement therapy, mortality, and total hospital length of stay. In this study, which included a total of 13 randomized, controlled trials ($n = 1{,}338$), use of NAC appeared to significantly lower (36%) the odds of developing postsurgery atrial fibrillation but had no significant impact on any of the other examined endpoints. Finally, in the meta-analysis by Adabag et al. [124] that included ten studies and 1,163 patients, the efficacy of NAC in preventing AKI was evaluated only in patients undergoing cardiac surgery (Fig. 5). Despite the overall negative result of this study concerning the major meta-analysis endpoints (AKI, need for dialysis, and death), the authors observed a trend toward

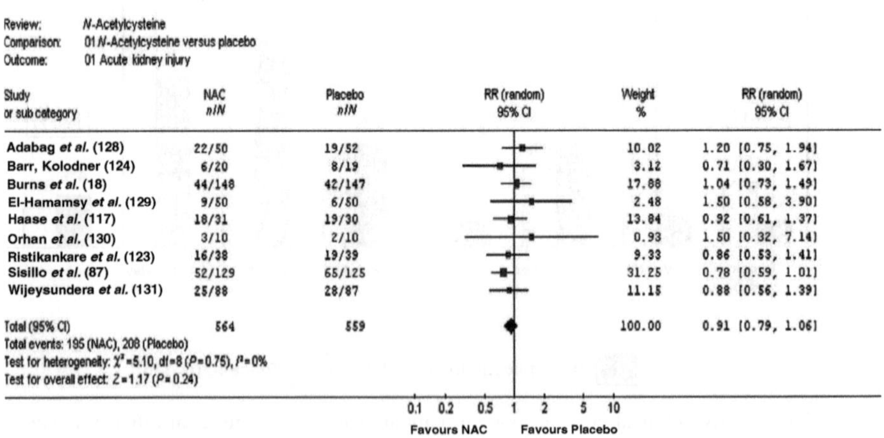

Fig. 5 Forest plot describing the relative risk of acute kidney injury among cardiac surgery patients randomized to *N*-acetylcysteine (NAC) vs. placebo. Modified from [124]

reduced AKI incidence among patients with preexisting CKD randomized to NAC (relative risk [RR] 0.86; 95% CI, 0.70–1.05; $P = 0.14$), particularly if NAC was administered intravenously (RR 0.80; 95% CI, 0.64–1.01; $P = 0.06$).

In summary, to date use of NAC for the prevention of postoperative AKI in all patients undergoing cardiac surgery is not supported by evidence. However, future trials are needed to determine if intravenous NAC may improve major clinical outcomes in patients with preexisting CKD undergoing cardiac surgery with CPB support. In addition, the most appropriate dose of NAC and the optimal length of therapy are still unknown. Thus, future studies should directly compare various doses of NAC and include different lengths of therapy to address these important issues.

4 Unresolved Issues

There have been some concerns that NAC may affect serum creatinine, the surrogate marker of glomerular filtration rate that is routinely employed in clinical practice, without affecting glomerular filtration rate. In an uncontrolled study of healthy volunteers, one group of investigators found that NAC reduced serum creatinine but not cystatin C, and speculated that NAC may have an effect on serum creatinine independent of glomerular filtration rate [129]. However, this NAC effect was never confirmed in the setting of AKI. Experimental administration of high doses of NAC (200 and 500 mg/kg) to protect against AKI showed a similar pattern of change in serum creatinine and cystatin C, indicating no influence of NAC on serum creatinine levels [113]. In another study [130] in which serum creatinine and cystatin C were both assayed, the NAC arm showed a significant correlation between creatinine and cystatin C at baseline and a better correlation 48 h after contrast exposure. In agreement with these results, a randomized trial designed to prevent AKI in CKD patients that employed high-dose NAC (300 mg/kg IV) showed no difference in the direction of creatinine and cystatin C response and ruled out a creatinine lowering effect or a difference in urinary creatinine excretion related to NAC [120, 131]. Finally, a more recent study in which a "double-dose" NAC was administered in the absence of iodinated contrast media to patients with stable CKD showed no effect of NAC on either serum creatinine or cystatin C levels [132].

N-acetylcysteine is generally considered to have an excellent side-effect profile and appears to be safe [18]. However, in a recent post-hoc analysis of a randomized controlled trial in patients with preexisting moderate CKD undergoing cardiac surgery, NAC was associated with increased blood loss and blood product transfusion [133]. This observation is consistent with recent data indicating that NAC impairs coagulation factors and platelet function [134]. Although a higher risk of surgical reexploration for bleeding or an increased need of blood transfusions was not observed in the meta-analysis by Ho and Morgan [122], future studies are warranted to elucidate the mechanism by which NAC may impair hemostasis in patients undergoing cardiac surgery and to assess the overall risk–benefit profile of NAC for perioperative kidney protection.

5 Conclusion and Future Directions

Acute kidney injury frequently causes prolongation of hospitalization, increase in cost of care, and need for dialysis in severe cases. However, the most disturbing consequence is poor long-term outcome, particularly in patients who develop AKI after contrast exposure and cardiac surgery. Abundant data suggest that development of AKI in these clinical settings is associated with increased short- and long-term mortality. *N*-acetylcysteine has been studied as a prophylactic agent because of its decreased effect on free radical formation, a potential pathogenetic mechanism of both CIN and postcardiac surgery AKI. Data regarding the efficacy of NAC are conflicting and prior meta-analyses demonstrated significant heterogeneity of NAC effect across studies. However, a recent meta-analysis of studies that used a dose higher than that originally employed by Tepel et al. [17] revealed no heterogeneity and suggested a benefit for the prevention of CIN [73]. Data on NAC use for the prevention of postoperative AKI in cardiac surgery patients cannot be considered conclusive and need further investigation.

Most studies that have investigated clinical outcomes related to prophylaxis of AKI assessed the incidence of acute postcontrast and postoperative renal failure. However, the question as to whether prevention of this clinical outcome leads to a decrease in mortality represents a very important and unresolved issue. The few studies that have analyzed a mortality endpoint were not planned or powered to detect the effect of prophylaxis on mortality. In this regard, there are data to support the assumption that NAC may have a positive impact on important clinical outcomes. Indeed, NAC has been shown to improve pulse pressure, endothelial function, and to reduce cardiovascular events in patients with advanced renal failure [135, 136]. Further research is needed to determine the effect of high-dose NAC employed for AKI prevention on the mortality of patients undergoing diagnostic and interventional procedures and cardiac surgery.

References

1. Maitino AJ, Levin DC, Parker L, et al. Nationwide trends in rates of utilization of noninvasive diagnostic imaging among the Medicare population between 1993 and 1999. Radiology 2003;227:113–117.
2. Nash K, Hafeez A, Hou S. Hospital-acquired renal insufficiency. Am J Kidney Dis 2002;39:930–936.
3. Chertow G, Levy E, Hammermeister K, Grover F, Daley J. Independent association between acute renal failure and mortality following cardiac surgery. Am J Med 1998;104:343–348.
4. Hobson CE, Yavas S, Segal MS, et al. Acute kidney injury is associated with increased long-term mortality after cardiothoracic surgery. Circulation 2009;119:2444–2453.
5. Wijeysundera DN, Rao V, Beattie WS, Ivanov J, Karkouti K. Evaluating surrogate measures of renal dysfunction after cardiac surgery. Anesth Analg 2003;96:1265–1273.
6. Lassnigg A, Schmidlin D, Mouhieddine M, et al. Minimal changes of serum creatinine predict prognosis in patients after cardiothoracic surgery: a prospective cohort study. J Am Soc Nephrol 2004;15:1597–1605.

7. Ryckwaert F, Boccara G, Frappier JM, Colson PH. Incidence, risk factors, and prognosis of a moderate increase in plasma creatinine early after cardiac surgery. Crit Care Med 2002;30:1495–1498.
8. Lameire N, van Biesen W, Hoste E, Vanholder R. The prevention of acute kidney injury: an in-depth narrative review. Nephrol Dial Transplant Plus 2009;1:1–10.
9. Baliga R, Ueda N, Walker PD, et al. Oxidant mechanisms in toxic acute renal failure. Am J Kidney Dis 1997;29:465–477.
10. Edelstein CL, Schrier RW. Pathophysiology of ischemic acute renal injury. In: Schrier RW (ed.). Diseases of the Kidney and Urinary Tract. Philadelphia: Lippincott Williams & Wilkins 2007;930–961.
11. Duru M, Nacar A, Yönden Z, et al. Protective effects of N-acetylcysteine on cyclosporine-A-induced nephrotoxicity. Ren Fail 2008;30:453–459.
12. Sheikh-Hamad D, Timmins K, Jalali Z. Cisplatin-induced renal toxicity: possible reversal by N-acetylcysteine treatment. J Am Soc Nephrol 1997;8:1640–1644.
13. Fernández-Fúnez A, Polo FJ, Broseta L, Valer J, Zafrilla L. Effects of N-acetylcysteine on myoglobinuric acute renal failure in rats. Ren Fail 2002;24:725–733.
14. Mazzon E, Britti D, De Sarro A, Caputi AP, Cuzzocrea S. Effect of N-acetylcysteine on gentamicin-mediated nephropathy in rats. Eur J Pharmacol 2001;424:75–83.
15. Schaller G, Pleiner J, Mittermayer F, Posch M, Kapiotis S, Wolzt M. Effects of N-acetylcysteine against systemic and renal hemodynamic effects of endotoxin in healthy humans. Crit Care Med 2007;35:1869–1875.
16. Hynninen MS, Niemi TT, Pöyhiä R, et al. N-acetylcysteine for the prevention of kidney injury in abdominal aortic surgery: a randomized, double-blind, placebo-controlled trial. Anesth Analg 2006;102:1638–1645.
17. Tepel M, van Del Giet M, Schwarzfeld NR, et al. Prevention of radiographic-contrast-agent-induced reductions in renal function by acetylcysteine. N Engl J Med 2000;343:180–184.
18. Burns KEA, Chu MWA, Novick RJ, et al. Perioperative N-acetylcysteine to prevent renal dysfunction in high-risk patients undergoing CABG surgery. A randomized controlled trial. JAMA 2005;294:342–350.
19. Barrett BJ, Parfrey PS. Preventing nephropathy induced by contrast medium. N Engl J Med 2006;354:379–386.
20. Bartorelli AL, Marenzi G. Contrast-induced nephropathy. J Interv Cardiol 2008;21:74–85.
21. McCullough PA, Dandberg KR. Epidemiology of contrast-induced nephropathy. Rev Cardiovasc Med 2003;4(Suppl 5):1172–1181.
22. McCullough PA, Wolyn R, Rocher LL, Levin RN, O'Neill WW. Acute renal failure after coronary intervention: incidence, risk factors and relationship to mortality. Am J Med 1997;103:368–375.
23. Bartholomew BA, Harjai KJ, Dukkipati S, et al. Impact of nephropathy after percutaneous coronary intervention and a method for risk stratification. Am J Cardiol 2004;93:1515–1519.
24. Rihal CS, Textor SC, Grill DE, et al. Incidence and prognostic importance of acute renal failure after percutaneous coronary intervention. Circulation 2002;105:2259–2264.
25. Gruber L, Mehran R, Dangas G, et al. Acute renal failure requiring dialysis after percutaneous coronary interventions. Catheter Cardiovasc Interv 2001;52:409–416.
26. Goldenberg I, Chonchol M, Guetta V. Reversible acute kidney injury following contrast exposure and the risk of long-term mortality. Am J Nephrol 2009;29:136–144.
27. Mehran R, Aymong ED, Nikolsky E, et al. A simple risk score for prediction of contrast-induced nephropathy after percutaneous coronary intervention. J Am Coll Cardiol 2004;44;1393–1399.
28. Cochran ST, Wong WS, Roe DJ. Predicting angiography-induced acute renal function impairment: clinical risk model. Am J Roentgenol 1983;141:1027–1033.
29. Rich MW, Crecelius CA. Incidence, risk factors and clinical course of acute renal insufficiency after cardiac catheterization in patients 70 years of age or older. A prospective study. Arch Intern Med 1990;150:1237–1242.

30. Briguori C, Marenzi G. Pharmacologic prophylaxis. Kidney Intern 2006;69:S30–S38.
31. Marenzi G, Bartorelli A. Recent advances in radiocontrast-induced nephropathy prevention. Curr Opin Crit Care 2004;10:505–509.
32. Reddan D, Laville M, Garovic VD. Contrast-induced nephropathy and its prevention: what do we really know from evidence-based findings? J Nephrol 2009;22:333–351.
33. Persson PB, Tepel M. Contrast medium-induced nephropathy: the pathophysiology. Kidney Int 2006;69(Suppl 100):S8–S10.
34. Tumlin J, Stacul F, Adam A, et al. Pathophysiology of contrast-induced nephropathy. Am J Cardiol 2006;98(Suppl):14K–20K.
35. Massy ZA, Nguyen-Khoa T. Oxidative stress and chronic renal failure: markers and management. J Nephrol 2002;15:336–341.
36. Scott JA, King GL. Oxidative stress and antioxidant treatment in diabetes. Ann N Y Acad Sci 2004;1031:204–213.
37. Yoshimasa Y, Fogo A, Beckman JD. Reduced activity of antioxidant enzymes underlies contrast media-induced renal injury in volume depletion. Kidney Int 1992;41:1008–1015.
38. Bakris GL, Lass N, Gaber AO, et al. Radiocontrast medium-induced declines in renal function: a role for oxygen free radicals. Am J Physiol Renal Physiol 1990;258:F115–F120.
39. Parvez Z, Rahman MA, Moncada R. Contrast media-induced lipid peroxidation in the rat kidney. Invest Radiol 1989;24:697–702.
40. Hanss BG, Valencia SH, Shah SV, Vari SC. The iron chelator deferoxamine prevents contrast media induced acute renal failure in the rabbit. J Am Soc Nephrol 1990;1:612.
41. Drager LF, Andrade L, Barros de Toledo JF, Laurindo FR, Machado César LA, Seguro AC. Renal effects of N-acetylcysteine in patients at risk for contrast nephropathy: decrease in oxidant stress-mediated renal tubular injury. Nephrol Dial Transplant 2004;19:1803–1807.
42. DiMari J, Megyesi J, Udvarbelyi N, et al. N-acetylcysteine ameliorates ischemic renal failure. Am J Physiol 1997;272:F292–F298.
43. Conesa EL, Valero F, Nadal JC, et al. N-acetyl-L-cysteine improves renal medullary hypoperfusion in acute renal failure. Am J Physiol 2001;281:R730–R737.
44. Forman MB, Puett DW, Cates CU, et al. Glutathione redox pathway and reperfusion injury. Effect of N-acetylcysteine on infarct size and ventricular function. Circulation 1988;78:202–213.
45. Lopez BL, Snyder JW, Birenbaum DS, Ma XI. N-acetylcysteine enhances endothelium-dependent vasorelaxation in the isolated rat mesenteric artery. Ann Emerg Med 1998;32:405–410.
46. Heyman SN, Goldfarb M, Shina A, et al. N-acetylcysteine ameliorates renal microcirculation: studies in rats. Kidney Int 2003;63:634–641.
47. Diaz-Sandoval LJ, Kosowsky BD, Losordo DW. Acetylcysteine to prevent angiography-related renal tissue injury (the APART trial). Am J Cardiol 2002;89:356–358.
48. Shyu KG, Cheng JJ, Kuan P. Acetylcysteine protects against acute renal damage in patients with abnormal renal function undergoing a coronary procedure. J Am Coll Cardiol 2002;40:1383–1388.
49. Kay J, Chow WH, Chan TM, et al. Acetylcysteine for prevention of acute deterioration of renal function following elective coronary angiography and intervention: a randomized controlled trial. JAMA 2003;289:553–558.
50. Briguori C, Manganelli F, Scarpato P, et al. Acetylcysteine and contrast-agent associated nephrotoxicity. J Am Coll Cardiol 2002;40:298–303.
51. Birck R, Krzossok S, Markowetz F, et al. Acetylcysteine for prevention of contrast nephropathy. Lancet 2003;362:598–603.
52. Isenbarger D, Kent S, O'Malley P. Meta-analysis of randomized clinical trials on the usefulness of acetylcysteine for prevention of contrast nephropathy. Am J Cardiol 2003;92:1454–1458.

53. Alonso A, Lau J, Jaber BL, et al. Prevention of radiocontrast nephropathy with N-acetylcysteine in patients with chronic kidney disease: a meta-analysis of randomized, controlled trials. Am J Kidney Dis 2004;43:1–9.
54. Kshirsagar A, Poole C, Mottl A, et al. N-acetylcysteine for the prevention of radiocontrast induced nephropathy: a meta-analysis of prospective controlled trials. J Am Soc Nephrol 2004;15:761–769.
55. Pannu N, Manns B, Lee H, Tonelli M. Systematic review of the impact of N-acetylcysteine on contrast nephropathy. Kidney Int 2004;65:1366–1374.
56. Guru V, Fremes S. The role of N-acetylcysteine in preventing radiographic contrast-induced nephropathy. Clin Nephrol 2004;62:77–83.
57. Bagshaw S, Ghali WA. Acetylcysteine for prevention of contrast-induced nephropathy: a systematic review and meta-analysis. BMC Med 2004;2:38.
58. Misra D, Leibowitz K, Gowda R, Shapiro M, Khan I. Role of N-acetylcysteine in prevention of contrast-induced nephropathy after cardiovascular procedures: a meta-analysis. Clin Cardiol 2004;27:607–610.
59. Nallamothu BK, Shojania KG, Saint S, et al. Is acetylcysteine effective in preventing contrast-related nephropathy? A meta-analysis. Am J Med 2004;117:938–947.
60. Liu R, Nair D, Ix J, Moore D, Bent S. N-acetylcysteine for prevention of contrast-induced nephropathy: a systematic review and meta-analysis. J Gen Intern Med 2005;20:193–200.
61. Duong M, MacKenzie T, Malenka D. N-acetylcysteine prophylaxis significantly reduces the risk of radiocontrast-induced nephropathy. Catheter Cardiovasc Interv 2005;64:471–479.
62. Tepel M, Aspelin P, Lameire N. Contrast-induced nephropathy: a clinical and evidence-based approach. Circulation 2006;113:1799–1806.
63. Kelly AM, Dwamena B, Cronin P, Bernstein SJ, Carlos RC. Meta-analysis: effectiveness of drugs for preventing contrast-induced nephropathy. Ann Intern Med 2008;148:284–294.
64. Eggebrecht H, Breuckmann F, Martini S, et al. Frequency and outcomes of acute renal failure following thoracic aortic stent-graft placement. Am J Cardiol 2006;98:458–463.
65. Baker CSR, Wragg A, Kumar S, et al. A rapid protocol for the prevention of contrast-induced renal dysfunction: the RAPPID study. J Am Coll Cardiol 2003;41:2114–2118.
66. Briguori C, Colombo A, Violante A, et al. Standard vs double dose of N-acetylcysteine to prevent contrast agent associated nephrotoxicity. Eur Heart J 2004;25:206–211.
67. Marenzi G, Assanelli E, Marana I, et al. N-acetylcysteine and contrast-induced nephropathy in primary angioplasty. N Engl J Med 2006; 354:2773–2782.
68. Marenzi G, Lauri G, Assanelli E, et al. Contrast-induced nephropathy in patients undergoing primary angioplasty for acute myocardial infarction. J Am Coll Cardiol 2004;44:1780–1785.
69. Arstall MA, Yang J, Stafford I, Betts WH, Horowitz JD. N-acetylcysteine in combination with nitroglycerin and streptokinase for the treatment of evolving acute myocardial infarction: safety and biochemical effects. Circulation 1995;92:2855–2862.
70. Sochman J, Kole J, Vrana M, Fabian J. Cardioprotective effects of N-acetylcysteine: the reduction in the extent of infarction and occurrence of reperfusion arrhythmias in the dog. Int J Cardiol 1990;28:191–196.
71. Anfossi G, Russo I, Massucco P, Mattiello L, Cavalot F, Trovati M. N-acetyl-L-cysteine exerts direct anti-aggregating effect on human platelets. Eur J Clin Invest 2001;31:452–461.
72. Recio-Majoral A, Chaparro M, Prado B, et al. The reno-protective effect of hydration with sodium bicarbonate plus N-acetylcysteine in patients undergoing emergency percutaneous coronary intervention. The RENO study. J Am Coll Cardiol 2007;49:1283–1288.
73. Trivedi H, Daram S, Szabo A, Bartorelli AL, Marenzi G. High-dose N-acetylcysteine for the prevention of contrast-induced nephropathy. Am J Med 2009;122:874.e9–874.e15.
74. Spargias K, Alexopoulos E, Kyrzopoulos S, et al. Ascorbic acid prevents contrast-mediated nephropathy in patients with renal dysfunction undergoing coronary angiography or intervention. Circulation 2004;110:2837–2842.
75. Onbasili AO, Yeniceriglu Y, Agaoglu P, et al. Trimetazidine in the prevention of contrast-induced nephropathy after coronary procedures. Heart 2007;93:698–702.

76. Khanal S, Attallah N, Smith DE, Kline-Rogers E, Share D, O'Donnell MJ, Moscucci M. Statin therapy reduces contrast-induced nephropathy: an analysis of contemporary percutaneous interventions. Am J Med 2005;118:843–849.
77. Mangano CM, Diamondstone LS, Ramsay JG, Aggarwal A, Herskowitz A, Mangano DT, The Multicenter Study of Perioperative Ischemia Research Group. Renal dysfunction after myocardial revascularization: risk factors, adverse outcomes, and hospital resources utilization. Ann Intern Med 1998;128:194–203.
78. Hix JK, Thakar CV, Katz EM, et al. Effect of off-pump coronary artery bypass graft surgery on postoperative acute kidney injury and mortality. Crit Care Med 2006;34:2979–2983.
79. Mangos GJ, Brown MA, Chan WY, et al. Acute renal failure following cardiac surgery: incidence, outcomes and risk factors. Aust NZ J Med 1995;25:284–289.
80. Chertow GM, Lazarus JM, Christiansen CL, et al. Preoperative renal risk stratification. Circulation 1997;95:878–884.
81. Llopart T, Lombardi R, Forselledo M, et al. Acute renal failure in open heart surgery. Ren Fail 1997;19:319–323.
82. Hein OV, Birnbaum J, Wernecke KD, et al. Three-year survival after four major postcardiac operative complications. Crit Care Med 2006;34:2729–2737.
83. Zanardo G, Michielon P, Paccagnella A, et al. Acute renal failure in the patient undergoing cardiac operation. Prevalence, mortality rate, and main risk factors. J Thorac Cardiovasc Surg 1994;107:1489–1495.
84. Novis BK, Roizen MF, Aronson S, Thisted RA. Association of preoperative risk factors with postoperative acute renal failure. Anesth Analg 1994;78:143–149.
85. Salis S, Mazzanti VV, Merli G, et al. Cardiopulmonary bypass duration is an independent predictor of morbidity and mortality after cardiac surgery. J Cardiothorac Vasc Anesth 2008;22:814–822.
86. Sisillo E, Ceriani R, Bortone F, et al. N-acetylcysteine for prevention of acute renal failure in patients with chronic renal insufficiency undergoing cardiac surgery. A prospective, randomized, clinical trial. Crit Care Med 2008;36:338–340.
87. Lock CE, Austin PC, Wuang H, Tu JV. Impact of renal insufficiency on short- and long-term outcomes after cardiac surgery. Am Heart J 2004;148:430–438.
88. Hillis GS, Croal BL, Buchan KG, et al. Renal function and outcome from coronary artery bypass grafting. Impact on mortality after a 2.3-year follow-up. Circulation 2006;113:1056–1062.
89. Cooper WA, O'Brien SM, Thourani VH, et al. Impact of renal dysfunction on outcomes of coronary artery bypass surgery. Results from the society of thoracic surgeons national adult cardiac database. Circulation 2006;113:1063–1070.
90. Anderson RJ, O'Brien M, MacWhinney S, et al. Renal failure predisposes patients to adverse outcome after coronary artery bypass surgery: VA Cooperative Study #5. Kidney Int 1999;55:1057–1062.
91. Hilberman M, Myers BD, Carrie BJ, Derby G, Jamison RL, Stinson EB. Acute renal failure following cardiac surgery. J Thorac Cardiovasc Surg 1979;77:880–888.
92. Tossios P, Bloch W, Huebner A, et al. N-acetylcysteine prevents reactive oxygen species-mediated myocardial stress in cardiac surgery patients: results of a randomized double-blinded placebo-controlled clinical trial. J Thorac Cardiovasc Surg 2003;126:1513–1520.
93. Andersen LW, Thiis J, Kharazmi A, et al. The role of N-acetylcysteine administration on the oxidative response of neutrophils during cardiopulmonary bypass. Perfusion 1995;10:21–26.
94. Starkopf J, Zilmer K, Vihalemm T, Kullisaar T, Zilmer M, Samarutel J. Time course of oxidative stress during open-heart surgery. Scand J Thorac Cardiovasc Surg 1995;29:181–186.
95. Paparella D, Yau TM, Young E. Cardiopulmonary bypass induced inflammation: pathophysiology and treatment. An update. Eur J Cardiothorac Surg 2002;21:232–244.
96. Chaney MA. Corticosteroids and cardiopulmonary bypass: a review of clinical investigations. Chest 2002;121:921–931.

97. Goettea A, Lendeckel U. Nonchannel drug targets in atrial fibrillation. Pharmacol Ther 2004;102:17–36.
98. Hilberman M, Derby GC, Spencer RJ. Sequential pathophysiological changes characterizing the progression from renal dysfunction to acute renal failure following cardiac operation. J Thorac Cardiovasc Surg 1980;79:838–844.
99. Mazzarella V, Gallucci MT, Toxxo C, et al. Renal function in patients undergoing cardiopulmonary bypass operations. J Thorac Cardiovasc Surg 1992;104:1625–1627.
100. Colson P, Ribstein J, Mimran A, et al. Effect of angiotensin converting enzyme inhibition on blood pressure and renal function during open heart surgery. Anesthesiology 1990;72:23–27.
101. Sreeram GM, Grocott HP, White WD, et al. Transcranial Doppler emboli count predicts rise in creatinine after coronary artery bypass graft surgery. J Cardiothorac Vasc Anesth 2004;18:548–551.
102. Bruins P, te Velthuis H, Yazdanbakhsh AP, et al. Activation of the complement system during and after cardiopulmonary bypass surgery: postsurgery activation involves C reactive protein and is associated with postoperative arrhythmia. Circulation 1997;96:3542–3548.
103. Gaudino M, Andreotti F, Zamparelli R, et al. The 174G/C interleukin-6 polymorphism influences postoperative interleukin-6 levels and postoperative atrial fibrillation. Is atrial fibrillation an inflammatory complication? Circulation 2003;108:195–199.
104. Baliga R, Ueda N,Walker PD, Shaw SW. Oxidant mechanisms in toxic acute renal failure. Am J Kidney Dis 1997;29:465–477.
105. Neuman RB, Bloom HL, Shukrullah I, et al. Oxidative stress markers are associated with persistent atrial fibrillation. Clin Chem 2007;53:1652–1657.
106. Ramlawi R, Out H, Mieno S, et al. Oxidative stress and atrial fibrillation after cardiac surgery: a case-control study. Ann Thorac Surg 2007;84:1166–1173.
107. McColl AJ, Keeble T, Hadjinikolaou L, et al. Plasma antioxidants: evidence for a protective role against reactive oxygen species following cardiac surgery. Ann Clin Biochem 1998;35:616–623.
108. Fischer UM, Tossios P, Huebner A, Geissler HJ, Bloch W, Mehlhorn U. Myocardial apoptosis prevention by radical scavenging in patients undergoing cardiac surgery. J Thorac Cardiovasc Surg 2004;128:103–108.
109. Sucu N, Cinel I, Unlu A, et al. N-acetylcysteine for preventing pump-induced oxidoinflammatory response during cardiopulmonary bypass. Surg Today 2004;34:237–242.
110. Baker WL, White CM, Kluger J, Denowitz A, Konecny CP, Coleman CI. Effect of perioperative corticosteroid use on the incidence of postcardiothoracic surgery atrial fibrillation and length of stay. Heart Rhythm 2007;4:461–468.
111. Lertsburapa K, White CM, Kluger J, Faheem O, Hammond J, Coleman CI. Preoperative statins for the prevention of atrial fibrillation after cardiothoracic surgery. J Thorac Cardiovasc Surg 2008;135:405–411.
112. Patti G, Chello M, Candura D, Pasceri V, D'Ambrosio A, Covino E, Di Sciascio G. Randomized trial of atorvastatin for reduction of postoperative atrial fibrillation in patients undergoing cardiac surgery. Circulation 2006;114:1455–1461.
113. Zhu J, Yin R, Shao H, Dong G, Luo L, Jing H. N-acetylcysteine to ameliorate acute renal injury in a rat cardiopulmonary bypass model. J Thorac Cardiovasc Surg 2007;133:696–703.
114. Ferguson TB Jr, Hammill BG, Peterson ED, DeLong ER, Grover FL, STS National Database Committee. Decade of change-risk profiles and outcomes for isolated coronary artery bypass grafting procedures, 1990–1999: a report from the STS National Database Committee and the Duke Clinical Research Institute. Ann Thorac Surg 2002;73:480–489.
115. Haase M, Haase-Fielitz A, Bagshaw SM, et al. Phase II, randomized, controlled trial of high-dose N-acetylcysteine in high-risk cardiac surgery patients. Crit Care Med 2007;35:1324–1331.
116. Royston D, Fleming JS, Desai JB, et al. Increased production of peroxidation products associated with cardiac operations. J Thorac Cardiovasc Surg 1986;91:759–766.
117. Pepper JR, Mumby S, Gutteridge JMC. Blood cardioplegia increases plasma iron overload and thiol levels during cardiopulmonary bypass. Ann Thorac Surg 1995;60:1735–1740.

118. Biglioli P, Cannata A, Alamanni F, et al. Biological effects of off-pump vs. on-pump coronary artery surgery: focus on inflammation, hemostasis and oxidative stress. Eur J Cardiothorac Surg 2003;24:260–269.
119. Fischer UM, Tossios P, Mehlhorn U. Renal protection by radical scavenging in cardiac surgery patients. Curr Med Res Opin 2005;21:1161–1164.
120. Ristikankare A, Kuitunen T, Kuitunen A, et al. Lack of renoprotective effect of IV N-acetylcysteine in patients with chronic renal failure undergoing cardiac surgery. Br J Anaesth 2006;97:611–616.
121. Barr LF, Kolodner K. N-acetylcysteine and fenoldopam protect the renal function of patients with chronic renal insufficiency undergoing cardiac surgery. Crit Care Med 2008;36:1427–1435.
122. Ho KM, Morgan DJR. Meta-analysis of N-acetylcysteine to prevent acute renal failure after major surgery. Am J Kidney Dis 2008;53:33–40.
123. Baker WL, Anglade MW, Baker EL, White CM, Kluger J, Coleman CI. Use of N-acetylcysteine to reduce post-cardiothoracic surgery complications: a meta-analysis. Eur J Cardiothorac Surg 2009;35:521–527.
124. Adabag AS, Ishani A, Bloomfield HE, Hgo AK, Wilt TJ. Efficacy of N-acetylcysteine in preventing renal injury after heart surgery: a systematic review of randomized trials. Eur Heart J 2009;30:1910–1917.
125. Adabag AS, Ishani A, Koneswaran S, et al. Utility of N-acetylcysteine to prevent acute kidney injury after cardiac surgery: a randomized controlled trial. Am Heart J 2008;155:1143–1149.
126. El-Hamamsy I, Stevens LM, Carrier M, et al. Effect of intravenous N-acetylcysteine on outcomes after coronary artery bypass surgery: a randomized, double-blind, placebo-controlled clinical trial. J Thorac Cardiovasc Surg 2007;133:7–12.
127. Orhan G, Yapici N, Yuksel M, et al. Effects of N-acetylcysteine on myocardial ischemia-reperfusion injury in bypass surgery. Heart Vessels 2006;21(1):42–47
128. Wijeysundera DN, Beattie WS, Rao V, Granton JT, Chan CT. N-acetylcysteine for preventing acute kidney injury in cardiac surgery patients with pre-existing moderate renal insufficiency. Can J Anaesth 2007;54(11):872–881
129. Hoffmann U, Fischereder M, Kruger B, et al. The value of N-acetyl-cysteine in the prevention of radiocontrast agent-induced nephropathy seems questionable. J Am Soc Nephrol 2004;15:407–410.
130. Poletti PA, Saudan P, Platon A, et al. IV N-acetylcysteine and emergency CT: use of serum creatinine and cystatin C as markers of radiocontrast nephrotoxicity. Am J Roentgenol 2007;189:687–692.
131. Haase M, Haase-Fielitz A, Ratnaike S, et al. N-acetylcysteine does not artifactually lower plasma creatinine concentration. Nephrol Dial Transplant 2008;23:1581–1587.
132. Rehman T, Fought J, Solomon R. N-acetylcysteine effect on serum creatinine and cystatin C levels in CKD patients. J Am Soc Nephrol 2008;3:1610–1614.
133. Wijeysundera DN, Karkouti K, Rao V, et al. N-acetylcysteine is associated with increased blood loss and blood product utilization during cardiac surgery. Crit Care Med 2009;37:1929–1934.
134. Niemi TT, Munsterhjelm E, Poyhia R, et al. The effect of N-acetylcysteine on blood coagulation and platelet function in patients undergoing open repair of abdominal aortic aneurysm. Blood Coagul Fibrinolysis 2006;17:29–34.
135. Scholze A, Rinder C, Beige J, Riezler R, Zidek W, Tepel M. Acetylcysteine reduces plasma homocysteine concentration and improves pulse pressure and endothelial function in patients with end-stage renal failure. Circulation 2004;109:369–374.
136. Tepel M, van der Giet M, Statz M, Jankowski J, Zidek W. The antioxidant acetylcysteine reduces cardiovascular events in patients with end-stage renal failure: a randomized, controlled trial. Circulation 2003;107:992–995.
137. Bagshaw SM, et al. Arch Intern Med 2006;166:161–166

Chapter 20
Advanced Glycation End Products Inhibitor

Takashi Dan, Charles van Ypersele de Strihou, and Toshio Miyata

Abstract Early intensive glycemic control in both type 1 and type 2 diabetes mellitus retards in the long term the development and progression of microvascular complications such as diabetic nephropathy, even despite a worsening glycemic control. This phenomenon is a so-called metabolic memory or legacy effect, partly ascribed to the advanced glycation end products (AGEs), whose formation is closely linked to the glycoxidative biology. In this chapter we summarize the role of AGEs and their receptor (receptor for AGEs: RAGE) in the onset and progression of diabetic vascular complications, especially in diabetic nephropathy, and review current and future treatment strategies targeting the AGE-RAGE system, including (a) AGE formation inhibitor (including angiotensin receptor blocker [ARB]), (b) AGE cross-link breaker, (c) RAGE antagonist, (d) AGE binder including sRAGE (soluble receptor for AGEs), and (e) hypoxia-inducible factor (HIF) activator. Several inhibitors of tissue accumulation of AGEs in diabetes have been clinically tested, including inhibitors of AGE formation, such as aminoguanidine, benfotiamine, and pyridoxamine, or AGE cross-link breakers, such as ALT-711. The clinical benefits of ARBs to protect the kidney are, at least in part, due to the inhibition of AGE formation. Recently antagonists of the peroxisome proliferator-activated receptor-γ (PPAR-γ), such as rosiglitazone and pioglitazone, which also provide renal benefits clinically, have been recognized as RAGE antagonists. Oral adsorbents, such as AST-120 (Kremezin) and Sevelamer carbonate, bind AGEs and reduce their plasma levels in experimental models. Several other approaches relying on the AGE hypothesis have been proposed to treat diabetic complications in experimental models, but their benefits are yet to be documented clinically. In summary, inhibition of the AGE-RAGE system confers some degree of experimental or clinical protection against diabetic microvascular complications, albeit to different degrees and by different mechanisms.

T. Dan (✉)
Division of Molecular Medicine and Therapy, United Centers for Advanced Research and Translational Medicine (ART), Tohoku University Graduate School of Medicine, 2-1 Seiryo-Machi, Aoba-ku, Sendai, Miyagi 980-8575, Japan
e-mail: dantks@m.tains.tohoku.ac.jp

T. Miyata et al. (eds.), *Studies on Renal Disorders*, Oxidative Stress in Applied Basic Research and Clinical Practice, DOI 10.1007/978-1-60761-857-7_20,

Keywords Advanced glycation end products (AGEs) · AGE-RAGE system · AGE inhibitor · AGE cross-link breaker · RAGE (receptor for AGE) antagonist · AGE binder · Hypoxia-inducible factor (HIF) activator · Diabetic microvascular complications

1 Introduction

Despite the availability of various insulin preparations or oral hypoglycemic agents, complete normalization of blood sugar and prevention of microvascular complications in diabetic patients are not always easy to accomplish. Diabetes mellitus has thus become an important contributor to the rising prevalence of diabetic microvascular complications (e.g., nephropathy, retinopathy, and neuropathy). The mechanisms of progression of diabetic microvascular complications as well as more innovative therapeutic approaches are therefore urgently needed.

Large-scale clinical studies, such as Diabetes Control and Complication Trial (DCCT) [1] and United Kingdom Prospective Diabetes Study (UKPDS) [2], have shown in the 1990s that the onset and progression of diabetic microvascular complications can be prevented by the strict control of blood sugar. Multivariate analyses from the large-scale clinical trials of DCCT's subsequent follow-up for another 10 years, in the Epidemiology of Diabetes Interventions and Complications (EDIC) study, have led to an interesting hypothesis concerning the mechanism of diabetic macrovascular and microvascular complications [3, 4], the so-called metabolic memory hypothesis. Strict control of blood sugar levels in the early stages of type 1 diabetes mellitus led to a significantly slower retinopathy progression and a decreased incidence of proliferative retinopathy, even if glycemic control worsened subsequently [5]. The difference in the prevalence of diabetic retinopathy between former intensive and conventional therapy persisted for at least 10 years and was attributed to metabolic memory. Early intensive therapy had the same benefits for nephropathy: an 86% reduction was found in the development of albuminuria (>300 mg/day) at year 4 of the EDIC study [6]. The metabolic memory was also maintained for neuropathy at year 8 of the EDIC study [7].

The best etiological hypothesis for the phenomenon of metabolic memory rests with the advanced glycation end products (AGEs), which are irreversibly generated and accumulate in type 1 diabetic patients in proportion to the degree and duration of glycemic control (so-called diabetic exposure). Glycation and subsequent AGEs formation in skin collagen predict progression of microvascular disease [4]. They may thus explain the risk of these complications and provide a rational basis for the phenomenon of metabolic memory.

The metabolic memory hypothesis [8], also known as the legacy effect [9], has been proposed subsequently by several researchers. Further studies indeed published in 2008 (ACCORD, VADT, UKPDS posttrial) [10, 11] highlight the existence of an AGE-related metabolic memory in type 2 diabetes.

In this chapter, we address the role of AGEs and their receptors (e.g., receptor for AGEs, RAGE) in the onset and progression of diabetic vascular complications, especially in diabetic nephropathy and retinopathy, and review established and new treatment strategies targeting the AGE-RAGE system.

2 Advanced Glycation End Products and Their Receptors

2.1 Advanced Glycation End Products

Advanced glycation end products are formed nonenzymatically between the protein amino group and reduced sugar carbonyl group, such as glucose and its metabolites, through the formation of Schiff base products and Amadori rearrangements followed by irreversible dehydration, condensation, and cross-linkage. This chain of events is called the Maillard reaction, as this reaction was first reported by French food chemist Louis Camille Maillard. Monnier and Cerami discovered in 1981 that, in humans, it progresses with aging [12]. Originally, AGEs shared both the physicochemical features of cross-linked molecules with characteristic fluorescence and browning: accumulation of browning products was correlated with the severity of diabetic complications such as retinopathy, nephropathy, and arterial and joint stiffness [13]. More recently, specific AGEs have been identified and used as surrogates for AGEs: pentosidine, crosslines, imidazolones, N^{ε}-(carboxymethyl)-lysine (CML) and pyrraline, although the latter two structures do not have the classical characteristics of AGEs. Still, they constitute only a minor fraction of all of the AGE structures present in the body, and it is not yet clear which AGE structures are concerned with the onset and progress of various chronic diseases. AGEs are also generated on proteins by reactions with reactive dicarbonyl compounds, such as glyoxal (GO), methylglyoxal (MGO), and 3-deoxyglucosone (3-DG), whose formation is dramatically accelerated by hyperglycemia, uremia, or oxidative stress [14].

In diabetic patients, the concentration of plasma and urine AGEs is correlated with the severity of retinopathy, neuropathy, and nephropathy as well as with macrovascular complications [15, 16]. Dicarbonyl compounds, such as GO, MGO, and 3-DG, also increase in diabetic blood but respond to glycemic control by insulin therapy [17]. Carbonyl stress, defined as excessive accumulation of reactive dicarbonyl compounds, further modifies proteins, lipids, and DNA, followed by oxidative stress and tissue damage [18–20]. AGE formation and its attendant cellular damage have thus been implicated in the genesis and progression of diabetic complications.

Oxidative stress contributes to AGE formation as shown by the correlation observed in diabetic and uremic serum between pentosidine (an AGE) and oxidative markers such as dehydroascorbate and advanced oxidation protein products (AOPP) [21, 22] as well as by the colocalization of both oxidation-dependent AGE structures and lipid peroxidation products observed in diabetic glomerular lesions [23, 24].

2.2 Receptors for AGEs

AGEs damage cells by two different mechanisms. First, they directly modify extracellular-matrix proteins, membrane proteins, or intracellular proteins, and alter their functions. Second, specific cellular receptors binding AGEs moieties induce intracellular signaling and cause broad cellular responses.

RAGE is a well-studied AGE receptor and was originally identified in cow lungs as a membrane protein of 35 kDa belonging to the immunoglobulin super family and able to bind AGEs [25]. Human RAGE incorporates a sugar chain and weights 55 kDa. Its extracellular portion includes three immunoglobulin-like domains and a variable region-like domain whose N-terminal binds an AGE ligand. Besides AGEs, RAGE binds several biomolecules, such as amyloid beta [26], HMGB-1/amphoterin [27], S100-/calgranulin [28], and transthyretin [29], suggesting the involvement of RAGE not only in the AGE disorders but also in various pathologies, such as Alzheimer's disease, cancer, and inflammation.

Several effects of AGEs on cells, such as mesangial cells, podocytes, endothelial cells, and pericytes, are mediated through an interaction with RAGE. RAGE expression indeed increases in mesangial cells and podocytes of human diabetic nephropathy [30]. The pivotal role of RAGE in diabetic renal injury has been demonstrated in RAGE transgenic or knockout mice [31, 32].

2.3 Soluble RAGE

Three alternative splicing forms of RAGE have been recently reported in humans: a full-length RAGE, a RAGE without one N-terminus immunoglobulin-like loop, and a soluble type RAGE lacking in the membrane domain [33]. The soluble form of RAGE, also named endogenous secretory RAGE (esRAGE), is actually secreted from the human endothelial cells and exists in the circulation. Since esRAGE has an AGE binding site, it potentially competes with membrane-spanning RAGE and captures circulating AGEs, suggesting that esRAGE might prevent the progression of AGE-induced events.

3 Inhibitors of the AGE-RAGE System

The AGE-RAGE system is a molecular target for the treatment of diabetic complications. Current and future therapeutic strategies interfering with the AGE-RAGE system include (a) AGE inhibitors (including angiotensin receptor blockers [ARB]), (b) AGE cross-link breakers, (c) RAGE antagonists, (d) AGE binders including esRAGE, and (e) hypoxia-inducible factor (HIF) activator (Fig. 1).

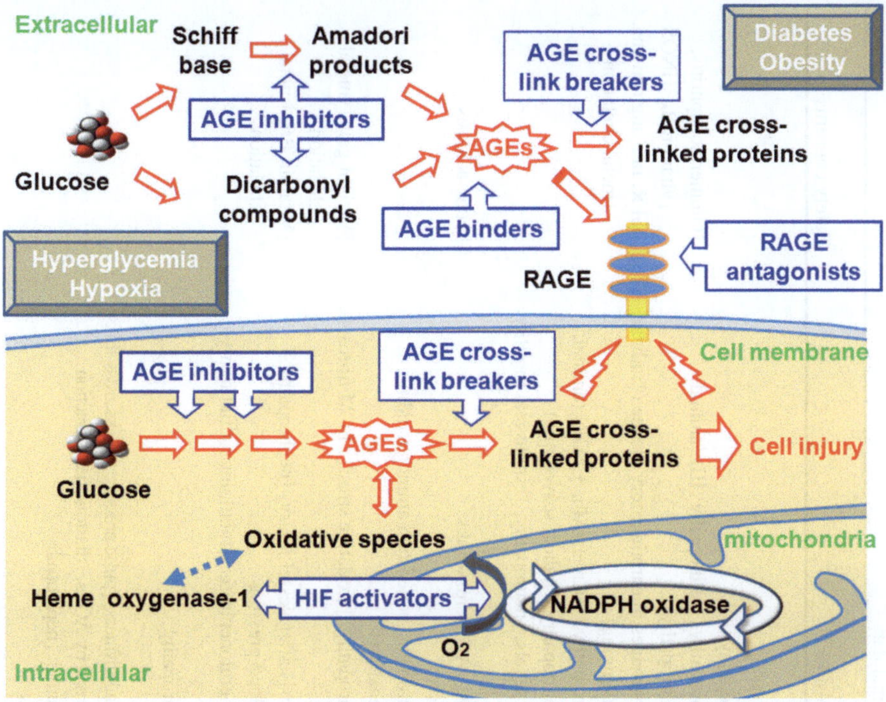

Fig. 1 Schematic representation of inhibitors of the AGE-RAGE system. Current and future therapeutic strategies interfering with the AGE-RAGE system include (a) AGE inhibitors (including angiotensin receptor blockers [ARB]), (b) AGE cross-link breakers, (c) RAGE antagonists, (d) AGE binders including esRAGE, and (e) hypoxia-inducible factor (HIF) activators

3.1 AGE Inhibitor

Aminoguanidine, pyridoxamine, and benfotiamine are known clinical AGE inhibitors (Table 1). They act through the covalent binding of toxic reactive carbonyl precursors for AGEs.

Aminoguanidine (Pimagedine) was the first AGE inhibitor studied [34]. In human clinical trials, it reduces the level of AGE-hemoglobin, independently of HbA1c [35]. Randomized, placebo-controlled trials have evaluated it in patients with type 1 and 2 diabetes [36, 37]. In a phase III clinical trial in type 1 diabetic patients, aminoguanidine significantly reduced proteinuria and progression of retinopathy, whereas the progression to overt nephropathy was not statistically improved [36]. Safety concerns and apparent lack of efficacy raised by the External Safety Monitoring Committee drastically limited the clinical evaluation of this agent in type 2 diabetic patients [38]. Clinical development of aminoguanidine was further hampered by its long-term toxicity, such as the development of myeloperoxidase and antineutrophil antibodies [37], glomerulonephritis [36] in some patients, and kidney tumors in diabetic rats [38].

Table 1 Therapeutic agents targeting the AGEs-RAGE system: human and animal data

Category of therapy	Most advanced stage relating to AGEs	Trial results	Safety concerns
AGE inhibitors 0(prevent formation of AGEs)			
Benfotiamine	Human, phase IV	↓Neuropathy	
Aminoguanidine	Human, phase III	↓Nephropathy, ↓retinopathy (III) (animal: ↓neuropathy)	↑Glomerulonephritis, ↓vitamin B_6, ↓iNOS
ARBs	Human, phase III	↓Macrophages in carotid artery plaque (↑radical scavenging)	↓GFR, rate angioedema, ↑metal chelation
Pyridoxamine	Human, phase II	↓Nephropathy (↓sCr, ±uAlb, ±AGE) (animal: ↓retinopathy, ↑radical scavenging)	
AR inhibitors (epalrestat, zopolrestat)	Human, phase, II	↓AGE levels, ↓neuropathy, ↑esophageal motility	
Metformin	Animal (diabetic rats)	↓AGEs, ↓AGE cross-links	Lactic acidosis
Aspirin	Animal (diabetic dogs)	↓AGEs	
ALT-946	Animal (diabetic rats)	↓Nephropathy better than aminoguanidine, ↓renal AGE	
LR-90	Animal (diabetic rats)	↓Nephropathy, ↓oxidative stress, ↓ECM fibrosis	↑Weight gain, ↑metal chelation
OPB-9195	Animal (diabetic rats)	↓Stenosis after vessel injury, ↓nephropathy, ↓blood pressure	↓Vitamin B6, ↑metal chelation
TM-2002	Animal (renal injury rats)	↓Glomerulonephritis, ↓tubulointerstitial injury	
R-147176	Animal (diabetic rats)	↓Nephropathy	
AGE cross-link breakers			
ALT-711 (alagebrium chloride)	Human, phase III	↓Arterial stiffness, ↓pulse pressure, breaks cross-links formed by AGEs, ↑diastolic heart function (animal: ↓nephropathy)	
PTB	Animal (diabetic rats)	↓AGEs	

RAGE antagonists			
ACE inhibitors	Human, phase II	↓RAGE levels	↓GFR, rare angioedema
Alpha-lipoic acid	Animal (diabetic rats)	↓AGE albumin-induced NF-κB	
TZDs rosiglitazone, pioglitazone	Animal (diabetic rats)	↓AGEs, ↓AGE cross-links	↑Hepatitis, ↑CHF if susceptible
Soluble RAGE	Animal (diabetic mice)	↓Stenosis after vessel injury, ↓neuropathy, ↓nephropathy	
AGE binders			
AST-120	Human, phase II	↓AGEs	
Sevelamer carbonate	Human, phase II		
Lysozyme	Animal (diabetic/apo E-knockout mice)	↓AGEs, ↓nephropathy, ↓atherosclerosis	
HIF activators			
Cobalt	Animal (diabetic rats)	↓AGE levels ↓nephropathy (proteinuria)	

Pyridoxamine, a vitamin B_6 derivative, prevents the formation of AGEs from Amadori intermediates [39] and cleaves 3-deoxyglucosone-reactive carbonyl intermediates [40]. It inhibits the progression of renal disease and decreases hyperlipidemia and apparent redox imbalances in type 1 diabetic rats [41]. In phase II studies in patients with diabetic nephropathy, pyridoxamine significantly reduced the change from baseline in serum creatinine, whereas no differences in urinary albumin excretion were seen [42]. It also protected peritoneal tissue [43].

Benfotiamine, a liposoluble derivative of vitamin B_1, has AGE lowering properties. In contrast with pyridoxamine, benfotiamine decreases the formation of reducing sugars and intermediates of the polyol pathway [44]. Benfotiamine as well as thiamine reduces diabetic nephropathy and retinopathy in experimental animal models [45]. Administration of benfotiamine to type 2 diabetic patients, on a high AGE content diet, reduced circulating AGE levels and markers of oxidative stress [46]. This benefit, however, has not been confirmed in type 1 diabetic patients [47].

Other drugs currently used in clinical medicine inhibit AGE formation. Metformin, a glucose lowering agent, traps reactive carbonyl compounds [48]. Aspirin also decreases AGE accumulation by targeting preformed intermediates, by chelation of copper and other transition metals contributing to oxidative stress, as well as by scavenging free carbonyls [49].

Aldose reductase inhibitors decrease AGE formation [50] by inhibiting the first rate-limiting steps in the polyol pathway. One such compound, epalrestat, reduces the production of a dicarbonyl intermediate, 3-deoxyglucosone [51]. These drugs improve nerve conduction velocity [52] and ameliorate the esophageal dysfunction of diabetic patients [53]. In a murine diabetic glomerulopathy model, the aldose reductase inhibitor zopolrestat suppresses the AGE-induced enhancement of transforming growth factor-β1 (TGF-β1) and type IV collagen expression [54].

ALT-946 therapy for 12 weeks reduces renal AGE accumulation and cortical tubular degeneration to a greater extent than aminoguanidine and, unlike aminoguanidine, reduces albumin excretion rate in the hypertensive transgenic (mRen-2) 27 rat with streptozotocin-induced diabetes [55]. In another study, ALT-946 reduced equally albuminuria when given to a streptozotocin-induced diabetic rat model either at the onset of diabetes or 16 weeks later [56]. No human data with this agent have been published.

LR-90 inhibits renal and circulating AGE accumulation through its potent metal chelating ability and its interaction with reactive carbonyl species [57]. Given to experimental models of both type 1 and type 2 diabetic nephropathy, it affords renoprotection such as improved albuminuria, glomerulosclerosis, and tubulointerstitial fibrosis, concomitantly with a reduction of renal AGEs, TGF-β1, connective tissue growth factor, fibronectin, and collagen IV deposition [58].

OPB-9195, a thiazolidine derivative, is an agonist of the peroxisome proliferator-activated receptor. It inhibits glycoxidation and lipoxidation thereby decreasing the formation of AGEs and dicarbonyl intermediates. OPB-9195 prevented the progression of diabetic glomerular sclerosis in OLETF (Otsuka Long Evans Tokushima Fatty) rats mainly by lowering serum levels of AGEs and attenuating AGE deposition in the glomeruli [59] but also, perhaps, by scavenging dicarbonyl intermediates [60].

In stroke-prone spontaneously hypertensive (SHRSP) rats, OPB-9195 also slowed the progression of nephropathy, lowered blood pressure, and reduced oxidative stress [61]. This compound unfortunately traps pyridoxal like aminoguanidine with an attendant toxicity, which has prevented its use in humans [62].

We screened a large chemical library (~1,300 compounds) and discovered that one compound, edaravone [63], a drug used to treat cerebral infarction, had in vitro AGE inhibitory activity. Unfortunately, edaravone also traps pyridoxal, thus limiting its clinical usefulness. Taking advantage of edaravone's structure, we synthesized a novel AGE inhibitor, TM-2002, that does not trap pyridoxal [64]. In vitro, TM-2002 is a powerful AGE inhibitor. Like ARBs it also inhibits efficiently markers of oxidation (i.e., o-tyrosine formation) and chelates transition metal but does not bind to the angiotensin II type 1 receptor. It is readily bioavailable and nontoxic. In vivo, TM-2002, given either acutely or for 8 weeks, has no adverse effects. In four different rat models of renal injury (anti-Thy1 and ischemia reperfusion) and of cardiovascular injury (carotid artery balloon injury and angiotensin II-induced cardiac fibrosis), it improved renal and cardiovascular lesions without modification of blood pressure. Interestingly, it also decreased significantly infarct volume in both transient and permanent focal ischemia rat models [65]. TM-2002 inhibition of advanced glycation and oxidative stress was confirmed by a significant reduction of the number of cells positive for AGEs and heme oxygenase-1 and by a reduced level of protein carbonyl formation in parts of the brain. Like ARBs, but unlike aminoguanidine, OPB-9195, pyridoxamine, and LR-90, TM-2002 does not trap carbonyl precursors and belongs thus to a new class of AGE inhibitors (Table 1).

3.2 Angiotensin Receptor Blocker

Renal protection of ARBs is usually explained exclusively by their blood pressure lowering action and by the inhibition of the renin angiotensin system. Indeed, angiotensin II concentration is significantly higher in the kidneys than in the systemic circulation [66]. Recently, however, other pleiotropic actions of ARBs have been invoked. For example, we and others [62, 67] reported that ARBs and angiotensin converting enzyme inhibitors (ACEIs) blocked the formation of AGEs through the reduction rather than through the entrapment of precursor carbonyl compounds. Interestingly, these inhibitory actions are not observed in other classes of antihypertensive agents, such as calcium channel blockers, diuretics, or beta-blockers. Three different types of antihypertensive agents – olmesartan (an ARB), nifedipine (a calcium channel blocker), atenolol (a beta-blocker) – were given for 20 weeks to SHR/NDmcr-cp, a hypertensive type 2 diabetes rat model, so as to depress blood pressure to the same extent. Only olmesartan strongly reduced proteinuria and prevented mesangial and interstitial injury [68]. This finding fits with clinical experience that inhibitors of the renin–angiotensin system (RAS) are better renoprotectors than other types of antihypertensive drugs [69]. In this model,

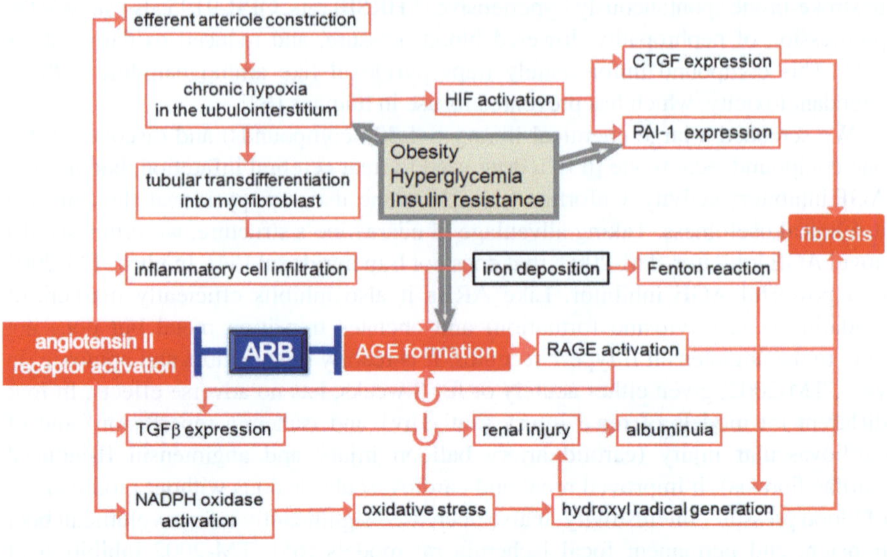

Fig. 2 Proposed renoprotective mechanism of ARB. Angiotensin receptor blockers (ARBs) have renal benefits (i.e., inhibition of the renin–angiotensin system [RAS], prevention of abnormal iron deposition in the interstitium, correction of chronic hypoxia, hydroxyl radical scavenging, reduction of expressions of heme oxygenase-1 and nicotinamide adenine dinucleotide phosphate [NADPH] oxidase, amelioration of inflammatory cell infiltration, and inhibition of pentosidine formation). These benefits of ARB may contribute to reduction of proteinuria and improvement of glomerular and tubulointerstitial damage

the renal benefits of ARB appear independent of the effects on systemic blood pressure and on metabolic abnormalities. Of note, the amount of the renal pentosidine fell only with ARBs and correlated with the inhibition of proteinuria. The expression of heme oxygenase-1 and p47phox (a subunit of nicotinamide adenine dinucleotide phosphate [NADPH] oxidase), markers of oxidative stress, decreased in parallel with the reduction of AGEs.

The rather heterogeneous list of potential determinants or mediators of diabetic nephropathy is tentatively integrated in the hypothetical scheme depicted in Fig. 2. Clearly, the interrelationships suggested between these elements preclude the identification of a single culprit in the genesis of diabetic kidney lesions. Whatever these hypotheses, it is now clear that ARBs and probably ACEIs have unique renoprotective properties, including certainly a decreased oxidative stress, the correction of chronic hypoxia, the inhibition of AGE formation, an abnormal iron deposition, PAI-1 activity, and inflammatory cell infiltration.

In order to dissect the mechanisms of ARBs' protective benefits, we synthesized a novel, nontoxic ARB derivative, R-147176, characterized by a very weak affinity for the angiotensin II type 1 receptor (AT_1) (6,700 times less effective than olmesartan in AT_1 binding inhibition), but a strong inhibition of oxidative stress and AGE formation [70]. Despite a minimal effect on blood pressure, this compound provides a significant renoprotection in two different experimental type 2 diabetic rat models, SHR/NDmcr-cp and Zucker diabetic fatty. Renal benefits of ARBs thus depend, at least partly, on their ability to inhibit oxidative stress and AGE formation. Of interest, R-147176, like ARBs, protected not only the kidney but also brain cells in an experimental rat stroke model [71].

3.3 AGE Breaker

N-phenacylthiazolium bromide (N-PTB) [72], known as an AGE breaker, cleaves nonreversible covalent cross-links, including those of AGEs, in tissue proteins and allows the elimination of glycated proteins via scavenger receptors and renal excretion. Unfortunately, it increases blood pressure [73], so that a more stable thiazolium derivative, ALT-711 (Alagebrium chloride) [74] was developed [75]. In a clinical trial [76], ALT-711 did not change ejection fraction, blood pressure, peak exercise oxygen consumption, or aortic distensibility, but proved able to decrease left ventricular mass and improve left ventricular diastolic filling and quality of life in patients with diastolic heart failure. Despite the therapeutic potential of ALT-711 for cardiovascular complications and dermatological alterations associated with diabetes, all ongoing clinical trials were terminated after review of the manufacturer's clinical development portfolio and financial status.

3.4 RAGE Antagonist

RAGE may prove another use if the signal transduction elicited by AGEs can be inhibited. This concept is supported by the observation that low molecular weight heparin functions as an antagonist of RAGE and prevents diabetic nephropathy [77].

Alpha-lipoic acid, known as an RAGE-post signaling blocker, inhibits the translocation of transcription nuclear factor κB (NFκB) from the cytoplasm to the nucleus. It thus reduces AGE albumin-induced NFκB-mediated transcription and expression of endothelial genes relevant in diabetes [78].

Thiazolidinediones (TZDs) are ligands of the peroxisome proliferator-activated receptor-γ (PPAR-γ). TZDs, such as rosiglitazone [79] and pioglitazone [80], have also been recently identified as RAGE antagonists. They exert beneficial effects in diabetic nephropathy independent of insulin sensitization. Interestingly, rosiglitazone administered to type 2 diabetic subjects increased the serum level of protective sRAGEs besides the decrease of circulating AGE levels [81].

Studies with RAGE knockout mice that do not express sRAGE or full-length RAGE suggest that sRAGE acts via inhibition of RAGE-dependent phenomena

[82]. Recently it has been postulated that ACE inhibition reduces the accumulation of AGE in diabetes partly through an increased production and secretion of sRAGE into the plasma [83]. It remains to be seen whether sRAGE acts as an antagonist inhibiting the RAGE-dependent signaling pathways or as a binder of various RAGE ligands such as AGEs. sRAGE or possibly a nonpeptide RAGE antagonist may thus become a future therapeutic target [84].

3.5 AGE Binder

A low-AGE diet administered for 6 weeks in a clinical trial lowered serum AGE levels and inflammatory markers such as C-reactive protein [85]. In an apolipoprotein E knockout diabetic mouse, a low-AGE diet reduced lesions at the aortic root [86]. An important link has therefore been suggested between dietary intake of preformed glycoxidation products, tissue-incorporated AGEs, and diabetes-accelerated atherosclerosis.

Inhibition of dietary AGEs absorption may be a novel means to reduce the deleterious effects of AGEs. AST-120 (Kremezin), an oral adsorbent, attenuates the progression of chronic kidney disease by removing uremic toxins. It also binds CML and decreases serum levels of AGEs in nondiabetic subjects with chronic kidney disease [87].

Patients with diabetic nephropathy were recruited for a ongoing clinical trial with sevelamer carbonate, a clinically used phosphate binder. It is argued that sevelamer might provide an advantage through a reduction of the levels of AGEs. An attendant reduction of both oxidative stress and inflammation may yield an independent benefit for the progression of chronic kidney and cardiovascular disease.

Lysozyme was also found to bind AGEs and improve their removal [88]. A lysozyme-linked AGE affinity column might thus serve as an efficient method for the removal of toxic AGEs from diabetic or uremic sera [89]. Additional studies in vitro and in vivo in diabetic mice demonstrated that lysozyme may accelerate renal AGE clearance, suppress macrophage and mesangial cell-specific gene activation, and improve albuminuria [90]. Lysozyme appears to confer resistance to AGE-induced oxidative species. It blocked cellular apoptosis in vitro, reduced mortality in vivo [91], and decreased atherosclerosis in apolipoprotein E knockout mice [92].

As previously mentioned, edaravone inhibits AGE formation more effectively than aminoguanidine and pyridoxamine. We therefore developed a new α-methylene-rich compound by binding several edaravone molecules with a polyethyleneimine linker, and coupled it to pyrazolinone-polyethyleneimine with cellulose beads (PPCB) to produce a novel, safe, reactive carbonyl compounds (RCO)-adsorption bead. This high-affinity device reduced the toxic RCO content and AGE formation of peritoneal dialysis fluid and might protect the peritoneal membrane of uremic patients on peritoneal dialysis [93].

3.6 HIF Activator

AGE reduction during renal hypoxia might rely on the activation of HIF [94]. In the kidney, the peritubular capillary plexus, fed by glomerular efferent arterioles, supplies oxygen and nutrients to tubular and interstitial cells. Diabetic glomerular damage decreases the number of peritubular capillaries and thus oxygen diffusion to tubulointerstitial cells, leading to tubular dysfunction and fibrosis [95].

HIF is crucial in the defense against hypoxia [14]. Its activation protects hypoxic tissues through the expression of a broad range of genes, erythropoietin, vascular endothelial growth factor (VEGF), heme oxygenase-1, glucose transporter (GLUT), all of which are critically involved in the defense against chronic hypoxia. HIF stability is drastically reduced by the oxygen-dependent hydroxylation of HIF proline residues by prolyl hydroxylase (PHD), a so-called oxygen sensor. Hydroxylated HIF recruits the von Hippel Lindau protein (pVHL), which in turn tags HIF with ubiquitin and targets it for degradation within the proteasome. Under hypoxic conditions HIF is not hydroxylated and is thus able to activate the genes mitigating the effects of hypoxia.

Cobalt inhibits HIF degradation by PHDs [96]. It was therefore given to hypertensive, type 2 diabetic rats with nephropathy, the SHR/NDmcr-cp strain [94]. Treatment initiated at the age of 13 weeks was continued for 26 weeks. Although it did not correct hypertension and metabolic abnormalities, cobalt reduced proteinuria as well as histological kidney injury and upregulated the renal expression of HIF and of the HIF-regulated genes, including erythropoietin, VEGF, and heme oxygenase-1. Furthermore, cobalt reduced the renal expression of TGF-β, the renal content of AGEs, and the expression of NADPH oxidase, a marker of oxidative stress. Renal AGEs correlated with proteinuria. Cobalt thus achieved renal protection independently of metabolic status and blood pressure through the upregulation of HIF and HIF-regulated genes and a mitigated advanced glycation and oxidative stress. Less toxic, more effective PHD inhibitors should provide renal benefits by improving not only local hypoxia but also local oxidative stress and advanced glycation.

4 Conclusion

A growing body of evidence demonstrates the role of AGEs in the development of diabetic complications, including diabetic nephropathy. Several agents interfering with the AGE-RAGE system have protected the kidney in experimental diabetic models. Some of them, tested in human clinical studies, open encouraging new therapeutic avenues.

References

1. Diabetes Control and Complications Trial Research Group. The effect of intensive therapy of diabetes on the development and progression of long-term complications in insulin-dependent diabetes mellitus. N Engl J Med 1993;329:977–986.
2. UK Prospective Diabetes Study (UKPDS) Group. Intensive blood-glucose control with sulphonylureas or insulin compared with conventional treatment and risk of complications in patients with type 2 diabetes (UKPDS 33). Lancet 1998;352:837–853.
3. Nathan DM, Cleary PA, Backlund JY, et al. Intensive diabetes treatment and cardiovascular disease in patients with type 1 diabetes. N Engl J Med 2005;353:2643–2653.
4. Genuth S, Sun W, Cleary P, et al. Glycation and carboxymethyllysine levels in skin collagen predict the risk of future 10-year progression of diabetic retinopathy and nephropathy in the diabetes control and complications trial and epidemiology of diabetes interventions and complications participants with type 1 diabetes. Diabetes 2005;54:3103–3111.
5. White NH, Sun W, Cleary PA, et al. Prolonged effect of intensive therapy on the risk of retinopathy complications in patients with type 1 diabetes mellitus: 10 years after the diabetes control and complications trial. Arch Ophthalmol 2008;126:1707–1715.
6. Writing Team for the Diabetes Control and Complications Trial/Epidemiology of Diabetes Interventions and Complications Research Group. Effect of intensive therapy on the microvascular complications of type 1 diabetes mellitus. JAMA 2002;287:2563–2569.
7. Martin CL, Albers J, Herman WH, et al. Neuropathy among the diabetes control and complications trail cohort 8 years after trial completion. Diabetes Care 2006;29:340–344.
8. Ceriello A, Ihnat MA, Thorpe JE. Clinical review 2: the "metabolic memory": is more than just tight glucose control necessary to prevent diabetic complications? J Clin Endocrinol Metab 2009;94:410–415.
9. Soldatos G, Cooper ME. Does intensive glycemic control for type 2 diabetes mellitus have long-term benefits for cardiovascular disease risk? Nat Clin Pract Endocrinol Metab 2009;5:138–139.
10. Chalmers J, Cooper ME. UKPDS and the legacy effect. N Engl J Med 2008;359:1618–1620.
11. Del Prato S. Megatrials in type 2 diabetes. From excitement to frustration? Diabetologia 2009;52:1219–1226.
12. Monnier VM, Cerami A. Nonenzymatic browning in vivo: possible process for aging of long-lived proteins. Science 1981;211:491–493.
13. Monnier VM, Vishwanath V, Frank KE, et al. Relation between complications of type I diabetes mellitus and collagen-linked fluorescence. N Engl J Med 1986;314:403–408.
14. Wells-Knecht KJ, Brinkmann E, Wells-Knecht MC, et al. New biomarkers of Maillard reaction damage to proteins. Nephrol Dial Transplant 1996;11:41–47.
15. Vlassara H, Brownlee M, Cerami A. Nonenzymatic glycosylation: role in the pathogenesis of diabetic complications. Clin Chem 1986;32:B37–B41.
16. Monnier VM, Sell DR, Genuth S. Glycation products as markers and predictors of the progression of diabetic complications. Ann N Y Acad Sci 2005;1043:567–581.
17. Ahmed N, Babaei-Jadidi R, Howell SK, et al. Glycated and oxidized protein degradation products are indicators of fasting and postprandial hyperglycemia in diabetes. Diabetes Care 2005;28:2465–2471.
18. Miyata T, Ueda Y, Yamada Y, et al. Accumulation of carbonyls accelerates the formation of pentosidine, an advanced glycation end product: carbonyl stress in uremia. J Am Soc Nephrol 1998;9:2349–2356.
19. Singh R, Barden A, Mori T, et al. Advanced glycation end-products: a review. Diabetologia 2001;44:129–146.
20. Miyata T, Ishikawa N, van Ypersele de Strihou C. Carbonyl stress and diabetic complications. Clin Chem Lab Med 2003;41:1150–1158.
21. Witko-Sarsat V, Friedlander M, Capeillère-Blandin C, et al. Advanced oxidation protein products as a novel marker of oxidative stress in uremia. Kidney Int 1996;49:1304–1313.

22. Miyata T, van Ypersele de Strihou C, Kurokawa K, et al. Alterations in nonenzymatic biochemistry in uremia: origin and significance of “carbonyl stress” in long-term uremic complications. Kidney Int 1999;55:389–399.
23. Horie K, Miyata T, Maeda K, et al. Immunohistochemical colocalization of glycoxidation products and lipid peroxidation products in diabetic renal glomerular lesions. Implication for glycoxidative stress in the pathogenesis of diabetic nephropathy. J Clin Invest 1997;100:2995–3004.
24. Uesugi N, Sakata N, Horiuchi S, et al. Glycoxidation-modified macrophages and lipid peroxidation products are associated with the progression of human diabetic nephropathy. Am J Kidney Dis 2001;38:1016–1025.
25. Neeper M, Schmidt A, Brett J, et al. Cloning and expression of a cell surface receptor for advanced glycosylation end products of proteins. J Biol Chem 1992;267:14998–15004.
26. Yan SD, Chen X, Fu J, et al. RAGE and amyloid-beta peptide neurotoxicity in Alzheimer’s disease. Nature 1996;382:685–691.
27. Hori O, Brett J, Slattery T, et al. The receptor for advanced glycation end product (RAGE) is a cellular binding site for amphoterin. Mediation of neurite outgrowth and co-expression of rage and amphoterin in the developing nervous system. J Biol Chem 1995;270:25752–25761.
28. Hofmann MA, Drury S, Fu C, et al. RAGE mediates a novel proinflammatory axis; a central cell surface receptor for S100/calgranulin polypeptides. Cell 1999;97:889–901.
29. Sousa MM, Yan SD, Stern D, et al. Interaction of the receptor for advanced glycation end products (RAGE) with transthyretin triggers nuclear transcription factor κB (NF-κB) activation. Lab Invest 2000;80:1101–1110.
30. Abel M, Ritthaler U, Zhang Y, et al. Expression of receptors for advanced glycosylated end-products in renal disease. Nephrol Dial Transplant 1995;10:1662–1667.
31. Yamamoto Y, Kato I, Doi T, et al. Development and prevention of advanced diabetic nephropathy in RAGE-overexpressing mice. J Clin Invest 2001;108:261–268.
32. Yonekura H, Yamamoto Y, Sakurai S, et al. Roles of the receptor for advanced glycation endproducts in diabetes-induced vascular injury. J Pharmacol Sci 2005;97:305–311.
33. Yonekura H, Yamamoto Y, Sakurai S, et al. Novel splice variants of the receptor for advanced glycation end-products expressed in human vascular endothelial cells and pericytes, and their putative roles in diabetes-induced vascular injury. Biochem J 2003;370:1097–1109.
34. Brownlee M, Vlassara H, Kooney A, et al. Aminoguanidine prevents diabetes- induced arterial wall protein cross-linking. Science 1986;232:1629–1632.
35. Bucala R, Vlassara H. Advanced glycosylation end products in diabetic renal and vascular disease. Am J Kidney Dis 1995;26:875–888.
36. Bolton WK, Cattran DC, Williams ME, et al. Randomized trial of an inhibitor of formation of advanced glycation end products in diabetic nephropathy. Am J Nephrol 2004;24:32–40.
37. Whittier F, Spinowitz B, Wuerth JP. Pimagedine safety profile in patients with type 1 diabetes. J Am Soc Nephrol 1999;10:184A (Abstract A0941)
38. Thornalley PJ. Use of aminoguanidine (Pimagedine) to prevent the formation of advanced glycation endproducts. Arch Biochem Biophys 2003;419:31–40.
39. Khalifah RG, Chen Y, Wassenberg JJ. Post-Amadori AGE inhibition as a therapeutic target for diabetic complications: a rational approach to second-generation Amadorin design. Ann N Y Acad Sci 2005;1043:793–806.
40. Chetyrkin SV, Zhang W, Hudson BG, et al. Pyridoxamine protects proteins from functional damage by 3-deoxyglucosone: mechanism of action of pyridoxamine. Biochemistry 2008;47:997–1006.
41. Degenhardt TP, Alderson NL, Arrington DD, et al. Pyridoxamine inhibits early renal disease and dyslipidemia in the streptozotocin-diabetic rat. Kidney Int 2002;61:939–950.
42. Williams ME, Bolton WK, Khalifah RG, et al. Effects of pyridoxamine in combined phase 2 studies of patients with type 1 and type 2 diabetes and overt nephropathy. Am J Nephrol 2007;27:605–614.

43. Kakuta T, Tanaka R, Satoh Y, et al. Pyridoxamine improves functional, structural, and biochemical alterations of peritoneal membranes in uremic peritoneal dialysis rats. Kidney Int 2005;68:1326–1336.
44. Berrone E, Beltramo E, Solimine C, et al. Regulation of intracellular glucose and polyol pathway by thiamine and benfotiamine in vascular cells cultured in high glucose. J Biol Chem 2006;281:9307–9313.
45. Babaei-Jadidi R, Karachalias N, Ahmed N, et al. Prevention of incipient diabetic nephropathy by high-dose thiamine and benfotiamine. Diabetes 2003;52:2110–2120.
46. Stirban A, Negrean M, Stratmann B, et al. Benfotiamine prevents macro- and microvascular endothelial dysfunction and oxidative stress following a meal rich in advanced glycation end products in individuals with type 2 diabetes. Diabetes Care 2006;29:2064–2071.
47. Du X, Edelstein D, Brownlee M. Oral benfotiamine plus alpha-lipoic acid normalizes complication-causing pathways in type 1 diabetes. Diabetologia 2008;51:1930–1932.
48. Beisswenger PJ, Howell SK, Touchette AD, et al. Metformine reduces systemic methylglyoxal levels in type 2 diabetes. Diabetes 1999;48:198–202.
49. Urios P, Grigorova-Borsos AM, Sternberg M. Aspirin inhibits the formation of pentosidine, a cross-linking advanced glycation end product, in collagen. Diabetes Res Clin Pract 2007;77:337–340.
50. Suarez G, Rajaram R, Bhuyan KC, et al. Administration of an aldose reductase inhibitor induces a decrease of collagen fluorescence in diabetic rats. J Clin Invest 1988;82:624–627.
51. Hamada Y, Nakamura J, Naruse K, et al. Epalrestat, an aldose reductase inhibitor, reduces the levels of Nε-(carboxymethyl)lysine protein adducts and their precursors in erythrocytes from diabetic patients. Diabetes Care 2000;23:1539–1544.
52. Brown MJ, Bird SJ, Watling S, et al. Natural progression of diabetic peripheral neuropathy in the Zenarestat study population. Diabetes Care 2004;27:1153–1159.
53. Kinekawa F, Kubo F, Matsuda K, et al. Effect of an aldose reductase inhibitor on esophageal dysfunction in diabetic patients. Hepatogastroenterology 2005;52:471–474.
54. Dan Q, Wong R, Chung SK, et al. Interaction between the polyol pathway and non-enzymatic glycation on mesangial cell gene expression. Nephron Exp Nephrol 2004;98:e89–e99.
55. Wilkinson-Berka JL, Kelly DJ, Koerner SM, et al. ALT-946 and aminoguanidine, inhibitors of advanced glycation, improve severe nephropathy in the diabetic transgenic (mREN-2)27 rat. Diabetes 2002;51:3283–3289.
56. Forbes JM, Soulis T, Thallas V, et al. Renoprotective effects of a novel inhibitor of advanced glycation. Diabetologia 2001;44:108–114.
57. Figarola JL, Scott S, Loera S, et al. LR-90 a new advanced glycation endproduct inhibitor prevents progression of diabetic nephropathy in streptozotocin-diabetic rats. Diabetologia 2003;46:1140–1152.
58. Figarola JL, Loera S, Weng Y, et al. LR-90 prevents dyslipidaemia and diabetic nephropathy in the Zucker diabetic fatty rat. Diabetologia 2008;51:882–891.
59. Nakamura S, Makita Z, Ishikawa S, et al. Progression of nephropathy in spontaneous diabetic rats is prevented by OPB-9195, a novel inhibitor of advanced glycation. Diabetes 1997;46:895–899.
60. Miyata T, Ishikawa S, Asahi K, et al. 2-Isopropylidenehydrazono-4-oxo-thiazolidin-5-ylacetanilide (OPB-9195) treatment inhibits the development of intimal thickening after balloon injury of rat carotid artery: role of glycoxidation and lipoxidation reactions in vascular tissue damage. FEBS Lett 1999;445:202–206.
61. Mizutani K, Ikeda K, Tsuda K, et al. Inhibitor for advanced glycation end products formation attenuates hypertension and oxidative damage in genetic hypertensive rats. J Hypertens 2002;20:1607–1614.
62. Miyata T, van Ypersele de Strihou C, Ueda Y, et al. Angiotensin II receptor antagonists and angiotensin converting enzyme inhibitors lower in vitro the formation of advanced glycation end products: biochemical mechanisms. J Am Soc Nephrol 2002;13:2478–2487.

63. Tsujita K, Shimomura H, Kawano H, et al. Effects of edaravone on reperfusion injury in patients with acute myocardial infarction. Am J Cardiol 2004;94:481–484.
64. Izuhara Y, Nangaku M, Takizawa S, et al. A novel class of advanced glycation inhibitors ameliorates renal and cardiovascular damage in experimental rat models. Nephrol Dial Transplant 2008;23:497–509.
65. Takizawa S, Izuhara Y, Kitao Y, et al. A novel inhibitor of advanced glycation and endoplasmic reticulum stress reduces infarct volume in rat focal cerebral ischemia. Brain Res 2007;1183:124–137.
66. Seikaly MG, Arant BS Jr, Seney BS Jr. Endogenous angiotensin concentrations in specific intrarenal fluid compartments of the rat. J Clin Invest 1990;86:1352–1357.
67. Forbes JM, Cooper ME, Thallas V, et al. Reduction of accumulation of advanced glycation end products by ACE inhibition in experimental diabetic nephropathy. Diabetes 2002;51:3274–3282.
68. Izuhara Y, Nangaku M, Inagi R, et al. Renoprotective properties of angiotensin receptor blockers beyond blood pressure lowering. J Am Soc Nephrol 2005;16:3631–3641.
69. Wolf G, Ritz E. Combination therapy with ACE inhibitors and angiotensin II receptor blockers to halt progression of chronic renal disease: pathophysiology and indications. Kidney Int 2005;67:799–812.
70. Izuhara Y, Sada T, Yanagisawa H, et al. A novel sartan derivative with very low angiotensin II type 1 receptor affinity protects the kidney in type 2 diabetic rats. Arterioscler Thromb Vasc Biol 2008;28:1767–1773.
71. Takizawa S, Dan T, Uesugi T, et al. A sartan derivative with a very low angiotensin II receptor affinity ameliorates ischemic cerebral damage. J Cereb Blood Flow Metab 2009;29:1665–1672.
72. Vasan S, Zhang X, Zhang X, et al. An agent cleaving glucose-derived protein crosslinks in vitro and in vivo. Nature 1996;382:275–278.
73. Cooper ME, Thallas V, Forbes J, et al. The cross-link breaker, N-phenacylthiazolium bromide prevents vascular advanced glycation endproduct accumulation. Diabetologia 2000;43:660–664.
74. Forbes JM, Thallas V, Thomas MC, et al. The breakdown of preexisting advanced glycation end products is associated with reduced renal fibrosis in experimental diabetes. FASEB J 2003;17:1762–1764.
75. Vasan S, Foiles P, Founds H. Therapeutic potential of breakers of advanced glycation end product-protein crosslinks. Arch Biochem Biophys 2003;419:89–96.
76. Little WC, Zile MR, Kitzman DW, et al. The effect of alagebrium chloride (ALT-711), a novel glucose cross-link breaker, in the treatment of elderly patients with diastolic heart failure. J Card Fail 2005;11:191–195.
77. Myint KM, Yamamoto Y, Doi T, et al. RAGE control of diabetic nephropathy in a mouse model: effects of RAGE gene disruption and administration of low-molecular weight heparin. Diabetes 2006;55:2510–2522.
78. Bierhaus A, Chevion S, Chevion M, et al. Advanced glycation end product-induced activation of NF-kappaB is suppressed by alpha-lipoic acid in cultured endothelial cells. Diabetes 1997;46:1481–1490.
79. Wang K, Zhou Z, Zhang M, et al. Peroxisome proliferator activated receptor gamma down-regulates receptor for advanced glycation end products and inhibits smooth muscle cell proliferation in a diabetic and nondiabetic rat carotid artery injury model. J Pharmacol Exp Ther 2006;317:37–43.
80. Hirasawa Y, Matsui Y, Yamane K, et al. Pioglitazone improves obesity type diabetic nephropathy: relation to the mitigation of renal oxidative reaction. Exp Anim 2008;57:423–432.
81. Tan KC, Chow WS, Tso AW, et al. Thiazolidinedione increases serum soluble receptor for advanced glycation end-products in type 2 diabetes. Diabetologia 2007;50:1819–1825.

82. Bierhaus A, Humpert PM, Stern DM, et al. Advanced glycation end product receptor-mediated cellular dysfunction. Ann N Y Acad Sci 2005;1043:676–680.
83. Forbes JM, Thorpe SR, Thallas-Bonke V, et al. Modulation of soluble receptor for advanced glycation end products by angiotensin-converting enzyme-1 inhibition in diabetic nephropathy. J Am Soc Nephrol 2005;16:2363–2372.
84. Goh SY, Cooper ME. The role of advanced glycation end products in progression and complications of diabetes. J Clin Endocrinol Metab 2008;93:1143–1152.
85. Vlassara H, Cai W, Crandall J, et al. Inflammatory mediators are induced by dietary glycotoxins, a major risk factor for diabetic angiopathy. Proc Natl Acad Sci U S A 2002;99:15596–15601.
86. Lin RY, Choudhury RP, Cai W, et al. Dietary glycotoxins promote diabetic atherosclerosis in apolipoprotein E deficient mice. Atherosclerosis 2003;168:213–220.
87. Yamagishi S, Nakamura K, Matsui T, et al. Oral administration of AST-120 (Kremezin) is a promising therapeutic strategy for advanced glycation end product (AGE)-related disorders. Med Hypotheses 2007;69:666–668.
88. Li YM, Tan AX, Vlassara H. Antibacterial activity of lysozyme and lactoferrin is inhibited by binding of advanced glycation-modified proteins to a conserved motif. Nat Med 1995;1:1057–1061.
89. Mitsuhashi T, Li YM, Fishbane S, et al. Depletion of reactive advanced glycation endproducts from diabetic uremic sera using a lysozyme-linked matrix. J Clin Invest 1997;100:847–854.
90. Zheng F, Cai W, Mitsuhashi T, et al. Lysozyme enhances renal excretion of advanced glycation endproducts in vivo and suppresses adverse age-mediated cellular effects in vitro: a potential AGE sequestration therapy for diabetic nephropathy? Mol Med 2001;7:737–747.
91. Liu H, Zheng F, Cao Q, et al. Amelioration of oxidant stress by the defensin lysozyme. Am J Physiol Endocrinol Metab 2006;290:E824–E832.
92. Liu H, Zheng F, Zhu L, Uribarri J, Tunstead JR, Ren B, Badimon J, Striker GE, Vlassara H. The immune defense protein lysozyme ameliorates acute vascular injury and atherosclerosis in hyperlipidemic mice. Am J Pathol 2005.
93. Liu H, Zheng F, Li Z, et al. Reduced acute vascular injury and atherosclerosis in hyperlipidemic mice transgenic for lysozyme. Am J Pathol. 2006;169:303–313.
94. Ohtomo S, Nangaku M, Izuhara Y, et al. Cobalt ameliorates renal injury in an obese, hypertensive type 2 diabetes rat model. Nephrol Dial Transplant 2007;23:1166–1172.
95. Nangaku M. Chronic hypoxia and tubulointerstitial injury: a final common pathway to end-stage renal failure. J Am Soc Nephrol 2006;17:17–25.
96. Salnikow K, Donald SP, Bruick RK, et al. Depletion of intracellular ascorbate by the carcinogenic metals nickel and cobalt results in the induction of hypoxic stress. J Biol Chem 2004;279:40337–40344.

Part IV
Hypoxia

Chapter 21
Involvement of Hypoxia-Inducible Factor 1 in Physiological and Pathological Responses to Continuous and Intermittent Hypoxia: Role of Reactive Oxygen Species

Gregg L. Semenza

Abstract The hypoxia-inducible factors (HIFs) are transcriptional activators that mediate homeostatic responses to hypoxia. At the cellular level, HIF-1 mediates adaptive metabolic responses to hypoxia that serve to maintain energy and redox homeostasis by reducing mitochondrial generation of reactive oxygen species (ROS). At the systemic level, HIFs control erythropoiesis and thereby maintain blood O_2-carrying capacity and delivery of O_2 to body tissues. In contrast to these adaptive responses, patients with obstructive sleep apnea are subjected to chronic intermittent hypoxia, a nonphysiological stimulus that induces HIF-1, which mediates a maladaptive response, systemic hypertension.

Keywords HIF-1 · Redox · Cytochrome-c oxidase

1 Introduction: Defining Hypoxia

The normal O_2 concentration to which cells in the human body are exposed varies from ~21% (corresponding to a partial pressure (PO_2) of ~150 mmHg at sea level) in the upper airway to ~1% at the corticomedullary junction of the kidney. Biologists usually maintain tissue culture cells in 20% O_2 (95% air and 5% CO_2) and refer to this concentration as normoxia despite the fact that most cells in the human body are exposed to much lower O_2 levels. Whatever the specific set point, complex

G.L. Semenza (✉)
Vascular Program, Institute for Cell Engineering;
McKusick-Nathans Institute of Genetic Medicine;
and Departments of Pediatrics, Medicine, Oncology, Radiation Oncology,
and Biological Chemistry, The Johns Hopkins University School of Medicine,
Baltimore, MD 21205, USA
e-mail: gsemenza@jhmi.edu

T. Miyata et al. (eds.), *Studies on Renal Disorders*, Oxidative Stress in Applied Basic Research and Clinical Practice, DOI 10.1007/978-1-60761-857-7_21,

homeostatic mechanisms serve to maintain the cellular O_2 concentration within a narrow range in vivo.

Hypoxia is defined as a reduction in the amount of O_2 available to a cell, tissue, or organism. As such, it is a relative term. Hypoxia can occur continuously (e.g., when individuals ascend to high altitude) or intermittently (e.g., in individuals with sleep apnea, in whom airway obstruction transiently blocks O_2 uptake, resulting in a rapid decline in blood PO_2 (hypoxemia), which causes the individual to awaken and resume breathing). Hypoxia can be divided into an acute phase, in which rapid but transient responses are mediated through the posttranslational modification of existing proteins, and a chronic phase, in which delayed but durable changes are mediated through altered gene transcription and protein synthesis. Finally, hypoxia can be systemic, as in the case of ascent to high altitude, or local, as in the case of myocardial ischemia associated with coronary artery disease.

Ultimately, hypoxia impacts the functioning of individual cells. Humans and other metazoan organisms are sustained by energy generated through the oxidative metabolism of glucose and fatty acids in the mitochondria, which results in the production of reducing equivalents that are used to maintain an electrochemical gradient that drives adenosine triphosphate (ATP) synthesis. This highly efficient mechanism for producing ATP is dependent upon the utilization of O_2 as the terminal electron acceptor at complex IV of the respiratory chain. When electrons react with O_2 prematurely (e.g., at complex III), reactive oxygen species (ROS) are generated. Tonic, low-level production of ROS represents a signal that mitochondrial function is intact, whereas increased ROS production, resulting from reduced or fluctuating O_2 availability, is a danger signal that the cell is at risk of oxidative damage and, if uncorrected, of death. Thus, many different adaptive responses are triggered by hypoxia, principally through the activity of hypoxia-inducible factor 1 (HIF-1), which is a transcription factor that functions as a master regulator of oxygen homeostasis.

2 Molecular Mechanisms of Oxygen Sensing: The PHD–VHL–HIF-1 Pathway

HIF-1 is a heterodimeric protein that is composed of a constitutively expressed HIF-1β subunit and an O_2-regulated HIF-1α subunit [1, 2]. Under normoxic conditions, the HIF-1α subunit is synthesized and subjected to hydroxylation on proline residue 402 or 564 by prolyl hydroxylase domain (PHD) proteins (principally PHD2) that use O_2 and α-ketoglutarate as substrates to catalyze a reaction in which one oxygen atom is inserted into the proline residue and the other oxygen atom is inserted into α-ketoglutarate (also known as 2-oxoglutarate) to form succinate and CO_2 [3]. Prolyl hydroxylation is required for the binding of the von Hippel–Lindau protein (VHL), which recruits a ubiquitin ligase complex [4–7]. Ubiquitination marks HIF-1α for degradation by the proteasome [8]. FIH-1 binds to HIF-1α and negatively regulates transactivation function [9] by hydroxylating asparagine residue 803, which blocks the interaction of the HIF-1α transactivation domain with the coactivator p300 or CBP [10]. Thus, both the stability and transcriptional activity of HIF-1 are negatively regulated by O_2-dependent hydroxylation.

When cells are acutely subjected to hypoxia, the hydroxylation reactions are inhibited as a result of substrate (O_2) deprivation or increased mitochondrial production of ROS, which may inhibit the hydroxylases by oxidizing a ferrous ion in the catalytic site [3, 11]. The loss of hydroxylase activity increases HIF-1α stability and transactivation function, leading to its dimerization with HIF-1β, binding of HIF-1 to its recognition sequence 5′-(A/G)CGTG-3′ [12] in target genes and increased transcription of target gene sequences into mRNA.

Using the HIF-1α DNA sequence to search databases, DNA sequences encoding a related protein, now designated HIF-2α, were identified [13–16]. HIF-2α is also expressed in an O_2-regulated manner and dimerizes with HIF-1β [16, 17]. HIF-1α and HIF-1β are ubiquitously expressed [18], whereas HIF-2α expression is restricted to a limited number of cell types, including cells of the developing lung, vascular endothelial cells, renal interstitial cells, hepatocytes, cardiomyocytes, and astrocytes [13–16]. Whereas HIF-1α homologues are present in all metazoan species studied (including *Caenorhabditis elegans*, which consists of only ~1,000 cells and contains no specialized systems for oxygen delivery), it appears that HIF-2α arose coincident with the evolution of complex respiratory and circulatory systems in vertebrate organisms.

3 Cellular Oxygen Homeostasis: Regulation of Glucose and Energy Metabolism

Individual cells must adapt to O_2 deprivation by reprogramming their metabolism. The metabolic alterations that are induced by hypoxia are profound. Perhaps the most subtle adaptation identified thus far is a subunit switch that occurs in cytochrome-c oxidase (COX; complex IV), in which the COX4-1 regulatory subunit is replaced by the COX4-2 isoform as a result of the HIF-1-mediated transcriptional activation of genes encoding COX4-2 and LON, a mitochondrial protease that is required for the hypoxia-induced degradation of COX4-1 [19]. This subunit switch serves to optimize the efficiency with which COX transfers electrons to O_2 under hypoxic conditions. Remarkably, the budding yeast *Saccharomyces cerevisiae* also switches COX subunits in response to hypoxia [20], but does so by a completely different molecular mechanism since yeast do not have a HIF-1 homologue. The similar regulation of COX activity in yeast and human cells indicate that the selection for O_2-dependent homeostatic regulation of mitochondrial respiration is ancient and likely to be shared by all eukaryotic organisms [19].

A more drastic alteration is the shunting of pyruvate away from the mitochondria by the HIF-1-mediated activation of the *PDK1* gene encoding pyruvate dehydrogenase (PDH) kinase 1 [21, 22], which phosphorylates the catalytic subunit of PDH, the enzyme that converts pyruvate into acetyl coenzyme A (AcCoA) for entry into the mitochondrial tricarboxylic acid cycle, which generates reducing equivalents that are donated to the electron transport chain. The reduced delivery of substrate to the mitochondria for oxidative phosphorylation results in reduced ATP synthesis, which must be compensated for by increased glucose uptake via glucose transporters and increased conversion of glucose to lactate by the activity of glycolytic enzymes and lactate dehydrogenase A, which are all encoded by HIF-1 target genes [23–28].

Induction of *PDK1* expression will inhibit the oxidative metabolism of AcCoA derived from glucose but will not affect the oxidative metabolism of AcCoA derived from fatty acids. The most dramatic response to persistent hypoxia is the active destruction of mitochondria by selective mitochondrial autophagy [29]. Remarkably, mouse embryo fibroblasts cultured at 1% O_2 reduce their mitochondrial mass by ~75% within 48 h through autophagy that is initiated by the HIF-1-dependent expression of BNIP3, a mitochondrial protein that competes with Beclin1 for binding to Bcl2, thereby freeing Beclin1 to trigger autophagy [29].

The adaptive significance of these metabolic responses to hypoxia were revealed by the finding that HIF-1α-deficient mouse embryo fibroblasts die when cultured under hypoxic conditions for 72 h, due to dramatically increased levels of ROS [21, 28]. The cells can be rescued by overexpression of PDK1 or BNIP3, or by treatment with free radical scavengers [21, 29]. It has long been known that mitochondrial production of ROS increases under hyperoxic conditions [30]. However, recent studies have demonstrated that acute hypoxia also leads to increased mitochondrial production of ROS, which is required for the inhibition of HIF-1α hydroxylase activity [11]. Exposure of wild-type mouse embryo fibroblasts to hypoxia for 48 h results in reduced levels of ROS, in contrast to HIF-1α-deficient in which the levels of ROS are markedly increased [21, 29].

The following conclusions can be drawn regarding the metabolic adaptation to hypoxia. The increase in glycolysis and decrease in respiration that occur in response to hypoxia do not represent a passive effect of substrate (O_2) deprivation but instead represent an active response of the cell to counteract the reduced efficiency of respiration under hypoxic conditions, which in the absence of adaptation results in the accumulation of toxic levels of ROS. These studies indicate that a major role of HIF-1 is to establish, at any O_2 concentration, the optimal balance between glycolytic and oxidative metabolism that maximizes ATP production without increasing levels of ROS. Finally, analysis of lung tissue from nonhypoxic Hif1a$^{+/-}$ mice, which are heterozygous for a HIF-1α null allele and thus partially HIF-1α deficient, revealed a ~50% decrease in mitochondrial mass compared to WT littermates [28]. This remarkable finding indicates that HIF-1 regulates mitochondrial metabolism even in the tissue exposed to the highest PO_2, indicating that HIF-1 performs this critical function over the entire range of physiological PO_2. Thus, HIF-1 maintains the metabolic/redox homeostasis that is essential metazoan cells to live with O_2.

4 Systemic Oxygen Homeostasis: Regulation of Erythropoiesis

We discovered HIF-1 in 1992 as a protein required for hypoxia-induced transcription of the human EPO gene encoding erythropoietin, which is the hormone that controls red blood cell production and thereby determines the O_2-carrying capacity of the blood [31]. Red blood cells function to deliver O_2 from the lungs to every cell in the body. Acute blood loss, ascent to high altitude, and pneumonia each results in a reduction in the blood O_2 content. The ensuing tissue hypoxia induces HIF-1 activity in cells throughout the body, including specialized cells in the kidney that produce erythropoietin, a glycoprotein hormone that is secreted into the blood and binds to its cognate receptor on erythroid progenitor cells, thereby stimulating their survival and differentiation [32]. Analysis of the cis-acting DNA sequences regulating hypoxia-induced EPO gene transcription (the hypoxia response element (HRE)) led to the discovery of HIF-1 as the transacting factor that bound to the HRE [31]. Subsequently, HIF-1 has been shown to orchestrate erythropoiesis by coordinately regulating the expression of multiple genes encoding proteins responsible for the intestinal uptake, tissue recycling, and delivery of iron to the bone marrow for its use in the synthesis of hemoglobin, including divalent metal transporter 1 [33], hepcidin [33], ceruloplasmin [34], transferrin [35], and transferrin receptor [37, 38]. Expression of the erythropoietin receptor is also regulated by HIF-1 [39].

Erythropoiesis is impaired in Hif1a$^{-/-}$ (homozygous HIF-1α-null) embryos, and the erythropoietic defects in HIF-1α-deficient erythroid colonies could not be corrected by cytokines, such as vascular endothelial growth factor or erythropoietin, but were ameliorated by administration of iron-salicylaldehyde isonicotinoylhydrazone, a compound that can deliver iron into cells independently of iron transport proteins, which was consistent with reduced levels of transferrin receptor in HIF-1α-deficient embryos and yolk sacs [40]. In this study, only yolk sac erythropoiesis could be studied because Hif1a$^{-/-}$ embryos arrest in their development on day 8.5 [26] prior to the establishment of definitive erythropoiesis in the liver or bone marrow. In contrast, deficiency of HIF-2α (which, like HIF-1α, is O_2-regulated, dimerizes with HIF-1β, and activates target gene expression) has a major effect on EPO production [41] and intestinal iron absorption [33] in adult mice.

In humans, familial erythrocytosis is an inherited disorder in which affected individuals produce excess red cells. The resulting increased blood viscosity can impair blood flow in cerebral vessels, leading to headache or stroke. Four types of familial erythrocytosis have been identified. Type 1 is inherited as an autosomal dominant trait and is due to heterozygosity for a mutation in the *EPOR* gene that results in increased erythropoietin receptor signaling, such as a frameshift that eliminates the last 64 amino acids of the protein [42]. Type 2 familial erythrocytosis, which is also known as Chuvash polycythemia, is inherited as an autosomal recessive trait and is due to homozygosity for a missense mutation that results in the substitution of tryptophan for arginine at codon 200 of VHL [43]. The mutant VHL protein binds to hydroxylated HIF-1α and HIF-2α with reduced affinity, leading to

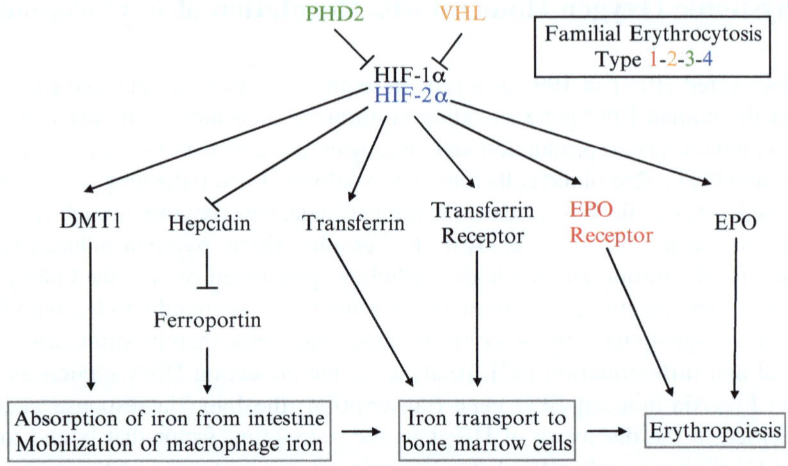

Fig. 1 Regulation of erythropoiesis by hypoxia-inducible factors (HIF)-1α and 2α. HIF-1α and HIF-2α control the expression of multiple genes encoding proteins required for iron absorption and transport and for the survival, proliferation, and differentiation of erythroid cells. Molecular defects in the four subtypes of familial erythrocytosis are *color-coded*. *DMT* divalent metal transporter; *EPO* erythropoietin

reduced ubiquitination of HIF-1α and HIF-2α, thereby increasing their steady state levels and the expression of HIF-1 target genes at any given O_2 concentration. Type 3 familial erythrocytosis is an autosomal dominant condition due to heterozygosity for a missense mutation in PHD2 that reduces hydroxylase activity [44]. Type 4 familial erythrocytosis is an autosomal dominant condition due to heterozygosity for a missense mutation in HIF-2α that reduces its hydroxylation [45]. These findings underscore the critical role of the PHD2–VHL–HIF-2α pathway in controlling erythropoiesis in the adult (Fig. 1).

5 Pathological Effects of Intermittent Hypoxia

Chronic intermittent hypoxia occurs in individuals with obstructive sleep apnea, in whom airway occlusion results in cessation of breathing leading to hypoxemia, which then arouses the individual to breathe. Obstructive sleep apnea may be a contributing factor in 30% of patients with essential hypertension [46]. The carotid body is a small chemosensory organ located at the bifurcation of the internal and external carotid arteries that senses arterial PO_2. Chronic intermittent hypoxia induces signaling from the carotid body that activates the sympathetic nervous system, leading to increased catecholamine secretion, which increases arterial tone, leading to hypertension [46, 47].

Whereas complete HIF-1α deficiency in Hif1a$^{-/-}$; mice results in embryonic lethality [25, 26], Hif1a$^{+/-}$ heterozygous-null mice develop normally but have impaired responses to hypoxia and ischemia [48–54]. Exposure of Hif1a$^{+/-}$ mice and their wild-type littermates to chronic intermittent hypoxia (15 s of hypoxia followed by 5 min of normoxia, 9 episodes per hour, 8 h/day) for 10 days results in marked increases in systolic and diastolic blood pressures and a significant elevation in plasma norepinephrine concentration in the wild-type mice, whereas their Hif1a$^{+/-}$ littermates are unaffected [52]. Remarkably, the carotid bodies of Hif1a$^{+/-}$ mice, although structurally and histologically normal, do not respond to hypoxia, although they respond normally to CO_2 and cyanide [49].

Chronic intermittent hypoxia induces increased production of ROS in rodents [54] and humans [56] and induces HIF-1α expression [52]. Administration of the superoxide scavenger manganese tetrakis(1-methyl-4-pyridyl)porphyrin pentachloride to wild-type mice reduces the levels of ROS that are generated by chronic intermittent hypoxia [57], blocks the development of hypertension [58], and inhibits the expression of HIF-1α [51]. Remarkably, in Hif1a$^{+/-}$ mice, neither HIF-1α expression, ROS generation, nor blood pressure are increased in response to chronic intermittent hypoxia [52]. These results indicate that the production of ROS is required for HIF-1α induction and that HIF-1α induction is required for the production of ROS, suggesting a feed-forward mechanism in which increased levels of ROS induce HIF-1α, which induces more ROS, leading to higher HIF-1α expression.

In contrast to the physiological response to continuous hypoxia observed in cultured mouse embryo fibroblasts (described above), in which HIF-1 activity ameliorates increases in ROS levels, the pathological response to chronic intermittent hypoxia is characterized by a HIF-1-dependent increase in the levels of ROS. Obstructive sleep apnea is a complication of obesity and, like other complications of obesity, has not been subject to evolutionary selection due to its recent origin. Thus, a nonphysiological stimulus (chronic intermittent hypoxia) elicits a maladaptive response (systemic hypertension) in which HIF-1 contributes to disease pathogenesis.

References

1. Wang GL, Semenza GL. Purification and characterization of hypoxia-inducible factor 1. J Biol Chem 1995; 270:1230–1237.
2. Wang GL, Jiang BH, Rue EA, et al. Hypoxia-inducible factor 1 is a basic-helix-loop-helix-PAS heterodimer regulated by cellular O_2 tension. Proc Natl Acad Sci USA 1995; 92:5510–5514.
3. Kaelin WG, Jr., Ratcliffe PJ. Oxygen sensing by metazoans: the central role of the HIF hydroxylase pathway. Mol Cell 2008; 30:393–402.
4. Maxwell PH, Wiesener MS, Chang GW, et al. The tumor suppressor protein VHL targets hypoxia-inducible factors for oxygen-dependent proteolysis. Nature 1999; 399:271–275.

5. Kamura T, Sato S, Iwai K, et al. Activation of HIF-1α ubiquitination by a reconstituted von Hippel-Lindau (VHL) tumor suppressor complex. Proc Natl Acad Sci U S A 2000; 97:10430–10435.
6. Ivan M, Kondo K, Yang H, et al. HIFα targeted for VHL-mediated destruction by proline hydroxylation: implications for O_2 sensing. Science 2001; 292:464–468.
7. Jaakkola P, Mole DR, Tian YM, et al. Targeting of HIF-α to the von Hippel-Lindau ubiquitylation complex by O_2-regulated prolyl hydroxylation. Science 2001; 292:468–472.
8. Salceda S, Caro J. Hypoxia-inducible factor 1α (HIF-1α) protein is rapidly degraded by the ubiquitin-proteasome system under normoxic conditions. Its stabilization by hypoxia depends on redox-induced changes. J Biol Chem 1997; 272:22642–22647.
9. Mahon PC, Hirota K, Semenza GL. FIH-1: a novel protein that interacts with HIF-1α and VHL to mediate repression of HIF-1 transcriptional activity. Genes Dev 2001; 15:2675–2686.
10. Lando D, Peet DJ, Gorman JJ, et al. FIH-1 is an asparaginyl hydroxylase enzyme that regulates the transcriptional activity of hypoxia-inducible factor. Genes Dev 2002; 16:1466–1471.
11. Guzy RD, Schumacker PT. Oxygen sensing by mitochondria at complex III: the paradox of increased reactive oxygen species during hypoxia. Exp Physiol 2006; 91:807–819.
12. Semenza GL, Jiang BH, Leung SW, et al. Hypoxia response elements in the aldolase A, enolase 1, and lactate dehydrogenase A gene promoters contain essential binding sites for hypoxia-inducible factor 1. J Biol Chem 1996; 271:32529–32537.
13. Ema M, Taya S, Yokotani N, et al. A novel bHLH-PAS factor with close sequence similarity to hypoxia-inducible factor 1α regulates the VEGF expression and is potentially involved in lung and vascular development. Proc Natl Acad Sci U S A 1997; 94:4273–4278.
14. Flamme I, Frohlich T, von Reutern M, et al. HRF, a putative basic helix-loop-helix-PAS-domain transcription factor is closely related to hypoxia-inducible factor-1α and developmentally expressed in blood vessels. Mech Dev 1997; 63:51–60.
15. Hogenesch JB, Chan WK, Jackiw VH, et al. Characterization of a subset of the basic-helix-loop-helix-PAS superfamily that interacts with components of the dioxin signaling pathway. J Biol Chem 1997; 272:8581–8593.
16. Tian H, McKnight SL, Russell DW. Endothelial PAS domain protein 1 (EPAS1), a transcription factor selectively expressed in endothelial cells. Genes Dev 1997; 11:72–82.
17. Wiesener MS, Turley H, Allen WE, et al. Induction of endothelial PAS domain protein-1 by hypoxia: characterization and comparison with hypoxia-inducible factor-1α. Blood 1998; 92:2260–2268.
18. Wiener CM, Booth G, Semenza GL. In vivo expression of mRNAs encoding hypoxia-inducible factor 1. Biochem Biophys Res Commun 1996; 225:485–488.
19. Fukuda R, Zhang H, Kim JW, et al. HIF-1 regulates cytochrome oxidase subunits to optimize efficiency of respiration in hypoxic cells. Cell 2007; 129:111–122.
20. Kwast KE, Burke PV, Poyton RO. Oxygen sensing and the transcriptional regulation of oxygen-responsive genes in yeast. J Exp Biol 1998; 201:1177–1195.
21. Kim JW, Tchernyshyov I, Semenza GL, Dang CV. HIF-1-mediated expression of pyruvate dehydrogenase kinase: a metabolic switch required for cellular adaptation to hypoxia. Cell Metab 2006; 3:177–185.
22. Papandreou I, Cairns RA, Fontana L, et al. HIF-1 mediates adaptation to hypoxia by actively downregulating mitochondrial oxygen consumption. Cell Metab 2006; 3:187–197.
23. Semenza GL, Roth PH, Fang HM, Wang GL. Transcriptional regulation of genes encoding glycolytic enzymes by hypoxia-inducible factor 1. J Biol Chem 1994; 269:23757–23763.
24. Ebert BL, Firth JD, Ratcliffe PJ. Hypoxia and mitochondrial inhibitors regulate expression of glucose transporter-1 via distinct cis-acting sequences. J Biol Chem 1995; 270:29083–29089.
25. Iyer NV, Kotch LE, Agani F, et al. Cellular and developmental control of O_2 homeostasis by hypoxia-inducible factor 1α. Genes Dev 1998; 12:149–162.
26. Ryan HE, Lo J, Johnson RS. HIF-1α is required for solid tumor formation and embryonic vascularization. EMBO J 1998; 17:3005–3015.

27. Seagroves TN, Ryan HE, Lu H, et al. Transcription factor HIF-1 is a necessary mediator of the Pasteur effect in mammalian cells. Mol Cell Biol 2001; 21:3436–3444.
28. Zhang H, Bosch-Marce M, Shimoda LA, et al. Mitochondrial autophagy is an HIF-1-dependent adaptive metabolic response to hypoxia. J Biol Chem 2008; 283:10892–10903.
29. Turrens JF, Freeman BA, Levitt JG, Crapo JD. The effect of hyperoxia on superoxide production by lung submitochondrial particles. Arch Biochem Biophys 1982; 217:401–410.
30. Semenza GL, Wang GL. A nuclear factor induced by hypoxia via de novo protein synthesis binds to the human erythropoietin gene enhancer at a site required for transcriptional activation. Mol Cell Biol 1992; 12:5447–5454.
31. Jelkmann W. Control of erythropoietin gene expression and its use in medicine. Methods Enzymol 2007; 435:179–197.
32. Mastrogiannaki M, Matak P, Keith B, et al. HIF-2α, but not HIF-1α, promotes iron absorption in mice. J Clin Invest 2009; 119:1159–1166.
33. Peyssonnaux C, Zinkernagel AS, Schuepbach RA, et al. Regulation of iron homeostasis by the hypoxia-inducible transcription factors (HIFs). J Clin Invest 2007; 117:1926–1932.
34. Mukhopadhyay CK, Mazumder B, Fox PL. Role of hypoxia-inducible factor 1 in transcriptional activation of ceruloplasmin by iron deficiency. J Biol Chem 2000; 275:21048–21054.
35. Rolfs A, Kvietikova I, Gassmann M, et al. Oxygen-regulated transferrin expression is mediated by hypoxia-inducible factor-1. J Biol Chem 1997; 272:20055–20062.
36. Lok CN, Ponka P. Identification of a hypoxia response element in the transferrin receptor gene. J Biol Chem 1999; 274:24147–24152.
37. Tacchini L, Bianchi L, Bernelli-Zazzera A, et al. Transferrin receptor induction by hypoxia. HIF-1-mediated transcriptional activation and cell-specific post-transcriptional regulation. J Biol Chem 1999; 274:24142–24146.
38. Manalo DJ, Rowan A, Lavoie T, et al. Transcriptional regulation of vascular endothelial cell responses to hypoxia by HIF-1. Blood 2005; 105:659–669.
39. Yoon D, Pastore YD, Divoky V, et al. Hypoxia-inducible factor 1 deficiency results in dysregulated erythropoiesis signaling and iron homeostasis in mouse development. J Biol Chem 2006; 281:25703–25711.
40. Gruber M, Hu CJ, Johnson RS, et al. Acute postnatal ablation of HIF-2α results in anemia. Proc Natl Acad Sci U S A 2007; 104:2301–2306.
41. Sokol L, Luhovy M, Guan Y, et al. Primary familial polycythemia: a frameshift mutation in the erythropoietin receptor gene and increased sensitivity of erythroid progenitors to erythropoietin. Blood 1995; 86:15–22.
42. Ang SO, Chen H, Hirota K, et al. Disruption of oxygen homeostasis underlies congenital Chuvash polycythemia. Nat Genet 2002; 32:614–621.
43. Percy MJ, Zhao Q, Flores A, et al. A family with erythrocytosis establishes a role for prolyl hydroxylase domain protein 2 in oxygen homeostasis. Proc Natl Acad Sci USA 2006; 103:654–659.
44. Percy MJ, Furlow PW, Lucas GS, et al. A gain-of-function mutation in the HIF2A gene in familial erythrocytosis. N Engl J Med 2008; 358:162–168.
45. Lesske J, Fletcher EC, Bao G, Unger T. Hypertension caused by chronic intermittent hypoxia – influence of chemoreceptors and sympathetic nervous system. J Hypertens 1997; 15:1593–1603.
46. Prabhakar NR, Dick TE, Nanduri J, Kumar GK. Systemic, cellular and molecular analysis of chemoreflex-mediated sympathoexcitation by chronic intermittent hypoxia. Exp Physiol 2007; 92:39–44.
47. Yu AY, Shimoda LA, Iyer NV, et al. Impaired physiological responses to chronic hypoxia in mice partially deficient for hypoxia-inducible factor 1α. J Clin Invest 1999; 103:691–696.
48. Kline DD, Peng YJ, Manalo DJ, et al. Defective carotid body function and impaired ventilatory responses to chronic hypoxia in mice partially deficient for hypoxia-inducible factor 1α. Proc Natl Acad Sci U S A 2002; 99:821–826.

49. Cai Z, Manalo DJ, Wei G, et al. Hearts from rodents exposed to intermittent hypoxia or erythropoietin are protected against ischemia-reperfusion injury. Circulation 2003; 108:79–85.
50. Li J, Bosch-Marce M, Nanayakkara A, et al. Altered metabolic responses to intermittent hypoxia in mice with partial deficiency of hypoxia-inducible factor 1α. Physiol Genomics 2006; 25:450–457.
51. Peng YJ, Yuan G, Ramakrishnan D, et al. Heterozygous HIF-1α deficiency impairs carotid body-mediated systemic responses and reactive oxygen species generation in mice exposed to intermittent hypoxia. J Physiol 2006; 577:705–716.
52. Bosch-Marce M, Okuyama H, Wesley JB, et al. Effects of aging and HIF-1 activity on angiogenic cell mobilization and recovery of perfusion following limb ischemia. Circ Res 2007; 101:1310–1318.
53. Cai Z, Zhong H, Bosch-Marce M, et al. Complete loss of ischemic preconditioning-induced cardioprotection in mice with partial deficiency of HIF-1α. Cardiovasc Res 2008; 77:463–470.
54. Prabhakar NR, Kumar GK, Nanduri J, Semenza GL. ROS signaling in systemic and cellular responses to chronic intermittent hypoxia. Antioxid Redox Signal 2007; 9:1397–1403.
55. Dyugovskaya L, Lavie P, Lavie L. Increased adhesion molecules expression and production of reactive oxygen species in leukocytes of sleep apnea patients. Am J Respir Crit Care Med 2002; 165:934–939.
56. Peng YJ, Overholt JL, Kline D, et al. Induction of sensory long-term facilitation in the carotid body by intermittent hypoxia: implications for recurrent apneas. Proc Natl Acad Sci USA 2003; 100:10073–10078.
57. Kumar GK, Rai V, Sharma SD, et al. Chronic intermittent hypoxia induces hypoxia-evoked catecholamine efflux in adult rat adrenal medulla via oxidative stress. J Physiol 2006; 575:229–239.

Chapter 22
Regulation of Oxygen Homeostasis by Prolyl Hydroxylase Domains

Kotaro Takeda and Guo-Hua Fong

Abstract Prolyl hydroxylase domain containing proteins (PHDs) are oxygen sensors critical for the adaptation of multicellular animals to fluctuating oxygen availability in the environment. A key function of PHDs is to catalyze oxygen-dependent prolyl hydroxylation of hypoxia-inducible factor (HIF)-α subunits, a modification that initiates HIF-α degradation. Because HIF-α proteins are transcription factors responsible for the expression of a large number of genes, oxygen regulated HIF-α abundance may enable cells to modify their gene expression programs in accordance to intracellular oxygen concentrations. In addition to HIF-α, the abundance or activity of several other proteins are also regulated by PHD-catalyzed hydroxylation, which suggests that these non-HIF proteins might also contribute to hypoxia responses. Although lower animals such as nematodes have only a single PHD isoform, higher animals such as mammals have multiple PHD or PHD-related proteins to regulate multiple physiological processes, such as angiogenesis, erythropoiesis, and energy metabolism. These features are now being explored to develop novel therapeutic strategies aimed at treating a wide range of diseases such as stroke, heart attack, anemia, inflammation, and cancer.

Keywords Prolyl hydroxylase domain containing proteins (PHDs) · Hypoxia · Hypoxia-inducible factors (HIFs) · Angiogenesis

1 Introduction

Essentially all eukaryotic cells generate the bulk of their adenosine triphosphate (ATP) supplies by oxidative phosphorylation in mitochondria, a complicated process that employs oxygen as the final electron acceptor. Not surprisingly,

K. Takeda (✉)
Department of Cardiology, Kyushu University of Medicine, 3-1-1 Maidashi, Higashi-ku, Fukuoka 812-8582, Japan
e-mail: ktakeda@cardiol.med.kyushu-u.ac.jp

T. Miyata et al. (eds.), *Studies on Renal Disorders*, Oxidative Stress in Applied Basic Research and Clinical Practice, DOI 10.1007/978-1-60761-857-7_22,

multicellular animals have evolved sophisticated mechanisms to deliver oxygen to different tissues and cells, most of which are not directly exposed to the atmosphere. On the other hand, since they first came into existence at about 550–600 million years ago, multicellular animals have lived under ever changing environmental conditions and have been often challenged with diminished oxygen accessibility (hypoxia). Even in the presence of adequate atmospheric oxygen, tissues within a multicellular organism may still encounter hypoxic conditions due to various pathophysiologic processes, such as hypoxia within and near infracted cardiac tissues due to heart attack. Hence, selective pressure during evolution has favored those species that are equipped with means of adapting to various forms of tissue hypoxia. It is now known that organisms as diverse as worms and humans all share a highly conserved oxygen sensing mechanism [1–4] and are capable of adapting to hypoxia by improving the efficiency of oxygen delivery and reducing oxygen consumption. Such adaptive processes are initiated by the accumulation of hypoxia-inducible factor (HIF)-α proteins, which are essential for the transcriptional activation of genes encoding hypoxia response proteins such as erythropoietin (EPO) and vascular endothelial growth factor (VEGF)-A [1, 2]. These protein molecules improve oxygen delivery by stimulating erythropoiesis and blood vessel growth. A direct link between HIF-α accumulation and hypoxia is provided by prolyl hydroxylase domain containing proteins (PHDs) [5–8]. When intracellular oxygen concentration is sufficiently high for normal cellular functions, PHDs trigger rapid HIF-α degradation through oxygen-dependent hydroxylation of specific HIF-α prolyl residues. Under hypoxia, HIF-α proteins escape the degradation pathway due to reduced prolyl hydroxylase activity and participate in the transcriptional activation of hypoxia response genes. In this chapter, we describe mechanisms of PHD-dependent regulation of HIF-α and discuss the associated biological and medical significances.

2 Milestones in the Field of Hypoxia Research

As early as over a century ago, it was already recognized that humans were capable of adapting to the low oxygen condition at high altitudes by increasing the number of red blood cells, a phenomenon now known as erythrocytosis [9]. By 1977, EPO was purified and identified as the main protein factor responsible for hypoxia-induced erythrocytosis [10]. A major breakthrough in the field of hypoxia research was achieved in 1991 when Semenza and coworkers [11, 12] discovered that HIF-1α was largely responsible for hypoxia-induced transcription of the *EPO* gene. Shortly after, HIF-1α was purified and cloned, and was shown to activate *EPO* gene transcription by binding to its enhancer at the 3′ end of the gene [13–15]. In the subsequent years, HIF-1α was found to activate the transcription of a large number of other target genes as well, such as *VEGF-A* [16–18], genes encoding glycolytic enzymes [19–22], and many more others involved in hypoxia adaptation [2, 23, 24]. In 1997, another HIF-α isoform, HIF-2α, was also cloned and was later found to be capable of activating the expression of many genes originally known as HIF-1α target genes, such as *EPO* and *VEGF-A* (HIF-2α is now considered to be the major isoform for EPO expression in vivo) [25–27]. However, while HIF-1α and HIF-2α share a long list of common target genes, there are genes that are exclusively activated by either, but not both, of them, such as genes encoding glycolytic enzymes, which are activated by HIF-1α, and several genes important for stem cell pluripotency, which are activated by HIF-2α [28–32].

Subsequent studies indicated that pVHL (von Hippel Lindau protein) dependent polyubiquitination by E3 ubiquitin ligase complex was essential for HIF-1α degradation in the proteasome [33]. This finding was further extended in 2001 by Ivan et al. [7] and Jaakkola et al. [8] who independently discovered that oxygen-dependent hydroxylation of specific HIF-1α proline residues was a key signal to polyubiquitination and proteasomal degradation of HIF-1α. While these findings represented significant progress in understanding the mechanism underlying oxygen-induced HIF-1α degradation, a major milestone was marked by the discovery in 2001–2002 that a subfamily of 2-oxoglutarate (2-OG)/Fe^{2+} dependent dioxygenases were the prolyl hydroxylases responsible for the hydroxylation of specific HIF-1α proline residues [5, 6, 34], a conclusion that was believed to be also true for HIF-2α [35, 36]. Several different names were given to these prolyl hydroxylases, including PHDs, based on a common sequence motif [6, 37], HIF-prolyl hydroxylases based on function [5], and EGLN proteins due to the fact that the prototype of these hydroxylase was originally identified as the as egg laying nine (EGL-9 or EGLN) in *Caenorhabditis elegans* [6].

3 Regulation of PHD Enzymatic Properties

3.1 Oxygen

PHDs belong to a subfamily of 2-OG/Fe^{2+}-dependent dioxygenases, which are highly conserved from *C. elegans* to *Homo sapiens* [5, 6, 38]. By using molecular oxygen as a donor to the oxygen atom in the hydroxyl group, PHDs hydroxylate two specific proline residues in the so-called oxygen dependent degradation (ODD) domain of HIF-α protein sequences (for example, P402 and P564 in human HIF-1α) [39, 40]. The specificity of proline residues is determined by their presence within a conserved LXXLAP sequence motif, which can be found at two separate locations of the ODD domain, commonly referred as N-terminal ODD (N-ODD) and C-terminal ODD (C-ODD).

PHDs can act as oxygen sensors largely because they require molecular oxygen as a substrate for hydroxylation reactions. If intracellular oxygen concentration is sufficiently high (normoxia), PHDs actively hydroxylate HIF-α proteins, which is a modification recognized by pVHL of the E3 ubiquitin ligase complex [5, 7, 8]. Once recruited to the E3 ubiquitin ligase complex by pVHL [41–43], HIF-α proteins are rapidly polyubiquitinated and routed to proteasomes where they are degraded [5, 6, 44, 45]. On the other hand, if intracellular oxygen concentrations reduce to levels insufficient for PHD activity (hypoxia), nonhydroxylated forms of HIF-α proteins are not recognized by pVHL, accumulate to high levels, and translocate to the nucleus where they heterodimerize with HIF-1β to form active transcription factors [13, 46]. In short, the evolutionary adaptation of PHDs to use oxygen as a substrate provides a primary mechanism that directly links PHD catalytic activity to intracellular oxygen concentration, therefore allowing PHDs to regulate HIF transcriptional activity in accordance to tissue oxygenation levels.

3.2 2-OG, Iron, Ascorbate, and Reactive Oxygen Species

Although a major role of HIF-α accumulation is to facilitate adaptation to hypoxia, it is now well appreciated that the roles of HIFs extend beyond the maintenance of oxygen homeostasis [47–49]. Many pathophysiological conditions, such as inflammation and tumor growth, are not only characterized by tissue hypoxia, but also bear multiple other abnormalities such as alterations in metabolism, cytokine secretion, and redox homeostasis [24, 50–52]. It is important to recognize that hypoxia and these other pathophysiological changes are intricately interrelated and can affect one another. In tumors, for example, hypoxia may have an impact on growth factor expression and glucose metabolism, whereas the latter changes may also have an effect on HIF-α expression.

PHDs are well adapted to the complexity of these conditions. Instead of being regulated by oxygen alone, PHD hydroxylase activities are under the control of multiple factors, such as 2-OG, iron (Fe^{2+}), and ascorbate. Among these, 2-OG and Fe^{2+} are cofactors directly required for PHD catalytic activities, whereas ascorbate is essential for the regeneration of Fe^{2+} from Fe^{3+} following a hydroxylation reaction [53]. The intracellular redox state also critically impacts on the activity of PHD hydroxylases, in part by regulating the oxidation status of iron. For example, reactive oxygen species (ROS) generated by oxidative stress strongly inhibits PHD activity, mostly by oxidizing Fe^{2+} to Fe^{3+} [54, 55]. Table 1 summarizes the factors that regulate PHD activity.

Regulation of PHDs by multiple factors provides a mechanism that allows these enzymes to integrate different pathophysiological signals to maximize the chance of animal survival. An excellent example is illustrated by succinate-mediated inhibition of PHD activity. Succinate is generated by decarboxylation of 2-OG and converted to fumarate by succinate dehydrogenase (SDH). Because the passage of succinate across mitochondrial membranes is dependent on translocation by membrane proteins, the amount of succinate in the cytoplasm is negligible under normal conditions. However, partial loss of function mutation in the *SDH* gene can cause significant build up of mitochondrial succinate levels due to inefficient conversion to fumarate, resulting in a corresponding increase in cytosolic succinate concentration [56]. Due to its structural similarity to 2-OG, which is a cofactor for

Table 1 Summary of regulatory mechanisms of PHD catalytic activities

Regulators	Positive/ negative	Mode of action	Major source	References
Oxygen	Positive	Substrate of PHD enzymatic reaction	Respiration/ circulation	[6, 34]
2-oxoglutarate (2-OG)	Positive	Substrate of PHD enzymatic reaction	TCA cycles	[6, 34]
Fe^{2+}	Positive	Cosubstrate of PHD enzymatic reaction	Absorption from stomach	[6, 34]
Ascorbate	Positive	Reduces Fe^{3+} to Fe^{2+}	Absorption from intestine	[6, 34]
NO (low concentration)	Positive	Redistribution of oxygen		[118]
NO (high concentrations)	Negative	Competitive inhibitor for oxygen		[119, 120]
ROS	Negative	Oxidizes Fe^{2+} to Fe^{3+}	Defective oxidative phosphorylation	[54, 55]
Succinate/fumarate	Negative	Competitive inhibition with 2-OG	Defective TCA cycles	[56]

PHD prolyl hydroxylase domain; *2-OG* 2-oxoglutarate; *ROS* reactive oxygen species

PHDs, succinate causes HIF-1α accumulation by acting as a competitive inhibitor [56, 57]. In this example, in spite of the fact that intracellular oxygen concentration is not directly affected, ATP production is still perturbed due to *SDH* mutation. Without succinate-mediated PHD inhibition and HIF-1α accumulation, such a mutation may very likely lead to lethality due to insufficient ATP production. Succinate-induced HIF-1α accumulation may help animal survival by modifying metabolic programs and activating angiogenesis, the latter of which facilitates the delivery of extra nutrition and oxygen to compensate for reduced efficiency of ATP production. However, such a survival strategy did not come without a price. Hypoxia-independent accumulation of HIF-1α is associated with tumorigenesis [56]. In the context of species survival, however, the evolutionary choice of death from cancer at a later stage of life does appear to be advantageous over death from defective energy metabolism early on in life.

4 Regulation of PHD Abundance and Functions

PHD protein levels are subject to regulation by multiple mechanisms. One important mechanism is hypoxia-dependent upregulation of PHD2 and PHD3 (Table 2) [58–60]. This feedback mechanism may prevent excessive HIF-α accumulation under hypoxia and prepare cells for efficient HIF-α degradation upon

Table 2 Comparison of different PHD isoforms and P4H-TM

Hydroxylases	PHD1 [6]	PHD2 [6]	PHD3 [6]	P4H-TM [65]
Alternative names	EGLN2 [6, 121], HPH3 [5], HIF-P4H1 [35], Folker [75]	EGLN1 [6, 121], HPH2 [5], HIF-P4H2 [35]	EGLN3 [6, 121], HPH1 [5], HIF-P4H3 [35], SM-20 (rat) [122]	PH-4 [66]
Tissues of expression	Testis, brain, liver, adrenal gland, skeletal muscle, adipose tissue, heart, kidney (human) [35, 66] Testis, liver, heart, brain, kidney (mice) [38]	Adipose tissue, heart, testis, kidney, brain, adrenal gland, liver (human) [35, 66] Liver, heart, kidney, brain, skeletal muscle, lung (mice) [38]	Heart, brain, adipose tissue, kidney, intestine (human) [35, 66] Heart, liver, brain, kidney, skeletal muscle, lung (mice) [38]	Brain, adrenal gland, kidney, testis, liver, skeletal muscle, lung, heart (human) [65, 66]
Relative substrate preferences	HIF-2α > HIF-1α [36] C-ODD ≫ N-ODD [35]	HIF-1α > HIF-2α [36] C-ODD > N-ODD [35]	HIF-2α > HIF-1α [36] C-ODD only [35, 123]	HIF-1α = HIF-2α [65, 66] C-ODD > N-ODD [65, 66]
K_m values for O_2	230 μM [35]	250 μM [35]	230 μM [35]	NA
Intracellular location	Nucleus [124]	Cytoplasm [124]	Cytoplasm, nucleus [124]	Endoplasmic reticulum [65]
Phenotypes in KO mice	Hypoxia tolerance in skeletal muscle [67]	Heart and placenta defect (fetus) [70] Angionenesis (adult) [72] Blood vessel maturation (adult) [108] Polycythemia (adult) [71]	Reduced neuronal apoptosis and systemic hypotension [125]	NA
Chromosomal locations	#19 (human) #7 (mice)	#1 (human) #8 (mice)	#14 (human) #12 (mice)	#3 (human)
Information of protein structure	Nuclear localization signal [126, 127], alternative translational initiation [128]	Nuclear export signal [127]	Lack of N-terminal half	Closely resembles collagen-P4H [65]
Inducibility by hypoxia	No [36, 58] or reduction [59]	Yes [36, 58]	Yes (strongly) [36, 58]	Yes [65]

reoxygenation [36, 60]. Interestingly, like HIF-α, PHD proteins are also degraded via the ubiquitin-mediated proteosomal degradation pathway, although in the latter case polyubiquitination is mediated by E3 ubiquitin ligases Siah1α/2 [61]. It is noteworthy that the expression of Siah2 itself is upregulated by hypoxia, which constitutes another level of feedback mechanism to suppress PHD1/3 protein levels under hypoxia [61]. The abundance of PHD2 is also regulated at the protein level, although trafficking of PHD2 to proteasomes is mediated by a ubiquitin-independent mechanism that involves interaction with FK506-binding protein (FKBP)-38 [62]. In short, PHD protein levels are positively regulated by hypoxia/HIF pathway and negatively regulated by Saih1α/2 or FKBP38. Such a dual regulatory mechanism is indicative of the need to fine tune PHD protein levels for optimal oxygen homeostasis.

In addition to regulation at the level of expression and protein stability, PHDs are also regulated at the functional level. Several proteins are known to physically associate with PHD to regulate HIF activity. For example, OS-9 enhances HIF-α hydroxylation by forming a multiprotein complex containing OS-9, HIF, and PHD [63]. Interestingly, inhibitor of growth-4 (ING4), which is a candidate tumor suppressor protein, was reported to regulate PHD2 function by a rather unusual mechanism [64]. Instead of regulating PHD2 hydroxylase activity, ING4 inhibits HIF transcriptional activity by forming a complex that contains HIF, PHD2, and ING4 itself [64].

5 PHD Isoforms

As animals evolved from relatively simple and small forms such as nematodes and flies into highly complex and large forms such as mammals, their tasks of handling hypoxia and other pathophysiological conditions also became increasingly complicated. Relative to nematodes and flies, mammals are generally much larger in size, contain many more cells and tissue types, and have much longer lifespans. Furthermore, different cells may respond to hypoxia in unique ways to maintain overall homeostasis of animal physiology. For example, the main response of renal interstitial cells to hypoxia is EPO expression and secretion, whereas cells of the vascular system respond to hypoxia by active angiogenesis.

Thus, it is not surprising that while *C. elegans* or *Drosophila melanogaster* has a single PHD isoform, mammals have acquired multiple HIF-α hydroxylases to take on more challenging tasks of maintaining oxygen homeostasis. The PHD subfamily of 2-OG/Fe^{2+} dependent dioxygenases includes three isoforms (PHD1, PHD2, and PHD3), all of which are soluble enzymes [5, 6]. Another proline 4-hydroxylase protein (PH-4) has also been reported and is referred to as P4H-TM based on the finding that it is a prolyl 4-hydroxylase possessing transmembrane domain [65, 66]. Different PHD isoforms have many features in common but also display some differences in terms of expression patterns, catalytic properties, and most notably physiological roles (see Table 2). We propose that both their commonality and differences may contribute to oxygen homeostasis in a context dependent manner.

Different PHD isoforms are expressed in overlapping although not identical tissue domains (see Table 2). For example, while PHD2 is broadly expressed in essentially all tissues examined, PHD1 expression partially overlaps with PHD2 in some tissues, including testis, followed by liver, heart, and brain [5, 6, 35, 38, 67]. At the functional level, all PHD isoforms display hydroxylase activities toward both HIF-1α and HIF-2α, and their catalytic activities are regulated rather similarly [37]. Such functional redundancy may be important for survival or reproduction of higher animals such as mammals. One relevant example is the nonessential role of PHD1 in testis. Even though PHD1 is highly expressed in the testis, PHD1 knockout does not have significant impact on male fertility [67]. Presumably, functional redundancy by other PHD isoforms, notably PHD2, which is also expressed in the testis, allows mice to maintain reproductive capacity in the absence of PHD1.

On the other hand, different PHD isoforms do display significantly different physiological roles, presumably due to a combination of the following mechanisms. First, although different PHD isoforms may be expressed in the same tissue and cell type, their relative abundances are often different and may vary depending on specific cell types. For example, PHD2 is expressed more abundantly than other isoforms in most tissues and cell lines, which explains for the most part why PHD2 knockdown in cultured cells most effectively led to HIF-1α accumulation [68]. Second, regulatory mechanisms may also differ to some extent. For example, PHD3 appears to be most robustly induced by hypoxia [36, 69], suggesting that PHD3 might act more efficiently in a feedback mechanism to prevent excessive HIF-α accumulation under hypoxia. Third, although all PHDs can hydroxylate both HIF-1α and HIF-2α, a certain degree of preference does exist, with PHD2 preferring HIF-1α over HIF-2α, and PHD1/PHD3 hydroxylating HIF-2α more efficiently (see Table 2) [36].

Different physiological roles of PHD isoforms are reflected in knockout phenotypes in mice. For example, germline *Phd2* knockout, but not *Phd1* or *Phd3* knockout, led to grossly defective placental and heart development and embryonic lethality by midgestation stages [70]. In adult mice, global *Phd2* knockout, but not *Phd1* or *Phd3* knockout, resulted in significantly increased vascular growth and polycythemia [71, 72]. On the other hand, double knockout of *Phd1* and *Phd3* led to moderate polycythemia without evidence of increased angiogenesis [71]. These differences suggest that different PHD isoforms may differentially contribute to oxygen homeostasis in different tissue environments.

6 Novel PHD Targets Other than HIF-α

PHDs were originally considered to be HIF-α specific prolyl hydroxylases, mostly based on the finding that they did not hydroxylate collagen [6]. However, subsequent studies demonstrated the existence of several other hydroxylation substrates. For instance, IκB kinase-β (IκKβ), which activates nuclear factor κB (NFκB) signaling by a kinase dependent mechanism, is also a PHD substrate [73]. The hydroxylation of proline residue (P191) of IκKβ by PHD1 reduces IκKβ activity, thus leading to the suppression of NFκβ pathway. Because NFκβ is known to mediate a HIF-independent mechanism of angiogenesis, regulation of NFκB activity by PHDs may provide an additional mechanism that links angiogenesis to tissue hypoxia [74]. Furthermore, since NFκB plays a critical role in cell growth, differentiation, and survival, its regulation by PHDs may provide a mechanism for cells to modify these properties in response to fluctuating oxygen availability.

Regulation of at least two other non-HIF-α proteins may also potentially contribute to HIF-independent adaptation to changes in oxygen tension. In one study, forced expression of C-terminal domain (i.e., the hydroxylase domain) of PHD1 accelerated cell proliferation, and a separate study showed that PHD1 promoted cell growth and tumorigenesis by hydroxylating proline residue (P1465) of the large subunit of RNA polymerase II, Rpb1 [75, 76]. Although tumorigenesis typically shortens lifespan, it may be a price that animals pay to avoid immediate death caused by cessation of cell growth.

PHD3-mediated hydroxylation of activating transcription factor-4 (ATF-4) illustrates yet another interesting example of how PHDs might be involved in HIF-independent mechanisms of hypoxia adaptation [77]. ATF-4 is a basic-leucine zipper transcription factor that regulates several genes, such as *CHOP/GADD153*, which are involved in stress regulatory pathways [78, 79]. ATF-4 has a unique ODD domain that is distinct from HIF-α ODD domain and contains five proline residues for hydroxylation [77]. It was found that all five of the proline residues had to be hydroxylated for degradation to occur. Because complete hydroxylation of all five proline residues may occur only in the presence of fairly high oxygen concentration, ATF-4 may accumulate to high levels under even moderate hypoxia, resulting in the expression of stress regulatory proteins such as CHOP/GADD153.

Hypoxia-induced expression of stress proteins may be another important survival strategy. While elevated ROS levels under hypoxia may interfere with normal protein folding, stress regulatory proteins may protect cells by facilitating degradation of misfolded proteins [80]. Furthermore, stress regulatory proteins may protect animals by triggering apoptosis of cells that have accumulated too many misfolded proteins [81].

7 PHDs as Potential Therapeutic Targets for Human Diseases

Various animal and clinical studies have raised the possibility that PHDs may be novel therapeutic targets. For example, increased erythropoiesis due to PHD deficiency suggests that pharmacological inhibition of PHD activities might be an effective way to treat anemia [71, 82, 83]. For these and other potential applications, tremendous efforts have been poured into developing pharmacological PHD inhibitors, which may activate HIF-regulated processes such as erythropoiesis and angiogenesis by stabilizing HIF-α proteins (hence also called HIF-α stabilizers) [84]. More recently, investigators are also gaining interest in developing PHD activators, which may have potential applications where decreased HIF abundance is beneficial, such as to inhibit tumor angiogenesis [85].

7.1 Ischemic Disorders

One of the most promising applications of PHD inhibitors is the treatment of ischemic diseases such as myocardial infarction and stroke [86, 87]. Although hypoxia in ischemic tissues may promote HIF-α stabilization, the extent of HIF-α accumulation is typically suboptimal for robust angiogenesis [88], presumably because hypoxia-induced PHD expression, coupled with residual PHD hydroxylase activities, can still cause HIF-α degradation at reduced rates. Thus, even under hypoxia, PHD inhibitors may further boost HIF activation and accelerate angiogenesis [86, 89].

PHD inhibition may protect against ischemia by several mechanisms. First, increased angiogenesis may improve perfusion and reduce cell death in ischemic tissues. In support of this mechanism, we have previously reported increased angiogenesis in PHD2 deficient mice [72]. Similarly, others have also found that HIF-1α overexpression stimulated angiogenesis and promoted the healing of ischemic tissues [90–92]. Second, a shift to anaerobic metabolism due to PHD1 inhibition may enhance hypoxia tolerance by reducing oxygen consumption. In PHD1 deficient mice, the expression level of pyruvate dehydrogenase kinase (PDK)1/4 is elevated, which restricts the entry of pyruvate into the Krebs cycle [67]. Interestingly, increased *Pdk4* expression in *Phd1* knockout mice is dependent on peroxisome proliferator-activated receptor alpha (PPARα) and HIF-2α. Third, PHD inhibition (particularly PHD2) increases the production of cardioprotective molecules including nitric oxide (NO) [93], carbon monoxide (CO) [87], and adenosine [94] by the induction of inducible NO synthase, heme oxygenase-1 and CD73 (an enzyme for adenosine generation). Since these factors are almost identical to the factors responsible for the ischemic preconditioning, administration of PHD inhibitor might mimic the effect of ischemic preconditioning [94, 95].

7.2 Anemia

In adult mice, global *Phd2* knockout increases EPO expression and induces dramatic erythrocytosis [71, 82]. In human, mutation of *PHD2* gene causes polycythemia [83, 96, 97].

Similarly, a novel PHD inhibitor FG-2216 increases EPO expression and erythrocytosis in rhesus macaques [98]. Thus, PHD inhibitors are potential drugs for the treatment of anemia.

PHD2 may also affect the bone marrow (BM) microenvironment because PHD2 deficient mice showed increased BM-derived hematopoietic stem cells (HSCs) [71]. Although it is also possible that increased HSCs in PHD2 deficient mice may be due to systemic PHD2 deficiency, it is known that local hypoxia in the bone marrow favors HSCs differentiation, proliferation, and survival [99]. Thus, PHD2 inhibition might be beneficial to anemic patients refractory to EPO by directly stimulating HSC proliferation.

7.3 *Cancer*

Rapid expansion of tumor tissues causes hypoxia, which induces tumor angiogenesis [100, 101]. Given the apparent importance of tumor vasculature for tumor survival and growth, antiangiogenesis therapies have attracted much attention. For instance, VEGF-A neutralizing antibodies or inhibitors of VEGF receptor tyrosine kinase-2 have been developed, with some success in clinical applications [102–105]. PHD activation may be an attractive addition to the currently available strategies. Indeed, a recent study showed that the administration of KRH102053, a novel PHD activator, reduced tumor growth and metastasis in animal models due to reduced HIF abundance and diminished expression of HIF target genes in tumor cells [85].

Besides antiangiogenic approach, methods aimed at enhancing vascular maturation are also being explored for cancer treatment [106]. Blood vessels in tumors are typically immature and leaky, conditions that favor metastasis by allowing easy passage of cancer cells through vascular walls [107]. A recently reported study demonstrated the feasibility of inhibiting tumor metastasis by promoting vascular maturation [108]. In mice with about 50% reduction in PHD2 protein level, maturation of tumor blood vessels was improved due to HIF-dependent upregulation of VEGF receptor-1 and VE-cadherin [108], resulting in reduced tumor cell intravasation and metastasis. Thus, in spite of the apparently counterintuitive nature, PHD inhibition could be an anticancer strategy. However, because PHD2 itself is now considered a tumor suppressor protein [96], systemic and long-term PHD inhibition may have dual effects: besides antitumorigenic effects, as discussed above, it may also accelerate tumor growth by other mechanisms such as increased angiogenesis or modified cell growth properties. Future studies should address these issues.

7.4 *Inflammation*

Recent evidence suggests that hypoxia accelerates inflammation [109–112], and that HIF-1α is a prerequisite for an innate immune reaction [112]. HIF-1α deficient myeloid cells show reduced inflammatory responses due to decreased glycolytic activity [109, 113], and mice carrying HIF-1α deficient myeloid cells produce several proinflammatory cytokines at reduced levels [114]. Moreover, PHD inhibitor mimosine boosts phagocytosis by human macrophages [115], and PHD1 knockdown leads to the activation of a key inflammatory transcriptional factor NFκB [73]. These data indicate that inflammatory responses are positively regulated by HIF-1α and negatively by PHDs. However, conflicting data also exist.

Several studies indicate that pharmacological inhibition of PHDs reduced inflammatory reactions in colitis and ischemic heart [87, 116, 117]. Therefore, PHDs might regulate inflammation either positively or negatively in a context-specific manner. Further studies are required to clarify the role of specific PHD isoforms and the effect of PHD inhibition in different types of inflammatory disorders.

8 Conclusion

PHDs are oxygen sensors and important negative regulators for the HIF pathway. A variety of biological processes upregulated by HIFs are inversely regulated by PHDs. Moreover, PHDs regulate non-HIF targets as well. Given the diversity of different PHD/HIF-α combinations, advanced animals such as mammals have a sophisticated oxygen-sensing network to fine tune adaptive responses. While PHDs are attractive targets for disease treatment, their maltifaceted roles raise a concern of potential side effects. However, it is hoped that isoform and tissue-specific PHD inhibition may offer novel opportunities to develop more specific strategies with minimal side effects.

Acknowledgments This work was supported in part by a Grant-in-Aid for Scientific Research from the Ministry of Education, Culture, Sports, Science and Technology of Japan (21790731) and Mitsubishi Pharma Research Foundation for Research Award on Cardiology (2008) to KT; and NIH grants 1R01EY019721-01 and 5R21HD052945-02 to G-HF.

References

1. Semenza GL. HIF-1, O(2), and the 3 PHDs: how animal cells signal hypoxia to the nucleus. Cell 2001; 107:1–3.
2. Semenza GL. HIF-1 and mechanisms of hypoxia sensing. Curr Opin Cell Biol 2001; 13:167–171.
3. Kaelin WG, Jr., Ratcliffe PJ. Oxygen sensing by metazoans: the central role of the HIF hydroxylase pathway. Mol Cell 2008; 30:393–402.
4. Schofield CJ, Ratcliffe PJ. Oxygen sensing by HIF hydroxylases. Nat Rev Mol Cell Biol 2004; 5:343–354.
5. Bruick RK, McKnight SL. A conserved family of prolyl-4-hydroxylases that modify HIF. Science 2001; 294:1337–1340.

6. Epstein AC, Gleadle JM, McNeill LA, et al. *C. elegans* EGL-9 and mammalian homologs define a family of dioxygenases that regulate HIF by prolyl hydroxylation. Cell 2001; 107:43–54.
7. Ivan M, Kondo K, Yang H, et al. HIFalpha targeted for VHL-mediated destruction by proline hydroxylation: implications for O_2 sensing. Science 2001; 292:464–468.
8. Jaakkola P, Mole DR, Tian YM, et al. Targeting of HIF-alpha to the von Hippel-Lindau ubiquitylation complex by O_2-regulated prolyl hydroxylation. Science 2001; 292:468–472.
9. Stockmann C, Fandrey J. Hypoxia-induced erythropoietin production: a paradigm for oxygen-regulated gene expression. Clin Exp Pharmacol Physiol 2006; 33:968–979.
10. Miyake T, Kung CK, Goldwasser E. Purification of human erythropoietin. J Biol Chem 1977; 252:5558–5564.
11. Semenza GL, Nejfelt MK, Chi SM, Antonarakis SE. Hypoxia-inducible nuclear factors bind to an enhancer element located 3′ to the human erythropoietin gene. Proc Natl Acad Sci U S A 1991; 88:5680–5684.
12. Semenza GL, Wang GL. A nuclear factor induced by hypoxia via de novo protein synthesis binds to the human erythropoietin gene enhancer at a site required for transcriptional activation. Mol Cell Biol 1992; 12:5447–5454.
13. Wang GL, Jiang BH, Rue EA, Semenza GL. Hypoxia-inducible factor 1 is a basic-helix-loop-helix-PAS heterodimer regulated by cellular O_2 tension. Proc Natl Acad Sci U S A 1995; 92:5510–5514.
14. Wang GL, Semenza GL. General involvement of hypoxia-inducible factor 1 in transcriptional response to hypoxia. Proc Natl Acad Sci U S A 1993; 90:4304–4308.
15. Wang GL, Semenza GL. Characterization of hypoxia-inducible factor 1 and regulation of DNA binding activity by hypoxia. J Biol Chem 1993; 268:21513–21518.
16. Levy AP, Levy NS, Wegner S, Goldberg MA. Transcriptional regulation of the rat vascular endothelial growth factor gene by hypoxia. J Biol Chem 1995; 270:13333–13340.
17. Liu Y, Cox SR, Morita T, Kourembanas S. Hypoxia regulates vascular endothelial growth factor gene expression in endothelial cells. Identification of a 5′ enhancer. Circ Res 1995; 77:638–643.
18. Forsythe JA, Jiang BH, Iyer NV, et al. Activation of vascular endothelial growth factor gene transcription by hypoxia-inducible factor 1. Mol Cell Biol 1996; 16:4604–4613.
19. Semenza GL, Roth PH, Fang HM, Wang GL. Transcriptional regulation of genes encoding glycolytic enzymes by hypoxia-inducible factor 1. J Biol Chem 1994; 269:23757–23763.
20. Graven KK, Yu Q, Pan D, Roncarati JS, Farber HW. Identification of an oxygen responsive enhancer element in the glyceraldehyde-3-phosphate dehydrogenase gene. Biochim Biophys Acta 1999; 1447:208–218.
21. Semenza GL, Jiang BH, Leung SW, et al. Hypoxia response elements in the aldolase A, enolase 1, and lactate dehydrogenase A gene promoters contain essential binding sites for hypoxia-inducible factor 1. J Biol Chem 1996; 271:32529–32537.
22. Firth JD, Ebert BL, Pugh CW, Ratcliffe PJ. Oxygen-regulated control elements in the phosphoglycerate kinase 1 and lactate dehydrogenase A genes: similarities with the erythropoietin 3′ enhancer. Proc Natl Acad Sci U S A 1994; 91:6496–6500.
23. Semenza GL. Hypoxia-inducible factor 1: master regulator of O_2 homeostasis. Curr Opin Genet Dev 1998; 8:588–594.
24. Semenza GL. Targeting HIF-1 for cancer therapy. Nat Rev Cancer 2003; 3:721–732.
25. Scortegagna M, Ding K, Zhang Q, et al. HIF-2alpha regulates murine hematopoietic development in an erythropoietin-dependent manner. Blood 2005; 105:3133–3140.
26. Gruber M, Hu CJ, Johnson RS, et al. Acute postnatal ablation of HIF-2alpha results in anemia. Proc Natl Acad Sci U S A 2007; 104:2301–2306.
27. Rankin EB, Biju MP, Liu Q, et al. Hypoxia-inducible factor-2 (HIF-2) regulates hepatic erythropoietin in vivo. J Clin Invest 2007; 117:1068–1077.
28. Tian H, McKnight SL, Russell DW. Endothelial PAS domain protein 1 (EPAS1), a transcription factor selectively expressed in endothelial cells. Genes Dev 1997; 11:72–82.

29. Ema M, Taya S, Yokotani N, et al. A novel bHLH-PAS factor with close sequence similarity to hypoxia-inducible factor 1alpha regulates the VEGF expression and is potentially involved in lung and vascular development. Proc Natl Acad Sci U S A 1997; 94:4273–4278.
30. Flamme I, Frohlich T, von Reutern M, et al. HRF, a putative basic helix-loop-helix-PAS-domain transcription factor is closely related to hypoxia-inducible factor-1 alpha and developmentally expressed in blood vessels. Mech Dev 1997; 63:51–60.
31. Hu CJ, Wang LY, Chodosh LA, Keith B, Simon MC. Differential roles of hypoxia-inducible factor 1alpha (HIF-1alpha) and HIF-2alpha in hypoxic gene regulation. Mol Cell Biol 2003; 23:9361–9374.
32. Covello KL, Kehler J, Yu H, et al. HIF-2alpha regulates Oct-4: effects of hypoxia on stem cell function, embryonic development, and tumor growth. Genes Dev 2006; 20:557–570.
33. Maxwell PH, Wiesener MS, Chang GW, et al. The tumour suppressor protein VHL targets hypoxia-inducible factors for oxygen-dependent proteolysis. Nature 1999; 399:271–275.
34. Ivan M, Haberberger T, Gervasi DC, et al. Biochemical purification and pharmacological inhibition of a mammalian prolyl hydroxylase acting on hypoxia-inducible factor. Proc Natl Acad Sci U S A 2002; 99:13459–13464.
35. Hirsila M, Koivunen P, Gunzler V, Kivirikko KI, Myllyharju J. Characterization of the human prolyl 4-hydroxylases that modify the hypoxia-inducible factor. J Biol Chem 2003; 278:30772–30780.
36. Appelhoff RJ, Tian YM, Raval RR, et al. Differential function of the prolyl hydroxylases PHD1, PHD2, and PHD3 in the regulation of hypoxia-inducible factor. J Biol Chem 2004; 279:38458–38465.
37. Huang J, Zhao Q, Mooney SM, Lee FS. Sequence determinants in hypoxia-inducible factor-1alpha for hydroxylation by the prolyl hydroxylases PHD1, PHD2, and PHD3. J Biol Chem 2002; 277:39792–39800.
38. Lieb ME, Menzies K, Moschella MC, Ni R, Taubman MB. Mammalian EGLN genes have distinct patterns of mRNA expression and regulation. Biochem Cell Biol 2002; 80:421–426.
39. Huang LE, Gu J, Schau M, Bunn HF. Regulation of hypoxia-inducible factor 1alpha is mediated by an O_2-dependent degradation domain via the ubiquitin-proteasome pathway. Proc Natl Acad Sci U S A 1998; 95:7987–7992.
40. Masson N, Willam C, Maxwell PH, Pugh CW, Ratcliffe PJ. Independent function of two destruction domains in hypoxia-inducible factor-alpha chains activated by prolyl hydroxylation. EMBO J 2001; 20:5197–5206.
41. Kamura T, Koepp DM, Conrad MN, et al. Rbx1, a component of the VHL tumor suppressor complex and SCF ubiquitin ligase. Science 1999; 284:657–661.
42. Kibel A, Iliopoulos O, DeCaprio JA, Kaelin WG, Jr. Binding of the von Hippel-Lindau tumor suppressor protein to elongin B and C. Science 1995; 269:1444–1446.
43. Pause A, Lee S, Worrell RA, et al. The von Hippel-Lindau tumor-suppressor gene product forms a stable complex with human CUL-2, a member of the Cdc53 family of proteins. Proc Natl Acad Sci U S A 1997; 94:2156–2161.
44. Ohh M, Park CW, Ivan M, et al. Ubiquitination of hypoxia-inducible factor requires direct binding to the beta-domain of the von Hippel-Lindau protein. Nat Cell Biol 2000; 2:423–427.
45. Tanimoto K, Makino Y, Pereira T, Poellinger L. Mechanism of regulation of the hypoxia-inducible factor-1 alpha by the von Hippel-Lindau tumor suppressor protein. EMBO J 2000; 19:4298–4309.
46. Jiang BH, Rue E, Wang GL, Roe R, Semenza GL. Dimerization, DNA binding, and transactivation properties of hypoxia-inducible factor 1. J Biol Chem 1996; 271:17771–17778.
47. Richard DE, Berra E, Pouyssegur J. Nonhypoxic pathway mediates the induction of hypoxia-inducible factor 1alpha in vascular smooth muscle cells. J Biol Chem 2000; 275:26765–26771.
48. Wolf G, Schroeder R, Stahl RA. Angiotensin II induces hypoxia-inducible factor-1 alpha in PC 12 cells through a posttranscriptional mechanism: role of AT2 receptors. Am J Nephrol 2004; 24:415–421.

49. McMahon S, Charbonneau M, Grandmont S, Richard DE, Dubois CM. Transforming growth factor beta 1 induces hypoxia-inducible factor-1 stabilization through selective inhibition of PHD2 expression. J Biol Chem 2006; 281:24171–24181.
50. Gottlieb E, Tomlinson IP. Mitochondrial tumour suppressors: a genetic and biochemical update. Nat Rev Cancer 2005; 5:857–866.
51. Mizukami Y, Jo WS, Duerr EM, et al. Induction of interleukin-8 preserves the angiogenic response in HIF-1alpha-deficient colon cancer cells. Nat Med 2005; 11:992–997.
52. Scortegagna M, Cataisson C, Martin RJ, et al. HIF-1alpha regulates epithelial inflammation by cell autonomous NFkappaB activation and paracrine stromal remodeling. Blood 2008; 111:3343–3354.
53. Page EL, Chan DA, Giaccia AJ, Levine M, Richard DE. Hypoxia-inducible factor-1alpha stabilization in nonhypoxic conditions: role of oxidation and intracellular ascorbate depletion. Mol Biol Cell 2008; 19:86–94.
54. Gerald D, Berra E, Frapart YM, et al. JunD reduces tumor angiogenesis by protecting cells from oxidative stress. Cell 2004; 118:781–794.
55. Pan Y, Mansfield KD, Bertozzi CC, et al. Multiple factors affecting cellular redox status and energy metabolism modulate hypoxia-inducible factor prolyl hydroxylase activity in vivo and in vitro. Mol Cell Biol 2007; 27:912–925.
56. Selak MA, Armour SM, MacKenzie ED, et al. Succinate links TCA cycle dysfunction to oncogenesis by inhibiting HIF-alpha prolyl hydroxylase. Cancer Cell 2005; 7:77–85.
57. MacKenzie ED, Selak MA, Tennant DA, et al. Cell-permeating alpha-ketoglutarate derivatives alleviate pseudohypoxia in succinate dehydrogenase-deficient cells. Mol Cell Biol 2007; 27:3282–3289.
58. Aprelikova O, Chandramouli GV, Wood M, et al. Regulation of HIF prolyl hydroxylases by hypoxia-inducible factors. J Cell Biochem 2004; 92:491–501.
59. Erez N, Stambolsky P, Shats I, et al. Hypoxia-dependent regulation of PHD1: cloning and characterization of the human PHD1/EGLN2 gene promoter. FEBS Lett 2004; 567:311–315.
60. D'Angelo G, Duplan E, Boyer N, Vigne P, Frelin C. Hypoxia up-regulates prolyl hydroxylase activity: a feedback mechanism that limits HIF-1 responses during reoxygenation. J Biol Chem 2003; 278:38183–38187.
61. Nakayama K, Frew IJ, Hagensen M, et al. Siah2 regulates stability of prolyl-hydroxylases, controls HIF1alpha abundance, and modulates physiological responses to hypoxia. Cell 2004; 117:941–952.
62. Barth S, Edlich F, Berchner-Pfannschmidt U, et al. Hypoxia-inducible factor prolyl-4-hydroxylase PHD2 protein abundance depends on integral membrane-anchoring of FKBP38. J Biol Chem 2009; 284(34):23046–23058.
63. Baek JH, Mahon PC, Oh J, et al. OS-9 interacts with hypoxia-inducible factor 1alpha and prolyl hydroxylases to promote oxygen-dependent degradation of HIF-1alpha. Mol Cell 2005; 17:503–512.
64. Ozer A, Wu LC, Bruick RK. The candidate tumor suppressor ING4 represses activation of the hypoxia inducible factor (HIF). Proc Natl Acad Sci U S A 2005; 102:7481–7486.
65. Koivunen P, Tiainen P, Hyvarinen J, et al. An endoplasmic reticulum transmembrane prolyl 4-hydroxylase is induced by hypoxia and acts on hypoxia-inducible factor alpha. J Biol Chem 2007; 282:30544–30552.
66. Oehme F, Ellinghaus P, Kolkhof P, et al. Overexpression of PH-4, a novel putative proline 4-hydroxylase, modulates activity of hypoxia-inducible transcription factors. Biochem Biophys Res Commun 2002; 296:343–349.
67. Aragones J, Schneider M, Van Geyte K, et al. Deficiency or inhibition of oxygen sensor PHD1 induces hypoxia tolerance by reprogramming basal metabolism. Nat Genet 2008; 40:170–180.
68. Berra E, Benizri E, Ginouves A, et al. HIF prolyl-hydroxylase 2 is the key oxygen sensor setting low steady-state levels of HIF-1alpha in normoxia. EMBO J 2003; 22:4082–4090.

69. Cioffi CL, Liu XQ, Kosinski PA, Garay M, Bowen BR. Differential regulation of HIF-1 alpha prolyl-4-hydroxylase genes by hypoxia in human cardiovascular cells. Biochem Biophys Res Commun 2003; 303:947–953.
70. Takeda K, Ho VC, Takeda H, et al. Placental but not heart defects are associated with elevated hypoxia-inducible factor alpha levels in mice lacking prolyl hydroxylase domain protein 2. Mol Cell Biol 2006; 26:8336–8346.
71. Takeda K, Aguila HL, Parikh NS, et al. Regulation of adult erythropoiesis by prolyl hydroxylase domain proteins. Blood 2008; 111:3229–3235.
72. Takeda K, Cowan A, Fong GH. Essential role for prolyl hydroxylase domain protein 2 in oxygen homeostasis of the adult vascular system. Circulation 2007; 116:774–781.
73. Cummins EP, Berra E, Comerford KM, et al. Prolyl hydroxylase-1 negatively regulates IkappaB kinase-beta, giving insight into hypoxia-induced NFkappaB activity. Proc Natl Acad Sci U S A 2006; 103:18154–18159.
74. Karashima T, Sweeney P, Kamat A, et al. Nuclear factor-kappaB mediates angiogenesis and metastasis of human bladder cancer through the regulation of interleukin-8. Clin Cancer Res 2003; 9:2786–2797.
75. Erez N, Milyavsky M, Goldfinger N, et al. Falkor, a novel cell growth regulator isolated by a functional genetic screen. Oncogene 2002; 21:6713–6721.
76. Mikhaylova O, Ignacak ML, Barankiewicz TJ, et al. The von Hippel-Lindau tumor suppressor protein and Egl-9-type proline hydroxylases regulate the large subunit of RNA polymerase II in response to oxidative stress. Mol Cell Biol 2008; 28:2701–2717.
77. Koditz J, Nesper J, Wottawa M, et al. Oxygen-dependent ATF-4 stability is mediated by the PHD3 oxygen sensor. Blood 2007; 110:3610–3617.
78. Hai T, Hartman MG. The molecular biology and nomenclature of the activating transcription factor/cAMP responsive element binding family of transcription factors: activating transcription factor proteins and homeostasis. Gene 2001; 273:1–11.
79. Ohoka N, Yoshii S, Hattori T, Onozaki K, Hayashi H. TRB3, a novel ER stress-inducible gene, is induced via ATF4-CHOP pathway and is involved in cell death. EMBO J 2005; 24:1243–1255.
80. Malhotra JD, Kaufman RJ. Endoplasmic reticulum stress and oxidative stress: a vicious cycle or a double-edged sword? Antioxid Redox Signal 2007; 9:2277–2293.
81. Maytin EV, Ubeda M, Lin JC, Habener JF. Stress-inducible transcription factor CHOP/GADD153 induces apoptosis in mammalian cells via p38 kinase-dependent and -independent mechanisms. Exp Cell Res 2001; 267:193–204.
82. Minamishima YA, Moslehi J, Bardeesy N, et al. Somatic inactivation of the PHD2 prolyl hydroxylase causes polycythemia and congestive heart failure. Blood 2008; 111:3236–3244.
83. Percy MJ, Zhao Q, Flores A, et al. A family with erythrocytosis establishes a role for prolyl hydroxylase domain protein 2 in oxygen homeostasis. Proc Natl Acad Sci U S A 2006; 103:654–659.
84. Nangaku M, Kojima I, Tanaka T, et al. Novel drugs and the response to hypoxia: HIF stabilizers and prolyl hydroxylase. Recent Pat Cardiovasc Drug Discov 2006; 1:129–139.
85. Choi HJ, Song BJ, Gong YD, Gwak WJ, Soh Y. Rapid degradation of hypoxia-inducible factor-1alpha by KRH102053, a new activator of prolyl hydroxylase 2. Br J Pharmacol 2008; 154:114–125.
86. Nangaku M, Izuhara Y, Takizawa S, et al. A novel class of prolyl hydroxylase inhibitors induces angiogenesis and exerts organ protection against ischemia. Arterioscler Thromb Vasc Biol 2007; 27:2548–2554.
87. Ockaili R, Natarajan R, Salloum F, et al. HIF-1 activation attenuates postischemic myocardial injury: role for heme oxygenase-1 in modulating microvascular chemokine generation. Am J Physiol Heart Circ Physiol 2005; 289:H542–H548.
88. Huang M, Chan DA, Jia F, et al. Short hairpin RNA interference therapy for ischemic heart disease. Circulation 2008; 118:S226–S233.

89. Milkiewicz M, Pugh CW, Egginton S. Inhibition of endogenous HIF inactivation induces angiogenesis in ischaemic skeletal muscles of mice. J Physiol 2004; 560:21–26.
90. Shyu KG, Wang MT, Wang BW, et al. Intramyocardial injection of naked DNA encoding HIF-1alpha/VP16 hybrid to enhance angiogenesis in an acute myocardial infarction model in the rat. Cardiovasc Res 2002; 54:576–583.
91. Vincent KA, Shyu KG, Luo Y, et al. Angiogenesis is induced in a rabbit model of hindlimb ischemia by naked DNA encoding an HIF-1alpha/VP16 hybrid transcription factor. Circulation 2000; 102:2255–2261.
92. Kido M, Du L, Sullivan CC, et al. Hypoxia-inducible factor 1-alpha reduces infarction and attenuates progression of cardiac dysfunction after myocardial infarction in the mouse. J Am Coll Cardiol 2005; 46:2116–2124.
93. Natarajan R, Salloum FN, Fisher BJ, Kukreja RC, Fowler AA, III. Hypoxia inducible factor-1 activation by prolyl 4-hydroxylase-2 gene silencing attenuates myocardial ischemia reperfusion injury. Circ Res 2006; 98:133–140.
94. Eckle T, Kohler D, Lehmann R, El Kasmi K, Eltzschig HK. Hypoxia-inducible factor-1 is central to cardioprotection: a new paradigm for ischemic preconditioning. Circulation 2008; 118:166–175.
95. Downey JM, Davis AM, Cohen MV. Signaling pathways in ischemic preconditioning. Heart Fail Rev 2007; 12:181–188.
96. Ladroue C, Carcenac R, Leporrier M, et al. PHD2 mutation and congenital erythrocytosis with paraganglioma. N Engl J Med 2008; 359:2685–2692.
97. Pappalardi MB, Martin JD, Jiang Y, et al. Biochemical characterization of human prolyl hydroxylase domain protein 2 variants associated with erythrocytosis. Biochemistry 2008; 47:11165–11167.
98. Hsieh MM, Linde NS, Wynter A, et al. HIF prolyl hydroxylase inhibition results in endogenous erythropoietin induction, erythrocytosis, and modest fetal hemoglobin expression in rhesus macaques. Blood 2007; 110:2140–2147.
99. Parmar K, Mauch P, Vergilio JA, Sackstein R, Down JD. Distribution of hematopoietic stem cells in the bone marrow according to regional hypoxia. Proc Natl Acad Sci U S A 2007; 104:5431–5436.
100. Carmeliet P. Angiogenesis in health and disease. Nat Med 2003; 9:653–660.
101. Carmeliet P. Manipulating angiogenesis in medicine. J Intern Med 2004; 255:538–561.
102. Ferrara N, Hillan KJ, Novotny W. Bevacizumab (Avastin), a humanized anti-VEGF monoclonal antibody for cancer therapy. Biochem Biophys Res Commun 2005; 333:328–335.
103. Hurwitz H, Fehrenbacher L, Novotny W, et al. Bevacizumab plus irinotecan, fluorouracil, and leucovorin for metastatic colorectal cancer. N Engl J Med 2004; 350:2335–2342.
104. Drevs J, Muller-Driver R, Wittig C, et al. PTK787/ZK 222584, a specific vascular endothelial growth factor-receptor tyrosine kinase inhibitor, affects the anatomy of the tumor vascular bed and the functional vascular properties as detected by dynamic enhanced magnetic resonance imaging. Cancer Res 2002; 62:4015–4022.
105. Nakamura K, Taguchi E, Miura T, et al. KRN951, a highly potent inhibitor of vascular endothelial growth factor receptor tyrosine kinases, has antitumor activities and affects functional vascular properties. Cancer Res 2006; 66:9134–9142.
106. Jain RK. Normalization of tumor vasculature: an emerging concept in antiangiogenic therapy. Science 2005; 307:58–62.
107. Carmeliet P, Jain RK. Angiogenesis in cancer and other diseases. Nature 2000; 407:249–257.
108. Mazzone M, Dettori D, Leite de Oliveira R, et al. Heterozygous deficiency of PHD2 restores tumor oxygenation and inhibits metastasis via endothelial normalization. Cell 2009; 136:839–851.
109. Cramer T, Yamanishi Y, Clausen BE, et al. HIF-1alpha is essential for myeloid cell-mediated inflammation. Cell 2003; 112:645–657.
110. Murdoch C, Muthana M, Lewis CE. Hypoxia regulates macrophage functions in inflammation. J Immunol 2005; 175:6257–6263.

111. Fraisl P, Aragones J, Carmeliet P. Inhibition of oxygen sensors as a therapeutic strategy for ischaemic and inflammatory disease. Nat Rev Drug Discov 2009; 8:139–152.
112. Rius J, Guma M, Schachtrup C, et al. NF-kappaB links innate immunity to the hypoxic response through transcriptional regulation of HIF-1alpha. Nature 2008; 453:807–811.
113. Peyssonnaux C, Datta V, Cramer T, et al. HIF-1alpha expression regulates the bactericidal capacity of phagocytes. J Clin Invest 2005; 115:1806–1815.
114. Peyssonnaux C, Cejudo-Martin P, Doedens A, et al. Cutting edge: essential role of hypoxia inducible factor-1alpha in development of lipopolysaccharide-induced sepsis. J Immunol 2007; 178:7516–7519.
115. Zinkernagel AS, Peyssonnaux C, Johnson RS, Nizet V. Pharmacologic augmentation of hypoxia-inducible factor-1alpha with mimosine boosts the bactericidal capacity of phagocytes. J Infect Dis 2008; 197:214–217.
116. Cummins EP, Seeballuck F, Keely SJ, et al. The hydroxylase inhibitor dimethyloxalylglycine is protective in a murine model of colitis. Gastroenterology 2008; 134:156–165.
117. Robinson A, Keely S, Karhausen J, et al. Mucosal protection by hypoxia-inducible factor prolyl hydroxylase inhibition. Gastroenterology 2008; 134:145–155.
118. Hagen T, Taylor CT, Lam F, Moncada S. Redistribution of intracellular oxygen in hypoxia by nitric oxide: effect on HIF1alpha. Science 2003; 302:1975–1978.
119. Metzen E, Zhou J, Jelkmann W, Fandrey J, Brune B. Nitric oxide impairs normoxic degradation of HIF-1alpha by inhibition of prolyl hydroxylases. Mol Biol Cell 2003; 14:3470–3481.
120. Sandau KB, Zhou J, Kietzmann T, Brune B. Regulation of the hypoxia-inducible factor 1alpha by the inflammatory mediators nitric oxide and tumor necrosis factor-alpha in contrast to desferroxamine and phenylarsine oxide. J Biol Chem 2001; 276:39805–39811.
121. Taylor MS. Characterization and comparative analysis of the EGLN gene family. Gene 2001; 275:125–132.
122. Wax SD, Rosenfield CL, Taubman MB. Identification of a novel growth factor-responsive gene in vascular smooth muscle cells. J Biol Chem 1994; 269:13041–13047.
123. Landazuri MO, Vara-Vega A, Viton M, Cuevas Y, del Peso L. Analysis of HIF-prolyl hydroxylases binding to substrates. Biochem Biophys Res Commun 2006; 351:313–320.
124. Metzen E, Berchner-Pfannschmidt U, Stengel P, et al. Intracellular localisation of human HIF-1 alpha hydroxylases: implications for oxygen sensing. J Cell Sci 2003; 116:1319–1326.
125. Bishop T, Gallagher D, Pascual A, et al. Abnormal sympathoadrenal development and systemic hypotension in PHD3−/− mice. Mol Cell Biol 2008; 28:3386–3400.
126. Steinhoff A, Pientka FK, Mockel S, et al. Cellular oxygen sensing: importins and exportins are mediators of intracellular localisation of prolyl-4-hydroxylases PHD1 and PHD2. Biochem Biophys Res Commun 2009; 387:705–711.
127. Yasumoto K, Kowata Y, Yoshida A, Torii S, Sogawa K. Role of the intracellular localization of HIF-prolyl hydroxylases. Biochim Biophys Acta 2009; 1793:792–797.
128. Tian YM, Mole DR, Ratcliffe PJ, Gleadle JM. Characterization of different isoforms of the HIF prolyl hydroxylase PHD1 generated by alternative initiation. Biochem J 2006; 397:179–186.

Chapter 23
Oxygen-Dependent Regulation of Erythropoiesis

Volker H. Haase

Abstract Oxygen-regulated erythropoiesis serves as a paradigm of systemic hypoxia responses. Efforts to understand its molecular basis have led to the identification of erythropoietin (EPO) and hypoxia-inducible factor (HIF). This chapter provides an overview of the most recent advances in oxygen-dependent erythropoiesis and discusses (a) how molecular oxygen-sensing mechanisms regulate and integrate cellular and systemic hypoxia responses that stimulate and support erythropoiesis and (b) how these mechanisms may be exploited pharmacologically for the treatment of anemias. Specifically this chapter addresses the role of the HIF pathway in regulating and coordinating tissue-dependent EPO production and iron metabolism, and discusses inherited forms of erythrocytosis that result from a disruption of molecular oxygen-sensing mechanisms.

Keywords Erythropoietin · HIF · pVHL · Iron · Oxygen sensing

1 Introduction

Tissue hypoxia is the result of a discrepancy between oxygen availability and demand and triggers systemic and cellular responses that are aimed at improving tissue oxygenation and thus cell survival. Hypoxia is a frequently encountered clinical problem in patients with pulmonary, cardiovascular, or hematological conditions that impair oxygen delivery, but also occurs under physiologic conditions during normal embryonic development, when alterations in tissue oxygen levels provide important microenvironmental signals that modulate cellular growth and differentiation. In order to adapt to low partial oxygen pressure following

V.H. Haase (✉)
Division of Nephrology and Hypertension, Vanderbilt University, C-3119A, MCN,
1161 21st Avenue, Nashville, TN 37232, USA
e-mail: volker.haase@vanderbilt.edu

T. Miyata et al. (eds.), *Studies on Renal Disorders*, Oxidative Stress in Applied Basic Research and Clinical Practice, DOI 10.1007/978-1-60761-857-7_23,

ascent to high altitude the human body undergoes multiple physiologic changes. These include an acute increase in ventilation, heart rate, and cardiac output. One of the more extensively studied systemic adaptations to high altitude is enhanced erythropoiesis, which is the result of a well-executed transcriptional response that is aimed at improving oxygen carrying capacity and blood oxygen content. It was the interest in understanding the physiologic and molecular basis of this erythropoietic response that not only led to the discovery of erythropoietin (EPO), the glycoprotein hormone, which is essential for the regulation of red blood cell mass, but more importantly, led to the identification of the molecular machinery, which senses oxygen and controls the wide spectrum of transcriptional responses to hypoxia. Oxygen-dependent induction of EPO therefore serves as the paradigm of hypoxic gene regulation, and the search for the transcription factor that mediates this increase resulted in the discovery of hypoxia-inducible factor (HIF) as the central mediator of cellular adaptation to low oxygen. There is now mounting experimental evidence that HIF not only stimulates red blood cell production by increasing serum EPO levels, but that it also promotes erythropoiesis through coordinated cell type–specific hypoxia responses that result in enhanced iron uptake and utilization, as well as favorable changes in the bone marrow microenvironment that facilitate erythroid progenitor maturation and proliferation. Because of this central role in the regulation of erythropoiesis, pharmacological targeting of the HIF oxygen-sensing pathway offers enormous potential for the treatment of anemias associated with inadequate EPO production. Recent insights into the molecular mechanisms that underlie the oxygen-dependent regulation of EPO synthesis, iron metabolism, and erythroid progenitor maturation and their relevance to clinical disorders are discussed in this chapter.

2 EPO Regulation as a Paradigm of Systemic Hypoxia Responses

Over 100 years ago Paul Bert, by many considered to be the founder of modern aerospace and high altitude medicine, and his collaborator the French physician Denis Jourdanet observed the association between reduced atmospheric oxygen pressure and elevated red blood cell numbers [1–3]. Jourdanet, who practiced medicine in Mexico and studied the clinical manifestations of chronic mountain sickness, noted that residing at high altitude was associated with increased blood viscosity, while Bert found elevated hemoglobin concentrations in animals living in the mountains of Bolivia and speculated that this increase in red cell mass might represent an inherited condition. During an expedition to the Peruvian Andes in 1890, Francois-Gilbert Viault discovered that high altitude, however, provided a direct stimulus for erythropoiesis and documented this acute elevation in red blood cell counts in his own blood and that of his companions [4]. In 1906, Paul Carnot and his assistant C. Deflandre postulated that serum contained an erythropoietic factor (hemopoietin) [5], which Bonsdorff and Jalavisto later called erythropoietin [6]. In their experiment, Carnot and Deflandre injected serum from mildly anemic rabbits into normal rabbits and found increased red blood cell counts [5]. However, it was not until almost 50 years later that their results could be reliably reproduced by Allan Erslev [7], finally proving the existence of EPO. Using biological assays, normal serum EPO concentrations were determined to be in the range of 5–30 mU/mL, whereby 1 U of EPO was defined as equivalent to the erythropoietic effects of 5 μmol of $CoCl_2$ [8]. The ability to quantify serum and urine EPO led to investigations into the dynamics of the hypoxia-induced EPO response, which is very rapid, pO_2-dependent, and results in exponential increases in serum EPO levels when hematocrit declines in a linear fashion [9–11]. Eventually human EPO was purified from large quantities of urine obtained from severely anemic patients by Miyake et al. [12]. This permitted determination of its protein sequence and cloning of the *EPO* gene [13, 14], ultimately resulting in the production and successful clinical use of recombinant EPO; EPO as a therapeutic agent was approved by the U.S. Food And Drug Administration in 1989. Important further steps in the molecular understanding of hypoxic EPO induction were reports by Nielsen et al. [15] and Goldberg et al. [16], who discovered that the human liver tumor cell lines Hep3B and HepG2 produced EPO in an oxygen-dependent manner. The availability of EPO-expressing cell lines permitted detailed studies of *EPO* gene regulation, which aimed at understanding the molecular basis of its hypoxia-inducibility (for a summary of studies see [17]). Gregg Semenza et al. from Johns Hopkins University used transgenic mice to identify and validate the hypoxia-sensitive regulatory elements in the *EPO* gene and then were able to purify HIF-1 as the transcription factor that binds to its 3′-hypoxia-enhancer, mediating its hypoxic induction [18–21].

EPO's major action is the prevention of apoptosis in EPO-dependent colony-forming unit-erythroid (CFU-E) cells and erythroblasts that have not begun hemoglobin synthesis. The human *EPO* gene encodes a glycoprotein hormone, which as a pro-hormone consists of 193 amino acids. Following removal of its peptide leader sequence and a carboxy-terminal arginine, it is reduced in size to 165 amino acids in its circulating form. Serum EPO is heavily glycosylated and amounts to a molecular mass of about 30 kDa, with approximately 40% being derived from its carbohydrate portion. Its receptor (EPO-R), which is also hypoxia-inducible [22–24], lacks intrinsic enzymatic function and associates with the tyrosine kinase Janus kinase 2 (JAK2). Upon ligand binding and homodimerization JAK2 phosphorylates EPO-R at multiple sites, thus providing docking sites for signal transducing molecules that contain src homology 2 (SH2) domains. Pathways through which EPO-R signaling occurs include the signal transduction and activator of transcription (STAT) 5 pathway, the phosphatidyl-inositol-3-kinase/protein kinase B (PI-3K/AKT), and mitogen-associated protein kinase/extracellular signal-related kinase (MAPK/ERK) pathways, as well as signaling through protein kinase C.

3 Regulation of EPO Synthesis by Molecular Oxygen

The primary physiologic stimulus of EPO production is tissue hypoxia, which can induce several 100-fold increases in circulating serum EPO levels, depending on the hypoxic condition [25]. Studies aimed at isolating transcription factors responsible for the hypoxia-mediated induction of EPO identified the heterodimeric basic helix-loop-helix (bHLH) transcription factor hypoxia-inducible factor-1 (HIF-1) [20, 21]. HIF-1 belongs to the PAS (PER/aryl-hydrocarbon-receptor nuclear translocator [ARNT]/single minded [SIM]) family of hypoxia-regulated transcription factors and consists of an oxygen-sensitive alpha subunit and the constitutively expressed beta-subunit ARNT [26–28]. Together with HIF-2α (also known as EPAS-1, HLF), HIF-1α facilitates oxygen delivery and cellular adaptation to hypoxia by stimulating erythropoiesis, angiogenesis, and anaerobic glucose metabolism [29]. HIFs regulate gene expression by binding to hypoxia-response elements (HREs), specific DNA recognition sequences located in the regulatory regions of HIF target genes (Fig. 1). Three HIF-α subunits, HIF-1α, HIF-2α, and HIF-3α, have been identified. Under normoxic conditions, all three homologues are targeted for rapid proteasomal degradation by the von Hippel-Lindau tumor suppressor pVHL, which acts as the substrate recognition component of an E3 ubiquitin-ligase complex [30, 31]. Whereas HIF-1α and HIF-2α heterodimers function as transcriptional activators, splice variants of HIF-3α have been shown to be inhibitory [32, 33]. Although HIF-1 and HIF-2 share many common transcriptional targets, they regulate unique targets and have specific biological functions. Anaerobic glycolysis, for example, appears to be predominantly controlled by HIF-1 [34], whereas HIF-2 has now emerged as the main regulator of EPO production in the adult [35–38] (Fig. 1). In addition to heterodimerization with HIF-β, HIF-α modulates cellular signaling pathways through functional interaction with proteins that do not contain PAS domains. These include, among others, tumor suppressor protein p53, the c-Myc proto-oncogene, and the Notch intracellular domain [39–42].

Under normal oxygen conditions HIF α-subunits are rapidly degraded following ubiquitylation by the pVHL-E3-ubiquitin ligase, precluding the formation of transcriptionally active heterodimers. pVHL-mediated polyubiquitylation of HIF-α requires hydroxylation of specific proline residues (Pro402 and Pro564 in human HIF-1α; Pro405 and Pro531 in human HIF-2α) within its oxygen-dependent degradation domain (ODD) [43–49]. Hydroxylation of HIF-α is carried out by three major 2-oxoglutarate-dependent dioxygenases (prolyl-4-hydroxylase domain [PHD] proteins), PHD1, PHD2 and PHD3, and requires molecular oxygen, ferrous iron, and ascorbate [50] (Fig. 1). To add complexity to this system, transcription of *PHD2* and *PHD3* is increased under hypoxia through HIF-dependent mechanisms, and PHD1 and PHD3 protein turnover under hypoxia is regulated by Siah proteins, which themselves are hypoxia-inducible [51, 52]. An additional hypoxic switch operates in the carboxy-terminal transactivation domain of HIF-α with oxygen-dependent asparagine hydroxylation by factor-inhibiting-HIF (FIH) blocking CBP/p300

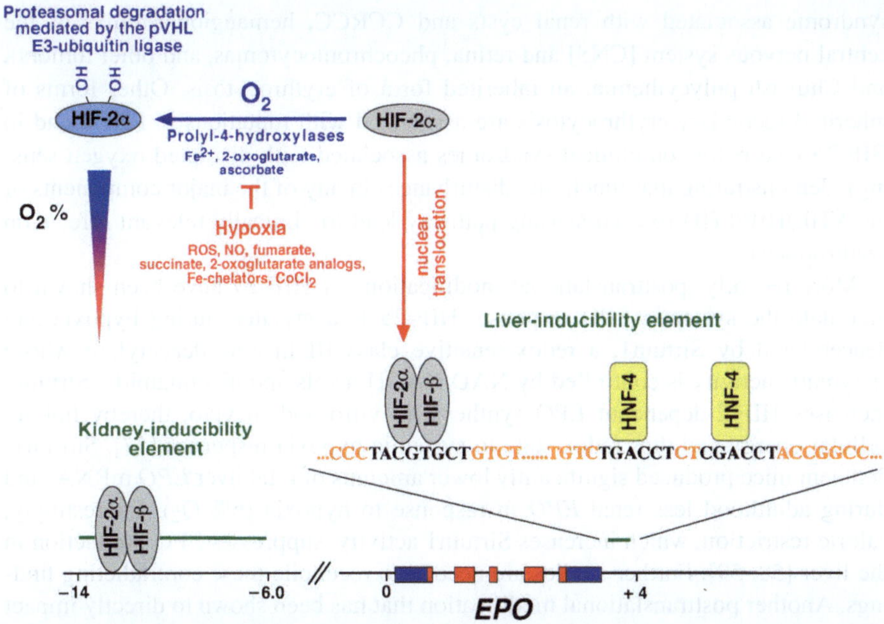

Fig. 1 HIF-2 regulates EPO in kidney and liver. Under normoxia, HIF-2**α** is hydroxylated by prolyl-4-hydroxylases and targeted for proteasomal degradation by the pVHL-E3-ubiquitin ligase complex. When prolyl-4-hydroxylation is inhibited (e.g., in the absence of molecular oxygen), HIF-2**α** is not degraded and translocates to the nucleus where it heterodimerizes with HIF-**β**. HIF-2**α**/**β** heterodimers bind to the HIF consensus binding site RCGTG and increase EPO transcription. The hypoxic induction of EPO in the liver is mediated by the liver inducibility element located in the 3′-end of the EPO gene and in renal interstitial fibroblast-like cells by the 5′ kidney inducibility element, which is located 6–14 kb upstream of its transcription start site. Nitric oxide, reactive oxygen species, Krebs cycle metabolites succinate and fumarate, cobalt chloride, and iron chelators such as desferrioxamine inhibit HIF prolyl-4-hydroxylases in the presence of oxygen. EPO mRNA is encoded by five exons, depicted by boxes, coding sequences are shown in red; untranslated regions are shown in blue; numbers indicate distance from the transcription start site in kb (not drawn to scale). Shown also are the binding sites of HNF-4 in the 3′ liver inducibility region. *HIF* hypoxia-inducible factor; *pVHL* von Hippel-Lindau protein; *CoCl$_2$* cobalt chloride; *EPO* erythropoietin; *Fe^{2+}* ferrous iron; *HNF-4* hepatocyte nuclear factor-4; *NO* nitric oxide; *ROS* reactive oxygen species

recruitment. Inhibition of FIH under hypoxic conditions facilitates CBP/p300 recruitment, resulting in increased HIF target gene expression in *VHL*-deficient cell lines or under pronounced hypoxia [53–56].

A defect in the ability to ubiquitylate HIF-α, for example as a consequence of mutated pVHL, results in HIF-α stabilization, increased HIF transcriptional activity, and upregulation of HIF target genes such as vascular endothelial growth factor (*VEGF*), glucose transporter 1 (*GLUT-1*), and *EPO* irrespective of oxygen levels. Clinically, mutations in pVHL are associated with the development of sporadic renal cell cancer of the clear cell type (CCRCC), VHL disease (a familial tumor syndrome associated with renal cysts and CCRCC, hemangioblastomas of the central nervous system [CNS] and retina, pheochromocytomas, and other tumors), and Chuvash polycythemia, an inherited form of erythrocytosis. Other forms of inherited secondary erythrocytosis are associated with mutations in PHD2 and in HIF-2α (see below on clinical syndromes associated with disrupted oxygen sensing), demonstrating that functional disturbances in any of the major components of the VHL/HIF/PHD oxygen-sensing pathway lead to clinically relevant effects on erythropoiesis.

More recently, posttranslational modifications of HIF-2α have been shown to modulate the systemic EPO response. HIF-2α is acetylated during hypoxia and deacetylated by Sirtuin1, a redox-sensitive class III histone deacetylase whose enzymatic activity is controlled by NAD/NADH levels and nicotinamide. Sirtuin1 increases HIF-2-dependent EPO synthesis in vitro and in vivo, thereby linking cellular energy and thus redox state to systemic hypoxia responses [57]. Sirtuin1-deficient mice produced significantly lower amounts of fetal liver *EPO* mRNA, and during adulthood less renal *EPO* in response to hypoxia (6% O_2). Interestingly, caloric restriction, which increases Sirtuin1 activity, suppresses EPO production in the liver [58, 59]. Further studies are needed to reconcile these contradicting findings. Another posttranslational modification that has been shown to directly impact EPO synthesis and erythropoiesis is SUMOylation. SUMO (small ubiquitin-like modifier) proteins are structurally similar to ubiquitin and are reversibly attached to target proteins by SUMOylating enzymes, thus altering their function and cellular localization. One of the enzymes that removes SUMO from target proteins is SENP (Sentrin/SUMO-specific protease). SENP1 knockout mice developed severe fetal anemia and died during midgestation [60]. In this model de-SUMOylation under hypoxic conditions did not occur in the absence of SENP and prevented the activation of HIF signaling in the nucleus, which resulted in a strong reduction of hepatic *EPO* mRNA levels. Instead, SUMOylated HIF-α was targeted for proteasomal degradation in a pVHL- and ubiquitin-dependent, but PHD-independent manner, as PHDs are not active in the absence of molecular oxygen [60]. Although SUMOylation of HIF-α was specifically investigated with regard to hypoxic HIF-1 signaling, the anemic phenotype suggests that, in light of HIF-2's function in fetal erythropoiesis, SENP's role in hypoxic signaling must extend to HIF-2α as well.

4 Tissue Sources of EPO

Animal experiments, in which various organs were resected, identified the kidney as the major site of EPO production [61], as nephrectomy led to anemia and to a severe blunting of the erythropoietic response to hypoxic stimuli. Despite initial uncertainties and debate, there is now strong experimental and morphological evidence that interstitial peritubular fibroblast-like cells located in the renal inner cortex and outer medulla, not tubular epithelial cells, are the cellular source of renal EPO [62–67]. *EPO* expression in other renal cell types, such as tubular epithelial cells, appears to be suppressed by inhibitory GATA transcription factors, in particular GATA-2 and GATA-3. Mutations of the GATA binding site in EPO-green fluorescent protein (GFP) transgenic animals resulted in GFP expression in distal tubular cells under normoxic and hypoxic conditions, suggesting that in the kidney, these factors are important in limiting the expression of EPO to peritubular fibroblast-like cells [67]. It is also interesting that the kidney does not respond to hypoxia by dynamically increasing the *EPO* message in individual cells, but rather by pO_2-dependent recruitment of additional cells that generate a fixed amount of EPO. In this regard, renal EPO production appears to be mechanistically different from EPO synthesis in hepatocytes, hepatoma, neuroblastoma, or melanoma cell lines [64, 67].

The isolation and establishment of EPO-producing cells from the kidney has been largely unsuccessful, although hypoxic induction of EPO synthesis was reported in 4E cells, a mesenchymal cell clone isolated from the adult kidney [68]. EPO-producing renal cells, however, have been successfully tagged with GFP [67], which should permit their isolation and purification by fluorescence-activated cell sorting (FACS) and facilitate the establishment of permanent cell lines in the near future. In their study, Obara et al. used homologous recombination to generate BAC transgenes, in which a segment of the murine *EPO* gene, spanning from exon 2 to the 3′-end of intron 4, was replaced with GFP cDNA, bringing GFP under the control of *EPO* regulatory elements [67]. GFP expression in transgenic mice was induced by anemia in renal peritubular interstitial cells specifically, and in hepatocytes surrounding the central vein, supporting the notion that these two cell types are the major sources of EPO production under hypoxic conditions. Obara et al. furthermore demonstrated that the number of GFP-expressing cells increased exponentially with a linear decline in hematocrit, supporting the notion that in the kidney, it is the number of EPO-producing cells and not cellular transcript abundance that changes in response to anemia. Renal EPO-producing cells, named REP cells by the authors, were unique in their morphologic appearance and possessed dendrite-like processes, which resembled those of neurons. Indeed, REP cells expressed neural-specific markers, such as microtubule-associated protein 2 (MAP2) and neurofilament protein, light polypeptide (NFL), suggesting that they may be derived from progenitor cells related to the neural lineage. In keeping with this notion, Frede et al. recently described the establishment of an EPO-producing renal tumor cell line with similar morphologic and molecular characteristics [69]. These exiting findings will certainly stimulate further interest and investigation into the histogenetic origin of renal EPO-producing cells.

While the kidney is the primary physiologic site of EPO synthesis in the adult, the liver is the main source during embryonic development. In the adult it is the major extrarenal source of EPO synthesis when stimulated by hypoxia. While hepatocytes have been identified as the primary cell type responsible for EPO synthesis in the liver, *EPO* can be detected in hepatic stellate cells, which are also known as ITO cells [70, 71]. The onset of the transition from liver to the kidney as the primary site of EPO production is species-dependent and usually occurs in late gestation or around birth [72–75]. Despite extensive efforts by many laboratories, the molecular mechanisms underlying this switch are still poorly understood and may involve transcriptional repression, or reduced expression of certain transcriptional activators such as transcription factor GATA-4 [76]. In the adult, liver *EPO* mRNA levels, which are very difficult to detect at baseline, rise substantially following hypoxic stimulation and account for most, if not all, physiologically relevant systemic EPO of extrarenal origin [77, 78]. While the liver responds poorly to mild hypoxia, hepatic EPO production is significantly stimulated by severe hypoxia (7.5% O_2) and may account for more than one third of total serum EPO, with the remainder being kidney-derived [73, 79]. Hypoxia-induced *EPO* transcription in the liver seems to be negatively affected by uremia in some studies, as demonstrated in bilaterally nephrectomized rats, in which liver *EPO* mRNA was reduced by almost 50% compared to controls [73]. Nevertheless, in stimulated uremic rats that had undergone partial nephrectomy, hepatic EPO accounted for more than 80% of an approximately tenfold increase in total serum EPO compared to 25–50% of an approximately 40-fold increase in control animals, illustrating the potential of the liver to produce large quantities of EPO under hypoxic conditions, despite the presence of uremia [80]. However, physiologic production of liver EPO under normoxic conditions is usually not adequate to compensate for the loss of renal EPO in the setting of nephrectomy or advanced renal failure. Because of the recent advances in understanding the molecular basis of HIF oxygen sensing, pharmacological stimulation of HIF signaling through prolyl-4-hydroxyase inhibition, with the goal stimulating hepatic EPO synthesis, offers great potential as a therapeutic strategy in the treatment of renal anemia or other conditions that are associated with inadequate EPO production (see Sect. 10).

The distinct regulatory elements that control HIF-mediated induction of EPO in liver and kidney are located on the opposite ends of the *EPO* gene, the kidney-inducibility element in the 5′-region and the liver-inducibility element in the 3′-region. Semenza et al. first identified these in transgenic mouse lines that expressed human *EPO* gene fragments of different sizes. The shortest of these transgenes, tgEPO4, contained, in addition to the *EPO* coding sequence, 0.4 kb of upstream and 0.7 kb of downstream regulatory sequence, sufficient for hypoxic transgene induction in the liver, but not in kidneys. The inclusion of additional 14 kb 5′-flanking sequence, tgEPO18, resulted in renal expression. Further in vivo mapping studies located the kidney-inducibility element to a region approximately 6–14 kb 5′ of the *EPO* transcription start site [18, 81, 82]. That hypoxic induction of EPO in the liver is mediated by the *EPO* 3′ HRE in vivo, was also shown by Suzuki et al., who, by mutating the 3′ HRE in an EPO-GFP transgenic mouse abolished hypoxia-inducibility of GFP in the liver, but not in kidneys, elegantly demonstrating that the *EPO* 3′ hypoxia enhancer is liver specific [83].

Aside from kidney and liver as the two major sources of synthesis, *EPO* mRNA expression has also been detected in the brain (neurons and astrocytes), lung, heart, bone marrow, spleen, hair follicles, and reproductive tract [75, 84–93]. However, it appears that these sites do not make a significant contribution to systemic EPO homeostasis and erythropoiesis, and that locally synthesized EPO acts, most likely, in an auto- or paracrine fashion, impacting, for example, regional angiogenesis and cell survival under hypoxic stress (for a overview of the nonhematopoietic actions of EPO see Jelkmann [94]).

5 Nonrenal Oxygen Sensing in EPO Homeostasis: Indirect Mechanisms

Experiments with isolated perfused kidneys and isolated hepatocytes provided the first direct evidence that kidneys and liver directly sense decreases in regional oxygen tension, to which they respond with increased EPO synthesis. In addition, it has also been shown that, under certain experimental conditions, EPO production in the kidney can be activated by extrinsic signals. Von Wussow et al. postulated the existence of an O_2-sensitive sensor in the brainstem, which triggers renal EPO production through the release of yet to be identified humoral factors (isolated brainstem hypoxia was generated by increased intracranial pressure) [95]. More recently, Boutin et al. demonstrated that the skin has the potential to modulate renal and hepatic EPO synthesis indirectly through HIF-1- and nitric oxide (NO)–mediated effects on dermal blood flow [96]. This, however, is dependent on the ability of the kidney and liver to perceive changes in tissue oxygen levels locally at the sites of EPO production. Keratinocyte-specific inactivation of HIF-1α, but not HIF-2α, resulted in a blunted renal EPO response to hypoxia (9% O_2 for 14 h), which was associated with increased blood flow to the kidney, presumably resulting in increased tissue oxygenation and thus suppression of EPO synthesis [96]. Conversely, activation of HIF signaling in the skin through genetic ablation of pVHL produced dermal vasodilatation in a NO-dependent manner, which redirected blood flow to the skin and away from the kidneys and liver, resulting in persistent tissue hypoxia, which in turn led to increased EPO production and the development of polycythemia (*EPO* mRNA in the skin was not detected in this study). It is of interest to mention here that experimental stenosis of the renal arteries causes only a small linear increase in serum EPO, as opposed to a large exponential increase in response to anemia [97]. This is most likely due to the fact that renal blood flow (RBF) and glomerular filtration rate (GFR) are functionally linked, that is, that the effects of decreased RBF and thus decreased O_2 delivery on renal tissue oxygenation are frequently offset by a simultaneous decrease in GFR and thus reduction in tubular workload and oxygen consumption. In light of these considerations additional studies are warranted to determine whether changes in RBF alone can completely account for renal tissue hypoxia that led to increased renal EPO synthesis in the skin knockout model. Boutin et al. furthermore proposed that the skin, independent of blood oxygen levels, senses differences in environmental oxygen to which it responds with changes in vascular tone, that is, vasoconstriction and a reduction in blood flow under conditions of acute hypoxia. This was demonstrated experimentally by delivering different oxygen concentrations to the lungs and skin of mice for a short period of time (5 h), that is, mice were breathing hypoxic gas while their body was exposed to either normoxic (21% O_2) or hypoxic gas (10% O_2). Mice breathing hypoxic gas with their skin exposed to normoxia had an enhanced renal EPO response, while the renal EPO response in mice whose lung and skin were exposed to the same degree of hypoxia was diminished. The latter response was associated with HIF-1-dependent dermal vasoconstriction and a relative increase in RBF. Given the finding of EPO synthesis in hair follicles [90], a direct contribution of dermal EPO to total serum EPO cannot be completely ruled out. Whether this is the case or not, the studies by Boutin et al. have certainly highlighted the importance of the skin in the regulation of EPO homeostasis and erythropoiesis.

6 HIF-1 Vs HIF-2 in EPO Synthesis

Although HIF-1 was the first HIF transcription factor that was isolated from human Hep3B cells utilizing the HIF binding sequence in the *EPO* 3′ hypoxia enhancer, there is now compelling experimental evidence that HIF-2 acts as the main mediator of the hypoxic induction of EPO in the adult kidney and liver. This notion is based on in vitro and in vivo studies using human cell lines and genetic mouse models, as well as on histological studies, which demonstrated that the location of HIF-2α-expressing renal cells coincided with the location of EPO-producing renal interstitial fibroblast-like cells [98].

The expression profile of HIF-2α was initially thought to be limited to endothelial cells. Further detailed analysis, however, demonstrated that HIF-2α was expressed in additional cell types, although not as widespread as HIF-1α, which is broadly expressed in all organs. Aside from endothelial cells, HIF-2α-expressing cells include hepatocytes, cardiomyocytes, glial cells in the central nervous system, type-II pneumocytes in the lung, and renal peritubular interstitial cells [98, 99].

First insights into the functional differences between HIF-1 and HIF-2 with regard to EPO regulation came from the analysis of HIF-1α and HIF-2α knockout mice. HIF-1- and HIF-2-deficient mice differ dramatically in their phenotypes. Mice that lack both copies of HIF-1α die in utero between embryonic day (E) 8 and E11 from neural tube defects, increased cell death in the cephalic mesenchyme, and cardiovascular malformations [100, 101]. Mice with homozygous deletion of HIF-2α die in utero or around birth, unless bred as heterozygotes in a mixed C57/BL6J and 129S6/SvEv genetic background [102]. Three different phenotypes have been described for homozygous HIF-2α germ line inactivation: (a) defective catecholamine synthesis in the organ of Zuckerkandl, leading to heart failure and midgestational death [103]; (b) abnormal VEGF-mediated lung maturation, resulting in perinatal death [104]; and (c) severe vascular defects in the yolk sac and embryo proper, resulting in death between E9.5 and E13.5 [105]. When bred in a mixed genetic background, HIF-2α knockout mice survived into adulthood but developed hepatic steatosis, skeletal myopathy, and cardiac hypertrophy, which were associated with mitochondrial abnormalities and deficiencies in reactive oxygen species (ROS) scavenging, resulting from an inadequate production of superoxide dismutase (SOD) [102]. Hematologic analysis revealed pancytopenia associated with hypocellularity of the bone marrow [106]. Since renal EPO synthesis was decreased and red blood cell counts were corrected with recombinant EPO [36], anemia in HIF-2-deficient mice resulted from inadequate renal EPO production and not from a significant cell-autonomous defect in erythroid precursor maturation. While Morita et al. had previously demonstrated a role for HIF-2 in the synthesis of retinal EPO [35], the study by Scortegagna et al. established for the first time that HIF-2 was essential in the regulation of systemic EPO homoeostasis [36, 106]. The relative contribution of HIF-2 to the hypoxic induction of EPO in the adult, however, remained unclear, since a direct comparison between HIF-1α and HIF-2α knockout mice was not possible due to the embryonic lethality that is associated with homozygous deletion of HIF-1α in the germ line. In order to study the effects of homozygous HIF-1 deficiency on EPO synthesis and erythropoiesis in vivo, Yoon et al. analyzed HIF-1-deficient embryos at E9.5 and were able to establish a role for HIF-1 in the regulation of embryonic EPO, erythropoiesis, and iron metabolism [24]. They found a reduction in myeloid multilineage and committed erythroid progenitor cells, while *EPO* mRNA levels were decreased in the embryo proper but not in yolk sac. *EPO-R* mRNA was decreased in both tissues. In adult mice with heterozygous HIF-1α deficiency, Yu et al. reported delayed erythrocytosis on exposure to chronic hypoxia (10% O_2 over a period of up to 6 weeks, serum EPO or tissue *EPO* mRNA levels were not measured) [108]. Given recent findings on the role of skin oxygen sensing in the regulation of EPO, it could be possible that in this study, reduced expression of dermal HIF-1α may have indirectly affected the renal EPO response through a redirection of blood flow [107]. Cai and colleagues observed that renal *EPO* mRNA did not rise following treatment with 1-h intermittent hypoxia, supporting a role for HIF-1 in the hypoxic induction of renal EPO in the adult [109]. However, the degree to which individual HIF-α subunits contributed to EPO production, that is, whether *EPO* was coregulated by HIF-1 and HIF-2, or whether it was preferentially regulated by one HIF or the other, could not be determined by these studies. Warnecke et al. used cell lines to address this issue and treated Hep3B and Kelly cells with small interfering RNAs (siRNA) directed against HIF-1α or HIF-2α. They found that the hypoxic induction of *EPO* was largely HIF-2-dependent, while HIF-1, in contrast to other HIF-regulated genes, played only a minor role in the hypoxic regulation of *EPO* [110]. The same was found in EPO-producing cultured cortical astrocytes [111], whereas studies in other cell lines suggested HIF-1 dependence [17].

The most compelling support for the notion that HIF-2 is the main regulator of EPO, at least in adults, comes from conditional knockout studies in mice. Gruber et al. observed that postnatal global ablation of HIF-2α, but not of HIF-1α, resulted in anemia, which, similar to the findings in mice with HIF-2α germ line deletion, was treatable with recombinant EPO [37]. This study made use of a tamoxifen-inducible and ubiquitously expressed Cre-recombinase transgene, which produced efficient, albeit not complete, recombination of HIF-α conditional alleles in every tissue. While the increase in renal EPO in response to phenylhydrazine treatment (phenylhydrazine induces hemolytic anemia) was blunted in HIF-2α ablated mice, postnatal deletion of HIF-1α had no effect on renal EPO production. Although the effects of chronic hypoxia on erythropoiesis were not examined in these mice, this study allowed direct comparison of HIF-1 to HIF-2, and did not find any obvious role for HIF-1 in the regulation of EPO homeostasis at baseline or in response to anemia [37].

Work from our laboratory has demonstrated that HIF-2, not HIF-1, regulates EPO in hepatocytes under different experimental conditions. Cell type-specific inactivation of pVHL in a subpopulation of hepatocytes (20–30%) resulted in HIF-2- but not HIF-1-dependent erythrocytosis, while pharmacological inhibition of HIF prolyl-4-hydroxylases caused an HIF-2-dependent increase in liver *EPO* mRNA levels. More importantly, we found that HIF-2, not HIF-1, induced liver EPO in response to anemia using hepatocyte-specific HIF-1α and HIF-2α knockout mice, and that inactivation of hepatic HIF-2 resulted in postnatal anemia, which resolved with maturation to adulthood (the liver is the main tissue source of EPO during embryonic development). Consistent with our findings are studies by Kim et al., who expressed a nondegradable form of HIF-1α or HIF-2α in hepatocytes [112]. Overexpression of HIF-2α in mouse livers produced erythrocytosis, whereas HIF-1α did not. In addition, our laboratory was able to demonstrate that inactivation of renal HIF-2α alone completely ablated the renal EPO response when mice were subjected to normobaric hypoxia (10% O_2 for 10 days), anemic hypoxia induced by phlebotomy, or when they were treated with an prolyl-4-hydroxylase inhibitor [113]. Taken together, conditional knockout studies support the concept that, in the adult, renal and hepatic EPO is regulated by HIF-2, not by HIF-1, identifying HIF-2 as a pharmacological target that, when activated, stimulates EPO synthesis.

To investigate the molecular basis for HIF-2-dependent regulation of EPO, we carried out chromatin immunoprecipitation (ChIP) assays in Hep3B cells and found that HIF-2, not HIF-1, associated with the endogenous *EPO* HRE element. In contrast, we found that HIF-1 preferentially bound to the unmodified *EPO* HRE fragment in vitro, consistent with the purification of HIF-1 from hypoxic Hep3B extracts using a 18-nucleotide fragment containing the *EPO* 3′ HRE [19]. Our findings suggested that in vivo, HIF-2 binding to the *EPO* HRE requires additional nuclear factors, which associate with the *EPO* gene. This notion is supported by the recent observation that HIF-2-mediated expression of an *EPO* HRE-luciferase construct required a 223-bp enhancer fragment, which contained additional transcription factor binding sites [19, 110]. One of these sites includes a DR-2 element, which is a binding site for members of the nuclear hormone receptor family. Hepatocyte nuclear factor-4 (HNF-4) binds to this element and has been proposed as a candidate factor that may cooperate with HIF-2 in the regulation of EPO [110]. Similar to HIF-2, HNF-4 expression coincides with sites of EPO production in the liver and renal cortex and is required for the hypoxic induction of *EPO* in Hep3B cells [110, 114, 115]. The notion that binding of HIF-2 to specific target genes depends on the availability and cooperation with other transcription factors bound to cognate elements in target gene regulatory sequences has been previously suggested; however, further studies are needed to isolate factors that interact with HIF-2 specifically and mediate HIF-2-dependent induction of *EPO* [116].

7 Oxygen-Dependent Regulation of Iron Homeostasis

The transition metal iron (Fe) is a component of hemoproteins, iron-sulfur proteins, and other metalloproteins, such as dioxygenases, which includes HIF prolyl-4-hydroxylases. Thus iron is critical for a multitude of biological processes that range from chemical catalysis and electron transfer to oxygen transport. In red blood cells O_2 is carried by hemoglobins, which are heme-associated tetrameric metalloproteins that are present in millimolar concentrations and contain slightly more than half of the human body's iron. Iron is therefore absolutely required for normal erythropoiesis, and a depletion of iron stores will inevitably lead to anemia. Recent studies have demonstrated that molecular oxygen-sensing mechanisms play a critical role in maintaining adequate serum iron levels and modulate the bone marrow microenvironment to facilitate erythroid maturation (Fig. 2).

The major control mechanisms that regulate serum iron levels involve intestinal absorption from diet, transport in blood, recycling of iron released from phagocytosed erythrocytes through splenic and hepatic macrophages, and release of iron from other tissue stores, such as the liver. Since little iron is lost on a daily basis (e.g., as a result of sloughing of epithelial surface cells or from menstruation), only relatively small amounts need to be replaced through dietary uptake, on average approximately 1–2 mg/day. Most of the iron used for normal erythropoiesis is recycled from phagocytosed erythrocytes (approximately 20 mg/day) (for a review see [117]). When erythropoiesis is stimulated by hypoxia, iron demand in the bone marrow increases. This necessitates increased intestinal iron uptake, an augmentation of serum iron-binding capacity, and enhanced mobilization of iron from internal stores. Therefore, it is not surprising that the expression levels of proteins with key roles in iron metabolism are oxygen-sensitive. HIF not only provides a strong stimulus for erythropoiesis through EPO, but it also regulates factors that indirectly support erythropoiesis. HIF-regulated proteins involved in iron metabolism include transferrin, which binds iron in its ferric form (Fe^{3+}) for transport to target organs, and its high affinity receptor, transferrin receptor-1 (TfR1) [118–120]; ceruloplasmin, which oxidizes Fe^{2+} to Fe^{3+} and is also important for iron transport [121]; the divalent metal transporter-1 (DMT1) [122, 123], which transports iron into the cytoplasm of cells; duodenal cytochrome b (DcytB) [122], which reduces ferric iron to its ferrous form (Fe^{2+}); heme-oxygenase-1 (HO1), which cleaves the heme ring of hemoglobin and produces iron, biliverdin or verdoglobin, and carbon monoxide and is important for the recycling of iron from phagocytosed erythrocytes [124]; hepcidin [125]; and others.

The intestinal uptake of dietary iron is hypoxia-sensitive and is mediated by DMT1, which moves iron across the cell membrane in its ferrous form. Before this

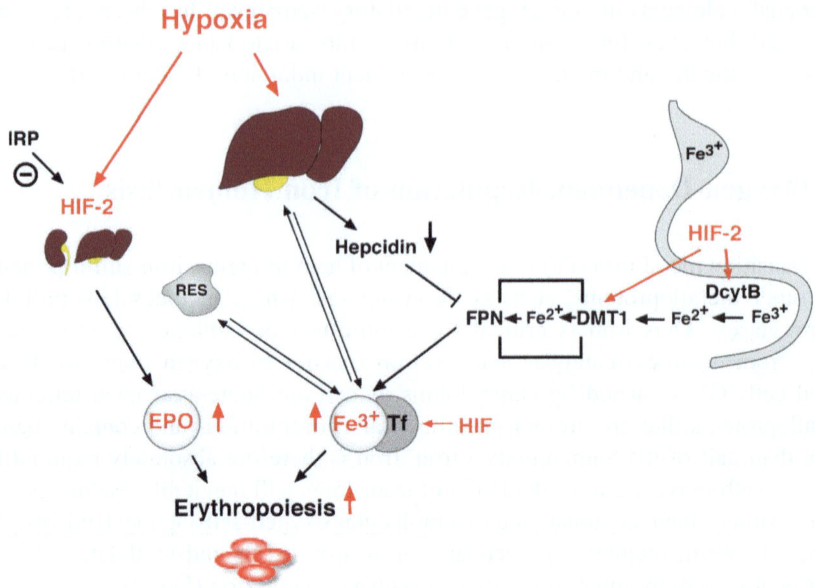

Fig. 2 The hypoxic EPO response is coordinated with enhanced iron metabolism. Shown is a simplified overview of hypoxic and hypoxia-inducible factor (HIF)-mediated effects on iron metabolism. HIF-2 induces renal and hepatic *EPO* transcription in response to hypoxia, which results in increased serum EPO levels (*circle*) and erythropoiesis, necessitating an adjustment of iron metabolism to satisfy the increased demand for iron in the bone marrow. In the duodenum, duodenal cytochrome b (DcytB) reduces ferric iron (Fe^{3+}) to ferrous iron (Fe^{2+}), which is then transported into the cytosol of enterocytes (*depicted by a square*) by divalent metal transporter-1 (DMT1). DcytB and DMT1 are hypoxia-inducible, which appears to be preferentially mediated by HIF-2. Ferroportin (FPN) exports iron, which, following oxidation, is transported in its ferric form to target organs (liver, reticuloendothelial [RES] cells, bone marrow) by transferrin (Tf). Tf is HIF-regulated and hypoxia increases its serum levels. Hepcidin synthesis in the liver, which is inhibited by low serum iron and by a putative factor that signals increased erythropoietic drive, is also hypoxia-sensitive, and hepcidin serum levels are low when erythropoiesis is stimulated by hypoxia. As a result, FPN degradation is diminished, and its cell-surface expression is increased and more iron is exported or released from enterocytes, hepatocytes, and RES cells. If intracellular iron is low, iron regulatory protein (IRP) inhibits HIF-2α translation, thereby hypoxia-induced erythropoiesis

occurs a reduction of dietary iron from its ferric form to Fe^{2+} is necessary, which is mediated by DcytB. HIF-2 increases the cellular expression of DMT-1 and DcytB and is capable of enhancing intestinal iron uptake [122, 123]. Following uptake by enterocytes, iron is then released into the circulation through the membrane protein ferroportin, which is the only known iron transporter that exports iron out of cells. It is also expressed in hepatocytes and macrophages. The number of ferroportin molecules on the cell surface determines how much iron is exported, that is, how much iron is either taken up from diet or released from iron stores in the liver or the reticuloendothelial system. Its surface expression is negatively regulated by hepcidin, a small polypeptide that is produced in the liver (25 amino acids in its active form). Elevated serum hepcidin levels result in low ferroportin cell surface expression, and thus in decreased iron release from enterocytes, hepatocytes, or macrophages, leading to hypoferremia. Since its synthesis is increased in inflammatory states (interleukin-6 induces *hepcidin* transcription through JAK/STAT), hepcidin has been proposed to have a key role in the development of anemia of chronic disease and inflammation [126]. Conversely, constitutively low expression of hepcidin is associated with the development of hemochromatosis [127]. Physiologically, hepcidin serum levels are low in hypoferremic states or when erythropoiesis is stimulated by an increase in serum EPO. This results in enhanced ferroportin-mediated intestinal iron uptake and mobilization of iron from intracellular stores in hepatocytes and macrophages, which in turn leads to increased iron availability for red cell production.

The regulation of hepcidin synthesis is complex and incompletely understood, but clearly dependent on (a) serum iron levels and (b) the degree of the underlying erythropoietic drive. In vitro and in vivo studies have demonstrated that molecular oxygen coregulates hepcidin synthesis [125, 128], and several mechanistic explanations have been proposed, the simplest model being that HIF-2 induces renal EPO, thereby indirectly suppressing hepcidin, because of the increase in erythropoietic drive, which would be signaled to the liver by yet to be identified humoral factors. A candidate factor is growth and differentiation factor 15 (GDF15), an iron- and oxygen-regulated (HIF-independent) member of the transforming growth factor-β (TGF-β) superfamily [129, 130]. Studies with liver-specific knockout mice suggested that HIF-1 contributes directly, albeit to a minor degree, to the hypoferremia-mediated repression of hepcidin synthesis by binding to an HRE located in the 5′-end of the hepcidin gene [125]. In this model liver hypoxia, as a result of hypoxemia or iron deficiency anemia would result in HIF activation and inhibition of hepcidin synthesis. However, the direct HRE-mediated regulation of hepcidin by HIF is controversial. Recent in vitro studies suggested that the hypoxic suppression of hepcidin involved HIF prolyl-4-hydroxylases but was independent of HIF [131]. Another model is based on the effects of hypoxia on the iron-dependent regulatory pathways that control hepcidin transcription. These involve HFE, a protein that is mutated in patients with hereditary hemochromatosis, TfR1 and TfR2 [127, 132], and possibly hemojuvelin (HJV), which acts as a coreceptor for bone morphogenetic protein-6 (BMP6) and induces hepcidin transcription in a SMAD-dependent fashion [133–135]. HIF-1 regulates furin, a proprotein convertase that cleaves HJV, generating soluble HJV (sHJV). sHJV lowers hepcidin synthesis by competing for binding to BMP6, thereby antagonizing signaling through membrane-bound HJV [136, 137].

The link between HIF and iron is bidirectional and complex. Iron is necessary for HIF prolyl-4-hydroxylation, and its oxidation state and abundance affects the activity of PHDs and the efficiency by which normoxic HIF degradation occurs (HIF prolyl-4-hydroxylation is dependent on ferrous iron). An additional feedback loop has been proposed that links intracellular iron levels to HIF-2 translation and would limit HIF-2-induced EPO synthesis when intracellular iron levels are low. *HIF-2α* mRNA contains an iron-regulatory element (IRE) in its 5′-untranslated region, which is an RNA stem loop structure that binds iron-regulatory protein

(IRP), when intracellular iron levels are low. IRPs (IRP1 and IRP2) function as intracellular iron sensors that control the expression of iron-regulated genes, such as transferrin and TfR [138, 139]. When abundant, iron is incorporated into an iron-sulphur cluster at the center of IRP1 and converts IRP1 to an enzyme with aconitase activity. In its aconitase form IRP1 does not bind to IREs. In contrast to IRP1, IRP2 is regulated via proteasomal degradation. Depending on the location of the IRE stem loop, IRP binding either inhibits translation (5′-IRE), as is the case for ferritin, or it has an mRNA stabilizing effect (3′-IRE), as in the case of TfR1 (*TfR1* mRNA levels increase when intracellular iron is low). In the case of HIF-2α, the IRE is located in the 5′-untranslated region, inhibiting translation in the presence of IRP. In the absence of adequate systemic iron, intracellular iron levels are expected to be low and would result in an inhibition of HIF-2α translation, thereby limiting EPO synthesis and adjusting the hypoxic induction of erythropoiesis to iron availability.

8 Oxygen and the Bone Marrow

Changes in oxygen levels have direct effects on erythroid progenitor maturation and proliferation and the bone marrow microenvironment [140, 141]. Hypoxia and anemia stimulates EPO receptor expression and regulates components of the hemoglobin synthesis pathway [22–24, 142]. It furthermore modulates interaction of erythroid cells with other cell types in the bone marrow. This interaction is important for stem cell maintenance, lineage differentiation, and maturation. Its importance is illustrated by recent observations, which demonstrate that endothelial HIF-2 facilitates hypoxia-stimulated erythropoiesis [141]. Yamashita et al. proposed that this occurs through HIF-2-dependent expression of vascular cell adhesion molecule-1 (VCAM1) in bone marrow endothelial cells [141]. VCAM1 is an integrin receptor that binds very late antigen-4 (VLA4), which is expressed on erythroblasts and supports erythroid maturation. In their studies, the investigators described an erythroid maturation defect in mice with globally decreased expression of HIF-2α (mutant mice had increased number of immature erythroblasts) that was dependent on endothelial HIF-2. These studies bring attention to an area of hypoxia research and HIF biology that is largely unexplored, and a better molecular understanding of erythroid cell-stromal interactions may lead to the development of novel therapeutic strategies for the treatment of anemias.

9 Erythrocytosis Associated with Genetic Alterations in the HIF Oxygen-Sensing Pathway

Clinically, even mild to moderate perturbations in the VHL/HIF/PHD oxygen-sensing pathway lead to the development of erythrocytosis. Erythrocytoses can be divided into primary and secondary forms. In primary erythrocytoses, the molecular defects reside within the erythroid progenitors themselves, most frequently affecting JAK2 and the EPO receptor [143, 144]. Whereas serum EPO levels are suppressed in primary erythrocytoses, secondary forms are typically the result of increased EPO production. They can occur as a response to chronic hypoxic conditions, such as chronic obstructive pulmonary disease, right-to-left cardiac shunts, and high altitude, or can be due to EPO-producing tumors or genetic alterations in the molecular machinery that controls EPO synthesis.

The initial observations that linked perturbations in the VHL/HIF/PHD oxygen-sensing pathway to the development of erythrocytosis were made in patients with Chuvash polycythemia, which is a rare autosomal recessive disease endemic to Chuvashia, a republic in central European Russia, and is associated with a homozygous mutation in the VHL tumor suppressor at codon 200 (C598T → R200W); compound heterozygotes have also been described [145–147]. Codon 200 is located in the C-terminal end of pVHL outside the central binding groove for hydroxylated HIF-α (β-domain core). C-terminal to the β-domain lies the α-domain (amino acid residues 157–189), a small α-helical region through which pVHL interacts with the E3-ligase component via Elongin C [148]. Genetic alterations in these two core regions are strongly associated with VHL disease, an inherited autosomal dominant tumor syndrome, where individuals who carry one defective *VHL* allele in their germ line (transmitted from an affected parent) are predisposed to the development of highly vascularized tumors that occur in multiple organs. These tumors include, among others, renal cell carcinoma, hemangioblastomas of the CNS and retina, and pheochromocytomas [149]. Tumor development is associated with inactivation or loss of the remaining *VHL* wild-type allele, which is consistent with Knudson's two-hit hypothesis of tumor development, that is, *VHL* behaves as a typical tumor suppressor gene, where the defective allele is transmitted through the germ line, while the second allele is somatically inactivated in affected tissues. Based on observations in knockout mice, homozygosity for mutations, which generate pVHL species that are incapable of capturing or ubiquitylating hydroxylated HIF-α, is predicted to result in embryonic lethality [150]. In contrast to patients with renal cancers and hemangioblastomas, who carry pVHL mutations affecting the α- or β-domain [151, 152], patients who are homozygous for the R200W mutation are not predisposed to the development of tumors. When expressed in murine embryonic stem cells or in *VHL*-defective renal carcinoma cells, pVHL-R200W targets HIF-α less efficiently for proteasomal degradation, affecting HIF-2α more than HIF-1α [153]. How this exactly occurs is unclear, but may involve changes in protein stability or conformational changes, which could impinge on the pVHL-HIF-α interaction. Although individuals with Chuvash polycythemia are not prone to VHL-associated tumor development, they suffer from premature morbidity and mortality due to pulmonary hypertension, cerebrovascular accidents, and vertebral hemangiomas [154, 155]. Following the initial discovery in Chuvash patients, the R200W mutation as well as other pVHL mutations have now been found in patients with idiopathic erythrocytosis that are ethnically distinct from this group [156, 157].

Further investigations into the role of the VHL/HIF/PHD oxygen-sensing pathway in the development of idiopathic erythrocytosis have identified families with heterozygous mutations in HIF-2α and PHD2. Interestingly, mutations in HIF-1α have not been found, underscoring the role of HIF-2 as the critical regulator of EPO synthesis in humans. The point mutations that were identified in HIF-2α change glycine at position 537 to tryptophan or arginine, and the methionine residue at position 535 to valine [158, 159]. Codons 537 and 535 are close to proline residue 531, which is one of the sites targeted for hydroxylation by PHDs (the other hydroxylation target is proline 405). Biochemical analysis of mutation G537W showed that this amino acid change impaired recognition, as well as the hydroxylation, of HIF-2α by PHD2, producing a partial gain of function mutation. Conversely, several mutations have been identified in PHD2, which result in diminished prolyl-hydroxylase activity and in increased red blood cell production that is associated with inappropriately normal serum EPO levels [160, 161]. These mutations include an amino acid change at position 317, which is in the direct vicinity of two iron-chelating amino acid residues (position 313 and 315) that are critical for PHD catalytic activity.

It is important to point out in this context that only approximately 5% of renal cancers are associated with erythrocytosis [162]. Most of these cancers are *VHL*-deficient and stabilize HIF-1α and HIF-2α constitutively, resulting in increased expression of many HIF-regulated genes; *EPO* transcription, however, is repressed, despite the very high levels of HIF-2 activity. The molecular basis for EPO repression in this setting is not clear, but may involve certain transcriptional regulators, such as GATA-2 and GATA-3 [67].

10 The VHL/HIF/PHD Axis as a Therapeutic Target

Pharmacological targeting of the VHL/HIF/PHD axis, with the goal of boosting synthesis of endogenous EPO by activating HIF-2, has the potential to benefit patients with anemias that result from inadequate EPO production. These include renal anemia, which is typically associated with advanced renal failure, independent of disease etiology. Mechanistically, renal anemia is not completely understood. Possible reasons for inappropriately low renal EPO production include, but are not limited to (a) decreased renal oxygen consumption as a consequence of reduced GFR and thus tubular workload, resulting in preservation of tissue oxygen levels despite anemia, (b) transdifferentiation of EPO-producing cells into myofibroblasts in areas of fibrosis, and (c) proinflammatory cytokine accumulation, impairing EPO production. As discussed above, the VHL/HIF/PHD oxygen-sensing pathway is still intact in diseased kidneys as renal EPO production increases in response to hypoxia, and the liver is capable of producing large quantities of EPO under hypoxic conditions or when HIF-2 is activated under normoxia (e.g., in the setting of *VHL*-deficiency) [38, 73, 80]. Thus, small molecule compounds that effectively inhibit HIF degradation in both organs have the potential to boost EPO synthesis in patients with chronic kidney disease and renal anemia.

In keeping with clinical observations in families with PHD2 mutations, inactivation of PHD enzymes in mouse models increases serum EPO levels and causes erythrocytosis, demonstrating that PHD inhibition is a feasible approach to boost erythropoiesis pharmacologically. More importantly, these studies, together with in vitro experiments, have also indicated that individual PHDs are functionally different with regard to EPO synthesis, and that targeting of individual PHDs may improve specificity. Global deletion of PHD2 in the adult, using inducible Cre/loxP-mediated gene targeting, results in severe erythrocytosis (HCT values >80%) and multiple other organ pathologies, while homozygous inactivation of PHD2 in the germ line is embryonically lethal, resembling the phenotype that resulted from pVHL inactivation, in particular when PHD3 is inactivated simultaneously [163–166]. *PHD1*- and *PHD3*-deficient mice, which survive into adulthood, develop erythrocytosis only when both enzymes are inactivated together. Erythrocytosis is moderate (HCT of 67% in mutants vs. 53% in controls), while serum EPO levels were suppressed compared to wild type. Taken together these in vivo findings illustrate that PHD2 is the most critical regulator of HIF proteolysis under normoxia, and that complete inactivation of PHD2 results in a major disruption of oxygen sensing and massive upregulation of HIF target genes, which leads to embryonic lethality, similar, but not identical, to pVHL inactivation [164]. PHD1 and PHD3, on the other hand, play distinct roles in oxygen-dependent HIF proteolysis and may be better suited as drug targets with regard to boosting EPO synthesis. Functional differences between individual PHDs are expected, because of differences in cellular localization, hypoxia-inducibility, and biochemical behavior. For example, only PHD2 and PHD3 are hypoxia-inducible, and PHD3 does not hydroxylate proline 402 in HIF-1α, which, although not directly proven, is most likely also the case for the corresponding residue in HIF-2α (for a review see [167, 168]). Furthermore, PHD1 and PHD3 appear to have a preference for HIF-2α in vitro and in vivo [163, 169]. PHD1/PHD3−/− mice upregulate HIF-2α but not HIF-1α in the liver [163], while PHD2 inactivation increases both HIFs [165]; a preference of PHD2 for HIF-1α, however, has been suggested by in vitro studies [169].

Analogues of 2-oxoglutarate, which is the substrate for PHDs and FIH, have been successfully used for the stimulation of endogenous EPO production in vivo, and furthermore have the potential to improve the clinical outcome of acute ischemic injuries as demonstrated in animal models [170–172]. Some compounds have entered clinical trials for the treatment of renal anemia and are awaiting safety and efficacy evaluations. One of these compounds was investigated in rhesus macaques by Hsieh et al., who demonstrated that oral administration of an PHD inhibitor resulted in transient activation of HIF and an induction of EPO synthesis and erythropoiesis [172]. The same compound resulted in an increase of serum EPO levels in patients on dialysis, including anephric patients [173]. Although pharmacological targeting of HIF may be an attractive alternative to treatment with recombinant EPO due to the predicted beneficial effects on iron metabolism, a word of caution is warranted. HIF transcription factors regulate a multitude of biological processes, and HIF activation over prolonged periods of time may lead to profound systemic and cellular changes affecting metabolism, growth, and differentiation. Whether long-term or intermittent treatment with HIF stabilizing compounds is safe will have to be carefully established.

Acknowledgments The author is supported by the Krick-Brooks chair in nephrology and grants from the National Cancer Institute (NCI), and the National Institutes of Diabetes, and Digestive and Kidney Diseases (NIDDK). The author declares that no conflicts of interest exist.

References

1. Bert P. Sur la richesse en hemoglobine du sang des animaux vivant sur les hauts lieux. C R Acad Sci Paris 1882; 94: 805–807.
2. Jourdanet D. Influence de la pression de l'air sur la vie de l'homme. G Masson, Paris, 1875.
3. Bert P. La pression barométrique. Recherches de physiologie expérimentale. G Masson, Paris, 1878.
4. Viault F. Sur laugmentation considérable du nombre des globules rouges dans le sang chez les inhabitants des hauts plateaux de l'Amérique du Sud. C R Acad Sci Paris 1890; 111: 917–918.
5. Carnot P, Deflandre C. Sur l''activité hémopoetique du sérum au course de la régéneration du sang. C R Acad Sci Paris 1906; 143: 384–386.
6. Bonsdorff E, Jalavisto E. A humoral mechanism in anoxic erythrocytosis. Acta Physiol Scand 1948; 16: 150–170.
7. Erslev A. Humoral regulation of red cell production. Blood 1953; 8: 349–357.
8. Koury MJ. Erythropoietin: the story of hypoxia and a finely regulated hematopoietic hormone. Exp Hematol 2005; 33: 1263–1270.
9. Abbrecht PH, Littell JK. Plasma erythropoietin in men and mice during acclimatization to different altitudes. J Appl Physiol 1972; 32: 54–58.
10. Faura J, Ramos J, Reynafarje C, et al. Effect of altitude on erythropoiesis. Blood 1969; 33: 668–676.
11. Erslev AJ, Caro J, Miller O, Silver R. Plasma erythropoietin in health and disease. Ann Clin Lab Sci 1980; 10: 250–257.
12. Miyake T, Kung CK, Goldwasser E. Purification of human erythropoietin. J Biol Chem 1977; 252: 5558–5564.
13. Jacobs K, Shoemaker C, Rudersdorf R, et al. Isolation and characterization of genomic and cDNA clones of human erythropoietin. Nature 1985; 313: 806–810.
14. Lin FK, Suggs S, Lin CH, et al. Cloning and expression of the human erythropoietin gene. Proc Natl Acad Sci U S A 1985; 82: 7580–7584.
15. Nielsen OJ, Costa-Giomi P, Weinmann R, Erslev AJ, Caro J. Erythropoietin-beta-D-galactosidase. The generation, purification and use of a fusion protein. J Immunol Methods 1988; 111: 1–9.
16. Goldberg MA, Glass GA, Cunningham JM, Bunn HF. The regulated expression of erythropoietin by two human hepatoma cell lines. Proc Natl Acad Sci USA 1987; 84: 7972–7976.
17. Fandrey J. Oxygen-dependent and tissue-specific regulation of erythropoietin gene expression. Am J Physiol Regul Integr Comp Physiol 2004; 286: R977–R988.
18. Semenza GL, Koury ST, Nejfelt MK, Gearhart JD, Antonarakis SE. Cell-type-specific and hypoxia-inducible expression of the human erythropoietin gene in transgenic mice. Proc Natl Acad Sci U S A 1991; 88: 8725–8729.

19. Semenza GL, Wang GL. A nuclear factor induced by hypoxia via de novo protein synthesis binds to the human erythropoietin gene enhancer at a site required for transcriptional activation. Mol Cell Biol 1992; 12: 5447–5454.
20. Wang GL, Semenza GL. Purification and characterization of hypoxia-inducible factor 1. J Biol Chem 1995; 270: 1230–1237.
21. Wang GL, Jiang BH, Rue EA, Semenza GL. Hypoxia-inducible factor 1 is a basic-helix-loop-helix-PAS heterodimer regulated by cellular O_2 tension. Proc Natl Acad Sci U S A 1995; 92: 5510–5514.
22. Chin K, Yu X, Beleslin-Cokic B, et al. Production and processing of erythropoietin receptor transcripts in brain. Brain Res Mol Brain Res 2000; 81: 29–42.
23. Manalo DJ, Rowan A, Lavoie T, et al. Transcriptional regulation of vascular endothelial cell responses to hypoxia by HIF-1. Blood 2005; 105: 659–669.
24. Yoon D, Pastore YD, Divoky V, et al. HIF-1alpha -deficiency results in dysregulated EPO signaling and iron homeostasis in mouse development. J Biol Chem 2006; 281: 25703–25711.
25. Ebert BL, Bunn HF. Regulation of the erythropoietin gene. Blood 1999; 94: 1864–1877.
26. Kewley RJ, Whitelaw ML, Chapman-Smith A. The mammalian basic helix-loop-helix/PAS family of transcriptional regulators. Int J Biochem Cell Biol 2004; 36: 189–204.
27. Semenza GL. Regulation of mammalian O2 homeostasis by hypoxia-inducible factor 1. Annu Rev Cell Dev Biol 1999; 15: 551–578.
28. Wenger RH, Stiehl DP, Camenisch G. Integration of oxygen signaling at the consensus HRE. Sci STKE 2005; 2005: re12.
29. Semenza GL. HIF-1 and mechanisms of hypoxia sensing. Curr Opin Cell Biol 2001; 13: 167–171.
30. Maxwell PH, Wiesener MS, Chang GW, et al. The tumour suppressor protein VHL targets hypoxia-inducible factors for oxygen-dependent proteolysis [see comments]. Nature 1999; 399: 271–275.
31. Maynard MA, Qi H, Chung J, et al. Multiple splice variants of the human HIF-3alpha locus are targets of the VHL E3 ubiquitin ligase complex. J Biol Chem 2003; 278: 1032–1040.
32. Makino Y, Cao R, Svensson K, et al. Inhibitory PAS domain protein is a negative regulator of hypoxia-inducible gene expression. Nature 2001; 414: 550–554.
33. Maynard MA, Evans AJ, Shi W, et al. Dominant-negative HIF-3 alpha 4 suppresses VHL-null renal cell carcinoma progression. Cell Cycle 2007; 6: 2810–2816.
34. Hu CJ, Wang LY, Chodosh LA, Keith B, Simon MC. Differential roles of hypoxia-inducible factor 1alpha (HIF-1alpha) and HIF-2alpha in hypoxic gene regulation. Mol Cell Biol 2003; 23: 9361–9374.
35. Morita M, Ohneda O, Yamashita T, et al. HLF/HIF-2alpha is a key factor in retinopathy of prematurity in association with erythropoietin. Embo J 2003; 22: 1134–1146.
36. Scortegagna M, Ding K, Zhang Q, et al. HIF-2alpha regulates murine hematopoietic development in an erythropoietin-dependent manner. Blood 2005; 105: 3133–3140.
37. Gruber M, Hu CJ, Johnson RS, et al. Acute postnatal ablation of Hif-2alpha results in anemia. Proc Natl Acad Sci USA 2007; 104: 2301–2306.
38. Rankin EB, Biju MP, Liu Q, et al. Hypoxia-inducible factor-2 (HIF-2) regulates hepatic erythropoietin in vivo. J Clin Invest 2007; 117: 1068–1077.
39. An WG, Kanekal M, Simon MC, et al. Stabilization of wild-type p53 by hypoxia-inducible factor 1alpha. Nature 1998; 392: 405–408.
40. Ravi R, Mookerjee B, Bhujwalla ZM, et al. Regulation of tumor angiogenesis by p53-induced degradation of hypoxia-inducible factor 1alpha. Genes Dev 2000; 14: 34–44.
41. Koshiji M, Kageyama Y, Pete EA, et al. HIF-1alpha induces cell cycle arrest by functionally counteracting Myc. Embo J 2004; 23: 1949–1956.
42. Gustafsson MV, Zheng X, Pereira T, et al. Hypoxia requires notch signaling to maintain the undifferentiated cell state. Dev Cell 2005; 9: 617–628.

43. Jaakkola P, Mole DR, Tian YM, et al. Targeting of HIF-alpha to the von Hippel-Lindau ubiquitylation complex by O_2-regulated prolyl hydroxylation. Science 2001; 292: 468–472.
44. Ivan M, Kondo K, Yang H, et al. HIFalpha targeted for VHL-mediated destruction by proline hydroxylation: implications for O2 sensing. Science 2001; 292: 464–468.
45. Epstein AC, Gleadle JM, McNeill LA, et al. C. elegans EGL-9 and mammalian homologs define a family of dioxygenases that regulate HIF by prolyl hydroxylation. Cell 2001; 107: 43–54.
46. Hon WC, Wilson MI, Harlos K, et al. Structural basis for the recognition of hydroxyproline in HIF-1 alpha by pVHL. Nature 2002; 417: 975–978.
47. Masson N, Willam C, Maxwell PH, Pugh CW, Ratcliffe PJ. Independent function of two destruction domains in hypoxia-inducible factor-alpha chains activated by prolyl hydroxylation. Embo J 2001; 20: 5197–5206.
48. Yu F, White SB, Zhao Q, Lee FS. HIF-1alpha binding to VHL is regulated by stimulus-sensitive proline hydroxylation. Proc Natl Acad Sci USA 2001; 98: 9630–9635.
49. Bruick RK, McKnight SL. A conserved family of prolyl-4-hydroxylases that modify HIF. Science 2001; 294: 1337–1340.
50. Schofield CJ, Ratcliffe PJ. Oxygen sensing by HIF hydroxylases. Nat Rev Mol Cell Biol 2004; 5: 343–354.
51. Nakayama K, Ronai Z. Siah: new players in the cellular response to hypoxia. Cell Cycle 2004; 3: 1345–1347.
52. Nakayama K, Qi J, Ronai Z. The ubiquitin ligase Siah2 and the hypoxia response. Mol Cancer Res 2009; 7: 443–451.
53. Mahon PC, Hirota K, Semenza GL. FIH-1: a novel protein that interacts with HIF-1alpha and VHL to mediate repression of HIF-1 transcriptional activity. Genes Dev 2001; 15: 2675–2686.
54. Lando D, Peet DJ, Gorman JJ, et al. FIH-1 is an asparaginyl hydroxylase enzyme that regulates the transcriptional activity of hypoxia-inducible factor. Genes Dev 2002; 16: 1466–1471.
55. Lando D, Peet DJ, Whelan DA, Gorman JJ, Whitelaw ML. Asparagine hydroxylation of the HIF transactivation domain a hypoxic switch. Science 2002; 295: 858–861.
56. Stolze IP, Tian YM, Appelhoff RJ, et al. Genetic analysis of the role of the asparaginyl hydroxylase factor inhibiting hypoxia-inducible factor (HIF) in regulating HIF transcriptional target genes. J Biol Chem 2004; 279: 42719–42725.
57. Dioum EM, Chen R, Alexander MS, et al. Regulation of hypoxia-inducible factor 2alpha signaling by the stress-responsive deacetylase sirtuin 1. Science 2009; 324: 1289–1293.
58. Kang MJ, Kim HJ, Kim HK, et al. The effect of age and calorie restriction on HIF-1-responsive genes in aged liver. Biogerontology 2005; 6: 27–37.
59. Cohen HY, Miller C, Bitterman KJ, et al. Calorie restriction promotes mammalian cell survival by inducing the SIRT1 deacetylase. Science 2004; 305: 390–392.
60. Cheng J, Kang X, Zhang S, Yeh ET. SUMO-specific protease 1 is essential for stabilization of HIF1alpha during hypoxia. Cell 2007; 131: 584–595.
61. Jacobson LO, Goldwasser E, Fried W, Plzak L. Role of the kidney in erythropoiesis. Nature 1957; 179: 633–634.
62. Lacombe C, Da Silva JL, Bruneval P, et al. Peritubular cells are the site of erythropoietin synthesis in the murine hypoxic kidney. J Clin Invest 1988; 81: 620–623.
63. Koury ST, Bondurant MC, Koury MJ. Localization of erythropoietin synthesizing cells in murine kidneys by in situ hybridization. Blood 1988; 71: 524–527.
64. Koury ST, Koury MJ, Bondurant MC, Caro J, Graber SE. Quantitation of erythropoietin-producing cells in kidneys of mice by in situ hybridization: correlation with hematocrit, renal erythropoietin mRNA, and serum erythropoietin concentration. Blood 1989; 74: 645–651.
65. Bachmann S, Le Hir M, Eckardt KU. Co-localization of erythropoietin mRNA and ecto-5'-nucleotidase immunoreactivity in peritubular cells of rat renal cortex indicates that fibroblasts produce erythropoietin. J Histochem Cytochem 1993; 41: 335–341.

66. Maxwell PH, Osmond MK, Pugh CW, et al. Identification of the renal erythropoietin-producing cells using transgenic mice. Kidney Int 1993; 44: 1149–1162.
67. Obara N, Suzuki N, Kim K, et al. Repression via the GATA box is essential for tissue-specific erythropoietin gene expression. Blood 2008; 111: 5223–5232.
68. Plotkin MD, Goligorsky MS. Mesenchymal cells from adult kidney support angiogenesis and differentiate into multiple interstitial cell types including erythropoietin-producing fibroblasts. Am J Physiol Renal Physiol 2006; 291: F902–F912.
69. Frede S, Geuting L, Freitag P, Fandrey J. Role of HIF-1 and HIF-2 for the oxygen-regulated expression of erythropoietin in human kidney cells (abstracts in issue 9). Am J Hematol 2009; 84: E3.
70. Koury ST, Bondurant MC, Koury MJ, Semenza GL. Localization of cells producing erythropoietin in murine liver by in situ hybridization. Blood 1991; 77: 2497–2503.
71. Maxwell PH, Ferguson DJ, Osmond MK, et al. Expression of a homologously recombined erythopoietin-SV40 T antigen fusion gene in mouse liver: evidence for erythropoietin production by Ito cells. Blood 1994; 84: 1823–1830.
72. Zanjani ED, Ascensao JL, McGlave PB, Banisadre M, Ash RC. Studies on the liver to kidney switch of erythropoietin production. J Clin Invest 1981; 67: 1183–1188.
73. Eckardt KU, Ratcliffe PJ, Tan CC, Bauer C, Kurtz A. Age-dependent expression of the erythropoietin gene in rat liver and kidneys. J Clin Invest 1992; 89: 753–760.
74. Moritz KM, Lim GB, Wintour EM. Developmental regulation of erythropoietin and erythropoiesis. Am J Physiol 1997; 273: R1829–R1844.
75. Dame C, Fahnenstich H, Freitag P, et al. Erythropoietin mRNA expression in human fetal and neonatal tissue. Blood 1998; 92: 3218–3225.
76. Dame C, Sola MC, Lim KC, et al. Hepatic erythropoietin gene regulation by GATA-4. J Biol Chem 2004; 279: 2955–2961.
77. Fried W, Kilbridge T, Krantz S, McDonald TP, Lange RD. Studies on extrarenal erythropoietin. J Lab Clin Med 1969; 73: 244–248.
78. Fried W. The liver as a source of extrarenal erythropoietin production. Blood 1972; 40: 671–677.
79. Tan CC, Eckardt KU, Firth JD, Ratcliffe PJ. Feedback modulation of renal and hepatic erythropoietin mRNA in response to graded anemia and hypoxia. Am J Physiol 1992; 263: F474–F481.
80. Tan CC, Eckardt KU, Ratcliffe PJ. Organ distribution of erythropoietin messenger RNA in normal and uremic rats. Kidney Int 1991; 40: 69–76.
81. Semenza GL, Traystman MD, Gearhart JD, Antonarakis SE. Polycythemia in transgenic mice expressing the human erythropoietin gene. Proc Natl Acad Sci USA 1989; 86: 2301–2305.
82. Semenza GL, Dureza RC, Traystman MD, Gearhart JD, Antonarakis SE. Human erythropoietin gene expression in transgenic mice: multiple transcription initiation sites and cis-acting regulatory elements. Mol Cell Biol 1990; 10: 930–938.
83. Suzuki N, Obara N, Yamamoto M. Use of gene-manipulated mice in the study of erythropoietin gene expression. Methods Enzymol 2007; 435: 157–177.
84. Marti HH, Wenger RH, Rivas LA, et al. Erythropoietin gene expression in human, monkey and murine brain. Eur J Neurosci 1996; 8: 666–676.
85. Marti HH, Gassmann M, Wenger RH, et al. Detection of erythropoietin in human liquor: intrinsic erythropoietin production in the brain. Kidney Int 1997; 51: 416–418.
86. Bernaudin M, Bellail A, Marti HH, et al. Neurons and astrocytes express EPO mRNA: oxygen-sensing mechanisms that involve the redox-state of the brain. Glia 2000; 30: 271–278.
87. Magnanti M, Gandini O, Giuliani L, et al. Erythropoietin expression in primary rat Sertoli and peritubular myoid cells. Blood 2001; 98: 2872–2874.
88. Dame C, Bartmann P, Wolber E, et al. Erythropoietin gene expression in different areas of the developing human central nervous system. Brain Res Dev Brain Res 2000; 125: 69–74.

89. Fandrey J, Bunn HF. In vivo and in vitro regulation of erythropoietin mRNA: measurement by competitive polymerase chain reaction. Blood 1993; 81: 617–623.
90. Bodo E, Kromminga A, Funk W, et al. Human hair follicles are an extrarenal source and a nonhematopoietic target of erythropoietin. Faseb J 2007; 21: 3346–3354.
91. Yasuda Y, Masuda S, Chikuma M, et al. Estrogen-dependent production of erythropoietin in uterus and its implication in uterine angiogenesis. J Biol Chem 1998; 273: 25381–25387.
92. Kobayashi T, Yanase H, Iwanaga T, Sasaki R, Nagao M. Epididymis is a novel site of erythropoietin production in mouse reproductive organs. Biochem Biophys Res Commun 2002; 296: 145–151.
93. Masuda S, Kobayashi T, Chikuma M, Nagao M, Sasaki R. The oviduct produces erythropoietin in an estrogen- and oxygen-dependent manner. Am J Physiol Endocrinol Metab 2000; 278: E1038–E1044.
94. Jelkmann W. Erythropoietin after a century of research: younger than ever. Eur J Haematol 2007; 78: 183–205.
95. von Wussow U, Klaus J, Pagel H. Is the renal production of erythropoietin controlled by the brain stem? Am J Physiol Endocrinol Metab 2005; 289: E82–E86.
96. Boutin AT, Weidemann A, Fu Z, et al. Epidermal sensing of oxygen is essential for systemic hypoxic response. Cell 2008; 133: 223–234.
97. Pagel H, Jelkmann W, Weiss C. A comparison of the effects of renal artery constriction and anemia on the production of erythropoietin. Pflugers Arch 1988; 413: 62–66.
98. Rosenberger C, Mandriota S, Jurgensen JS, et al. Expression of hypoxia-inducible factor-1alpha and -2alpha in hypoxic and ischemic rat kidneys. J Am Soc Nephrol 2002; 13: 1721–1732.
99. Wiesener MS, Jurgensen JS, Rosenberger C, et al. Widespread hypoxia-inducible expression of HIF-2alpha in distinct cell populations of different organs. Faseb J 2003; 17: 271–273.
100. Ryan HE, Lo J, Johnson RS. HIF-1 alpha is required for solid tumor formation and embryonic vascularization. Embo J 1998; 17: 3005–3015.
101. Iyer NV, Kotch LE, Agani F, et al. Cellular and developmental control of O2 homeostasis by hypoxia-inducible factor 1 alpha. Genes Dev 1998; 12: 149–162.
102. Scortegagna M, Ding K, Oktay Y, et al. Multiple organ pathology, metabolic abnormalities and impaired homeostasis of reactive oxygen species in Epas1-/- mice. Nat Genet 2003; 35: 331–340.
103. Tian H, Hammer RE, Matsumoto AM, Russell DW, McKnight SL. The hypoxia-responsive transcription factor EPAS1 is essential for catecholamine homeostasis and protection against heart failure during embryonic development. Genes Dev 1998; 12: 3320–3324.
104. Compernolle V, Brusselmans K, Acker T, et al. Loss of HIF-2alpha and inhibition of VEGF impair fetal lung maturation, whereas treatment with VEGF prevents fatal respiratory distress in premature mice. Nat Med 2002; 8: 702–710.
105. Peng J, Zhang L, Drysdale L, Fong GH. The transcription factor EPAS-1/hypoxia-inducible factor 2alpha plays an important role in vascular remodeling. Proc Natl Acad Sci USA 2000; 97: 8386–8391.
106. Scortegagna M, Morris MA, Oktay Y, Bennett M, Garcia JA. The HIF family member EPAS1/HIF-2alpha is required for normal hematopoiesis in mice. Blood 2003; 102: 1634–1640.
107. Jelkmann W, Wagner K. Beneficial and ominous aspects of the pleiotropic action of erythropoietin. Ann Hematol 2004; 83: 673–686.
108. Yu AY, Shimoda LA, Iyer NV, et al. Impaired physiological responses to chronic hypoxia in mice partially deficient for hypoxia-inducible factor 1alpha. J Clin Invest 1999; 103: 691–696.
109. Cai Z, Manalo DJ, Wei G, et al. Hearts from rodents exposed to intermittent hypoxia or erythropoietin are protected against ischemia-reperfusion injury. Circulation 2003; 108: 79–85.

110. Warnecke C, Zaborowska Z, Kurreck J, et al. Differentiating the functional role of hypoxia-inducible factor (HIF)-1alpha and HIF-2alpha (EPAS-1) by the use of RNA interference: erythropoietin is a HIF-2alpha target gene in Hep3B and Kelly cells. Faseb J 2004; 18: 1462–1464.
111. Chavez JC, Baranova O, Lin J, Pichiule P. The transcriptional activator hypoxia inducible factor 2 (HIF-2/EPAS-1) regulates the oxygen-dependent expression of erythropoietin in cortical astrocytes. J Neurosci 2006; 26: 9471–9481.
112. Kim WY, Safran M, Buckley MR, et al. Failure to prolyl hydroxylate hypoxia-inducible factor alpha phenocopies VHL inactivation in vivo. Embo J 2006; 25: 4650–4662.
113. Kapitsinou PP, Liu Q, Unger TL, et al. Hepatic HIF-2 regulates erythropoietic responses to hypoxia in level aueuria. Blood 2010; 13 [Epub ahead of print].
114. Blanchard KL, Acquaviva AM, Galson DL, Bunn HF. Hypoxic induction of the human erythropoietin gene: cooperation between the promoter and enhancer, each of which contains steroid receptor response elements. Mol Cell Biol 1992; 12: 5373–5385.
115. Galson DL, Tsuchiya T, Tendler DS, et al. The orphan receptor hepatic nuclear factor 4 functions as a transcriptional activator for tissue-specific and hypoxia-specific erythropoietin gene expression and is antagonized by EAR3/COUP-TF1. Mol Cell Biol 1995; 15: 2135–2144.
116. Aprelikova O, Wood M, Tackett S, Chandramouli GV, Barrett JC. Role of ETS transcription factors in the hypoxia-inducible factor-2 target gene selection. Cancer Res 2006; 66: 5641–5647.
117. Ganz T. Molecular control of iron transport. J Am Soc Nephrol 2007; 18: 394–400.
118. Rolfs A, Kvietikova I, Gassmann M, Wenger RH. Oxygen-regulated transferrin expression is mediated by hypoxia-inducible factor-1. J Biol Chem 1997; 272: 20055–20062.
119. Lok CN, Ponka P. Identification of a hypoxia response element in the transferrin receptor gene. J Biol Chem 1999; 274: 24147–24152.
120. Tacchini L, Bianchi L, Bernelli-Zazzera A, Cairo G. Transferrin receptor induction by hypoxia. HIF-1-mediated transcriptional activation and cell-specific post-transcriptional regulation. J Biol Chem 1999; 274: 24142–24146.
121. Mukhopadhyay CK, Mazumder B, Fox PL. Role of hypoxia-inducible factor-1 in transcriptional activation of ceruloplasmin by iron deficiency. J Biol Chem 2000; 275: 21048–21054.
122. Shah YM, Matsubara T, Ito S, Yim SH, Gonzalez FJ. Intestinal hypoxia-inducible transcription factors are essential for iron absorption following iron deficiency. Cell Metab 2009; 9: 152–164.
123. Mastrogiannaki M, Matak P, Keith B, et al. HIF-2alpha, but not HIF-1alpha, promotes iron absorption in mice. J Clin Invest 2009; 119: 1159–1166.
124. Lee PJ, Jiang BH, Chin BY, et al. Hypoxia-inducible factor-1 mediates transcriptional activation of the heme oxygenase-1 gene in response to hypoxia. J Biol Chem 1997; 272: 5375–5381.
125. Peyssonnaux C, Zinkernagel AS, Schuepbach RA, et al. Regulation of iron homeostasis by the hypoxia-inducible transcription factors (HIFs). J Clin Invest 2007; 117: 1926–1932.
126. Ganz T. Hepcidin, a key regulator of iron metabolism and mediator of anemia of inflammation. Blood 2003; 102: 783–788.
127. Andrews NC. Forging a field: the golden age of iron biology. Blood 2008; 112: 219–230.
128. Nicolas G, Chauvet C, Viatte L, et al. The gene encoding the iron regulatory peptide hepcidin is regulated by anemia, hypoxia, and inflammation. J Clin Invest 2002; 110: 1037–1044.
129. Lakhal S, Talbot NP, Crosby A, et al. Regulation of growth differentiation factor 15 expression by intracellular iron. Blood 2009; 113: 1555–1563.
130. Tanno T, Bhanu NV, Oneal PA, et al. High levels of GDF15 in thalassemia suppress expression of the iron regulatory protein hepcidin. Nat Med 2007; 13: 1096–1101.
131. Braliou GG, Verga Falzacappa MV, Chachami G, et al. 2-Oxoglutarate-dependent oxygenases control hepcidin gene expression. J Hepatol 2008; 48: 801–810.

132. Gao J, Chen J, Kramer M, et al. Interaction of the hereditary hemochromatosis protein HFE with transferrin receptor 2 is required for transferrin-induced hepcidin expression. Cell Metab 2009; 9: 217–227.
133. Babitt JL, Huang FW, Wrighting DM, et al. Bone morphogenetic protein signaling by hemojuvelin regulates hepcidin expression. Nat Genet 2006; 38: 531–539.
134. Andriopoulos B, Jr., Corradini E, Xia Y, et al. BMP6 is a key endogenous regulator of hepcidin expression and iron metabolism. Nat Genet 2009; 41: 482–487.
135. Meynard D, Kautz L, Darnaud V, et al. Lack of the bone morphogenetic protein BMP6 induces massive iron overload. Nat Genet 2009; 41: 478–481.
136. Silvestri L, Pagani A, Camaschella C. Furin-mediated release of soluble hemojuvelin: a new link between hypoxia and iron homeostasis. Blood 2008; 111: 924–931.
137. McMahon S, Grondin F, McDonald PP, Richard DE, Dubois CM. Hypoxia-enhanced expression of the proprotein convertase furin is mediated by hypoxia-inducible factor-1: impact on the bioactivation of proproteins. J Biol Chem 2005; 280: 6561–6569.
138. Sanchez M, Galy B, Muckenthaler MU, Hentze MW. Iron-regulatory proteins limit hypoxia-inducible factor-2alpha expression in iron deficiency. Nat Struct Mol Biol 2007; 14: 420–426.
139. Zimmer M, Ebert BL, Neil C, et al. Small-molecule inhibitors of HIF-2a translation link its 5'UTR iron-responsive element to oxygen sensing. Mol Cell 2008; 32: 838–848.
140. Adelman DM, Maltepe E, Simon MC. Multilineage embryonic hematopoiesis requires hypoxic ARNT activity. Genes Dev 1999; 13: 2478–2483.
141. Yamashita T, Ohneda O, Sakiyama A, et al. The microenvironment for erythropoiesis is regulated by HIF-2alpha through VCAM-1 in endothelial cells. Blood 2008; 112: 1482–1492.
142. Liu YL, Ang SO, Weigent DA, Prchal JT, Bloomer JR. Regulation of ferrochelatase gene expression by hypoxia. Life Sci 2004; 75: 2035–2043.
143. Gordeuk VR, Stockton DW, Prchal JT. Congenital polycythemias/erythrocytoses. Haematologica 2005; 90: 109–116.
144. Levine RL. Janus kinase mutations. Semin Oncol 2009; 36: S6–S11.
145. Ang SO, Chen H, Hirota K, et al. Disruption of oxygen homeostasis underlies congenital Chuvash polycythemia. Nat Genet 2002; 32: 614–621.
146. Liu E, Percy MJ, Amos CI, et al. The worldwide distribution of the VHL 598C>T mutation indicates a single founding event. Blood 2004; 103: 1937–1940.
147. Percy MJ, McMullin MF, Jowitt SN, et al. Chuvash-type congenital polycythemia in 4 families of Asian and Western European ancestry. Blood 2003; 102: 1097–1099.
148. Stebbins CE, Kaelin WG, Jr., Pavletich NP. Structure of the VHL-ElonginC-ElonginB complex: implications for VHL tumor suppressor function. Science 1999; 284: 455–461.
149. Lonser RR, Glenn GM, Walther M, et al. von Hippel-Lindau disease. Lancet 2003; 361: 2059–2067.
150. Gnarra JR, Ward JM, Porter FD, et al. Defective placental vasculogenesis causes embryonic lethality in VHL-deficient mice. Proc Natl Acad Sci USA 1997; 94: 9102–9107.
151. Clifford SC, Cockman ME, Smallwood AC, et al. Contrasting effects on HIF-1alpha regulation by disease-causing pVHL mutations correlate with patterns of tumourigenesis in von Hippel-Lindau disease. Hum Mol Genet 2001; 10: 1029–1038.
152. Hoffman MA, Ohh M, Yang H, et al. von Hippel-Lindau protein mutants linked to type 2C VHL disease preserve the ability to downregulate HIF. Hum Mol Genet 2001; 10: 1019–1027.
153. Hickey MM, Lam JC, Bezman NA, Rathmell WK, Simon MC. von Hippel-Lindau mutation in mice recapitulates Chuvash polycythemia via hypoxia-inducible factor-2alpha signaling and splenic erythropoiesis. J Clin Invest 2007; 117(12): 3879–3889.
154. Gordeuk VR, Sergueeva AI, Miasnikova GY, et al. Congenital disorder of oxygen sensing: association of the homozygous Chuvash polycythemia VHL mutation with thrombosis and vascular abnormalities but not tumors. Blood 2004; 103: 3924–3932.

155. Bushuev VI, Miasnikova GY, Sergueeva AI, et al. Endothelin-1, vascular endothelial growth factor and systolic pulmonary artery pressure in patients with Chuvash polycythemia. Haematologica 2006; 91: 744–749.
156. Perrotta S, Nobili B, Ferraro M, et al. Von Hippel-Lindau-dependent polycythemia is endemic on the island of Ischia: identification of a novel cluster. Blood 2006; 107: 514–519.
157. Cario H, Schwarz K, Jorch N, et al. Mutations in the von Hippel-Lindau (VHL) tumor suppressor gene and VHL-haplotype analysis in patients with presumable congenital erythrocytosis. Haematologica 2005; 90: 19–24.
158. Percy MJ, Furlow PW, Lucas GS, et al. A gain-of-function mutation in the HIF2A gene in familial erythrocytosis. N Engl J Med 2008; 358: 162–168.
159. Percy MJ, Beer PA, Campbell G, et al. Novel exon 12 mutations in the HIF2A gene associated with erythrocytosis. Blood 2008; 111: 5400–5402.
160. Percy MJ, Zhao Q, Flores A, et al. A family with erythrocytosis establishes a role for prolyl hydroxylase domain protein 2 in oxygen homeostasis. Proc Natl Acad Sci U S A 2006; 103: 654–659.
161. Percy MJ, Furlow PW, Beer PA, et al. A novel erythrocytosis-associated PHD2 mutation suggests the location of a HIF binding groove. Blood 2007; 110: 2193–2196.
162. Kazal LA, Erslev AJ. Erythropoietin production in renal tumors. Ann Clin Lab Sci 1975; 5: 98–109.
163. Takeda K, Cowan A, Fong GH. Essential role for prolyl hydroxylase domain protein 2 in oxygen homeostasis of the adult vascular system. Circulation 2007; 116: 774–781.
164. Takeda K, Ho VC, Takeda H, et al. Placental but not heart defects are associated with elevated hypoxia-inducible factor alpha levels in mice lacking prolyl hydroxylase domain protein 2. Mol Cell Biol 2006; 26: 8336–8346.
165. Minamishima YA, Moslehi J, Bardeesy N, et al. Somatic inactivation of the PHD2 prolyl hydroxylase causes polycythemia and congestive heart failure. Blood 2008; 111: 3236–3244.
166. Minamishima YA, Moslehi J, Padera RF, et al. A feedback loop involving the Phd3 prolyl hydroxylase tunes the mammalian hypoxic response in vivo. Mol Cell Biol 2009; 29(21): 5729–5741.
167. Bruick RK. Oxygen sensing in the hypoxic response pathway: regulation of the hypoxia-inducible transcription factor. Genes Dev 2003; 17: 2614–2623.
168. Kaelin WG, Jr., Ratcliffe PJ. Oxygen sensing by metazoans: the central role of the HIF hydroxylase pathway. Mol Cell 2008; 30: 393–402.
169. Appelhoff RJ, Tian YM, Raval RR, et al. Differential function of the prolyl hydroxylases PHD1, PHD2, and PHD3 in the regulation of hypoxia-inducible factor. J Biol Chem 2004; 279: 38458–38465.
170. Ivan M, Haberberger T, Gervasi DC, et al. Biochemical purification and pharmacological inhibition of a mammalian prolyl hydroxylase acting on hypoxia-inducible factor. Proc Natl Acad Sci U S A 2002; 99: 13459–13464.
171. Bernhardt WM, Campean V, Kany S, et al. Preconditional activation of hypoxia-inducible factors ameliorates ischemic acute renal failure. J Am Soc Nephrol 2006; 17: 1970–1978.
172. Hsieh MM, Linde NS, Wynter A, et al. HIF prolyl hydroxylase inhibition results in endogenous erythropoietin induction, erythrocytosis, and modest fetal hemoglobin expression in rhesus macaques. Blood 2007; 110: 2140–2147.
173. Bernhardt W, Wiesener M, Schmieder R, Guenzler V, Eckardt KU. The prolylhydroxyalse inhibitor FG2216 stimulates EPO production in nephric and anephric dialysis patients: evidence for an underutilized production capacity in liver and kidneys. ASN annual meeting, San Francisco, CA. J Am Soc Nephrol 2007, pp SA-PO784.

Chapter 24
Intricate Link between Hypoxia and Oxidative Stress in Chronic Kidney Disease

Tetsuhiro Tanaka

Abstract Hypoxia in the tubulointerstitium has been implicated in a number of progressive renal diseases. Following the initial hypothesis of chronic hypoxia raised about a decade ago, much effort has been made to substantiate hypoxia in the tubulointerstitial compartment in various models of progressive renal disease. The discovery of hypoxia-inducible factor (HIF) further accelerated research in this field. Expression and the functional operation of this transcription factor have been extensively characterized in pathophysiological contexts, and there is accumulating evidence that HIF may either protect kidneys from injury or aggravate damage, depending on the context. On the other hand, hypoxia per se, or subtle changes in tissue oxygenation, result in mitochondrial generation of reactive oxygen species (ROS) in various organs including the kidney. Depending on the kind of ROS and the quantity generated, ROS either serve to transduce cellular signaling or cause cellular injury and death. Notably, there is experimental evidence that ROS generated by complex III of the electron transport chain stabilize HIF-α, suggesting that mitochondria serve as oxygen sensors and provide a link between hypoxia and oxidative stress. This chapter focuses on the significance of hypoxia and oxidative stress in understanding the pathogenesis of renal diseases and provides an insight into how they are mutually related and cooperate to modify disease progression.

Keywords HIF · Tubulointerstitial injury · Reactive oxygen species · Prolyl hydroxylase · Mitochondria · Redox

T. Tanaka (✉)
Division of Nephrology and Endocrinology, University of Tokyo School of Medicine, 7-3-1 Hongo, Bunkyo-ku, 113-8655 Tokyo, Japan
e-mail: tetsu-tky@umin.ac.jp

T. Miyata et al. (eds.), *Studies on Renal Disorders*, Oxidative Stress in Applied Basic Research and Clinical Practice, DOI 10.1007/978-1-60761-857-7_24,

1 Introduction

Human beings consume as much as 500–600 L of oxygen per day to meet the metabolic demand. Under circumstances in which ambient oxygen levels are below optimal, individuals try to transport up more oxygen, by increasing respiratory depth and frequency, increasing heart rate, and raising the number of erythrocytes. However, when this compensation is less than enough and the ambient O_2 level falls below 8–10%, humans experience severe neurological insufficiency and are no longer able to survive. Therefore, oxygen utilization is extremely important in our daily lives.

At the cellular level, nearly 90% of available oxygen is consumed by mitochondria, to generate adenosine triphosphate (ATP) in the most efficient way. In the steady state of normoxia, glucose is converted to pyruvate by the glycolytic pathway, which occurs in the cytoplasm. Pyruvate then enters mitochondria and is converted to acetyl CoA to initiate the tricarboxylic acid (TCA) cycle. In the TCA cycle, the electron carrier NADH (nicotinamide adenine dinucleotide plus hydrogen) is generated, and its electrons are further passed to the process of oxidative phosphorylation. Oxidative phosphorylation takes place in complexes I–IV, in which electrons are passed through cytochromes to molecular oxygen, the final acceptor. During this process, a proton gradient is generated in the inner mitochondrial membrane, which serves as a driving force to synthesize ATP. The entire process is extremely efficient and one molecule of glucose results in generation of as many as 38 molecules of ATP, which are then utilized to maintain cellular function. Importantly, however, this process does not use all the available oxygen, and regulated proton leakage in the electron transport results in generation of reactive oxygen species (ROS). This is even more evident when the balance between oxygen demand and supply is distorted or fluctuates, temporarily. Therefore, changes in cellular oxygenation status are closely related to ROS production.

2 Tubulointerstitial Hypoxia in the Kidney

During evolution, kidneys in land mammals, including humans, developed a mechanism with which to reabsorb sodium and water as efficiently as possible, by maximally concentrating urine. This is accomplished by constructing a countercurrent system of blood vessels and tubules in the medulla. As a consequence, renal medulla is reasonably placed at the borderline of hypoxia in the physiological state [1].

In the pathological state, however, hypoxic areas extend to the cortical tubulointerstitium as well. In many types of chronic renal diseases, whether glomerular, vascular, or metabolic in origin, tubulointerstitial hypoxia is reported as one of the common mediators. This is supported by observation of human kidneys, indicating that the loss of the peritubular capillaries in the renal cortex correlates with the degree of residual renal function in various types of glomerular diseases [2, 3].

About a decade ago, a hypothesis on "chronic hypoxia in the tubulointerstitium" was proposed by Fine et al. [4]. In glomerulopathies, for example, the initial insult to the glomeruli damages local endothelial structures and occludes postglomerular capillary blood flow. Ischemia in the affected region then triggers several phenotypic changes in tubular cells, including abnormal proliferation, epithelial–mesenchymal transition (EMT), and cell death. Hypoxic tubular cells in turn serve as a source of key mediators involved in macrophage infiltration and tubulointerstitial fibrosis. Interstitial accumulation of extracellular matrix further impairs local oxygenation by impeding oxygen diffusion, thus accelerating regional hypoxia and creating a vicious cycle. At the cellular level, hypoxia serves as a driving force for tubular epithelium to alter the rate of cell proliferation [5], experience EMT [6], or undergo cell death [7]. These lines of experimental evidence in cell culture studies further raise the possibility that hypoxia is a critical mediator in progression of chronic renal diseases.

These hypotheses, although highly plausible in theory, were initially met with some skepticism because evidence was lacking that the tubulointerstitial compartment experienced hypoxia during the pathogenesis of different diseases. Direct demonstration of tubular hypoxia is still awaited.

In subsequent studies, a lot of efforts have been made to substantiate tubulointerstitial hypoxia in experimental models of progressive renal diseases, using various tools. Pimonidazole is a chemical hypoxia probe that allows detection of hypoxia at the cellular level. With this probe, hypoxia was demonstrated in the diseased tubulointerstitium of the rat progressive Thy-1.1 nephritis model, which was associated with loss of peritubular capillaries or stagnant capillary blood flow [8]. A similar method was applied in the rat remnant kidney, and hypoxic tubules were observed at day 7, a time point before any pathological signs of tubulointerstitial injury were to be observed. Tubular hypoxia in this model was linked to narrowing and distortion of neighboring cortical peritubular capillaries [9]. However, this method is potentially unreliable and based on the assumption that the chemical adduct is evenly distributed in target organs. Moreover, it only allows detection of hypoxic cells at oxygen tension below 10 mmHg. This degree of hypoxia may be too severe to occur during the course of many diseases, and we may potentially have missed detection of hypoxic tubulointerstitium, which is of a milder degree and still pathophysiologically more relevant.

Blood oxygen level-dependent (BOLD) magnetic resonance imaging (MRI) is a noninvasive method to measure oxygen bioavailability in organs. Changes in blood concentrations of deoxyhemoglobin result in increased magnetic spin dephasing of blood water protons and decreased signal intensity on T2-weighted MRI sequences. Therefore, the apparent relaxation rate (1/T2) is directly proportional to tissue deoxyhemoglobin levels and the increased 1/T2 implies decreased oxygen bioavailability in the tissue. Conceptually, this may be more physiologically relevant, and by using this method, hypoxia was detected in the outer stripe of outer medulla, the part most susceptible to ischemic injury, in streptozotocin (STZ)-induced type 1 diabetes model in the rat [10]. In summary, results of these studies have provided further evidence that the tubulointerstitial compartment is hypoxic during the pathogenesis of certain types of progressive renal diseases.

As exemplified above, tubular hypoxia is most frequently associated with stagnant capillary flow or rarefaction of peritubular capillary network. However, this is probably not the only cause of regional hypoxia, given that hypoxia is determined by the balance between oxygen supply and demand. In this regard, many studies have focused on the possible role of alterations in oxygen consumption. Conceptually, oxygen consumption in the kidney is primarily determined by tubular uptake of electrolytes, mainly sodium, which correlates with the amount filtered in the glomerulus [11]. However, recent studies suggest that the linear relationship between oxygen consumption (QO_2) and sodium reabsorption (T_{Na}) may be variable, and increases in QO_2/T_{Na} are observed in spontaneously hypertensive rats and in a number of high angiotensin II and low nitric oxide conditions [12], suggesting a disproportionate increase in oxygen utilization under these conditions, which, at least in theory, may contribute to net hypoxia.

3 Hypoxia-Inducible Factors and Their Role in Renal Disorders

Hypoxia-inducible factor (HIF) is a transcription factor that plays a central role in cellular response to hypoxia. It is a heterodimer that belongs to the basic helix-loop-helix per-arnt-sim (bHLH-PAS) family, composed of an oxygen-dependent α subunit and constitutively expressed β subunit (also referred to as aryl hydrocarbon receptor nuclear translocator [ARNT]). The expression of the α subunit is under tight regulation by oxygen at the posttranslational level. In normoxia, the α subunit is hydroxylated at two conserved proline residues (Pro402 and Pro564 of human HIF-1α) by members of the prolyl hydroxylase domain (PHD) family (also referred to as the EglN family), which allows recruitment of the von Hippel-Lindau protein (pVHL), a recognition component of the E3-ubiquitin ligase complex, and promotes proteasomal degradation [13–15]. In hypoxia, the α subunit escapes degradation, translocates into the nucleus, forms a heterodimer with the β subunit, and can transactivate 100–200 target genes, including erythropoietin (EPO),vascular endothelial growth factor (VEGF), and glycolytic enzymes. In addition, factor-inhibiting HIF-1 (FIH-1) mediates oxygen-dependent hydroxylation at the asparagine residue in the C-terminal transactivator domain (C-TAD), which inhibits recruitment of the coactivators CBP/p300 and subsequently transactivation. Therefore, hypoxia regulates the expression and functional operation of HIF through similar mechanisms of hydroxylation.

Earlier immunohistochemical studies characterized HIF-α expression in the ischemic kidney [16]. After a number of ischemic challenges, such as ambient hypoxia, functional anemia, and renal artery ligation, HIF-1α was expressed in tubular and glomerular epithelial cells, as well as in papillary interstitial cells. Likewise, HIF-2α expression was identified in endothelial and interstitial cells.

In addition, renal expression of functional HIF was confirmed using transgenic rats carrying the HRE-driven luciferase vector (HRE-luc transgenic rats) [17]. In these rats, an increase in transgene expression, reflecting HIF activation, was observed in the cortical tubular epithelium in proteinuric renal disease (puromycin nephropathy) and glomerular hypertension (remnant kidney). This provides further evidence that HIF is activated and likely involved in the pathogenesis of chronic renal diseases.

Demonstration of HIF expression in renal diseases, in turn, raised an important question of what role it plays in renal pathology. To address this question, we pharmacologically overexpressed HIF in the rat using cobalt and tested for its function. Cobalt reduces PHD activity by serving as a chelator of iron or through depletion of intracellular ascorbate, thus allowing HIF-α protein to accumulate. Administration of cobalt in rats protected kidneys from injury in a number of acute and chronic renal diseases, through induction of a set of protective HIF-target genes [18].

Kidneys facing ischemic insult, whether acute or chronic by nature, experience bursts of ROS. The simplest example will be ischemia–reperfusion injury, such as it occurs in kidney transplants in the clinical setting. Rapid reintroduction of oxygen and nutrients in areas in which blood flow was initially occluded triggers ROS generation and contributes to tissue injury. Importantly, sustained hypoxia also appears to generate ROS in ischemic organs, although the amounts of ROS measured during ischemia are considerably less and appear more transient as compared to reperfusion, as seen, for example, in the heart [19]. Measurement of such small and transient signals is challenging with the methods available to date, and, therefore, there is no firm conclusion to draw on this issue for now. Nevertheless, hypoxic production of ROS is likely and is to be envisaged by the incessant stagnation/occlusion and reflow/reopening of blood flow at the microvascular level in the chronically ischemic kidney, which leads to changes in tissue oxygenation status and triggers ROS. Sources of ROS could be either resident cells or infiltrating inflammatory cells.

4 Reactive Oxygen Species in Hypoxia

As discussed above, mitochondria bind oxygen at cytochrome oxidase, and they represent the primary site of oxygen consumption [20]. Oxygen is utilized as a final acceptor of electrons during oxidative phosphorylation to generate maximal amounts of ATP. Although this entire process is extremely efficient and sufficient to fuel individual cells under normal conditions, it is not completely efficient, and about 20% of protons undergo regulated proton leak during oxidative phosphorylation, which yields mitochondrial ROS. In terms of quantity, the mitochondrial electron transport chain is by far the most important source of ROS.

When oxygen demand exceeds its supply (hypoxia), there is an increase in glycolytic ATP production mediated by transcriptional upregulation of genes involved in glycolysis (Pasteur effect), while the relative contribution of ATP production through the TCA cycle and subsequent oxidative phosphorylation becomes less relevant. HIF-1 plays a central role in this metabolic switch [21]. In hypoxia, HIF-1 upregulates a set of genes involved in glycolysis. At the same time, HIF-1 induces pyruvate dehydrogenase kinase 1 (PDK1) to cause a decrease in the amount of pyruvate/acetyl-CoA entering the TCA cycle and lactate dehydrogenase A (LDHA) to convert pyruvate to lactate, thus switching the machinery of ATP production from oxidative phosphorylation to glycolytic metabolism. From the viewpoint of mitochondrial ROS, this switch could be regarded as a defense mechanism to protect cells from excessive production of ROS, which potentially damages cellular function.

With respect to mitochondrial ROS generation in a hypoxic milieu, there is controversy regarding whether hypoxia causes changes in the amount of ROS. While several reports describe a decease in ROS decrease in hypoxia, others documented an increase in ROS when cells experience insufficient oxygen supply [22–24]. Possible explanations for this apparent discrepancy include cellular background, basal cell metabolism, and the degree and duration of hypoxia experienced, as well as methods used for the detection of ROS.

Hypoxic induction of ROS in mitochondria may sound somewhat paradoxical at first glance, because, as stated above, ROS is produced during the process of oxidative phosphorylation through the electron transport chain, in which oxygen represents a final acceptor of electrons. Lack of molecular oxygen would therefore make it unlikely for mitochondria to generate ROS. Nevertheless, hypoxic cells do appear to generate ROS utilizing oxygen. In fact, there is evidence that considerable molecular oxygen is apparently available for reaction in most conditions of hypoxia. On the basis of solubility of oxygen in water and lipids, even at 1 mmHg partial pressure of oxygen (PO_2), 1.3 and 6.5 μM O_2, respectively, are available for reactions [25]. In addition, it is reported that hypoxia or ischemia alters mitochondrial inner membrane fluidity [26], and such changes allow the ubisemiquinone, an electron donor in complex III that mediates conversion of O_2 to O_2^-, to gain access for longer periods, necessary for oxygen reduction [27]. Taken together, it is likely that there is still a sufficient amount of oxygen available for mitochondrial ROS production, even under pathological conditions in which oxygen availability is limited.

One caveat in substantiating hypoxic production of mitochondrial ROS is the limitation of available methods to reliably measure ROS. One of the most widely used would be the one based on oxidation of dichlorofluorescein to the fluorescent product. Using this method, several groups observed that the intracellular fluorescence increases when gas bubbling in the media is switched to a hypoxic mixture, indicating that hypoxia indeed induces ROS in the cell [28–30]. In cardiac myocytes, the extent of fluorescence intensity was found to be proportional to the severity of hypoxia, and the largest ROS signal was observed at approximately 7 mmHg PO_2 in the perfusate [31]. These changes were attenuated by rotenone and thenoyltrifluoroacetone, inhibitors that block the electron transfer at complex I and II, respectively. In addition, myxothiazol, a specific inhibitor of complex III at the binding site of ubiquinol to Q_O that prevents generation of ubisemiquinone, also inhibited hypoxic induction of ROS, suggesting that hypoxia generates ROS at the Q cycle in complex III [32]. But again, these findings had to be viewed with caution, partly because they relied heavily on pharmacological tools, which may have effects other than those related to their class effect. Obviously, demonstration of hypoxic ROS by approaches using nonchemical tools, such as genetic manipulation, are still awaited.

5 A Role of Mitochondrial ROS in Hypoxia

Mitochondria have been implicated in a number of hypoxic responses, many of which take place through the hypoxic induction of mitochondrial ROS. Such responses include myocyte contraction [30, 31], adipose differentiation [33], Na/K-ATPase activation [34], and glutathione depletion [35]. One of the most extensively studied and relevant with respect to renal diseases would be the impact of ROS on HIF-α expression, which will be discussed below. Details on the role of ROS in acute kidney injury, as well as chronic kidney disorders, will be reviewed separately in other chapters.

The role of cellular redox status on hypoxic gene induction was initially investigated in cell culture models of EPO production, yielding conflicting results. In one study, hydrogen peroxide (H_2O_2) inhibited hypoxic induction of EPO, which was reversed by catalase [36]. In another study, ectopic administration of H_2O_2 at concentrations of 0.1–1.0 mM resulted in a decrease in HIF-1α protein, with parallel impairment of its DNA binding and a resultant dose-dependent decrease in EPO mRNA expression [37]. On the other hand, there is experimental evidence that H_2O_2– and ROS-generating systems increased hypoxic EPO production, while ROS scavengers showed the opposite effect in EPO-producing renal carcinoma cells [38] and Hep3B cells [39].

In the midst of these existing controversies, the first report concerning the positive regulation of HIF-1α expression by mitochondrial ROS appeared in 1988, describing that ROS generated by mitochondria in hypoxic cells were necessary and sufficient to accumulate HIF-1α [28]. But again, these studies relied heavily on pharmacological approaches, and their proposal was challenged by reports demonstrating that cells lacking the mitochondrial respiratory chain (ρ^0 cells) were still able to accumulate HIF-1α [40, 41]. Those apparent discrepancies on the effect of ROS, in hindsight, might have been reconciled by differences in hypoxic conditions. In contrast to the subnormal range of oxygen tensions used in the former study, the latter group [40] employed near anoxic conditions. Anoxic accumulation of HIF-α occurs even without functional mitochondrial respiratory chains and ROS generated therein [42], likely because HIF-hydroxylases play a dominant role under these conditions.

6 Regulation of HIF-α by PHDs

In terms of regulation of HIF-α proteins, PHDs, especially PHD-2, play a dominant role [43, 44]. In higher mammals, three PHD paralogs have been reported, and they belong to the Fe(II) and 2-oxoglutarate-dependent oxygenase superfamily, the enzymatic activity of which is completely dependent on oxygen. Measurement of values for the apparent K_M for oxygen (the concentration of oxygen that supports a half-maximal initial catalytic rate) that use HIF-α polypeptides suggests that the PHD oxygen K_M values are approximately 100 μM [45]. Given that oxygen concentrations in tissues are typically 10–30 μM and are always below the K_M value for oxygen, it follows that the enzymatic activity is modulated by molecular oxygen availability over the entire physiological range. This explains why PHDs are excellent candidates to operate as an oxygen sensor. However, several other factors participate in HIF-α hydroxylation by influencing PHD enzymatic activity. In fact, these enzymes require Fe^{2+}, 2-oxoglutarate, and ascorbate, in addition to molecular oxygen, and there is experimental evidence that PHD-2 activity is also influenced by changes in intracellular succinate (converted from 2-oxoglutarate) and redox states [46, 47].

7 Mitochondria as an Oxygen Sensor

Several lines of evidence now support the view that ROS also plays a role in the hypoxic induction of HIF-α. In 2005, three groups reported the role of mitochondrial ROS in HIF-α expression simultaneously, using novel approaches to overcome technical limitations.

First, Guzy et al. [48] used a novel ROS-sensitive fluorescence resonance energy transfer (FRET) probe and confirmed hypoxic induction of mitochondrial ROS. Then, using siRNA techniques to suppress expression of the Rieske iron-sulfur protein of complex III, they proved that ROS, generated by complex III of the electron transport, is required for the hypoxic stabilization of HIF-1α and HIF-2α, under physiological hypoxia.

Second, Mansfield et al. [49] demonstrated that murine embryonic cells, lacking cytochrome c and mitochondrial activity, were unable to accumulate HIF-1α and HIF-2α, and that ectopic administration of H_2O_2 in these cells was sufficient to stabilize HIF-α, suggesting that ROS generated by mitochondrial respiration play a critical role for HIF-α accumulation.

Third, Brunelle et al. [50] used fibroblasts from a patient with Leigh's syndrome, a disease in which the Surf-1 gene product is disrupted, thus displaying improper assembly of cytochrome c oxidase (COX) and residual levels of electron transport activity. They demonstrated that these cells were able to accumulate HIF-1α, indicating that oxidative phosphorylation is not required. Instead, H_2O_2 produced at complex III was suggested as the responsible signaling molecule, because hypoxic stabilization of HIF-1α was abolished when glutathione peroxidase 1 (GPX1) or catalase was overexpressed, while overexpression of copper zinc superoxide dismutase (SOD1) or manganese superoxide dismutase (SOD2) had no clear effect. Results from these genetic approaches have provided more solid evidence that ROS generated by mitochondria do contribute to hypoxic induction of HIF-1α and HIF-2α.

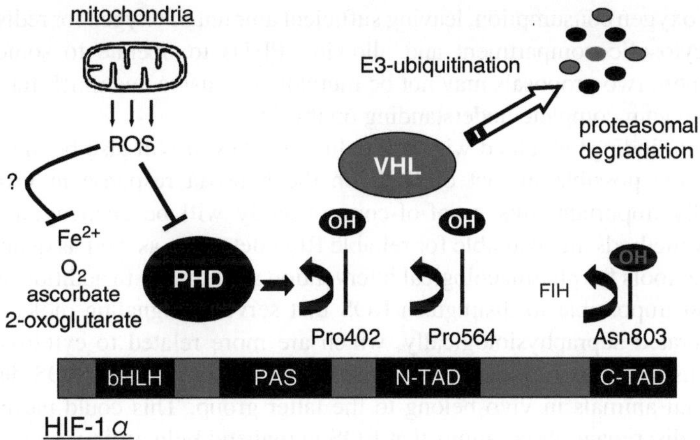

Fig. 1 Reactive oxygen species (ROS)-mediated stabilization of hypoxia-inducible factor (HIF)-1α. Proposed mechanisms to date are shown in schematic diagrams. Expression of HIF-1α is largely controlled by mechanisms of proteasomal degradation, which are initiated by prolyl hydroxylation and mediated by prolyl hydroxylase domain (PHDs). Since PHDs require Fe^{2+}, ascorbate, and 2-oxoglutarate in addition to O_2 for enzymatic activity, they fail to hydroxylate HIF-1α and allow its expression in hypoxia. In addition, hypoxia triggers bursts of ROS production by complex III of the mitochondrial electron transport chain, which inhibits PHD enzymatic activity through redox changes in PHD-bound iron or yet to be identified mechanisms

In contrast to the accumulating evidence that mitochondrial ROS are required for hypoxic induction of HIF-α, it remains unclear how generated ROS participate in the process of HIF-α accumulation. One of the plausible explanations is that ROS dampens enzymatic activity of PHDs, thereby preventing the interaction of HIF-α with pVHL and the subsequent proteasomal degradation (Fig. 1). Indeed, the loss of JunD, a member of the AP-1 family of transcription factors, increases oxidative stress in cells and results in normoxic accumulation of HIF-1 and its target gene VEGF, by reducing enzymatic activity of PHDs [46]. It is also likely that ROS in hypoxia could affect the oxidation status of PHD-bound iron, thereby modifying its enzymatic activity. The pathological relevance of ROS in HIF signaling is highlighted by the observation that radiation-induced reoxygenation of tumor cells induces ROS, which activate HIF-1 and contribute to resistance against radiation-induced endothelial cell injury [51]. Conversely, there is experimental evidence that antioxidants contribute to inhibition of tumor growth by diminishing HIF-1α expression [52].

In addition to the previously described role of ROS, mitochondria also participate in hypoxia sensing by redistributing intracellular oxygen through respiration [53]. This theory is based on the low K_M of cytochrome oxidase for oxygen (less than 1 μM) as compared to that of HIF hydroxylases. According to this theory, the majority of oxygen consumed by mitochondria leaves little oxygen available for enzymatic activities to take place in the cytosol. The inhibition of mitochondrial function in hypoxia increases the apparent K_M of cytochrome oxidase and results in decreased oxygen consumption, leaving sufficient amount of oxygen for redistribution into the cytosolic compartment and allowing PHDs to operate to some degree. Notably, these two proposals may not be mutually exclusive and much further work is needed to gain complete understanding on this issue.

Based on findings obtained with cell culture studies, it will also be interesting to determine the possible impact of ROS on the hypoxia response in vivo. While undoubtedly important, this proof-of-concept study will be challenging, because only a few methods are available for reliable ROS detection as well as quantification in vivo, and tools for pharmacological intervention are limited. In addition, above all, it is almost impossible to distinguish ROS that serve as signaling molecules from those generated supraphysiologically, which are more related to cytotoxicity in a general sense. In many cases, with the methods available to date, ROS detected in experimental animals in vivo belong to the latter group. This could partly explain apparently discrepant observations that ROS in diabetic kidneys in vivo dampen the hypoxia response through dysfunction of HIF-1, rather than facilitating its expression [54, 55]. Although a detailed explanation for this discrepancy remains to be elucidated, it may be due to a biphasic effect of oxidative stress, depending on the kind and amounts. Again, while physiological amounts of ROS serve as a mediator of signal transduction pathways, supraphysiological amounts of ROS are definitely cytotoxic.

8 Conclusion

During the past decade, much effort has been made to substantiate chronic hypoxia in the diseased tubulointerstitium of the kidney. Discovery of HIF accelerated this issue and contributed greatly to a better understanding of renal pathology mediated by hypoxia. On the other hand, accumulating evidence indicates that hypoxia and hypoxia/reoxygenation, which diseased kidneys often encounter, lead to ROS production. The mutual relationship between hypoxia and ROS is still not completely understood, but recent discoveries with respect to the role of hypoxia on mitochondrial ROS generation, and conversely, the impact of ROS on hypoxia sensing, would certainly compose an intricate, but physiologically relevant, link in multiple organs, including the kidney. Together with PHDs, ROS may thus add an additional layer of complexity to hypoxia sensing. Understanding this link could ultimately offer a new strategy to prevent or minimize injury and treat both acute and chronic renal diseases.

References

1. Brezis M, Rosen S. Hypoxia of the renal medulla – its implications for disease. N Engl J Med 1995; 332: 647–655.
2. Bohle A, Mackensen-Haen S, Wehrmann M. Significance of postglomerular capillaries in the pathogenesis of chronic renal failure. Kidney Blood Press Res 1996; 19: 191–195.

3. Bohle A, Muller GA, Wehrmann M, et al. Pathogenesis of chronic renal failure in the primary glomerulopathies, renal vasculopathies, and chronic interstitial nephritides. Kidney Int Suppl 1996; 54: S2–S9.
4. Fine LG, Bandyopadhay D, Norman JT. Is there a common mechanism for the progression of different types of renal diseases other than proteinuria? Towards the unifying theme of chronic hypoxia. Kidney Int Suppl 2000; 75: S22–S26.
5. Sahai A, Mei C, Schrier RW, Tannen RL. Mechanisms of chronic hypoxia-induced renal cell growth. Kidney Int 1999; 56: 1277–1281.
6. Manotham K, Tanaka T, Matsumoto M, et al. Transdifferentiation of cultured tubular cells induced by hypoxia. Kidney Int 2004; 65: 871–880.
7. Saikumar P, Dong Z, Weinberg JM, Venkatachalam MA. Mechanisms of cell death in hypoxia/reoxygenation injury. Oncogene 1998; 17: 3341–3349.
8. Matsumoto M, Tanaka T, Yamamoto T, et al. Hypoperfusion of peritubular capillaries induces chronic hypoxia before progression of tubulointerstitial injury in a progressive model of rat glomerulonephritis. J Am Soc Nephrol 2004; 15: 1574–1581.
9. Manotham K, Tanaka T, Matsumoto M, et al. Evidence of tubular hypoxia in the early phase in the remnant kidney model. J Am Soc Nephrol 2004; 15: 1277–1288.
10. Ries M, Basseau F, Tyndal B, et al. Renal diffusion and BOLD MRI in experimental diabetic nephropathy. Blood oxygen level–dependent. J Magn Reson Imaging 2003; 17: 104–113.
11. Lassen NA, Munck O, Thaysen JH. Oxygen consumption and sodium reabsorption in the kidney. Acta Physiol Scand 1961; 51: 371–384.
12. Welch WJ. Intrarenal oxygen and hypertension. Clin Exp Pharmacol Physiol 2006; 33: 1002–1005.
13. Maxwell PH, Wiesener MS, Chang GW, et al. The tumour suppressor protein VHL targets hypoxia-inducible factors for oxygen-dependent proteolysis. Nature 1999; 399: 271–275.
14. Ivan M, Kondo K, Yang H, et al. HIF-alpha targeted for VHL-mediated destruction by proline hydroxylation: implications for O2 sensing. Science 2001; 292: 464–468.
15. Jaakkola P, Mole DR, Tian YM, et al. Targeting of HIF-alpha to the von Hippel-Lindau ubiquitylation complex by O2-regulated prolyl hydroxylation. Science 2001; 292: 468–472.
16. Rosenberger C, Mandriota S, Jurgensen JS, et al. Expression of hypoxia-inducible factor-1alpha and -2alpha in hypoxic and ischemic rat kidneys. J Am Soc Nephrol 2002; 13: 1721–1732.
17. Tanaka T, Miyata T, Inagi R, et al. Hypoxia in renal disease with proteinuria and/or glomerular hypertension. Am J Pathol 2004; 165: 1979–1992.
18. Nangaku M, Inagi R, Miyata T, Fujita T. Hypoxia and hypoxia-inducible factor in renal disease. Nephron Exp Nephrol 2008; 110: e1–e7.
19. Stoner JD, Clanton TL, Aune SE, Angelos MG. O_2 delivery and redox state are determinants of compartment-specific reactive O_2 species in myocardial reperfusion. Am J Physiol Heart Circ Physiol 2007; 292: H109–H116.
20. Bunn HF, Poyton RO. Oxygen sensing and molecular adaptation to hypoxia. Physiol Rev 1996; 76: 839–885.
21. Semenza GL. Oxygen-dependent regulation of mitochondrial respiration by hypoxia-inducible factor 1. Biochem J 2007; 405: 1–9.
22. Pouyssegur J, Mechta-Grigoriou F. Redox regulation of the hypoxia-inducible factor. Biol-Chem 2006; 387: 1337–1346.
23. Chandel NS, Budinger GR. The cellular basis for diverse responses to oxygen. Free Radic Biol Med 2007; 42: 165–174.
24. Moudgil R, Michelakis ED, Archer SL. Hypoxic pulmonary vasoconstriction. J Appl Physiol 2005; 98: 390–403.
25. Snedden W, LeDez K, Manson HJ. A new method for the measurement of gas solubility. J Appl Physiol 1996; 80: 1371–1378.
26. Irazu CE, Rajagopalan PR, Orak JK, et al. Mitochondrial membrane fluidity changes in renal ischemia. J Exp Pathol 1990; 5: 1–6.

27. Gille L, Nohl H. The ubiquinol/bc1 redox couple regulates mitochondrial oxygen radical formation. Arch Biochem Biophys 2001; 388: 34–38.
28. Chandel NS, Maltepe E, Goldwasser E, et al. Mitochondrial reactive oxygen species trigger hypoxia-induced transcription. Proc Natl Acad Sci U S A 1998; 95: 11715–11720.
29. Vanden Hoek TL, Becker LB, Shao Z, et al. Reactive oxygen species released from mitochondria during brief hypoxia induce preconditioning in cardiomyocytes. J Biol Chem 1998; 273: 18092–18098.
30. Waypa GB, Marks JD, Mack MM, et al. Mitochondrial reactive oxygen species trigger calcium increases during hypoxia in pulmonary arterial myocytes. Circ Res 2002; 91: 719–726.
31. Duranteau J, Chandel NS, Kulisz A, et al. Intracellular signaling by reactive oxygen species during hypoxia in cardiomyocytes. J Biol Chem 1998; 273: 11619–11624.
32. Guzy RD, Schumacker PT. Oxygen sensing by mitochondria at complex III: the paradox of increased reactive oxygen species during hypoxia. Exp Physiol 2006; 91: 807–819.
33. Carriere A, Carmona MC, Fernandez Y, et al. Mitochondrial reactive oxygen species control the transcription factor CHOP-10/GADD153 and adipocyte differentiation: a mechanism for hypoxia-dependent effect. J Biol Chem 2004; 279: 40462–40469.
34. Dada LA, Chandel NS, Ridge KM, et al. Hypoxia-induced endocytosis of Na,K-ATPase in alveolar epithelial cells is mediated by mitochondrial reactive oxygen species and PKC-zeta. J Clin Invest 2003; 111: 1057–1064.
35. Mansfield KD, Simon MC, Keith B. Hypoxic reduction in cellular glutathione levels requires mitochondrial reactive oxygen species. J Appl Physiol 2004; 97: 1358–1366.
36. Fandrey J, Frede S, Jelkmann W. Role of hydrogen peroxide in hypoxia-induced erythropoietin production. Biochem J 1994; 303 (Pt 2): 507–510.
37. Huang LE, Arany Z, Livingston DM, Bunn HF. Activation of hypoxia-inducible transcription factor depends primarily upon redox-sensitive stabilization of its alpha subunit. J Biol Chem 1996; 271: 32253–32259.
38. Ueno M, Brookins J, Beckman BS, Fisher JW. Effects of reactive oxygen metabolites on erythropoietin production in renal carcinoma cells. Biochem Biophys Res Commun 1988; 154: 773–780.
39. Ndele JK, Yoshioka K, Fisher JW. Hydrogen peroxide in the regulation of erythropoietin (Epo) gene expression in hepatocellular carcinoma cells. East Afr Med J 1996; 73: 143–146.
40. Vaux EC, Metzen E, Yeates KM, Ratcliffe PJ. Regulation of hypoxia-inducible factor is preserved in the absence of a functioning mitochondrial respiratory chain. Blood 2001; 98: 296–302.
41. Srinivas V, Leshchinsky I, Sang N, et al. Oxygen sensing and HIF-1 activation does not require an active mitochondrial respiratory chain electron-transfer pathway. J Biol Chem 2001; 276: 21995–21998.
42. Schroedl C, McClintock DS, Budinger GR, Chandel NS. Hypoxic but not anoxic stabilization of HIF-1alpha requires mitochondrial reactive oxygen species. Am J Physiol Lung Cell Mol Physiol 2002; 283: L922–L931.
43. Schofield CJ, Ratcliffe PJ. Oxygen sensing by HIF hydroxylases. Nat Rev Mol Cell Biol 2004; 5: 343–354.
44. Kaelin WG, Jr., Ratcliffe PJ. Oxygen sensing by metazoans: the central role of the HIF hydroxylase pathway. Mol Cell 2008; 30: 393–402.
45. Koivunen P, Hirsila M, Kivirikko KI, Myllyharju J. The length of peptide substrates has a marked effect on hydroxylation by the hypoxia-inducible factor prolyl 4-hydroxylases. J Biol Chem 2006; 281: 28712–28720.
46. Gerald D, Berra E, Frapart YM, et al. JunD reduces tumor angiogenesis by protecting cells from oxidative stress. Cell 2004; 118: 781–794.
47. Selak MA, Armour SM, MacKenzie ED, et al. Succinate links TCA cycle dysfunction to oncogenesis by inhibiting HIF-alpha prolyl hydroxylase. Cancer Cell 2005; 7: 77–85.
48. Guzy RD, Hoyos B, Robin E, et al. Mitochondrial complex III is required for hypoxia-induced ROS production and cellular oxygen sensing. Cell Metab 2005; 1: 401–408.

49. Mansfield KD, Guzy RD, Pan Y, et al. Mitochondrial dysfunction resulting from loss of cytochrome c impairs cellular oxygen sensing and hypoxic HIF-alpha activation. Cell Metab 2005; 1: 393–399.
50. Brunelle JK, Bell EL, Quesada NM, et al. Oxygen sensing requires mitochondrial ROS but not oxidative phosphorylation. Cell Metab 2005; 1: 409–414.
51. Moeller BJ, Cao Y, Li CY, Dewhirst MW. Radiation activates HIF-1 to regulate vascular radiosensitivity in tumors: role of reoxygenation, free radicals, and stress granules. Cancer Cell 2004; 5: 429–441.
52. Gao P, Zhang H, Dinavahi R, et al. HIF-dependent antitumorigenic effect of antioxidants in vivo. Cancer Cell 2007; 12: 230–238.
53. Hagen T, Taylor CT, Lam F, Moncada S. Redistribution of intracellular oxygen in hypoxia by nitric oxide: effect on HIF1alpha. Science 2003; 302: 1975–1978.
54. Katavetin P, Miyata T, Inagi R, et al. High glucose blunts vascular endothelial growth factor response to hypoxia via the oxidative stress-regulated hypoxia-inducible factor/hypoxia-responsible element pathway. J Am Soc Nephrol 2006; 17: 1405–1413.
55. Rosenberger C, Khamaisi M, Abassi Z, et al. Adaptation to hypoxia in the diabetic rat kidney. Kidney Int 2008; 73: 34–42.

Chapter 25
RNA Interference and the Regulation of Renal Gene Expression in Hypoxia

Carsten C. Scholz, Colin R. Lenihan, Cormac T. Taylor, and Ulrike Bruning

Abstract Hypoxia is a common occurrence both in health and disease, and it contributes to both physiologic and pathophysiologic processes through the regulation of gene expression. At the cellular level, highly conserved signaling pathways that activate transcription factors, including (but not limited to) the hypoxia-inducible factor (HIF) and nuclear factor κB (NFκB) demonstrate hypoxic sensitivity and modulate gene expression in response to changes in microenvironmental oxygen concentrations. Recent evidence has revealed that posttranscriptional mechanisms based on RNA interference are also critically important in the regulation of global gene expression. In this chapter, we will discuss the potential importance of RNA interference in the regulation of gene expression in response to hypoxia with a particular emphasis on renal disease.

Keywords HIF · NFκB · miRNA · siRNA · shRNA

1 Clinical Perspective: Hypoxia and Renal Disease

Tissue hypoxia occurs when the metabolic demand for oxygen exceeds the vascular supply. Hypoxia is implicated in the underlying pathology of multiple renal diseases including ischemic acute kidney injury (AKI) and chronic kidney disease (CKD). It is surprising that an organ receiving 20% of cardiac output should be vulnerable to hypoxic injury, however the process of renal counter-current oxygen exchange results in a decline in the partial pressure of oxygen (pO_2) from

C.T. Taylor (✉)
Conway Institute of Biomolecular and Biomedical Research, School of Medicine and Medical Science, College of Life Sciences, University College Dublin, Belfield, Dublin 4, Ireland
e-mail: cormac.taylor@ucd.ie

T. Miyata et al. (eds.), *Studies on Renal Disorders*, Oxidative Stress in Applied Basic Research and Clinical Practice, DOI 10.1007/978-1-60761-857-7_25,

the well-oxygenated cortex to the borderline hypoxic medulla. Because of their high metabolic activity, the S3 segment of the proximal tubule and the medullary thick ascending limb are particularly prone to hypoxic injury following reductions in renal perfusion [1].

AKI is a syndrome defined by an abrupt decline in glomerular filtration rate (GFR), with consequent retention of nitrogenous waste products and loss of homeostasis. AKI occurs in up to 7% of all hospitalized patients [2] and is a major cause of morbidity and mortality. The mortality rate in AKI patients associated with multiorgan failure is greater than 50% [3]. Management of AKI is expensive and resource intensive when compared to other common medical conditions [4]. Incomplete renal recovery may necessitate long-term renal replacement therapy or result in CKD [5, 6]. The current management of AKI is limited to supportive measures, including dialysis.

Acute tubular necrosis (ATN) is the most common form of intrinsic AKI and is most frequently caused by a renal hypoxic/ischemic insult [6]. Three main factors contribute to the reduction in GFR associated with ATN: hypoxic tubular injury, vascular dysfunction, and intrarenal inflammation. Hypoxic tubular injury gives rise to intratubular obstruction, causing back-leak of glomerular filtrate and renal insufficiency. Profound vasoconstriction directly lowers GFR and exacerbates hypoxic injury. An intense inflammatory reaction within the renal parenchyma is also associated with ischemic AKI [7].

Chronic kidney disease is a major public health problem with a rising rate of incidence [8]. CKD management, particularly dialysis and renal transplantation, is extremely expensive [9]. End-stage renal disease is characterized by inflammation, capillary rarefaction, glomerulosclerosis, and tubulointerstitial fibrosis [10]. The chronic hypoxia hypothesis proposes that hypoxia mediates the progression of CKD to end-stage renal failure [11]. This hypothesis suggests that an initial disease specific insult results in postglomerular peritubular capillary injury, which in turn leads to hypoxia of surrounding tissue. Hypoxia-induced renal fibrosis accelerates capillary destruction and impairs tissue oxygen diffusion, creating a cycle of fibrosis and hypoxia that is independent of the initiating disease aetiology. Multiple experimental studies confirm the presence of hypoxia in animal and human CKD [12–16]. The case for tubulointerstitial hypoxia as a driver of CKD progression is furthered by evidence that it precedes the development of overt histological injury [17]. Hypoxia induces many of the key pathophysiological events that cause renal fibrogenesis including epithelial to mesenchymal transition (EMT), fibroblast activation, collagen matrix deposition, and inflammation [18–20].

Hypoxia induces an extensive transcriptional response with a change in expression of a large number of genes involved in a range of processes, including cell survival and adaptation [21, 22]. A number of transcriptional regulators have been shown to demonstrate responsiveness to hypoxia [22]. Cellular adaptation to hypoxia is largely mediated by a transcription factor known as hypoxia-inducible factor (HIF), which is the best understood of the hypoxia-responsive transcriptional regulators. The regulatory subunits, HIF-1α or HIF-2α, are degraded in normoxia [23] in a manner dependent upon a family of oxygen-sensitive HIF hydroxylases, including three prolyl hydroxylases, PHD1, PHD2, and PHD3, and one asparagine hydroxylase, the factor inhibiting HIF (FIH) [24]. In hypoxia, HIF-hydroxylase activity is inhibited, leading to accumulation and nuclear localization of HIF-α, dimerization with HIF-β subunits and activation of HIF-dependent gene expression. Active HIF binds to hypoxia response elements (HREs) on target genes, causing increased expression of genes involved in glycolysis, angiogenesis, and vasodilatation [25].

HIF-1 accumulates in renal tubular, interstitial, and vascular endothelial cells exposed to hypoxia [26]. In experimental AKI, there is a positive correlation between HIF-1 staining and cell survival, suggesting a protective role for HIF-1 [26]. In addition some HIF-1-induced gene products (erythropoietin [EPO] and heme oxygenase-1 [HO-1]) have been shown to ameliorate ischemic renal injury [27, 28]. HIF-1 is also likely to play a role in ischemic preconditioning. Indeed in AKI models, preactivation of the HIF pathways, using either hypoxia or pharmacological hydroxylase inhibition, results in the attenuation of renal injury in response to subsequent ischemic insult [29]. Therefore, therapies that promote HIF-induced hypoxic adaptation are a potentially attractive prophylaxis for AKI. In contrast, however, in the setting of the sustained hypoxia associated with CKD the activation of HIF may contribute to disease progression by inducing EMT and the transcription of pro-fibrotic gene products [30].

The inflammatory response in AKI is maladaptive and worsens renal injury. In animal models, renal damage is attenuated by various interventions that interrupt the inflammatory process, including depletion/blockade of various leukocyte subtypes [31], complement depletion [32], the addition of anti-inflammatory cytokines [33], and treatment with monoclonal antibody against the pro-inflammatory cytokine interleukin-6 (IL-6) [34]. Additionally, in patients with AKI, increased plasma cytokine levels are associated with higher mortality [35]. Nuclear factor κB (NFκB) is central to the function of the human immune system and regulates the expression of cytokines, growth factors, and immune-effector enzymes in response to cellular signals [36]. NFκB is rapidly activated in experimental models of AKI [37]. Experimental reduction of NFκB activity ameliorates functional and histological renal injury. This is demonstrated indirectly via the experimental silencing of Toll-like receptor-2 (TLR-2) and directly with reduction of NFκB activity using decoy oligonucleotide treatment [38, 39]. Hypoxia activates NFκB in cultured cells, at least in part through decreased activity of PHD1 and PHD2 [40].

In summary, hypoxia is an important mediator of injury in both acute and CKD. Transcriptional regulators of the response to hypoxia include HIF and NFκB, which, depending on the context, may be adaptive or maladaptive in nature. A further mechanism of regulating gene expression that has been recently recognized is through RNA interference (RNAi). This will be discussed in the next section.

2 RNA Interference

In 1990 Napoli et al. introduced a chimeric gene for chalcone synthase (CHS) in petunia to overexpress the enzyme in an attempt to intensify the violet color of the flowers [41]. However, counter to expectations, the flowers color was reduced or absent (white flowers occurred). This was associated with a reduced abundance of the chalcone synthase mRNA. In fact, the expression of both the introduced and the endogenous gene was suppressed in a phenomenon the authors termed "co-suppression." The mechanism, however, remained unclear. Two years later, in 1992, Romano and Macino demonstrated a transient inactivation of gene expression in *Neurospora crassa* (a type of red bread mold of the phylum *Ascomycota*) by transformation with homologous DNA sequences in a response that the authors termed "quelling" [41a]. The mechanism behind these findings remained unknown until 1998, when Fire et al. injected double-stranded RNA (dsRNA) into the nematode *Caenorhabditis elegans* and were able to show an efficient and sequence-dependent inhibition of the gene expression caused by the introduced dsRNAs [42]. This landmark study established the concept of RNAi. In 2006 Andrew Fire and Craig Mello were awarded the Nobel Prize of Physiology and Medicine for their discovery.

RNAi is an evolutionarily conserved process that controls sequence-dependent silencing of genes via small dsRNAs. These RNAs can inhibit the gene activity by posttranscriptional gene silencing (PTGS) [43] and transcriptional gene silencing (TGS) [44, 45]. Small silencing RNAs can be separated into different classes including microRNAs (miRNAs) and small interfering RNAs (siRNAs), which will be discussed in more detail in the following sections.

3 The Biology of Micro RNA

Micro RNAs (miRNAs) are approximately 22 nucleotides in length and have recently been recognized to play an important role in the regulation of protein expression at the PTGS level. The first miRNA *lin-4* was discovered in *C. elegans* in 1993 and was found to be complementary to the 3′ untranslated region (UTR) of the *lin-14* mRNA [46]. The translational repression of the *lin-14* mRNA by *lin-4* was demonstrated in the same year [47] and subsequently demonstrated to be important in temporal regulation of larval development. These studies represented the first functional characterization of miRNAs.

Currently, it is thought that a third of all genes in mammals may be regulated by miRNAs [48, 49] and hundreds of miRNAs have been identified in different species [50, 51]. The structure of miRNAs includes both complementary regions and characteristic mismatches, resulting in bulges and stem loops. Genes for miRNAs can be included in either coding regions or introns of their pre-mRNA host genes. In addition, there are miRNA genes under the control of their own promoter region, and several miRNA genes form characteristic clusters on chromosomes [52]. Primary miRNA transcripts (pri-miRNAs) have a length of up to 2,000 nucleotides and are predominantly transcribed by RNA polymerase II with a 5′ cap and 3′ polyadenylation [53]. In the nucleus the ribonuclease type-III enzyme Drosha associates with the dsRNA binding protein DGCR8 to cleave pri-miRNA, liberating the 60–70 nucleotide stem loop intermediate, known as the miRNA precursor (pre-miRNA). The pre-miRNAs, which feature characteristic 5′ phosphate and a two nucleotide 3′ overhang, are transported into the cytoplasm by the Ran GTP/GDP shuttle system via the transport factor Exportin 5. The maturation of pre-miRNAs in the cytoplasm is completed by another RNase type-III enzyme called Dicer, which is complexed with TAR RNA binding protein (TRBP) and results in double-stranded mature miRNA after cleavage. The mature miRNA duplex enters the multi-protein complex, which contains one of several Argonaute proteins (AGO1–4). Argonaute 2 (AGO2) was identified to have endonuclease activity, whereas the actual function of other Argonaute proteins remains unclear [54, 55]. One strand of the miRNA duplex binds to target mRNA in the RISC, while the other strand is degraded. The miRNA nucleotides 2 to 8 are known as the seed region and are responsible for sequence specific mRNA target binding [56]. The presence of multiple conserved miRNA binding sites on the 3′UTR of a given mRNA strongly suggests that its parent gene is a bona fide miRNA target [57]. In animals, miRNAs generally have imperfect complementarity to their target sequences. Bulges or mismatches in the central region of the miRNA–mRNA duplex prevent AGO2-mediated endonucleolytic cleavage of mRNA. Instead, miRNAs inhibit protein expression by sequestering mRNA in the RISC, halting the translation machinery [58]. mRNA degradation by partially complementary miRNAs in animals may additionally occur through the removal of the poly(A) tail by the deadenylase Ccr/Not [59, 60]. In plants but rarely in animals, miRNAs bind with total complementarity to their target sequence, leading to AGO2-mediated endonucleolytic cleavage of the miRNA:mRNA duplex between the nucleotides pairing the miRNA residues 10 and 11 [61]. Recent studies suggest greater complexity in the regulatory action of miRNAs, raising the possibility that more than one mechanism may be involved [62]. At least one group has shown that miRNA can upregulate translation [63]. Specialized protein complexes in the cytoplasm, known as the P-bodies, were identified as the place of mRNA decay and translational repression.

The regulatory role of miRNA covers all spheres of physiological processes, including tissue and organ development, proliferation, death, and metabolism. Furthermore, altered expression of miRNAs is often associated with pathological features such as tumorigenic transformation or chronic inflammation [64–69]. Each miRNA has a large number of gene targets, so it is not surprising that miRNA activity leads to complex posttranscriptional fine tuning of protein expression. In mammals the average number of predicted conserved target genes per miRNA is approximately 200 [70]. Bioinformatic algorithms such as MiRanda, TargetScan, and PicTar are useful for individual miRNA target prediction [57, 70, 71], however, few predicted miRNA targets have been experimentally validated.

4 Hypoxia and miRNAs

As well as being a physiologic stimulus, hypoxia is a microenvironmental feature of neoplastic and chronically inflamed tissues. Recent studies have investigated the impact of hypoxia upon miRNA expression. A number of hypoxia-regulated miRNAs have been recently identified [72]. A subgroup of these miRNAs (including miR-26, miR-107, and miR-210) have been predicted to play a role in cell survival under hypoxic conditions. miR-210 is generally upregulated in hypoxia in a manner that is not cell-type dependent. In endothelial cells, the receptor tyrosine kinase ligand ephrin A3 was validated as a bona fide miR-210 target, and its inhibition was shown to be necessary for stimulation of tubulogenesis and cell chemotaxis in hypoxia [73]. Another target of miR-210 is the e2f transcription factor 3 (e2f3), which plays a role in cell cycle regulation in low-oxygen environment [74]. Gene copy deletion of miR-210 is associated with ovarian cancer [75] and breast cancer [76], whereas other studies have shown overexpression of miR-210 in cancer tissues [72]. Further functional aspects of hypoxia regulated miRNAs are in the early stages of investigation, and several miRNAs have been shown to regulate expression of the hypoxia-dependent protein vascular endothelial growth factor (VEGF) under hypoxic conditions in a nasopharyngeal carcinoma cell line [77]. HIF-1α was found to be critical in the hypoxic induction of miR-210 and miR-373, and both miRNAs seem to play a role in DNA repair regulation [78]. In one study HIF-1α was investigated as a target of the miR-17-92 cluster in lung cancer cells [79]. One recent study has demonstrated a high degree of complexity in mRNA regulation by miRNAs in hypoxia [80]. In this approach the pattern of hypoxia-regulated miRNA expression did not match the predicted expression pattern of target genes. These results raise the possibility of combinatorial actions of several miRNAs in the regulation of specific target genes. As outlined above, PHDs are inactive under hypoxic conditions, leading to an accumulation and nuclear translocation of HIF-1α [81]. A key question that remains as yet unanswered is the role of miRNAs in the expression of PHDs and HIF under conditions of normoxia and hypoxia (Fig. 1).

5 Inflammation and miRNAs in the Immune System

Several miRNAs were found as key molecules in immune cell development and hematopoiesis, including miR-155, miR-150, miR-223, the miR-17-92 cluster, and the miR-146 family [82]. miRNA-155 has been shown to be involved in the inflammatory response of macrophages, and it appears that the c-Jun N-terminal kinases (JNK) pathway is involved in this [83]. In the bone marrow miR-155 induction after lipopolysaccharide (LPS) treatment leads to an expansion in hematopoietic stem cell proliferation, indicating a potential role in inflammatory stresses and pathological states such as acute myeloid leukemia (AML) [84]. In an LPS-stimulated human acute monocytic leukemia cell line, miR-146a/b was upregulated

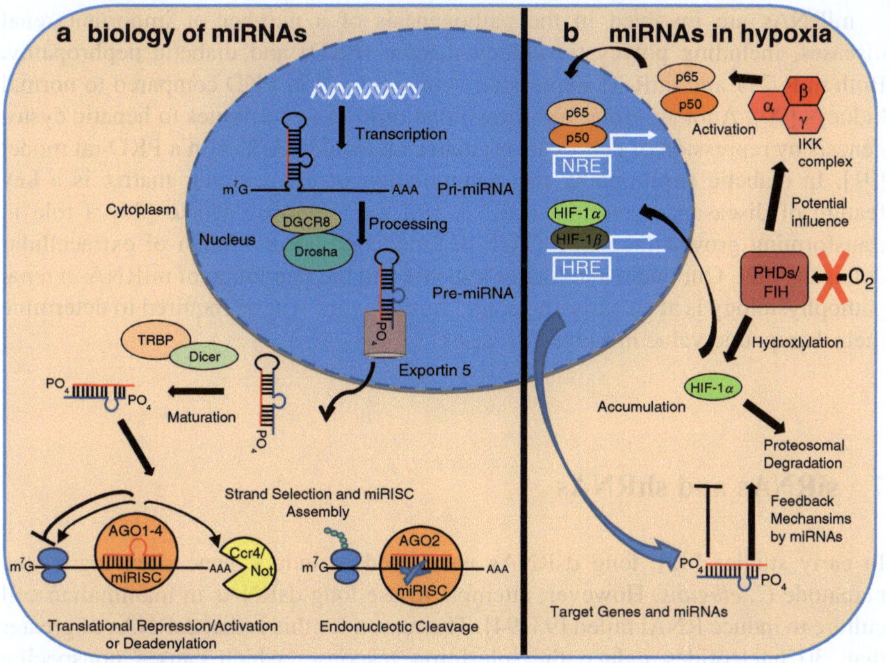

Fig. 1 (**a**) The miRNA pathway which describes the processing, maturation, and mRNA silencing activity of miRNA. (**b**) Potential roles for miRNA in the regulation of the cellular response to hypoxia

in a NFκB-dependent manner and functioned as a negative feedback on NFκB signaling by targeting IL-1 receptor-associated kinase 1 (IRAK1) and tumor necrosis factor (TNF) receptor-associated factor 6 (TRAF6) [85].

6 MicroRNA Expression and Regulation in the Kidney

Microarray and bioinformatic analyses were used in many studies to investigate expression patterns of relevant miRNAs in the kidney. For instance, one group has shown the differential expression of miRNAs in the renal medulla and renal cortex in rats where several reciprocal pairs of miRNAs and their predicted target proteins could be identified [86]. The importance of functional miRNAs in the kidney was demonstrated in studies on conditional (podocyte-specific) Dicer knockout mice. The loss of function of this key molecule in the miRNA biogenesis demonstrated a critical role of miRNAs in maintaining the glomerular filtration barrier [87–89].

miRNAs are involved in the pathogenesis of a number of important renal diseases, including polycystic kidney disease (PKD) and diabetic nephropathy. Both miR-217 and miR-31 expression is upregulated in PKD compared to normal kidneys [90]. Another group has shown that miR-15a contributes to hepatic cystogenesis by repression of cell-cycle regulator molecule Cdc25A in a PKD rat model [91]. In diabetic nephropathy the accumulation of extracellular matrix is a key feature of disease progression. miRNA-192 was demonstrated to play a role in transforming growth factor-β1 (TGF-β1)-mediated accumulation of extracellular collagens [92]. Our understanding of hypoxia and the regulation of miRNA in renal pathophysiology is at an early stage, and further work will be required to determine their therapeutic value in kidney diseases.

7 siRNAs and shRNAs

In early studies [42], long dsRNAs were used to induce gene silencing in the nematode *C. elegans*. However, attempts to use long dsRNAs in mammalian cell culture to induce RNAi failed [93, 94]. The reason for this is that dsRNAs of greater than 30 nucleotides induce the interferon response, which causes nonspecific destruction of RNA and inhibition of protein synthesis [95, 96]. Three years after the discovery of RNAi Elbashir et al. found a solution to this problem by introducing 21 nucleotide dsRNAs into mammalian cells [95]. These small dsRNAs were able to specifically suppress gene expression in the transfected cells and were termed siRNAs. siRNAs can either be generated within a mammalian cell by Dicer mediated cleavage of long dsRNAs [97, 98] or be introduced into the cell in their biologically active form.

Another important step in developing our understanding of siRNA was the demonstration that siRNAs can be delivered within plasmids or viral vector genomes as gene expression cassettes and can be expressed in the intracellular compartment (reviewed in [99]). These small silencing RNAs are expressed as miRNAs or short hairpins (shRNAs) and subsequently processed into ~21–25 nucleotide RNAs by Drosha and Dicer (miRNAs) or Dicer alone (shRNAs). The advantage of this approach is the long-lasting gene silencing in comparison to transfected siRNAs.

siRNAs and shRNAs are usually designed to be 100% complementary to their targeted mRNA. This mediates PTGS by inducing cleavage and subsequent degradation of its target. siRNAs contain a two-nucleotide overhang at the 3′-end, a phosphate group at the 5’-end, and have an overall length of approximately 21 nucleotides [95, 100], mimicking Dicer derived cleavage products. Each siRNA duplex consists of an “antisense” strand, which contains the nucleotide sequence complementary to its mRNA target and a corresponding “sense” strand. The antisense strand is incorporated into AGO2 and serves as a template to bind

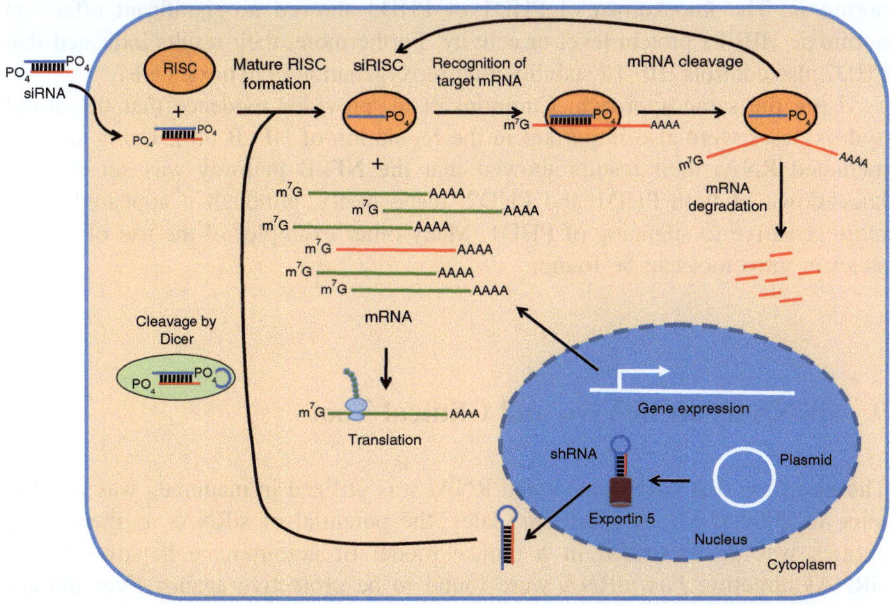

Fig. 2 mRNA processing by exogenously applied siRNA or plasmid-coded shRNA

sequence-specific mRNA targets, while its paired sense strand is broken down. After binding the target mRNA to the antisense strand-protein complex (siRISC), AGO2 catalyzes its cleavage adjacent to guide strand nucleotide 10 and 11 and releases the products [101]. Subsequently, the cleaved mRNA is degraded and the siRISC can go on to process further targets [61] (Fig. 2).

8 siRNA-Mediated RNAi as an In Vitro Tool

Soon after the discovery of RNAi in *C. elegans* [42] and its first successful induction in mammalian cells [95], siRNA-mediated RNAi became a widely used in vitro tool for functional gene analysis in cell culture because of its high efficiency and specificity. Here, we will use two examples pertaining to hypoxia-elicited gene expression to visualize the usefulness of siRNAs as an in vitro tool.

In 2003, three prolyl hydroxylases (PHD1–3) were identified in mammals and were implicated in the regulation of HIF. The level of contribution of each of the PHDs to the HIF regulation was not known. Berra et al. utilized siRNAs against PHD1, PHD2, and PHD3 to demonstrate that PHD2 was the vital oxygen sensor responsible for the low HIF-1α protein levels in normoxia [102]. They showed that siRNA-mediated knockdown of PHD2 led to an increase of HIF-1α protein level and activity in normoxia. The knockdown of PHD1 or PHD3 showed no significant effect on normoxic HIF-1α protein level or activity. Furthermore, their results indicated that PHD2 also controls HIF-1α stability after reoxygenation of hypoxic cells.

Using the same approach, Cummins et al. provided evidence that the prolyl hydroxylases were also important in the regulation of NFκB [40]. Using siRNA-mediated RNAi their results showed that the NFκB pathway was sensitive to knockdown of both PHD1 and PHD2, respectively, although it appeared to be more sensitive to silencing of PHD1. Many other examples of the use of siRNA as an in vitro tool can be found.

9 siRNA as an In Vivo and Clinical Tool

The first time that siRNA-mediated RNAi was utilized in mammals was in adult mice in 2002 [103]. A short time later, the potential of siRNAs as therapeutic entities was demonstrated in a mouse model of autoimmune hepatitis where siRNAs targeting *Fas* mRNA were found to be protective against liver fibrosis [104]. In 2004, phase I of the first clinical trial of siRNA-based human therapy was undertaken in patients with wet age-related macular degeneration.

Another example of the in vivo use of siRNAs is the work of Yuan et al. [105]. It was previously shown that the 12/15-lipooxygenase (12/15-LO) pathway of arachidonate acid metabolism is important in several incidences related to diabetic nephropathy (DN). Therefore, the authors investigated the impact of siRNAs targeting 12/15-LO in a mouse model of streptozotocin-induced type 1 diabetes. For a higher delivery rate of the siRNAs to the kidney in vivo, the oligonucleotides were conjugated with cholesterol, which increases the uptake of oligonucleotides by distinct tissues [106]. This study demonstrated that cholesterol conjugated siRNAs directed against 12/15-LO were renoprotective in diabetic nephropathy [105].

A further example for the use of siRNAs in vivo is the study of Molitoris et al. [107]. Their results show that naked siRNAs are rapidly taken up by the proximal tubule cells (PTCs) of the kidney following intravenous injection in rats. The renal proximal tubules are prone to ischemic and nephrotoxic injury. The rapid uptake of naked siRNAs by PTCs was therefore utilized by the authors to investigate the efficacy of siRNAs directed against p53 (an important protein in apoptosis) on ischemic and cisplatin-induced AKI. The siRNA treatment effectively reduced apoptosis in PTCs by knocking down p53, indicating a potential therapeutic benefit of these siRNAs in ischemic and nephrotoxic kidney injury.

The most advanced RNAi-based therapeutic in clinical trials is an siRNA against the VEGF, Bevasiranib, which is used for the treatment of wet age-related macular degeneration. A more detailed overview on the current RNAi-based therapeutics in clinical trials is given by Castanotto and Rossi [108].

10 Side Effects and Problems with siRNA- or shRNA-Mediated RNAi

Despite great advances with RNAi in vitro and in vivo and several promising clinical trials, there are various reports, which underline the need for further improvements in understanding and utilization of the RNAi mechanism. One of the biggest problems is the so-called "off-target" effect. siRNAs may induce unwanted knockdowns through their interactions with partially compatible mRNA sequences in a manner similar to that described for miRNA:mRNA interaction. Unfortunately, these effects are hard to predict and to eliminate [109, 110]. Furthermore, passenger strands may also mediate RNAi and can "misguide" the RISC [111, 112]. Great efforts are being made to overcome these problems by optimizing the siRNA designs (e.g., with asymmetric interfering RNAs [aiRNAs] or chemical modifications of siRNAs) [96, 101, 113, 114].

Another important issue is the capacity of siRNAs to stimulate an immune response. Initially, this problem seemed to be avoided by using RNAs with a length of ~21 nucleotides, circumventing the recognition of long dsRNAs by cellular sensors [95]. It is now evident that other receptors also detect foreign RNA via several mechanisms. These include the Toll-like-receptors TLR-3, -7, and -8, which detect sequence elements within siRNAs, or the helicase RIG-1 (retinoic acid-inducible gene-1), which recognizes distinct structural features [115–117]. A recent example of siRNA-induced immune stimulation was demonstrated by Kleinman et al. in an investigation of the mechanism of action of siRNAs against VEGF or its receptor VEGFR1 (also called FLT1), both of which are currently in clinical trials for the treatment of blinding choroidal neovascularization (CNV) from wet age-related macular degeneration [118]. The results show that the effect of the siRNAs on CNV is due to nonspecific inhibition of angiogenesis by activation of TLR-3. Active TLR-3 signaling leads to activation of IL-12 and interferon-γ, which then downregulates VEGF [118].

In 2006 Grimm et al. reported that expression of shRNAs in the liver of mice could result in early liver damage, organ failure, and death [119]. Their results indicated that this problem occurred due to oversaturation of the miRNA pathway. The exact mechanism is still under investigation, but the results of Grimm et al. suggested that saturation of Exportin 5 was crucial. Another report demonstrated that high doses of siRNAs could also interfere with the endogenous RNAi machinery [120]. Exogenously introduced siRNAs do not need Exportin 5 to be biologically active [121]. Therefore, these results indicate that other parts of the machinery play an important role in the competition of exogenous with endogenous small silencing RNAs. Overall, these results highlight the importance of delivering the right amount of si/shRNA to targeted cells in order to knockdown the desired mRNA without interfering with the endogenous RNAi machinery.

A further major issue in the use of siRNAs and shRNAs in animal and human studies is the specific and efficient delivery of the nucleic acids. Naked siRNAs are not readily taken up by most of the cells because of their size and negative charge and are rapidly degraded in serum. Furthermore, small RNAs are rapidly excreted via the kidneys or cleared by the liver (for more detailed information's regarding the pharmacokinetics and biodistribution of oligonucleotides as individual molecules see the study by Juliano et al. [122]). The delivery of shRNAs by viral vectors is also problematic, for instance even nonpathogenic viruses are potentially immunogenic. Additionally, a viral vector genome mutation could result in aberrant gene expression or insertional mutagenesis [108].

11 Conclusion

Hypoxia is a key event in a number of renal pathologies, and a major component of this occurs via the regulation of gene expression. While transcription factors including HIF and NFκB (and likely several others) are important in regulating this response, recent evidence has indicated that RNAi through hypoxia-regulated miRNAs also plays a key role in shaping the cellular response to hypoxia. Furthermore, manipulation of this response through the introduction of therapeutic siRNA has heralded a new and exciting therapeutic approach in a range of disorders where hypoxia plays a role in disease development.

References

1. Brezis M, Rosen S. Hypoxia of the renal medulla – its implications for disease. N Engl J Med 1995;332(10):647–55.
2. Nash K, Hafeez A, Hou S. Hospital-acquired renal insufficiency. Am J Kidney Dis 2002;39 (5):930–6.
3. Brivet FG, Kleinknecht DJ, Loirat P, Landais PJ. Acute renal failure in intensive care units – causes, outcome, and prognostic factors of hospital mortality; a prospective, multicenter study. French Study Group on Acute Renal Failure. Crit Care Med 1996;24(2):192–8.
4. Fischer MJ, Brimhall BB, Lezotte DC, Glazner JE, Parikh CR. Uncomplicated acute renal failure and hospital resource utilization: a retrospective multicenter analysis. Am J Kidney Dis 2005;46(6):1049–57.
5. Shaw NJ, Brocklebank JT, Dickinson DF, Wilson N, Walker DR. Long-term outcome for children with acute renal failure following cardiac surgery. Int J Cardiol 1991;31(2):161–5.
6. Star RA. Treatment of acute renal failure. Kidney Int 1998;54(6):1817–31.
7. Devarajan P. Update on mechanisms of ischemic acute kidney injury. J Am Soc Nephrol 2006;17(6):1503–20.
8. K/DOQI clinical practice guidelines for chronic kidney disease: evaluation, classification, and stratification. Am J Kidney Dis 2002;39(2 Suppl 1):S1–266.
9. Foley RN, Collins AJ. End-stage renal disease in the United States: an update from the United States Renal Data System. J Am Soc Nephrol 2007;18(10):2644–8.
10. Eddy AA. Progression in chronic kidney disease. Adv Chronic Kidney Dis 2005;12 (4):353–65.
11. Fine LG, Orphanides C, Norman JT. Progressive renal disease: the chronic hypoxia hypothesis. Kidney Int Suppl 1998;65:S74–8.

12. Palm F, Cederberg J, Hansell P, Liss P, Carlsson PO. Reactive oxygen species cause diabetes-induced decrease in renal oxygen tension. Diabetologia 2003;46(8):1153–60.
13. Rosenberger C, Pratschke J, Rudolph B, Heyman SN, Schindler R, Babel N, et al. Immunohistochemical detection of hypoxia-inducible factor-1alpha in human renal allograft biopsies. J Am Soc Nephrol 2007;18(1):343–51.
14. Bernhardt WM, Wiesener MS, Weidemann A, Schmitt R, Weichert W, Lechler P, et al. Involvement of hypoxia-inducible transcription factors in polycystic kidney disease. Am J Pathol 2007;170(3):830–42.
15. Rosenberger C, Khamaisi M, Abassi Z, Shilo V, Weksler-Zangen S, Goldfarb M, et al. Adaptation to hypoxia in the diabetic rat kidney. Kidney Int 2008;73(1):34–42.
16. Ries M, Basseau F, Tyndal B, Jones R, Deminiere C, Catargi B, et al. Renal diffusion and BOLD MRI in experimental diabetic nephropathy. Blood oxygen level-dependent. J Magn Reson Imaging 2003;17(1):104–13.
17. Matsumoto M, Tanaka T, Yamamoto T, Noiri E, Miyata T, Inagi R, et al. Hypoperfusion of peritubular capillaries induces chronic hypoxia before progression of tubulointerstitial injury in a progressive model of rat glomerulonephritis. J Am Soc Nephrol 2004;15(6):1574–81.
18. Manotham K, Tanaka T, Matsumoto M, Ohse T, Inagi R, Miyata T, et al. Transdifferentiation of cultured tubular cells induced by hypoxia. Kidney Int 2004;65(3):871–80.
19. Norman JT, Clark IM, Garcia PL. Hypoxia promotes fibrogenesis in human renal fibroblasts. Kidney Int 2000;58(6):2351–66.
20. Kong T, Eltzschig HK, Karhausen J, Colgan SP, Shelley CS. Leukocyte adhesion during hypoxia is mediated by HIF-1-dependent induction of beta2 integrin gene expression. Proc Natl Acad Sci USA 2004;101(28):10440–5.
21. Leonard MO, Cottell DC, Godson C, Brady HR, Taylor CT. The role of HIF-1 alpha in transcriptional regulation of the proximal tubular epithelial cell response to hypoxia. J Biol Chem 2003;278(41):40296–304.
22. Cummins EP, Taylor CT. Hypoxia-responsive transcription factors. Pflugers Arch 2005;450 (6):363–71.
23. Bunn HF, Poyton RO. Oxygen sensing and molecular adaptation to hypoxia. Physiol Rev 1996;76(3):839–85.
24. Kaelin WG, Jr., Ratcliffe PJ. Oxygen sensing by metazoans: the central role of the HIF hydroxylase pathway. Mol Cell 2008;30(4):393–402.
25. Huang LE, Pete EA, Schau M, Milligan J, Gu J. Leu-574 of HIF-1alpha is essential for the von Hippel-Lindau (VHL)-mediated degradation pathway. J Biol Chem 2002;277 (44):41750–5.
26. Rosenberger C, Heyman SN, Rosen S, Shina A, Goldfarb M, Griethe W, et al. Up-regulation of HIF in experimental acute renal failure: evidence for a protective transcriptional response to hypoxia. Kidney Int 2005;67(2):531–42.
27. Sharples EJ, Patel N, Brown P, Stewart K, Mota-Philipe H, Sheaff M, et al. Erythropoietin protects the kidney against the injury and dysfunction caused by ischemia-reperfusion. J Am Soc Nephrol 2004;15(8):2115–24.
28. Blydt-Hansen TD, Katori M, Lassman C, Ke B, Coito AJ, Iyer S, et al. Gene transfer-induced local heme oxygenase-1 overexpression protects rat kidney transplants from ischemia/ reperfusion injury. J Am Soc Nephrol 2003;14(3):745–54.
29. Bernhardt WM, Campean V, Kany S, Jurgensen JS, Weidemann A, Warnecke C, et al. Preconditional activation of hypoxia-inducible factors ameliorates ischemic acute renal failure. J Am Soc Nephrol 2006;17(7):1970–8.
30. Higgins DF, Kimura K, Iwano M, Haase VH. Hypoxia-inducible factor signaling in the development of tissue fibrosis. Cell Cycle 2008;7(9):1128–32.
31. Friedewald JJ, Rabb H. Inflammatory cells in ischemic acute renal failure. Kidney Int 2004;66(2):486–91.

32. Thurman JM, Ljubanovic D, Edelstein CL, Gilkeson GS, Holers VM. Lack of a functional alternative complement pathway ameliorates ischemic acute renal failure in mice. J Immunol 2003;170(3):1517–23.
33. Deng J, Kohda Y, Chiao H, Wang Y, Hu X, Hewitt SM, et al. Interleukin-10 inhibits ischemic and cisplatin-induced acute renal injury. Kidney Int 2001;60(6):2118–28.
34. Patel NS, Chatterjee PK, Di Paola R, Mazzon E, Britti D, De Sarro A, et al. Endogenous interleukin-6 enhances the renal injury, dysfunction, and inflammation caused by ischemia/reperfusion. J Pharmacol Exp Ther 2005;312(3):1170–8.
35. Simmons EM, Himmelfarb J, Sezer MT, Chertow GM, Mehta RL, Paganini EP, et al. Plasma cytokine levels predict mortality in patients with acute renal failure. Kidney Int 2004;65 (4):1357–65.
36. Karin M, Greten FR. NF-kappaB: linking inflammation and immunity to cancer development and progression. Nat Rev Immunol 2005;5(10):749–59.
37. Gueler F, Rong S, Park JK, Fiebeler A, Menne J, Elger M, et al. Postischemic acute renal failure is reduced by short-term statin treatment in a rat model. J Am Soc Nephrol 2002;13 (9):2288–98.
38. Leemans JC, Stokman G, Claessen N, Rouschop KM, Teske GJ, Kirschning CJ, et al. Renal-associated TLR2 mediates ischemia/reperfusion injury in the kidney. J Clin Invest 2005;115 (10):2894–903.
39. Cao CC, Ding XQ, Ou ZL, Liu CF, Li P, Wang L, et al. In vivo transfection of NF-kappaB decoy oligodeoxynucleotides attenuate renal ischemia/reperfusion injury in rats. Kidney Int 2004;65(3):834–45.
40. Cummins EP, Berra E, Comerford KM, Ginouves A, Fitzgerald KT, Seeballuck F, et al. Prolyl hydroxylase-1 negatively regulates IkappaB kinase-beta, giving insight into hypoxia-induced NFkappaB activity. Proc Natl Acad Sci USA 2006;103(48):18154–9.
41. Napoli C, et al. Introduction of a Chimeric Chalcone Synthase Gene into Petunia Results in Reversible Co-Suppression of Homologous Genes in trans. Plant cell 1990;2(4):279–89.
41a. Romano N, Macino G. Quelling: transient inactivation of gene expression in Neurospora crassa by transformation with homologous sequences. Mol Microbiol 1992;6(22):3343–53.
42. Fire A, Xu S, Montgomery MK, Kostas SA, Driver SE, Mello CC. Potent and specific genetic interference by double-stranded RNA in *Caenorhabditis elegans*. Nature 1998;391 (6669):806–11.
43. Zamore PD, Tuschl T, Sharp PA, Bartel DP. RNAi: double-stranded RNA directs the ATP-dependent cleavage of mRNA at 21 to 23 nucleotide intervals. Cell 2000;101(1):25–33.
44. Matzke MA, Birchler JA. RNAi-mediated pathways in the nucleus. Nat Rev Genet 2005;6 (1):24–35.
45. Wassenegger M. The role of the RNAi machinery in heterochromatin formation. Cell 2005;122(1):13–6.
46. Lee RC, Feinbaum RL, Ambros V. The *C. elegans* heterochronic gene *lin-4* encodes small RNAs with antisense complementarity to *lin-14*. Cell 1993;75(5):843–54.
47. Wightman B, Ha I, Ruvkun G. Posttranscriptional regulation of the heterochronic gene *lin-14* by *lin-4* mediates temporal pattern formation in *C. elegans*. Cell 1993;75(5):855–62.
48. Lewis BP, Burge CB, Bartel DP. Conserved seed pairing, often flanked by adenosines, indicates that thousands of human genes are microRNA targets. Cell 2005;120(1):15–20.
49. Xie X, Lu J, Kulbokas EJ, Golub TR, Mootha V, Lindblad-Toh K, et al. Systematic discovery of regulatory motifs in human promoters and 3′ UTRs by comparison of several mammals. Nature 2005;434(7031):338–45.
50. Lagos-Quintana M, Rauhut R, Lendeckel W, Tuschl T. Identification of novel genes coding for small expressed RNAs. Science 2001;294(5543):853–8.
51. Lagos-Quintana M, Rauhut R, Meyer J, Borkhardt A, Tuschl T. New microRNAs from mouse and human. RNA 2003;9(2):175–9.
52. Bartel DP. MicroRNAs: genomics, biogenesis, mechanism, and function. Cell 2004;116 (2):281–97.

53. Lee Y, Kim M, Han J, Yeom KH, Lee S, Baek SH, et al. MicroRNA genes are transcribed by RNA polymerase II. Embo J 2004;23(20):4051–60.
54. Song JJ, Smith SK, Hannon GJ, Joshua-Tor L. Crystal structure of Argonaute and its implications for RISC slicer activity. Science 2004;305(5689):1434–7.
55. Liu J, Carmell MA, Rivas FV, Marsden CG, Thomson JM, Song JJ, et al. Argonaute2 is the catalytic engine of mammalian RNAi. Science 2004;305(5689):1437–41.
56. Carthew RW, Sontheimer EJ. Origins and Mechanisms of miRNAs and siRNAs. Cell 2009;136(4):642–55.
57. Lewis BP, Shih IH, Jones-Rhoades MW, Bartel DP, Burge CB. Prediction of mammalian microRNA targets. Cell 2003;115(7):787–98.
58. Williams AE. Functional aspects of animal microRNAs. Cell Mol Life Sci 2008;65 (4):545–62.
59. Wu L, Fan J, Belasco JG. MicroRNAs direct rapid deadenylation of mRNA. Proc Natl Acad Sci U S A 2006;103(11):4034–9.
60. Giraldez AJ, Mishima Y, Rihel J, Grocock RJ, Van Dongen S, Inoue K, et al. Zebrafish miR-430 promotes deadenylation and clearance of maternal mRNAs. Science 2006;312 (5770):75–9.
61. Hutvagner G, Zamore PD. A microRNA in a multiple-turnover RNAi enzyme complex. Science 2002;297(5589):2056–60.
62. Wu L, Belasco JG. Let me count the ways: mechanisms of gene regulation by miRNAs and siRNAs. Mol Cell 2008;29(1):1–7.
63. Vasudevan S, Tong Y, Steitz JA. Switching from repression to activation: microRNAs can up-regulate translation. Science 2007;318(5858):1931–4.
64. Wu F, Zikusoka M, Trindade A, Dassopoulos T, Harris ML, Bayless TM, et al. MicroRNAs are differentially expressed in ulcerative colitis and alter expression of macrophage inflammatory peptide-2 alpha. Gastroenterology 2008;135(5):1624–35.e24.
65. Calin GA, Ferracin M, Cimmino A, Di Leva G, Shimizu M, Wojcik SE, et al. A microRNA signature associated with prognosis and progression in chronic lymphocytic leukemia. N Engl J Med 2005;353(17):1793–801.
66. Yanaihara N, Caplen N, Bowman E, Seike M, Kumamoto K, Yi M, et al. Unique microRNA molecular profiles in lung cancer diagnosis and prognosis. Cancer Cell 2006;9(3):189–98.
67. Iorio MV, Ferracin M, Liu CG, Veronese A, Spizzo R, Sabbioni S, et al. MicroRNA gene expression deregulation in human breast cancer. Cancer Res 2005;65(16):7065–70.
68. Lu J, Getz G, Miska EA, Alvarez-Saavedra E, Lamb J, Peck D, et al. MicroRNA expression profiles classify human cancers. Nature 2005;435(7043):834–8.
69. Volinia S, Calin GA, Liu CG, Ambs S, Cimmino A, Petrocca F, et al. A microRNA expression signature of human solid tumors defines cancer gene targets. Proc Natl Acad Sci U S A 2006;103(7):2257–61.
70. Krek A, Grun D, Poy MN, Wolf R, Rosenberg L, Epstein EJ, et al. Combinatorial microRNA target predictions. Nat Genet 2005;37(5):495–500.
71. Griffiths-Jones S, Grocock RJ, van Dongen S, Bateman A, Enright AJ. miRBase: microRNA sequences, targets and gene nomenclature. Nucleic Acids Res 2006;34(Database issue): D140–4.
72. Kulshreshtha R, Davuluri RV, Calin GA, Ivan M. A microRNA component of the hypoxic response. Cell Death Differ 2008;15(4):667–71.
73. Fasanaro P, D'Alessandra Y, Di Stefano V, Melchionna R, Romani S, Pompilio G, et al. MicroRNA-210 modulates endothelial cell response to hypoxia and inhibits the receptor tyrosine kinase ligand Ephrin-A3. J Biol Chem 2008;283(23):15878–83.
74. Ivan M, Harris AL, Martelli F, Kulshreshtha R. Hypoxia response and microRNAs: no longer two separate worlds. J Cell Mol Med 2008;12(5A):1426–31.
75. Giannakakis A, Sandaltzopoulos R, Greshock J, Liang S, Huang J, Hasegawa K, et al. miR-210 links hypoxia with cell cycle regulation and is deleted in human epithelial ovarian cancer. Cancer Biol Ther 2008;7(2):255–64.

76. Camps C, Buffa FM, Colella S, Moore J, Sotiriou C, Sheldon H, et al. hsa-miR-210 is induced by hypoxia and is an independent prognostic factor in breast cancer. Clin Cancer Res 2008;14(5):1340–8.
77. Hua Z, Lv Q, Ye W, Wong CK, Cai G, Gu D, et al. miRNA-directed regulation of VEGF and other angiogenic factors under hypoxia. PLoS One 2006;1:e116.
78. Crosby ME, Kulshreshtha R, Ivan M, Glazer PM. MicroRNA regulation of DNA repair gene expression in hypoxic stress. Cancer Res 2009;69(3):1221–9.
79. Taguchi A, Yanagisawa K, Tanaka M, Cao K, Matsuyama Y, Goto H, et al. Identification of hypoxia-inducible factor-1 alpha as a novel target for miR-17-92 microRNA cluster. Cancer Res 2008;68(14):5540–5.
80. Guimbellot JS, Erickson SW, Mehta T, Wen H, Page GP, Sorscher EJ, et al. Correlation of microRNA levels during hypoxia with predicted target mRNAs through genome-wide microarray analysis. BMC Med Genomics 2009;2:15.
81. Semenza GL. Hydroxylation of HIF-1: oxygen sensing at the molecular level. Physiology (Bethesda) 2004;19:176–82.
82. Baltimore D, Boldin MP, O'Connell RM, Rao DS, Taganov KD. MicroRNAs: new regulators of immune cell development and function. Nat Immunol 2008;9(8):839–45.
83. O'Connell RM, Taganov KD, Boldin MP, Cheng G, Baltimore D. MicroRNA-155 is induced during the macrophage inflammatory response. Proc Natl Acad Sci USA 2007;104 (5):1604–9.
84. O'Connell RM, Rao DS, Chaudhuri AA, Boldin MP, Taganov KD, Nicoll J, et al. Sustained expression of microRNA-155 in hematopoietic stem cells causes a myeloproliferative disorder. J Exp Med 2008;205(3):585–94.
85. Taganov KD, Boldin MP, Chang KJ, Baltimore D. NF-kappaB-dependent induction of microRNA miR-146, an inhibitor targeted to signaling proteins of innate immune responses. Proc Natl Acad Sci USA 2006;103(33):12481–6.
86. Tian Z, Greene AS, Pietrusz JL, Matus IR, Liang M. MicroRNA-target pairs in the rat kidney identified by microRNA microarray, proteomic, and bioinformatic analysis. Genome Res 2008;18(3):404–11.
87. Ho J, Ng KH, Rosen S, Dostal A, Gregory RI, Kreidberg JA. Podocyte-specific loss of functional microRNAs leads to rapid glomerular and tubular injury. J Am Soc Nephrol 2008;19(11):2069–75.
88. Harvey SJ, Jarad G, Cunningham J, Goldberg S, Schermer B, Harfe BD, et al. Podocyte-specific deletion of dicer alters cytoskeletal dynamics and causes glomerular disease. J Am Soc Nephrol 2008;19(11):2150–8.
89. Shi S, Yu L, Chiu C, Sun Y, Chen J, Khitrov G, et al. Podocyte-selective deletion of dicer induces proteinuria and glomerulosclerosis. J Am Soc Nephrol 2008;19(11):2159–69.
90. Pandey P, Brors B, Srivastava PK, Bott A, Boehn SN, Groene HJ, et al. Microarray-based approach identifies microRNAs and their target functional patterns in polycystic kidney disease. BMC Genomics 2008;9:624.
91. Lee SO, Masyuk T, Splinter P, Banales JM, Masyuk A, Stroope A, et al. MicroRNA15a modulates expression of the cell-cycle regulator Cdc25A and affects hepatic cystogenesis in a rat model of polycystic kidney disease. J Clin Invest 2008;118(11):3714–24.
92. Kato M, Zhang J, Wang M, Lanting L, Yuan H, Rossi JJ, et al. MicroRNA-192 in diabetic kidney glomeruli and its function in TGF-beta-induced collagen expression via inhibition of E-box repressors. Proc Natl Acad Sci USA 2007;104(9):3432–7.
93. Caplen NJ, Fleenor J, Fire A, Morgan RA. dsRNA-mediated gene silencing in cultured Drosophila cells: a tissue culture model for the analysis of RNA interference. Gene 2000;252 (1–2):95–105.
94. Ui-Tei K, Zenno S, Miyata Y, Saigo K. Sensitive assay of RNA interference in Drosophila and Chinese hamster cultured cells using firefly luciferase gene as target. FEBS Lett 2000;479(3):79–82.

95. Elbashir SM, Harborth J, Lendeckel W, Yalcin A, Weber K, Tuschl T. Duplexes of 21-nucleotide RNAs mediate RNA interference in cultured mammalian cells. Nature 2001;411(6836):494–8.
96. Pei Y, Tuschl T. On the art of identifying effective and specific siRNAs. Nat Methods 2006;3 (9):670–6.
97. Bernstein E, Caudy AA, Hammond SM, Hannon GJ. Role for a bidentate ribonuclease in the initiation step of RNA interference. Nature 2001;409(6818):363–6.
98. Zhang H, Kolb FA, Brondani V, Billy E, Filipowicz W. Human Dicer preferentially cleaves dsRNAs at their termini without a requirement for ATP. EMBO J 2002;21(21):5875–85.
99. Paddison PJ, Hannon GJ. RNA interference: the new somatic cell genetics? Cancer Cell 2002;2(1):17–23.
100. Meister G, Tuschl T. Mechanisms of gene silencing by double-stranded RNA. Nature 2004;431(7006):343–9.
101. Rana TM. Illuminating the silence: understanding the structure and function of small RNAs. Nat Rev Mol Cell Biol 2007;8(1):23–36.
102. Berra E, Benizri E, Ginouves A, Volmat V, Roux D, Pouyssegur J. HIF prolyl-hydroxylase 2 is the key oxygen sensor setting low steady-state levels of HIF-1alpha in normoxia. EMBO J 2003;22(16):4082–90.
103. McCaffrey AP, Meuse L, Pham TT, Conklin DS, Hannon GJ, Kay MA. RNA interference in adult mice. Nature 2002;418(6893):38–9.
104. Song E, Lee SK, Wang J, Ince N, Ouyang N, Min J, et al. RNA interference targeting Fas protects mice from fulminant hepatitis. Nat Med 2003;9(3):347–51.
105. Yuan H, Lanting L, Xu ZG, Li SL, Swiderski P, Putta S, et al. Effects of cholesterol-tagged small interfering RNAs targeting 12/15-lipoxygenase on parameters of diabetic nephropathy in a mouse model of type 1 diabetes. Am J Physiol Renal Physiol 2008;295(2):F605–17.
106. Cheng K, Ye Z, Guntaka RV, Mahato RI. Enhanced hepatic uptake and bioactivity of type alpha1(I) collagen gene promoter-specific triplex-forming oligonucleotides after conjugation with cholesterol. J Pharmacol Exp Ther 2006;317(2):797–805.
107. Molitoris BA, Dagher PC, Sandoval RM, Campos SB, Ashush H, Fridman E, et al. siRNA targeted to p53 attenuates ischemic and cisplatin-induced acute kidney injury. J Am Soc Nephrol 2009;20(8):1754–64.
108. Castanotto D, Rossi JJ. The promises and pitfalls of RNA-interference-based therapeutics. Nature 2009;457(7228):426–33.
109. Jackson AL, Burchard J, Schelter J, Chau BN, Cleary M, Lim L, et al. Widespread siRNA "off-target" transcript silencing mediated by seed region sequence complementarity. RNA 2006;12(7):1179–87.
110. Jackson AL, Bartz SR, Schelter J, Kobayashi SV, Burchard J, Mao M, et al. Expression profiling reveals off-target gene regulation by RNAi. Nat Biotechnol 2003;21(6):635–7.
111. Elbashir SM, Lendeckel W, Tuschl T. RNA interference is mediated by 21- and 22-nucleotide RNAs. Genes Dev 2001;15(2):188–200.
112. Wei JX, Yang J, Sun JF, Jia LT, Zhang Y, Zhang HZ, et al. Both strands of siRNA have potential to guide posttranscriptional gene silencing in mammalian cells. PLoS One 2009;4 (4):e5382.
113. Behlke MA. Chemical modification of siRNAs for in vivo use. Oligonucleotides 2008;18 (4):305–19.
114. Grimm D. Small silencing RNAs: state-of-the-art. Adv Drug Deliv Rev 2009;61 (9):672–703.
115. Schlee M, Hornung V, Hartmann G. siRNA and isRNA: two edges of one sword. Mol Ther 2006;14(4):463–70.
116. Robbins M, Judge A, MacLachlan I. siRNA and innate immunity. Oligonucleotides 2009;19 (2):89–102.
117. Judge A, MacLachlan I. Overcoming the innate immune response to small interfering RNA. Hum Gene Ther 2008;19(2):111–24.

118. Kleinman ME, Yamada K, Takeda A, Chandrasekaran V, Nozaki M, Baffi JZ, et al. Sequence- and target-independent angiogenesis suppression by siRNA via TLR3. Nature 2008;452(7187):591–7.
119. Grimm D, Streetz KL, Jopling CL, Storm TA, Pandey K, Davis CR, et al. Fatality in mice due to oversaturation of cellular microRNA/short hairpin RNA pathways. Nature 2006;441 (7092):537–41.
120. Castanotto D, Sakurai K, Lingeman R, Li H, Shively L, Aagaard L, et al. Combinatorial delivery of small interfering RNAs reduces RNAi efficacy by selective incorporation into RISC. Nucleic Acids Res 2007;35(15):5154–64.
121. Yi R, Qin Y, Macara IG, Cullen BR. Exportin-5 mediates the nuclear export of pre-microRNAs and short hairpin RNAs. Genes Dev 2003;17(24):3011–6.
122. Juliano R, Bauman J, Kang H, Ming X. Biological barriers to therapy with antisense and siRNA oligonucleotides. Mol Pharm 2009;6(3):686–95.

Part V
Hypoxia Pathology in Renal Disorders

Chapter 26
Cardio-Renal Connection: The Role of Hypoxia and Oxidative Stress

Carsten Willam, Tilmann Ditting, Roland Veelken, and Johannes Jacobi

Abstract In recent years the interference of heart and kidney disease has increasingly found attention in basic science and clinical research. A new classification of cardiorenal syndromes has now been proposed to address acute and chronic heart or kidney disease with its consequences on the respective other organ. However, it also becomes clear that the different clinical cardiorenal syndromes share common pathways. These basically include activation of the renin angiotensin system and of the sympathetic nervous system, but also inflammation, fibrosis, and accelerated atherosclerosis. These factors eventually support generation of oxidative stress and tissue hypoxia in the kidney and heart, culminating in irreversible tissue damage and aggravation of organ ischemia. This chapter summarizes aspects of the cardiorenal connection and oxygenation imbalances.

Keywords Renin · Angiotensin · Cardiorenal syndrome · Chronic kidney disease · Hypoxia · HIF · Acute kidney injury

1 Introduction: The Concept of the Cardiorenal Connection

1.1 The Association of Cardiovascular Disease in Chronic Kidney Disease Patients

Chronic kidney injury is clearly associated with increased cardiovascular morbidity and mortality [1]. In particular diabetes and hypertension are the main contributors of an increasing number of patients with chronic kidney disease (CKD) that ultimately require renal replacement therapy. The enormous economic burden of

C. Willam (✉)
Department of Nephrology and Hypertension,
Friedrich-Alexander-University Erlangen, Krankenhausstrasse 12,
D-91054 Erlangen, Germany
e-mail: carsten.willam@uk-erlangen.de

T. Miyata et al. (eds.), *Studies on Renal Disorders*, Oxidative Stress in Applied Basic Research and Clinical Practice, DOI 10.1007/978-1-60761-857-7_26,

end stage renal disease (ESRD) for health care systems throughout the world is substantial. About 13% of the U.S. population suffers from CKD, with 341,000 on chronic dialysis and 140,000 with kidney transplants [2, 3]. Epidemiologic studies indicate that CKD patients have a markedly higher mortality rate due to cardiovascular events compared to patients with normal kidney function [4, 5]. Cardiovascular events increase inversely with declining glomerular filtration rate (GFR) and account for 45% of mortality of patients on dialysis [5]. This led to a report of the National Kidney Foundation in 1998 emphasizing the risk of cardiovascular disease in CKD patients, with a 10–30 times higher mortality of CKD patients compared to others [6]. If myocardial infarction occurs in CKD patients, their prognosis is poor. A large population-based study demonstrated a 1- and a 5-year mortality rate of 59 and 90%, respectively [7]. On the other hand, progressive renal impairment has clearly been identified as a major risk factor in patients with acute or chronic heart failure. Mortality increases incrementally by approximately 7% for each decline of 10 mL/min in eGFR [8]. These clinical observations led to the conclusion that renal function and cardiovascular morbidity are closely related to each other and that renal function is a major risk factor for cardiovascular death, which eventually led to the new paradigm of the cardiorenal connection.

1.2 The Acute or Severe Cardiorenal Syndrome

Although many aspects of the association of heart and kidney failure were intensively analyzed within the past two decades [9], the term "cardiorenal syndrome" was coined primarily with respect to *acute heart failure* and its impact on renal function. About one fourth of patients who are hospitalized for acute heart failure are at risk to suffer a decline in renal function [10, 11]. So far cardiorenal syndrome has been identified as the clinical condition where (a) heart failure is associated with (b) a decline in renal function and (c) a diuretic resistance, which can lead to a vicious cycle of sodium retention, volume overload, and progressive congestive heart failure [10]. The underlying pathophysiology was attributed to a reduction of effective circulating arterial blood volume and the activation of the neurohormonal axis. Accordingly, Bongartz et al. readopted Guyton's description of the physiological framework of the relationship of cardiac output, extracellular volume control, and blood pressure to describe the severe cardiorenal syndrome (CRS) [12] and related its pathophysiology to inflammation, the renin angiotensin aldosterone system (RAAS), sympathetic nervous system (SNS), and the balance of nitric oxide (NO) and reactive oxygen species (ROS) generation.

1.3 The New Proposed Classification for CRS

Regarding the effects of acute heart failure on kidney function, it became clear that the relationship of heart and kidney is more complex and also includes kidney diseases affecting the cardiovascular system. This led to a recent proposal by a consensus conference under the auspices of the Acute Dialysis Quality Initiative in 2008 to establish a new classification, including five types of CRS [13, 14]. The acute CRS (type 1) includes the classical concept of acute heart failure and the rapidly decreasing kidney function before and during cardiac therapy. Increasing doses of diuretics are needed to resolve peripheral and lung edema, while creatinine rises and diuresis often declines. The common concept to explain CRS1 is that heart failure is followed by decreased renal perfusion, which stimulates renin release, sympathetic tone, and sodium retention [10]. Vasoconstrictive effects of vasopressin and the sympathetic tone hereby support renal ischemia. Ischemia leads to acute energy deprivation in particular of the tubular epithelium and acute tubular necrosis, which leads to acute kidney injury (AKI) [10]. This concept primarily includes patients with severe heart failure such as patients in cardiogenic shock who develop AKI in about 70% of cases [15]. In cardiogenic shock an acute decline in blood and oxygen delivery usually occurs and can lead to hypoxia and ischemia of the kidney.

The chronic CRS (type 2) is characterized by chronic heart dysfunction and progressive chronic kidney disease. In theory, a decrease of cardiac output leads to critical organ supply of blood and oxygen, which compromises kidney oxygenation. Intriguingly, the overall correlation of cardiac output and kidney function in chronic CRS is rather poor. Thus, analysis of cardiac function parameters in the ESCAPE study, which compared clinical evaluation and use of the pulmonary artery catheter to diagnose and treat heart failure [16], revealed that only the right arterial pressure correlated – albeit weakly – to increases in creatinine. There was no correlation to the pulmonary capillary wedge pressure (PCWP), the cardiac index, or the vascular systemic resistance, which could have indicated a reduced blood supply to the kidney [17]. Accordingly, Damman et al. analyzed 2,557 patients undergoing right heart catheterization and found a significant association of increased central venous pressure (and thus elevated venous pressure on the kidney) and glomerular filtration rate [18]. The conclusion would be that the venous congestion is potentially a central pathophysiological event for CRS. Hereby higher venous pressure reduces effective arterial perfusion pressure and thus worsens oxygen supply. Activation of the RAAS and SNS system in the chronic CRS are essential mediators that maintain renal perfusion; on the other hand, angiotensin II (Ang II) and the SNS induce oxidative stress and contribute to a progressive inflammatory and pro-fibrotic state, which eventually injures heart and kidney tissues.

In contrast to the acute and chronic CRS, where the heart leads to kidney dysfunction, in types 3 and 4 CRS kidney injury affects heart function. In this context, acute renocardiac CRS (type 3) is characterized by acute kidney injury, which leads to acute cardiac malfunction due to volume overload, arrhythmia, acid base disorders, or uremic pericarditis. AKI has been recently redefined according to the Acute Kidney Injury Network consensus criteria 1–3. These criteria allow clinical evaluation, make studies comparable, and enable cardiovascular and mortality risk stratifications [19]. The chronic renocardiac CRS (type 4) is characterized by cardiovascular alterations that follow CKD. This includes accelerated atherosclerosis, which is common in patients with CKD, hypertension, and renal anemia, possibly leading to left ventricular hypertrophy and cardiac microangiopathy.

Finally, the secondary CRS (type 5) includes clinical situation where a disease induces renal and cardiac dysfunction. In acute disease the septic AKI is probably the most important secondary CRS. The common concept of septic AKI is that inflammatory mediators lead to acute systemic vasodilatation. Baroreceptors, Ang II, and the SNS are activated to counterbalance decreases in vascular organ blood supply. When renal perfusion drops below the autoregulatory threshold, endogenous vasoconstrictors like angiotensin and the SNS increase the afferent arteriolar resistance, which aggravates acute renal ischemia and leads to acute tubular necrosis [20]. Ischemic tubular injury predominantly occurs in the outer medulla, where the S3 segment of the proximal tubule and the thick ascending limb are the prevailing cellular structures. A further characteristic is vascular congestion in the deep medullary segments, which has been associated with an endothelial damage and disturbed vascular blood transport, potentially due to vasoconstrictive effects in the renal vasculature [21]. However, this concept has been challenged, since several investigations failed to observe a decline in renal blood flow (RBF) in septic AKI [22, 23]. In a pig model of hyperdynamic sepsis, Ravikant and Lucas [24] instead found an increase in RBF, and Brenner, who used a thermodilution catheter in critically ill patients, measured unaltered RBF [25], which has been supported by newer findings by Bellomo and coworkers [26–28]

2 Pathophysiological Paradigms in CRS

A major criticism of the CRS nomenclature is that, albeit the sequelae of clinical events allows classification, the underlying pathophysiological processes essentially overlap [29]. Indeed, common paradigms for the chronic kidney–heart and heart–kidney connection could be delineated, which includes the accelerated atherosclerosis, the RAAS, and SNS activation and an inflammatory state.

2.1 RAAS Activation

Oxidative stress is a key regulator of accelerated atherosclerosis, development of fibrosis, and organ dysfunction. Coronary calcification, microangiopathy, and left ventricular hypertrophy result in reduced cardiac output. The baroreceptors in the aortic arch and coronary sinus sense a decreased cardiac output and attenuate the tonic inhibition of the afferent parasympathetic signals to the CNS. This activates the sympathetic tone and supports vasoconstriction and sodium retention and promotes oxidative stress on a cellular level (discussed below). Furthermore, the reduced cardiac output and the increasing venous congestion, which reduce effective perfusion pressure, lead to hypoperfusion of the kidney with activation of counter-regulatory mechanisms, mainly the release of renin and Ang II. Ang II leads to arteriolar constriction in the glomeruli, leading to a decrease in renal blood flow. Acute or chronic infusions of Ang II [30–32] increase vascular resistance and impair renal blood and oxygen supply [33]. Video microscopy studies have impressively shown that infusion of Ang II leads to a rapid decline of renal blood flow by 30–40% due to an increased vascular resistance [34]. Also in blood oxygen level-dependent magnetic resonance imaging, Ang II infusion was associated with a decreased renal oxygenation in humans [35]. Conversely, angiotensin receptor blockers improve tissue oxygenation [32]. In an experimental setting, increased sodium transport after Ang II administration enhances renal oxygen consumption and is associated with increased ROS production [36]. Sodium retention and fluid overload can vice versa aggravate cardiac congestion and worsen heart failure [37]. This can lead to a vicious cycle resulting in tissue hypoxia and organ failure [38]. Ang II has been shown to induce reactive oxygen species that interfere with protein synthesis, cell hypertrophy, inflammation, and cellular injury [39–43]. Ang II-induced release of oxygen free radicals can directly contribute to reduced renal oxygenation, as evidenced by direct measurements using intrarenal oxygen probes. The application of antioxidant scavengers was protective under these conditions [31].

2.2 Sympathetic Nervous System

Numerous studies have shown that CKD as well as chronic heart failure (CHF) are characterized by an increased activity of the SNS. Increased SNS activity mediates high catecholamine-levels and -spillover and a high sympathetic nerve activity (SNA), which both have been shown to be clearly associated with increased mortality and morbidity [44–47].

The basal sympathetic tone is generated by premotor neurons in the rostral ventrolateral medulla oblongata (RVLM), which receives input from a number of hypothalamic areas, the caudal ventrolateral medulla (CVLM), the caudal pressure area (CPA), as well as from the nucleus tractus solitarii (NTS), which receives input from numerous afferent pathways controlling the activity of the SNS [48]. These include vascular and cardiac mechano- and chemoreceptors and renal and further visceral afferent peptidergic nerve fibers. The term "peptidergic" refers to the fact that in addition to electrical signal transmission the afferents are able to release different neurokinins at their terminals, such as calcitonin gene related peptide (CGRP), substance P, and others, which are involved in the regulation of local perfusion and immune response. Release and afferent trafficking is mediated mainly via transient receptor potential vallinoid type 1 (TRPV1) channels.

Cardiac sympathetic efferents act positively chronotropic, inotropic, and dromotropic, while parasympathetic efferents act in the opposite way. Interestingly, the kidney only has sympathetic innervation, no parasympathetic components have been described. Efferent renal sympathetic nerve activity (RSNA) is involved in renin release from the juxtaglomerular granular cells via alpha1- and beta1-adrenergic receptor mechanisms [48, 49], thus activating the RAAS. Furthermore, RSNA is involved in tubular sodium and water reabsorption (salt retention) [50] and renal hemodynamics (cortical vasoconstriction). These effects are dependent on characteristic renal nerve discharge patterns and contribute to the long-term regulation of blood pressure and volume homeostasis [48, 51–53]. These effects also lead to increased oxidative stress, ROS formation, volume overload, hypertension, cardiac hypertrophy, and pro-inflammatory signaling due to noradrenaline-mediated cytokine production [12]. Several mechanisms have been identified that are involved in the increased SNS activity in different types of the cardiorenal syndrome. Ang II is one of the most important mediators in the cardiorenal interplay. On the one hand, RSNA can activate the RAAS and thus increases Ang II levels. On the other hand, Ang II leads to sympathoexcitation via peripheral and central AT1-receptor mechanisms. Interestingly, at least in the central nervous system, high levels of Ang II induce upregulation of AT1 receptors, which is part of a positive feedback mechanism [45].

A further important aspect of reno-cardial sympathoexcitation is related to NO-dependent mechanisms. CKD is characterized by decreased NO availability, due to decreased NO precursors, decreased NOS activity, and accumulation of natural nitric oxide synthase (NOS) inhibitors such as asymmetric dimethyl arginine (ADMA). Probably more important are the central effects of NO deficiency [44]. Neuronal NOS seems to be critically involved in transduction pathways that tonically inhibit sympathetic outflow from the brainstem [54]. Sympathetic outflow is strictly controlled by reflex mechanisms whose afferent pathways originate from the great vessels (arterial baroreceptor reflex [ABR]) the lungs and the heart [46]. These reflexes mainly act as inhibitory feedback loops: for example, increases in blood pressure or cardiac filling pressure lead to sympathoinhibition, which might be drastically impaired in various disease states like CKD and CHF [55]. A further inhibitory chemosensitive reflex is the Bezold-Jarisch reflex mediated by the 5HT3 system and many others. However, the polymodality of some afferent fibers [56] might lead to a paradoxical SNS disinhibition. For example, a subthreshold 5-hydroxytryptamine activation (serotonin) has been shown to desensitize cardiac mechanoreceptors [57, 58] in rats. Furthermore, there are sympathoexcitatory reflex mechanisms (cardiac afferent sympathetic reflex), which might be activated by some inflammatory mediators (e.g., bradykinin, ROS, and others) that play a role in the setting of acute cardiac ischemia [58, 59].

There is strong evidence in humans [60, 61] and animals [62] that the diseased kidney conveys some sympathoexcitatory afferent signals to the brainstem. However, neither the signal nor the exact role of the peptidergic renal afferent nerve fibers that innervate all functionally relevant structures in the kidney together with sympathetic efferents [63, 64] are as yet clearly determined. Renal ischemia might be one of the factors contributing to increased SNS activity; however, this issue is still under investigation.

2.3 Uremic Calcification and Coronary Heart Disease

Higher cardiovascular morbidity and mortality in patients with renal impairment are related to excess cardiovascular complications, such as left ventricular hypertrophy, cardiac microangiopathy, and most importantly accelerated atherosclerosis[65].

Lindner et al.[66] have shown that 60% of a cohort of patients undergoing dialysis for 6.5 years died, and cardiovascular events as a consequence of severe atherosclerosis were the most common causes of death. Overt atherosclerosis usually becomes evident at a GFR of 50–60 mL/min and below [5, 67] and is often not symptomatic until complications occur [68]. Atherosclerosis is not only more frequent in patients with CKD, but the manifestation of vascular changes is also different. One of the characteristic findings in patients with kidney disease is excessive vascular calcification. Possible primary reasons for this phenomenon are (a) mineral and bone disorder due to secondary hyperparathyroidism, (b) systemic inflammation and increased C-reactive protein (CRP) levels, (c) hyperhomocysteinemia, (d) increased oxidative stress with the accumulation of endogenous inhibitors of NOS such as ADMA, (e) renal anemia due to erythropoietin (EPO) deficiency, and (f) yet poorly defined disturbances in endogenous activators and inhibitors of calcification. Importantly, in uremic patients the secondary hyperparathyroidism, which develops due to a reduced capacity to hydroxylate the vitamin D precursor, together with elevated phosphate levels markedly contributes to the pro-inflammatory and atherosclerotic phenotype. Although many CKD patients are diabetic, hypertensive, and less frequently physically active, they have some protective factors (less obesity, lower cholesterol) that are considered to be favorable in the general population. Nonetheless, these so-called protective factors do *not* translate into a better outcome in CKD patients, a phenomenon that has been termed *reverse epidemiology* [69]. The arterial calcification in CKD differs from the normal atherosclerotic plaque in patients with intact kidney function. Much of the calcification resides in the media of the artery and does not directly lead to narrowing of the vessel lumen [70]. Calcification scores obtained in computed tomography of CKD patients correlate with the severity of cardiovascular disease and indicate a poor prognosis [71]. Accelerated atherosclerosis results in vessels calcification with reduced vascular compliance due to an increased stiffness of large arteries, a reduced pulse wave velocity, and an increased cardiac afterload [65]. Together these factors aggravate coronary heart disease and left ventricular remodeling. Vascular insufficiency in turn leads to tissue ischemia and induction of hypoxic gene regulation. This includes activation of neoangiogenesis essentially by the HIF pathway (as discussed below) in order to counteract chronic oxygen deficiency. However, if lack of oxygen prevails, heart and kidney tissues undergo a fibrotic state with ultimate loss of organ function.

2.4 Inflammation

Chronic kidney disease has been shown to be associated with a chronic inflammatory state [72]. In CKD patients elevated levels of proinflammatory cytokines like tumor necrosis factor α (TNF-α) or interleukin (IL) 1β are present [73]. A pathophysiological link may be the fact that T-cells from CKD patients demonstrate higher levels of stimulation by increased surface marker expression and cytokine production in vitro [74]. Moreover, increased expression of adhesion molecules on endothelial cells support tissue infiltration of inflammatory cells. Elevated CRP levels correlate with increased mortality (see discussion below).

A few models have been proposed to identify factors that contribute to the increased cytokine, adhesion molecule, and T-cell response as factors of the chronic inflammatory state of CKD. Besides uremic toxins, malnutrition, hypoalbuminemia, and atherosclerotic processes play important roles in the chronic inflammatory state.

Circulating and membrane-bound factors that have been associated with inflammation in CKD include leptin, the peroxisome-proliferators activated receptor, fetuin-A, lipopolysaccharides, beta-glycans, L-fucose, decreased glutathione levels, and others (reviewed in [72]). Kimmel et al. found that reduced albumin levels as a marker of malnutrition are associated with increased levels of IL-13 and TNF-α, but inversely correlate with IL-2 and T-cell counts and function. These effects were accompanied by an increased cardiovascular mortality [75]. In diabetes advanced glycation end products support inflammation and contribute to the generation of diabetic retinopathy, nephropathy, or neuropathy [76]. Inflammation contributes to the generation of oxidative stress (discussed below). Inflammation and oxidative stress are centrally involved in the onset of endothelial dysfunction, which is regarded as an early step of atherosclerotic vascular disease.

The four major pathophysiological events – RAAS and SNS activation, inflammation, and uremic calcification – have diverse effects on cellular integrity and are the key events in the cardiorenal connection. Its numerous aspects are probably still not very well understood in all their complexities. However, oxidative stress and tissue hypoxia turn out to be centrally involved in the generation of organ dysfunction and cellular injury following the cardiorenal pathophysiological sequelae. Next, the major aspects of hypoxia and, later, of oxidative stress in the cardiorenal connection will be further discussed.

3 Hypoxia and Its Role in CRS

3.1 The Kidney and Hypoxia

About 20% of the cardiac output is supplied to the kidneys. Nevertheless, the prevailing oxygen tension in the medullary parts is about as low as 10–15 mmHg even under normal physiological conditions [77]. This paradox can best be explained by the specific vascular architecture of the kidney. In the cortex, high oxygen tensions of about 50–60 mmHg can be found, where blood supplies the glomeruli by the interlobular and arcuatae arteries. The efferent vessels that emerge from the cortical glomeruli then supply the cortical tubules through the peritubular capillary network, before the blood drains into the venous system. In comparison the medullary tubules are supplied with blood by the efferent vessels, which leave the juxtamedullary glomeruli.

The drop in oxygen tension in the renal medulla has been explained by an oxygen countercurrent mechanism of its arterial and venous vessels [78]. Oxygen shunting leads to a comparable low oxygen extraction in the kidney (10–15%) compared to the heart (45%) [79]. Correspondingly, oxygen extraction remains relatively constant over the physiological blood flow range [80, 81]. Using ultrasound flowmetry for blood flow and optical oxygen probes for tissue oxygenation, Leong et al. [81] observed that fractional extraction of oxygen and tissue oxygen tension remain relatively constant despite changes in renal blood flow within a physiological range. Adaption in oxygen extraction can best be explained by increased preglomerular diffusional shunting of oxygen between the arterial and venous circulation, and it has been calculated that only about 10% of the renal blood flow enters the medulla [82]. Thus, autoregulation can maintain renal oxygenation over a certain range in situations of prerenal volume depletion, whereas a drop in systemic oxygen delivery cannot be compensated by reduced oxygen shunting or higher oxygen tissue extraction, thus leading to a rapid decline in oxygenation particularly of the outer medulla [83]. Correspondingly, Leong et al. [81] saw a 60–70% reduction of cortical and medullary partial pressure of oxygen (pO_2) when rabbits were ventilated with 10% oxygen.

The specific vascular architecture of the kidney also implies that injuries of the cortical vasculature can lead to a critical decrease of oxygen and to hypoxia in the deeper regions of the kidney. Matsumoto from Nangaku's group [270] elegantly demonstrated a reduced capillary blood flow by intravital microscopy and tissue hypoxia in a glomerulonephritis model, implying an impaired blood supply to the postglomerular capillary network in the medulla, even before tubulointerstitial fibrosis occurs.

Interestingly, in most experimental settings, the tubules of the inner medulla with low metabolic demands seem to be well adapted to hypoxia with high capacities to produce adenosine triphosphate (ATP) by lactate synthesis [84–86]. In contrast, the outer medulla is very susceptible to ischemic injury. In particular, the S3 segment of the proximal tubule, which is located in the outer medulla, is prone to ischemia and acute tubular necrosis under conditions like severe hypotension, shock, or hemorrhage. This has been explained by a restricted glycolytic capacity (S3 proximal tubule) and by the high oxygen-consuming ATP demand due to the extensive transtubular transport activities (mTAL) [85, 87]. The mTAL, which is also located in the outer medulla, has the highest rate of oxygen consumption compared to other epithelia of the nephron [88] and possesses a high mitochondrial density similar to cardiac myocytes [89].

Taken together, acute tubular injury and necrosis of the S3-proximal tubule and the mTAL in the outer medulla can occur because of (a) the relatively low fraction of renal blood flow entering the medulla (10%), (b) a further postglomerular diffusional oxygen shunting in the vessels running parallel to the U-shaped loop of Henle, and (c) the high metabolic demand in S3-proximal tubules and in the mTAL.

3.2 *Hypoxia and Hypoxic Gene Regulation in the Kidney and the Heart*

Hypoxia has long been regarded as a state of oxygen and energy deprivation. Research over the past two decades has unraveled the complex genetic and physiological regulatory mechanisms, how the organism responds to hypoxia and can switch on protective pathways to adapt and survive. These hypoxic responses are largely mediated by the two hypoxia-inducible transcription factors (HIF-1α and HIF-2β), which are stabilized under low oxygen tensions and can initiate transcription of more than 100 genes, which are involved in glycolysis, angiogenesis, erythropoiesis, and cell cycle regulation or cell survival [90]. In the kidney HIF-1α has been localized to the tubular epithelia, whereas HIF-2α is predominantly located in endothelial, glomerular, and interstitial cells [91–93]. Concordantly, HIF-2α appears to be the HIF variant controlling EPO production in peritubular interstitial cells [94, 95]. In the presence of oxygen, two proline residues within the oxygen-dependent degradation domain of the HIFα chain are hydroxylated [96, 97] by HIF-prolyl hydroxylases (prolyl hydroxylase domain proteins [PHD]), of which three functional different forms have been identified (PHD 1–3) [98]. Hydroxylation of the proline residues leads to recognition by the pVHL ubiquitin ligase and subsequent ubiquitination, which targets HIF-α for proteasomal degradation. When molecular oxygen is not available for hydroxylation, HIF can accumulate in the cell and can bind to specific hypoxia response elements (HREs) in the promoter regions of its target genes (review in [99]).

Kidney ischemia-reperfusion experiments in rats have demonstrated an induction of HIF-1 in tubules. These HIF signals correlated well to the hypoxia marker pimonidazole, which accumulates in regions of low oxygen tension and can be detected by a specific antibody [91, 100]. Also in segmental infarction models of the kidney [101], in a multihit injury model, including contrast media application [102], and in human specimens of kidney transplant rejections [100], HIF was significantly induced. Complexity of the renal HIF system rises with a specific tissue expression pattern of the PHDs. PHD 1–3 are predominantly found in distal tubular cells and podocytes and appear to be present where HIF accumulation occurs [103].

In the heart both HIF variants, HIF-1α and HIF-2α, have been found in experimental or clinical states associated with oxygen deprivation [104]. In myocardial infarction models both isoforms localize to the perinfarcted, still viable regions of the infarct [105]. HIF correlated well with increased PHD-2 and -3 levels in infarct borders, implicating a feedback mechanism to restrict an uncontrolled HIF activation and target gene induction [106]. Myocytes have a high capacity to use fatty acids as an energy substrate as long as oxygen is available [107]. Hypoxia leads to a shift to glucose metabolism to maintain the basic energy supply [108]. A major event in more severe hypoxia is the induction of the HIF target gene pyruvate dehydrogenase kinase 1 (*pdk1*), which is a negative regulator of the pyruvate dehydrogenase complex. Pyruvate dehydrogenase complex controls entry of pyruvate into the oxygen-dependent oxidative phosphorylation by the citric cycle and the mitochondrial respiration chain. Blockade induces lactate generation and ATP generation by anaerobic glycolysis[109, 110]. This effect was attributed to HIF-1α since elimination only of HIF-1α [111] but not HIF-2α muted the myocardial glycolytic response [112]. Experimentally HIF-2α supports HIF-1α for normal vascular development and is likely involved in vascular repair after ischemic injury of the heart.

There are numerous associations of hypoxia, HIF regulation, and oxidative stress [113, 114]. A direct link of HIF regulation and the cellular redox system has been demonstrated, since hydroxylation and thus degradation of HIF is dependent on ascorbate to prevent Fe^{2+} oxidation, which is needed to utilize oxygen as the hydroxylation substrate [115]. The interaction of both systems as well as its linkage to Ang II [116] are still under intensive research.

Some of the identified HIF target genes play important roles particular in the kidney and in renal physiology. These include hemoxygenase-1, adrenomedulin, iNOS, endothelin, and erythropoietin. Hemoxygenase-1 is a HIF target gene that catalyzes the degradation of heme to biliverdin/bilirubin, iron, and carbon monoxide and thus has protective effects on the endothelium and reduces reactive oxygen species (reviewed in [117]). It is known to reduce oxidative stress by reducing heme and to generate anti-inflammatory and antioxidant metabolites like bilirubin. Hemoxygenase-1 is evaluated as a therapeutic target, since it can attenuate Ang II-induced, ischemic or inflammatory tissue injury [118]. Experimentally, mice that have a targeted deletion of the HO-1 gene are characterized not only by higher creatinine levels and a higher mortality but also exhibit higher systemic levels of IL-6 in serum, heart, and lung compared to wild-type mice after ischemia reperfusion injury of the kidney [119].

The HIF target adrenomedullin [120] is a known potent vasodilator and is largely expressed in the heart and lung, followed by the adrenal, kidney, and the central nervous system [121, 122]. In the normal kidney it causes diuresis and natriuresis probably via renal vasodilatation and increases in renal blood flow reviewed in [123]. In fact, it has been reported that adrenomedullin plasma levels are elevated in the context in AKI [124] or heart failure [125] and may be part of the cardiorenal connection and response. Overall, adrenomedullin has been associated with organ- and tissue-protective effects (vasodilatation, natriuresis, and anti-inflammation) and is thus regarded as a potential therapeutic target [123].

At least one NOS, the inducible NOS (iNOS), has been shown to be an HIF target gene [126]. Production of NO has not only potent vasodilatory effects but has numerous other biological consequences. Importantly, NOS itself is involved in the generation of oxidative stress and has effects on mitochondrial oxygen consumption (discussed below).

In contrast, endothelin is a known HIF-1 target gene [127, 128] that induces vasoconstriction and is involved in hypertension and heart failure. Endothelin induction by chronic or intermittent hypoxia has been identified to contribute to the development of systemic pulmonary hypertension (reviewed in [129, 130]). High ET1 acts synergistically with Ang II and increases vascular resistance and cardiac workload.

The net effect of the induction of the interplay of the diverse vasodilators and vasoconstrictors on regional and systemic vascular tone is difficult to assess and still a matter of debate. In clinical practice acute hypoxia leads to vasodilation as long as the SNS does not counteract hypotension. The NO system and the SNS are important factors for the instantaneous vasodilatory effects of acute hypoxia. On the other hand, chronic tissue hypoxia maintained by endothelial damage, fibrosis, and decreased systemic oxygen supply (anemia, low cardiac output) is likely to participate in the generation of a high vascular tone. Here HIF target genes like endothelin and systemic activation of Ang II and the SNS are centrally involved.

3.3 Acute Kidney Injury and the Concept of Distant Organ Damage

Hypoxia and the HIF/PHD system not only act locally in the kidney or in the heart but also have systemic effects, which importantly participate in the cardiorenal connection. Abundant evidence indicates that AKI rapidly induces distant organ damage and dysfunction. Rabb and coworkers (reviewed in [131, 132]) and others [133, 134] have shown that shortly after induction of an ischemic injury of one kidney infiltration of T-cells and macrophages occurs in the contralateral kidney, the lung, and the cardiac interstitium. Overall induction of AKI leads to various derangements in lung and heart, including (a) a reduced antioxidative defense, (b) higher levels of inflammatory cytokines and enhanced infiltration of inflammatory cells, and (c) a reduced clearance of fluids thereby causing edema [131]. The latter can be explained by changes in the expression of water channels, such as aquaporin-5 (AQP5), or sodium channels, such as the epithelial sodium channel (ENaC). Forty-eight hours after bilateral renal ischemia in rats, both channels were markedly reduced in pulmonary tissues, which leads to decreased transport of fluids from the alveolar space into the circulation and thus supports the development of edema [135]. Increased systemic levels of IL-1 and TNF-α and markers of increased apoptosis were observed in cardiac myocytes only 1 h after renal ischemia in the rat, which was paralleled by left ventricular dilatation and a reduced shortening fraction in the echocardiogram [136]. These experimental findings correlate in a striking way to day-to-day clinical observations, where AKI (cardiorenal syndrome type 3) is associated with a pro-inflammatory state, the involvement of other organ dysfunctions, the development of edema, and an increased cardiovascular morbidity and mortality.

3.4 Chronic Hypoxia and Kidney Fibrosis

The vascular supply of the kidney is not only crucial for acute kidney failure and the ischemic injury of the S3 segment of the outer medulla, but it also plays a major role in slowly developing renal malfunction in patients with CKD. Here, a renal injury triggers a fibrotic process of the tubulointerstitium, which is paralleled by a rarefaction of the tubular capillary network (reviewed in [137–139]). Kang et al. [140–142] have shown that the renal peritubular capillary network decreased with progressive tubulointerstitial fibrosis [143]. And it has been convincingly shown that interstitial fibrosis and the decreasing capillary network correlate better to the declining renal function than the glomerular injury [144–146]. Capillary loss has been associated with an imbalance of angiogenic and antiangiogenic factors and endothelial injury [147]. In particular induction of the antiangiogenic factor thrombospondin-1 and reduced expression of the vascular growth factor (VEGF), which is important for endothelial growth and survival, may play a central role [140, 148]. The capillary rarefaction then leads to a reduced blood supply and to hypoxia. To prove renal ischemia and hypoxia the hypoxia marker pimonidazole [149] and a transgenic rat have been used. In the latter the DNA binding site of HIF, the HRE of HIF-target genes, has been used to control a FLAG-tag luciferase transgene [150]. A similar strategy was applied by Safran et al. [151] in transgenic mice, which demonstrated a five- to tenfold increase in HRE-driven reporter gene activity under systemic hypoxia. Noteworthy, exclusively the kidney appeared to have HRE activity already under room air and showed a markedly higher response in animals being exposed to 8% oxygen for 4 h compared to other organs.

Epithelial tubular cells are the second cell type in the kidney that is affected by hypoxia besides endothelial and vascular cells. In epithelial cells renal hypoxia is involved in the epithelial to mesenchymal transition (EMT), where the epithelial cell loses its polarity, disorganizes its cell-to-cell junctions, and reorganizes to mesenchymal cells, which can be analyzed by detection of mesenchymal markers such as α-smooth muscle actin, fibroblast-specific protein-1, and N-cadherin (reviewed in [152]).

Third, renal interstitial fibroblasts play a central role in the process of renal scaring because they present the main source for extracellular matrix synthesis [153]. In CKD fibroblasts proliferate and differentiate into myofibroblasts and increase ECM synthesis. The increased ECM production makes their isolation for in vitro analysis technically difficult.

EMT, hypoxic induction of pro-fibrotic genes, and recruitment of inflammatory cells all contribute to renal fibrosis. Renal hypoxia, resulting from the reduced number of capillaries in the kidney, is further aggravated by the extended oxygen diffusion distance in the fibrotic tissues. This may lead to a vicious cycle, when hypoxia itself stimulates fibrotic processes in the kidney, which then aggravate ischemia. To this end, many aspects of the chronic hypoxia hypothesis that have been proposed by Fine et al. [38, 154] have been experimentally confirmed [92, 137, 138]. Overall, chronic hypoxia and fibrosis appear to importantly aggravate CKD and its cardiorenal complications.

3.5 Anemia and Left Ventricular Hypertrophy

Another potential mediator of the cardiorenal connection is EPO, which a 165 amino acid glycoprotein, type 1 cytokine hormone that acts via the EPO receptor on erythroid progenitors to stimulate erythropoiesis in the bone marrow. The strongest inducers of EPO synthesis are hypoxia and anemia. EPO expression occurs in the fetal liver and then after birth for about 90% in the adult kidney. In the kidney interstitial fibroblasts have been identified to synthesize EPO [155]. So far the hypoxic gradient in the kidney and disturbances in renal oxygen delivery are directly linked to each other and play a central role in the cardiorenal connection. In CKD anemia typically develops due to a reduced renal capacity to synthesize EPO. Anemia is further aggravated by inflammatory cytokines like IL-6 and TNF-α, which inhibit EPO synthesis, iron uptake and release from the reticuloendothelial system through hepcidin induction [156], which can directly diminish red cell production in the bone marrow, and which reduce red blood cell survival [157].

It has been postulated that anemia, together with expansion of the extracellular volume, and arteriovenous shunting in dialysis fistulas of CKD patients lead to cardiac volume overload and thereby promote eccentric cardiac hypertrophy. Arterial hypertension, on the other hand, is accompanied by increased pressure, which is associated with concentric hypertrophy. The prevalence of left ventricular hypertrophy (LVH) clearly correlates with the degree of anemia [158, 159], and more than 75% of patients entering dialysis have LVH [160]. Epidemiologic studies implied that LVH and anemia increase cardiac failure and mortality [158, 161]. Experimental studies have shown that anemia is compensated by increased oxygen extraction within tissues and by activation of the SNS with a rise in heart rate and elevation in stroke volume [162, 163]. LVH leads to impaired contractile function and to a marked interstitial fibrosis in the myocardium [164, 165]. Importantly, myocardial fibrosis even progresses in uremia alone, independent of blood pressure and LVH [166]. Myocardial fibrosis – in analogy to the kidney – impairs oxygen diffusion capacity in the tissue. Moreover, capillary growth in uremic hearts does not keep pace with myocyte hypertrophy [167], both supporting hypoxia in the myocardium.

In clinical nephrology recombinant EPO and iron treatment are cornerstones in the management of patients with CKD. However, the debate regarding the optimal hemoglobin (Hb) target level is still ongoing and controversial. In a retrospective analysis of about 22,000 patients on dialysis, patients with a Hb < 8 g/dL had a twofold higher mortality compared to patients having an Hb of 10–11 g/dL [168]. As a matter of fact EPO substitution is associated with a reduction in cardiac workload and with a regression of LVH in these patients [169, 170]. However, in up to one third of cases EPO treatment was reported to induce or aggravate hypertension [171]. Albeit smaller studies showed positive effects of EPO and implicated that correction of anemia reduces morbidity and mortality, although recent trials could not confirm positive effects of Hb correction in CKD patients: The Correction of Hemoglobin and Outcomes in Renal Insufficiency (CHOIR) study enrolled 1,432 patients and showed a higher incidence of death, myocardial infarction, and hospitalization due to CHF in the higher target Hb group (13.5 vs. 11.3 g/dL Hb) [172]. Also the study of Cardiovascular Risk Reduction by Early Anemia Treatment with

Epoetin Beta (CREATE) with 603 patients did not show a decrease in cardiovascular events in the group targeted to a Hb of 13–15 g/dL as compared to a lower Hb target with later onset of therapy [173].

Anemia, as one of the possible components of the cardiorenal connection, not only occurs in CKD patients but is also regularly found in patients with chronic heart failure (reviewed in [174–176]). Anemia has a prevalence of 9.1, 19.2, 52.6, and 79.1% in NYHA (New York Heart Association) stages I, II, III, and IV, respectively [177]. The reasons for anemia in CHF patients are inflammation, progressive CKD, myelosuppressive effects of uremic toxins, iron deficiency, and the use of angiotensin-converting enzyme (ACE) inhibitors (reviewed in [175]). ACE inhibitors have been identified to aggravate anemia, since in the Studies of Left Ventricular Dysfunction (SOLVD) enalapril increased the risk for developing anemia by 56% at 1 year [178]. In analogy with the treatment of renal anemia, recombinant EPO has been applied in anemic CHF patients to test for improved outcome. Here some trials saw beneficial effects of applying EPO (and iron) [177, 179, 180], but larger studies are still ongoing.

Experimental findings imply that EPO may exert protective effects not only by correcting hematocrit but also by other effects mediated by the EPO receptor. Besides the bone marrow, EPO receptors have been postulated in many cells of different organs, like neurons [181], smooth muscle cells [182], endothelial cells, and cardiomyocytes [181, 183, 184]. Here EPO has been shown to contribute to cellular effects, which can be protective and adaptive, for example, in the brain or the kidney itself [185]. EPO exerts its protective effects as a survival factor and is important for growth, angiogenesis, and repair [186]. Protective effects have been evaluated, for example, for wound healing [187], ocular hypertension [188], hepatic ischemia, and in many other experimental settings. Concerning the myocardium, EPO reduced apoptosis in isolated myocytes [189, 190]. In murine models of myocardial infarction EPO reduced fibrosis and improved myocardial function [189, 191] and had positive effects on ventricular remodeling [192]. Some experimental findings imply a role of EPO in upregulating eNOS and a consequent inhibition of nicotinamide adenine dinucleotide phosphate (NADPH) oxidase, thus limiting oxidative stress [193, 194]. Either application of L-NAME (N^G-nitro-L-arginine methyl ester) or eNOS knockouts abrogated the protective effect of EPO [195]. Experimentally and clinically in CHF patients, EPO given for longer times led to endothelial growth and enhanced angiogenesis [196, 197].

However, the results of more stringent randomized controlled trials are supposed to assess the possible beneficial effects of EPO, including the Reduction of Events With Darbepoietin Alfa in Heart Failure (RED-HF) [198] or the Rationale-Trial to Reduce Cardiovascular Events with Aranesp Therapy (TREAT). As a matter of fact, recent publication of the results of the TREAT trial suggested no benefit of EPO treatment for cardiovascular events in patients with diabetes, chronic kidney disease, or anemia [199].

4 Oxidative Stress and Its Role in CRS

4.1 *CRP and Oxidative Stress*

The paradigm shift of atherosclerosis as a state of chronic vascular inflammation in response to various stimuli challenged the traditional concepts of vascular pathobiology and paved the way for an entirely novel field of cardiovascular research [200]. Advances in basic science within the past decade indicate a fundamental role of inflammation in all stages of atherosclerotic vascular disease and have helped to link cardiovascular risk factors to mechanisms of disease progression [201]. A wide range of pro-inflammatory molecules have been associated with atherosclerosis, and the list of potential cardiovascular biomarkers is steadily growing [202]. The most appealing marker identified so far is CRP [203]. CRP, a member of the pentraxin family of highly conserved cyclic pentameric proteins, is an acute phase protein that is predominantly synthesized in the liver in response to inflammation [204]. It was originally identified as a protein that could precipitate the C-polysaccharide of pneumococcal cell walls [205]. Biosynthesis is not restricted to the liver; extrahepatic local generation within atherosclerotic foci has been demonstrated in humans [206–208].

Several epidemiologic studies support the notion of CRP as a useful biomarker in the setting of primary prevention in cardiovascular disease, even in low-risk category patients [209–211]. Whether CRP is merely a marker of underlying atherosclerosis or itself plays a causal role as a culprit molecule in vascular disease remains controversial. In a recently published large observational study CRP polymorphisms were associated with an increase in CRP levels of up to 64%, an increase that should have translated into a theoretically predicted increased risk of up to 32% for ischemic heart disease and up to 25% for ischemic cerebrovascular disease [212]. However, this was not the case since genotype combinations were not associated with an increased risk of ischemic vascular disease. Even more intriguingly, mice overexpressing CRP are resistant to endotoxemia, pointing to an anti-inflammatory role of CRP, which has been explained by its ability to inhibit neutrophil chemotaxis [213]. The pro- and anti-inflammatory actions of CRP have in part been related to distinct species of CRP formed during inflammation. Thus, pentameric or native CRP (nCRP) can be dissociated into free subunits. These conformationally rearranged subunits are also referred to as modified or monomeric CRP (mCRP). Interestingly nCRP and mCRP seem to exert opposite effects in atherosclerosis. In apolipoprotein (apo) E-deficient mice, nCRP led to a dramatic increase in vascular plaque burden, whereas mrCRP treatment was associated with decreased atherosclerosis [214]. The latter effect may be explained by reduced uptake of acetylated LDL (acLDL) by mCRP as shown in vitro [215].

As a pro-inflammatory molecule CRP has been linked with oxidative stress. In coronary atherectomy specimens CRP co-localized with p22phox of NADH/NADPH oxidase, an important source of ROS [208]. In addition, CRP impairs endothelium-dependent NO-mediated vasodilation via activation of p38 kinase

[216]. CRP may further bind oxidized LDL (oxLDL) and thereby promote oxLDL uptake via macrophages [217]. However, the precise mechanisms of an association between oxidative stress and elevated levels of CRP are not firmly established (i.e., direct evidence in humans is sparse). To date, most studies in humans simply show an association between markers of oxidative stress and CRP levels, which does not ultimately imply a direct interaction [218–220]. Despite these limitations the widespread use of CRP as a clinical biomarker in cardiovascular disease is an indicator of its broad acceptance in the medical community.

4.2 Angiotensin II and Oxidative Stress

Ang II has long been identified as a major source of ROS formation. In vitro studies in Ang II–treated vascular smooth muscle cells revealed marked production of superoxide anions [221, 222]. Ang II–mediated ROS formation is largely driven by NADPH oxidase. This membrane-bound enzyme complex consists of six subunits: a Rho GTPase and five phagocytic oxidase (phox) units (gp91phox, p22phox, p40phox, p47phox, p67phox). NADPH oxidase generates superoxide by transferring electrons from NADPH inside the cell across the membrane and coupling these to molecular oxygen to produce superoxide anion. Ang II stimulates oxidative stress by upregulation of all known subunits of NADPH oxidase [223–230]. All these effects are mediated by the AT1 receptor. In contrast, stimulation of AT2 receptors seems to be vasoprotective via downregulation of Nox-1, p22phox, and p67phox [231, 232]. Inhibition of the renin angiotensin system with ACE inhibitors or AT1 receptor antagonists effectively inhibits NADPH oxidase activation [233]. In addition specific NADPH oxidase inhibitors, such as apocynin (prevents the assembly of subunits) or diphenylene iodonium, have shown promising results as potential therapeutic interventions, although their safety and effectiveness in treating vascular disease need to be determined [233–236]. While NADPH oxidase is the predominant source of Ang II–induced ROS formation, other pathways have been described. These include activation of mitogen-activated protein kinases (MAPK) as well as involvement of transcription factors such as hypoxia-inducible factor-1α (HIF-1α) [116, 228, 237–239].

4.3 Nitric Oxide Synthase and Oxidative Stress

Endothelial NOS is the key enzyme responsible for NO synthesis. Under normal conditions a functional NOS transfers electrons from the C-terminally bound NADPH reductase domain to the heme center in the amino-terminal oxygenase domain, where L-arginine is oxidized to L-citrulline and NO. Lack of substrate (L-arginine) or enzyme cofactors such as tetrahydrobiopterin (BH_4) may lead to uncoupling of oxygen reduction and arginine oxidation, a state referred to as NOS uncoupling. In an uncoupled state electrons flowing from the reductase domain to the oxygenase domain are diverted to molecular oxygen rather than to L-arginine, resulting in production of the free radical superoxide anion ($\bullet O_2^-$). NADPH oxidase-derived superoxide interacts with eNOS-derived NO to form peroxynitrite ($ONOO^-$). Peroxynitrite is a highly reactive oxidant and nitrating agent. Due to its oxidizing properties peroxynitrite avidly oxidizes the eNOS cofactor tetrahydrobiopterin (BH_4) to cofactor inactive molecules [240]. In addition, peroxynitrite can interfere with NOS directly by oxidizing the zinc-thiolate cluster, causing the release of zinc from the enzyme and leading to eNOS uncoupling [241]. Thus, a dysfunctional NOS enzyme may become the major source of ROS formation. Other enzymes that contribute to oxidative stress include xanthine oxidase, NADPH oxidases, and enzymes of the mitochondrial respiratory chain.

Evidence for eNOS uncoupling stems from in vitro studies in endothelial cells treated with oxidized LDL [242] as well as in vivo studies in Ang II or DOCA salt-induced hypertension models in rodents [243, 244]. NOS inhibitors such as L-NAME significantly attenuated ROS formation in these studies, pointing to the crucial role of NOS uncoupling as a major source of oxidative stress [243]. On the contrary endogenous NOS inhibitors such as N^G-monomethyl L-arginine (L-NMMA) or N^G,N^G dimethyl L-arginine (ADMA) may also facilitate or contribute to NOS uncoupling. Thus, enhanced superoxide production stimulates protein arginine methyltransferases (PRMTs), thereby increasing the generation of both L-NMMA and ADMA. Furthermore, oxidative stress reduces L-NMMA and ADMA degradation via impaired expression or activity of the metabolizing enzyme dimethylarginine dimethylaminohydrolase (DDAH) [245]. The resulting net increase of intracellular L-NMMA and ADMA concentration diminishes endogenous NO production via competition with L-arginine for NOS enzyme. Both endogenous NOS inhibitors may also shift the NO/ROS balance toward ROS via induction of eNOS uncoupling [246].

Various cardiovascular risk factors such as hypercholesterolemia [247], diabetes [248], and essential hypertension [249] are associated with eNOS uncoupling in humans.

Studies in rats indicate that within the kidneys NADPH oxidase and uncoupled NOS are major sources of glomerular superoxide generation [250]. Blockade of the renin-angiotensin system with angiotensin receptor blockers improves NOS uncoupling in diabetic nephropathy by increasing BH_4 bioavailability [251]. In addition, oral administration of NOS cofactor BH_4 or substrate supply with L-arginine ameliorates endothelial function in rats suffering from chronic renal failure [252].

4.4 The Sympathetic Nervous System and Oxidative Stress

There are multiple links between the sympathetic nervous system and oxidative stress (ROS), which mostly act as positive feedback loops. On the one hand, the SNS is able to induce or perpetuate oxidative stress via Ang II induction and its downstream effects; on the other hand, oxidative stress is able to further activate the SNS, which culminates in a viscous circle.

The most obvious connection between oxidative stress and SNS can be related to the RSNA, which directly controls renin release from juxtaglomerular granular cells via adrenergic mechanisms [48]. To this end, all the mechanisms described above for the renal RAAS can be more or less analyzed in the context of the renal sympathetic innervation as well. However, research of the past decade has emphasized the *systemic*, the *afferent nerve*, and the *CNS effects* of ROS on the SNS, which will be outlined next.

Systemic effects of ROS on SNS: hydroxy-2,2,6,6-tetramethylpiperidinyloxy (TEMPOL), a superoxide dismutase (SOD) mimetic has been shown to lower both blood pressure and RSNA in normotensive rats [253], as well as in DOCA- and spontaneously hypertensive rats (SHR) [254, 255]. This seems to be a systemic effect because intracerebroventricular injection of TEMPOL did not have any effect in SHR and the effects of TEMPOL were markedly reduced after pharmacological blockade of sympathetic ganglia by hexamethonium in these three experimental models [253–255]. Moreover, the effects on RSNA were not due to increased NO availability, since they persisted after NOS inhibition [253, 255]. In an ischemia reperfusion model of the kidney TEMPOL significantly reduced renal dysfunction and injury. Similar observations were made in remnant kidney model (5/6 nephrectomy model). However, the precise role of the SNS in these kidney injury models still remains unclear.

Afferent effects of ROS on SNS: Oxidative stress certainly plays a pivotal role in myocardial ischemia and reperfusion where large amounts of ROS are produced. Some animal studies have been conducted to address this issue. Activation of the sympatho-sympathetic cardiac sympathoexcitatory afferent reflex (CSAR) and pain during ischemia could be shown to be at least in part due to direct activation of a subgroup of cardiac sympathetic C- and Aδ-fibers by ROS, especially ^{-}OH [256], which finally leads to RSNA increase with consecutive activation of the RAAS. However, vagal fibers are also activated by ROS in the setting of ischemia and reperfusion, mediating a cardiodepressory and sympathoinhibitory reflex (Bezold-Jarisch reflex) [257, 258]. It was demonstrated that ischemia-induced fiber activation could be blocked by indomethacin, whereas the reperfusion-derived activation was unaffected, which suggests differential ROS-mediated activation pathways [258]. The net effect of the CSAR and Bezold-Jarisch reflex due to ROS-mediated stimulation, however, seems to be sympathoexcitatory. Furthermore, many other chemosensory pathways that originate in the heart play an important role in ischemia induced sympathoexcitation [259].

Central actions of ROS on SNS: The RVLM harbors the sympathetic premotor neurons, which are pivotal in the regulation of central sympathetic outflow. Several studies have shown that ROS stimulate central sympathetic outflow when acting in the RVLM: SOD injected into the RVLM decreased sympathetic nerve activity in swine [260]. TEMPOL was given in the lateral ventricle, leading to reduced norepinephrine secretion from the posterior hypothalamus (PH) and RSNA. It seems that ROS may raise blood pressure via activation of the SNS.

This activation may be mediated in part by downregulation of nNOS and NO production and in part by mechanisms independent of NO [261, 262].

Some studies have investigated the role of ROS in the nucleus tractus solitarius (NTS), a major hindbrain area involved in cardiovascular regulation, which receives primary afferent fibers from peripheral mechano- and chemoreceptors. Interestingly, in contrast to the action in the RVLM, ROS injected into the NTS induces hypotension and bradycardia probably due to glutaminergic mechanisms.

Importantly, there is evidence from numerous studies for a link between the generation of central ROS and Ang II and the AT1 receptor in normal and pathological states [263–265]. For example, in a rabbit model of chronic cardiac failure a chronic upregulation of AT1 receptors along with various subunits of the NADPH oxidase subunits in the RVLM was found [266].

In summary, oxidative stress activates the sympathetic nervous system. Due to the link between ROS, AT1-receptors, and Ang II, many studies point to positive-feedback mechanisms by which oxidative stress increases sympathetic activity, which in turn further aggravates oxidative stress via Ang II–mediated effects. However, some studies point to the existence of ROS defense mechanisms, such as the Bezold-Jarisch reflex, which is mediated via vagal afferent cardiac fibers and the sympathoinhibitory ROS-induced actions of the NTS. Furthermore, there is increasing evidence for a protective action of peptidergic visceral afferent fibers, which seems to be mediated via TRPV1 receptors. This has been shown in rats with destruction of sensory nerves due to postnatal capsaicin treatment, in which enhanced O_2^- could be measured in the cortex and medulla of the kidney if these animals where on a high-salt diet [267]. Activation of TRPV1 receptors leads to release of potent microvascular vasodilator neuropeptides such as calcitonin gene–related peptide and substance P, and it was recently shown that ROS seem to be a mediator of such a neurogenic vasodilation [268]. The exact impact of this finding and the role of other members of the TRP receptor family need further investigation.

5 Perspectives: Therapeutic Options

Until now the pathophysiological cascades that are in involved in the complex cardiorenal interaction are far from being fully understood. However, it is clear that major players such as oxidative stress and SNS play an integral part in this interaction. Factors known to be involved in the cardiorenal interaction give rise to hope that beneficial therapeutic strategies that are in part already established or are under current investigation find their way into the clinical routine. Among these, blockade of the RAAS seems to be most promising. ACE inhibitors (ACEI) or angiotensin receptor blockers (ARBs) have already been shown to reduce oxidative stress, tissue hypoxia, and fibrosis. Blockade of the SNS has a long history in cardiology and nephrology and may also be a promising tool to counteract cardiorenal complications. Antioxidant therapy in acute and chronic disease is evaluated in several clinical trials, but at present the results are conflicting and difficult to interpret [269]. EPO treatment has been established for almost two decades, but the risks and benefits still have to be better defined.

References

1. Sarnak MJ. Cardiovascular complications in chronic kidney disease. Am J Kidney Dis 2003; 41:11–17.
2. Coresh J, Astor BC, Greene T, Eknoyan G, Levey AS. Prevalence of chronic kidney disease and decreased kidney function in the adult US population: Third National Health and Nutrition Examination Survey. Am J Kidney Dis 2003; 41:1–12.
3. Hage FG, Venkataraman R, Zoghbi GJ, Perry GJ, DeMattos AM, Iskandrian AE. The scope of coronary heart disease in patients with chronic kidney disease. J Am Coll Cardiol 2009; 53:2129–2140.
4. Tonelli M, Wiebe N, Culleton B, House A, Rabbat C, Fok M, McAlister F, Garg AX. Chronic kidney disease and mortality risk: a systematic review. J Am Soc Nephrol 2006; 17:2034–2047.
5. Go AS, Chertow GM, Fan D, McCulloch CE, Hsu CY. Chronic kidney disease and the risks of death, cardiovascular events, and hospitalization. N Engl J Med 2004; 351:1296–1305.
6. Levey AS. Controlling the epidemic of cardiovascular disease in chronic renal disease: where do we start? Am J Kidney Dis 1998; 32:S5–S13.
7. Herzog CA, Ma JZ, Collins AJ. Poor long-term survival after acute myocardial infarction among patients on long-term dialysis. N Engl J Med 1998; 339:799–805.
8. Smith GL, Lichtman JH, Bracken MB, Shlipak MG, Phillips CO, DiCapua P, Krumholz HM. Renal impairment and outcomes in heart failure: systematic review and meta-analysis. J Am Coll Cardiol 2006; 47:1987–1996.
9. Sarnak MJ, Levey AS, Schoolwerth AC, Coresh J, Culleton B, Hamm LL, McCullough PA, Kasiske BL, Kelepouris E, Klag MJ, Parfrey P, Pfeffer M, Raij L, Spinosa DJ, Wilson PW. Kidney disease as a risk factor for development of cardiovascular disease: a statement from the American Heart Association Councils on Kidney in Cardiovascular Disease, High Blood Pressure Research, Clinical Cardiology, and Epidemiology and Prevention. Hypertension 2003; 42:1050–1065.
10. Liang KV, Williams AW, Greene EL, Redfield MM. Acute decompensated heart failure and the cardiorenal syndrome. Crit Care Med 2008; 36:S75–S88.
11. Owan TE, Hodge DO, Herges RM, Jacobsen SJ, Roger VL, Redfield MM. Secular trends in renal dysfunction and outcomes in hospitalized heart failure patients. J Card Fail 2006; 12:257–262.
12. Bongartz LG, Cramer MJ, Doevendans PA, Joles JA, Braam B. The severe cardiorenal syndrome: "Guyton revisited." Eur Heart J 2005; 26:11–17.
13. Ronco C, Haapio M, House AA, Anavekar N, Bellomo R. Cardiorenal syndrome. J Am Coll Cardiol 2008; 52:1527–1539.
14. Ronco C, House AA, Haapio M. Cardiorenal and renocardiac syndromes: the need for a comprehensive classification and consensus. Nat Clin Pract Nephrol 2008; 4:310–311.
15. Jose P, Skali H, Anavekar N, Tomson C, Krumholz HM, Rouleau JL, Moye L, Pfeffer MA, Solomon SD. Increase in creatinine and cardiovascular risk in patients with systolic dysfunction after myocardial infarction. J Am Soc Nephrol 2006; 17:2886–2891.
16. Binanay C, Califf RM, Hasselblad V, O'Connor CM, Shah MR, Sopko G, Stevenson LW, Francis GS, Leier CV, Miller LW. Evaluation study of congestive heart failure and pulmonary artery catheterization effectiveness: the ESCAPE trial. JAMA 2005; 294:1625–1633.

17. Nohria A, Hasselblad V, Stebbins A, Pauly DF, Fonarow GC, Shah M, Yancy CW, Califf RM, Stevenson LW, Hill JA. Cardiorenal interactions: insights from the ESCAPE trial. J Am Coll Cardiol 2008; 51:1268–1274.
18. Damman K, van Deursen VM, Navis G, Voors AA, van Veldhuisen DJ, Hillege HL. Increased central venous pressure is associated with impaired renal function and mortality in a broad spectrum of patients with cardiovascular disease. J Am Coll Cardiol 2009; 53:582–588.
19. Mehta RL, Kellum JA, Shah SV, Molitoris BA, Ronco C, Warnock DG, Levin A. Acute Kidney Injury Network: report of an initiative to improve outcomes in acute kidney injury. Crit Care 2007; 11:R31.
20. Schrier RW, Wang W. Acute renal failure and sepsis. N Engl J Med 2004; 351:159–169.
21. Sutton TA, Fisher CJ, Molitoris BA. Microvascular endothelial injury and dysfunction during ischemic acute renal failure. Kidney Int 2002; 62:1539–1549.
22. Wan L, Bagshaw SM, Langenberg C, Saotome T, May C, Bellomo R. Pathophysiology of septic acute kidney injury: what do we really know? Crit Care Med 2008; 36:S198–S203.
23. Langenberg C, Wan L, Egi M, May CN, Bellomo R. Renal blood flow and function during recovery from experimental septic acute kidney injury. Intensive Care Med 2007; 33:1614–1618.
24. Ravikant T, Lucas CE. Renal blood flow distribution in septic hyperdynamic pigs. J Surg Res 1977; 22:294–298.
25. Brenner M, Schaer GL, Mallory DL, Suffredini AF, Parrillo JE. Detection of renal blood flow abnormalities in septic and critically ill patients using a newly designed indwelling thermodilution renal vein catheter. Chest 1990; 98:170–179.
26. Bellomo R, Wan L, Langenberg C, May C. Septic acute kidney injury: new concepts. Nephron Exp Nephrol 2008; 109:e95–e100.
27. Langenberg C, Bagshaw SM, May CN, Bellomo R. The histopathology of septic acute kidney injury: a systematic review. Crit Care 2008; 12:R38.
28. Langenberg C, Bellomo R, May C, Wan L, Egi M, Morgera S. Renal blood flow in sepsis. Crit Care 2005; 9:R363–R374.
29. van der Putten K, Bongartz LG, Braam B, Gaillard CA. The cardiorenal syndrome a classification into 4 groups? J Am Coll Cardiol 2009; 53:1340; author reply 1340–1341.
30. Nobes MS, Harris PJ, Yamada H, Mendelsohn FA. Effects of angiotensin on renal cortical and papillary blood flows measured by laser-Doppler flowmetry. Am J Physiol 1991; 261: F998–F1006.
31. Welch WJ, Blau J, Xie H, Chabrashvili T, Wilcox CS. Angiotensin-induced defects in renal oxygenation: role of oxidative stress. Am J Physiol Heart Circ Physiol 2005; 288:H22–H28.
32. Manotham K, Tanaka T, Matsumoto M, Ohse T, Miyata T, Inagi R, Kurokawa K, Fujita T, Nangaku M. Evidence of tubular hypoxia in the early phase in the remnant kidney model. J Am Soc Nephrol 2004; 15:1277–1288.
33. Norman JT, Stidwill R, Singer M, Fine LG. Angiotensin II blockade augments renal cortical microvascular pO2 indicating a novel, potentially renoprotective action. Nephron Physiol 2003; 94:p39–p46.
34. Kondo N, Kiyomoto H, Yamamoto T, Miyatake A, Sun GP, Rahman M, Hitomi H, Moriwaki K, Hara T, Kimura S, Abe Y, Kohno M, Nishiyama A. Effects of calcium channel blockade on angiotensin II-induced peritubular ischemia in rats. J Pharmacol Exp Ther 2006; 316:1047–1052.
35. Schachinger H, Klarhofer M, Linder L, Drewe J, Scheffler K. Angiotensin II decreases the renal MRI blood oxygenation level-dependent signal. Hypertension 2006; 47:1062–1066.
36. Silva GB, Garvin JL. Angiotensin II-dependent hypertension increases Na transport-related oxygen consumption by the thick ascending limb. Hypertension 2008; 52:1091–1098.
37. Nangaku M, Fujita T. Activation of the renin-angiotensin system and chronic hypoxia of the kidney. Hypertens Res 2008; 31:175–184.

38. Fine LG, Norman JT. Chronic hypoxia as a mechanism of progression of chronic kidney diseases: from hypothesis to novel therapeutics. Kidney Int 2008; 74:867–872.
39. Gorin Y, Kim NH, Feliers D, Bhandari B, Choudhury GG, Abboud HE. Angiotensin II activates Akt/protein kinase B by an arachidonic acid/redox-dependent pathway and independent of phosphoinositide 3-kinase. Faseb J 2001; 15:1909–1920.
40. Gorin Y, Ricono JM, Kim NH, Bhandari B, Choudhury GG, Abboud HE. Nox4 mediates angiotensin II-induced activation of Akt/protein kinase B in mesangial cells. Am J Physiol Renal Physiol 2003; 285:F219–F229.
41. Hannken T, Schroeder R, Zahner G, Stahl RA, Wolf G. Reactive oxygen species stimulate p44/42 mitogen-activated protein kinase and induce p27(Kip1): role in angiotensin II-mediated hypertrophy of proximal tubular cells. J Am Soc Nephrol 2000; 11:1387–1397.
42. Lassegue B, Sorescu D, Szocs K, Yin Q, Akers M, Zhang Y, Grant SL, Lambeth JD, Griendling KK. Novel gp91(phox) homologues in vascular smooth muscle cells: nox1 mediates angiotensin II-induced superoxide formation and redox-sensitive signaling pathways. Circ Res 2001; 88:888–894.
43. Shenberger JS, Adams MH, Zimmer SG. Oxidant-induced hypertrophy of A549 cells is accompanied by alterations in eukaryotic translation initiation factor 4E and 4E-binding protein-1. Am J Respir Cell Mol Biol 2002; 27:250–256.
44. Schlaich MP, Socratous F, Hennebry S, Eikelis N, Lambert EA, Straznicky N, Esler MD, Lambert GW. Sympathetic activation in chronic renal failure. J Am Soc Nephrol 2009; 20:933–939.
45. Zucker IH. Novel mechanisms of sympathetic regulation in chronic heart failure. Hypertension 2006; 48:1005–1011.
46. Floras JS. Sympathetic nervous system activation in human heart failure: clinical implications of an updated model. J Am Coll Cardiol 2009; 54:375–385.
47. Koomans HA, Blankestijn PJ, Joles JA. Sympathetic hyperactivity in chronic renal failure: a wake-up call. J Am Soc Nephrol 2004; 15:524–537.
48. DiBona GF. Physiology in perspective: the wisdom of the body. Neural control of the kidney. Am J Physiol Regul Integr Comp Physiol 2005; 289:R633–R641.
49. Blair ML, Chen YH, Izzo JL, Jr. Influence of renal perfusion pressure on alpha- and beta-adrenergic stimulation of renin release. Am J Physiol 1985; 248:E317–E326.
50. DiBona GF. Neural control of renal function: role of renal alpha adrenoceptors. J Cardiovasc Pharmacol 1985; 7 (Suppl 8):S18–S23.
51. DiBona GF. Nervous kidney. Interaction between renal sympathetic nerves and the renin-angiotensin system in the control of renal function. Hypertension 2000; 36:1083–1088.
52. DiBona GF, Kopp UC. Neural control of renal function. Physiol Rev 1997; 77:75–197.
53. Veelken R. Neural control of kidney function. Kidney Blood Press Res 1998; 21:249–252.
54. Campese VM, Ye S, Zhong H. Downregulation of neuronal nitric oxide synthase and interleukin-1beta mediates angiotensin II-dependent stimulation of sympathetic nerve activity. Hypertension 2002; 39:519–524.
55. DiBona GF, Sawin LL. Reflex regulation of renal nerve activity in cardiac failure. Am J Physiol 1994; 266:R27–R39.
56. Veelken R, Stetter A, Dickel T, Hilgers KF. Bimodality of cardiac vagal afferent Cfibres in the rat. Pflugers Arch 2003; 446(5):516–522.
57. Linz P, Veelken R. Serotonin 5-HT(3) receptors on mechanosensitive neurons with cardiac afferents. Am J Physiol Heart Circ Physiol 2002; 282:H1828–H1835.
58. Veelken R, Hilgers KF, Ditting T, Fierlbeck W, Geiger H, Schmieder RE. Subthreshold stimulation of a serotonin 5-HT3 reflex attenuates cardiovascular reflexes. Am J Physiol 1996; 271:R1500–R1506.
59. Schultz HD, Ustinova EE. Capsaicin receptors mediate free radical-induced activation of cardiac afferent endings. Cardiovasc Res 1998; 38:348–355.

60. Hausberg M, Kosch M, Harmelink P, Barenbrock M, Hohage H, Kisters K, Dietl KH, Rahn KH. Sympathetic nerve activity in end-stage renal disease. Circulation 2002; 106:1974–1979.
61. Converse RL, Jr., Jacobsen TN, Toto RD, Jost CM, Cosentino F, Fouad-Tarazi F, Victor RG. Sympathetic overactivity in patients with chronic renal failure. N Engl J Med 1992; 327:1912–1918.
62. Campese VM. A new model of neurogenic hypertension caused by renal injury: pathophysiology and therapeutic implications. Clin Exp Nephrol 2003; 7:167–171.
63. Ditting T, Tiegs G, Rodionova K, Reeh PW, Neuhuber W, Freisinger W, Veelken R. Do distinct populations of dorsal root ganglion neurons account for the sensory peptidergic innervation of the kidney? Am J Physiol Renal Physiol 2009; 297(5):F1427–F1434.
64. Veelken R, Vogel EM, Hilgers K, Amann K, Hartner A, Sass G, Neuhuber W, Tiegs G. Autonomic renal denervation ameliorates experimental glomerulonephritis. J Am Soc Nephrol 2008; 19:1371–1378.
65. Berl T, Henrich W. Kidney-heart interactions: epidemiology, pathogenesis, and treatment. Clin J Am Soc Nephrol 2006; 1:8–18.
66. Lindner A, Charra B, Sherrard DJ, Scribner BH. Accelerated atherosclerosis in prolonged maintenance hemodialysis. N Engl J Med 1974; 290:697–701.
67. Anavekar NS, McMurray JJ, Velazquez EJ, Solomon SD, Kober L, Rouleau JL, White HD, Nordlander R, Maggioni A, Dickstein K, Zelenkofske S, Leimberger JD, Califf RM, Pfeffer MA. Relation between renal dysfunction and cardiovascular outcomes after myocardial infarction. N Engl J Med 2004; 351:1285–1295.
68. deFilippi C, Wasserman S, Rosanio S, Tiblier E, Sperger H, Tocchi M, Christenson R, Uretsky B, Smiley M, Gold J, Muniz H, Badalamenti J, Herzog C, Henrich W. Cardiac troponin T and C-reactive protein for predicting prognosis, coronary atherosclerosis, and cardiomyopathy in patients undergoing long-term hemodialysis. JAMA 2003; 290:353–359.
69. Levin NW, Handelman GJ, Coresh J, Port FK, Kaysen GA. Reverse epidemiology: a confusing, confounding, and inaccurate term. Semin Dial 2007; 20:586–592.
70. Moe SM, O'Neill KD, Duan D, Ahmed S, Chen NX, Leapman SB, Fineberg N, Kopecky K. Medial artery calcification in ESRD patients is associated with deposition of bone matrix proteins. Kidney Int 2002; 61:638–647.
71. Haydar AA, Hujairi NM, Covic AA, Pereira D, Rubens M, Goldsmith DJ. Coronary artery calcification is related to coronary atherosclerosis in chronic renal disease patients: a study comparing EBCT-generated coronary artery calcium scores and coronary angiography. Nephrol Dial Transplant 2004; 19:2307–2312.
72. Silverstein DM. Inflammation in chronic kidney disease: role in the progression of renal and cardiovascular disease. Pediatr Nephrol 2009; 24:1445–1452.
73. Goldstein SL, Leung JC, Silverstein DM. Pro- and anti-inflammatory cytokines in chronic pediatric dialysis patients: effect of aspirin. Clin J Am Soc Nephrol 2006; 1:979–986.
74. Costa E, Lima M, Alves JM, Rocha S, Rocha-Pereira P, Castro E, Miranda V, do SF, Loureiro A, Quintanilha A, Belo L, Santos-Silva A. Inflammation, T-cell phenotype, and inflammatory cytokines in chronic kidney disease patients under hemodialysis and its relationship to resistance to recombinant human erythropoietin therapy. J Clin Immunol 2008; 28:268–275.
75. Kimmel PL, Phillips TM, Simmens SJ, Peterson RA, Weihs KL, Alleyne S, Cruz I, Yanovski JA, Veis JH. Immunologic function and survival in hemodialysis patients. Kidney Int 1998; 54:236–244.
76. Yan SF, Ramasamy R, Schmidt AM. The receptor for advanced glycation endproducts (RAGE) and cardiovascular disease. Expert Rev Mol Med 2009; 11:e9.
77. Brezis M, Rosen S. Hypoxia of the renal medulla – its implications for disease. N Engl J Med 1995; 332:647–655.
78. Schurek HJ, Jost U, Baumgartl H, Bertram H, Heckmann U. Evidence for a preglomerular oxygen diffusion shunt in rat renal cortex. Am J Physiol 1990; 259:F910–F915.

79. Persson PB, Ehmke H, Kirchheim HR, Janssen B, Baumann JE, Just A, Nafz B. Autoregulation and non-homeostatic behaviour of renal blood flow in conscious dogs. J Physiol 1993; 462:261–273.
80. Levy MN. Effect of variations of blood flow on renal oxygen extraction. Am J Physiol 1960; 199:13–18.
81. Leong CL, Anderson WP, O'Connor PM, Evans RG. Evidence that renal arterial-venous oxygen shunting contributes to dynamic regulation of renal oxygenation. Am J Physiol Renal Physiol 2007; 292:F1726–F1733.
82. Pallone TL, Robertson CR, Jamison RL. Renal medullary microcirculation. Physiol Rev 1990; 70:885–920.
83. Dyson A, Stidwill R, Taylor V, Singer M. Tissue oxygen monitoring in rodent models of shock. Am J Physiol Heart Circ Physiol 2007; 293:H526–H533.
84. Lee JB, Peter HM. Effect of oxygen tension on glucose metabolism in rabbit kidney cortex and medulla. Am J Physiol 1969; 217:1464–1471.
85. Bagnasco S, Good D, Balaban R, Burg M. Lactate production in isolated segments of the rat nephron. Am J Physiol 1985; 248:F522–F526.
86. Thomas SR. Inner medullary lactate production and accumulation: a vasa recta model. Am J Physiol Renal Physiol 2000; 279:F468–F481.
87. Vandewalle A, Wirthensohn G, Heidrich HG, Guder WG. Distribution of hexokinase and phosphoenolpyruvate carboxykinase along the rabbit nephron. Am J Physiol 1981; 240: F492–F500.
88. Katz AI, Doucet A, Morel F. Na-K-ATPase activity along the rabbit, rat, and mouse nephron. Am J Physiol 1979; 237:F114–F120.
89. Abrahams S, Greenwald L, Stetson DL. Contribution of renal medullary mitochondrial density to urinary concentrating ability in mammals. Am J Physiol 1991; 261:R719–R726.
90. Semenza GL. Hypoxia-inducible factor 1 (HIF-1) pathway. Sci STKE 2007; 2007:cm8.
91. Rosenberger C, Mandriota S, Jurgensen JS, Wiesener MS, Horstrup JH, Frei U, Ratcliffe PJ, Maxwell PH, Bachmann S, Eckardt KU. Expression of hypoxia-inducible factor-1alpha and -2alpha in hypoxic and ischemic rat kidneys. J Am Soc Nephrol 2002; 13:1721–1732.
92. Eckardt KU, Bernhardt W, Willam C, Wiesener M. Hypoxia-inducible transcription factors and their role in renal disease. Semin Nephrol 2007; 27:363–372.
93. Eckardt KU, Bernhardt WM, Weidemann A, Warnecke C, Rosenberger C, Wiesener MS, Willam C. Role of hypoxia in the pathogenesis of renal disease. Kidney Int Suppl 2005; 99: S46–S51.
94. Rankin EB, Biju MP, Liu Q, Unger TL, Rha J, Johnson RS, Simon MC, Keith B, Haase VH. Hypoxia-inducible factor-2 (HIF-2) regulates hepatic erythropoietin in vivo. J Clin Invest 2007; 117:1068–1077.
95. Warnecke C, Zaborowska Z, Kurreck J, Erdmann VA, Frei U, Wiesener M, Eckardt KU. Differentiating the functional role of hypoxia-inducible factor (HIF)-1alpha and HIF-2alpha (EPAS-1) by the use of RNA interference: erythropoietin is a HIF-2alpha target gene in Hep3B and Kelly cells. Faseb J 2004; 18:1462–1464.
96. Bruick RK, McKnight SL. A conserved family of prolyl-4-hydroxylases that modify HIF. Science 2001; 294:1337–1340.
97. Jaakkola P, Mole DR, Tian YM, Wilson MI, Gielbert J, Gaskell SJ, Kriegsheim A, Hebestreit HF, Mukherji M, Schofield CJ, Maxwell PH, Pugh CW, Ratcliffe PJ. Targeting of HIF-alpha to the von Hippel-Lindau ubiquitylation complex by O2-regulated prolyl hydroxylation. Science 2001; 292:468–472.
98. Epstein AC, Gleadle JM, McNeill LA, Hewitson KS, O'Rourke J, Mole DR, Mukherji M, Metzen E, Wilson MI, Dhanda A, Tian YM, Masson N, Hamilton DL, Jaakkola P, Barstead R, Hodgkin J, Maxwell PH, Pugh CW, Schofield CJ, Ratcliffe PJ. *C. elegans* EGL-9 and mammalian homologs define a family of dioxygenases that regulate HIF by prolyl hydroxylation. Cell 2001; 107:43–54.

99. Kaelin WG, Jr., Ratcliffe PJ. Oxygen sensing by metazoans: the central role of the HIF hydroxylase pathway. Mol Cell 2008; 30:393–402.
100. Rosenberger C, Pratschke J, Rudolph B, Heyman SN, Schindler R, Babel N, Eckardt KU, Frei U, Rosen S, Reinke P. Immunohistochemical detection of hypoxia-inducible factor-1alpha in human renal allograft biopsies. J Am Soc Nephrol 2007; 18:343–351.
101. Rosenberger C, Griethe W, Gruber G, Wiesener M, Frei U, Bachmann S, Eckardt KU. Cellular responses to hypoxia after renal segmental infarction. Kidney Int 2003; 64:874–886.
102. Rosenberger C, Heyman SN, Rosen S, Shina A, Goldfarb M, Griethe W, Frei U, Reinke P, Bachmann S, Eckardt KU. Up-regulation of HIF in experimental acute renal failure: evidence for a protective transcriptional response to hypoxia. Kidney Int 2005; 67:531–542.
103. Schodel J, Klanke B, Weidemann A, Buchholz B, Bernhardt W, Bertog M, Amann K, Korbmacher C, Wiesener M, Warnecke C, Kurtz A, Eckardt KU, Willam C. HIF-prolyl hydroxylases in the rat kidney: physiologic expression patterns and regulation in acute kidney injury. Am J Pathol 2009; 174:1663–1674.
104. Shohet RV, Garcia JA. Keeping the engine primed: HIF factors as key regulators of cardiac metabolism and angiogenesis during ischemia. J Mol Med 2007; 85:1309–1315.
105. Jurgensen JS, Rosenberger C, Wiesener MS, Warnecke C, Horstrup JH, Grafe M, Philipp S, Griethe W, Maxwell PH, Frei U, Bachmann S, Willenbrock R, Eckardt KU. Persistent induction of HIF-1alpha and -2alpha in cardiomyocytes and stromal cells of ischemic myocardium. Faseb J 2004; 18:1415–1417.
106. Willam C, Maxwell PH, Nichols L, Lygate C, Tian YM, Bernhardt W, Wiesener M, Ratcliffe PJ, Eckardt KU, Pugh CW. HIF prolyl hydroxylases in the rat; organ distribution and changes in expression following hypoxia and coronary artery ligation. J Mol Cell Cardiol 2006; 41:68–77.
107. Nieth H, Schollmeyer P. Substrate-utilization of the human kidney. Nature 1966; 209:1244–1245.
108. Bolukoglu H, Goodwin GW, Guthrie PH, Carmical SG, Chen TM, Taegtmeyer H. Metabolic fate of glucose in reversible low-flow ischemia of the isolated working rat heart. Am J Physiol 1996; 270:H817–H826.
109. Kim JW, Tchernyshyov I, Semenza GL, Dang CV. HIF-1-mediated expression of pyruvate dehydrogenase kinase: a metabolic switch required for cellular adaptation to hypoxia. Cell Metab 2006; 3:177–185.
110. Papandreou I, Cairns RA, Fontana L, Lim AL, Denko NC. HIF-1 mediates adaptation to hypoxia by actively downregulating mitochondrial oxygen consumption. Cell Metab 2006; 3:187–197.
111. Seagroves TN, Ryan HE, Lu H, Wouters BG, Knapp M, Thibault P, Laderoute K, Johnson RS. Transcription factor HIF-1 is a necessary mediator of the Pasteur effect in mammalian cells. Mol Cell Biol 2001; 21:3436–3444.
112. Hu CJ, Wang LY, Chodosh LA, Keith B, Simon MC. Differential roles of hypoxia-inducible factor 1alpha (HIF-1alpha) and HIF-2alpha in hypoxic gene regulation. Mol Cell Biol 2003; 23:9361–9374.
113. Simon MC. Mitochondrial reactive oxygen species are required for hypoxic HIF alpha stabilization. Adv Exp Med Biol 2006; 588:165–170.
114. Zou AP, Cowley AW, Jr. Reactive oxygen species and molecular regulation of renal oxygenation. Acta Physiol Scand 2003; 179:233–241.
115. Pan Y, Mansfield KD, Bertozzi CC, Rudenko V, Chan DA, Giaccia AJ, Simon MC. Multiple factors affecting cellular redox status and energy metabolism modulate hypoxia-inducible factor prolyl hydroxylase activity in vivo and in vitro. Mol Cell Biol 2007; 27:912–925.
116. Page EL, Chan DA, Giaccia AJ, Levine M, Richard DE. Hypoxia-inducible factor-1alpha stabilization in nonhypoxic conditions: role of oxidation and intracellular ascorbate depletion. Mol Biol Cell 2008; 19:86–94.
117. Abraham NG, Cao J, Sacerdoti D, Li X, Drummond GS. Heme oxygenase: the key to renal function regulation. Am J Physiol Renal Physiol 2009; 297(5):F1137–F1152.

118. Dawn B, Bolli R. HO-1 induction by HIF-1: a new mechanism for delayed cardioprotection? Am J Physiol Heart Circ Physiol 2005; 289:H522–H524.
119. Tracz MJ, Juncos JP, Croatt AJ, Ackerman AW, Grande JP, Knutson KL, Kane GC, Terzic A, Griffin MD, Nath KA. Deficiency of heme oxygenase-1 impairs renal hemodynamics and exaggerates systemic inflammatory responses to renal ischemia. Kidney Int 2007; 72:1073–1080.
120. Sandner P, Hofbauer KH, Tinel H, Kurtz A, Thiesson HC, Ottosen PD, Walter S, Skott O, Jensen BL. Expression of adrenomedullin in hypoxic and ischemic rat kidneys and human kidneys with arterial stenosis. Am J Physiol Regul Integr Comp Physiol 2004; 286: R942–R951.
121. Owji AA, Smith DM, Coppock HA, Morgan DG, Bhogal R, Ghatei MA, Bloom SR. An abundant and specific binding site for the novel vasodilator adrenomedullin in the rat. Endocrinology 1995; 136:2127–2134.
122. Juaneda C, Dumont Y, Chabot JG, Fournier A, Quirion R. Adrenomedullin receptor binding sites in rat brain and peripheral tissues. Eur J Pharmacol 2003; 474:165–174.
123. Ishimitsu T, Ono H, Minami J, Matsuoka H. Pathophysiologic and therapeutic implications of adrenomedullin in cardiovascular disorders. Pharmacol Ther 2006; 111:909–927.
124. Ishimitsu T, Nishikimi T, Saito Y, Kitamura K, Eto T, Kangawa K, Matsuo H, Omae T, Matsuoka H. Plasma levels of adrenomedullin, a newly identified hypotensive peptide, in patients with hypertension and renal failure. J Clin Invest 1994; 94:2158–2161.
125. Nishikimi T, Saito Y, Kitamura K, Ishimitsu T, Eto T, Kangawa K, Matsuo H, Omae T, Matsuoka H. Increased plasma levels of adrenomedullin in patients with heart failure. J Am Coll Cardiol 1995; 26:1424–1431.
126. Palmer LA, Semenza GL, Stoler MH, Johns RA. Hypoxia induces type II NOS gene expression in pulmonary artery endothelial cells via HIF-1. Am J Physiol 1998; 274: L212–L219.
127. Earley S, Nelin LD, Chicoine LG, Walker BR. Hypoxia-induced pulmonary endothelin-1 expression is unaltered by nitric oxide. J Appl Physiol 2002; 92:1152–1158.
128. Hu J, Discher DJ, Bishopric NH, Webster KA. Hypoxia regulates expression of the endothelin-1 gene through a proximal hypoxia-inducible factor-1 binding site on the antisense strand. Biochem Biophys Res Commun 1998; 245:894–899.
129. Semenza GL. O2-regulated gene expression: transcriptional control of cardiorespiratory physiology by HIF-1. J Appl Physiol 2004; 96:1173–1177; discussion 1170–1172.
130. Belaidi E, Joyeux-Faure M, Ribuot C, Launois SH, Levy P, Godin-Ribuot D. Major role for hypoxia inducible factor-1 and the endothelin system in promoting myocardial infarction and hypertension in an animal model of obstructive sleep apnea. J Am Coll Cardiol 2009; 53:1309–1317.
131. Feltes CM, Van Eyk J, Rabb H. Distant-organ changes after acute kidney injury. Nephron Physiol 2008; 109:p80–p84.
132. Ko GJ, Rabb H, Hassoun HT. Kidney-lung crosstalk in the critically ill patient. Blood Purif 2009; 28:75–83.
133. Miyazawa S, Watanabe H, Miyaji C, Hotta O, Abo T. Leukocyte accumulation and changes in extra-renal organs during renal ischemia reperfusion in mice. J Lab Clin Med 2002; 139:269–278.
134. Tokuyama H, Kelly DJ, Zhang Y, Gow RM, Gilbert RE. Macrophage infiltration and cellular proliferation in the non-ischemic kidney and heart following prolonged unilateral renal ischemia. Nephron Physiol 2007; 106:p54–p62.
135. Rabb H, Wang Z, Nemoto T, Hotchkiss J, Yokota N, Soleimani M. Acute renal failure leads to dysregulation of lung salt and water channels. Kidney Int 2003; 63:600–606.
136. Kelly KJ. Distant effects of experimental renal ischemia/reperfusion injury. J Am Soc Nephrol 2003; 14:1549–1558.
137. Nangaku M. Hypoxia and tubulointerstitial injury: a final common pathway to end-stage renal failure. Nephron Exp Nephrol 2004; 98:e8–e12.

138. Nangaku M. Chronic hypoxia and tubulointerstitial injury: a final common pathway to end-stage renal failure. J Am Soc Nephrol 2006; 17:17–25.
139. Nangaku M, Inagi R, Miyata T, Fujita T. Hypoxia and hypoxia-inducible factor in renal disease. Nephron Exp Nephrol 2008; 110:e1–e7.
140. Kang DH, Anderson S, Kim YG, Mazzalli M, Suga S, Jefferson JA, Gordon KL, Oyama TT, Hughes J, Hugo C, Kerjaschki D, Schreiner GF, Johnson RJ. Impaired angiogenesis in the aging kidney: vascular endothelial growth factor and thrombospondin-1 in renal disease. Am J Kidney Dis 2001; 37:601–611.
141. Kang DH, Hughes J, Mazzali M, Schreiner GF, Johnson RJ. Impaired angiogenesis in the remnant kidney model: II. Vascular endothelial growth factor administration reduces renal fibrosis and stabilizes renal function. J Am Soc Nephrol 2001; 12:1448–1457.
142. Kang DH, Joly AH, Oh SW, Hugo C, Kerjaschki D, Gordon KL, Mazzali M, Jefferson JA, Hughes J, Madsen KM, Schreiner GF, Johnson RJ. Impaired angiogenesis in the remnant kidney model: I. Potential role of vascular endothelial growth factor and thrombospondin-1. J Am Soc Nephrol 2001; 12:1434–1447.
143. Norman JT, Fine LG. Intrarenal oxygenation in chronic renal failure. Clin Exp Pharmacol Physiol 2006; 33:989–996.
144. Mackensen-Haen S, Bader R, Grund KE, Bohle A. Correlations between renal cortical interstitial fibrosis, atrophy of the proximal tubules and impairment of the glomerular filtration rate. Clin Nephrol 1981; 15:167–171.
145. Risdon RA, Berry CL, Chrispin AR. Urographic changes in acute papillary necrosis in the rat. Br Med J 1970; 3:263–265.
146. Schainuck LI, Striker GE, Cutler RE, Benditt EP. Structural-functional correlations in renal disease. II. The correlations. Hum Pathol 1970; 1:631–641.
147. Kang DH, Kanellis J, Hugo C, Truong L, Anderson S, Kerjaschki D, Schreiner GF, Johnson RJ. Role of the microvascular endothelium in progressive renal disease. J Am Soc Nephrol 2002; 13:806–816.
148. Hugo C, Shankland SJ, Pichler RH, Couser WG, Johnson RJ. Thrombospondin 1 precedes and predicts the development of tubulointerstitial fibrosis in glomerular disease in the rat. Kidney Int 1998; 53:302–311.
149. Rosenberger C, Rosen S, Paliege A, Heyman SN. Pimonidazole adduct immunohistochemistry in the rat kidney: detection of tissue hypoxia. In: Hewitson TD, Becker GJ (eds). Kidney Research. Experimental Protocols. Humana Press: New York, 2009.
150. Tanaka T, Miyata T, Inagi R, Fujita T, Nangaku M. Hypoxia in renal disease with proteinuria and/or glomerular hypertension. Am J Pathol 2004; 165:1979–1992.
151. Safran M, Kim WY, O'Connell F, Flippin L, Gunzler V, Horner JW, Depinho RA, Kaelin WG, Jr. Mouse model for noninvasive imaging of HIF prolyl hydroxylase activity: assessment of an oral agent that stimulates erythropoietin production. Proc Natl Acad Sci U S A 2006; 103:105–110.
152. Haase VH. Oxygen regulates epithelial-to-mesenchymal transition: insights into molecular mechanisms and relevance to disease. Kidney Int 2009; 76(5):492–499.
153. Liu Y. Epithelial to mesenchymal transition in renal fibrogenesis: pathologic significance, molecular mechanism, and therapeutic intervention. J Am Soc Nephrol 2004; 15:1–12.
154. Fine LG, Orphanides C, Norman JT. Progressive renal disease: the chronic hypoxia hypothesis. Kidney Int Suppl 1998; 65:S74–S78.
155. Eckardt KU, Ratcliffe PJ, Tan CC, Bauer C, Kurtz A. Age-dependent expression of the erythropoietin gene in rat liver and kidneys. J Clin Invest 1992; 89:753–760.
156. Deicher R, Horl WH. Hepcidin: a molecular link between inflammation and anaemia. Nephrol Dial Transplant 2004; 19:521–524.
157. Means RT, Jr. Advances in the anemia of chronic disease. Int J Hematol 1999; 70:7–12.
158. Silberberg JS, Barre PE, Prichard SS, Sniderman AD. Impact of left ventricular hypertrophy on survival in end-stage renal disease. Kidney Int 1989; 36:286–290.

159. London GM, Fabiani F, Marchais SJ, de Vernejoul MC, Guerin AP, Safar ME, Metivier F, Llach F. Uremic cardiomyopathy: an inadequate left ventricular hypertrophy. Kidney Int 1987; 31:973–980.
160. Foley RN, Parfrey PS, Harnett JD, Kent GM, Murray DC, Barre PE. The impact of anemia on cardiomyopathy, morbidity, and mortality in end-stage renal disease. Am J Kidney Dis 1996; 28:53–61.
161. Ritz E, McClellan WM. Overview: increased cardiovascular risk in patients with minor renal dysfunction: an emerging issue with far-reaching consequences. J Am Soc Nephrol 2004; 15:513–516.
162. Neill WA, Oxendine JM, Moore SC. Acute and chronic cardiovascular adjustments to induced anemia in dogs. Am J Physiol 1969; 217:710–714.
163. Schwartz S, Frantz RA, Shoemaker WC. Sequential hemodynamic and oxygen transport responses in hypovolemia, anemia, and hypoxia. Am J Physiol 1981; 241:H864–H871.
164. Mall G, Rambausek M, Neumeister A, Kollmar S, Vetterlein F, Ritz E. Myocardial interstitial fibrosis in experimental uremia – implications for cardiac compliance. Kidney Int 1988; 33:804–811.
165. Mall G, Huther W, Schneider J, Lundin P, Ritz E. Diffuse intermyocardiocytic fibrosis in uraemic patients. Nephrol Dial Transplant 1990; 5:39–44.
166. Amann K, Ritz E. Cardiovascular abnormalities in ageing and in uraemia – only analogy or shared pathomechanisms? Nephrol Dial Transplant 1998; 13 (Suppl 7):6–11.
167. Amann K, Breitbach M, Ritz E, Mall G. Myocyte/capillary mismatch in the heart of uremic patients. J Am Soc Nephrol 1998; 9:1018–1022.
168. Madore F, Lowrie EG, Brugnara C, Lew NL, Lazarus JM, Bridges K, Owen WF. Anemia in hemodialysis patients: variables affecting this outcome predictor. J Am Soc Nephrol 1997; 8:1921–1929.
169. Eckardt KU. Cardiovascular consequences of renal anaemia and erythropoietin therapy. Nephrol Dial Transplant 1999; 14:1317–1323.
170. Radermacher J, Koch KM. Treatment of renal anemia by erythropoietin substitution. The effects on the cardiovascular system. Clin Nephrol 1995; 44 (Suppl 1):S56–S60.
171. Mann JF. Hypertension and cardiovascular effects – long-term safety and potential long-term benefits of r-HuEPO. Nephrol Dial Transplant 1995; 10 (Suppl 2):80–84.
172. Singh AK, Szczech L, Tang KL, Barnhart H, Sapp S, Wolfson M, Reddan D. Correction of anemia with epoetin alfa in chronic kidney disease. N Engl J Med 2006; 355:2085–2098.
173. Drueke TB, Locatelli F, Clyne N, Eckardt KU, Macdougall IC, Tsakiris D, Burger HU, Scherhag A. Normalization of hemoglobin level in patients with chronic kidney disease and anemia. N Engl J Med 2006; 355:2071–2084.
174. Smith K, Semple D, Bhandari S, Seymour AM. Cellular basis of uraemic cardiomyopathy: arole for erythropoietin? Eur J Heart Fail 2009; 11:732–738.
175. Pagourelias ED, Koumaras C, Kakafika AI, Tziomalos K, Zorou PG, Athyros VG, Karagiannis A. Cardiorenal anemia syndrome: do erythropoietin and iron therapy have a place in the treatment of heart failure? Angiology 2009; 60:74–81.
176. Jie KE, Verhaar MC, Cramer MJ, van der Putten K, Gaillard CA, Doevendans PA, Koomans HA, Joles JA, Braam B. Erythropoietin and the cardiorenal syndrome: cellular mechanisms on the cardiorenal connectors. Am J Physiol Renal Physiol 2006; 291:F932–F944.
177. Silverberg DS, Wexler D, Blum M, Keren G, Sheps D, Leibovitch E, Brosh D, Laniado S, Schwartz D, Yachnin T, Shapira I, Gavish D, Baruch R, Koifman B, Kaplan C, Steinbruch S, Iaina A. The use of subcutaneous erythropoietin and intravenous iron for the treatment of the anemia of severe, resistant congestive heart failure improves cardiac and renal function and functional cardiac class, and markedly reduces hospitalizations. J Am Coll Cardiol 2000; 35:1737–1744.
178. Ishani A, Weinhandl E, Zhao Z, Gilbertson DT, Collins AJ, Yusuf S, Herzog CA. Angiotensin-converting enzyme inhibitor as a risk factor for the development of anemia, and the

impact of incident anemia on mortality in patients with left ventricular dysfunction. J Am Coll Cardiol 2005; 45:391–399.
179. Mancini DM, Katz SD, Lang CC, LaManca J, Hudaihed A, Androne AS. Effect of erythropoietin on exercise capacity in patients with moderate to severe chronic heart failure. Circulation 2003; 107:294–299.
180. Silverberg DS, Wexler D, Sheps D, Blum M, Keren G, Baruch R, Schwartz D, Yachnin T, Steinbruch S, Shapira I, Laniado S, Iaina A. The effect of correction of mild anemia in severe, resistant congestive heart failure using subcutaneous erythropoietin and intravenous iron: a randomized controlled study. J Am Coll Cardiol 2001; 37:1775–1780.
181. Brines M, Grasso G, Fiordaliso F, Sfacteria A, Ghezzi P, Fratelli M, Latini R, Xie QW, Smart J, Su-Rick CJ, Pobre E, Diaz D, Gomez D, Hand C, Coleman T, Cerami A. Erythropoietin mediates tissue protection through an erythropoietin and common beta-subunit heteroreceptor. Proc Natl Acad Sci U S A 2004; 101:14907–14912.
182. Ammarguellat F, Gogusev J, Drueke TB. Direct effect of erythropoietin on rat vascular smooth-muscle cell via a putative erythropoietin receptor. Nephrol Dial Transplant 1996; 11:687–692.
183. Depping R, Kawakami K, Ocker H, Wagner JM, Heringlake M, Noetzold A, Sievers HH, Wagner KF. Expression of the erythropoietin receptor in human heart. J Thorac Cardiovasc Surg 2005; 130:877–878.
184. Weidemann A, Johnson RS. Nonrenal regulation of EPO synthesis. Kidney Int 2009; 75:682–688.
185. Merchionne F, Dammacco F. Biological functions and therapeutic use of erythropoiesis-stimulating agents: perplexities and perspectives. Br J Haematol 2009; 146:127–141.
186. Arcasoy MO. The non-haematopoietic biological effects of erythropoietin. Br J Haematol 2008; 141:14–31.
187. Haroon ZA, Amin K, Jiang X, Arcasoy MO. A novel role for erythropoietin during fibrin-induced wound-healing response. Am J Pathol 2003; 163:993–1000.
188. Fu QL, Wu W, Wang H, Li X, Lee VW, So KF. Up-regulated endogenous erythropoietin/erythropoietin receptor system and exogenous erythropoietin rescue retinal ganglion cells after chronic ocular hypertension. Cell Mol Neurobiol 2008; 28:317–329.
189. Calvillo L, Latini R, Kajstura J, Leri A, Anversa P, Ghezzi P, Salio M, Cerami A, Brines M. Recombinant human erythropoietin protects the myocardium from ischemia-reperfusion injury and promotes beneficial remodeling. Proc Natl Acad Sci U S A 2003; 100:4802–4806.
190. Tramontano AF, Muniyappa R, Black AD, Blendea MC, Cohen I, Deng L, Sowers JR, Cutaia MV, El-Sherif N. Erythropoietin protects cardiac myocytes from hypoxia-induced apoptosis through an Akt-dependent pathway. Biochem Biophys Res Commun 2003; 308:990–994.
191. Li Y, Takemura G, Okada H, Miyata S, Maruyama R, Li L, Higuchi M, Minatoguchi S, Fujiwara T, Fujiwara H. Reduction of inflammatory cytokine expression and oxidative damage by erythropoietin in chronic heart failure. Cardiovasc Res 2006; 71:684–694.
192. van der Meer P, Groenveld HF, Januzzi JL, Jr., van Veldhuisen DJ. Erythropoietin treatment in patients with chronic heart failure: a meta-analysis. Heart 2009; 95:1309–1314.
193. Razavi HM, Hamilton JA, Feng Q. Modulation of apoptosis by nitric oxide: implications in myocardial ischemia and heart failure. Pharmacol Ther 2005; 106:147–162.
194. Smith RS, Jr., Agata J, Xia CF, Chao L, Chao J. Human endothelial nitric oxide synthase gene delivery protects against cardiac remodeling and reduces oxidative stress after myocardial infarction. Life Sci 2005; 76:2457–2471.
195. Burger D, Lei M, Geoghegan-Morphet N, Lu X, Xenocostas A, Feng Q. Erythropoietin protects cardiomyocytes from apoptosis via up-regulation of endothelial nitric oxide synthase. Cardiovasc Res 2006; 72:51–59.
196. Nishiya D, Omura T, Shimada K, Matsumoto R, Kusuyama T, Enomoto S, Iwao H, Takeuchi K, Yoshikawa J, Yoshiyama M. Effects of erythropoietin on cardiac remodeling after myocardial infarction. J Pharmacol Sci 2006; 101:31–39.

197. Westenbrink BD, Lipsic E, van der Meer P, van der Harst P, Oeseburg H, Du Marchie Sarvaas GJ, Koster J, Voors AA, van Veldhuisen DJ, van Gilst WH, Schoemaker RG. Erythropoietin improves cardiac function through endothelial progenitor cell and vascular endothelial growth factor mediated neovascularization. Eur Heart J 2007; 28:2018–2027.
198. McMurray JJ, Anand IS, Diaz R, Maggioni AP, O'Connor C, Pfeffer MA, Polu KR, Solomon SD, Sun Y, Swedberg K, Tendera M, van Veldhuisen DJ, Wasserman SM, Young JB. Design of the reduction of events with Darbepoetin alfa in Heart Failure (RED-HF): a phase III, anaemia correction, morbidity-mortality trial. Eur J Heart Fail 2009; 11:795–801.
199. Pfeffer MA, Burdmann EA, Chen CY, Cooper ME, de Zeeuw D, Eckardt KU, Feyzi JM, Ivanovich P, Kewalramani R, Levey AS, Lewis EF, McGill JB, McMurray JJ, Parfrey P, Parving HH, Remuzzi G, Singh AK, Solomon SD, Toto R. A trial of darbepoetin alfa in type 2 diabetes and chronic kidney disease. N Engl J Med 2009; 361:2019–2032.
200. Ross R. Atherosclerosis – an inflammatory disease. N Engl J Med 1999; 340:115–126.
201. Libby P, Ridker PM, Maseri A. Inflammation and atherosclerosis. Circulation 2002; 105:1135–1143.
202. Zethelius B, Berglund L, Sundstrom J, Ingelsson E, Basu S, Larsson A, Venge P, Arnlov J. Use of multiple biomarkers to improve the prediction of death from cardiovascular causes. N Engl J Med 2008; 358:2107–2116.
203. Willerson JT, Ridker PM. Inflammation as a cardiovascular risk factor. Circulation 2004; 109:II2–II10.
204. Pepys MB, Hirschfield GM. C-reactive protein: a critical update. J Clin Invest 2003; 111:1805–1812.
205. Gotschlich EC, Liu TY, Oliveira E. Binding of C-reactive protein to C-carbohydrate and PC-substituted protein. Ann N Y Acad Sci 1982; 389:163–171.
206. Calabro P, Willerson JT, Yeh ET. Inflammatory cytokines stimulated C-reactive protein production by human coronary artery smooth muscle cells. Circulation 2003; 108:1930–1932.
207. Jabs WJ, Theissing E, Nitschke M, Bechtel JF, Duchrow M, Mohamed S, Jahrbeck B, Sievers HH, Steinhoff J, Bartels C. Local generation of C-reactive protein in diseased coronary artery venous bypass grafts and normal vascular tissue. Circulation 2003; 108:1428–1431.
208. Kobayashi S, Inoue N, Ohashi Y, Terashima M, Matsui K, Mori T, Fujita H, Awano K, Kobayashi K, Azumi H, Ejiri J, Hirata K, Kawashima S, Hayashi Y, Yokozaki H, Itoh H, Yokoyama M. Interaction of oxidative stress and inflammatory response in coronary plaque instability: important role of C-reactive protein. Arterioscler Thromb Vasc Biol 2003; 23:1398–1404.
209. Koenig W, Sund M, Frohlich M, Fischer HG, Lowel H, Doring A, Hutchinson WL, Pepys MB. C-reactive protein, a sensitive marker of inflammation, predicts future risk of coronary heart disease in initially healthy middle-aged men: results from the MONICA (Monitoring Trends and Determinants in Cardiovascular Disease) Augsburg Cohort Study, 1984 to 1992. Circulation 1999; 99:237–242.
210. Ridker PM, Danielson E, Fonseca FA, Genest J, Gotto AM, Jr., Kastelein JJ, Koenig W, Libby P, Lorenzatti AJ, MacFadyen JG, Nordestgaard BG, Shepherd J, Willerson JT, Glynn RJ. Rosuvastatin to prevent vascular events in men and women with elevated C-reactive protein. N Engl J Med 2008; 359:2195–2207.
211. Ridker PM, Rifai N, Rose L, Buring JE, Cook NR. Comparison of C-reactive protein and low-density lipoprotein cholesterol levels in the prediction of first cardiovascular events. N Engl J Med 2002; 347:1557–1565.
212. Zacho J, Tybjaerg-Hansen A, Jensen JS, Grande P, Sillesen H, Nordestgaard BG. Genetically elevated C-reactive protein and ischemic vascular disease. N Engl J Med 2008; 359:1897–1908.
213. Xia D, Samols D. Transgenic mice expressing rabbit C-reactive protein are resistant to endotoxemia. Proc Natl Acad Sci U S A 1997; 94:2575–2580.

214. Schwedler SB, Amann K, Wernicke K, Krebs A, Nauck M, Wanner C, Potempa LA, Galle J. Native C-reactive protein increases whereas modified C-reactive protein reduces atherosclerosis in apolipoprotein E-knockout mice. Circulation 2005; 112:1016–1023.
215. Schwedler SB, Hansen-Hagge T, Reichert M, Schmiedeke D, Schneider R, Galle J, Potempa LA, Wanner C, Filep JG. Monomeric C-reactive protein decreases acetylated LDL uptake in human endothelial cells. Clin Chem 2009; 55:1728–1731.
216. Qamirani E, Ren Y, Kuo L, Hein TW. C-reactive protein inhibits endothelium-dependent NO-mediated dilation in coronary arterioles by activating p38 kinase and NAD(P)H oxidase. Arterioscler Thromb Vasc Biol 2005; 25:995–1001.
217. Chang MK, Binder CJ, Torzewski M, Witztum JL. C-reactive protein binds to both oxidized LDL and apoptotic cells through recognition of a common ligand: phosphorylcholine of oxidized phospholipids. Proc Natl Acad Sci U S A 2002; 99:13043–13048.
218. Hulthe J, Fagerberg B. Circulating oxidized LDL is associated with subclinical atherosclerosis development and inflammatory cytokines (AIR Study). Arterioscler Thromb Vasc Biol 2002; 22:1162–1167.
219. Pihl E, Zilmer K, Kullisaar T, Kairane C, Pulges A, Zilmer M. High-sensitive C-reactive protein level and oxidative stress-related status in former athletes in relation to traditional cardiovascular risk factors. Atherosclerosis 2003; 171:321–326.
220. Yasunari K, Maeda K, Nakamura M, Yoshikawa J. Oxidative stress in leukocytes is a possible link between blood pressure, blood glucose, and C-reacting protein. Hypertension 2002; 39:777–780.
221. Griendling KK, Minieri CA, Ollerenshaw JD, Alexander RW. Angiotensin II stimulates NADH and NADPH oxidase activity in cultured vascular smooth muscle cells. Circ Res 1994; 74:1141–1148.
222. Zafari AM, Ushio-Fukai M, Akers M, Yin Q, Shah A, Harrison DG, Taylor WR, Griendling KK. Role of NADH/NADPH oxidase-derived H2O2 in angiotensin II-induced vascular hypertrophy. Hypertension 1998; 32:488–495.
223. Cifuentes ME, Rey FE, Carretero OA, Pagano PJ. Upregulation of p67(phox) and gp91 (phox) in aortas from angiotensin II-infused mice. Am J Physiol Heart Circ Physiol 2000; 279:H2234–H2240.
224. Lavigne MC, Malech HL, Holland SM, Leto TL. Genetic demonstration of p47phox-dependent superoxide anion production in murine vascular smooth muscle cells. Circulation 2001; 104:79–84.
225. Matsuno K, Yamada H, Iwata K, Jin D, Katsuyama M, Matsuki M, Takai S, Yamanishi K, Miyazaki M, Matsubara H, Yabe-Nishimura C. Nox1 is involved in angiotensin II-mediated hypertension: a study in Nox1-deficient mice. Circulation 2005; 112:2677–2685.
226. Pagano PJ, Chanock SJ, Siwik DA, Colucci WS, Clark JK. Angiotensin II induces p67phox mRNA expression and NADPH oxidase superoxide generation in rabbit aortic adventitial fibroblasts. Hypertension 1998; 32:331–337.
227. Pagano PJ, Clark JK, Cifuentes-Pagano ME, Clark SM, Callis GM, Quinn MT. Localization of a constitutively active, phagocyte-like NADPH oxidase in rabbit aortic adventitia: enhancement by angiotensin II. Proc Natl Acad Sci U S A 1997; 94:14483–14488.
228. Ushio-Fukai M, Alexander RW, Akers M, Griendling KK. p38 Mitogen-activated protein kinase is a critical component of the redox-sensitive signaling pathways activated by angiotensin II. Role in vascular smooth muscle cell hypertrophy. J Biol Chem 1998; 273:15022–15029.
229. Zimmerman MC, Dunlay RP, Lazartigues E, Zhang Y, Sharma RV, Engelhardt JF, Davisson RL. Requirement for Rac1-dependent NADPH oxidase in the cardiovascular and dipsogenic actions of angiotensin II in the brain. Circ Res 2004; 95:532–539.
230. Touyz RM, Chen X, Tabet F, Yao G, He G, Quinn MT, Pagano PJ, Schiffrin EL. Expression of a functionally active gp91phox-containing neutrophil-type NAD(P)H oxidase in smooth muscle cells from human resistance arteries: regulation by angiotensin II. Circ Res 2002; 90:1205–1213.

231. Chabrashvili T, Kitiyakara C, Blau J, Karber A, Aslam S, Welch WJ, Wilcox CS. Effects of Ang II type 1 and 2 receptors on oxidative stress, renal NADPH oxidase, and SOD expression. Am J Physiol Regul Integr Comp Physiol 2003; 285:R117–R124.
232. Mehta PK, Griendling KK. Angiotensin II cell signaling: physiological and pathological effects in the cardiovascular system. Am J Physiol Cell Physiol 2007; 292:C82–C97.
233. Williams HC, Griendling KK. NADPH oxidase inhibitors: new antihypertensive agents? J Cardiovasc Pharmacol 2007; 50:9–16.
234. Liu F, Wei CC, Wu SJ, Chenier I, Zhang SL, Filep JG, Ingelfinger JR, Chan JS. Apocynin attenuates tubular apoptosis and tubulointerstitial fibrosis in transgenic mice independent of hypertension. Kidney Int 2009; 75:156–166.
235. Miller AA, Drummond GR, Schmidt HH, Sobey CG. NADPH oxidase activity and function are profoundly greater in cerebral versus systemic arteries. Circ Res 2005; 97:1055–1062.
236. Virdis A, Neves MF, Amiri F, Touyz RM, Schiffrin EL. Role of NAD(P)H oxidase on vascular alterations in angiotensin II-infused mice. J Hypertens 2004; 22:535–542.
237. Chen TH, Wang JF, Chan P, Lee HM. Angiotensin II stimulates hypoxia-inducible factor 1alpha accumulation in glomerular mesangial cells. Ann N Y Acad Sci 2005; 1042:286–293.
238. Peng YJ, Yuan G, Ramakrishnan D, Sharma SD, Bosch-Marce M, Kumar GK, Semenza GL, Prabhakar NR. Heterozygous HIF-1alpha deficiency impairs carotid body-mediated systemic responses and reactive oxygen species generation in mice exposed to intermittent hypoxia. J Physiol 2006; 577:705–716.
239. Taniyama Y, Ushio-Fukai M, Hitomi H, Rocic P, Kingsley MJ, Pfahnl C, Weber DS, Alexander RW, Griendling KK. Role of p38 MAPK and MAPKAPK-2 in angiotensin II-induced Akt activation in vascular smooth muscle cells. Am J Physiol Cell Physiol 2004; 287:C494–C499.
240. Forstermann U. Janus-faced role of endothelial NO synthase in vascular disease: uncoupling of oxygen reduction from NO synthesis and its pharmacological reversal. Biol Chem 2006; 387:1521–1533.
241. Zou MH, Shi C, Cohen RA. Oxidation of the zinc-thiolate complex and uncoupling of endothelial nitric oxide synthase by peroxynitrite. J Clin Invest 2002; 109:817–826.
242. Fleming I, Mohamed A, Galle J, Turchanowa L, Brandes RP, Fisslthaler B, Busse R. Oxidized low-density lipoprotein increases superoxide production by endothelial nitric oxide synthase by inhibiting PKCalpha. Cardiovasc Res 2005; 65:897–906.
243. Landmesser U, Dikalov S, Price SR, McCann L, Fukai T, Holland SM, Mitch WE, Harrison DG. Oxidation of tetrahydrobiopterin leads to uncoupling of endothelial cell nitric oxide synthase in hypertension. J Clin Invest 2003; 111:1201–1209.
244. Mollnau H, Wendt M, Szocs K, Lassegue B, Schulz E, Oelze M, Li H, Bodenschatz M, August M, Kleschyov AL, Tsilimingas N, Walter U, Forstermann U, Meinertz T, Griendling K, Munzel T. Effects of angiotensin II infusion on the expression and function of NAD(P)H oxidase and components of nitric oxide/cGMP signaling. Circ Res 2002; 90:E58–E65.
245. Boger RH, Sydow K, Borlak J, Thum T, Lenzen H, Schubert B, Tsikas D, Bode-Boger SM. LDL cholesterol upregulates synthesis of asymmetrical dimethylarginine in human endothelial cells: involvement of S-adenosylmethionine-dependent methyltransferases. Circ Res 2000; 87:99–105.
246. Sydow K, Munzel T. ADMA and oxidative stress. Atheroscler Suppl 2003; 4:41–51.
247. Stroes E, Kastelein J, Cosentino F, Erkelens W, Wever R, Koomans H, Luscher T, Rabelink T. Tetrahydrobiopterin restores endothelial function in hypercholesterolemia. J Clin Invest 1997; 99:41–46.
248. Heitzer T, Krohn K, Albers S, Meinertz T. Tetrahydrobiopterin improves endothelium-dependent vasodilation by increasing nitric oxide activity in patients with type II diabetes mellitus. Diabetologia 2000; 43:1435–1438.
249. Higashi Y, Sasaki S, Nakagawa K, Fukuda Y, Matsuura H, Oshima T, Chayama K. Tetrahydrobiopterin enhances forearm vascular response to acetylcholine in both normotensive and hypertensive individuals. Am J Hypertens 2002; 15:326–332.

250. Satoh M, Fujimoto S, Haruna Y, Arakawa S, Horike H, Komai N, Sasaki T, Tsujioka K, Makino H, Kashihara N. NAD(P)H oxidase and uncoupled nitric oxide synthase are major sources of glomerular superoxide in rats with experimental diabetic nephropathy. Am J Physiol Renal Physiol 2005; 288:F1144–F1152.
251. Satoh M, Fujimoto S, Arakawa S, Yada T, Namikoshi T, Haruna Y, Horike H, Sasaki T, Kashihara N. Angiotensin II type 1 receptor blocker ameliorates uncoupled endothelial nitric oxide synthase in rats with experimental diabetic nephropathy. Nephrol Dial Transplant 2008; 23:3806–3813.
252. Yamamizu K, Shinozaki K, Ayajiki K, Gemba M, Okamura T. Oral administration of both tetrahydrobiopterin and L-arginine prevents endothelial dysfunction in rats with chronic renal failure. J Cardiovasc Pharmacol 2007; 49:131–139.
253. Xu H, Fink GD, Chen A, Watts S, Galligan JJ. Nitric oxide-independent effects of tempol on sympathetic nerve activity and blood pressure in normotensive rats. Am J Physiol Heart Circ Physiol 2001; 281:H975–H980.
254. Shokoji T, Nishiyama A, Fujisawa Y, Hitomi H, Kiyomoto H, Takahashi N, Kimura S, Kohno M, Abe Y. Renal sympathetic nerve responses to tempol in spontaneously hypertensive rats. Hypertension 2003; 41:266–273.
255. Xu H, Fink GD, Galligan JJ. Nitric oxide-independent effects of tempol on sympathetic nerve activity and blood pressure in DOCA-salt rats. Am J Physiol Heart Circ Physiol 2002; 283:H885–H892.
256. Huang HS, Pan HL, Stahl GL, Longhurst JC. Ischemia- and reperfusion-sensitive cardiac sympathetic afferents: influence of H2O2 and hydroxyl radicals. Am J Physiol 1995; 269: H888–H901.
257. Ustinova EE, Schultz HD. Activation of cardiac vagal afferents by oxygen-derived free radicals in rats. Circ Res 1994; 74:895–903.
258. Ustinova EE, Schultz HD. Activation of cardiac vagal afferents in ischemia and reperfusion. Prostaglandins versus oxygen-derived free radicals. Circ Res 1994; 74:904–911.
259. Longhurst JC, Tjen ALSC, Fu LW. Cardiac sympathetic afferent activation provoked by myocardial ischemia and reperfusion. Mechanisms and reflexes. Ann N Y Acad Sci 2001; 940:74–95.
260. Zanzinger J, Czachurski J. Chronic oxidative stress in the RVLM modulates sympathetic control of circulation in pigs. Pflugers Arch 2000; 439:489–494.
261. Campese VM, Ye S, Zhong H, Yanamadala V, Ye Z, Chiu J. Reactive oxygen species stimulate central and peripheral sympathetic nervous system activity. Am J Physiol Heart Circ Physiol 2004; 287:H695–H703.
262. Han Y, Zhang Y, Wang HJ, Gao XY, Wang W, Zhu GQ. Reactive oxygen species in paraventricular nucleus modulates cardiac sympathetic afferent reflex in rats. Brain Res 2005; 1058:82–90.
263. Campese VM, Shaohua Y, Huiquin Z. Oxidative stress mediates angiotensin II-dependent stimulation of sympathetic nerve activity. Hypertension 2005; 46:533–539.
264. Gao L, Wang W, Li YL, Schultz HD, Liu D, Cornish KG, Zucker IH. Sympathoexcitation by central Ang II: roles for AT1 receptor upregulation and NAD(P)H oxidase in RVLM. Am J Physiol Heart Circ Physiol 2005; 288:H2271–H2279.
265. Zimmerman MC, Lazartigues E, Lang JA, Sinnayah P, Ahmad IM, Spitz DR, Davisson RL. Superoxide mediates the actions of angiotensin II in the central nervous system. Circ Res 2002; 91:1038–1045.
266. Gao L, Wang W, Li YL, Schultz HD, Liu D, Cornish KG, Zucker IH. Superoxide mediates sympathoexcitation in heart failure: roles of angiotensin II and NAD(P)H oxidase. Circ Res 2004; 95:937–944.
267. Wang Y, Chen AF, Wang DH. Enhanced oxidative stress in kidneys of salt-sensitive hypertension: role of sensory nerves. Am J Physiol Heart Circ Physiol 2006; 291: H3136–H3143.

268. Starr A, Graepel R, Keeble J, Schmidhuber S, Clark N, Grant A, Shah AM, Brain SD. A reactive oxygen species-mediated component in neurogenic vasodilatation. Cardiovasc Res 2008; 78:139–147.
269. Koyner JL, Sher Ali R, Murray PT. Antioxidants. Do they have a place in the prevention or therapy of acute kidney injury? Nephron Exp Nephrol 2008; 109:e109–e117.
270. Matsumoto M, Tanaka T, Yamamoto T, Noiri E, Miyata T, Inagi R, Fujita T, Nangaku M. Hypoperfusion of peritubular capillaries induces chronic hypoxia before progression of tubulointerstitial injury in a progressive model of rat glomerulonephritis. J Am Soc Nephrol 2004; 15:1574-1581.

Chapter 27
Hypoxia-Inducible Factors in Acute Kidney Injury: From Pathophysiology to a Novel Approach of Organ Protection

Wanja M. Bernhardt, Carsten Willam, and Kai-Uwe Eckardt

Abstract Acute kidney injury remains a major clinical problem associated with poor patient outcome and therapeutic options are still limited. The transcription of hypoxia-inducible factors (HIF) is a key regulator of the cellular response to hypoxia. Far more than 100 HIF target genes have been identified, including genes that have been demonstrated to protect the kidney. Although renal cells have a broad capacity to activate HIF, HIF is only sporadically stabilized in cases of acute renal injury and therefore protective effects may be limited. Newly identified inhibitors of the oxygen-sensing HIF-prolyl-hydroxylases now offer the possibility to pharmacologically stabilize HIF independent of oxygen prior to a renal insult in order to achieve protection. Thus, HIF activation appears as a promising therapeutic target for the prevention of acute kidney injury.

Keywords Renin · Angiotensin · Cardiorenal syndrome · Chronic kidney disease

1 Introduction

Following the discovery of hypoxia-inducible factors (HIF) as a main regulator of oxygen-dependent erythropoietin production, the HIF transcription factors have been recognized as a master switch of cellular adaptation to reduce oxygen availability [1–3]. The functional consequences of HIF induction during health and disease are increasingly recognized as complex and context dependent. From a clinical perspective HIF induction can probably be either adaptive or maladaptive in certain disease states and thus either limit or aggravate alterations in tissue structure and function. The net effect of HIF induction may depend on

W.M. Bernhardt (✉)
Department of Nephrology and Hypertension, Friedrich-Alexander-University, Erlangen-Nuremberg, Krankenhausstrasse 10–12, D-91054 Erlangen, Germany
e-mail: wanja.bernhardt@uk-erlangen.de

T. Miyata et al. (eds.), *Studies on Renal Disorders*, Oxidative Stress in Applied Basic Research and Clinical Practice, DOI 10.1007/978-1-60761-857-7_27,

circumstances and cells or tissues, but probably also on the process of disease. Currently available evidence suggests that in the short term HIF induction mainly induces protective effects, while the consequences of long-term activation are more difficult to assess. Increasing understanding of the molecular mechanisms of HIF regulation has started to provide opportunities for therapeutic intervention. While interfering with the HIF system may be clinically meaningful with respect to the variety of organ systems, most comprehensive data about beneficial effects of HIF induction have so far been generated from studies of the kidney.

2 Renal Oxygenation in Acute Kidney Injury

Renal arteries normally receive approximately 25% of the total cardiac output blood supply. Nevertheless, oxygen tension within certain areas of the kidney is low and remains usually below the pO_2 in the renal vein [4]. This pattern has been attributed to the vascular architecture of the kidney, with arterial and venous vessels running in parallel and close proximity. The countercurrent flow allows oxygen to diffuse from arterial into venous branches before it has reached the peritubular capillary bed [5]. Oxygen tensions are particularly low in the renal medulla, but marked variability has also been measured in the renal cortex [6]. Tubular sodium reabsorption varies with glomerular filtration rate and accounts for most oxygen consumption in the kidney. Changes in oxygen delivery due to changes in renal blood flow may thus provide an adaptive mechanism in situations with changes in renal oxygen demand. In addition, the above mentioned diffusional arteriovenous oxygen shunting provides a mechanism by which blood and oxygen can bypass the renal microcirculation, thus preventing development of tissue hyperoxia or hyperperfusion. Actual concepts of renal oxygenation assume that renal blood flow-dependent changes in oxygen shunt flow help to maintain constant renal oxygen tensions [7]. However, the comparatively low oxygen tension under normoxic conditions is considered a main reason for the extraordinary susceptibility of the kidney to acute hypoxic injury [8]. In case of an acute renal injury, the breakdown of the glomerular filtration rate may reduce renal oxygen consumption, but a number of pathophysiological mechanisms, such as endothelial injury with microvascular dysfunction or local inflammation, lead to a net aggravation of renal hypoxia [9]. Activation of hypoxia-inducible transcription factors may help to adapt renal cells to low oxygen tensions and thus protect these cells from hypoxic injury.

3 The Kidney Has a Widespread Capacity to Induce HIF

Several studies have shown that genetic deficiency in one of the components of the HIF system impairs normal development [10–14]. While these findings indicate that HIF expression is an important signal for normal tissue development, the early mortality of HIF-1 knockout animals made assessment of the specific role of HIF in renal development impossible. Kidney development still proceeds after embryonic day (E) 10.5, where embryonic lethality in HIF-deficient animals occurred [10]. However, descriptive analysis of HIF expression in rodent and human fetal kidneys revealed a widespread activation, which ceases with determination of kidney development [15, 16]. These data suggest that HIF does play a role in nephron formation. When renal development is completed, little or no HIF expression can be detected in the kidney under baseline conditions, despite well-known marked oxygen gradients and persistently low oxygen tensions in the renal medulla [17, 18]. However, exposure to systemic hypoxia reveals a widespread capability of tubular and nonparenchymal cells in the kidney to induce HIF [17, 18]. Interestingly HIF-1α and HIF-2α, the two most relevant oxygen regulated HIF isoforms, are expressed in different renal cell populations. While HIF-1α is mainly expressed in tubular cells, HIF-2α can be found in glomerular cells and interstitial cells, including peritubular endothelial cells and interstitial fibroblasts [17, 18]. Demonstration of HIF-2α in the latter cell type provided the first hint that HIF-2α, rather than HIF-1α, is the critical regulator of erythropoietin synthesis that occurs in the cell population [17–19]. Meanwhile, the critical role of HIF-2α in erythropoietin regulation has been supported by both in vitro studies [20] and several genetic approaches [21–23].

Within tubular cells, HIF-1α is not uniformly expressed along the nephron. Some segments, for instance S1 and S2, as well as a collecting duct reveal much stronger HIF accumulation than the remaining parts of the nephron [17]. The reason for this nonhomogenous expression remains poorly understood. The HIF prolyl-hydroxylases (PHDs), the enzymes responsible for HIF hydroxylation in the presence of oxygen that serve as gatekeepers of HIF degradation [24–27], are also not uniformly expressed along the nephron [28]. PHD2 protein, probably the most relevant PHD enzyme, was predominantly expressed in distal tubular cells, in particular in the medullary thick ascending limb, the distal convoluted tubule, and the collecting duct. Overall, PHD2 protein expression gradually increased from the cortical toward the medullary region of the kidney. Here, higher PHD2 expression levels could potentially lead to a control of HIF accumulation and target gene activation under normal conditions in the medulla, despite the prevailing low physiological oxygen tensions [28].

4 HIF Is Upregulated in Different Forms of Acute Kidney Injury

Similar to systemic hypoxia, acute regional hypoxia in the kidney regularly induces HIF, albeit in a much more restricted fashion. The regional expression pattern in relation to areas of severe hypoxia can most easily be assessed in infarct models [29]. A band of cells at the border of renal infarct were found to upregulate HIF, presumably cells that are affected by reduced oxygen availability in the area supplied by an obstructed vessel, but still viable enough to mount an HIF response. A similar "border zone" expression pattern of HIF can also be observed in more complex models of renal injury, in which hypoxia plays a role. For instance, in a model of radio-contrast nephropathy in rats, tubules adjacent to the most hypoxic segments in the outer medulla were found to show strongest upregulation of HIF[30].

Further insight into the mechanisms regulating HIF was obtained from maneuvers modulating oxygen consumption. It is well established that renal oxygen tensions depend on the balance between oxygen supply and consumption. Under experimental conditions renal damage can be mitigated by inhibition of oxygen consumption, for instance by the use of diuretics that block active sodium readsorption. Both in vivo and in the isolated perfused rat kidney a reduction in tissue damage achieved by blocking sodium readsorption was found to be associated with increased HIF accumulation [30, 31]. It is tempting to speculate that by reducing oxygen consumption and thereby shifting the ratio between oxygen supply and consumption into more positive range, cells are may be moved back into the operational range of HIF expression. Moreover, it appears likely that the increased HIF expression seen under these conditions contributes to mitigation of renal injury.

A further aspect important for the functional consequences of HIF activation is time. For instance, in a model of myoglobulinuric acute kidney injury tissue hypoxia as assessed by staining for the bioreductive dye pimonidazole persisted for up to 24 h, whereas HIF was induced by 3 h but had virtually disappeared after 6–24 h [32]. The reasons for this rapid downregulation of the HIF system during persistent hypoxia are not fully understood, but may include HIF-dependent induction of PHDs [33, 34]. The expression of HIF target genes is characterized by a delay of several hours and was found to persist for up to 24 h, and thus beyond demonstrable HIF activation [35]. In terms of potential intervention these data suggest that several hours may be needed to induce a maximal HIF target gene response.

4.1 *Partial Deficiency of HIF Worsens Renal Injury*

Because systemic HIF-1 knockout leads to embryonic lethality, regional and partial knockout of the HIF variants were experimentally tested and gave new insights into the adaptive role of HIF activation in acute renal injury. Kojima et al. demonstrated that ischemia reperfusion injury of the kidney was aggravated in HIF-2α knockout mice [36]. Similarly, Hill et al. showed that in mice, which were heterozygous for either HIF-1α or HIF-2α, the injury following ischemia reperfusion was more severe than in control animals [37]. These data were corroborated by a number of experiments showing that, conversely, HIF induction ameliorated acute kidney injury.

4.2 HIF Induction Protects Against Acute Renal Injury

Pathophysiological characteristics of acute kidney injury include changes in tubular integrity, such as tubular necrosis and flattening, microvascular changes due endothelial injury, local inflammation, and leukocyte infiltration. Via the activation of its target genes, HIF stabilization has the potential to protect the kidney by inducing adaptive mechanisms. In a simplified scheme Fig. 1 summarizes those adaptive mechanisms by which HIF may contribute to renal protection.

In order to test the effect of HIF induction on the sensitivity toward subsequent ischemic injury, rats were preexposed to systemic hypoxia prior to induction of renal ischemia reperfusion injury [38]. Despite the much more prolonged reduction in oxygen supply, renal structure and function were much better preserved in those animals previously exposed to systemic hypoxia [38]. In addition to hypoxia, inhibitors of the HIF-prolyl- and asparaginyl hydroxylases, which stabilize HIF and subsequently activate HIF targets, can be used to mimic the hypoxic response. Correspondingly, cobalt chloride, which most likely substitutes the active iron in the catalytic domain of PHD and thus inhibits its enzymatic activity, has been demonstrated to protect the rat kidney from acute ischemic injury [39].

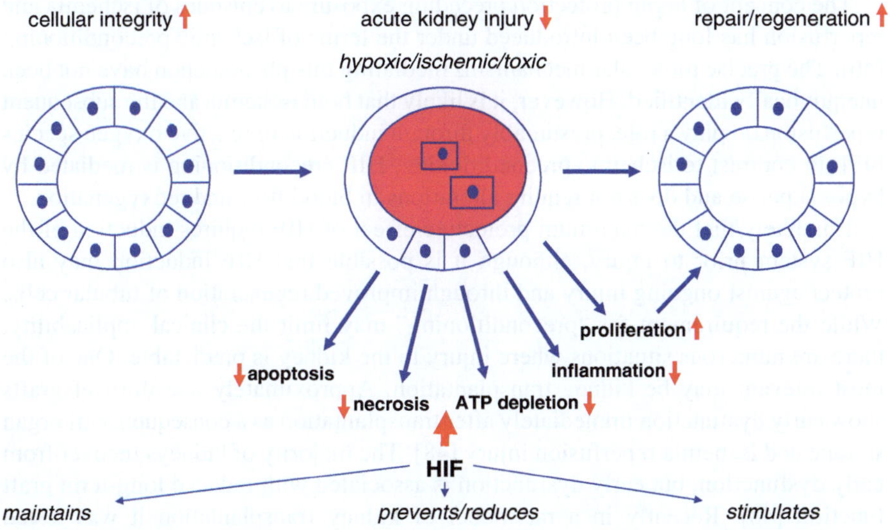

Fig. 1 A simplified schematic view with tubular cells representing pathophysiological changes in acute kidney injury. Beside other mechanisms, the consequences of an acute renal insult are characterized by a loss of cellular integrity, necrosis, and apoptosis of tubular epithelial cells. The recovery phase is characterized by tubular regeneration. Via the activation of a whole array of target genes, Hypoxia-inducible factor (HIF) may ameliorate acute kidney injury by reducing cellular necrosis, apoptosis, inflammation and adenosine triphosphate depletion and may improve renal regeneration and repair by increasing tubular proliferation. In summary, HIF activation may help to maintain cellular integrity, to prevent a more severe renal damage, and to accelerate renal regeneration

Because the PHDs use 2-oxoglutarate as a substrate to hydroxylate the HIF proteins, 2-oxoglutarate analogues have been used to inhibit the PHDs competitively even in the presence of oxygen, and the protective effect of systemic hypoxia could be mimicked by a single dose of a small molecule PHD inhibitor, which is able to stabilize HIF and activate HIF target genes [38]. Experiments in the isolated perfused kidney have demonstrated that these protective mechanisms reside in the kidney itself [40].

Besides hypoxia and pharmacological inhibition of PHDs, HIF can also be activated by other stimuli. In a recent study, preconditional xenon treatment protected mice kidneys from ischemia reperfusion injury via HIF activation in a mammalian target of rapamycin dependent manner [41]. In models of renal ischemia reperfusion, HIF regulation has been shown to be modulated by p53, where inhibition of p53 enhanced HIF accumulation [42].

Renal protection has not only been demonstrated in ischemic models but it has also been shown in toxic models of acute kidney injury. Activation of HIF has demonstrated that hypoxia can be protective in models of cisplatinum-induced renal tubular cell injury [45, 43, 44]. However, there is still debate whether in cisplatinum nephropathy protective effects of HIF can be explained due to protection from a concomitant renal tissue ischemia or by counteracting apoptosis and direct cell damage [45, 43, 44].

The concept of organ protection preceding exposure to episodes of ischemia and reperfusion has long been introduced under the terms of ischemic preconditioning [46]. The precise molecular mechanisms mediating this phenomenon have not been unequivocally identified. However, it is likely that both ischemia and the subsequent reperfusion do play a role, presumably through induction of reactive oxygen species [47]. In contrast to ischemic preconditioning, HIF preconditioning is mediated by hypoxia per se and does not require alterations in blood flow and reoxygenation.

It is likely that the maximum protective effect of HIF requires induction of the HIF system prior to injury, although it is possible that HIF induction may also protect against ongoing injury and through improved regeneration of tubular cells. While the requirement for "preconditioning" may limit the clinical applicability, there are numerous situations where injury to the kidney is predictable. One of the most relevant may be kidney transplantation. Approximately one third of grafts show early dysfunction immediately after transplantation as a consequence of organ storage and ischemia reperfusion injury [48]. The majority of kidneys recover from early dysfunction, but early dysfunction is associated with reduced long-term graft function [48]. Recently in a rat model of kidney transplantation it was tested whether HIF induction in organ donors following the application of a PHD inhibitor can improve transplant outcome [35]. In an acute model, in which the recipient animal was bilaterally nephrectomized and thus entirely dependent on the function of the graft, donor treatment with the PHD inhibitor reduced early mortality and improved function in surviving animals [35]. In a modified setting, in which the contralateral nephrectomy was postponed in order to allow animals to survive an early period of graft nonfunction, donor pretreatment with a PHD inhibitor was also found to improve long-term outcome [35].

5 The Perspective of HIF as a Therapeutic Target

Since the HIF system and the mechanism of its regulation have been unraveled, our knowledge of interventional procedures to stabilize HIF independent of prevailing oxygen tensions have remarkably expanded. Nowadays, the use of new pharmacological approaches to modulate the HIF system has become increasingly an option in humans for therapeutic purposes. Oxoglutarate analogues, which had been previously developed as potential antifibrotic agents [49, 50], are currently being explored as HIF stabilizers. This approach focuses on induction of endogenous erythropoietin formation as a novel strategy for the management of anemia [51]. Based on preclinical data, such as those described above, organ protection is another potentially rewarding application of interfering with HIF stability. In contrast to the induction of erythropoietin formation, organ protection reflects an unmet medical need. Beyond organ protection in the context of transplantation, several other indications probably deserve exploration. These include other forms of acute kidney injury, including radio-contrast-induced nephropathy or acute kidney injury associated with cardiac surgery as well as protective effects on other organ systems.

References

1. Gleadle JM. Review article: how cells sense oxygen: lessons from and for the kidney. Nephrology (Carlton) 2009;14(1):86–93.
2. Kaelin WG, Jr., Ratcliffe PJ. Oxygen sensing by metazoans: the central role of the HIF hydroxylase pathway. Mol Cell 2008;30(4):393–402.
3. Nangaku M, Inagi R, Miyata T, Fujita T. Hypoxia and hypoxia-inducible factor in renal disease. Nephron Exp Nephrol 2008;110(1):e1–7.
4. Lubbers DW, Baumgartl H. Heterogeneities and profiles of oxygen pressure in brain and kidney as examples of the pO_2 distribution in the living tissue. Kidney Int 1997;51(2):372–80.
5. Schurek HJ, Jost U, Baumgartl H, Bertram H, Heckmann U. Evidence for a preglomerular oxygen diffusion shunt in rat renal cortex. Am J Physiol 1990;259(6 Pt 2):F910–15.
6. Liss P, Nygren A, Revsbech NP, Ulfendahl HR. Intrarenal oxygen tension measured by a modified Clark electrode at normal and low blood pressure and after injection of X-ray contrast media. Pflugers Arch 1997;434(6):705–11.
7. Legrand M, Mik EG, Johannes T, Payen D, Ince C. Renal hypoxia and dysoxia after reperfusion of the ischemic kidney. Mol Med 2008;14(7–8):502–16.
8. Rosen S, Epstein FH, Brezis M. Determinants of intrarenal oxygenation: factors in acute renal failure. Ren Fail 1992;14(3):321–5.
9. Evans RG, Gardiner BS, Smith DW, O'Connor PM. Intrarenal oxygenation: unique challenges and the biophysical basis of homeostasis. Am J Physiol Renal Physiol 2008;295(5): F1259–70.
10. Iyer NV, Kotch LE, Agani F, Leung SW, Laughner E, Wenger RH, Gassmann M, Gearhart JD, Lawler AM, Yu AY, Semenza GL. Cellular and developmental control of O_2 homeostasis by hypoxia-inducible factor 1 alpha. Genes Dev 1998;12(2):149–62.
11. Ryan HE, Lo J, Johnson RS. HIF-1 alpha is required for solid tumor formation and embryonic vascularization. EMBO J 1998;17(11):3005–15.

12. Kotch LE, Iyer NV, Laughner E, Semenza GL. Defective vascularization of HIF-1alpha-null embryos is not associated with VEGF deficiency but with mesenchymal cell death. Dev Biol 1999;209(2):254–67.
13. Compernolle V, Brusselmans K, Acker T, Hoet P, Tjwa M, Beck H, Plaisance S, Dor Y, Keshet E, Lupu F, Nemery B, Dewerchin M, Van VP, Plate K, Moons L, Collen D, Carmeliet P. Loss of HIF-2alpha and inhibition of VEGF impair fetal lung maturation, whereas treatment with VEGF prevents fatal respiratory distress in premature mice. Nat Med 2002;8(7):702–10.
14. Scortegagna M, Ding K, Oktay Y, Gaur A, Thurmond F, Yan LJ, Marck BT, Matsumoto AM, Shelton JM, Richardson JA, Bennett MJ, Garcia JA. Multiple organ pathology, metabolic abnormalities and impaired homeostasis of reactive oxygen species in Epas1−/− mice. Nat Genet 2003;35(4):331–40.
15. Bernhardt WM, Schmitt R, Rosenberger C, Munchenhagen PM, Grone HJ, Frei U, Warnecke C, Bachmann S, Wiesener MS, Willam C, Eckardt KU. Expression of hypoxia-inducible transcription factors in developing human and rat kidneys. Kidney Int 2006;69 (1):114–22.
16. Freeburg PB, Robert B, St JP, Abrahamson DR. Podocyte expression of hypoxia-inducible factor (HIF)-1 and HIF-2 during glomerular development. J Am Soc Nephrol 2003;14 (4):927–38.
17. Rosenberger C, Mandriota S, Jorgensen JS, Wiesener MS, Hörstrup JH, Frei U, Ratcliffe PJ, Maxwell PH, Bachmann S, Eckardt KU. Expression of hypoxia-inducible factor -1alpha and -2alpha in hypoxic and ischemic rat kidneys. J Am Soc Nephrol 2002;13(7):1721-32.
18. Wiesener MS, Jurgensen JS, Rosenberger C, Scholze CK, Horstrup JH, Warnecke C, Mandriota S, Bechmann I, Frei UA, Pugh CW, Ratcliffe PJ, Bachmann S, Maxwell PH, Eckardt KU. Widespread hypoxia-inducible expression of HIF-2alpha in distinct cell populations of different organs. FASEB J 2003;17(2):271–3.
19. Paliege A, Rosenberger C, Bondke A, Sciesielski L, Shina A, Heyman SN, Flippin LA, Arend M, Klaus SJ, Bachmann S. Hypoxia-inducible factor-2alpha-expressing interstitial fibroblasts are the only renal cells that express erythropoietin under hypoxia-inducible factor stabilization. Kidney Int 2009;77(4):312–8.
20. Warnecke C, Zaborowska Z, Kurreck J, Erdmann VA, Frei U, Wiesener M, Eckardt KU. Differentiating the functional role of hypoxia-inducible factor (HIF)-1alpha and HIF-2alpha (EPAS-1) by the use of RNA interference: erythropoietin is a HIF-2alpha target gene in Hep3B and Kelly cells. FASEB J 2004;18(12):1462–4.
21. Rankin EB, Biju MP, Liu Q, Unger TL, Rha J, Johnson RS, Simon MC, Keith B, Haase VH. Hypoxia-inducible factor-2 (HIF-2) regulates hepatic erythropoietin in vivo. J Clin Invest 2007;117(4):1068–77.
22. Scortegagna M, Morris MA, Oktay Y, Bennett M, Garcia JA. The HIF family member EPAS1/HIF-2alpha is required for normal hematopoiesis in mice. Blood 2003;102 (5):1634–40.
23. Scortegagna M, Ding K, Zhang Q, Oktay Y, Bennett MJ, Bennett M, Shelton JM, Richardson JA, Moe O, Garcia JA. HIF-2alpha regulates murine hematopoietic development in an erythropoietin-dependent manner. Blood 2005;105(8):3133–40.
24. Epstein AC, Gleadle JM, McNeill LA, Hewitson KS, O'Rourke J, Mole DR, Mukherji M, Metzen E, Wilson MI, Dhanda A, Tian YM, Masson N, Hamilton DL, Jaakkola P, Barstead R, Hodgkin J, Maxwell PH, Pugh CW, Schofield CJ, Ratcliffe PJ. *C. elegans* EGL-9 and mammalian homologs define a family of dioxygenases that regulate HIF by prolyl hydroxylation. Cell 2001;107(1):43–54.
25. Ivan M, Kondo K, Yang H, Kim W, Valiando J, Ohh M, Salic A, Asara JM, Lane WS, Kaelin WG, Jr. HIFalpha targeted for VHL-mediated destruction by proline hydroxylation: implications for O_2 sensing. Science 2001;292(5516):464–8.
26. Bruick RK, McKnight SL. A conserved family of prolyl-4-hydroxylases that modify HIF. Science 2001;294(5545):1337–40.

27. Lando D, Peet DJ, Gorman JJ, Whelan DA, Whitelaw ML, Bruick RK. FIH-1 is an asparaginyl hydroxylase enzyme that regulates the transcriptional activity of hypoxia-inducible factor. Genes Dev 2002;16(12):1466–71.
28. Schodel J, Klanke B, Weidemann A, Buchholz B, Bernhardt W, Bertog M, Amann K, Korbmacher C, Wiesener M, Warnecke C, Kurtz A, Eckardt KU, Willam C. HIF-prolyl hydroxylases in the rat kidney: physiologic expression patterns and regulation in acute kidney injury. Am J Pathol 2009;174(5):1663–74.
29. Rosenberger C, Griethe W, Gruber G, Wiesener M, Frei U, Bachmann S, Eckardt KU. Cellular responses to hypoxia after renal segmental infarction. Kidney Int 2003;64(3):874–86.
30. Rosenberger C, Heyman SN, Rosen S, Shina A, Goldfarb M, Griethe W, Frei U, Reinke P, Bachmann S, Eckardt KU. Up-regulation of HIF in experimental acute renal failure: evidence for a protective transcriptional response to hypoxia. Kidney Int 2005;67(2):531–42.
31. Rosenberger C, Rosen S, Shina A, Bernhardt W, Wiesener MS, Frei U, Eckardt KU, Heyman SN. Hypoxia-inducible factors and tubular cell survival in isolated perfused kidneys. Kidney Int 2006;70(1):60–70.
32. Rosenberger C, Goldfarb M, Shina A, Bachmann S, Frei U, Eckardt KU, Schrader T, Rosen S, Heyman SN. Evidence for sustained renal hypoxia and transient hypoxia adaptation in experimental rhabdomyolysis-induced acute kidney injury. Nephrol Dial Transplant 2008;23(4):1135–43.
33. D'Angelo G, Duplan E, Boyer N, Vigne P, Frelin C. Hypoxia up-regulates prolyl hydroxylase activity: a feedback mechanism that limits HIF-1 responses during reoxygenation. J Biol Chem 2003;278(40):38183–7.
34. Stiehl DP, Wirthner R, Koditz J, Spielmann P, Camenisch G, Wenger RH. Increased prolyl 4-hydroxylase domain proteins compensate for decreased oxygen levels. Evidence for an autoregulatory oxygen-sensing system. J Biol Chem 2006;281(33):23482–91.
35. Bernhardt WM, Gottmann U, Doyon F, Buchholz B, Campean V, Schodel J, Reisenbuechler A, Klaus S, Arend M, Flippin L, Willam C, Wiesener MS, Yard B, Warnecke C, Eckardt KU. Donor treatment with a PHD-inhibitor activating HIFs prevents graft injury and prolongs survival in an allogenic kidney transplant model. Proc Natl Acad Sci U S A 2009;106 (50):21276–81.
36. Kojima I, Tanaka T, Inagi R, Kato H, Yamashita T, Sakiyama A, Ohneda O, Takeda N, Sata M, Miyata T, Fujita T, Nangaku M. Protective role of hypoxia-inducible factor-2alpha against ischemic damage and oxidative stress in the kidney. J Am Soc Nephrol 2007;18 (4):1218–26.
37. Hill P, Shukla D, Tran MG, Aragones J, Cook HT, Carmeliet P, Maxwell PH. Inhibition of hypoxia inducible factor hydroxylases protects against renal ischemia-reperfusion injury. J Am Soc Nephrol 2008;19(1):39–46.
38. Bernhardt WM, Campean V, Kany S, Jurgensen JS, Weidemann A, Warnecke C, Arend M, Klaus S, Gunzler V, Amann K, Willam C, Wiesener MS, Eckardt KU. Preconditional activation of hypoxia-inducible factors ameliorates ischemic acute renal failure. J Am Soc Nephrol 2006;17(7):1970–8.
39. Matsumoto M, Makino Y, Tanaka T, Tanaka H, Ishizaka N, Noiri E, Fujita T, Nangaku M. Induction of renoprotective gene expression by cobalt ameliorates ischemic injury of the kidney in rats. J Am Soc Nephrol 2003;14(7):1825–32.
40. Rosenberger C, Rosen S, Shina A, Frei U, Eckardt KU, Flippin LA, Arend M, Klaus SJ, Heyman SN. Activation of hypoxia-inducible factors ameliorates hypoxic distal tubular injury in the isolated perfused rat kidney. Nephrol Dial Transplant 2008;23(11):3472–8.
41. Ma D, Lim T, Xu J, Tang H, Wan Y, Zhao H, Hossain M, Maxwell PH, Maze M. Xenon preconditioning protects against renal ischemic-reperfusion injury via HIF-1alpha activation. J Am Soc Nephrol 2009;20(4):713–20.
42. Sutton TA, Wilkinson J, Mang HE, Knipe NL, Plotkin Z, Hosein M, Zak K, Wittenborn J, Dagher PC. p53 regulates renal expression of HIF-1alpha and pVHL under physiological

conditions and after ischemia-reperfusion injury. Am J Physiol Renal Physiol 2008;295(6): F1666–77.
43. Tanaka T, Kojima I, Ohse T, Inagi R, Miyata T, Ingelfinger JR, Fujita T, Nangaku M. Hypoxia-inducible factor modulates tubular cell survival in cisplatin nephrotoxicity. Am J Physiol Renal Physiol 2005;289(5):F1123–33.
44. Wang J, Biju MP, Wang MH, Haase VH, Dong Z. Cytoprotective effects of hypoxia against cisplatin-induced tubular cell apoptosis: involvement of mitochondrial inhibition and p53 suppression. J Am Soc Nephrol 2006;17(7):1875–85.
45. Weidemann A, Bernhardt WM, Klanke B, Daniel C, Buchholz B, Campean V, Amann K, Warnecke C, Wiesener MS, Eckardt KU, Willam C. HIF activation protects from acute kidney injury. J Am Soc Nephrol 2008;19(3):486–94.
46. Murry CE, Jennings RB, Reimer KA. Preconditioning with ischemia: a delay of lethal cell injury in ischemic myocardium. Circulation 1986;74(5):1124–6.
47. Kim J, Jang HS, Park KM. Reactive oxygen species generated by renal ischemia and reperfusion trigger protection against subsequent renal ischemia and reperfusion injury in mice. Am J Physiol Renal Physiol 2010;298(1):F158-F166.
48. Shoskes DA, Cecka JM. Effect of delayed graft function on short- and long-term kidney graft survival. Clin Transpl 1997;297–303.
49. Gunzler V, Brocks D, Henke S, Myllyla R, Geiger R, Kivirikko KI. Syncatalytic inactivation of prolyl 4-hydroxylase by synthetic peptides containing the unphysiologic amino acid 5-oxaproline. J Biol Chem 1988;263(36):19498–504.
50. Baader E, Tschank G, Baringhaus KH, Burghard H, Gunzler V. Inhibition of prolyl 4-hydroxylase by oxalyl amino acid derivatives in vitro, in isolated microsomes and in embryonic chicken tissues. Biochem J 1994;300 (Pt 2):525–30.
51. Hsieh MM, Linde NS, Wynter A, Metzger M, Wong C, Langsetmo I, Lin A, Smith R, Rodgers GP, Donahue RE, Klaus SJ, Tisdale JF. HIF prolyl hydroxylase inhibition results in endogenous erythropoietin induction, erythrocytosis, and modest fetal hemoglobin expression in rhesus macaques. Blood 2007;110(6):2140–7.

Chapter 28
Hypoxia in Chronic Kidney Disease: The Final Common Pathway to End Stage Renal Disease

Masaomi Nangaku

Abstract Because of the presence of arterial to venous oxygen shunt diffusion between arterial and venous vessels, renal tissue oxygen tensions are comparatively low, and accumulating evidence emphasizes chronic hypoxia in the tubulointerstitium as the final common pathway to end stage kidney disease.

Chronic hypoxia in the kidney is multifactorial and occurs via several mechanisms acting in concert. At a late stage of chronic kidney disease (CKD), as renal scarring evolves, the involvement of progressive tissue hypoxia secondary to obliteration and loss of the interstitial capillary bed predominates. Further, interstitial fibrosis impairs tubular oxygen supply because of the reduced efficiency of oxygen diffusion. Even at an early stage, kidneys suffer from hypoxia via functional mechanisms, including decreased peritubular capillary flow due to imbalances in vasoactive substances and inefficient oxygen utilization due to oxidative stress.

Studies verified a pathogenic role of hypoxia in various animal models of CKD as well as in the aged kidney. Furthermore, recent studies showed that glomerular cells are the target of hypoxia in addition to tubulointerstitial cells. Hypoxic cells suffer from energy depletion and oxidative stress, and defensive mechanisms against hypoxia are the sophisticated network of adaptation to oxygen depletion and reduction of oxidative stress. Sitting at the center of this web of cellular responses to hypoxia is hypoxia-inducible factor (HIF) and the concept that activation of HIF is effective in CKD has been validated in a variety of animal models of chronic as well as acute kidney injury.

Keywords HIF · Prolyl hydroxylase · Renal failure · Final common pathway · Tubulointerstitial injury

---

M. Nangaku (✉)
Division of Nephrology and Endocrinology,
University of Tokyo School of Medicine, 7-3-1 Hongo, Bunkyo-ku, Tokyo 113–8655, Japan
e-mail: mnangaku-tky@umin.ac.jp

T. Miyata et al. (eds.), *Studies on Renal Disorders*, Oxidative Stress in Applied Basic Research and Clinical Practice, DOI 10.1007/978-1-60761-857-7_28,

1 Introduction

Chronic kidney disease (CKD) is a precursor to end stage kidney disease (ESKD) and is associated with an increased risk of cardiovascular death. The early identification of subjects with CKD and development of therapeutic approaches to halt or retard the progression of CKD have therefore been the subject of intense investigation.

Once renal damage reaches a certain threshold, the progression of renal disease is consistent, irreversible, and largely independent of the initial insult. The final common pathway in this process has been closely studied in the hope that therapeutic approaches that target it may stop or slow the progression of CKD, whatever its etiology. Close pathological analysis shows that functional impairment of the kidney is better correlated with the degree of tubulointerstitial damage than with that of glomerular injury, and this finding has in turn led to broad recognition that the final common pathway of kidney failure operates principally in the tubulointerstitium. Accumulating evidence emphasizes chronic hypoxia at the center of tubulointerstitial injury and ESKD.

2 Anatomical Basis of Hypoxia in the Kidney

Although blood flow to the kidney is high, accounting for 20% of cardiac output, renal tissue oxygen tensions are in fact comparatively low because of the presence of arterial to venous (AV) oxygen shunt diffusion between arterial and venous vessels that run in close parallel contact. The existence of AV diffusional shunt for oxygen that bypasses the circulation was shown by measurement of the respective transit times for oxygen and erythrocytes, which revealed that oxygen transit was always faster than that of erythrocytes [1]. Furthermore, Welch and colleagues demonstrated that oxygen tension in the renal vein exceeded that in the efferent arteriole [2], while Schurek and colleagues observed that oxygen tension in the renal vein exceeded that in glomerular capillaries [3]. These studies clearly proved the existence of renal arterial-to-venous oxygen shunting. In fact, the kidney extracts only 10% of the oxygen delivered in the renal artery.

The kidney consumes oxygen mainly to generate chemical potential (adenosine triphosphate [ATP]), which is needed for sodium reabsorption [4]. Approximately two-thirds of this sodium reabsorption normally occurs in the proximal tubule. In contrast to other vascular beds, in which hypoxia causes vasodilation and hyperoxia induces vasoconstriction, renal blood flow is relatively insensitive to the effects of hypoxia [5]. This allows the control of renal perfusion to be dominated by the need to regulate renal excretory function, and not by energy demand.

As a consequence, the kidney is somewhat sensitive to changes in oxygen delivery. While this sensitivity has the merit of facilitating the kidneys in their adjustment of erythropoietin (EPO) production to changes in oxygen supply, it also renders them prone to hypoxic injury.

The chronic hypoxia hypothesis, proposed by Fine et al., emphasizes chronic ischemic damage in the tubulointerstitium as the final common pathway in end stage kidney injury [6]. Since its introduction, this fascinating hypothesis has been intensively investigated and validated.

3 Mechanisms of Hypoxia in Chronic Kidney Disease

Chronic hypoxia in the kidney is multifactorial and occurs via several mechanisms acting in concert.

In human kidney biopsy specimens, the degree of tubulointerstitial damage correlates with the loss of peritubular capillaries [7]. Studies in 110 patients with CKD showed that capillary density was significantly inversely correlated with an index of chronic damage, which incorporated global glomerular sclerosis, interstitial fibrosis, and atrophic tubules with thickened basement membranes, or large tubules with thin epithelium among others [8]. Furthermore, loss of peritubular capillaries predicted disease progression. A reduction in peritubular capillary density was also shown to be associated with the decrease of renal vascular endothelial growth factor (VEGF)-A expression in kidney samples of human diabetic nephropathy [9]. Recent studies also showed marked narrowing and a significant reduction in peritubular capillary density of the kidney in patients with congenital nephrotic syndrome of the Finnish type (NPHS1), which is associated with the rapid development of tubulointerstitial fibrosis [10]. There was increased expression (up to 43-fold) of hypoxia-inducible factor-1alpha (HIF-1α), suggesting tubulointerstitial hypoxia. Thus, as renal scarring evolves, the involvement of progressive tissue hypoxia secondary to obliteration and loss of the interstitial capillary bed predominate. Further, interstitial fibrosis impairs tubular oxygen supply, because the extended distance between the capillaries and tubular cells reduces the efficiency of oxygen diffusion.

Even at an early stage before the development of structural changes in the tubulointerstitium, kidneys suffer from hypoxia via functional mechanisms. Vasoconstriction of renal arterioles due to imbalances in vasoactive substances functionally decreases postglomerular peritubular capillary blood flow, leading to a decrease in oxygenation of the corresponding compartment. Among various vasoactive substances, local activation of the renin-angiotensin system (RAS) is an important contributor to the pathogenesis of chronic renal injury [11]. We observed the initial hypoxia in the remnant kidney model, which was associated with decreased perfusion status of the postglomerular peritubular capillary network, as evaluated by lectin perfusion and Hoechst 33342 diffusion techniques [12]. We found that RAS activation leads to disturbances in postglomerular capillary flow and precedes histological injury, demonstrating that these findings were dependent on the activation of the RAS but antedated any histologic evidence of tubulointerstitial damage.

There is evidence that renal oxygen consumption is increased in the remnant kidney model, and that this increase is associated with angiotensin II activity but not blood pressure or glomerular filtration rate [13]. Oxygen consumption in the kidney is determined by energy demand and oxygen utilization efficiency. Oxidative stress induced by the activation of RAS may impair oxygen utilization efficiency by exaggerating mitochondrial respiration and thereby induce relative hypoxia in the kidney. In addition, intact remnant nephrons need to deal with the compensatory increase in transport burden following the loss of nephrons, while the increased demand for energy required for active transport means that oxygen consumption is increased in remnant nephrons, resulting in a decrease in local oxygen tensions.

Further, renal anemia hinders oxygen delivery due to the decrease in oxygen-carrying capacity as well as the increase in oxygen consumption [14].

4 Animal Models of Chronic Hypoxia in the Kidney

Hypoxia of the kidney has been demonstrated in a variety of animal models. Using a pimonidazole adduct, which binds to hypoxic cells in vivo, we showed tubulointerstitial hypoxia in a model of chronic glomerulonephritis induced by uninephrectomy combined with repeated injection of anti-Thy-1.1 antibody in rats [15]. In diseased tubulointerstitium, cortical tubular cells were exposed to hypoxia, which was associated with stagnant regional blood flow as estimated by intravital microscopy analysis in rats. A decrease in cortical oxygenation in association with narrowing and distortion of cortical peritubular capillaries was also demonstrated in the remnant kidney model utilizing pimonidazole in rats [12].

Cellular metabolism depends on the availability of oxygen. The major regulator of oxygen homeostasis is HIF, a highly conserved transcription factor that plays an essential role in cellular and systemic homeostatic responses to hypoxia. Under hypoxic conditions, HIF is stabilized and translocates to the nucleus where it binds to hypoxia response elements (HREs) and activates the expression of target genes implicated in angiogenesis, cell growth, and survival. To develop a novel method of identifying and quantifying tissue oxygenation at the cellular level, we developed a hypoxia-responsive reporter vector using a HRE of the 5′ VEGF untranslated region and generated a hypoxia-sensing transgenic rat. This genetically engineered animal enabled us to identify diffuse cortical hypoxia in puromycin aminonucleoside-induced nephrotic syndrome and focal and segmental hypoxia in the remnant kidney model even at an early stage of disease [16].

Hypoxia in the tubulointerstitial compartment is not limited to glomerulopathy, but applies to nephropathy associated with metabolic disease. Chronic hypoxia in diabetic nephropathy is discussed in detail in the next chapter (see Chap. 29).

Johnson and colleagues proposed that hypertension is initiated by agents that cause systemic and intrarenal vasoconstriction [17]. Intrarenal injury develops with microvascular disease, maintaining intrarenal vasoconstriction and hypoxia with a change in local vasoconstrictor-vasodilator balance favoring sodium retention. As blood pressure rises, renal ischemia is ameliorated and sodium balance is restored completely (in salt-resistant) or partially (in salt-sensitive) for hypertension, but at the expense of a rightward shift in the pressure natriuresis curve and persistent hypertension. Thus, it is feasible that chronic hypoxia in the kidney plays a pathogenic role in hypertension and hypertensive kidney injury. To support this notion, recent studies utilizing transfection of HIF decoy oligodeoxynucleotides into the renal medulla in uninephrectomized rats showed that HIF-mediated gene activation serves as an antihypertensive pathway by regulating renal medullary function and sodium excretion [18]. Among oxygen-sensitive genes regulated by the HIF-HRE pathway are nitric oxide synthase, cyclooxygenase-2, and heme oxygenase-1. The natriuretic responses and increases in renal medullary blood flow that respond to the elevations of renal perfusion pressure were significantly blunted in decoy rats: in these rats, high salt intake caused a greater positive sodium balance with subsequent elevation of systemic blood pressure.

5 Chronic Hypoxia in the Aging Kidney

Aging is associated with a progressive decline in renal function and the development of glomerulosclerosis, interstitial fibrosis, and loss of peritubular capillaries [19]. Impaired angiogenesis and subsequent loss of renal microvasculature associated with alterations in renal expression of VEGF may induce tubulointerstitial injury via chronic hypoxia of the corresponding region [20]. Using two distinct methods, pimonidazole immunostaining and expression of the "hypoxia-responsive" reporter of transgenic rats, we identified the age-related expansion of hypoxia in all areas of the kidney [21]. Clusters of hypoxic tubules were observed in the superficial cortical zones and areas adjacent to the outer nephrons and increased in size with age in the medullary rays. The degree of hypoxia was positively correlated with the age-related tubulointerstitial injury. These findings point to the involvement of hypoxia and highlight the pathological relevance of HIF and its target genes in the aging kidney.

Aging is also associated with an increased susceptibility to hypoxic injury. When renal ischemia was induced in young and old rats by renal arterial occlusion, the resulting elevation of blood urea nitrogen (BUN) and serum creatinine levels as well as histological renal tubular damage were more severe in older animals [22]. Studies utilizing renal slices subjected to hypoxic stress showed that slices from aged rats were more susceptible to injury than young counterparts [23]. Importantly, caloric restriction of rats attenuated the increased susceptibility to hypoxic injury. An attempt to identify the molecular pathway(s) underlying this response by microarray analysis showed that the expression of 92 genes was changed during aging and attenuated by caloric restriction, including claudin-7, kidney injury molecule-1 (Kim-1), and matrix metalloproteinase-7 (MMP-7). A previously reported age-dependent decrease in HIF-1α expression in the kidney may be associated with the decreased adaptation to hypoxia found in aged animals and humans [24]. With increasing age, decreased expression of HIF-1α, which correlated to an increased expression of prolyl-4-hydroxylase domain (PHD3), was also found in mouse and human heart [25]. These data may explain the reduction in HIF-1α and HIF-1 target genes, such as the VEGF, in aging tissue.

6 Molecular Pathophysiology of Chronic Hypoxia in the Kidney

To elucidate the molecular pathophysiology of chronic hypoxia in the kidney, we studied expression profiles of genes and proteins in the kidney under chronic hypoxic conditions. We first induced chronic hypoxia in the rat kidney by constricting the left main renal artery and then performed microarray analysis and proteomic studies with the kidneys. The renal artery stenosis model is a useful tool for studying hypoxia of the kidney because it lacks changes in urinary protein excretion or other confounding factors at an early stage. We confirmed hypoxia of the kidney with renal artery stenosis utilizing pimonidazole and our hypoxia-sensing transgenic rats. Induction of chronic hypoxia in this model has also been confirmed by direct measurement of oxygen tensions via microelectrode [26]. In spite of decreases in oxygen consumption, tissue oxygenation of the kidney decreased in a pig model of renal artery stenosis due to a decrease in oxygen delivery [27].

Although the notion of oxidative stress under hypoxic conditions sounds paradoxical, hypoxic cells do in fact suffer from energy depletion and oxidative stress (see Chap. 24). We identified a paradoxical decrease in Cu/Zn superoxide dismutase (SOD1) in the chronic hypoxic kidney with renal artery stenosis. The tubulointerstitial injury in this model was ameliorated by TEMPOL (4-Hydroxy-2,2,6,6-tetramethylpiperidinoxyl), an SOD mimetic, in association with a decrease in levels of oxidative stress markers such as 4-hydroxyl-2-nonenal and nitrotyrosine. Our in vitro studies utilizing cultured tubular cells revealed a role for tumor necrosis factor alpha (TNF-α) in the downregulation of Cu/Zn SOD. Since the administration of anti-TNF-α antibody ameliorated Cu/Zn SOD suppression, TNF-α appears to be a suppressant of Cu/Zn SOD. These results revealed that maladaptive suppression of Cu/Zn SOD was a mediator of the vicious cycle of oxidative stress and the subsequent renal injury induced by chronic hypoxia [28].

Among the various genes upregulated in the kidney by chronic hypoxia in our microarray analysis was metallothionein, which is known to provide protection against metal toxicity, play a role in the regulation of physiological metals (Zn and Cu), and provide protection against oxidative stress. In addition to these physiological functions, we found that metallothionein stimulates the HIF-HRE pathway through the ERK/mTOR pathway [29]. Hypoxia and exogenous metallothionein increased HIF-1α protein in cultured tubular cells without changes in its mRNA levels, suggesting protein stabilization.

We also found the endogenous expression of hemoglobin in the kidney. This unexpected finding was confirmed by several techniques, including manual dissection and laser capture microdissection. Our overexpression studies showed that hemoglobin decreased the production of hydrogen peroxide–induced intracellular radical oxygen species and enhanced cell viability against oxidative stress in cultured kidney cells [30] (see Chap. 6).

7 Consequences of Chronic Hypoxia

Chronic hypoxia in the affected region then triggers several phenotypic changes in tubular cells, such as proliferation, epithelial-mesenchymal transdifferentiation (EMT), and cell death. We previously demonstrated that exposure of cultured tubular cells to chronic hypoxia induced EMT [31]. Tubular cells cultured under hypoxic conditions showed morphological changes, enhanced cell motility, increased collagen I production, and expression of alpha-smooth muscle actin. Furthermore, we observed EMT in the chronic hypoxic kidney due to renal artery stenosis. Recent studies showed the HIF-dependent involvement of Twist, a basic helix-loop-helix transcription factor, in promoting EMT of human tubular cells under hypoxic conditions [32].

Hypoxic tubular cells in turn serve as a source of key mediators involved in macrophage infiltration and tubulointerstitial fibrosis. Interstitial accumulation of extracellular matrix components further impairs local oxygenation by blocking oxygen diffusion, thus accelerating regional hypoxia.

Connective tissue growth factor (CTGF) is associated with increased synthesis of extracellular matrix and fibrosis and is upregulated in kidney diseases of different etiology. One study showed upregulation of CTGF by hypoxia, which was dependent on transcriptional activation of HRE in the mouse CTGF promoter by HIF-1 [33]. However, hypoxic regulation of CTGF remains controversial. The HRE are not conserved in the human CTGF promoter, and downregulation rather than upregulation of CTGF mRNA was observed in microarray analyses of human proximal tubule cells exposed to hypoxia [34]. In addition, exposure of two human proximal tubular epithelial cell lines to hypoxia reduced cellular as well as secreted CTGF protein synthesis [35]. Studies of HIF-1α knockdown by siRNA showed that hypoxic repression of CTGF synthesis was dependent on HIF-1. These results suggested that CTGF is suppressed by target genes and activated by HIF under hypoxic conditions in humans.

8 Hypoxia and Glomerular Injury

While most studies have focused on chronic hypoxia in the tubulointerstitium as the final common pathway to ESKD, hypoxia also has an adverse effect on glomeruli.

Mesangial cells are the target of injury in many forms of glomerulonephritis and other glomerular diseases. Hypoxia also stimulates the production of extracellular matrix such as type IV collagen and fibronectin by mesangial cells as well as mesangial proliferation [36, 37].

Glomerular endothelial cells play a key role in the preservation and reconstruction of the glomerular capillary network following injury and thereby to maintain tissue oxygenation. Accumulating evidence has shown that the failure to maintain microcirculation leads to irreversible glomerular injury and glomerular sclerosis. We exposed cultured glomerular endothelial cells to hypoxia and found that hypoxic treatment induced apoptosis in association with a decrease in mitochondrial membrane potentials and an increase in caspase-9 activity [38]. It was likely that changes in the quantity and localization of Bcl2 and Bax contributed to this hypoxia-mediated apoptosis of glomerular endothelial cells.

Podocytes are terminally differentiated, and a decline in podocyte number is crucial to the progression to glomerulosclerosis. Podocytes exposed to low-oxygen conditions in diseases that affect primarily the vascular compartment of the glomerulus and hypoxia may contribute to the progression of human glomerular disease by altering podocyte metabolism and survival [39]. HIF is essentially a protective factor that induces broad and coordinated adaptive responses to hypoxia. However, HIF activation may have deleterious effects on podocytes under some conditions. Deletion of the Von Hippel–Lindau gene from podocytes of mice initiated a necrotizing crescentic glomerulonephritis due to the stabilization of HIFs and de novo expression of the HIF target gene *Cxcr4*, which encodes a chemokine receptor in podocytes [40]. A pathogenic role of *CXCR4* in podocytes was highlighted by a recent observation of prominent induction of *CXCR4* in podocytes of human nephrosclerosis [41]. In addition to *CXCR4*, inappropriate expression of VEGF-164 plays a pathogenic role in glomerular diseases. Immunohistochemistry revealed that kidneys from HIV-1 transgenic mice and from patients with collapsing focal segmental glomerulosclerosis (FSGS) of HIV-associated nephropathy (HIVAN) had greater expression of both VEGF and its transcriptional regulator, HIF-2 [42]. Furthermore, expression of VEGF and HIF-2 were increased in HIV-infected podocytes in vitro.

9 Therapeutic Approaches Targeting Hypoxia in Chronic Kidney Disease

To economize energy use under hypoxic conditions, cells suppress protein synthesis by activating the unfolded protein response in response to endoplasmic reticulum stress (see Chap. 33). At the same time as this pathway is bringing about a general inhibition of translation, however, the cell is also activating other pathways essential to survival, which result in the expression of genes involved in adaptation to hypoxia. This multiactor response occurs via the activation of a "master gene" switch, which results in a broad and coordinated downstream reaction. Sitting at the center of this web of cellular responses to hypoxia is HIF (see Chap. 21).

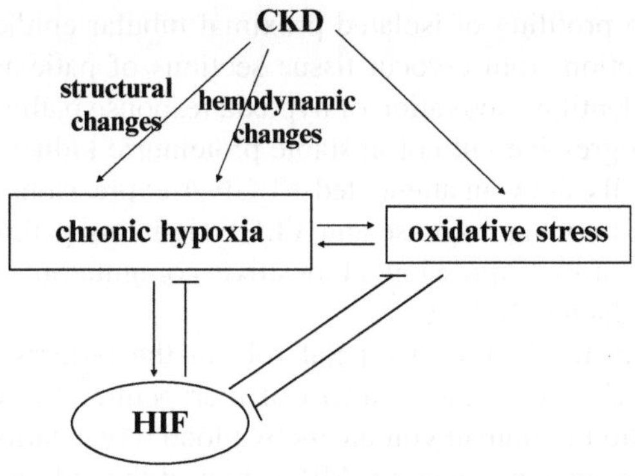

Fig. 1 Chronic hypoxia in the kidney is multifactorial and occurs via both structural and functional mechanisms acting in concert. Hypoxic cells suffer from energy depletion and oxidative stress, and oxidative stress aggravates hypoxia via reduction of oxygen utilization efficiency. Sitting at the center of this web of cellular responses to hypoxia is hypoxia-inducible factor (HIF). Under certain conditions, oxidative stress suppresses HIF activation, and HIF leads to upregulation of adaptive genes to hypoxic conditions as well as antioxidative enzymes

If HIF activation were already maximal under pathological conditions, therapeutic approaches that target HIF might be of little use. However, accumulating evidence suggests that HIF activation is suboptimal in CKD (Fig. 1). In a model of diabetic nephropathy, oxidative stress hinders activation of HIF [43, 44]. Proteinuric states may also inhibit HIF activation [45]. Furthermore, optimal HIF activation requires a moderate and sublethal decrease in oxygen tension. If hypoxia is too severe, cells suffer from suppression or disappearance of HIF and induction of apoptotic cell death. This "window of opportunity" for HIF was demonstrated by studies utilizing an ex vivo model of isolated perfused rat kidney with controlled oxygen consumption [46].

HIF level is determined by hydroxylation-induced degradation (see Chap. 22), bringing enthusiasm to the development of prolyl hydroxylase inhibitors as a therapeutic approach (see Chap. 32). The concept that inhibition of prolyl hydroxylase is effective in CKD has been validated in a variety of animal models of chronic as well as acute kidney injury by our group and others.

10 Findings in Humans

Demonstration of chronic hypoxia in human kidneys has been hampered by technical difficulties in measuring oxygen tensions in human organs. Indirect evidence of chronic hypoxia was provided by the demonstration of hemodynamic changes in the kidneys of patients with CKD. Measurement of intrarenal vascular resistance by Doppler ultrasonography showed an increase in renal resistive index (RRI) even at an early stage of disease in patients with tubulointerstitial injury [47].

Gene-expression profiling of isolated proximal tubular epithelial cells by laser capture microdissection from cryocut tissue sections of patients with proteinuric glomerulopathies identified activation of hypoxia response pathways in progressive patients [48]. In progressive but not in stable proteinuric kidney disease, proximal tubular epithelial cells show an attenuated VEGF-A expression, despite the activation of intracellular hypoxia response and VEGF signaling pathways, which might be due to the decreased expression of positive coregulators, such as EGF and insulin-like growth factor (IGF)-1.

Activation of the RAS plays a crucial role in the progression of CKD, and blockade of the RAS is the gold standard of therapeutic approaches. Estimation of kidney oxygenation in human volunteers by blood oxygenation level-dependent (BOLD) magnetic resonance imaging (MRI) showed that administration of angiotensin II caused an immediate decrease in kidney oxygenation [49]. It is thus likely that conditions associated with local activation of the RAS are associated with hypoxia of the kidney. Studies utilizing Doppler ultrasonography showed that treatment of chronic proteinuric patients with RAS inhibitors decreased resistive index (RI) and pulsatility index (PI), suggesting local hemodynamic improvement by RAS inhibitors [50].

The ability to estimate the oxygenation status of the kidney at bedside would have an enormous impact on renal care. While BOLD-MRI is promising, hurdles in daily use need to be overcome (see Chap. 30). Oxygen-specific markers may also be useful, but many good candidates, such as KIM-1 and N-GAL, were found to be upregulated by various insults and are not hypoxia specific. In addition, HIF, an oxygen-dependent molecule, is difficult to detect because of its short half-life.

11 Conclusion

Chronic hypoxia is the final common pathway to ESKD. Hypoxia of the kidney may develop in the tubulointerstitium before structural microvasculature damage in the corresponding region and play a pathogenic role from an early stage of CKD. Given expectations of efficacy in a broad range of renal diseases, therapeutic approaches against this final common pathway are eagerly awaited.

Acknowledgments This work was in part supported by Grant-in-Aid (21390360) from Japan Society for the Promotion of Science.

References

1. Levy MN, Sauceda G. Diffusion of oxygen from arterial to venous segments of renal capillaries. Am J Physiol 1959; 196: 1336–39.
2. Welch WJ, Baumgärtl H, Lubbers D, Wilcox CS. Nephron pO_2 and renal oxygen usage in the hypertensive rat kidney. Kidney Int 2001; 59: 230–7.

3. Schurek HJ, Jost U, Baumgärtl H, Bertram H, Heckmann U. Evidence for a preglomerular oxygen diffusion shunt in rat renal cortex. Am J Renal Fluid Electrolyte Physiol 1990; 259: F910–5.
4. Deng A, Miracle CM, Lortie M, Satriano J, Gabbai FB, Munger KA, Thomson SC, Blantz RC. Kidney oxygen consumption, carbonic anhydrase, and proton secretion. Am J Physiol Renal Physiol 2006; 290: F1009–15.
5. Evans RG, Gardiner BS, Smith DW, O'Connor PM. Intrarenal oxygenation: unique challenges and the biophysical basis of homeostasis. Am J Physiol Renal Physiol 2008; 295: F1259–70.
6. Fine LG, Bandyopadhay D, Norman JT. Is there a common mechanism for the progression of different types of renal diseases other than proteinuria? Towards the unifying theme of chronic hypoxia. Kidney Int 2000; 75: S22–6.
7. Bohle A, Mackensen-Haen S, Wehrmann M. Significance of postglomerular capillaries in the pathogenesis of chronic renal failure. Kidney Blood Press Res 1996; 19: 191–5.
8. Eardley KS, Kubal C, Zehnder D, Quinkler M, Lepenies J, Savage CO, Howie AJ, Kaur K, Cooper MS, Adu D, Cockwell P. The role of capillary density, macrophage infiltration and interstitial scarring in the pathogenesis of human chronic kidney disease. Kidney Int 2008; 74: 495–504.
9. Lindenmeyer MT, Kretzler M, Boucherot A, Berra S, Yasuda Y, Henger A, Eichinger F, Gaiser S, Schmid H, Rastaldi MP, Schrier RW, Schlondorff D, Cohen CD. Interstitial vascular rarefaction and reduced VEGF-A expression in human diabetic nephropathy. J Am Soc Nephrol 2007; 18: 1765–76.
10. Kaukinen A, Lautenschlager I, Helin H, Karikoski R, Jalanko H. Peritubular capillaries are rarefied in congenital nephrotic syndrome of the Finnish type. Kidney Int 2009; 75: 1099–108.
11. Kobori H, Nangaku M, Navar LG, Nishiyama A. The intrarenal renin-angiotensin system: from physiology to the pathobiology of hypertension and kidney disease. Physiol Rev 2007; 59: 251–87.
12. Manotham K, Tanaka T, Matsumoto M, Ohse T, Miyata T, Inagi R, Kurokawa K, Fujita T, Nangaku M. Evidence of tubular hypoxia in the early phase in the remnant kidney model. J Am Soc Nephrol 2004; 15: 1277–88.
13. Deng A, Tang T, Singh P, Wang C, Satriano J, Thomson SC, Blantz RC. Regulation of oxygen utilization by angiotensin II in chronic kidney disease. Kidney Int 2009; 75: 197–204.
14. Johannes T, Miki EG, Nohe B, Unertl KE, Ince C. Acute decrease in renal microvascular pO_2 during acute normovolemic hemodilution. Am J Physiol Renal Physiol 2007; 292: F796–803.
15. Matsumoto M, Tanaka T, Yamamoto T, Noiri E, Miyata T, Inagi R, Fujita T, Nangaku M. Hypoperfusion of peritubular capillaries induces chronic hypoxia before progression of tubulointerstitial injury in a progressive model of rat glomerulonephritis. J Am Soc Nephrol 2004; 15: 1574–81.
16. Tanaka T, Miyata T, Inagi R, Fujita T, Nangaku M. Hypoxia in renal disease with proteinuria and/or glomerular hypertension. Am J Pathol 2004; 165: 1979–92.
17. Johnson RJ, Feig DI, Nakagawa T, Sanchez-Lozada LG, Rodriguez-Iturbe B. Pathogenesis of essential hypertension: historical paradigms and modern insights. J Hypertens 2008; 26: 381–91.
18. Li N, Chen L, Yi F, Xia M, Li PL. Salt-sensitive hypertension induced by decoy of transcription factor hypoxia-inducible factor-1alpha in the renal medulla. Circ Res 2008; 102: 1101–8.
19. Thomas SE, Anderson S, Gordon KL, Oyama TT, Shankland SJ, Johnson RJ. Tubulointerstitial disease in aging: evidence for underlying peritubular capillary damage, a potential role for renal ischemia. J Am Soc Nephrol 1998; 9: 231–42.
20. Kang DH, Anderson S, Kim YG, Mazzalli M, Suga S, Jefferson JA, Gordon KL, Oyama TT, Hughes J, Hugo C, Kerjaschki D, Schreiner GF, Johnson RJ. Impaired angiogenesis in the aging kidney: vascular endothelial growth factor and thrombospondin-1 in renal disease. Am J Kidney Dis 2001; 37: 601–11.

21. Tanaka T, Kato H, Kojima I, Ohse T, Son D, Tawakami T, Yatagawa T, Inagi R, Fujita T, Nangaku M. Hypoxia and expression of hypoxia-inducible factor in the aging kidney. J Gerontol A Biol Sci Med Sci 2006; 61: 795–805.
22. Miura K, Goldstein RS, Morgan DG, Pasino DA, Hewitt WR, Hook JB. Age-related differences in susceptibility to renal ischemia in rats. Toxicol Appl Pharmacol 1987; 87: 284–96.
23. Chen G, Bridenbaugh EA, Akintola AD, Catania JM, Vaidya VS, Bonventre JV, Dearman AC, Sampson HW, Zawieja DC, Burghardt RC, Parrish AR. Increased susceptibility of aging kidney to ischemic injury: identification of candidate genes changed during aging, but corrected by caloric restriction. Am J Physiol Renal Physiol 2007; 293: F1272–81.
24. Frenkel-Denkberg G, Gershon D, Levy AP. The function of hypoxia-inducible factor 1 (HIF-1) is impaired in senescent mice. FEBS Lett 1999; 462: 341–4.
25. Rohrbach S, Simm A, Pregla R, Franke C, Katschinski DM. Age-dependent increase of prolyl-4-hydroxylase domain (PHD) 3 expression in human and mouse heart. Biogerontology 2005; 6: 165–71.
26. Palm F, Connors SG, Mendonca M, Welch WJ, Wilcox CS. Angiotensin II type 2 receptors and nitric oxide sustain oxygenation in the clipped kidney of early Goldblatt hypertensive rats. Hypertension 2008; 51: 345–51.
27. Warner L, Gomez SI, Bolterman R, Haas JA, Bentley MD, Lerman LO, Romero JC. Regional decreases in renal oxygenation during graded acute renal arterial stenosis: a case for renal ischemia. Am J Physiol Regul Integr Comp Physiol 2009; 296: R67–71.
28. Son D, Kojima I, Inagi R, Matsumoto M, Fujita T, Nangaku M. Chronic hypoxia aggravates renal injury via suppression of Cu/Zn-SOD: a proteomic analysis. Am J Physiol Renal Physiol 2008; 294: F62–72.
29. Kojima I, Tanaka T, Inagi R, Nishi H, Aburatani H, Kato H, Miyata T, Fujita T, Nangaku M. Metallothionein is upregulated by hypoxia and stabilizes hypoxia-inducible factor in the kidney. Kidney Int 2009; 75: 268–77.
30. Nishi H, Inagi R, Kato H, Tanemoto M, Kojima I, Son D, Fujita T, Nangaku M. Hemoglobin is expressed by mesangial cells and reduces oxidant stress. J Am Soc Nephrol 2008; 19: 1500–8.
31. Manotham K, Tanaka T, Matsumoto M, Ohse T, Inagi R, Miyata T, Kurokawa K, Fujita T, Ingelfinger JR, Nangaku M. Transdifferentiation of cultured tubular cells induced by hypoxia. Kidney Int 2004; 65: 871–80.
32. Sun S, Ning X, Zhang Y, Lu Y, Nie Y, Han S, Liu L, Du R, Xia L, He L, Fan D. Hypoxia-inducible factor-1alpha induces Twist expression in tubular epithelial cells subjected to hypoxia, leading to epithelial-to-mesenchymal transition. Kidney Int 2009; 75: 1278–87.
33. Higgins DF, Biju MP, Akai Y, Wutz A, Johnson RS, Haase VH. Hypoxic induction of Ctgf is directly mediated by HIF-1. Am J Physiol Renal Physiol 2004; 287: F1223–32.
34. Chi JT, Wang Z, Nuyten DS, Rodriguez EH, Schaner ME, Salim A, Wang Y, Kristensen GB, Helland A, Børresen-Dale AL, Giaccia A, Longaker MT, Hastie T, Yang GP, van de Vijver MJ, Brown PO. Gene expression programs in response to hypoxia: cell type specificity and prognostic significance in human cancers. PLoS Med 2006; 3: e47.
35. Kroening S, Neubauer E, Wessel J, Wiesener M, Goppelt-Struebe M. Hypoxia interferes with connective tissue growth factor (CTGF) gene expression in human proximal tubular cell lines. Nephrol Dial Transplant 2009; 24(11): 3319–25.
36. Sodhi CP, Phadke SA, Batlle D, Sahai A. Hypoxia and high glucose cause exaggerated mesangial cell growth and collagen synthesis: role of osteopontin. Am J Physiol Renal Physiol 2001; 280: F667–74.
37. Kim SB, Kang SA, Park JS, Lee JS, Hong CD. Effects of hypoxia on the extracellular matrix production of cultured rat mesangial cells. Nephron 1996; 72: 275–80.
38. Tanaka T, Miyata T, Inagi R, Kurokawa K, Adler S, Fujita T, Nangaku M. Hypoxia-induced apoptosis in cultured glomerular endothelial cells: involvement of mitochondrial pathways. Kidney Int 2003; 64: 2020–32.
39. Brukamp K, Jim B, Moeller MJ, Haase VH. Hypoxia and podocyte-specific Vhlh deletion confer risk of glomerular disease. Am J Physiol Renal Physiol 2007; 293: F1397–407.

40. Ding M, Cui S, Li C, Jothy S, Haase V, Steer BM, Marsden PA, Pippin J, Shankland S, Rastaldi MP, Cohen CD, Kretzler M, Quaggin SE. Loss of the tumor suppressor Vhlh leads to upregulation of Cxcr4 and rapidly progressive glomerulonephritis in mice. Nat Med 2006; 12: 1081–7.
41. Neusser MA, Lindenmeyer MT, Moll AG, Segerer S, Edenhofer I, Sen K, Stiehl DP, Kretzler M, Grone HJ, Schlondorff D, Cohen CD. Human nephrosclerosis triggers a hypoxia-related glomerulopathy. Am J Pathol 2009; 176(2): 594–607.
42. Korgaonkar SN, Feng X, Ross MD, Lu TC, D'Agati V, Iyengar R, Klotman PE, He JC. HIV-1 upregulates VEGF in podocytes. J Am Soc Nephrol 2008; 19: 877–83.
43. Katavetin P, Miyata T, Inagi R, Tanaka T, Sassa R, Ingelfinger JR, Fujita T, Nangaku M. High glucose blunts vascular endothelial growth factor response to hypoxia via the oxidative stress-regulated hypoxia-inducible factor/hypoxia-responsible element pathway. J Am Soc Nephrol 2006; 17: 1405–13.
44. Rosenberger C, Khamaisi M, Abassi Z, Shilo V, Weksler-Zangen S, Goldfarb M, Shina A, Zibertrest F, Eckardt KU, Rosen S, Heyman SN. Adaptation to hypoxia in the diabetic rat kidney. Kidney Int 2008; 73: 34–42.
45. Katavetin P, Inagi R, Miyata T, Tanaka T, Sassa R, Ingelfinger JR, Fujita T, Nangaku M. Albumin suppresses vascular endothelial growth factor via alteration of hypoxia-inducible factor/hypoxia-responsive element pathway. Biochem Biophys Res Commun 2008; 367: 305–10.
46. Rosenberger C, Rosen S, Shina A, Bernhardt W, Wiesener MS, Frei U, Eckardt KU, Heyman SN. Hypoxia-inducible factors and tubular cell survival in isolated perfused kidneys. Kidney Int 2006; 70: 60–70.
47. Boddi M, Cecioni I, Poggesi L, Fiorentino F, Olianti K, Berardino S, La Cava G, Gensini G. Renal resistive index early detects chronic tubulointerstitial nephropathy in normo- and hypertensive patients. Am J Nephrol 2006; 26: 16–21.
48. Rudnicki M, Perco P, Enrich J, Eder S, Heininger D, Bernthaler A, Wiesinger M, Sarközi R, Noppert SJ, Schramek H, Mayer B, Oberbauer R, Mayer G. Hypoxia response and VEGF-A expression in human proximal tubular epithelial cells in stable and progressive renal disease. Lab Invest 2009; 89: 337–46.
49. Schachinger H, Klarhöfer M, Linder L, Drewe J, Scheffler K. Angiotensin II decreases the renal MRI blood oxygenation level-dependent signal. Hypertension 2006; 47: 1062–6.
50. Loriga G, Vidili G, Ruggenenti P, Faedda R, Sanna M, Satta AE. Renal hemodynamics and renoprotection. Nephron Clin Pract 2008; 110: c213–9.

Chapter 29
Oxidative Stress and Hypoxia in the Pathogenesis of Diabetic Nephropathy

Fredrik Palm, Lina Nordquist, Christopher S. Wilcox, and Peter Hansell

Abstract The etiology of diabetic nephropathy is presently not fully understood, and several pathophysiological mechanisms are likely to be involved. However, there is a fair amount of support for a central role of oxidative stress and subsequent tissue hypoxia for the development of diabetes-induced kidney damage. This chapter will summarize some of the major biochemical pathways activated in the diabetic kidney and discuss how these are related to increased oxidative stress, tissue hypoxia, and the progression of diabetic nephropathy.

Keywords Reactive oxygen species · Oxygen consumption · Polyol · Nitric oxide · Advanced glycation end product

1 Introduction

Diabetic nephropathy can be defined as a progressive impairment of kidney function, which usually shows increased serum creatinine or calculated glomerular filtration rate (GFR) and increased excretion of albumin and protein (24-h urinary albumin excretion $\geq$300 mg). There are characteristic morphological changes that include Kimmelstiel-Wilson syndrome (intercapillary glomerulonephritis), mesangial expansion, and ultimately diffuse glomerulosclerosis, interstitial fibrosis, and extracellular matrix accumulation [1]. Hypertension and cardiovascular morbidity are commonly associated with the progression of diabetes and correlate with angiopathy of the glomerular capillaries. However, diabetic nephropathy usually is delayed 12–15 years from diagnosis. Therefore, metabolic alterations occur before the morphological

---

F. Palm (✉)
Department of Medical Cell Biology, Uppsala University,
Box 571751 23 Uppsala, Sweden
e-mail: Fredrik.Palm@mcb.uu.se

T. Miyata et al. (eds.), *Studies on Renal Disorders*, Oxidative Stress in Applied Basic Research and Clinical Practice, DOI 10.1007/978-1-60761-857-7_29,

changes. At this entry stage, the kidney can demonstrate functional defects including increased GFR and oxygen consumption (QO_2), resulting in decreased oxygen tension (pO_2) in the diabetic kidney [2–7]. The increased GFR is caused by increased glomerular capillary pressure [8]. This increases the tubular electrolyte load and the tubular Na^+ transport (T_{Na}) in order to maintain electrolyte homeostasis. The increased T_{Na} inevitably increases the oxygen (O_2) demand and thereby renal QO_2. Indeed, studies have shown that diabetic kidneys have reduced tissue pO_2, which may be linked in part to increased QO_2 [5, 9, 10]. However, it should be noted that the diabetes-induced increase in renal QO_2 cannot be explained fully by increased active T_{Na}. Although animal models can reproduce many of the functional defects and the presence of proteinuria, the special histological features are seen in full only in human diabetic nephropathy.

Although accounting for only about 0.5% of total body mass, the kidney consumes approximately 10% of the body O_2. About 80% of the renal QO_2 is attributed to active transport of electrolytes by tubular cells [11], while the remaining 20% is due to basal metabolism [12]. The major renal consumer of energy is Na^+/K^+ adenosine triphosphatase (ATPase), located in the basolateral membrane of most tubular cells. The metabolism in the renal cortex is mainly aerobic and therefore highly dependent on pO_2. In the renal medulla, on the other hand, glucose oxidation is relatively high compared to in the cortex, indicating a high glycolytic rate. The medulla has a relatively low QO_2 and anaerobic metabolism [12]. As a consequence, the cells in the renal medulla function under close to hypoxic conditions even in normal physiology [13]. The energy metabolism within the renal medulla is also highly heterogeneous. Glucose oxidation and QO_2 are higher in the outer part of the medulla, while the deeper, inner medulla is more dependent on anaerobic metabolism, with high glycolytic rate, but almost undetectable QO_2 [14].

Approximately 25% of the cardiac output is directed through the kidneys. Most of this perfusion is directed toward the renal cortex, while only about 2.5% of cardiac output perfuses the O_2-poor medullary structures. The low medullary blood flow is required to maintain an osmotic gradient, thereby optimizing urinary concentration capacity [15]. Blood flow to the renal medulla is derived through the closely situated ascending and descending vasa recta, creating a counter current system where electrolytes are recirculated from the ascending to the descending vessel. This counter-current system is necessary for maintaining the high osmotic gradient used for concentrating the urine. However, the high pO_2 of blood entering the vasa recta ensures that O_2 is shunted in the opposite direction, leading to a low O_2 delivery to the medulla [16, 17]. Clearly, shunting of O_2 could influence medullary pO_2 [18], possibly leading to further reduced pO_2, tubular damage, and reduced kidney function. Several factors have been incriminated in the genesis of diabetic nephropathy, and various defense mechanisms have been targeted for prevention of ischemic and hypoxic kidney injury [19, 20]. In this chapter, these factors and mechanisms will be discussed and the interplay between them critically summarized.

2 The Role of Hypoxia and Oxidative Stress

Approximately 30% (3.1% per year) of all type 1 diabetic patients will eventually develop diabetic nephropathy [21, 22], and 20% will develop end stage renal failure [23, 24]. Diabetes is the most ubiquitous reason patients require dialysis or kidney transplant [25, 26]. Diabetic nephropathy develops as a result of genetic susceptibility interacting with environmental factors. Increased plasma glucose is accompanied by activation of numerous cellular pathways, and the exact mechanisms that account for the deterioration in renal function have not yet been elucidated. However, it has been suggested that a reduced renal parenchymal pO_2 may accelerate the development of nephropathy during several pathological conditions [27–29].

Diabetes is closely associated with increased oxidative stress in many tissues, including the kidneys. Five major metabolic pathways have been reported to induce oxidative stress in hyperglycemia and diabetes: increased polyol pathway; activation of protein kinase C (PKC); glucose-derived advanced glycation end products (AGE); induction of the hexosamine pathway; and angiotensin II (Ang II)-induced activation of the nicotinamide adenine dinucleotide phosphate (NADPH) oxidase.

2.1 Diabetes-Induced Decreases in Kidney Oxygen Tension

The pO_2 in any tissue will be the result of the O_2 delivery and the QO_2. The majority of QO_2 in the kidney is related to T_{Na}, which itself correlates with the filtered load [30]. Therefore, the diabetes-induced increase in kidney QO_2 has been ascribed to the effect of glomerular hyperfiltration that increases proximal tubular reabsorption [5, 31]. However, although in the early stages of diabetes, glomerular hyperfiltration and concomitant increases in transport-dependent QO_2 are common [9], QO_2 is also elevated in tubular cells from diabetic rats with normal GFR [10], suggesting additional mechanisms for the increased QO_2 in diabetic kidneys (Fig. 1). In diabetes, excess formation of reactive oxygen species (ROS) can derive from the electron transport chain located in the inner mitochondria membrane or the NADPH oxidase [32–34]. Remarkably, correcting the diabetes-induced ROS formation completely prevents the decreased pO_2 in the diabetic kidney [9]. In addition, a diabetes-induced decrease in interstitial renomedullary pH [10] would increase the shunting of O_2 from the arterial blood in descending vasa recta to the venous blood in ascending vasa recta due to the Bohr effect. The net result will be a reduced O_2 delivery to renal medulla and hence reduced pO_2. A decreased pO_2 could be involved in progression of nephropathy [27].

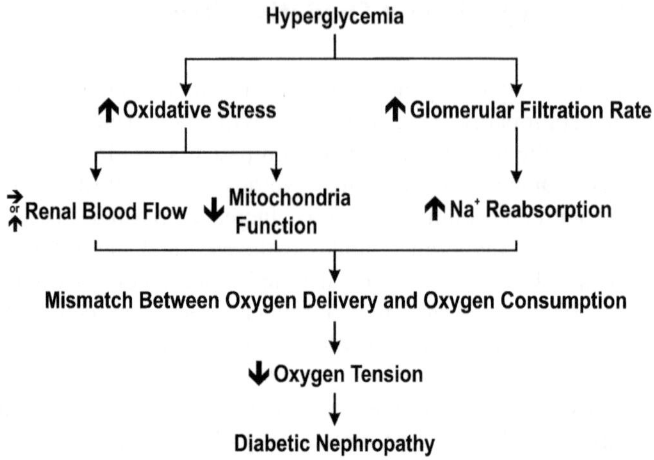

Fig. 1 Schematic view of mechanisms linking hyperglycemia to reduced kidney oxygenation and the development of diabetic nephropathy

2.2 *Diabetes Induces Increases in Kidney Oxygen Consumption*

Since QO_2 has been found to be increased in diabetic cells from the kidney, there must be impaired effects on O_2 metabolism in diabetes that are independent of hemodynamic changes such as hyperfiltration [9, 10]. The increased cellular QO_2 has been linked to increased Na^+/K^+-ATPase activity [5, 35]. Increased activity of the Na^+/glucose-linked transporters (SGLT) as a result of the increased filtered load of glucose can increase Na^+/K^+-ATPase activity [5]. Moreover, hyperglycemia enhances the expression of uncoupling protein-2 and increases gluconeogenesis and fatty acid metabolism, thus inducing a further increased QO_2 [36, 37]. Nitric oxide (NO) inhibits mitochondrial QO_2 reversibly and dose dependently. Thus, the ROS-induced increase in cellular QO_2 in the diabetic kidney is at least partly due to increased degradation of NO [38].

2.2.1 Activation of the Polyol Pathway and Pseudohypoxia

With their report on polyol pathway in *Science* in 1966, Gabbay et al. described the first pathological mechanism in basic diabetes research [39]. The diabetes-induced increase in intracellular glucose concentrations activated the polyol pathway in tissues with insulin-independent glucose uptake [40, 41]. The NADPH-dependent enzyme aldose reductase not only inactivates toxic aldehydes by reducing them to alcohols but also metabolizes glucose. Within the renal tissue, aldose reductase was found predominantly in the medulla, whereas low enzymatic activity was detected in the kidney cortex [42]. Aldose reductase has a low affinity for glucose (K_m = 70mmol/L), which accounts for polyol pathway as being only a minor contributor to the total glucose usage during normoglycemia. However, during manifest diabetes with excessive intracellular glucose levels the activity of the polyol pathway increased substantially.

The initial step of the polyol pathway (i.e., aldose reductase) entails a reduction of glucose to sorbitol, consuming NADPH. Since NADPH is required to regenerate the antioxidant capacity of tissues, this will predispose to oxidative damage. Moreover, the sorbitol formed by aldose reductase is further oxidized to fructose, thereby reducing NAD^+ to NADH and altering the intracellular redox-state [43]. This can be detected as an increased lactate/pyruvate ratio, mainly due to increased lactate concentration [40]. In diabetes, the lactate/pyruvate ratio increases in the renal cortex as well as in the medulla, but inhibition of polyol pathway prevented the increase in the medulla only [10], corresponding to the medulla being the main location of aldose reductase [44]. An increased $NADH/NAD^+$ ratio can mimic cellular abnormalities induced by hypoxia, even though the available O_2 is well above hypoxic threshold. Thus, this state is commonly referred to as "pseudohypoxia" [40]. However, an increased activity of the polyol pathway also will reduce renal pO_2 in the diabetic kidney [10]. Glucose-induced activation of polyol pathway has been demonstrated in numerous tissues affected by diabetes, including renal medulla [10, 45–47]. Activation of the polyol pathway may induce pathology by activation of the pentose phosphate pathway, diacylglycerol (DAG), and PKC [40]. Additionally, the protein expression, as well as the activity of the aldose reductase, is increased during manifest diabetes mellitus [41].

Presently, there are no clinical treatments directed against the polyol pathway to test the impact of this mechanism. However, a polymorphism of the aldose reductase gene (*ALR2*) is associated with decreased risk of microvascular diabetic complications and thus could be a therapeutic target [48]. Since NO inhibits aldose reductase [49], strategies to restore NO may be effective in part by reducing this pathway. Aldose reductase inhibition for 5 years preserved nerve conduction velocity in diabetic dogs [50]. However, a similar study showed no beneficial effects of aldose reductase inhibition for albuminuria or renal morphological changes [51]. It should be noted that inhibiting excessive ROS formation at the level of mitochondrial membrane totally inhibited the activation of the polyol pathway [33]. Thus, beneficial effects of antioxidant strategies might be mediated in part by reduced activity of the polyol pathway.

2.3 Alterations in the Nitric Oxide System

NO in the kidney acts as a vasodilator, and it regulates GFR, tubuloglomerular feedback, proton secretion, Na^+ reabsorption, and pO_2 by affecting both O_2 delivery and QO_2 [52–57]. In proximal tubular cells, NO inhibits QO_2 both by inhibiting T_{Na}, and fluid transport and by direct inhibition of the mitochondrial respiration [56–58]. On the other hand, it should be noted that in pathological states such as diabetes nitric oxide synthase (NOS) may be the source of superoxide through "NOS uncoupling," linked to increased cytotoxic peroxynitrite formation and a

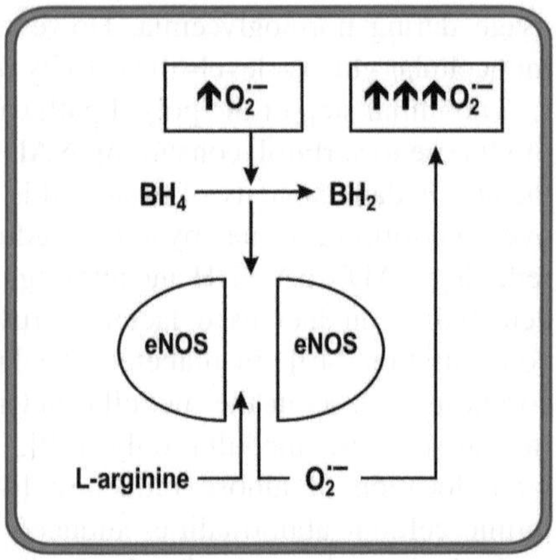

Fig. 2 Elevated superoxide levels causes oxidation of tetrahydrobiopterin (BH_4) to dihydrobiopterin (BH_2), which causes uncoupling of endothelial and neuronal nitric oxide synthase (eNOS and nNOS, respectively). The result is that the NOS switch from producing nitric oxide to superoxide radicals, which further increases the oxidative stress

reduced bioavailability of tetrahydrobiopterin (BH_4) [59, 60] (Fig. 2). Peroxynitrite in diabetic mice increased renal eNOS monomer expression and decreased renal eNOS dimer expression, which is predicted to switch eNOS from producing NO to superoxide radicals [60–62]. Additionally, peroxynitrite may cause irreversible damage to the mitochondria, which may induce further oxidative stress and inflammation [63].

There are at least three known isoforms of NOS: constitutive endothelial NOS (eNOS), neuronal NOS (nNOS), and inducible NOS (iNOS). There are reports that diabetic patients with overt nephropathy have altered allele frequencies in the gene expressing eNOS compared to normoalbuminuric diabetic patients [64], but there is controversy regarding the nature of NOS alterations in diabetes. The expression of eNOS in diabetic rats is often enhanced [65–67], but the renal cortex from diabetic rats displays a 50% reduction in total NOS activity [68]. Both the renal density and the distribution of NOS binding sites are decreased in diabetic rabbits [69]. Also, it should be noted that increased NOS expression does not necessarily result in increased NO release [66]. Hyperglycemia reduced eNOS activity in bovine aortic endothelial cells [70]. Komers et al. showed that eNOS expression was decreased in cortical membrane preparations, whereas the expression in cytosolic fractions was increased [71]. Since eNOS is a membrane-bound enzyme, this may imply decreased NOS activity. However, the nNOS isoform has been assigned a special role in the development of diabetic nephropathy [72]. The expression of nNOS is decreased in the macula densa and in the renal cortex of diabetic rats [73]. There may be a relationship between decreased expression of nNOS and diabetic nephropathy [74]. However, there are also reports of increased nNOS in diabetes [75]. Moreover, inhibition of nNOS with S-methyl-L-thiocitrulline (SMTC) has been shown to be renoprotective [76]. Hyperglycemia-induced alterations in NOS expression and activity can be restored fully by intense insulin treatment [68, 71, 77]. The iNOS isoform is soluble. It has little constitutive expression but is induced by immunostimulants such as infections, bacteria, and cytokines. The involvement of iNOS in diabetic nephropathy is controversial [67], but cortical iNOS is induced during diabetes [78], and NO synthesis is increased during the development of nephropathy [79]. Ketoacidosis is thought to reduce NO production via reduction of iNOS activity [80]. An 80% reduction of iNOS activity has been reported in mesangial cells during low pH, despite close to normal iNOS mRNA and protein levels. This is restored by normalizing the pH, but not by L-arginine supplementation. It is not clear yet how this relates to the effects of ketoacidosis.

Some studies in diabetes have reported increased levels of NO [81, 82] and an increased urinary excretion of NO metabolites [83], whereas others have reported decreased NO bioavailability [69, 84, 85]. These results may not be in conflict since the renal tissues become unresponsive to the vasodilatory effect of NO in diabetes [72, 86] and may therefore have increased susceptibility to vasoconstriction [87]. Oxidative stress promotes vasoconstriction and scavenges NO, which add further to the overall hypoperfusion of the kidney during the more advanced state of the diabetic nephropathy [34, 88]. Schnackenberg and Wilcox demonstrated that the response to acetylcholine in isolated afferent arterioles was shifted from a profound vasodilatation in vessels from normoglycemic rabbits to a profound vasoconstriction in diabetic vessels [89] (Fig. 3). Importantly, acute treatment with the antioxidant tempol (4-hydroxy tempo; 1 mmol/L) in the bath shifted the response back to a vasodilatation, demonstrating a pivotal involvement of oxidative stress. It is well-established that NO deficiency eventually results in hypertension and the development of progressive kidney dysfunction [90].

There are at least two additional mechanisms that may account for decreased NO activity in the diabetic kidney. Diabetic animals have increased NOS uncoupling and reduced plasma L-arginine concentration, both of which will limit NO production [84]. The administration of L-arginine induces a pronounced increase in

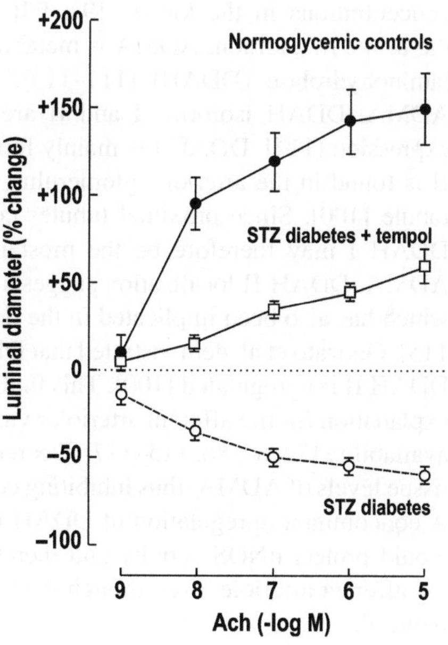

Fig. 3 The response of isolated afferent arterioles from normoglycemic control and streptozotocin (STZ)-diabetic rabbits to acetylcholine (Ach). Ach caused a vasodilation in normal arterioles, whereas it induced vasoconstriction in afferent arterioles from diabetic rabbits. The vasodilation was partly restored by acute administration of tempol (1 mmol/L) to the isolated vessels. Data from Schnackenberg and Wilcox [89]

bioavailable NO specifically in diabetic animals [84] and causes vasodilatation, an effect that is enhanced by insulin [91]. Romero et al. suggested further that there was an augmented intracellular arginase activity in diabetes, which would further limit L-arginine levels inside the NO-producing cells [92]. This may be a mechanism that will limit the efficiency of L-arginine supplementation to diabetics.

2.3.1 Dimethyl Arginines

Endogenously produced NOS inhibitors may regulate NO generation in a number of disease states [93]. Among the most important is the L-arginine analogue asymmetric dimethyl arginine (ADMA), which is a competitive inhibitor of NOS isoforms [94, 95], whereas its enantiometric form symmetric dimethyl arginine (SDMA) has no direct inhibitory effect on NOS activity. However, SDMA may compete with L-arginine for cellular uptake by cationic amino acid transporters and thereby interfering with NO production [96]. Augmented plasma ADMA concentrations have been reported in diabetic animal models, as well as in patients with chronic renal disease, diabetic neuropathy and retinopathy, patients with insulinopenic diabetes, type 2 diabetes, or insulin resistance syndromes [97–107]. Plasma concentrations of ADMA in patients with end stage renal disease strongly predict both cardiovascular disease and overall mortality [108, 109]. Intracellular accumulation of ADMA inhibits NO production and may be a causative mechanism involved in the development of renal and cardiovascular failure.

Cellular ADMA originates from turnover of arginine residues within proteins that have been posttranslationally methylated by protein arginine *N*-methyltransferase (PRMT) class I [95, 110]. ADMA is an important regulator of NO production and NO concentrations in the kidney [94, 95]. ADMA and SDMA are both excreted in the urine. In addition, ADMA is metabolized by N^G, N^G-dimethylarginine dimethylaminohydrolase (DDAH) [111–113]. The kidneys are crucial for metabolizing ADMA. DDAH isoforms I and II are expressed in the kidney at sites of NOS expression [114]. DDAH I is mainly found in the proximal tubules, whereas DDAH II is found in the arteriole, glomerulus, macula densa, renal vasculature, and distal tubule [100]. Since proximal tubules comprise the major part of the renal cortex, DDAH I may therefore be the most important isoform for renal metabolism of ADMA. DDAH II localization suggests that it may regulate kidney hemodynamics, which has also been implicated in the development of diabetic nephropathy [72, 86, 115]. Onozato et al. demonstrated that DDAH I is downregulated in diabetes, whereas DDAH II is upregulated [100]. This finding of upregulated DDAH II may provide an explanation for the afferent arteriolar vasodilation seen in states of reduced renal NO availability [72, 84, 86, 115–117]. Decreased DDAH I in diabetes could increase renal tissue levels of ADMA, thus inhibiting cortical NOS and reducing NO bioavailability. A concomitant upregulation of DDAH II in the macula densa and afferent arteriole would protect nNOS activity and thereby cause a NO-dependent vasodilatation of the afferent arteriole, even though the total NO availability in the diabetic kidney is reduced.

2.4 Proinsulin C-Peptide

Other factors besides hyperglycemia are implicated in the development of diabetic nephropathy. Proinsulin C-peptide is secreted from the pancreatic islets of Langerhans together with insulin. When, as in type 1 diabetes mellitus, insulin synthesis is impaired, there is a corresponding reduction in the synthesis and secretion of C-peptide. C-peptide has been shown to improve function in many of the tissues commonly affected by diabetes complications [118–123]. This provides some insight into the finding that pancreas transplantation reduces diabetic lesions after 10 years of normoglycemia [124]. C-peptide had renoprotective effects both in diabetic patients and in animal models [120, 125]. C-peptide acutely decreases hyperfiltration in diabetic patients [118, 125, 126] and improved renal function when administered chronically to diabetic patients [127]. In experimental diabetes mellitus, C-peptide reduced renal hypertrophy, proteinuria, and albuminuria [120, 128, 129].

C-peptide inhibited Na^+/K^+-ATPase in isolated proximal tubular cells from diabetic animals [130]. This normalized baseline proximal tubular QO_2 [131]. The C-peptide–induced inhibition of Na^+/K^+-ATPase in diabetes is supported by the in vivo observation that C-peptide increases both urinary lithium clearance and fractional Na^+ excretion, which are the results of inhibited T_{Na} in diabetic rat kidney [130].

C-peptide has been shown to affect O_2 transport, uptake, and consumption in several tissues, including the kidney, which in part is accomplished by inducing NO release [131]. In the kidney, C-peptide administration likely possesses dual actions: the peptide constricted the afferent arteriole but induced a compensatory dilation of the efferent arteriole, which maintained renal blood flow [130, 132]. Similar effects have been reported for adenosine, as well as for insulin [133–136]. C-peptide also improved diabetes-induced erythrocyte deformability, which has the potential to improve O_2 uptake in capillaries and thus tissue pO_2 [131].

It appears that C-peptide possesses state-specific effects; although generally reported to be a stimulator of eNOS, C-peptide counteracts diabetes-induced increases in renal eNOS expression [137, 138]. In normoglycemic rat proximal tubular cells, C-peptide stimulated Na^+/K^+-ATPase [139, 140], whereas, it had the opposite effect in isolated proximal tubular cells from diabetic animals [130] and normalized baseline proximal tubular QO_2 [130, 131].

2.5 Adenosine

In addition to regulating blood flow by modulating vascular resistance, adenosine is also coupled to energy metabolism and T_{Na}. The renal effects of adenosine are more pronounced in the renal medulla than in the cortex. Adenosine increased medullary pO_2 but decreased cortical pO_2 [141]. Adenosine caused vasoconstriction in the renal cortex via acting on the adenosine A1-receptor (A1AR), which directly reduced GFR [142]. Furthermore, selective A1AR blockade increased cortical blood flow by 15%, whereas medullary blood flow remained unchanged in normoglycemic control rats [143]. However, the same study reported that medullary pO_2 decreased more than 40% after A1AR blockade. Thus, it is not yet fully elucidated whether adenosine is protective or damaging to the diabetic kidney [144, 145]. It is clear that the vascular effects of adenosine are both concentration and time dependent. Adenosine caused renal vasoconstriction in short time and at low doses, and renal vasodilation later on and in higher concentrations [142, 146–148]. The vasoconstrictor effect of adenosine is enhanced in diabetic kidneys [149], perhaps via interaction with Ang II [146].

3 Sources and Molecular Effects of Oxidative Stress

Diabetes is closely associated with increased oxidative stress in the kidneys [9]. This may be offset by upregulation of the antioxidant enzyme hemeoxygenase-1 [150]. Treatment with antioxidants (e.g., α-tocopherol, pycnogenol, beta-carotene, and alpha-lipoic acid) have reduced oxidative stress in diabetic rats [151]. Antioxidant treatment reduced the diabetes-induced increase of the hemeoxygenase-1 and increased the activity of glutathione and glutathione redox enzymes [152, 153]. We have previously shown that treating diabetic animals with the free radical scavenger α-tocopherol fully prevented the diabetes-induced decrease in kidney pO_2 and normalized the cellular QO_2 [9]. Prevention of the early metabolic alterations occurring in diabetic kidney is likely to significantly improve or delay the onset of the structural and functional changes commonly associated with the progression of diabetic nephropathy.

3.1 Reactive Oxygen Species: The Unifying Mechanism

Oxidative stress is the result of an imbalance between ROS formation and the activity of the antioxidant defense system (e.g., superoxide dismutase [SOD]). Increased production of ROS during diabetes is accomplished by several different pathways, for example, increased NADPH oxidase activity, uncoupled NOS, and mitochondrial superoxide production. ROS are implicated in the development of diabetic nephropathy [9, 34, 154–157]. ROS can damage enzymes and DNA and regulate transcription factors and the activity of various O_2-sensitive genes. Hypoxia-inducible factor (HIF)-1 has emerged as an important mediator for medullary adaptation to low pO_2 under physiological conditions [158]. A unifying mechanism has been reported whereby hyperglycemia-induced ROS activated several pathways known to participate in the development of diabetic complications, namely the hexosamine pathway, glucose flux through the polyol pathway, formation of AGE, and activation of PKC [33, 157, 159]. However, this theory does not specify where the ROS are produced, only that the normalization of the superoxide concentrations at the levels of the mitochondria membrane prevents activation of these pathways. In the view of the excessive capability of NADPH oxidase to produce superoxide anions as a result of AT_1-receptor activation, it seems likely that superoxide anions produced in the cytosol could influence mitochondria function and potentially activate these pathways.

3.1.1 Activation of Protein Kinase C

PKC is activated via synthesis of DAG within a few days of hyperglycemia [160, 161]. Increased PKC activity is linked to numerous diabetic complications, including nephropathy [157, 162]. PKC mediated alterations in renal blood flow [163], possibly by decreasing the production of NO [164]. PKC is also an activator of the mitogen-activated protein kinase (MAPK) pathway, a regulator of cell growth, and is pivotal for development of hypertrophy in the diabetic kidney [165]. Superoxide production by the medullary thick asending loop of Henle (mTAL) has recently been shown to be partially PKC dependent [166]. mTALs from diabetic rats display increased superoxide production that can be abolished by specific inhibition of PKCα or PKCδ. Inhibition of PKC decreased mesangial expansion, albuminuria, and GFR and increased pro-inflammatory gene expression and vascular permeability in several models of experimental diabetes [157, 163, 167].

3.1.2 Advanced Glycation End Products

Posttranslationally modified proteins, such as glucose-derived AGE, are increased in diabetic glomeruli and urine [156, 168]. AGEs are formed when glucose auto-oxidizes to dicarbonyls, which thereafter react with protein amino groups and modify proteins, alter extracellular matrix function and stimulate the production of cytokines and ROS-specific receptors [169, 170]. AGE induce mitochondria-induced oxidative stress in the kidney [171], and AGE-modified plasma proteins caused vascular pathology [170]. A polymorphism in an AGE receptor (RAGE) was associated with a weak protective effect and a longer duration of nephropathy-free diabetes [172]. Recently, the renoprotective effects of angiotensin receptor blocker (ARB), calcium channel blocker, lipid- (bezafibrate) or glucose-lowering (pioglitazone) agents, and the HIF activator cobalt chloride in a model of type 2 diabetes were related to a reduction in the intrarenal AGE accumulation [173]. Taken together with reports of inhibition of AGE that prevented manifestations of diabetic nephropathy [174–176], this suggests that AGE play an important role for progression of diabetic nephropathy.

3.1.3 The Hexosamine Pathway

Shunting of excess intracellular glucose increased the delivery of substrates to the hexosamine pathway [177] whose end product (UDP-*N*-acetylglucosamine) causes glycosylation and affects several transcription factors [178–180]. The rate limiting step in the formation of hexosamines is glutamine:fructose-6-phosphate amidotransferase (GFPT), and increase in mesangial GFPT activity increased cytokine levels [181]. Intracellular glucose that enters the hexosamine pathway is metabolized to glycolipids, proteoglycans, and glycoproteins [157], which in turn increase the transcription of pro-sclerotic transforming growth factor and plasminogen activator inhibitor-1 (PAI-1) [179, 182], which cause microvascular dysfunction. The hexosamine pathway may also increase the expression of growth factors and leptin [183]. However, activation of the hexosamine pathway is dependent on excessive ROS levels at the level of the mitochondrial membrane [33].

3.2 Hypoxia and Oxidative Stress in Mitochondria

The mitochondria require a proton gradient across the inner membrane to produce ATP, which is accomplished by the transfer of electrons along the electron transfer chain. Complexes I, III, and IV are all coupled to proton transport, and for every two electrons passing through the electron transport chain, ten protons are pumped into the mitochondrial intermembrane space. This creates a proton gradient across the inner membrane, which is used by ATP-synthase to produce ATP from adenosine diphosphate (ADP) and inorganic phosphate (P_i). Mitochondria generate ROS even during normal respiration, which may account for approximately 0.2% of total normal cellular QO_2 [184, 185]. However, mitochondrial ROS generation is elevated in diabetes and aging [186, 187]. An abnormally high proton gradient is created across the inner mitochondrial membrane during sustained hyperglycemia, causing increased ROS formation [162, 188].

Several aspects of mitochondria function are altered in the diabetic kidney. Within the kidney, mitochondrial QO_2 is normally coupled to active transport (i.e., it depends on the availability of ADP and the subsequent ATP production) [189]. In brown adipose tissue, however, mitochondria display no such relationship due to the presence of uncoupling protein (UCP)-1, which is responsible for nonshivering thermogenesis. UCPs belong to a mitochondria anion carrier family and have an uncoupling function (i.e., uncoupling QO_2 from ATP production). This is accomplished by allowing protons to leak back across the inner membrane without passing through the ATP synthase [33, 188, 190–192]. Even minor UCP-related uncoupling will cause a mild depolarization, thus reducing the mitochondrial membrane potential [191].

Recently, UCP-2 was identified in both rodent and human kidneys [193–197]. Proximal tubular cells as well as cells of the medullary thick ascending loop of

Henle showed heavy staining for UCP-2 in the rat kidney [195]. It was also shown that mitochondria isolated from kidney cortex of diabetic rats display glutamate-stimulated QO_2 during inhibition of ATP synthase, indicating mitochondria uncoupling. This was further confirmed by addition of the UCP specific blocker guanosine diphosphate (GDP), which blocked all uncoupled QO_2 [196, 198]. Although the increased mitochondria uncoupling likely has a protective role against excessive ROS formation in the diabetic kidney, it will increase the QO_2 needed to sustain ATP production [190, 199–201]. This may provide an explanation for the lack of relationship between the diabetes-induced increase in renal QO_2 and active transport of electrolytes, and explain the increased electrolyte transport-independent QO_2 occurring in these kidneys [9, 10]. It should be noted that intense insulin treatment abolished the increased UCP-2 protein expression and the increased cellular QO_2 in the diabetic kidney [196].

3.3 NADPH and the Pentose Phosphate Shunt

NADPH is the main intracellular reductant and regenerator of the antioxidants glutathione and lipoic acid. The entire antioxidant system is dependent on a sufficient supply of NADPH since glutathione and lipoic acid in turn exert important intracellular antioxidant activities by reacting with ROS and organic peroxides [202]. NADPH is formed during glycolysis, oxidative phosphorylation, and by the pentose phosphate shunt, which controls the synthesis of sorbitol and glutathione. Binding of glucose to mitochondria leads to the formation of both ADP and glucose-6-phosphate (G6P). ADP that enter the mitochondria will stimulate oxidative phosphorylation, whereas G6P acts as the switch between glycolysis, glycogen synthesis, and the pentose phosphate shunt and can induce NADPH production. Since production of NADPH is dependent on G6P dehydrogenase, a decrease in G6P dehydrogenase activity results in decreased NADPH levels, increasing the risk for oxidant damage [203, 204].

The enzymes of the renal pentose phosphate shunt are activated in kidneys from both animal models of experimental diabetes and in suboptimally treated diabetic patients with chronic metabolic acidosis [205, 206]. These alterations are likely of relevance in development of renal hypertrophy [206].

3.4 NADPH Oxidase

Several studies have reported a role for NADPH oxidase in the development of experimental hypertension and nephropathy [207, 208]. NADPH oxidase includes the membrane-associated subunits Nox-2 (originally referred to as $gp91^{phox}$), $p22^{phox}$, and the regulatory cytosolic proteins $p47^{phox}$, $p67^{phox}$, $p40^{phox}$, and Rac1 [209]. The cytosolic elements all assemble and bind to subunits Nox-2 and $p22^{phox}$ to form an active superoxide producing enzyme [210, 211].

All components of the NADPH oxidase, as well as isozymes Nox-1 and -4, are expressed in the kidney [212]. The main locations correlate with the sites of ROS production. The sites of intrarenal expression include renal vessels, glomeruli, podocytes, thick ascending limbs of the loop of Henle (TAL), macula densa, distal tubules, collecting ducts, and cortical interstitial fibroblasts.

Renal NADPH oxidase produces a significant amount of superoxide radicals [213], which can react with NO to form peroxynitrite, which is a potent oxidant and nitrosylating agent. This reaction coincidentally causes NO deficiency. Since NO normally regulates tubuloglomerular feedback and enhances renal blood flow, with some relevance for Na^+ excretion, NO deficiency could result in hypertension, glomerulosclerosis, proteinuria, and eventually the development of renal failure [90]. Furthermore, several vasoconstrictors are activated by oxidative stress [90]. Thus, oxidative stress has been increasingly implicated in the pathogenesis of diabetic renal injury [214]. A high salt intake increased renal superoxide generation by diminishing the renal expression of SOD and increased NADPH oxidase activity [215]. Nox4 is induced by activation of the renin-angiotensin system (RAS) [216]. Nox1 activity and $p22^{phox}$ are also upregulated by activation of RAS and by infusion of Ang II [216, 217]. Interventions directed against the RAS normalized $p47^{phox}$ protein expression [34]. However, the mechanisms of NADPH oxidase activation in the diabetic kidney are not fully elucidated [218], but the protein expressions of several subtypes increased shortly after induction of diabetes and correlated with proteinuria [34, 154, 207, 219]. Translocation of $p47^{phox}$ to the membrane resulted in increased oxidative stress in diabetic kidneys [154]. NADPH oxidase expressions were upregulated in glomeruli of type 1 diabetic mice [220]. Streptozotocin-induced diabetic rats had increased mRNA for $p47^{phox}$, $p22^{phox}$, and Nox4 in both tubular cells and glomerular podocytes [154, 221]. Exposure of mesangial cells to advanced oxidation protein products led to membrane-associated phosphorylation of PKCα, which caused phosphorylation of cytosolic $p47^{phox}$ and led to membrane translocation that activated of NADPH oxidase [222]. The NADPH oxidase inhibitor apocynin decreased expression of the subunit $gp91^{phox}$ and repressed activation of $p47^{phox}$ in diabetic rat kidneys [154].

Nox4 is believed to be the major renal source of ROS in diabetic animals. Renal oxidative stress is paralleled by an increase in Nox4 and decreased superoxide dismutase activity in type 1 as well as in type 2 diabetes [223, 224]. Knockdown of Nox4 by siRNA reduced NADPH oxidase activity and blocked glucose-induced mitochondrial superoxide generation [223]. Part of the diabetes-induced renal mitochondrial oxidative stress was due to increased mitochondrial membrane permeability and to protein modifications by nitrotyrosine generation [224].

3.5 Angiotensin II

The RAS plays a key role in the development of diabetic nephropathy, and intrarenal Ang II is increased in insulinopenic diabetes [100]. Through its action on Ang II

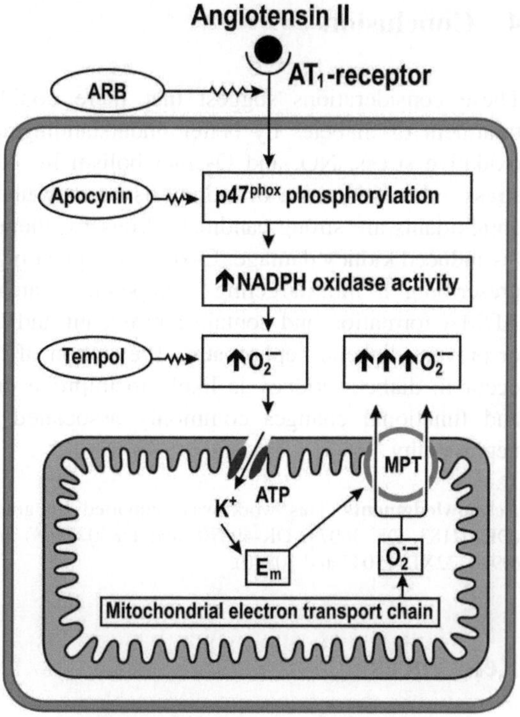

Fig. 4 Angiotensin II activates the nicotinamide adenine dinucleotide phosphate (NADPH) oxidase to produce superoxide radicals via AT_1-receptors, which induces a further increase in superoxide production by the mitochondrial electron transport chain. *ARB* angiotensin II AT_1-receptor blocker; *MPT* mitochondrial permeability transition pore

receptor subtype 1 (AT_1), Ang II increased NADPH oxidase activity, which stimulated the formation of superoxide anions [34, 215, 217, 225, 226] (Fig. 4). Ang II modulated the NO pathway in the diabetic kidney [115], partly by directly regulating the intrarenal ADMA levels [100]. Additionally, Ang II activated PKC [227]. Early diabetes mellitus in rats was associated with a transient increase in plasma renin activity [228], causing an increase in Ang II formation. A renin gene polymorphism may contribute to diabetic nephropathy through increased renin gene expression [229]. In addition, certain polymorphisms of angiotensin converting enzymes (ACE) are associated with increased susceptibility to diabetic nephropathy [64].

Inhibition of RAS activity has a renoprotective effect during diabetes [65, 230–233]. ARB and ACE inhibition have beneficial effects on albuminuria in both animal models of experimental diabetes and diabetic patients [234–236]. RAS inhibition prevented diabetes-induced eNOS expression in rats with streptozotocin-induced diabetes [34, 65], reduced diabetes-induced endothelin-1 activation [237], and inhibited NADPH oxidase activity [238, 239]. Onozato et al. reported that diabetes-induced increases in renal Ang II and ADMA both are normalized by an ARB, which increased the renal expression of ADMA metabolizing enzyme DDAH I, decreased the expression of the ADMA synthesizing enzyme PRMT-1, and resulted in increased intrarenal NO production [100]. These novel results provide at least a partial explanation for why patients with advanced nephropathy likely will be those who benefit the most from blockade of the RAS [240].

4 Conclusion

These considerations suggest that there could be substantial advances in the treatment of diabetes by better understanding of the complex interplay between oxidative stress, NO, and O_2 metabolism in the diabetic kidney. Since oxidative stress-induced hypoxia of kidney tissue may underlie diabetic nephropathy, potent antioxidants are strong candidate drugs for therapeutic strategies to prevent diabetes-induced kidney damage. Even more appealing may be combining treatments, by prescribing a multitargeting therapeutic strategy that inhibits RAS, AGE, and ADMA formation and contain antioxidant and C-peptide supplementation to treat or prevent diabetic nephropathy. Prevention of the early metabolic alterations that occur in diabetic kidney is likely to improve or delay the onset of the structural and functional changes commonly associated with the progression of diabetic nephropathy.

Acknowledgments This work was supported by grants from NHLBI (HL-68686), NIDDK (DK-07183, DK-36079, DK-49870, and DK-077858) and from the Swedish Research Council (9940, 72XD-15043 and 10840).

References

1. Najafian B, Mauer M. Progression of diabetic nephropathy in type 1 diabetic patients. Diabetes Res Clin Pract 2009; 83:1–8.
2. Mogensen CE. Glomerular filtration rate and renal plasma flow in normal and diabetic man during elevation of blood sugar levels. Scand J Clin Lab Invest 1971; 28:177–182.
3. Palm F, Liss P, Fasching A, Hansell P, Carlsson PO. Transient glomerular hyperfiltration in the streptozotocin-diabetic Wistar Furth rat. Ups J Med Sci 2001; 106:175–182.
4. Edlund J, Hansell P, Fasching A, Liss P, Weis J, Glickson JD, Palm F. Reduced oxygenation in diabetic rat kidneys measured by T2* weighted magnetic resonance micro-imaging. Adv Exp Med Biol 2009; 645:199–204.
5. Korner A, Eklof AC, Celsi G, Aperia A. Increased renal metabolism in diabetes. Mechanism and functional implications. Diabetes 1994; 43:629–633.
6. dos Santos EA, Li LP, Ji L, Prasad PV. Early changes with diabetes in renal medullary hemodynamics as evaluated by fiberoptic probes and BOLD magnetic resonance imaging. Invest Radiol 2007; 42:157–162.
7. Rosenberger C, Khamaisi M, Abassi Z, Shilo V, Weksler-Zangen S, Goldfarb M, Shina A, Zibertrest F, Eckardt KU, Rosen S, Heyman SN. Adaptation to hypoxia in the diabetic rat kidney. Kidney Int 2008; 73:34–42.
8. O'Bryan GT, Hostetter TH. The renal hemodynamic basis of diabetic nephropathy. Semin Nephrol 1997; 17:93–100.
9. Palm F, Cederberg J, Hansell P, Liss P, Carlsson PO. Reactive oxygen species cause diabetes-induced decrease in renal oxygen tension. Diabetologia 2003; 46:1153–1160.
10. Palm F, Hansell P, Ronquist G, Waldenstrom A, Liss P, Carlsson PO. Polyol-pathway-dependent disturbances in renal medullary metabolism in experimental insulin-deficient diabetes mellitus in rats. Diabetologia 2004; 47:1223–1231.
11. Thaysen JH, Lassen NA, Munck O. Sodium transport and oxygen consumption in the mammalian kidney. Nature 1961; 190:919–921.

12. Cohen J, Kamm D. Renal metabolism: Relation to renal function. In: Brenner B, Rector F, eds. The Kidney. Philadelphia: W.B. Saunders, 1985:144–248.
13. Leonhardt KO, Landes RR. Oxygen tension of the urine and renal structures. Preliminary report of clinical findings. N Engl J Med 1963; 269:115–121.
14. Klahr S, Hammarman M. Renal metabolism. In Seldin DW, Giebisch G, eds. The Kidney: Physiology and Pathophysiology. New York: Raven Press, 1985:699–718.
15. Chou SY, Porush JG, Faubert PF. Renal medullary circulation: hormonal control. Kidney Int 1990; 37:1–13.
16. Aukland K, Krog J. Renal oxygen tension. Nature 1960; 188:671.
17. Levy MN, Imperial ES. Oxygen shunting in renal cortical and medullary capillaries. Am J Physiol 1961; 200:159–162.
18. Aukland K. Studies on intrarenal circulation with special reference to gas exchange. J Oslo City Hosp 1964; 14:115–146.
19. Brezis M, Heyman SN, Dinour D, Epstein FH, Rosen S. Role of nitric oxide in renal medullary oxygenation. Studies in isolated and intact rat kidneys. J Clin Invest 1991; 88:390–395.
20. Epstein FH. Oxygen and renal metabolism. Kidney Int 1997; 51:381–385.
21. Schjoedt KJ, Hansen HP, Tarnow L, Rossing P, Parving HH. Long-term prevention of diabetic nephropathy: an audit. Diabetologia 2008; 51:956–961.
22. Krolewski M, Eggers PW, Warram JH. Magnitude of end-stage renal disease in IDDM: a 35 year follow-up study. Kidney Int 1996; 50:2041–2046.
23. Nishimura R, Dorman JS, Bosnyak Z, Tajima N, Becker DJ, Orchard TJ. Incidence of ESRD and survival after renal replacement therapy in patients with type 1 diabetes: a report from the Allegheny County Registry. Am J Kidney Dis 2003; 42:117–124.
24. Anon. The effect of intensive treatment of diabetes on the development and progression of long-term complications in insulin-dependent diabetes mellitus. The Diabetes Control and Complications Trial Research Group. N Engl J Med 1993; 329:977–986.
25. Berg UB, Torbjornsdotter TB, Jaremko G, Thalme B. Kidney morphological changes in relation to long-term renal function and metabolic control in adolescents with IDDM. Diabetologia 1998; 41:1047–1056.
26. Yip JW, Jones SL, Wiseman MJ, Hill C, Viberti G. Glomerular hyperfiltration in the prediction of nephropathy in IDDM: a 10-year follow-up study. Diabetes 1996; 45:1729–1733.
27. Brezis M, Rosen S. Hypoxia of the renal medulla – its implication for disease. N Engl J Med 1995; 332:647–655.
28. Fine LG, Orphanides C, Norman JT. Progressive renal disease: the chronic hypoxia hypothesis. Kidney Int Suppl 1998; 65:S74–S78.
29. Palm F. Intrarenal oxygen in diabetes and a possible link to diabetic nephropathy. Clin Exp Pharmacol Physiol 2006; 33:997–1001.
30. Lassen NA, Munck O, Thaysen JH. Oxygen consumption and sodium reabsorption in the kidney. Acta Physiol Scand 1961; 51:371–384.
31. Thomson SC, Deng A, Bao D, Satriano J, Blantz RC, Vallon V. Ornithine decarboxylase, kidney size, and the tubular hypothesis of glomerular hyperfiltration in experimental diabetes. J Clin Invest 2001; 107:217–224.
32. Reis JS, Bosco AA, Veloso CA, Mattos RT, Purish S, Nogueira-Machado JA. Oxidizing and reducing responses in type 1 diabetic patients determined up to 5 years after the clinical onset of the disease. Acta Diabetol 2008; 45:221–224.
33. Nishikawa T, Edelstein D, Du XL, Yamagishi S, Matsumura T, Kaneda Y, Yorek MA, Beebe D, Oates PJ, Hammes HP, Giardino I, Brownlee M. Normalizing mitochondrial superoxide production blocks three pathways of hyperglycaemic damage. Nature 2000; 404:787–790.
34. Onozato ML, Tojo A, Goto A, Fujita T, Wilcox CS. Oxidative stress and nitric oxide synthase in rat diabetic nephropathy: Effects of ACEI and ARB. Kidney Int 2002; 61:186–194.
35. Baines A, Ho P. Glucose stimulates O2 consumption, NOS, and Na/H exchange in diabetic rat proximal tubules. Am J Physiol Renal Physiol 2002; 283:F286–F293.

36. Peroni O, Large V, Diraison F, Beylot M. Glucose production and gluconeogenesis in postabsorptive and starved normal and streptozotocin-diabetic rats. Metabolism 1997; 46:1358–1363.
37. Guder WG, Schmolke M, Wirthensohn G. Carbohydrate and lipid metabolism of the renal tubule in diabetes mellitus. Eur J Clin Chem Clin Biochem 1992; 30:669–674.
38. Schnackenberg CG. Physiological and pathophysiological roles of oxygen radicals in the renal microvasculature. Am J Physiol Regul Integr Comp Physiol 2002; 282:R335–R342.
39. Gabbay KH, Merola LO, Field RA. Sorbitol pathway: presence in nerve and cord with substrate accumulation in diabetes. Science 1966; 151:209–210.
40. Williamson JR, Chang K, Frangos M, Hasan KS, Ido Y, Kawamura T, Nyengaard JR, van den Enden M, Kilo C, Tilton RG. Hyperglycemic pseudohypoxia and diabetic complications. Diabetes 1993; 42:801–813.
41. Shah VO, Dorin RI, Sun Y, Braun M, Zager PG. Aldose reductase gene expression is increased in diabetic nephropathy. J Clin Endocrinol Metab 1997; 82:2294–2298.
42. Terubayashi H, Sato S, Nishimura C, Kador PF, Kinoshita JH. Localization of aldose and aldehyde reductase in the kidney. Kidney Int 1989; 36:843–851.
43. Tilton RG, Baier LD, Harlow JE, Smith SR, Ostrow E, Williamson JR. Diabetes-induced glomerular dysfunction: links to a more reduced cytosolic ratio of NADH/NAD+. Kidney Int 1992; 41:778–788.
44. Dorin RI, Shah VO, Kaplan DL, Vela BS, Zager PG. Regulation of aldose reductase gene expression in renal cortex and medulla of rats. Diabetologia 1995; 38:46–54.
45. Pugliese G, Tilton RG, Williamson JR. Glucose-induced metabolic imbalances in the pathogenesis of diabetic vascular disease. Diabetes Metab Rev 1991; 7:35–59.
46. Stevens MJ, Dananberg J, Feldman EL, Lattimer SA, Kamijo M, Thomas TP, Shindo H, Sima AA, Greene DA. The linked roles of nitric oxide, aldose reductase and, (Na+, K+)-ATPase in the slowing of nerve conduction in the streptozotocin diabetic rat. J Clin Invest 1994; 94:853–859.
47. Van den Enden MK, Nyengaard JR, Ostrow E, Burgan JH, Williamson JR. Elevated glucose levels increase retinal glycolysis and sorbitol pathway metabolism. Implications for diabetic retinopathy. Invest Ophthalmol Vis Sci 1995; 36:1675–1685.
48. Lajer M, Tarnow L, Fleckner J, Hansen BV, Edwards DG, Parving HH, Boel E. Association of aldose reductase gene Z+2 polymorphism with reduced susceptibility to diabetic nephropathy in Caucasian type 1 diabetic patients. Diabet Med 2004; 21:867–873.
49. Srivastava SK, Ramana KV, Chandra D, Srivastava S, Bhatnagar A. Regulation of aldose reductase and the polyol pathway activity by nitric oxide. Chem Biol Interact 2003; 143–144:333–340.
50. Engerman RL, Kern TS, Larson ME. Nerve conduction and aldose reductase inhibition during 5 years of diabetes or galactosaemia in dogs. Diabetologia 1994; 37:141–144.
51. Kern TS, Engerman RL. Aldose reductase and the development of renal disease in diabetic dogs. J Diabetes Complications 1999; 13:10–16.
52. Wilcox CS, Welch WJ, Murad F, Gross SS, Taylor G, Levi R, Schmidt HH. Nitric oxide synthase in macula densa regulates glomerular capillary pressure. Proc Natl Acad Sci USA 1992; 89:11993–11997.
53. Ortiz PA, Garvin JL. Role of nitric oxide in the regulation of nephron transport. Am J Physiol Renal Physiol 2002; 282:F777–F784.
54. Tojo A, Tisher CC, Madsen KM. Angiotensin II regulates H(+)-ATPase activity in rat cortical collecting duct. Am J Physiol 1994; 267:F1045–F1051.
55. Welch WJ, Tojo A, Wilcox CS. Roles of NO and oxygen radicals in tubuloglomerular feedback in SHR. Am J Physiol Renal Physiol 2000; 278:F769–F776.
56. Koivisto A, Matthias A, Bronnikov G, Nedergaard J. Kinetics of the inhibition of mitochondrial respiration by NO. FEBS Lett 1997; 417:75–80.
57. Koivisto A, Pittner J, Froelich M, Persson AE. Oxygen-dependent inhibition of respiration in isolated renal tubules by nitric oxide. Kidney Int 1999; 55:2368–2375.

58. Palm F, Teerlink T, Hansell P. Nitric oxide and kidney oxygenation. Curr Opin Nephrol Hypertens 2009; 18:68–73.
59. Bevers LM, Braam B, Post JA, van Zonneveld AJ, Rabelink TJ, Koomans HA, Verhaar MC, Joles JA. Tetrahydrobiopterin, but not L-arginine, decreases NO synthase uncoupling in cells expressing high levels of endothelial NO synthase. Hypertension 2006; 47:87–94.
60. Zou MH, Shi C, Cohen RA. Oxidation of the zinc-thiolate complex and uncoupling of endothelial nitric oxide synthase by peroxynitrite. J Clin Invest 2002; 109:817–826.
61. Hink U, Li H, Mollnau H, Oelze M, Matheis E, Hartmann M, Skatchkov M, Thaiss F, Stahl RA, Warnholtz A, Meinertz T, Griendling K, Harrison DG, Forstermann U, Munzel T. Mechanisms underlying endothelial dysfunction in diabetes mellitus. Circ Res 2001; 88:E14–E22.
62. Cai S, Khoo J, Channon KM. Augmented BH4 by gene transfer restores nitric oxide synthase function in hyperglycemic human endothelial cells. Cardiovasc Res 2005; 65:823–831.
63. Ghafourifar P, Cadenas E. Mitochondrial nitric oxide synthase. Trends Pharmacol Sci 2005; 26:190–195.
64. Shestakova MV, Vikulova OK, Gorashko NM, Voronko OE, Babunova NB, Nosikov VV, Dedov II. The relationship between genetic and haemodynamic factors in diabetic nephropathy (DN): Case-control study in type 1 diabetes mellitus (T1DM). Diabetes Res Clin Pract 2006; 74(Suppl 1):S41–S50.
65. Li B, Yao J, Kawamura K, Oyanagi-Tanaka Y, Hoshiyama M, Morioka T, Gejyo F, Uchiyama M, Oite T. Real-time observation of glomerular hemodynamic changes in diabetic rats: effects of insulin and ARB. Kidney Int 2004; 66:1939–1948.
66. Hoshiyama M, Li B, Yao J, Harada T, Morioka T, Oite T. Effect of high glucose on nitric oxide production and endothelial nitric oxide synthase protein expression in human glomerular endothelial cells. Nephron Exp Nephrol 2003; 95:e62–e68.
67. Veelken R, Hilgers KF, Hartner A, Haas A, Bohmer KP, Sterzel RB. Nitric oxide synthase isoforms and glomerular hyperfiltration in early diabetic nephropathy. J Am Soc Nephrol 2000; 11:71–79.
68. Khamaisi M, Keynan S, Bursztyn M, Dahan R, Reinhartz E, Ovadia H, Raz I. Role of renal nitric oxide synthase in diabetic kidney disease during the chronic phase of diabetes. Nephron Physiol 2006; 102:p72–p80.
69. Mumtaz FH, Dashwood MR, Khan MA, Thompson CS, Mikhailidis DP, Morgan RJ. Down-regulation of nitric oxide synthase in the diabetic rabbit kidney: potential relevance to the early pathogenesis of diabetic nephropathy. Curr Med Res Opin 2004; 20:1–6.
70. Du XL, Edelstein D, Dimmeler S, Ju Q, Sui C, Brownlee M. Hyperglycemia inhibits endothelial nitric oxide synthase activity by posttranslational modification at the Akt site. J Clin Invest 2001; 108:1341–1348.
71. Komers R, Schutzer WE, Reed JF, Lindsley JN, Oyama TT, Buck DC, Mader SL, Anderson S. Altered endothelial nitric oxide synthase targeting and conformation and caveolin-1 expression in the diabetic kidney. Diabetes 2006; 55:1651–1659.
72. Komers R, Lindsley JN, Oyama TT, Allison KM, Anderson S. Role of neuronal nitric oxide synthase (NOS1) in the pathogenesis of renal hemodynamic changes in diabetes. Am J Physiol Renal Physiol 2000; 279:F573–F583.
73. Yagihashi N, Nishida N, Seo HG, Taniguchi N, Yagihashi S. Expression of nitric oxide synthase in macula densa in streptozotocin diabetic rats. Diabetologia 1996; 39:793–799.
74. Keynan S, Hirshberg B, Levin-Iaina N, Wexler ID, Dahan R, Reinhartz E, Ovadia H, Wollman Y, Chernihovskey T, Iaina A, Raz I. Renal nitric oxide production during the early phase of experimental diabetes mellitus. Kidney Int 2000; 58:740–747.
75. Maeda M, Yabuki A, Suzuki S, Matsumoto M, Taniguchi K, Nishinakagawa H. Renal lesions in spontaneous insulin-dependent diabetes mellitus in the nonobese diabetic mouse: acute phase of diabetes. Vet Pathol 2003; 40:187–195.
76. Komers R, Lindsley JN, Oyama TT, Anderson S. Effects of long-term inhibition of neuronal nitric oxide synthase (NOS1) in uninephrectomized diabetic rats. Nitric Oxide 2004; 11:147–155.

77. Ding Y, Vaziri ND, Coulson R, Kamanna VS, Roh DD. Effects of simulated hyperglycemia, insulin, and glucagon on endothelial nitric oxide synthase expression. Am J Physiol Endocrinol Metab 2000; 279:E11–E17.
78. Chen H, Brahmbhatt S, Gupta A, Sharma AC. Duration of streptozotocin-induced diabetes differentially affects p38-mitogen-activated protein kinase (MAPK) phosphorylation in renal and vascular dysfunction. Cardiovasc Diabetol 2005; 4:3.
79. Mensah-Brown EP, Obineche EN, Galadari S, Chandranath E, Shahin A, Ahmed I, Patel SM, Adem A. Streptozotocin-induced diabetic nephropathy in rats: the role of inflammatory cytokines. Cytokine 2005; 31:180–190.
80. Prabhakar SS. Inhibition of mesangial iNOS by reduced extracellular pH is associated with uncoupling of NADPH oxidation. Kidney Int 2002; 61:2015–2024.
81. Nakanishi K, Onuma S, Higa M, Nagai Y, Inokuchi T. The intrarenal blood flow distribution and role of nitric oxide in diabetic rats. Metabolism 2005; 54:788–792.
82. Bank N, Aynedjian HS. Role of EDRF (nitric oxide) in diabetic renal hyperfiltration. Kidney Int 1993; 43:1306–1312.
83. Komers R, Allen TJ, Cooper ME. Role of endothelium-derived nitric oxide in the pathogenesis of the renal hemodynamic changes of experimental diabetes. Diabetes 1994; 43:1190–1197.
84. Palm F, Buerk DG, Carlsson PO, Hansell P, Liss P. Reduced nitric oxide concentration in the renal cortex of streptozotocin-induced diabetic rats: effects on renal oxygenation and microcirculation. Diabetes 2005; 54:3282–3287.
85. Wang YX, Brooks DP, Edwards RM. Attenuated glomerular cGMP production and renal vasodilation in streptozotocin-induced diabetic rats. Am J Physiol 1993; 264:R952–R956.
86. Komers R, Anderson S. Paradoxes of nitric oxide in the diabetic kidney. Am J Physiol Renal Physiol 2003; 284:F1121–F1137.
87. Pflueger AC, Larson TS, Hagl S, Knox FG. Role of nitric oxide in intrarenal hemodynamics in experimental diabetes mellitus in rats. Am J Physiol 1999; 277:R725–R733.
88. Tesfamariam B. Free radicals in diabetic endothelial cell dysfunction. Free Radic Biol Med 1994; 16:383–391.
89. Schnackenberg CG, Wilcox CS. The SOD mimetic tempol restores vasodilation in afferent arterioles of experimental diabetes. Kidney Int 2001; 59:1859–1864.
90. Modlinger PS, Wilcox CS, Aslam S. Nitric oxide, oxidative stress, and progression of chronic renal failure. Semin Nephrol 2004; 24:354–365.
91. Dallinger S, Sieder A, Strametz J, Bayerle-Eder M, Wolzt M, Schmetterer L. Vasodilator effects of L-arginine are stereospecific and augmented by insulin in humans. Am J Physiol Endocrinol Metab 2003; 284:E1106–E1111.
92. Romero MJ, Iddings J, Platt D, Caldwell RB, Caldwell RW. Diabetes-induced arginase activity contributes to vascular and renal fibrosis and dysfunction. FASEB J 2008; 22:643.
93. Vallance P, Leone A, Calver A, Collier J, Moncada S. Accumulation of an endogenous inhibitor of nitric oxide synthesis in chronic renal failure. Lancet 1992; 339:572–575.
94. Baylis C. Arginine, arginine analogs and nitric oxide production in chronic kidney disease. Nat Clin Pract Nephrol 2006; 2:209–220.
95. Durban E, Lee HW, Kim S, Paik WK. Purification and characterization of protein methylase I (S-adenosylmethionine: protein-arginine methyltransferase; EC 2.1.1.23) from calf brain. Methods Cell Biol 1978; 19:59–67.
96. Closs EI, Basha FZ, Habermeier A, Forstermann U. Interference of L-arginine analogues with L-arginine transport mediated by the y+ carrier hCAT-2B. Nitric Oxide 1997; 1:65–73.
97. Abhary S, Kasmeridis N, Burdon KP, Kuot A, Whiting MJ, Yu WP, Petrovsky N, Craig JE. Diabetic retinopathy is associated with elevated serum asymmetric and symmetric dimethylarginines. Diabetes Care 2009; 32(11):2084–2086.
98. Kielstein JT, Boger RH, Bode-Boger SM, Frolich JC, Haller H, Ritz E, Fliser D. Marked increase of asymmetric dimethylarginine in patients with incipient primary chronic renal disease. J Am Soc Nephrol 2002; 13:170–176.

99. Tarnow L, Hovind P, Teerlink T, Stehouwer CD, Parving HH. Elevated plasma asymmetric dimethylarginine as a marker of cardiovascular morbidity in early diabetic nephropathy in type 1 diabetes. Diabetes Care 2004; 27:765–769.
100. Onozato ML, Tojo A, Leiper J, Fujita T, Palm F, Wilcox CS. Expression of NG,NG-dimethylarginine dimethylaminohydrolase and protein arginine N-methyltransferase isoforms in diabetic rat kidney: effects of angiotensin II receptor blockers. Diabetes 2008; 57:172–180.
101. Chan NN, Chan JC. Asymmetric dimethylarginine (ADMA): a potential link between endothelial dysfunction and cardiovascular diseases in insulin resistance syndrome? Diabetologia 2002; 45:1609–1616.
102. Sydow K, Mondon CE, Schrader J, Konishi H, Cooke JP. Dimethylarginine dimethylaminohydrolase overexpression enhances insulin sensitivity. Arterioscler Thromb Vasc Biol 2008; 28:692–697.
103. Xiong Y, Lei M, Fu S, Fu Y. Effect of diabetic duration on serum concentrations of endogenous inhibitor of nitric oxide synthase in patients and rats with diabetes. Life Sci 2005; 77:149–159.
104. Lin KY, Ito A, Asagami T, Tsao PS, Adimoolam S, Kimoto M, Tsuji H, Reaven GM, Cooke JP. Impaired nitric oxide synthase pathway in diabetes mellitus: role of asymmetric dimethylarginine and dimethylarginine dimethylaminohydrolase. Circulation 2002; 106:987–992.
105. Mittermayer F, Mayer BX, Meyer A, Winzer C, Pacini G, Wagner OF, Wolzt M, Kautzky-Willer A. Circulating concentrations of asymmetrical dimethyl-L-arginine are increased in women with previous gestational diabetes. Diabetologia 2002; 45:1372–1378.
106. Sydow K, Mondon CE, Cooke JP. Insulin resistance: potential role of the endogenous nitric oxide synthase inhibitor ADMA. Vasc Med 2005; 10(Suppl 1):S35–S43.
107. Hanai K, Babazono T, Nyumura I, Toya K, Tanaka N, Tanaka M, Ishii A, Iwamoto Y. Asymmetric dimethylarginine is closely associated with the development and progression of nephropathy in patients with type 2 diabetes. Nephrol Dial Transplant 2009; 24:1884–1888.
108. Lajer M, Tarnow L, Jorsal A, Teerlink T, Parving HH, Rossing P. Plasma concentration of asymmetric dimethylarginine (ADMA) predicts cardiovascular morbidity and mortality in type 1 diabetic patients with diabetic nephropathy. Diabetes Care 2008; 31:747–752.
109. Zoccali C, Bode-Boger S, Mallamaci F, Benedetto F, Tripepi G, Malatino L, Cataliotti A, Bellanuova I, Fermo I, Frolich J, Boger R. Plasma concentration of asymmetrical dimethylarginine and mortality in patients with end-stage renal disease: a prospective study. Lancet 2001; 358:2113–2117.
110. Anthony S, Leiper J, Vallance P. Endogenous production of nitric oxide synthase inhibitors. Vasc Med 2005; 10(Suppl 1):S3–S9.
111. Ogawa T, Kimoto M, Sasaoka K. Occurrence of a new enzyme catalyzing the direct conversion of NG,NG-dimethyl-L-arginine to L-citrulline in rats. Biochem Biophys Res Commun 1987; 148:671–677.
112. Ogawa T, Kimoto M, Sasaoka K. Purification and properties of a new enzyme, NG, NG-dimethylarginine dimethylaminohydrolase, from rat kidney. J Biol Chem 1989; 264:10205–10209.
113. Ogawa T, Kimoto M, Watanabe H, Sasaoka K. Metabolism of NG,NG-and NG, N'G-dimethylarginine in rats. Arch Biochem Biophys 1987; 252:526–537.
114. Leiper JM, Santa Maria J, Chubb A, MacAllister RJ, Charles IG, Whitley GS, Vallance P. Identification of two human dimethylarginine dimethylaminohydrolases with distinct tissue distributions and homology with microbial arginine deiminases. Biochem J 1999; 343 (Pt 1):209–214.
115. Komers R, Oyama TT, Chapman JG, Allison KM, Anderson S. Effects of systemic inhibition of neuronal nitric oxide synthase in diabetic rats. Hypertension 2000; 35:655–661.
116. Thomson SC, Deng A, Komine N, Hammes JS, Blantz RC, Gabbai FB. Early diabetes as a model for testing the regulation of juxtaglomerular NOS I. Am J Physiol Renal Physiol 2004; 287:F732–F738.

117. O'Byrne S, Forte P, Roberts LJ, 2nd, Morrow JD, Johnston A, Anggard E, Leslie RD, Benjamin N. Nitric oxide synthesis and isoprostane production in subjects with type 1 diabetes and normal urinary albumin excretion. Diabetes 2000; 49:857–862.
118. Johansson BL, Borg K, Fernqvist-Forbes E, Kernell A, Odergren T, Wahren J. Beneficial effects of C-peptide on incipient nephropathy and neuropathy in patients with type 1 diabetes mellitus. Diabet Med 2000; 17:181–189.
119. Forst T, Kunt T. Effects of C-peptide on microvascular blood flow and blood hemorheology. Exp Diabesity Res 2004; 5:51–64.
120. Samnegard B, Jacobson SH, Jaremko G, Johansson BL, Sjoquist M. Effects of C-peptide on glomerular and renal size and renal function in diabetic rats. Kidney Int 2001; 60:1258–1265.
121. Hansen A, Johansson BL, Wahren J, von Bibra H. C-peptide exerts beneficial effects on myocardial blood flow and function in patients with type 1 diabetes. Diabetes 2002; 51:3077–3082.
122. Forst T, Kunt T, Pohlmann T, Goitom K, Engelbach M, Beyer J, Pfutzner A. Biological activity of C-peptide on the skin microcirculation in patients with insulin-dependent diabetes mellitus. J Clin Invest 1998; 101:2036–2041.
123. Johansson BL, Sjoberg S, Wahren J. The influence of human C-peptide on renal function and glucose utilization in type 1 (insulin-dependent) diabetic patients. Diabetologia 1992; 35:121–128.
124. Fioretto P, Steffes MW, Sutherland DE, Goetz FC, Mauer M. Reversal of lesions of diabetic nephropathy after pancreas transplantation. N Engl J Med 1998; 339:69–75.
125. Huang DY, Richter K, Breidenbach A, Vallon V. Human C-peptide acutely lowers glomerular hyperfiltration and proteinuria in diabetic rats: a dose-response study. Naunyn Schmiedebergs Arch Pharmacol 2002; 365:67–73.
126. Stridh S, Sallstrom J, Friden M, Hansell P, Nordquist L, Palm F. C-peptide normalizes glomerular filtration rate in hyperfiltrating conscious diabetic rats. Adv Exp Med Biol 2009; 645:219–225.
127. Forst T, Kunt T, Pfutzner A, Beyer J, Wahren J. New aspects on biological activity of C-peptide in IDDM patients. Exp Clin Endocrinol Diabetes 1998; 106:270–276.
128. Samnegard B, Jacobson SH, Johansson BL, Ekberg K, Isaksson B, Wahren J, Sjoquist M. C-peptide and captopril are equally effective in lowering glomerular hyperfiltration in diabetic rats. Nephrol Dial Transplant 2004; 19:1385–1391.
129. Rebsomen L, Pitel S, Boubred F, Buffat C, Feuerstein JM, Raccah D, Vague P, Tsimaratos M. C-peptide replacement improves weight gain and renal function in diabetic rats. Diabetes Metab 2006; 32:223–228.
130. Nordquist L, Brown RD, Fasching A, Persson P, Palm F. Proinsulin C-peptide reduces diabetes-induced glomerular hyperfiltration via efferent arteriole dilation and inhibition of tubular sodium reabsorption. Am J Physiol Renal Physiol 2009; 297(5):F1265–F1272.
131. Nordquist L, Stridh S. Effects of proinsulin C-peptide on oxygen transport, uptake and utilization in insulinopenic diabetic subjects – a review. Adv Exp Med Biol 2009; 645:193–198.
132. Nordquist L, Lai EY, Sjoquist M, Patzak A, Persson AE. Proinsulin C-peptide constricts glomerular afferent arterioles in diabetic mice. A potential renoprotective mechanism. Am J Physiol Regul Integr Comp Physiol 2008; 294:R836–R841.
133. Al-Mashhadi RH, Skott O, Vanhoutte PM, Hansen PB. Activation of A(2) adenosine receptors dilates cortical efferent arterioles in mouse. Kidney Int 2009; 75:793–799.
134. Murray RD, Churchill PC. Effects of adenosine receptor agonists in the isolated, perfused rat kidney. Am J Physiol 1984; 247:H343–H348.
135. Juncos LA, Ito S. Disparate effects of insulin on isolated rabbit afferent and efferent arterioles. J Clin Invest 1993; 92:1981–1985.
136. Holz FG, Steinhausen M. Renovascular effects of adenosine receptor agonists. Ren Physiol 1987; 10:272–282.

137. Kamikawa A, Ishii T, Shimada K, Makondo K, Inanami O, Sakane N, Yoshida T, Saito M, Kimura K. Proinsulin C-peptide abrogates type-1 diabetes-induced increase of renal endothelial nitric oxide synthase in rats. Diabetes Metab Res Rev 2008; 24:331–338.
138. Wahren J, Shafqat J, Johansson J, Chibalin A, Ekberg K, Jornvall H. Molecular and cellular effects of C-peptide – new perspectives on an old peptide. Exp Diabesity Res 2004; 5:15–23.
139. Johansson J, Ekberg K, Shafqat J, Henriksson M, Chibalin A, Wahren J, Jornvall H. Molecular effects of proinsulin C-peptide. Biochem Biophys Res Commun 2002; 295:1035–1040.
140. Ohtomo Y, Aperia A, Sahlgren B, Johansson BL, Wahren J. C-peptide stimulates rat renal tubular Na+, K(+)-ATPase activity in synergism with neuropeptide Y. Diabetologia 1996; 39:199–205.
141. Dinour D, Brezis M. Effects of adenosine on intrarenal oxygenation. Am J Physiol 1991; 261:F787–F791.
142. Gabriels G, Endlich K, Rahn KH, Schlatter E, Steinhausen M. In vivo effects of diadenosine polyphosphates on rat renal microcirculation. Kidney Int 2000; 57:2476–2484.
143. Liss P, Carlsson PO, Palm F, Hansell P. Adenosine A1 receptors in contrast media-induced renal dysfunction in the normal rat. Eur Radiol 2004; 14:1297–1302.
144. Awad AS, Huang L, Ye H, Duong ET, Bolton WK, Linden J, Okusa MD. Adenosine A2A receptor activation attenuates inflammation and injury in diabetic nephropathy. Am J Physiol Renal Physiol 2006; 290:F828–F837.
145. Valladares D, Quezada C, Montecinos P, Concha, II, Yanez AJ, Sobrevia L, San Martin R. Adenosine A(2B) receptor mediates an increase on VEGF-A production in rat kidney glomeruli. Biochem Biophys Res Commun 2008; 366:180–185.
146. Lai EY, Patzak A, Steege A, Mrowka R, Brown R, Spielmann N, Persson PB, Fredholm BB, Persson AE. Contribution of adenosine receptors in the control of arteriolar tone and adenosine-angiotensin II interaction. Kidney Int 2006; 70:690–698.
147. Nakamoto H, Ogasawara Y, Kajiya F. Visualisation of the effects of dilazep on rat afferent and efferent arterioles in vivo. Hypertens Res 2008; 31:315–324.
148. Eppel GA, Ventura S, Evans RG. Regional vascular responses to ATP and ATP analogues in the rabbit kidney in vivo: roles for adenosine receptors and prostanoids. Br J Pharmacol 2006; 149:523–531.
149. Pflueger AC, Schenk F, Osswald H. Increased sensitivity of the renal vasculature to adenosine in streptozotocin-induced diabetes mellitus rats. Am J Physiol 1995; 269: F529–F535.
150. Di Noia MA, Van Driesche S, Palmieri F, Yang LM, Quan S, Goodman AI, Abraham NG. Heme oxygenase-1 enhances renal mitochondrial transport carriers and cytochrome C oxidase activity in experimental diabetes. J Biol Chem 2006; 281:15687–15693.
151. Berryman AM, Maritim AC, Sanders RA, Watkins JB, 3rd. Influence of treatment of diabetic rats with combinations of pycnogenol, beta-carotene, and alpha-lipoic acid on parameters of oxidative stress. J Biochem Mol Toxicol 2004; 18:345–352.
152. Bach FH. Heme oxygenase-1 as a protective gene. Wien Klin Wochenschr 2002; 114 (Suppl 4):1–3.
153. Koya D, Hayashi K, Kitada M, Kashiwagi A, Kikkawa R, Haneda M. Effects of antioxidants in diabetes-induced oxidative stress in the glomeruli of diabetic rats. J Am Soc Nephrol 2003; 14:S250–S253.
154. Asaba K, Tojo A, Onozato ML, Goto A, Quinn MT, Fujita T, Wilcox CS. Effects of NADPH oxidase inhibitor in diabetic nephropathy. Kidney Int 2005; 67:1890–1898.
155. Baynes JW. Role of oxidative stress in development of complications in diabetes. Diabetes 1991; 40:405–412.
156. Horie K, Miyata T, Maeda K, Miyata S, Sugiyama S, Sakai H, van Ypersole de Strihou C, Monnier VM, Witztum JL, Kurokawa K. Immunohistochemical colocalization of glycoxidation products and lipid peroxidation products in diabetic renal glomerular lesions.

Implication for glycoxidative stress in the pathogenesis of diabetic nephropathy. J Clin Invest 1997; 100:2995–3004.
157. Brownlee M. The pathobiology of diabetic complications: a unifying mechanism. Diabetes 2005; 54:1615–1625.
158. Zou AP, Cowley AW, Jr. Reactive oxygen species and molecular regulation of renal oxygenation. Acta Physiol Scand 2003; 179:233–241.
159. Nishikawa T, Edelstein D, Brownlee M. The missing link: a single unifying mechanism for diabetic complications. Kidney Int Suppl 2000; 77:S26–S30.
160. Derubertis FR, Craven PA. Activation of protein kinase C in glomerular cells in diabetes. Mechanisms and potential links to the pathogenesis of diabetic glomerulopathy. Diabetes 1994; 43:1–8.
161. Ishii H, Koya D, King GL. Protein kinase C activation and its role in the development of vascular complications in diabetes mellitus. J Mol Med 1998; 76:21–31.
162. Brownlee M. A radical explanation for glucose-induced beta cell dysfunction. J Clin Invest 2003; 112:1788–1790.
163. Ishii H, Jirousek MR, Koya D, Takagi C, Xia P, Clermont A, Bursell SE, Kern TS, Ballas LM, Heath WF, Stramm LE, Feener EP, King GL. Amelioration of vascular dysfunctions in diabetic rats by an oral PKC beta inhibitor. Science 1996; 272:728–731.
164. Craven PA, Studer RK, DeRubertis FR. Impaired nitric oxide-dependent cyclic guanosine monophosphate generation in glomeruli from diabetic rats. Evidence for protein kinase C-mediated suppression of the cholinergic response. J Clin Invest 1994; 93:311–320.
165. Haneda M, Koya D, Kikkawa R. Cellular mechanisms in the development and progression of diabetic nephropathy: activation of the DAG-PKC-ERK pathway. Am J Kidney Dis 2001; 38:S178–S181.
166. Yang J, Lane PH, Pollock JS, Carmines PK. PKC-dependent superoxide production by the renal medullary thick ascending limb from diabetic rats. Am J Physiol Renal Physiol 2009; 297(5): F1220–F1228.
167. Kikkawa R, Koya D, Haneda M. Progression of diabetic nephropathy. Am J Kidney Dis 2003; 41:S19–S21.
168. Brownlee M, Vlassara H, Cerami A. Measurement of glycosylated amino acids and peptides from urine of diabetic patients using affinity chromatography. Diabetes 1980; 29:1044–1047.
169. Brownlee M. The pathological implications of protein glycation. Clin Invest Med 1995; 18:275–281.
170. Brownlee M. Glycation products and the pathogenesis of diabetic complications. Diabetes Care 1992; 15:1835–1843.
171. Rosca MG, Mustata TG, Kinter MT, Ozdemir AM, Kern TS, Szweda LI, Brownlee M, Monnier VM, Weiss MF. Glycation of mitochondrial proteins from diabetic rat kidney is associated with excess superoxide formation. Am J Physiol Renal Physiol 2005; 289:F420–F430.
172. Poirier O, Nicaud V, Vionnet N, Raoux S, Tarnow L, Vlassara H, Parving HH, Cambien F. Polymorphism screening of four genes encoding advanced glycation end-product putative receptors. Association study with nephropathy in type 1 diabetic patients. Diabetes 2001; 50:1214–1218.
173. Miyata T, Izuhara Y. Inhibition of advanced glycation end products: an implicit goal in clinical medicine for the treatment of diabetic nephropathy? Ann N Y Acad Sci 2008; 1126:141–146.
174. Nakamura S, Makita Z, Ishikawa S, Yasumura K, Fujii W, Yanagisawa K, Kawata T, Koike T. Progression of nephropathy in spontaneous diabetic rats is prevented by OPB-9195, a novel inhibitor of advanced glycation. Diabetes 1997; 46:895–899.
175. Bolton WK, Cattran DC, Williams ME, Adler SG, Appel GB, Cartwright K, Foiles PG, Freedman BI, Raskin P, Ratner RE, Spinowitz BS, Whittier FC, Wuerth JP. Randomized trial of an inhibitor of formation of advanced glycation end products in diabetic nephropathy. Am J Nephrol 2004; 24:32–40.
176. Brownlee M, Vlassara H, Kooney A, Ulrich P, Cerami A. Aminoguanidine prevents diabetes-induced arterial wall protein cross-linking. Science 1986; 232:1629–1632.

177. Kolm-Litty V, Sauer U, Nerlich A, Lehmann R, Schleicher ED. High glucose-induced transforming growth factor beta1 production is mediated by the hexosamine pathway in porcine glomerular mesangial cells. J Clin Invest 1998; 101:160–169.
178. Yang X, Su K, Roos MD, Chang Q, Paterson AJ, Kudlow JE. O-linkage of N-acetylglucosamine to Sp1 activation domain inhibits its transcriptional capability. Proc Natl Acad Sci USA 2001; 98:6611–6616.
179. Du XL, Edelstein D, Rossetti L, Fantus IG, Goldberg H, Ziyadeh F, Wu J, Brownlee M. Hyperglycemia-induced mitochondrial superoxide overproduction activates the hexosamine pathway and induces plasminogen activator inhibitor-1 expression by increasing Sp1 glycosylation. Proc Natl Acad Sci USA 2000; 97:12222–12226.
180. Sayeski PP, Kudlow JE. Glucose metabolism to glucosamine is necessary for glucose stimulation of transforming growth factor-alpha gene transcription. J Biol Chem 1996; 271:15237–15243.
181. Weigert C, Friess U, Brodbeck K, Haring HU, Schleicher ED. Glutamine:fructose-6-phosphate aminotransferase enzyme activity is necessary for the induction of TGF-beta1 and fibronectin expression in mesangial cells. Diabetologia 2003; 46:852–855.
182. Ziyadeh FN, Sharma K, Ericksen M, Wolf G. Stimulation of collagen gene expression and protein synthesis in murine mesangial cells by high glucose is mediated by autocrine activation of transforming growth factor-beta. J Clin Invest 1994; 93:536–542.
183. Schleicher ED, Weigert C. Role of the hexosamine biosynthetic pathway in diabetic nephropathy. Kidney Int Suppl 2000; 77:S13–S18.
184. Staniek K, Nohl H. Are mitochondria a permanent source of reactive oxygen species? Biochim Biophys Acta 2000; 1460:268–275.
185. St-Pierre J, Buckingham JA, Roebuck SJ, Brand MD. Topology of superoxide production from different sites in the mitochondrial electron transport chain. J Biol Chem 2002; 277:44784–44790.
186. Balaban RS, Nemoto S, Finkel T. Mitochondria, oxidants, and aging. Cell 2005; 120:483–495.
187. Brand MD, Buckingham JA, Esteves TC, Green K, Lambert AJ, Miwa S, Murphy MP, Pakay JL, Talbot DA, Echtay KS. Mitochondrial superoxide and aging: uncoupling-protein activity and superoxide production. Biochem Soc Symp 2004; (71):203–213.
188. Korshunov SS, Skulachev VP, Starkov AA. High protonic potential actuates a mechanism of production of reactive oxygen species in mitochondria. FEBS Lett 1997; 416:15–18.
189. Balaban RS, Mandel LJ. Coupling of aerobic metabolism to active ion transport in the kidney. J Physiol 1980; 304:331–348.
190. Brand MD. Uncoupling to survive? The role of mitochondrial inefficiency in ageing. Exp Gerontol 2000; 35:811–820.
191. Dulloo AG, Samec S. Uncoupling proteins: their roles in adaptive thermogenesis and substrate metabolism reconsidered. Br J Nutr 2001; 86:123–139.
192. Miwa S, St-Pierre J, Partridge L, Brand MD. Superoxide and hydrogen peroxide production by Drosophila mitochondria. Free Radic Biol Med 2003; 35:938–948.
193. Jezek P, Zackova M, Rehakova Z, Ruzicka M, Borecky J, Skobisova E, Brucknerova J, Garlid KD, Gimeno RE, Tartaglia LA. Existence of uncoupling protein-2 antigen in isolated mitochondria from various tissues. FEBS Lett 1999; 455:79–82.
194. Jezek P, Garlid KD. Mammalian mitochondrial uncoupling proteins. Int J Biochem Cell Biol 1998; 30:1163–1168.
195. Friederich M, Nordquist L, Olerud J, Johansson M, Hansell P, Palm F. Identification and distribution of uncoupling protein isoforms in the normal and diabetic rat kidney. Adv Exp Med Biol 2009; 645:205–212.
196. Friederich M, Fasching A, Hansell P, Nordquist L, Palm F. Diabetes-induced up-regulation of uncoupling protein-2 results in increased mitochondrial uncoupling in kidney proximal tubular cells. Biochim Biophys Acta 2008; 1777:935–940.

197. Fleury C, Neverova M, Collins S, Raimbault S, Champigny O, Levi-Meyrueis C, Bouillaud F, Seldin MF, Surwit RS, Ricquier D, Warden CH. Uncoupling protein-2: a novel gene linked to obesity and hyperinsulinemia. Nat Genet 1997; 15:269–272.
198. Friederich M, Olerud J, Fasching A, Liss P, Hansell P, Palm F. Uncoupling protein-2 in diabetic kidneys: increased protein expression correlates to increased non-transport related oxygen consumption. Adv Exp Med Biol 2008; 614:37–43.
199. Skulachev VP. Uncoupling: new approaches to an old problem of bioenergetics. Biochim Biophys Acta 1998; 1363:100–124.
200. Pi J, Bai Y, Daniel KW, Liu D, Lyght O, Edelstein D, Brownlee M, Corkey BE, Collins S. Persistent oxidative stress due to absence of uncoupling protein 2 associated with impaired pancreatic beta-cell function. Endocrinology 2009; 150:3040–3048.
201. Negre-Salvayre A, Hirtz C, Carrera G, Cazenave R, Troly M, Salvayre R, Penicaud L, Casteilla L. A role for uncoupling protein-2 as a regulator of mitochondrial hydrogen peroxide generation. FASEB J 1997; 11:809–815.
202. Packer L, Witt EH, Tritschler HJ. Alpha-lipoic acid as a biological antioxidant. Free Radic Biol Med 1995; 19:227–250.
203. Gerbitz KD, Gempel K, Brdiczka D. Mitochondria and diabetes. Genetic, biochemical, and clinical implications of the cellular energy circuit. Diabetes 1996; 45:113–126.
204. Peragon J, Aranda F, Garcia-Salguero L, Vargas AM, Lupianez JA. Long-term adaptive response to dietary protein of hexose monophosphate shunt dehydrogenases in rat kidney tubules. Cell Biochem Funct 1990; 8:11–17.
205. Peragon J, Aranda F, Garcia-Salguero L, Corpas FJ, Lupianez JA. Stimulation of rat-kidney hexose monophosphate shunt dehydrogenase activity by chronic metabolic acidosis. Biochem Int 1989; 18:1041–1050.
206. Peragon J, Aranda F, Garcia-Salguero L, Lupianez JA. Influence of experimental diabetes on the kinetic behaviour of renal cortex hexose monophosphate dehydrogenases. Int J Biochem 1989; 21:689–694.
207. Haque MZ, Majid DS. Assessment of renal functional phenotype in mice lacking gp91PHOX subunit of NAD(P)H oxidase. Hypertension 2004; 43:335–340.
208. Rey FE, Cifuentes ME, Kiarash A, Quinn MT, Pagano PJ. Novel competitive inhibitor of NAD(P)H oxidase assembly attenuates vascular O(2)(-) and systolic blood pressure in mice. Circ Res 2001; 89:408–414.
209. Leusen JH, Verhoeven AJ, Roos D. Interactions between the components of the human NADPH oxidase: intrigues in the phox family. J Lab Clin Med 1996; 128:461–476.
210. Lambeth JD. NOX enzymes and the biology of reactive oxygen. Nat Rev Immunol 2004; 4:181–189.
211. Sumimoto H, Miyano K, Takeya R. Molecular composition and regulation of the Nox family NAD(P)H oxidases. Biochem Biophys Res Commun 2005; 338:677–686.
212. Gill PS, Wilcox CS. NADPH oxidases in the kidney. Antioxid Redox Signal 2006; 8:1597–1607.
213. Chabrashvili T, Tojo A, Onozato ML, Kitiyakara C, Quinn MT, Fujita T, Welch WJ, Wilcox CS. Expression and cellular localization of classic NADPH oxidase subunits in the spontaneously hypertensive rat kidney. Hypertension 2002; 39:269–274.
214. Mollsten A, Marklund SL, Wessman M, Svensson M, Forsblom C, Parkkonen M, Brismar K, Groop PH, Dahlquist G. A functional polymorphism in the manganese superoxide dismutase gene and diabetic nephropathy. Diabetes 2007; 56:265–269.
215. Kitiyakara C, Chabrashvili T, Chen Y, Blau J, Karber A, Aslam S, Welch WJ, Wilcox CS. Salt intake, oxidative stress, and renal expression of NADPH oxidase and superoxide dismutase. J Am Soc Nephrol 2003; 14:2775–2782.
216. Wingler K, Wunsch S, Kreutz R, Rothermund L, Paul M, Schmidt HH. Upregulation of the vascular NAD(P)H-oxidase isoforms Nox1 and Nox4 by the renin-angiotensin system in vitro and in vivo. Free Radic Biol Med 2001; 31:1456–1464.

217. Chabrashvili T, Kitiyakara C, Blau J, Karber A, Aslam S, Welch WJ, Wilcox CS. Effects of ANG II type 1 and 2 receptors on oxidative stress, renal NADPH oxidase, and SOD expression. Am J Physiol Regul Integr Comp Physiol 2003; 285:R117–R124.
218. Doi K, Noiri E, Tokunaga K. The association of NAD(P)H oxidase p22phox with diabetic nephropathy is still uncertain: response to Hodgkinson, Millward, and Demaine. Diabetes Care 2004; 27:1518–1519; author reply 1519.
219. Vaziri ND, Dicus M, Ho ND, Boroujerdi-Rad L, Sindhu RK. Oxidative stress and dysregulation of superoxide dismutase and NADPH oxidase in renal insufficiency. Kidney Int 2003; 63:179–185.
220. Eid AA, Gorin Y, Fagg BM, Maalouf R, Barnes JL, Block K, Abboud HE. Mechanisms of podocyte injury in diabetes: role of cytochrome P450 and NADPH oxidases. Diabetes 2009; 58:1201–1211.
221. Etoh T, Inoguchi T, Kakimoto M, Sonoda N, Kobayashi K, Kuroda J, Sumimoto H, Nawata H. Increased expression of NAD(P)H oxidase subunits, NOX4 and p22phox, in the kidney of streptozotocin-induced diabetic rats and its reversibility by interventive insulin treatment. Diabetologia 2003; 46:1428–1437.
222. Wei XF, Zhou QG, Hou FF, Liu BY, Liang M. Advanced oxidation protein products induce mesangial cell perturbation through PKC-dependent activation of NADPH oxidase. Am J Physiol Renal Physiol 2009; 296:F427–F437.
223. Block K, Gorin Y, Abboud HE. Subcellular localization of Nox4 and regulation in diabetes. Proc Natl Acad Sci U S A 2009; 106:14385–14390.
224. Kartha GK, Moshal KS, Sen U, Joshua IG, Tyagi N, Steed MM, Tyagi SC. Renal mitochondrial damage and protein modification in type-2 diabetes. Acta Diabetol 2008; 45:75–81.
225. Touyz RM, Yao G, Viel E, Amiri F, Schiffrin EL. Angiotensin II and endothelin-1 regulate MAP kinases through different redox-dependent mechanisms in human vascular smooth muscle cells. J Hypertens 2004; 22:1141–1149.
226. Wilcox CS. Oxidative stress and nitric oxide deficiency in the kidney: a critical link to hypertension? Am J Physiol Regul Integr Comp Physiol 2005; 289:R913–R935.
227. Craven PA, Studer RK, Negrete H, DeRubertis FR. Protein kinase C in diabetic nephropathy. J Diabetes Complications 1995; 9:241–245.
228. Brands MW, Fitzgerald SM. Arterial pressure control at the onset of type I diabetes: the role of nitric oxide and the renin-angiotensin system. Am J Hypertens 2001; 14:126S–131S.
229. Deinum J, Tarnow L, van Gool JM, de Bruin RA, Derkx FH, Schalekamp MA, Parving HH. Plasma renin and prorenin and renin gene variation in patients with insulin-dependent diabetes mellitus and nephropathy. Nephrol Dial Transplant 1999; 14:1904–1911.
230. Kasiske BL, Kalil RS, Ma JZ, Liao M, Keane WF. Effect of antihypertensive therapy on the kidney in patients with diabetes: a meta-regression analysis. Ann Intern Med 1993; 118:129–138.
231. Brenner BM, Cooper ME, de Zeeuw D, Keane WF, Mitch WE, Parving HH, Remuzzi G, Snapinn SM, Zhang Z, Shahinfar S. Effects of losartan on renal and cardiovascular outcomes in patients with type 2 diabetes and nephropathy. N Engl J Med 2001; 345:861–869.
232. Parving HH, Lehnert H, Brochner-Mortensen J, Gomis R, Andersen S, Arner P. The effect of irbesartan on the development of diabetic nephropathy in patients with type 2 diabetes. N Engl J Med 2001; 345:870–878.
233. Lewis EJ, Hunsicker LG, Clarke WR, Berl T, Pohl MA, Lewis JB, Ritz E, Atkins RC, Rohde R, Raz I. Renoprotective effect of the angiotensin-receptor antagonist irbesartan in patients with nephropathy due to type 2 diabetes. N Engl J Med 2001; 345:851–860.
234. Nishiyama A, Nakagawa T, Kobori H, Nagai Y, Okada N, Konishi Y, Morikawa T, Okumura M, Meda I, Kiyomoto H, Hosomi N, Mori T, Ito S, Imanishi M. Strict angiotensin blockade prevents the augmentation of intrarenal angiotensin II and podocyte abnormalities in type 2 diabetic rats with microalbuminuria. J Hypertens 2008; 26:1849–1859.
235. ACE Inhibitors in Diabetic Nephropathy Trialist Group. Should all patients with type 1 diabetes mellitus and microalbuminuria receive angiotensin-converting enzyme inhibitors? A meta-analysis of individual patient data. Ann Intern Med 2001; 134:370–379.

236. Ravid M, Brosh D, Levi Z, Bar-Dayan Y, Ravid D, Rachmani R. Use of enalapril to attenuate decline in renal function in normotensive, normoalbuminuric patients with type 2 diabetes mellitus. A randomized, controlled trial. Ann Intern Med 1998; 128:982–988.
237. Benigni A, Colosio V, Brena C, Bruzzi I, Bertani T, Remuzzi G. Unselective inhibition of endothelin receptors reduces renal dysfunction in experimental diabetes. Diabetes 1998; 47:450–456.
238. Rajagopalan S, Kurz S, Munzel T, Tarpey M, Freeman BA, Griendling KK, Harrison DG. Angiotensin II-mediated hypertension in the rat increases vascular superoxide production via membrane NADH/NADPH oxidase activation. Contribution to alterations of vasomotor tone. J Clin Invest 1996; 97:1916–1923.
239. Pagano PJ, Chanock SJ, Siwik DA, Colucci WS, Clark JK. Angiotensin II induces p67phox mRNA expression and NADPH oxidase superoxide generation in rabbit aortic adventitial fibroblasts. Hypertension 1998; 32:331–337.
240. Thorp ML. Diabetic nephropathy: common questions. Am Fam Physician 2005; 72:96–99.

Chapter 30
Estimation of Kidney Oxygenation by Blood Oxygenation Level Dependent Magnetic Resonance Imaging

Lu-Ping Li and Pottumarthi V. Prasad

Abstract Presence of renal hypoxia and its consequences to renal pathophysiology is well accepted now. Most data on renal oxygenation available today are based on animal models, and an ability to translate the findings to humans was highly desired. Although, several novel methodologies are being pursued, to date blood oxygenation level dependent (BOLD) magnetic resonance imaging (MRI) is the only known technique available to evaluate renal oxygenation in humans. The technique is noninvasive, based on an endogenous contrast mechanism, and can be applied to both animal models and humans. The ability to evaluate relative renal oxygenation status in both health and disease could be useful in better understanding the pathophysiology and allowing for monitoring of potential novel interventions.

In this chapter, we provide an overview of the principles involved and the implementation and various applications that investigators around the world have pursued to date.

Keywords Hypoxia · Magnetic resonance imaging · Diabetes · Hypertension · Transplantation

Abbreviations

ATN	Acute tubular necrosis
BOLD	Blood oxygenation level dependent
CAN	Chronic allograft nephropathy
CsA-ME	Closporine micro-emulsion
EPR	Electron paramagnetic resonance

P.V. Prasad (✉)
Department of Radiology, Evanston Northwestern Healthcare,
Suite G507, Evanston Hospital, 2650 Ridge Avenue, Evanston, IL 60201, USA
e-mail: pprasad@enh.org

T. Miyata et al. (eds.), *Studies on Renal Disorders*, Oxidative Stress in Applied Basic Research and Clinical Practice, DOI 10.1007/978-1-60761-857-7_30,

GFR	Glomerular filtration rate
HIF	Hypoxia-inducible factors
LiPc	Lithium phthalocyanine
L-NAME	N (G)-nitro-L-arginine methyl ester
MRI	Magnetic Resonance Imaging
NADPH	Nicotinamide adenine dinucleotide phosphate
NO	Nitric oxide
NOS	Nitric oxide snythase
NOSi	Nitric oxide synthase inhibition
NSAID	Non-steroidal anti-inflammatory drugs
pO2	Partial oxygen pressure
R_2*,R2*	The observed rate constant of the FID due to loss of phase coherence, MRI parameter, (R2* = 1/T2*)
RAS	Renal artery stenosis
RBF	Renal blood flow
RCM	Radio-contrast media
STZ	Streptozotocin, a naturally occurring chemical that is particularly toxic to the insulin-producing beta cells of the pancreas in mammals
T1	Longitudinal relaxation time, MRI parameter
T2	Transverse relaxation time, MRI parameter
T2*	The observed time constant of the FID due to loss of phase coherence, MRI parameter
TE	Echo time, MRI parameter
TR	Repetition time, MRI parameter
UUO	Unilateral ureteral obstruction

1 Introduction

Kidney is a highly perfused organ and has the least arteriovenous difference in oxygen saturation, suggesting minimal oxygen consumption [1, 2] and hence well-oxygenated tissue. In fact, it is estimated that the ratio of oxygen consumption to delivery is on the order of 7% for kidney as a whole [3]. However, it is now well known that kidney is highly susceptible to ischemic injury [4]. This apparent contradiction was resolved when invasive microprobe measurements became available and it was shown that the hemodynamics were significantly different in the renal cortex and medulla [5, 6]. This led to the theory of medullary hypoxia and its potential implications to kidney disease, especially acute ischemic injury [4]. Hypoxia is also becoming widely accepted as a key proponent of progression in chronic kidney disease [7, 8].

While early estimates were based on whole organ measurements [9], the more widely accepted data on renal oxygenation and blood flow are based on invasive microprobe measurements. These have led to a better understanding of the regional differences in these measurements within the kidney and have allowed for the understanding of several molecular mechanisms that influence them [1]. However, the invasive nature left many of the findings not being translated to humans. The first demonstration of the ability to translate some of these findings to humans was done in mid-1990s using blood oxygenation level dependent (BOLD) magnetic resonance imaging (MRI) [10]. The study for the first time demonstrated that renal medulla in humans is hypoxic and that it can be influenced by maneuvers such as administration of furosemide and water diuresis.

In this chapter we will review the basic principles involved in BOLD MRI, the different techniques available, the many applications pursued by several different groups around the world, and finally summarize the advantages and pitfalls of the method.

1.1 Renal Medullary Hypoxia

Since this topic is more extensively dealt by other chapters, we will restrict the discussion here so as to put the rest of the chapter in context. The magnitude of medullary hypoxia can be explained by a combination of (a) low blood flow to the region, (b) presence of oxygen shunting between the arterial and venous sections of the vasa recta, and (c) relatively higher oxygen consumption to support the sodium reabsorption along the medullary thick ascending limbs. The renal medulla has a significantly low ambient partial oxygen pressure (pO_2) (<20 mmHg) [5, 11] compared to even systemic venous blood (<40 mmHg). Sufficient evidence has now been obtained to support a pathogenic role of hypoxia in a number of diseases, such as ischemic acute renal failure [4], in the development of nephrotoxic acute kidney injury [12, 13] and in radiocontrast nephropathy [14, 15]. Furthermore, accumulating evidence suggests that chronic hypoxia is a final common pathway to end stage kidney failure in chronic kidney disease [8, 16].

1.2 Renal Oxygenation Measurements

The most widely applied method to study renal medullary hypoxia uses invasive microelectrodes to measure tissue pO_2 [5, 6, 17–23]. Recently a novel laser-based probe for pO_2 measurements has become available [24–27]. With the probe size on the order of a couple of hundred micrometers, as opposed to tens of micrometers with conventional microelectrodes, there are some fundamental differences in the data obtained. However, considering the extreme fragility of microelectrodes, the laser probes may have practical advantages. A histological method involving the reduction of pimonidazole by nicotinamide adenine dinucleotide phosphate (NADPH) has been used to map hypoxic regions [28]. Pimonidazole mapping is not a quantitative technique and is typically sensitive only at pO_2 less than

10 mmHg. Another objection to pimonidazole is that it may be reduced by an increase in NADPH produced by increased synthesis rather than hypoxia. Recently, the endogenous marker of hypoxia, upregulation of hypoxia-inducible factors (HIF) has been demonstrated [11, 29]. Electron paramagnetic resonance (EPR) has been shown to be useful in measuring changes related to tissue oxygenation [30, 31]. It necessitates introducing lithium phthalocyanine (LiPc) crystals in to the tissue, from which the EPR signals can then be monitored. The line width of the LiPc signals is directly related to surrounding tissue pO_2. EPR imaging may be used in concert with suitable spin probes [32]. Oximetry based on ^{19}F MRI has been demonstrated [33, 34]. None of these methods are readily applicable to humans today.

BOLD MRI is a noninvasive technique that has been shown to be sensitive to tissue oxygenation. The technique has been widely used in the brain [35, 36]. It has also been evaluated in the heart [37, 38], kidney [10, 39], muscle [40, 41], and tumors [33, 42, 43] in both animal models and in humans.

For a more exhaustive review of the methods to measure renal oxygenation, please refer to a recent review by Evans et al. [44].

1.2.1 Magnetic Resonance Imaging Technique

A detailed discussion on the principles of MRI is beyond the scope of the present review. However, those interested are referred to a recent chapter by Storey [45].

MRI utilizes the large number of hydrogen nuclei present within the body (water and fat molecules mostly). The hydrogen nuclei behave like tiny rotating magnets (and hence referred to as spins) and are generally incoherent in terms of their orientation, resulting in a zero net magnetization. When the body is placed in a large constant homogenous magnetic field B_0, these small spins align along or against the field, resulting in a net magnetization along the direction of B_0. An electromagnetic pulse applied at the resonant frequency of the rotating spins is used to change the orientation of hydrogen nuclei. Once the pulse is turned off, the spins relax back to the equilibrium status, depending on two characteristic relaxation times: T_1 and T_2. Even though the magnetic field within the scanner is highly homogenous, in the presence of inhomogenous objects, such as a human body, the uniformity is lost. This results in an apparent enhancement in T_2 relaxation process and is usually termed T_2* ($T_2^* \leq T_2$). Repetition time (TR) refers to the time between the excitation pulses, and echo time (TE) refers to the time between the excitation pulse and when the signal is acquired. These two usually control the amount of T_1 and T_2 weighting on the image intensity.

1.2.2 BOLD MRI Principle

As the name BOLD suggests, the technique uses changes in blood oxygenation levels to modulate the observed MRI signal. Deoxyhemoglobin is paramagnetic, while oxyhemoglobin is diamagnetic. Changes in the oxygenation status within the blood vessels lead to changes in regional magnetic field homogeneity, resulting in a change in regional T_2*. Thus, an area with more deoxygenated blood will show up relatively dark on a T_2*-weighted image compared to region with well-oxygenated blood around. The ratio of oxyhemoglobin to deoxyhemoglobin is related to the pO_2 of blood. Since pO_2 of capillary blood is thought to be in a dynamic equilibrium with that of surrounding tissue, changes estimated by BOLD MRI can be interpreted as changes in tissue pO_2 [10].

While changes in T_2* weighted signal intensity can be used to perform BOLD MRI measurements [10], it is desirable to use T_2* or R_2* (=1/T_2*) as a BOLD MRI parameter. This is because signal intensity will depend on many settings not related to T_2*. Signal intensity measurements are preferred when fast temporal response is desired or when the magnitude of change in T_2* is small, such as in brain functional MRI [46]. Since most applications in the kidney do not need fast temporal resolution and the main interest is to follow the oxygenation status in the renal medulla, which is relatively hypoxic at baseline, it is feasible to use R_2* as a BOLD MRI parameter.

1.2.3 R_2* as a BOLD MRI Index

Calculation of R_2* involves acquiring images with different echo times, as illustrated in Fig. 1. The initial study utilized echo planar imaging technique to acquire images with each echo time during a different breathhold interval [10]. Due to potential misregistration artifacts between each of these images, no attempt to create an R_2* map was made. Later a more robust and simpler acquisition sequence was used to acquire multiple gradient echo images within a single breathhold to allow for inline R_2* mapping [47]. This is the method that is being widely used for renal BOLD MRI.

BOLD MRI acquisition typically would include two or more images with different echo times to provide different amounts of T_2* weighting. These images are then analyzed on a pixel-by-pixel basis to construct an R_2* map. R_2* is calculated by fitting an exponential decay function to the signal intensity vs. echo time data. Figure 2 is an example of multiple echo BOLD MRI images and the R_2* map generated from those images. Owing to the relative simplicity of fitting to a linear function, some prefer to use a linear fit to Ln(signal intensity) vs. echo time data [47]. While both approaches provide similar results, the potential differences may be when the signal at longer echo times reaches the noise levels. One guiding principle to avoid this condition is to use the longest echo time to be equal to the T_2* of the tissue of interest. At 1.5 T, T_2* of medulla is approximately 50 ms, and so we usually limit the maximum TE to be around 50 ms.

The R_2* maps can be displayed either in gray scale [10] or color format [48, 49]. Given the magnitude of changes observed with pharmacological maneuvers, such as furosemide, nitric oxide synthase (NOS) inhibition, gray scale display of R_2* maps is adequate. Figure 3 is an example of human kidney R_2* map with the indication of medulla and cortex. However, in situations where more subtle changes

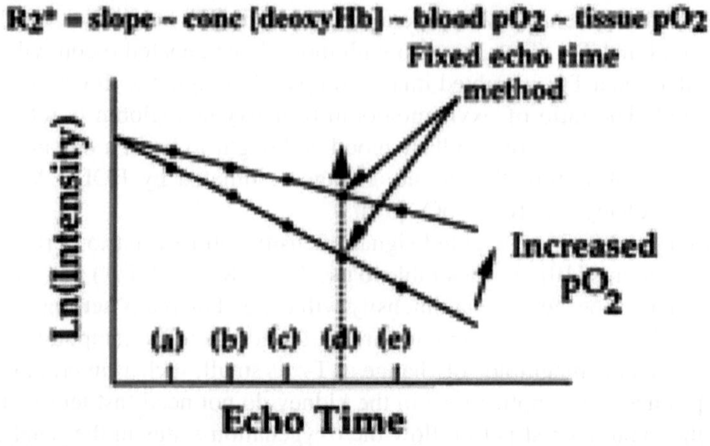

Fig. 1 Blood oxygenation level–dependent (BOLD) magnetic resonance imaging (MRI) changes with partial pressure of oxygen (pO_2). The deoxygenation of hemoglobin changes its magnetic characteristics, leading to changes in a parameter of magnetic resonance called R_2* (apparent spin-spin relaxation rate). R_2* can be estimated from signal intensity measurements made at several different echo times (*a–e*). The slope of Ln (intensity) vs. echo time determines R_2* and is directly related to the amount of deoxygenated blood. A decrease in the slope implies an increase in the pO_2 of blood. We can either measure the slope or obtain intensity measurements at a single echo time (e.g., *d*) to detect a difference in pO_2. Because blood pO_2 is thought to be in rapid equilibrium with tissue pO_2, changes in BOLD signal intensity or R_2* should reflect changes in the pO_2 of the tissue. Reproduced from [10]

are expected, visualization in color may be advantageous [10, 49]. Here it is important to realize that color scales need to be empirically determined. While the color display allows for potentially better visualization, it has no influence over the regional quantitative measurements.

1.2.4 Practical Logistical Issues When Using BOLD MRI

When scanning human patients, it is preferred to ask them to fast overnight (i.e., without food and drink) and perform the study in the morning. This is a way to reduce the variability introduced by the hydration status. Our own studies have indicated the effects of waterloading [50–52]. On the other hand, consumption of coffee and alcohol lead to dehydration and could adversely influence renal oxygenation level. Consumption of certain types of food could result in production of excess gas, leading to bulk susceptibility artifacts on the T_2* weighted images.

Patients should be covered with sheets to keep them at a comfortable temperature. In our experience, breathholding provides the ideal solution to minimizing motion artifacts. For this reason, we prefer to acquire a single slice during a breathhold so as to minimize the breathhold interval (typically 12–15 s). This is important especially in patients who may not be able to hold their breath for longer durations.

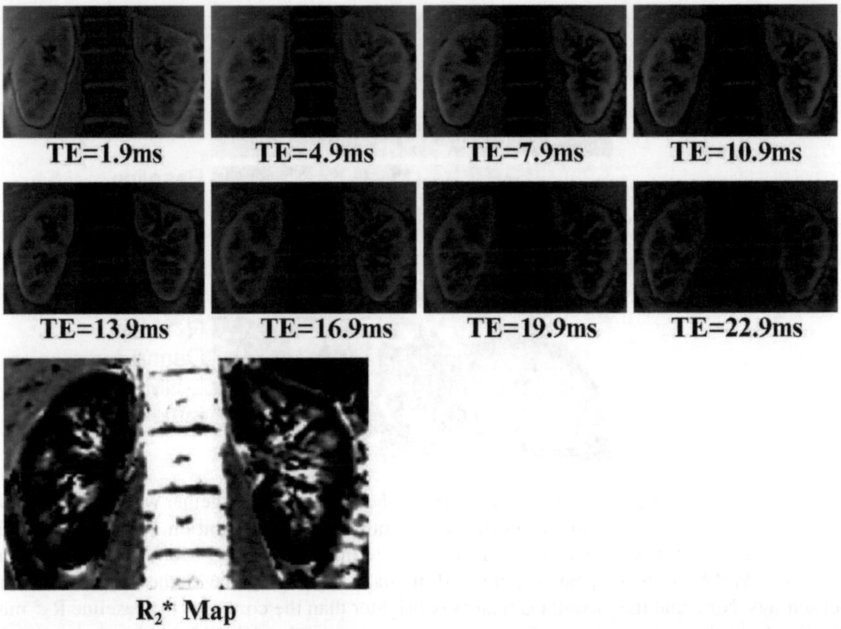

Fig. 2 Shown is a typical set of multiple gradient echo (mGRE) images with different echo times along with the corresponding R2* map, which is generated by fitting the signal intensity in each echo time image to a single exponential decay function on a pixel-by-pixel basis

For rodent studies, imaging can be performed either on dedicated small bore scanners or using the scanners designed for human use. Studies are performed in anesthetized animals and the type of anesthesia could potentially influence renal physiology. Most widely used agent is inactin. An inherent advantage of this agent is that the animals stay under its influence for a sufficiently long time with no need for redosing. On the flip side, the animals do not recover, so the experiments are terminal. For longitudinal studies, alternate short-acting anesthesia should be used. Inhalant anesthesia such as isoflurane is commonly used for small animal scanning.

The use of whole body scanners designed for humans have certain practical advantages when being used for rodent studies: (a) more widespread availability of scanners, especially in large academic medical centers; (b) more open access to the animal during the experiments for monitoring; and (c) easy translation to larger animal models and humans. In addition to these general advantages, BOLD MRI methods may have fewer limitations due to bulk susceptibility artifacts at the clinical field strengths, especially in the abdomen. While studies in rat kidneys can be performed using standard coils, for mouse studies custom-designed coils may be necessary [53].

Minimizing bulk susceptibility artifacts, primarily due to bowel gas, is a major concern. Shimming is essential prior to performing BOLD MRI measurements. In our experience, overnight fasting and positioning the animals on their sides help to reduce the effects of bowel gas. Others have used MRI-compatible kidney isolator [54]. It is not known if use of such a device has any effect on renal hemodynamics.

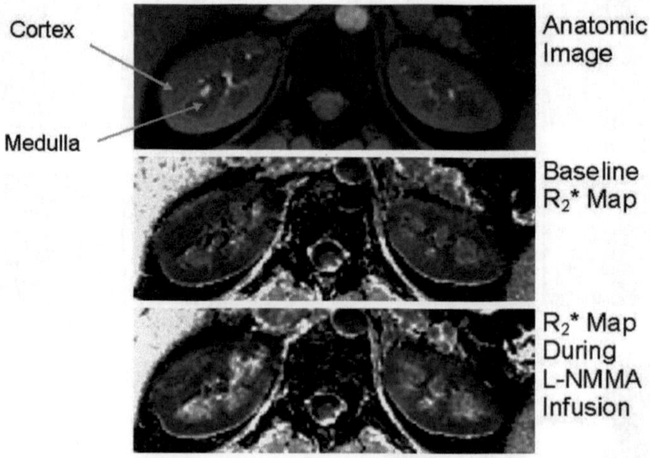

Fig. 3 A representative data set: anatomic image of the kidney (*top*), together with R_2* maps from the same slice, showing the effect of nitric oxide synthase (NOS) inhibition. At the center is the baseline R_2* map, at the bottom the R_2* map acquired 18 min after the NG-Monomethyl-L-arginine acetate (L-NMMA) infusion (peak response). Both maps are displayed with the same window and level settings. Note that the medulla is relatively brighter than the cortex on the baseline R_2* map, indicating lower basal oxygenation. Its brightness increases with L-NMMA administration, implying a further reduction in oxygenation in response to NOS inhibition. Reproduced from [26]

1.2.5 Effect of Field Strength

Because the susceptibility affects scale with field strength, sensitivity to BOLD increases with higher field strength. Most reports to date on renal BOLD applications in humans have been at 1.5 T. However, 3.0 T scanners are becoming increasingly available, and preliminary evaluation has demonstrated almost doubling of the R_2* values at 3.0 T compared to 1.5 T. Changes in R_2* in response to furosemide also scaled similarly [55]. At the same time the severity of bulk susceptibility-induced artifacts were not much different from those at 1.5 T. These observations suggest that measurements at 3.0 T would be preferred when a choice is available.

2 Experimental Studies

2.1 Renal BOLD MRI Validation Studies

There have been few studies to date where BOLD MRI measurements have been correlated with microprobe measurements. The initial BOLD MRI studies evaluating the effects of furosemide and acetazolamide in rat kidneys [56] showed a good correspondence with previous reports using microprobes [5, 6]. Pedersen et al. [57]

performed studies in swine where blood pO_2 was modulated by varying the breathing gas. They obtained BOLD MRI measurements in one kidney and correlated them with microelectrode measurements performed in the contralateral kidney. They did not use the electrodes in the same kidney to avoid susceptibility artifacts due to the electrode. More recently, two studies have performed BOLD MRI and microprobe measurements in the same animal models [25, 26]. One study reported an increasing level of hypoxia over the first 4 weeks after administration of streptozotocin (STZ) to later develop type 1 diabetes [25]. Simultaneous blood flow measurements performed in the microprobe group showed that the increase in hypoxia was not related to reduced blood flow (Fig. .4). On the other hand, when evaluating the effects of L-NAME (N^G-nitro-L-arginine methyl ester) [26], it was found there was a dose-dependent increase in the level of hypoxia that correlated with the reduction in blood flow (Fig. 4). These later studies were performed with Oxylite/Oxyflo probes, which are MRI compatible. However, with the lack of MRI compatible probe holders, simultaneous measurements have not yet been performed.

2.1.1 Reproducibility Study

Over the past decade since the introduction of renal BOLD MRI, there have been a number of studies performed by different investigators throughout the world that have shown consistent findings [48, 52, 54, 57–73]. These studies were performed on different vendor platforms and at different field strengths. Response to furosemide has been repeated in different species [47, 53, 56, 67] by different groups. In human studies, the baseline values and the change following administration of furosemide were comparable. In addition, two studies reported on reproducibility specifically [70, 74]. One looked at short-term reproducibility (i.e., during the same study session) [70, 73], and the other studied the long-term reproducibility [74]. Both studies showed the R_2* measurements were reproducible within 12%. We have shown measurements with a waterloading paradigm in three studies in humans [50–52] and also reported limited reproducibility in one study [74]. Overall we believe that waterloading may be more subject dependent compared to use of furosemide.

2.2 *Renal BOLD MRI Applications*

2.2.1 Physiological/Pharmacological Induced Changes in Intra-Renal Oxygenation

BOLD MRI is most effective in monitoring changes induced by pharmacological or physiological maneuvers [10, 66]. The most widely applied maneuver is administration of furosemide [55, 64, 71, 75, 76]. Waterload is another simple and effective

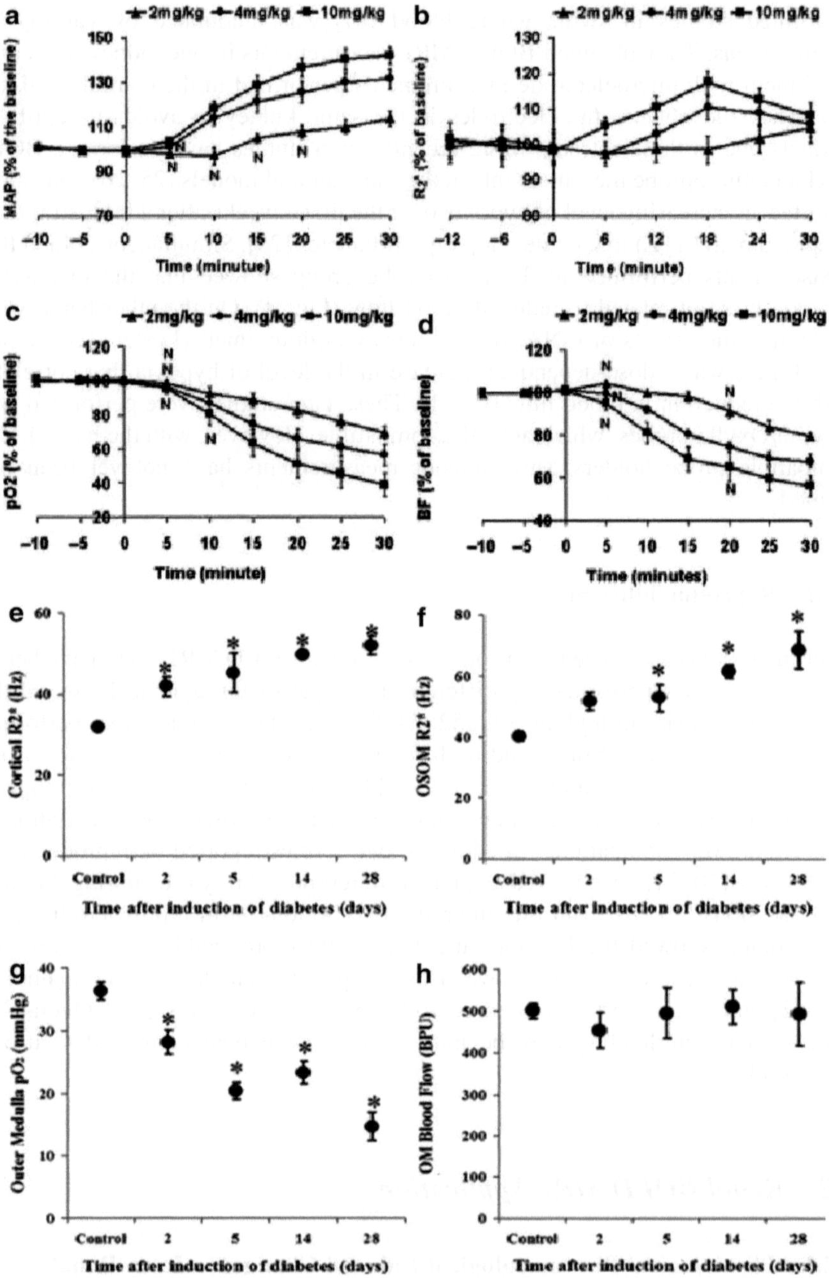

Fig. 4 Summary of blood oxygenation level dependent (BOLD) magnetic resonance imaging (MRI) measurements vs. direct partial pressure of oxygen (pO_2) and blood flow measurement using OxyLite/OxyFlo in L-NAME (N^G-nitro-L-arginine methyl ester) induced hypertension model and streptozotocin (STZ)-induced diabetes model. Blood pressure (**a**), medullary R2* (**b**), renal medullary pO_2 (**c**), and blood flow (**d**) data obtained in six rats at different dose of

maneuver to acutely change medullary oxygenation [10, 50–52]. The observed improvement in medullary oxygenation (lower R_2*) has been shown to be related to endogenous prostaglandin production in waterloading studies [51, 52, 77]. This maneuver has also been shown to differentiate responses in the elderly [51] and patients with diabetes [50] compared to healthy young subjects.

Vasoactive substances influence intrarenal oxygenation, and hence their actions could be monitored using BOLD MRI. Angiotensin II is known to diminish renal perfusion [78] and is thus expected to reduce renal oxygenation. This was recently confirmed by the BOLD MRI measurement in healthy subjects where angiotensin II caused a shortening of BOLD T_2* in renal cortex [59].

Nitric oxide (NO) is a soluble gas that is continuously synthesized by the endothelium [79] and has a wide range of biological properties including relaxation of vascular tone. In rat kidneys, administration of L-NAME (a nitric oxide synthase inhibitor) resulted in a further increase in medullary R_2*, suggesting enhanced hypoxia [39, 80]. A dose-dependent response to L-NAME in medullary R_2* has been found in rats and be confirmed by invasive pO_2 and blood flow measurement [26]. Interestingly, such an increase was absent in a genetic model of hypertension [39] and was shown to be restored when treated with an antioxidant [81] (Fig. 5). Preliminary data in a small number of healthy young human subjects with NOS inhibition have been reported [26].

Iodinated radio-contrast media (RCM) iopromide produced an increase in medullary and cortical R_2* values in humans 20 min after administration [66]. Calcineurin inhibitors, cyclosporine microemulsion (CsA-ME), and tacrolimus are currently the most widely used baseline immunosuppressants for prevention of acute rejection following kidney transplantation. A significant reduction in medullary R_2* values (suggesting improvement in oxygenation) was observed 2 h after CsA-ME administration in healthy subjects [66]. This is in apparent contradiction to the previously reported decrease in renal blood flow related to afferent arteriolar vasoconstriction [82]. However, the study also suggested reduction in glomerular filtration rate (GFR), and it is possible that there is an associated reduction in oxygen consumption related to reduced sodium reabsorption in the medulla. Tacrolimus had

Fig. 4 (continued) l-NAME. L-NAME infusion started at time 0. L-NAME infusion resulted in increased mean arterial pressure (MAP) and decreased renal medullary pO_2 and blood flow. All measured parameters show a dose response. Although medullary R2* increased consistently with the reduced medullary pO_2 and blood flow, a trend toward baseline was observed even during the infusion of L-NAME. All time points are statistically significant compared with the baseline based on Student *t*-test, except the few time points marked with "N." Error bars represent standard errors. Effect of diabetes on the renal R_2* signal in the cortex (**e**) and outer medulla (**f**). Note the progressive increase in R_2* values (suggesting increasing levels of hypoxia) in both cortex and medulla during the 4-week period. Error bars show the standard error over different animals. Effect of induced diabetes on the renal oxygenation (**g**) and blood flow (**h**) in rats (control, $n = 6$; 2 days after inducting, $n = 6$; 5 days, $n = 6$; 14 days, $n = 6$; 28 days, $n = 5$). The outer medulla pO_2 was measured with OxyLite. The outer medulla blood flow was measured with OxyFlo. Error bars show the standard error over different animals

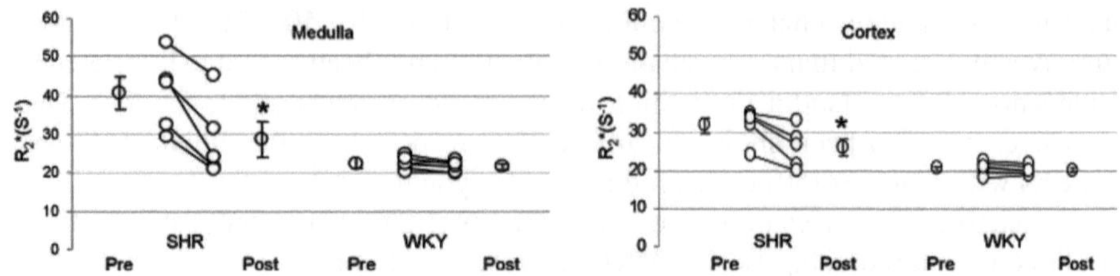

Fig. 5 Illustration of individual changes post-tempol in SHR and WKY rats. Average (mean ± SE) of all points acquired at least 20 min after tempol administration was used as post-tempol R_2*. Mean R_2* values pre- and post-tempol in the renal medulla and cortex were averaged over all rats of each strain. SHRs show a significant response to tempol in medulla and cortex as evaluated by blood oxygenation level dependent (BOLD) magnetic resonance imaging (MRI) measurements. WKY rats did not show significant response to tempol. Reproduced from [81]

no significant effect on R_2* values for medulla or cortex in healthy subjects [66], even though the nephrotoxic effects are known to be similar to CsA. Further studies are necessary to fully evaluate the significance of these observations.

Indomethacin has been shown to reduce medullary oxygenation by microelectrodes [67] and BOLD MRI [83] in rat kidneys. However, it did not induce a significant change in renal medullary R_2* in healthy subjects [66]. This might indicate that a single dose of indomethacin as prescribed routinely does not significantly influence renal oxygenation in humans. Similarly, another common nonsteroidal anti-inflammatory drug (NSAID), ibuprofen, did not change baseline renal medullary and cortical R_2* [75]. However, administration of ibuprofen (and similarly naproxen) significantly reduced the response to waterload [51, 52, 77]. Naproxen abolished the improvement in medullary oxygenation during waterloading in both humans [52] and animal models [77]. On the other hand, naproxcinod, a nitric oxide-donating anti-inflammatory compound, left intact the improvement in oxygenation in renal medulla during waterload based on BOLD MRI measurement [77]. This implies that naproxcinod may have less nephrotoxicity, and that the NO-donating moiety partially compensates for the hemodynamic effects of prostaglandin inhibition by naproxen. This may also suggest that use of a provocative maneuver such as waterloading may be necessary to evaluate the effects of prostaglandin inhibition.

2.2.2 Renal BOLD MRI in Disease

Diabetes Mellitus

A recent animal study using invasive microelectrodes has shown that the pO_2 in chronic diabetic rats is decreased throughout the renal parenchyma [17]. Ries et al. [54] and Edlund et al. [63] observed in an animal model that diabetic kidney had significantly lower oxygenation level in renal medulla compared to a control group using BOLD MRI at 5 days and 14 days postinduction [54, 63]. A similar study [25] with BOLD MRI, along with invasive blood flow and oxygenation measurement by optical fiber probes, showed that the pO_2 was considerably lower in diabetic rats as early as 2 days after induction. Further, the oxygenation continued to decrease at 5, 14, and 28 days following induction of diabetes compared to control rats. No blood flow changes were observed in both diabetes and normal groups over this time period, suggesting that the reduced oxygenation was related to increased consumption probably related to hyperfiltration, known to be associated with early diabetes.

A recent study showed that renal oxygenation in mice with type 2 diabetes is significantly lower than their normal littermates by BOLD MRI [84]. There is also evidence that acute hyperglycemia reduces renal oxygenation [85, 86].

BOLD MRI has also been used to evaluate diabetic human subjects. A study of 18 patients (nine healthy nondiabetics and nine with mild, controlled diabetes) [50] showed that in the healthy, water diuresis led to a significant increase in the oxygenation of the renal medulla, but not in the patients with diabetes. These results suggest that even patients with mild diabetes already show signs of renal injury long before the onset of symptoms that usually accompany kidney disease and a likely deficiency in the synthesis of endogenous vasodilator substances, such as prostaglandin or NO. BOLD MRI may provide important insight into the pathophysiology of renal injury at early stages in diabetes and allow for the means to evaluate novel drug interventions, especially those targeting renal hypoxia.

Hypertension

A study using BOLD MRI technique showed that medullary R_2* increased significantly in control rats in response to NOS inhibition, while hypertensive rats exhibited a minimal change [39]. This is consistent with the known deficiency of NO in hypertension rats [39] and previous reports based on invasive blood flow measurement [87, 88]. Oxidative stress is known to play a key role in the development of hypertension [89]. Tempol (4-hydroxy-2,2,6,6-tetramethyl piperidinoyl), a superoxide scavenger, is known to reduce reactive oxygen species (ROS) and hence improve NO bioavailability [90]. Tempol showed no effect on the R_2* in normal rats but significantly decreased in hypertensive rats evaluated by BOLD MRI [81]. The degree of R_2* changes is in qualitative agreement with the observed medullary blood flow and mean arterial pressure (MAP) changes induced by tempol administration assessed by invasive measurement [91]. These studies, combined with the report on angiotensin II [59], support a role for BOLD MRI in the understanding of pathophysiology of hypertension and evaluation of novel drug interventions.

Renal Artery Stenosis (RAS)

RAS is a common cause of ischemia and hence has consequences to intrarenal oxygenation. Juillard et al. [67] used a well-designed preclinical model to test if

BOLD can detect the presence of renal hypoxia induced by RAS. They found that R_2* increased continuously and progressively in parallel with the decrease in renal blood flow (RBF) in response to increasing levels of stenosis, suggesting evolving hypoxia in both the medulla and cortex. Alford et al. [61] also documented an increase in R_2* following acute occlusion of renal artery. Contralateral kidney showed no such change. They also demonstrated that the R_2* returned to baseline values upon releasing the occlusion. Recently, Textor et al. [71] reported on measurements in human subjects with RAS. In normal-sized kidneys downstream of high-grade RAS, R_2* was elevated at baseline (suggesting enhanced hypoxia) and fell after administration of furosemide. This was true even when the GFR was significantly reduced. These results are supported by previous reports of preserved cortical tissue volume in poststenotic kidneys, despite reduced function as measured by isotope renography [92]. These in turn may suggest that GFR might be recoverable in such cases and that nonfiltering kidney tissue represents a form of "hibernation" in the kidney with the potential for restoring kidney function after restoring blood flow [92]. On the other hand, atrophic kidneys beyond totally occluded renal arteries demonstrated low levels of R_2* (improved oxygenation) that did not change after administration of furosemide [71] (Fig. 6). This may suggest a nonfunctioning kidney with limited or no oxygen consumption. That study also includes an example where a kidney with multiple arteries showed different R_2* values in regions supplied by a stenosed renal artery.

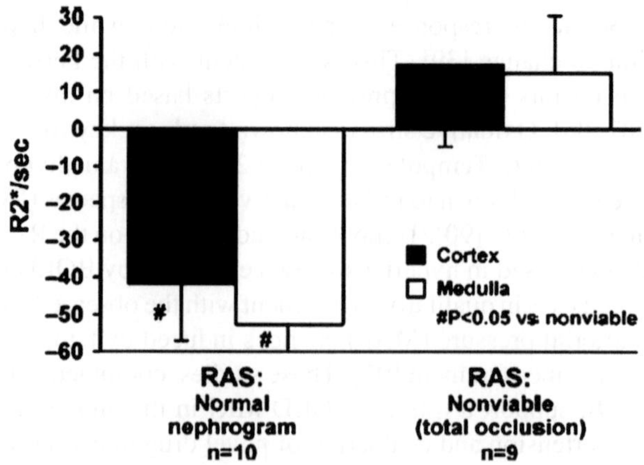

Fig. 6 Changes in blood oxygenation level dependent (BOLD) magnetic resonance imaging (MRI) measurements (Delta R_2*/s) before and after furosemide in patients with atherosclerotic renal artery stenosis (RAS) with "normal" kidneys as compared with "nonviable" kidneys with total occlusion. Nonfunctioning kidneys demonstrated no R_2* response after intravenous furosemide, whereas poststenotic kidneys otherwise had consistent falls in R_2* after furosemide, particularly in medullary regions. Reproduced from [71]

Unilateral Ureteral Obstruction

Pedersen et al. [57] demonstrated changes on BOLD MRI in a pig model of unilateral ureteral obstruction (UUO). Twenty-four hours of UUO was associated with an increased R_2* in the cortex and a decreased R_2* in medulla, as compared with the baseline that indicated pO_2 levels were reduced in the cortex and increased in the medulla during and after release of obstruction. Similar results were reported in a small number of patients with distal unilateral urethral calculus [72]. All patients had significantly lower medullary and cortical R_2* values in the obstructed kidney compared to the contralateral nonobstructed kidney and literature values in healthy subjects. The increase in oxygen content in medulla may indicate decreased oxygen consumption as a result of reduced GFR in the obstructed kidney [72].

Renal Transplants

Several groups have evaluated the feasibility of BOLD MR imaging in patients with renal transplants. Thoeny et al. [73] compared the BOLD index in transplanted kidneys with those in kidneys of healthy volunteers. The medullary R_2* was found to be lower in transplant patients compared to healthy volunteers ($P < 0.004$), implying a relatively improved oxygenation in transplanted kidney. This could be explained as the result of reduced tubular fractional reabsorption of sodium and increased blood flow due to allograft denervation. We believe that these observations may also be influenced by the time of the study following transplantation. Sadowski et al. [93] evaluated 20 patients who had recently received renal transplants in an attempt to obtain preliminary data on potential differences between normal functioning transplants and those experiencing acute rejection and acute tubular necrosis (ATN). Six patients had clinically normal functioning transplants, eight had biopsy-proved rejection, and six had biopsy-proven ATN. Their results showed that R_2* measurements in the medullary regions of transplanted kidneys with acute rejection were significantly lower than those in normally functioning transplants or transplants with ATN. Similar findings were reported by a more recent study in a much larger number of subjects ($n = 82$) including biopsy-proven acute rejection and ATN (Fig. 7) [65].

Djamali et al. applied BOLD-MRI to discriminate among different types of rejection early after kidney transplantation [94]. Twenty-three patients underwent imaging in the first 4 months posttransplant. Five had normal functioning transplants and 18 had biopsy-proven acute allograft dysfunction, acute tubular necrosis, and acute rejection, including borderline rejection ($n = 3$; IA rejection: $n = 4$; IIA rejection: $n = 6$: C4d(+) rejection: $n = 9$). Their results in general agreed with those of Sadowski et al. [93] in that medullary R_2* levels were higher (increased local deoxyhemoglobin concentration) in normal functioning allografts (24.3/s ± 2.3) compared to acute rejection (16.0/s ± 2.1) and ATN (20.9/s ± 1.8) ($P < 0.05$). There was no statistically significant difference in cortical R_2*. Medullary R_2* was the lowest in acute rejection with a vascular component (i.e., IIA

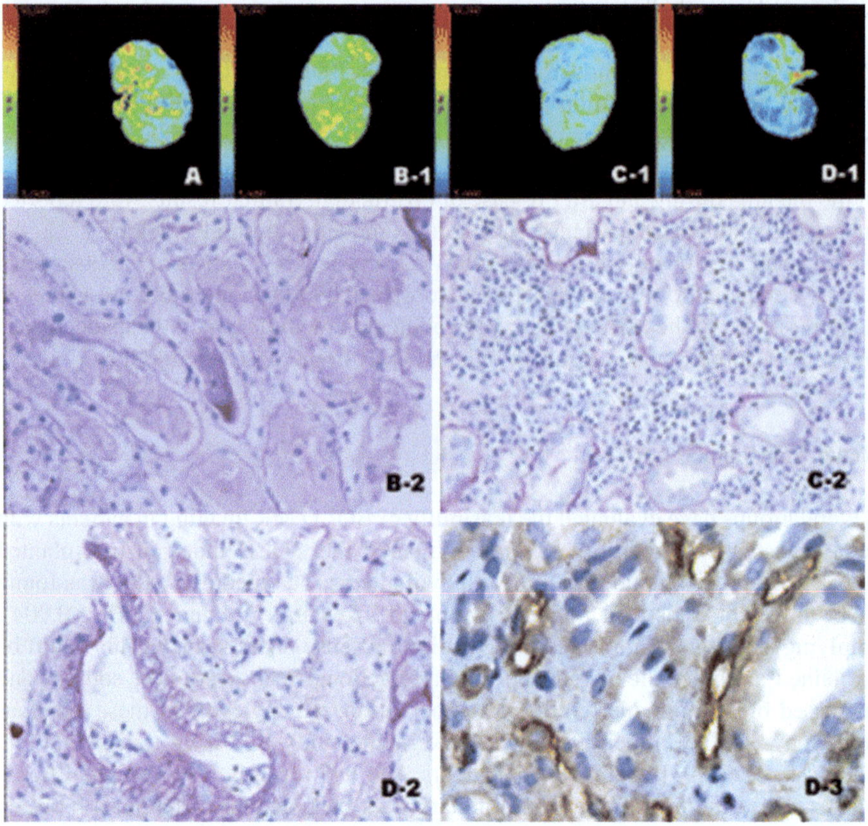

Fig. 7 The color blood oxygenation level dependent (BOLD) magnetic resonance imaging (MRI) R_2^* maps of coronal transplant sections and corresponding pathological photos. Blue represents the lowest R_2^* level, indicating the lowest tissue deoxyhemoglobin concentration, and green and orange distinctively represent increasing R_2^* levels, showing the increase of tissue deoxyhemoglobin concentration. In a normal functioning allograft, the cortex is blue with areas of green, and the medulla is green and orange reflecting the decrease in tissue oxygen bioavailability from outer medulla to inner medulla. In acute rejection (AR) allograft, the color gradient disappears with increased blue areas in the medulla. Conversely, acute tubular necrosis (ATN) allograft shows increased green areas in the cortex and more orange areas in the medulla, reflecting the decrease in tissue oxygen bioavailability both in the cortex and in the medulla. (**a**) The color R_2^* map of normal functioning transplant. (**b**) The color R_2^* map and pathological photo of ATN transplant shows more green areas in the cortex and more orange areas in the medulla in B-1, and pathologically tubular epithelial cell necrosis in B-2 (×200). (**c**) The color R_2^* map and pathological photo of T-cell-mediated (TMR) transplant shows the color gradient disappears with increasing blue areas in the medulla in C-1, and tubulitis with interstitial inflammation in C-2 (×200). (**d**) The color R_2^* map and pathological photos of antibody-mediated rejection (AMR) transplant shows more blue areas in the medulla in D-1, mild intimal arteritis in D-2 (×200), and C4d staining (+) in peritubular capillary in D-3 (×400). Reproduced from [65]

Table 1 Correlation between intrarenal oxygenation and oxidative stress in chronic allograft nephropathy

Patient characteristics	Multiple liner regression analyses	R^2	P
BOLD			
Medullary R2* (M R2*)	6.923 + (0.954 × C R2*) + (0.0000815 × UF_2*) − (0.00146 × UNO)	0.97	<0.001
Cortical R2* (C R2*)	19.472−(0.00518 × UNO*) − (0.564 × UH_2O_2*)	0.71	<0.03
MCR2*	1.313 + (0.00000851 × UF_2*) + (0.0257 × UH_2O_2*) − (0.0143 × S H_2O_2)	0.84	<0.03
OS			
Urine F_2-isoprostances (F_2)	−37191.367 + (10174.393 × MR2*)−11132.035 × CR2*)	0.83	<0.001
Urine nitric oxide	2975.627−(85.026 × MR2*) − (97.771 × UH_2O_2*) − (101.880 × UHSP27*)	0.72	<0.05
Urine HSP27 (UHSP27)	−0.446 + (1.466 × CR2*) − (1.106 × MR2*) − (1.946 × Fe_{Na}*) + (3.271 × albumin)	0.69	<0.04
Urine H_2O_2 (UH_2O_2)	19.112−(0.723 × CR2*) − (0.00572 × UNO*) − (0.694 × UHSP27*)	0.76	<0.05
MAP, mmHg	217.902−(37.299 × MCR2*) − (2.325 × Ht*) + (0.04482 × SF_2*) + (2.490 × UH_2O_2*)	0.92	<0.03

Reproduced from [62]

and C4d(+) compared to IA and "borderline" rejection). Receiver operator characteristic (ROC) curve analyses suggested that the medullary R_2* and medullary-cortical ratios could accurately discriminate acute rejection in the early posttransplant period.

Djamali et al. [62] also used BOLD MRI to evaluate patients with chronic allograft nephropathy (CAN) and looked for correlations with other conventional biomarkers of oxidative stress. Similar to previous reports on acute rejection, those with CAN showed lower medullary and cortical R_2* values. More importantly, they observed that intrarenal oxygenation as evaluated by BOLD MRI showed a high level of correlation with serum and urine biomarkers of oxidative stress (Table 1). They concluded, "this pilot study is provocative in suggesting that oxygenation patterns are different in CAN and, moreover, are strongly associated with oxidative stress. Our therapeutics to date have not used oxygen delivery as an outcome of therapy, but it may well be the case that optimal tissue oxygenation, not hypoxia nor hyperoxia, is a target of therapy. The association in CAN between aberrant kidney oxygenation and oxidative stress is important and may provide leads as to how to slow loss of transplant function."

3 Renal BOLD MRI: Limitations

Several limitations of BOLD MR imaging technique have to be considered. BOLD signal is influenced by oxygen supply, oxygen consumption, blood flow/blood volume [95, 96] hematocrit [97], and pO_2 [98]. Moreover, changes in the oxygen-hemoglobin dissociation curve may be influenced by factors such as pH and temperature [99]. In addition, R_2* is influenced by the vessel geometry and applied pulse-sequence parameters. Therefore, the absolute magnitude of R_2* is less reliable in practice than the relative changes observed. For the same reason, a direct calibration of R_2* vs. pO_2 has to be viewed with caution.

Susceptibility artifacts caused by bowel gas [100] are sometimes marked and at times lead to uninterpretable observations. Motion artifacts caused by breathing should also be carefully monitored. Use of a respiratory monitor could minimize errors because of improper breathholding.

Because hydration status can significantly influence the renal BOLD MRImeasurements, it is preferred to perform studies following 12-h fasting (overnight). This would facilitate combining data from different individual subjects and comparison of different groups of subjects.

4 Conclusion

We have provided a comprehensive review of renal BOLD MRI in this chapter. Based on collective experience to date, BOLD MRI offers a unique opportunity to study renal oxygenation status in humans. The technique is translational in that measurements in preclinical models can be carried out in the same way as in humans. While the technique is relatively simple in implementation, interpretation of BOLD MRI is complicated by the fact that R_2* is influenced by many factors other than blood oxygenation. But with a combination of appropriate validation studies and choice of physiological or pharmacological maneuvers, BOLD MRI could offer a way to better understand human renal physiology and pathophysiology. Although not yet extensively investigated, renal BOLD MRI measurements should be inherently sensitive to oxidative stress.

References

1. O'Connor PM. Renal oxygen delivery: matching delivery to metabolic demand. Clin Exp Pharmacol Physiol 2006; 33: 961–967.
2. O'Connor PM, Anderson WP, Kett MM, et al. Renal preglomerular arterial-venous O_2 shunting is a structural anti-oxidant defence mechanism of the renal cortex. Clin Exp Pharmacol Physiol 2006; 33: 637–641.
3. Epstein FH. Oxygen and renal metabolism. Kidney Int 1997; 51: 381–385.
4. Brezis M, Rosen S. Hypoxia of the renal medulla – its implications for disease. N Engl J Med 1995; 332: 647–655.
5. Brezis M, Agmon Y, Epstein FH. Determinants of intrarenal oxygenation. I. Effects of diuretics. Am J Physiol 1994; 267: F1059–F1062.
6. Brezis M, Heyman SN, Epstein FH. Determinants of intrarenal oxygenation. II. Hemodynamic effects. Am J Physiol 1994; 267: F1063–F1068.
7. Fine LG, Norman JT. Chronic hypoxia as a mechanism of progression of chronic kidney diseases: from hypothesis to novel therapeutics. Kidney Int 2008; 74: 867–872.
8. Nangaku M. Chronic hypoxia and tubulointerstitial injury: a final common pathway to end-stage renal failure. J Am Soc Nephrol 2006; 17: 17–25.
9. Clark JK, Barker HG. Studies of renal oxygen consumption in man. I. The effect of tubular loading (PAH), water diuresis and osmotic (mannitol) diuresis. J Clin Invest 1951; 30: 745–750.
10. Prasad PV, Edelman RR, Epstein FH. Noninvasive evaluation of intrarenal oxygenation with BOLD MRI. Circulation 1996; 94: 3271–3275.
11. Nangaku M, Eckardt KU. Hypoxia and the HIF system in kidney disease. J Mol Med 2007; 85: 1325–1330.
12. Klenzak J, Himmelfarb J. Sepsis and the kidney. Crit Care Clin 2005; 21: 211–222.
13. Issa N, Messer J, Paganini EP. Renal assist device and treatment of sepsis-induced acute kidney injury in intensive care units. Contrib Nephrol 2007; 156: 419–427.
14. Heyman SN, Reichman J, Brezis M. Pathophysiology of radiocontrast nephropathy: a role for medullary hypoxia. Invest Radiol 1999; 34: 685–691.
15. Heyman SN, Rosenberger C, Rosen S. Regional alterations in renal haemodynamics and oxygenation: a role in contrast medium-induced nephropathy. Nephrol Dial Transplant 2005; 20(Suppl 1): i6–i11.
16. Fine LG, Orphanides C, Norman JT. Progressive renal disease: the chronic hypoxia hypothesis. Kidney Int Suppl 1998; 65: S74–S78.
17. Palm F, Cederberg J, Hansell P, et al. Reactive oxygen species cause diabetes-induced decrease in renal oxygen tension. Diabetologia 2003; 46: 1153–1160.
18. Brezis M, Heyman SN, Dinour D, et al. Role of nitric oxide in renal medullary oxygenation. Studies in isolated and intact rat kidneys. J Clin Invest 1991; 88: 390–395.

19. Dinour D, Brezis M. Effects of adenosine on intrarenal oxygenation. Am J Physiol 1991; 261: F787–F791.
20. Heyman SN, Kaminski N, Brezis M. Dopamine increases renal medullary blood flow without improving regional hypoxia. Exp Nephrol 1995; 3: 331–337.
21. Leong CL, O'Connor PM, Eppel GA, et al. Measurement of renal tissue oxygen tension: systematic differences between fluorescence optode and microelectrode recordings in anaesthetized rabbits. Nephron Physiol 2008; 108: p11–p17.
22. Liss P, Aukland K, Carlsson PO, et al. Influence of iothalamate on renal medullary perfusion and oxygenation in the rat. Acta Radiol 2005; 46: 823–829.
23. Liss P, Nygren A, Erikson U, et al. Injection of low and iso-osmolar contrast medium decreases oxygen tension in the renal medulla. Kidney Int 1998; 53: 698–702.
24. Baudelet C, Gallez B. Effect of anesthesia on the signal intensity in tumors using BOLD-MRI: comparison with flow measurements by Laser Doppler flowmetry and oxygen measurements by luminescence-based probes. Magn Reson Imaging 2004; 22: 905–912.
25. dos Santos EA, Li LP, Ji L, et al. Early changes with diabetes in renal medullary hemodynamics as evaluated by fiberoptic probes and BOLD magnetic resonance imaging. Invest Radiol 2007; 42: 157–162.
26. Li LP, Ji L, Santos EA, et al. Effect of nitric oxide synthase inhibition on intrarenal oxygenation as evaluated by blood oxygenation level-dependent magnetic resonance imaging. Invest Radiol 2009; 44: 67–73.
27. O'Connor PM, Anderson WP, Kett MM, et al. Simultaneous measurement of pO_2 and perfusion in the rabbit kidney in vivo. Adv Exp Med Biol 2007; 599: 93–99.
28. Manotham K, Tanaka T, Matsumoto M, et al. Evidence of tubular hypoxia in the early phase in the remnant kidney model. J Am Soc Nephrol 2004; 15: 1277–1288.
29. Heyman SN, Khamaisi M, Rosen S, et al. Renal parenchymal hypoxia, hypoxia response and the progression of chronic kidney disease. Am J Nephrol 2008; 28: 998–1006.
30. Hirayama A, Nagase S, Ueda A, et al. In vivo imaging of oxidative stress in ischemia-reperfusion renal injury using electron paramagnetic resonance. Am J Physiol Renal Physiol 2005; 288: F597–F603.
31. Subramanian S, Yamada K, Irie A, et al. Noninvasive in vivo oximetric imaging by radiofrequency FT EPR. Magn Reson Med 2002; 47: 1001–1008.
32. James PE, Goda F, Grinberg OY, et al. Intrarenal pO2 measured by EPR oximetry and the effects of bacterial endotoxin. Adv Exp Med Biol 1997; 411: 557–568.
33. Zhao D, Jiang L, Hahn EW, et al. Comparison of (1)H blood oxygen level-dependent (BOLD) and (19)F MRI to investigate tumor oxygenation. Magn Reson Med 2009; 62: 357–364.
34. Zimmermann U, Noth U, Grohn P, et al. Non-invasive evaluation of the location, the functional integrity and the oxygen supply of implants: 19F nuclear magnetic resonance imaging of perfluorocarbon-loaded Ba2+-alginate beads. Artif Cells Blood Substit Immobil Biotechnol 2000; 28: 129–146.
35. Ndubuizu O, LaManna JC. Brain tissue oxygen concentration measurements. Antioxid Redox Signal 2007; 9: 1207–1219.
36. van Eijsden P, Hyder F, Rothman DL, et al. Neurophysiology of functional imaging. Neuroimage 2009; 45: 1047–1054.
37. Fieno DS, Shea SM, Li Y, et al. Myocardial perfusion imaging based on the blood oxygen level-dependent effect using T2-prepared steady-state free-precession magnetic resonance imaging. Circulation 2004; 110: 1284–1290.
38. Klocke FJ, Li D. Testing coronary flow reserve without a provocative stress. A "BOLD" approach. J Am Coll Cardiol 2003; 41: 841–842.
39. Li L, Storey P, Kim D, et al. Kidneys in hypertensive rats show reduced response to nitric oxide synthase inhibition as evaluated by BOLD MRI. J Magn Reson Imaging 2003; 17: 671–675.

40. Huegli RW, Schulte AC, Aschwanden M, et al. Effects of percutaneous transluminal angioplasty on muscle BOLD-MRI in patients with peripheral arterial occlusive disease: preliminary results. Eur Radiol 2009; 19: 509–515.
41. Ledermann HP, Schulte AC, Heidecker HG, et al. Blood oxygenation level-dependent magnetic resonance imaging of the skeletal muscle in patients with peripheral arterial occlusive disease. Circulation 2006; 113: 2929–2935.
42. Hoskin PJ, Carnell DM, Taylor NJ, et al. Hypoxia in prostate cancer: correlation of BOLD-MRI with pimonidazole immunohistochemistry-initial observations. Int J Radiat Oncol Biol Phys 2007; 68: 1065–1071.
43. Sanghvi DA. Recent advances in imaging of brain tumors. Indian J Cancer 2009; 46: 82–87.
44. Evans RG, Gardiner BS, Smith DW, et al. Methods for studying the physiology of kidney oxygenation. Clin Exp Pharmacol Physiol 2008; 35: 1405–1412.
45. Storey P. Introduction to magnetic resonance: imaging and spectroscopy. In: Magnetic Resonance Imaging: Methods and Biologic Applications, edited by Prasad PV, Totowa, Humana Press, 2005, pp 3–58.
46. Caria A, Veit R, Sitaram R, et al. Regulation of anterior insular cortex activity using real-time fMRI. Neuroimage 2007; 35: 1238–1246.
47. Prasad PV, Chen Q, Goldfarb JW, et al. Breath-hold R2* mapping with a multiple gradient-recalled echo sequence: application to the evaluation of intrarenal oxygenation. J Magn Reson Imaging 1997; 7: 1163–1165.
48. Sadowski EA, Djamali A, Wentland AL, et al. Blood oxygen level-dependent and perfusion magnetic resonance imaging: detecting differences in oxygen bioavailability and blood flow in transplanted kidneys. Magn Reson Imaging 2010; 28: 56–64.
49. Zuo CS, Rofsky NM, Mahallati H, et al. Visualization and quantification of renal R2* changes during water diuresis. J Magn Reson Imaging 2003; 17: 676–682.
50. Epstein FH, Veves A, Prasad PV. Effect of diabetes on renal medullary oxygenation during water diuresis. Diabetes Care 2002; 25: 575–578.
51. Prasad PV, Epstein FH. Changes in renal medullary pO_2 during water diuresis as evaluated by blood oxygenation level-dependent magnetic resonance imaging: effects of aging and cyclooxygenase inhibition. Kidney Int 1999; 55: 294–298.
52. Tumkur SM, Vu AT, Li LP, et al. Evaluation of intra-renal oxygenation during water diuresis: a time-resolved study using BOLD MRI. Kidney Int 2006; 70: 139–143.
53. Li LP, Ji L, Lindsay S, et al. Evaluation of intrarenal oxygenation in mice by BOLD MRI on a 3.0 T human whole-body scanner. J Magn Reson Imaging 2007; 25: 635–638.
54. Ries M, Basseau F, Tyndal B, et al. Renal diffusion and BOLD MRI in experimental diabetic nephropathy. Blood oxygen level-dependent. J Magn Reson Imaging 2003; 17: 104–113.
55. Li LP, Vu AT, Li BS, et al. Evaluation of intrarenal oxygenation by BOLD MRI at 3.0 T. J Magn Reson Imaging 2004; 20: 901–904.
56. Priatna A, Epstein FH, Spokes K, et al. Evaluation of changes in intrarenal oxygenation in rats using multiple gradient-recalled echo (mGRE) sequence. J Magn Reson Imaging 1999; 9: 842–846.
57. Pedersen M, Dissing TH, Morkenborg J, et al. Validation of quantitative BOLD MRI measurements in kidney: application to unilateral ureteral obstruction. Kidney Int 2005; 67: 2305–2312.
58. Gomez SI, Warner L, Haas JA, et al. Increased hypoxia and reduced renal tubular response to furosemide detected by BOLD magnetic resonance imaging in swine renovascular hypertension. Am J Physiol Renal Physiol 2009; 297: F981–F986.
59. Schachinger H, Klarhofer M, Linder L, et al. Angiotensin II decreases the renal MRI blood oxygenation level-dependent signal. Hypertension 2006; 47: 1062–1066.
60. Malvezzi P, Bricault I, Terrier N, et al. Evaluation of intrarenal oxygenation by blood oxygen level-dependent magnetic resonance imaging in living kidney donors and their recipients: preliminary results. Transplant Proc 2009; 41: 641–644.

61. Alford SK, Sadowski EA, Unal O, et al. Detection of acute renal ischemia in swine using blood oxygen level-dependent magnetic resonance imaging. J Magn Reson Imaging 2005; 22: 347–353.
62. Djamali A, Sadowski EA, Muehrer RJ, et al. BOLD-MRI assessment of intrarenal oxygenation and oxidative stress in patients with chronic kidney allograft dysfunction. Am J Physiol Renal Physiol 2007; 292: F513–F522.
63. Edlund J, Hansell P, Fasching A, et al. Reduced oxygenation in diabetic rat kidneys measured by T2* weighted magnetic resonance micro-imaging. Adv Exp Med Biol 2009; 645: 199–204.
64. Gloviczki ML, Glockner J, Gomez SI, et al. Comparison of 1.5 and 3 T BOLD MR to study oxygenation of kidney cortex and medulla in human renovascular disease. Invest Radiol 2009; 44: 566–571.
65. Han F, Xiao W, Xu Y, et al. The significance of BOLD MRI in differentiation between renal transplant rejection and acute tubular necrosis. Nephrol Dial Transplant 2008; 23: 2666–2672.
66. Hofmann L, Simon-Zoula S, Nowak A, et al. BOLD-MRI for the assessment of renal oxygenation in humans: acute effect of nephrotoxic xenobiotics. Kidney Int 2006; 70: 144–150.
67. Juillard L, Lerman LO, Kruger DG, et al. Blood oxygen level-dependent measurement of acute intra-renal ischemia. Kidney Int 2004; 65: 944–950.
68. Mason RP. Non-invasive assessment of kidney oxygenation: a role for BOLD MRI. Kidney Int 2006; 70: 10–11.
69. Boss A, Martirosian P, Jehs M, et al. Influence of oxygen and carbogen breathing on renal oxygenation measured by BOLD-imaging at 3.0 Tesla. In ISMRM 16th Scientific Meeting and Exhibition, edited by ISMRM, Toronto, Canada, ISMRM 2008.
70. Simon-Zoula SC, Hofmann L, Giger A, et al. Non-invasive monitoring of renal oxygenation using BOLD-MRI: a reproducibility study. NMR Biomed 2006; 19: 84–89.
71. Textor SC, Glockner JF, Lerman LO, et al. The use of magnetic resonance to evaluate tissue oxygenation in renal artery stenosis. J Am Soc Nephrol 2008; 19: 780–788.
72. Thoeny HC, Kessler TM, Simon-Zoula S, et al. Renal oxygenation changes during acute unilateral ureteral obstruction: assessment with blood oxygen level-dependent MR imaging – initial experience. Radiology 2008; 247: 754–761.
73. Thoeny HC, Zumstein D, Simon-Zoula S, et al. Functional evaluation of transplanted kidneys with diffusion-weighted and BOLD MR imaging: initial experience. Radiology 2006; 241: 812–821.
74. Li LP, Storey P, Pierchala L, et al. Evaluation of the reproducibility of intrarenal R2* and DeltaR2* measurements following administration of furosemide and during waterload. J Magn Reson Imaging 2004; 19: 610–616.
75. Epstein FH, Prasad P. Effects of furosemide on medullary oxygenation in younger and older subjects. Kidney Int 2000; 57: 2080–2083.
76. Tumkur S, Vu A, Li L, et al. Evaluation of intrarenal oxygenation at 3.0 T using 3-dimensional multiple gradient-recalled echo sequence. Invest Radiol 2006; 41: 181–184.
77. Ji L, Li L-P, Schnitzer T, et al. Effects of naproxen and naproxcinod on intra-renal oxygenation during water-loading as evaluated by BOLD MRI. In ISMRM 15th Scientific Meeting amd Exhibition (vol. 1), edited by Medicine ISfMRi, Berlin, Germany, International Society for Magnetic Resonance in Medicine, 2007, p 3051.
78. Hall JE, Granger JP. Renal hemodynamic actions of angiotensin II: interaction with tubuloglomerular feedback. Am J Physiol 1983; 245: R166–R173.
79. Palmer RM, Ashton DS, Moncada S. Vascular endothelial cells synthesize nitric oxide from L-arginine. Nature 1988; 333: 664–666.
80. Li L-P, Elizabete Santos LJ, Pierchala L, Dunkle E, Fogelson L, Prasad PV. Effect of NOS inhibition by L-NAME on intra-renal BOLD MRI: dose response in rat and human kidneys.

In ISMRM 13rd Scientific Meeting and Exhibition (vol. 17), edited by ISMRM, Miami, USA, ISMRM, 2005, p 559.
81. Li LP, Li BS, Storey P, et al. Effect of free radical scavenger (tempol) on intrarenal oxygenation in hypertensive rats as evaluated by BOLD MRI. J Magn Reson Imaging 2005; 21: 245–248.
82. Klein IH, Abrahams A, van Ede T, et al. Different effects of tacrolimus and cyclosporine on renal hemodynamics and blood pressure in healthy subjects. Transplantation 2002; 73: 732–736.
83. Prasad PV, Priatna A, Spokes K, et al. Changes in intrarenal oxygenation as evaluated by BOLD MRI in a rat kidney model for radiocontrast nephropathy. J Magn Reson Imaging 2001; 13: 744–747.
84. Li L-P, Halter S, Cabry J, et al. Evaluation of intra-renal oxygenation in db/db mice by BOLD MRI. In ISMRM 17th Scientific Meeting and Exhibition, edited by ISMRM, Honolulu, Hawaii, USA, ISMRM, 2009, p 4109.
85. Li L-P, Ji L, Carbray J, et al. Effect of acute hyperglycemia on intra-renal oxygenation as estimated by BOLD MRI in rats. In 16th Scientific Meeting and Exhibition, edited by ISMRM, Toronto, Canada, ISMRM, 2008, p 2692.
86. Li L-P, Carbray J, Halter S, et al. Effect of acute hyperglycemia with octreotide on intra-renal oxygenation as estimated by BOLD MRI in rats. In 17th Scientific Meeting & Exhibition, edited by ISMRM, Honolulu, USA, ISMRM, 2009, p 4140.
87. Mattson DL, Lu S, Nakanishi K, et al. Effect of chronic renal medullary nitric oxide inhibition on blood pressure. Am J Physiol 1994; 266: H1918–H1926.
88. Nakanishi K, Mattson DL, Cowley AW, Jr. Role of renal medullary blood flow in the development of L-NAME hypertension in rats. Am J Physiol 1995; 268: R317–R323.
89. Touyz RM. Reactive oxygen species, vascular oxidative stress, and redox signaling in hypertension: what is the clinical significance? Hypertension 2004; 44: 248–252.
90. Schnackenberg CG, Wilcox CS. The SOD mimetic tempol restores vasodilation in afferent arterioles of experimental diabetes. Kidney Int 2001; 59: 1859–1864.
91. Schnackenberg CG, Welch WJ, Wilcox CS. Normalization of blood pressure and renal vascular resistance in SHR with a membrane-permeable superoxide dismutase mimetic: role of nitric oxide. Hypertension 1998; 32: 59–64.
92. Cheung CM, Shurrab AE, Buckley DL, et al. MR-derived renal morphology and renal function in patients with atherosclerotic renovascular disease. Kidney Int 2006; 69: 715–722.
93. Sadowski EA, Fain SB, Alford SK, et al. Assessment of acute renal transplant rejection with blood oxygen level-dependent MR imaging: initial experience. Radiology 2005; 236: 911–919.
94. Djamali A, Sadowski EA, Samaniego-Picota M, et al. Noninvasive assessment of early kidney allograft dysfunction by blood oxygen level-dependent magnetic resonance imaging. Transplantation 2006; 82: 621–628.
95. Shen Q, Ren H, Duong TQ. CBF, BOLD, CBV, and CMRO(2) fMRI signal temporal dynamics at 500-msec resolution. J Magn Reson Imaging 2008; 27: 599–606.
96. Wu G, Luo F, Li Z, et al. Transient relationships among BOLD, CBV, and CBF changes in rat brain as detected by functional MRI. Magn Reson Med 2002; 48: 987–993.
97. Zhao JM, Clingman CS, Narvainen MJ, et al. Oxygenation and hematocrit dependence of transverse relaxation rates of blood at 3T. Magn Reson Med 2007; 58: 592–597.
98. Spees WM, Yablonskiy DA, Oswood MC, et al. Water proton MR properties of human blood at 1.5 Tesla: magnetic susceptibility, T(1), T(2), T*(2), and non-Lorentzian signal behavior. Magn Reson Med 2001; 45: 533–542.
99. Hess W. [Affinity of oxygen for hemoglobin – its significance under physiological and pathological conditions]. Anaesthesist 1987; 36: 455–467.
100. Grenier N, Basseau F, Ries M, et al. Functional MRI of the kidney. Abdom Imaging 2003; 28: 164–175.

Chapter 31
Anemia and Progression of Chronic Kidney Disease

Danilo Fliser and Ferdinand H. Bahlmann

Abstract Tubulointerstitial fibrosis that results from renal tissue hypoxia is thought to be a key element of progressive chronic kidney diseases (CKD). Findings from epidemiological studies suggest that anemia in patients with CKD due to inadequate erythropoietin (EPO) secretion may contribute to progression. However, results of prospective controlled studies that evaluated the effect of recombinant human erythropoietin (rHuEPO) on the course of CKD were inconsistent. Nevertheless, slowing CKD progression with rHuEPO (and correction of anemia) may be achieved with a better understanding of the processes involved in the damage caused by renal tissue hypoxia.

Keywords Erythropoietin · Recombinant human erythropoietin · Darbepoetin · Continuous erythropoietin receptor activator · Hypoxia-inducible factor

Abbreviations

BOLD-MRI	Blood oxygenation level-dependent magnetic resonance imaging
C.E.R.A.	Continuous erythropoietin receptor activator
CKD	Chronic kidney disease
EPO	Erythropoietin
EPOR	Erythropoietin receptor
eGFR	Estimated glomerular filtration rate
GFR	Glomerular filtration rate
HIF	Hypoxia-inducible factor
NO	Nitric oxide

D. Fliser (✉)
Department of Internal Medicine – Renal and Hypertensive Diseases
Saarland University Medical Centre, Kirrberger Strasse, 66421 Homburg/Saar, Germany
e-mail: indfli@uks.eu

T. Miyata et al. (eds.), *Studies on Renal Disorders*, Oxidative Stress in Applied Basic Research and Clinical Practice, DOI 10.1007/978-1-60761-857-7_31,

pO_2	Partial oxygen pressure
rHuEPO	Recombinant human erythropoietin
ROS	Reactive oxygen species
RRT	Renal replacement therapy
STAT	Signal transducers and activators of transcription
VEGF	Vascular endothelial growth factor

1 The Progressive Nature of Chronic Kidney Disease

Until the 1990s, the "hyperfiltration" hypothesis put forward by Brenner et al. [1] advocated altered glomerular hemodynamics as the key event that governs progression of chronic kidney disease (CKD): initial reduction in renal tissue causes glomerular hypertension, hyperfiltration, and hypertrophy of functioning glomeruli, leading to glomerular sclerosis and obsolescence, and finally into an ongoing vicious circle of progressive renal damage. However, subsequent clinical work revealed that in CKD patients the extent of kidney dysfunction is rather poorly associated with glomerular pathology, but instead correlates well with chronic tubulointerstitial injury and fibrosis, shifting the interest to other progression promoters [2–4]. An increasing body of evidence from experimental and clinical studies highlights the role of renal tissue hypoxia for the progressive nature of CKD [5].

In their seminal paper, Fine et al. [6] set the basis for the concept of renal tissue hypoxia in CKD progression, proposing that kidney injury leads to a vicious circle of tubulointerstitial fibrosis, pursuant obliteration of the renal microvasculature with consecutive tissue hypoxia, and perpetuation of tissue injury. Indeed, analyses of kidney biopsies obtained from patients with CKD and from animals with various experimental kidney diseases have shown that, besides tubular destruction, interstitial fibrosis is associated with a decrease in the number of peritubular capillaries, and that there is a good inverse correlation between renal function, the severity of interstitial fibrosis, and the number of peritubular capillaries [2, 7, 8]. However, it also became clear that the ischemic lesions of tubular cells may also be worsened by an increased consumption of oxygen by the remaining tubular cells, and this is likely to be one of the mechanisms by which proteinuria accelerate the progression of CKD [3]. Finally, hypoxia and reactive oxygen species (ROS) overproduction are two factors of overriding importance that promote the release of proinflammatory and pro-fibrotic molecules and can directly enhance the synthesis of extracellular matrix by fibroblastic cells [9–12].

2 Renal Tissue Hypoxia and CKD Progression

2.1 Methods to Assess Oxygenation of Renal Tissue

In the past decade the use of modern technologies have brought much insight into renal parenchymal tissue oxygenation and regional hypoxia, for example, needle-laser Doppler flow probes, oxygen microelectrodes, immunostaining for pimonidazole (a molecular "hypoxia probe," which, following injection, binds to tissues with oxygen tension below 10 mmHg), and immunostaining for hypoxia-inducible factors (HIFs) [13]. Moreover, blood oxygenation level-dependent magnetic resonance imaging (BOLD MRI), a noninvasive tool to measure dynamic changes in the distribution and extent of deoxygenated hemoglobin within renal tissue, is now available for experimental and clinical research [14]. These methods even permit the quantification of renal tissue hypoxia in progressive CKD.

Appropriate oxygenation of renal parenchyma reflects the balance between regional oxygen supply and consumption, mostly as a result of tubular solute transport (mainly sodium). Oxygen supply of renal tissue depends on the oxygen-carrying capacity of the blood and on factors that control intrarenal blood flow and distribution such as the structure of the renal microcirculation. Experimental studies have revealed that efferent arteriolar partial oxygen pressure (pO_2) is lower than renal venous pO_2, implying cortical precapillary shunting that renders the renal parenchymal at the edge of hypoxia [13]. Moreover, renal parenchymal pO_2 sharply declines at the cortico-medullary junction, reaching levels far below 25 mmHg within the renal medulla even under normal physiologic conditions [15]. This regional "physiologic hypoxia" of the renal medulla may further worsen under pathophysiologic conditions that reduce perfusion of medullary tissue. The medulla receives about 10% of the total renal blood flow, originating from efferent arterioles of deep juxtamedullary glomeruli. These blood vessels merge to form vasa recta, a central component of the renal concentrating machinery. Importantly, the countercurrent blood flow system in the vasa recta facilitates oxygen diffusion from descending to ascending blood vessels, and this shift of oxygen reduces its availability to the medulla. Complex mechanisms participate in the preservation of renal medullary oxygen supply, and several mediators such as nitric oxide (NO), adenosine, and others act in concert to maintain this oxygenation balance by enhancement of medullary blood flow and control of tubular transport.

2.2 Response to Renal Tissue Hypoxia

Chronic renal tissue hypoxia invokes a hypoxic response, mediated to a large extent by HIFs [16, 17], as explained in detail elsewhere in this book. These ubiquitous key transcription factors consist of α- and β-subunits. The former is rapidly degraded and removed under normal ambient oxygenation by specific oxygen-sensitive prolyl hydroxylases. Hypoxia inactivates the degradation of HIF-α by these enzymes, and it accumulates and binds with the β-subunit. The formed heterodimer translocates into the nucleus, binds to hypoxia-response elements, and activates a host of HIF-mediated genes that transcript for protective factors that participate in cell adaptation to hypoxia such as erythropoietin (EPO) and vascular endothelial growth factor (VEGF). These in turn may promote angiogenesis and improve tissue oxygenation and cell survival. Indeed, rats with chronic tubulointerstitial disease or diabetes have shown unexpected resistance to acute hypoxic insults, conceivably through hypoxia tolerance induced by HIF-mediated pathways [18]. However, diabetic kidneys subjected to extreme hypoxic conditions, such as ex vivo perfusion with red cell–free perfusate, are highly vulnerable to hypoxic damage, suggesting that HIF-mediated reno-protection is ineffective beyond a "window of opportunity" [18, 19].

There are at least two HIF-α isoforms: renal HIF is cell-type specific with HIF-1α isoform principally expressed in tubular segments (mostly in collecting ducts), while HIF-2α is mainly detected in vascular endothelial cells. Similarly, genes activated by HIF are also cell specific, and EPO is selectively induced by interstitial cells at the cortico-medullary junction [20]. Not surprisingly, this region at the verge of hypoxia serves in the control of the blood oxygen-carrying capacity by regulating EPO production. Using different techniques such as immunostaining for expression of HIF and HIF-mediated genes, the extent and distribution of renal hypoxia under different pathologic settings have been mapped [20, 21].

2.3 Renal Tissue Hypoxia in CKD

In patients with CKD increased concentration of deoxygenated hemoglobin within the renal medulla (i.e., intensified regional hypoxia) was found by using BOLD MRI, as compared to kidneys of healthy subjects [22]. Administration of an angiotensin II receptor blocker attenuated hypoxia, implicating an important role for angiotensin II in the generation of renal tissue hypoxia in CKD patients [23]. Importantly, BOLD MRI may also distinguish between severely compromised but viable renal parenchymal tissue that is characterized by high basal deoxygenated hemoglobin values that decreases after the administration of furosemide, and nonfunctional scar tissue with low basal deoxygenated hemoglobin levels (unaffected by furosemide administration) [24]. Basile et al. [25] showed that in an animal model of chronic tubulointerstitial disease accompanied by rarefaction of peritubular capillaries, the perfusion index falls in all renal regions along time, resulting in less oxygen supply and tissue hypoxia. As the peritubular capillaries originate from efferent glomerular arterioles, it is noteworthy that severe glomerular diseases with loss of functioning nephrons also compromise downstream peritubular microcirculation and intensify parenchymal hypoxia [26, 27].

In an experimental model of chronic tubulointerstitial disease induced by ischemia-reperfusion injury, regions with depleted renal microvasculature and expanded extracellular matrix display deposition of pimonidazole, indicating critical hypoxia, as well as HIF expression, reflecting a cellular response to hypoxia [28]. Similar indices of evolving hypoxia were documented in other animal models of hypoxic/nephrotoxic tubulointerstitial disease induced by adriamycin or cyclosporine A [29–31]. Thus, the distorted anatomy of CKD further augments hypoxia. Hypoxia may also exist in transplanted kidneys, especially during rejection episodes [32], but studies using BOLD MRI have revealed conflicting results [33].

2.4 Other Clinical Conditions Resulting in Significant Renal Tissue Hypoxia

Not unexpectedly, aging is associated with intensified renal tissue hypoxia. As shown by BOLD MRI, ambient pO_2 is some 20 mmHg lower both in the medulla in otherwise healthy aged individuals as compared with young subjects [34]. Hypoxia and hypoxia adaptation were also noted in aged rat kidneys, and this may also reflect the reduced capacity of the aged kidney to generate vasoactive substances such as NO [35, 36].

Another important condition related to renal parenchymal hypoxia is diabetes. As shown by Palm et al. [37] using oxygen microelectrodes in rats, the decline in parenchymal oxygenation in the diabetic kidney is most pronounced in the medulla, and this has also been documented using other techniques including BOLD MRI [38–40]. With the control of hyperglycemia, medullary hypoxia disappears. Moreover, renal oxygenation is restored by the administration of antioxidants, which scavenge ROS. Interestingly, though hypoxia is ameliorated, HIF expression intensifies with the addition of a ROS scavenger, suggesting that ROS suppress hypoxic adaptive responses [41]. These findings are in agreement with studies in cultured proximal tubular cells and in diabetic rats, showing oxidative stress-regulated attenuation of the HIF–VEGF pathway [42]. Collectively, the role of renal tissue hypoxia in the progression diabetic kidney injury is now well acknowledged.

Besides hyperglycemia, hypertension may also lead to renal tissue hypoxia via enhanced oxygen demand due to defective coupling of tubular transport and oxygen consumption mediated by ROS [43]. In this respect it is of interest that chronic salt consumption and volume depletion may invert the normal renal oxygenation gradient: cortical pO_2 declines, whereas medullary oxygenation paradoxically improves [44]. The attenuated medullary hypoxia conceivably reflects decreased tubular (sodium) transport. In summary, renal "physiologic" hypoxia, particularly of the medulla, is aggravated in clinical conditions that predispose to the evolution of CKD [5, 13, 45].

3 Anemia and CKD Progression

Low hematocrit levels due to anemia result in decreased oxygen supply to the kidney, and this was documented by the use of oxygen microelectrodes inserted into the kidney during acute decrements in hematocrit; both cortical and medullary pO_2 decline significantly [46]. Whereas the relationship between chronically low hematocrit (hemoglobin) levels, renal tissue hypoxia, and progression of CKD has not been extensively explored in experimental studies, there is ample evidence from observational trials and also from interventional trials that (renal) anemia in patients with CKD is related to progression [47–51]. For example, a post-hoc analysis of the RENAAL (Reduction in Endpoints with the Angiotensin Antagonist Losartan) study, which explored the reno-protective effects of angiotensin II-receptor blockade with Losartan in patients with type 2 diabetes mellitus and diabetic kidney injury, revealed a clear-cut relationship between the level of hemoglobin and progression of renal disease [48]. This relationship was independent from the randomization to the treatment intervention. Moreover, epidemiologic trials in patients with primary and secondary (e.g., diabetic nephropathy) CKD have confirmed a close correlation between hemoglobin levels and progression of kidney injury [49–51]. The question arose, however, if this was a causative relationship: Does anemia contributes to CKD progression or is it merely a marker of progressive CKD? This question can be properly answered only in intervention trials in which anemia is treated with recombinant human erythropoietin (rHuEPO).

4 Renal Anemia, EPO, and rHuEPO

In the hematopoietic system, the principal function of EPO is the regulation of red blood cell production. Consequently, following the cloning of the EPO gene, rHuEPO forms, such as EPOetin-α and EPOetin-β and the long-acting analogs darbepoetin α and continuous erythropoietin receptor activator (C.E.R.A.), have been widely used for treatment of anemia in patients with CKD.

4.1 Tissue Expression of EPO and the Erythropoietin Receptor

In humans, EPO gene expression primarily occurs in the fetal liver and adult kidney under the tightly control of an oxygen-sensing, HIF-dependent mechanism [52, 53]. Additionally, recent studies could identify EPO expression in several extrarenal tissues and cell types, including astrocytes, neurons, the female genital tract, male reproductive organs, mammary glands, placental trophoblasts, bone marrow macrophages, and erythroid progenitors [54–62]. The expression of erythropoietin receptor (EPOR) in nonerythroid tissues such as the brain, retina, heart, kidney, smooth muscle cells, myoblasts, and vascular endothelium has been associated with the discovery of novel biological functions of endogenous EPO signaling in nonhematopoietic tissues [63–67]. For example, targeted disruption of either EPO or EPOR in mice leads to in utero death between embryonic days (E) 11.5 and 13.5 because of lack of definitive erythropoiesis in the fetal liver and, interestingly, defects in the onset of the circulatory system, indicating the vital role for EPO–EPOR signaling during vascular development [68].

The EPO molecule is a glycoprotein with a molecular weight of 30.4 kDa, and its tertiary structure is defined by four antiparallel α-helices. Binding of a single molecule to two adjacent EPOR on the membrane of target cells leads to homodimerization of the EPOR and the triggering of different intracellular signaling cascades (Fig. 1). The major mechanism for degradation of EPO in the body occurs in cells that express the EPOR, through receptor-mediated endocytosis of EPO, followed by degradation in lysosomes [69]. However, these signaling cascades are highly complex and their relevance for the cellular effects of EPO is not completely understood so far. For example, by using EPOR fusion proteins, it has been shown that distinct conformations of the EPOR exist that may activate different

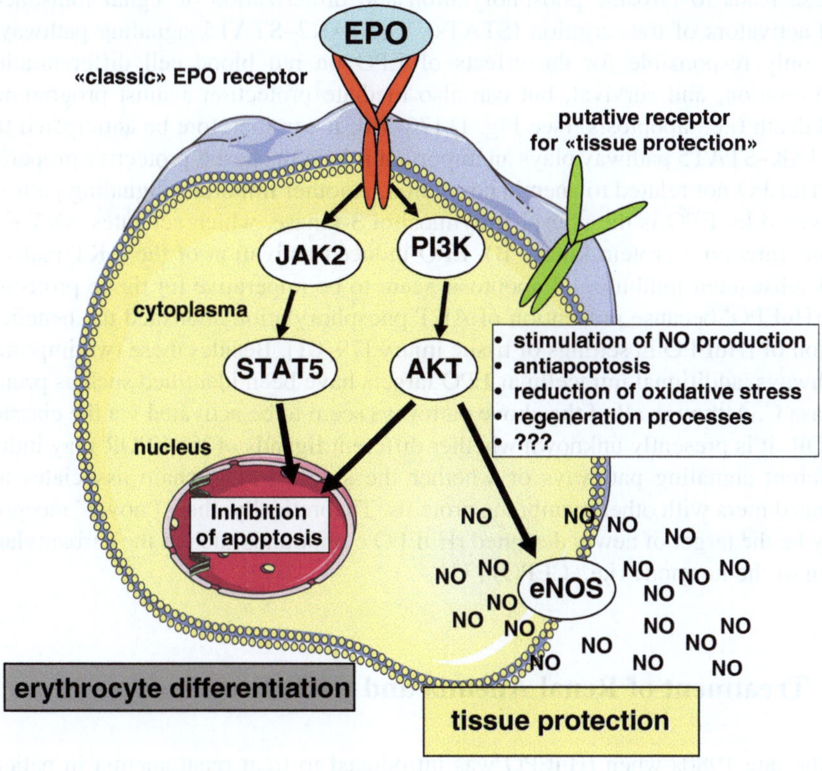

Fig. 1 Cellular and molecular actions of erythropoietin (EPO). *NO* nitric oxide; *STAT* signal transducers and activators of transcription; *eNOS* endothelial NO synthase

intracellular pathways. Thus, it may depend on the extracellular binding site of EPO, from which receptor conformation is achieved and through which the signaling pathway is subsequently activated [70, 71]. The tissue-protective effects of EPO appear to require the expression of CD131 and a low-affinity, heterodimeric EPOR–CD131 receptor [72], which might exert a different signaling behavior than the "classical" EPOR [73]. In addition, different classes of EPOR with respect to receptor affinity for EPO binding have been described: high (KD90–900 pM) and low (KD20–9,000 pM) affinity EPOR [74, 75]. The former mediate the well-known hematopoietic effects, whereas the latter seem to be involved in tissue protection by EPO. Collectively, these data indicate that the interaction of EPO with its receptor may be far more complicated than previously believed.

4.2 Molecular Actions of EPO

The activated EPOR exhibits more than 40 binding sites, and a pivotal molecule that induces intracellular signaling is JAK2 tyrosine kinase. Activation of this kinase leads to tyrosine phosphorylation and dimerization of signal transducers and activators of transcription (STATs). The JAK2–STAT5 signaling pathway is not only responsible for the effects of EPO on red blood cell differentiation, proliferation, and survival, but can also mediate protection against programmed cell death (i.e., apoptosis) (see Fig. 1) [76–78]. It can therefore be anticipated that the JAK–STAT5 pathway plays an important role in the tissue-protective properties of rHuEPO not related to anemia correction. Another important signaling pathway triggered by EPO is the phosphatidylinositol 3-kinase, which activates AKT (i.e., serine/threonine protein kinase B). EPO-induced activation of the AKT pathway and subsequent inhibition of apoptosis seem to be imperative for tissue protection by rHuEPO, because prevention of AKT phosphorylation abolished the beneficial action of rHuEPO in settings of tissue injury [79–81]. Besides these two important pathways, additional intracellular EPO targets have been identified such as protein kinase C. Although all of the above pathways seem to be activated via the classical EPOR, it is presently unknown whether different ligands of the EPOR may induce different signaling pathways or whether the single EPOR chain associates and forms dimers with other membrane proteins. Theoretically, these "novel" receptors may be the target of newly designed rHuEPO compounds such as the carbamylated form of the hormone (i.e., CEPO) [73].

5 Treatment of Renal Anemia and CKD Progression

In the late 1980s when rHuEPO was introduced to treat renal anemia in patients with CKD who were not on renal replacement therapy (RRT), a major concern was that it could accelerate progression. This fear was nurtured by the observation of

Garcia et al. [82] who found pronounced tissue injury and faster progression in the classical 5/6 nephrectomy model in the rat (remnant kidney model) when, in anemic animals, an overcorrection of hemoglobin levels was achieved with rHuEPO. The fact that ameliorating anemia does not accelerate the progression of CKD in humans has been shown later in clinical studies and it is no longer questioned [83]. Nevertheless, the question of whether correction of anemia with rHuEPO slows the progression remained unanswered for more than a decade. Results of several retrospective as well as smaller prospective clinical studies supported the hypothesis that treatment of anemia in CKD patients with rHuEPO favorably influences progression [84–90], but some trials did not confirm this (Table 1)

Table 1 Clinical studies on chronic kidney disease (CKD) progression with recombinant human erythropoietin (rHuEPO)

References	Type and duration of the study	Number of patients	Primary (P) or secondary (S) study parameter and outcome
[84]	Multicenter/prospective/randomized/open label/48 weeks	83	Glomerular filtration rate (GFR) with iothalamate clearance (P) *difference between treatments*
[85]	Monocenter/prospective/randomized/open label	73	Doubling of serum-creatinine (P) *retardation of progression*
[86]	Monocenter/retrospective	63	Creatinine clearance *retardation of progression*
[87]	Monocenter/retrospective	18	1/Serum creatinine *retardation of progression*
[88]	Monocenter/prospective/randomized/early vs. late treatment/22.5 months	88	Doubling of serum creatinine or hemodialysis (P) *retardation of progression*
[89]	Monocenter/prospective/randomized/high vs. low hemoglobin/36 months	241	GFR with iohexol clearance (P) *difference between treatment arms*
[91]	Monocenter/prospective/randomized/high vs. low hemoglobin	19	GFR with iohexol clearance (P) *retardation of progression*
[92]	Multicenter/prospective/randomized/normal vs. subnormal hemoglobin/48–76 weeks	416	Serum creatinine (S) *no effect*
[93]	Monocenter/prospective/randomized/high vs. low hemoglobin/24 months	155	GFR with iothalamate clearance (S) *no effect*
[94]	Multicenter/prospective/randomized/high vs. low hemoglobin	172	Serum creatinine (S) *no effect*
[95]	Multicenter/prospective/randomized/high vs. low hemoglobin/36 months	603	Estimated glomerular filtration rate (eGFR) or hemodialysis (S) *no effect*
[96]	Multicenter/prospective/randomized/open label/high vs. low hemoglobin/16 months	1,432	Doubling of serum creatinine or hemodialysis (S) *no effect*

[91–93]. Data from recently published larger trials with rHuEPO in patients with CKD and anemia (e.g., the ACORD [Anemia Correction in Diabetes], CREATE [Cardiovascular risk Reduction by Early Anemia Treatment with Epoetin], and CHOIR [Correction of Hemoglobin Outcomes in Renal Insufficiency] studies) revealed no beneficial effect on progression [94–96].

For example, Ritz et al. [94] have studied 172 patients with type 1 and type 2 diabetes mellitus and nephropathy (ACORD study). Patients had mild to moderate anemia were randomly assigned to attain a target hemoglobin level of either 13–15 g/dL or 10.5–11.5 g/dL. The primary endpoint was a change in left ventricular mass, and secondary endpoints included renal function, quality of life, and safety. At study's end, mean Hb levels were 13.5 and 12.1 g/dL, respectively ($p < 0.001$), but no significant differences were observed in median left ventricular mass index between study groups. Furthermore, anemia correction had no effect on the rate of decrease in creatinine clearance, whereas it resulted in significantly improved quality of life in the group with higher hemoglobin levels ($p = 0.04$). In the CREATE trial [95] 603 patients with an estimated glomerular filtration rate (eGFR) of 15.0–35.0 mL/min/1.73 m^2, and mild-to-moderate anemia (hemoglobin level 11.0–12.5 g/dL) were randomized to a target hemoglobin value in the normal range (13.0–15.0 g/dL) or in the subnormal range (10.5–11.5 g/dL). Treatment with rHuEPO was initiated either at randomization (group 1) or only after the hemoglobin level fell below 10.5 g/dL (group 2). The primary endpoint was a composite of eight cardiovascular events, and secondary endpoints included quality-of-life scores and CKD progression. During the 3-year study, complete correction of anemia did not affect the likelihood of a first cardiovascular event ($p = 0.20$), and the decrease in mean eGFR was comparable in both groups (3.6 and 3.1 mL/min/year; $p = 0.40$). In addition, dialysis was required in more patients in group 1 than in group 2 (127 vs. 111; $p = 0.03$). However, general health and physical function was significantly better in the group with hemoglobin normalization, despite hypertensive episodes and headaches being more prevalent.

Finally, in the open-label CHOIR trial [96] 1,432 patients with CKD were studied, 715 of whom were randomly assigned to receive a dose of rHuEPO targeted to achieve a hemoglobin level of 13.5 g/dL and 717 of whom were assigned to receive a dose targeted to achieve a level of 11.3 g/dL. During a median follow-up of 16 months 222 composite events (death, myocardial infarction, hospitalization for congestive heart failure, and stroke) occurred: 125 events in the high-hemoglobin group, as compared with 97 events in the low-hemoglobin group (hazard ratio, 1.34; 95% confidence interval, 1.03–1.74; $p = 0.03$). In addition, no difference was observed with respect to progression of CKD.

In the above studies, CKD progression was a secondary endpoint, and (almost) complete correction of anemia was accomplished within weeks using high doses of rHuEPO. For example, the initial weekly starting dose in the CHOIR study was 10,000 IU of rHuEPO [96]. Such an abrupt increase in the hematocrit level in patients with vascular problems might have mitigated putative beneficial effects of rHuEPO on the kidney, particularly if one assumes that much lower doses could be adequate for tissue protection (discussed next). Indeed, reanalysis of the CHOIR data indicated that a beneficial effect of rHuEPO on cardiovascular outcome is present only in patient who responded to the therapy, irrespective of the randomization to high or low hemoglobin target levels [97]. It is not clear whether this also holds true for renal endpoints.

6 Other Actions of rHuEPO and Its Derivates to Prevent Renal Tissue Injury

Several recent studies investigated the ability of rHuEPO to modulate organ function and cellular responses to diverse types of injury. Thus, in addition to its essential role in the regulation of erythropoiesis, EPO signaling, activated either by exogenous EPO or by endogenous EPO in an autocrine or paracrine fashion, has emerged as a major tissue-protective survival factor in various nonhematopoietic organs [98].

6.1 Tissue Protection by Nonhematologic Effects of rHuEPO

Recent in vivo data have revealed that the antiapoptotic effects of EPO can also be observed with other cell types that express EPOR, such as neuronal cells, and that systemic administration of rHuEPO can decrease the consequences of various neurological experimental injuries [99, 100]. In vitro, rHuEPO has also been shown to protect endothelial cells or vascular smooth muscle cells against apoptosis [79, 101]. One can thus speculate that this effect also exists for other cells that express EPOR, such as proximal tubular cells and renal endothelial cells. Furthermore, in vitro and in vivo experiments suggest that EPO may also have pro-angiogenic properties [102, 103]. The antiapoptotic effects of EPO could also be beneficial for the progression of CKD because apoptosis has been implicated in the progressive loss of tubular cells observed during CKD [104]. For example, apoptosis appears to play an important role in the progression of tubular lesions observed in rats submitted to subtotal nephrectomy or to experimental antiglomerular basement membrane nephritis [105]. The mechanisms underlying the increased apoptosis of tubular cells are still poorly understood, but ROS could play a role in this process. Thus, treatment with rHuEPO may have beneficial effects not only by protecting tubular cells against apoptosis but also by decreasing the production of ROS (i.e., protection against oxidative stress). The links between oxidative stress and anemia come mostly from the fact that erythrocytes represent a major antioxidant component of the blood [106]. Their antioxidant effects are mediated through the glutathione system, enzymes such as superoxide dismutase or catalase, and cellular proteins that are devoid of enzymatic activity but can react with ROS, such as the low-molecular weight proteins of the erythrocyte membrane, vitamin E, or coenzyme Q. Furthermore, erythrocytes can regenerate consumed redox equivalents through the pentose phosphate pathway and through reduction of oxidized glutathione by glutathione reductase. In addition to increasing the number of red blood cells, EPO may also reduce oxidative stress by increasing the antioxidant potential of erythrocytes. Experimental data have shown that the binding of EPO to its receptor activates nuclear factor κB (NFκB), which in turn enhances the expression of genes encoding proteins, such as superoxide dismutase or glutathione, that have antioxidant properties [107].

7 Renoprotection by Nonhematologic Effects of rHuEPO

As previously outlined, hypoxia of renal tissue appears to be an important factor for interstitial fibrosis. Thus, preventing hypoxia with EPO should decrease tissue injury and ultimately protect against nephron loss. Since EPO ameliorates cell and tissue hypoxia by direct mechanisms that are unrelated to its hematological action, renoprotection could be achieved with lower doses of rHuEPO than those used for treatment of anemia [108, 109].

We used a hematologically noneffective dose of the long-acting rHuEPO analog darbepoetin (i.e., a dose that did not affect hematocrit levels in treated animals) in the established 5/6 nephrectomy remnant kidney model in the rat [110]. This model features progressive injury to the renal microvascular endothelium, leading to glomerular sclerosis and ischemia-induced tubulointerstitial damage. We could demonstrate that chronic treatment with darbepoetin conferred renal vascular and tissue protection, preserved renal function, and improved survival in this experimental setting of chronic renal ischemia (Fig. 2). However, we could also document that escalating doses of darbepoetin mitigate the protective effects on the remnant kidney tissue and even worsen glomerulosclerosis [111].

We and others recently expanded these findings in experimental models of diabetic, toxic, and immunologic kidney injury [112–116]. Chronic administration of C.E.R.A. had beneficial dose-dependent effects on molecular pathways of diabetic kidney damage and on loss of podocyte in vivo [112, 113]. However, only the nonhematologically effective (low) dose was also clinically renoprotective, whereas high-dose C.E.R.A. aggravated albuminuria in this experimental setting, despite clear-cut beneficial molecular effects. Interestingly, phlebotomy in high-dose C.E.R.A.-treated mice preserved its tissue-protective effect. Eto et al. [114] examined the mechanisms by which darbepoetin confers renoprotection in a puromycin aminonucleoside-induced model of nephrotic syndrome in the rat. They found that the decrease in proteinuria was correlated with the immunohistochemical disappearance of the podocyte injury markers desmin and the immune costimulator molecule B7.1, with the reappearance of nephrin expression in the slit diaphragm. Podocyte foot process retraction and effacement along with actin filament rearrangement were reversed by darbepoetin treatment. Furthermore, puromycin treatment of rat podocytes in culture caused actin cytoskeletal reorganization along

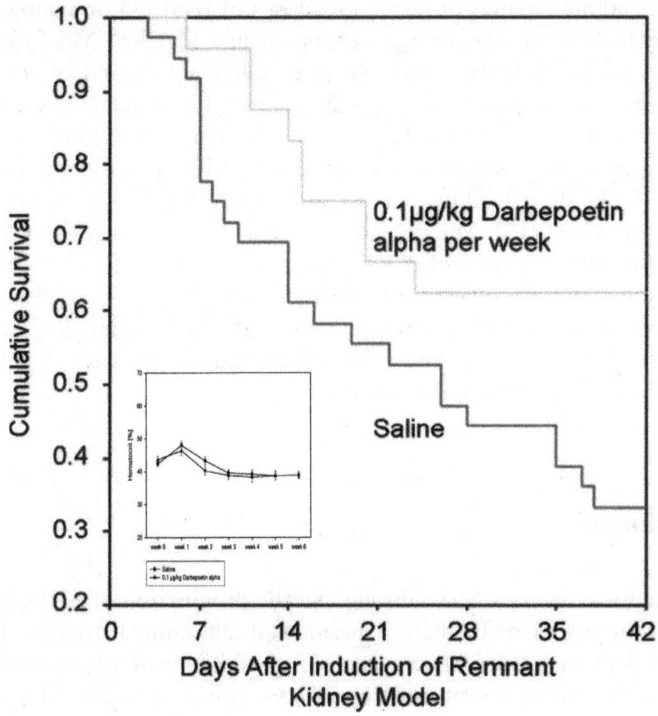

Fig. 2 Kaplan–Meier survival curves of remnant kidney model rats treated with either saline or 0.1 µg/kg body weight darbepoetin. Survival of darbepoetin-treated animals was significantly better ($p < 0.05$) than in the saline-treated group. Note that hematocrit values were not different between groups over time (*inset*)

with deranged nephrin distribution. All these effects in vitro were reversed by darbepoetin. The same authors also investigated the role of rHuEPO in the regulation of heme oxygenase-1, an antioxidative stress protein in an animal model of chronic tubulointerstitial injury [115]. Treatment with rHuEPO reduced proteinuria and renal injury, including peritubular capillaries rarefaction. This renoprotection was associated with upregulation of heme oxygenase-1 in kidney tissue. Finally, Logar et al. [116] tested the effects of darbepoetin in preventing podocyte apoptosis in cultured immortalized mouse podocytes treated with low-dose ultraviolet-C irradiation to induce apoptosis. Darbepoetin pretreatment significantly reduced podocyte apoptosis, with this effect involving intact JAK2 and AKT signaling pathways. Moreover, in mice with antiglomerular antibody–induced glomerulonephritis, chronic darbepoetin treatment significantly reduced podocyte apoptosis, glomerulosclerosis, and proteinuria.

Collectively, these observations could be of considerable clinical relevance, because “low dose” rHuEPO treatment may be a safe strategy to avoid potential adverse effects of high-dose therapy (i.e., doses that cause a large increase in hematocrit with accompanying changes in rheology and activate thrombocytes).

Thus, earlier administration of rather low doses of rHuEPO or analogs may be a feasible way to limit renal tissue damage in patients with CKD. Moreover, rHuEPO analogs that maintain tissue-protective effects but are devoid of the action on erythropoiesis (and thrombopoiesis?) may represent a valuable alternative. Such molecules like the carbamylated form of the hormone (CEPO) have already been tested in experimental studies, revealing tissue-protective properties comparable with that of classic rHuEPO, but without any effect on hematocrit or pro-coagulative activity [117]. The potential of such rHuEPO analogs to prevent loss of renal tissue in patients with CKD awaits further investigation, however. In addition, better characterization of the nonerythropoietic biological effects of EPO and understanding of the mechanisms of EPO–EPOR signaling activation in nonhematopoietic organs and cell types are critical to the future development of novel applications for EPO and its derivatives.

8 Conclusion

Renal tissue hypoxia does occur throughout the progression of CKD, as has been shown by a variety of methods in experimental and clinical studies. The kidney responds to declining ambient oxygenation by the activation of genes providing adaptation and endurance, such as EPO. Since anemia was found to be an independent predictor not only of cardiovascular morbidity but also of progression in patients with CKD, treatment with rHuEPO was expected not only to improve clinical symptoms of anemia, but also to reduce cardiovascular sequel and progression in this population. Indeed, data from epidemiologic studies suggest that progression in CKD patients is faster with lower hemoglobin levels, but so far intervention studies did not reveal a definite beneficial effect of rHuEPO treatment on progression. It is conceivable that the development of renal anemia in CKD patients is accompanied by renal tissue hypoxia and progression, but these two conditions may not be as closely related as suggested. Thus, in clinical studies treatment of anemia with rHuEPO may not have the same outcome as treatment of renal tissue hypoxia.

References

1. Brenner BM, Meyer TW, Hostetter TH (1982) Dietary protein intake and the progressive nature of kidney disease: the role of hemodynamically mediated glomerular injury in the pathogenesis of progressive glomerular sclerosis in aging, renal ablation, and intrinsic renal disease. N Engl J Med 307:652–659
2. Bohle A, Müller GA, Wehrmann M et al (1996) Pathogenesis of chronic renal failure in the primary glomerulopathies, renal vasculopathies, and chronic interstitial nephritides. Kidney Int 49[Suppl 54]:S2–S9

3. Remuzzi G, Bertani T (1998) Pathophysiology of progressive nephropathies. N Engl J Med 339:1448–1456
4. De Heer E, Sijpkens YW, Verkade M et al (2000) Morphometry of interstitial fibrosis. Nephrol Dial Transplant 15[Suppl 6]:72–73
5. Brezis M, Rosen S (1995) Hypoxia of the renal medulla – its implications for disease. N Engl J Med 332:647–655
6. Fine LG, Bandyopadhay D, Norman JT (2000) Is there a common mechanism for the progression of different types of renal diseases other than proteinuria? Towards the unifying theme of chronic hypoxia. Kidney Int 57[Suppl 75]:S22–S26
7. Kang DH, Kanellis J, Hugo C et al (2002) Role of the microvascular endothelium in progressive renal disease. J Am Soc Nephrol 13:806–816
8. Seron D, Alexopulos E, Raftery MJ et al (1990) Number of interstitial capillary cross-section assessed by monoclonal antibodies: relation to interstitial damage. Nephrol Dial Transplant 5:889–893
9. Orphanides C, Fine LG, Norman JT (1997) Hypoxia stimulates proximal tubular cell matrix production via a TGF-beta1-independent mechanism. Kidney Int 52:637–647
10. Morigi M, Macconi D, Zoja C et al (2002) Protein overload-induced NF-kappaβ activation in proximal tubular cells requires H_2O_2 through a PKC-dependent pathway. J Am Soc Nephrol 13:1179–1189
11. Nath KA, Grande J, Croatt A et al (1998) Redox regulation of renal DNA synthesis, transforming growth factor-beta 1 and collagen gene expression. Kidney Int 53:367–381
12. Norman JT, Clark IM, Garcia PL (2000) Hypoxia promotes fibrogenesis in human renal fibroblasts. Kidney Int 58:2351–2366
13. Heyman SN, Khamaisi M, Rosen S et al (2008) Renal parenchymal hypoxia, hypoxia response and the progression of chronic kidney disease. Am J Nephrol 28:998–1006
14. Pedersen M, Dissing TH, Morkenborg J et al (2005) Validation of quantitative BOLD MRI measurements in kidney: application to unilateral ureteral obstruction. Kidney Int 67:2305–2312
15. Welch WJ, Baumgartl H, Lubbers D et al (2001) Nephron pO_2 and renal oxygen usage in the hypertensive rat kidney. Kidney Int 59:230–237
16. Haase VH (2006) Hypoxia-inducible factors in the kidney. Am J Physiol Renal Physiol 291: F271–F281
17. Rosenberger C, Rosen S, Heyman S (2005) Current understanding of HIF in renal disease. Kidney Blood Press Res 28:325–340
18. Rosenberger C, Shina A, Rosen S et al (2006) Hypoxia inducible factors and tubular cell survival in isolated perfused kidneys. Kidney Int 70:60–70
19. Rosenberger C, Goldfarb M, Khamaisi M et al (2008) Acute kidney injury in the diabetic rat: studies in the isolated perfused and intact kidney. Am J Nephrol 28:831–839
20. Rosenberger C, Mandriota S, Jurgensen JS et al (2002) Expression of hypoxia-inducible factor-1alpha and -2alpha in hypoxic and ischemic rat kidneys. J Am Soc Nephrol 13:1721–1732
21. Rosenberger C, Heyman SN, Rosen S et al (2005) Up-regulation of HIF in experimental acute renal failure: evidence for a protective transcriptional response to hypoxia. Kidney Int 67:531–542
22. Manotham K, Ongvilawan B, Urusopone P et al (2006) Intra-renal hypoxia in CKD patients, a BOLD MRI study. J Am Soc Nephrol 17:164A
23. Norman JT, Stidwill R, Singer M et al (2003) Angiotensin II blockade augments renal cortical microvascular pO_2 indicating a novel, potentially renoprotective action. Nephron Physiol 94:39–46
24. Textor SC, Glockner JF, Lerman LO et al (2008) The use of magnetic resonance to evaluate tissue oxygenation in renal artery stenosis. J Am Soc Nephrol 19:780–788

25. Basile DP, Donohoe D, Roethe K et al (2001) Renal ischemic injury results in permanent damage to peritubular capillaries and influences long-term function. Am J Physiol Renal Physiol 281:F887–F899
26. Matsumoto M, Tanaka T, Yamamoto T et al (2004) Hypoperfusion of peritubular capillaries induces chronic hypoxia before progression of tubulointerstitial injury in a progressive model of rat glomerulonephritis. J Am Soc Nephrol 15:1574–1581
27. Manotham K, Tanaka T, Matsumoto M et al (2004) Evidence of tubular hypoxia in the early phase of the remnant kidney model. J Am Soc Nephrol 15:1277–1288
28. Goldfarb M, Rosenberger C, Abassi Z et al (2006) Acute-on-chronic renal failure in the rat: functional compensation and hypoxia tolerance. Am J Nephrol 26:22–33
29. Kairaitis LK, Wang Y, Gassman M et al (2005) HIF-1α expression follows microvascular loss in advanced murine adriamycin nephrosis. Am J Physiol Renal Physiol 288:F198–F206
30. Zhong Z, Arteel GE, Connor HD et al (1998) Cyclosporin A increases hypoxia and free radical production by rat kidneys: prevention by dietary glycine. Am J Physiol 275: F595–F604
31. Yuan H-T, Li X-Z, Pitera JE et al (2003) Peritubular capillary loss after mouse acute nephrotoxicity correlates with down-regulation of vascular endothelial growth factor A and hypoxia-inducible factor-1alpha. Am J Pathol 163:2289–2301
32. Rosenberger C, Pratschke J, Rudolph B et al (2007) Immunohistochemical detection of hypoxia-inducible factor-1alpha in human renal allograft biopsies. J Am Soc Nephrol 18:343–351
33. Sadowski EA, Fain SB, Alford SK et al (2005) Assessment of acute renal transplant rejection with blood oxygen level-dependent MR imaging: initial experience. Radiology 236:911–919
34. Epstein FH, Prasad P (2000) Effects of furosemide on medullary oxygenation in younger and older subjects. Kidney Int 57:2080–2083
35. Tanaka T, Kato H, Kojima I et al (2006) Hypoxia and expression of hypoxia-inducible factor in the aging kidney. J Gerontol A Biol Sci Med Sci 61:795–805
36. Adler S, Huang H, Wolin MS et al (2004) Oxidant stress leads to impaired regulation of renal cortical oxygen consumption by nitric oxide in the aging kidney. J Am Soc Nephrol 15:52–60
37. Palm F, Cederberg J, Hansell P (2003) Reactive oxygen species cause diabetes-induced decrease in renal oxygen tension. Diabetologia 46:1153–1160
38. Ries M, Basseau F, Tyndal B et al (2003) Renal diffusion and BOLD MRI in experimental diabetic nephropathy. Blood oxygen leveled pendent. J Magn Reson Imaging 17:104–113
39. Epstein FH, Veves A, Prasad PV (2002) Effect of diabetes on renal medullary oxygenation during water diuresis. Diabetes Care 25:575–578
40. Rosenberger C, Khamaisi M, Abassi Z et al (2008) Adaptation to hypoxia in the diabetic rat kidney. Kidney Int 73:34–42
41. Katavetin P, Miyata T, Inagi R et al (2006) High glucose blunts vascular endothelial growth factor response to hypoxia via the oxidative stress-regulated hypoxia-inducible factor/hypoxia-responsible element pathway. J Am Soc Nephrol 17:1405–1413
42. Palm F, Friederich M, Carlsson PO et al (2008) Reduced nitric oxide in diabetic kidneys due to increased hepatic arginine metabolism: implications for renomedullary oxygen availability. Am J Physiol Renal Physiol 294:F30–F37
43. Welch WJ (2006) Intra-renal oxygen and hypertension. Clin Exp Pharmacol Physiol 33:1002–1005
44. Stillman LE, Brezis M, Heyman SN et al (1994) Effects of salt depletion on the kidney: changes in medullary oxygenation and thick ascending limb size. J Am Soc Nephrol 4:1538–1545
45. Nangaku M (2006) Chronic hypoxia and tubulo-interstitial injury: a final common pathway to end-stage renal failure. J Am Soc Nephrol 17:17–25
46. Johannes T, Mik EG, Nohe B et al (2007) Acute decrease in renal microvascular pO_2 during acute normovolemic hemodilution. Am J Physiol Renal Physiol 292:F796–F803

47. Rossert J, Fouqueray B, Boffa JJ (2003) Anemia management and delay of chronic renal failure. J Am Soc Nephrol 14:S173–S177
48. Mohanram A, Zhang Z, Shahinfar S et al (2004) Anemia and end-stage renal disease in patients with type 2 diabetes and nephropathy. Kidney Int 66:1183–1189
49. Ravani P, Tripepi G, Malberti F et al (2005) Asymmetrical dimethylarginine predicts progression to dialysis and death in patients with chronic kidney disease: a competing risks modeling approach. J Am Soc Nephrol 16:2449–2455
50. Furth SL, Cole SR, Fadrowski JJ et al; Council of Pediatric Nephrology and Urology, New York/New Jersey; Kidney and Urology Foundation of America (2007) The association of anemia and hypoalbuminemia with accelerated decline in GFR among adolescents with chronic kidney disease. Pediatr Nephrol 22:265–271
51. Johnson ES, Thorp ML, Platt RW et al (2008) Predicting the risk of dialysis and transplant among patients with CKD: a retrospective cohort study. Am J Kindey Dis 52:653–660
52. Zanjani ED, Poster J, Burlington H et al (1977) Liver as the primary site of erythropoietin formation in the fetus. J Lab Clin Med 89:640–644
53. Rankin EB, Biju MP, Liu Q et al (2007) Hypoxia-inducible factor-2 (HIF-2) regulates hepatic erythropoietin in vivo. J Clin Invest 117:1068–1077
54. Juul SE, Yachnis AT, Rojiani AM et al (1999) Immunohistochemical localization of erythropoietin and its receptor in the developing human brain. Pediatr Dev Pathol 2:148–158
55. Masuda S, Kobayashi T, Chikuma M et al (2000) The oviduct produces erythropoietin in an estrogen- and oxygen-dependent manner. Am J Physiol Endocrinol Metab 278: E1038–E1044
56. Magnanti M, Gandini O, Giuliani L et al (2001) Erythropoietin expression in primary rat Sertoli and peritubular myoid cells. Blood 98:2872–2874
57. Conrad KP, Benyo DF, Westerhausen-Larsen A et al (1996) Expression of erythropoietin by the human placenta. FASEB J 10:760–768
58. Vogt C, Pentz S, Rich IN (1989) A role for the macrophage in normal hemopoiesis: III. In vitro and in vivo erythropoietin gene expression in macrophages detected by in situ hybridization. Exp Hematol 17:391–397
59. Stopka T, Zivny JH, Stopkova P, Prchal JF, Prchal JT (1998) Human hematopoietic progenitors express erythropoietin. Blood 91:3766–3772
60. Sato T, Maekawa T, Watanabe S et al (2000) Erythroid progenitors differentiate and mature in response to endogenous erythropoietin. J Clin Invest 106:263–270
61. Liu C, Shen K, Liu Z et al (1997) Regulated human erythropoietin receptor expression in mouse brain. J Biol Chem 272:32395–32400
62. Grimm C, Wenzel A, Groszer M et al (2002) HIF-1-induced erythropoietin in the hypoxic retina protects against light-induced retinal degeneration. Nat Med 8:718–724
63. Wu H, Lee SH, Gao J et al (1999) Inactivation of erythropoietin leads to defects in cardiac morphogenesis. Development 126:3597–3605
64. Westenfelder C, Biddle DL, Baranowski RL (1999) Human, rat, and mouse kidney cells express functional erythropoietin receptors. Kidney Int 55:808–820
65. Ammarguellat F, Gogusev J, Drueke TB (1996) Direct effect of erythropoietin on rat vascular smooth-muscle cell via a putative erythropoietin receptor. Nephrol Dial Transplant 11:687–692
66. Ogilvie M, Yu X, Nicolas-Metral V et al (2000) Erythropoietin stimulates proliferation and interferes with differentiation of myoblasts. J Biol Chem 275:39754–39761
67. Anagnostou A, Liu Z, Steiner M et al (1994) Erythropoietin receptor mRNA expression in human endothelial cells. Proc Natl Acad Sci U S A 91:3974–3978
68. Kertesz N, Wu J, Chen TH et al (2004) The role of erythropoietin in regulating angiogenesis. Dev Biol 276:101–110
69. Gross AW, Lodish HF (2006) Cellular trafficking and degradation of erythropoietin and novel erythropoiesis stimulating protein (NESP). J Biol Chem 281:2024–2032

70. Frank SJ (2002) Receptor dimerization in GH and erythropoietin action – it takes two to tango, but how? Endocrinology 143:2–10
71. McCaffery PJ, Fraser JK, Lin FK (1989) Subunit structure of the erythropoietin receptor. J Biol Chem 264:10507–10512
72. Brines M, Grasso G, Fiordaliso F et al (2004) Erythropoietin mediates tissue protection through an erythropoietin and common beta-subunit heteroreceptor. Proc Natl Acad Sci U S A 101:14907–14912
73. Leist M, Ghezzi P, Grasso G et al (2004) Derivatives of erythropoietin that are tissue protective but not erythropoietic. Science 305:239–242
74. Sawada K, Krantz SB, Sawyer ST et al (1998) Quantitation of specific binding of erythropoietin to human erythroid colony-forming cells. J Cell Physiol 137:337–345
75. Fraser JK, Lin FK, Berridge MV (1988) Expression of high affinity receptors for erythropoietin on human bone marrow cells and on the human erythroleukemic cell line, HEL. Exp Hematol 16:836–842
76. Byts N, Samoylenko A, Fasshauer T et al (2008) Essential role for STAT5 in the neurotrophic but not in the neuroprotective effect of erythropoietin. Cell Death Differ 15:783–792
77. Socolovsky M, Fallon AE, Wang S et al (1999) Fetal anemia and apoptosis of red cell progenitors in Stat5a−/−5b−/− mice: a direct role for STAT5 in Bcl-X(L) induction. Cell 98:181–191
78. Haseyama Y, Sawada K, Oda A et al (1999) Phosphatidylinositol 3-kinase is involved in the protection of primary cultured human erythroid precursor cells from apoptosis. Blood 94:1568–1577
79. Akimoto T, Kusano E, Inaba T et al (2000) Erythropoietin regulates vascular smooth muscle cell apoptosis by a phosphatidylinositol 3 kinase-dependent pathway. Kidney Int 58:269–282
80. Um M, Lodish HF (2006) Antiapoptotic effects of erythropoietin in differentiated neuroblastoma SH-SY5Y cells require activation of both the STAT5 and AKT signaling pathways. J Biol Chem 281:5648–5656
81. Um M, Gross AW, Lodish HF (2007) A “classical” homodimeric erythropoietin receptor is essential for the antiapoptotic effects of erythropoietin on differentiated neuroblastoma SH-SY5Y and pheochromocytoma PC-12 cells. Cell Signal 19:634–645
82. Garcia DL, Anderson S, Rennke HG et al (1988) Anemia lessens and its prevention with recombinant human erythropoietin worsens glomerular injury and hypertension in rats with reduced renal mass. Proc Natl Acad Sci U S A 85:6142–6146
83. Rossert J, Froissart M (2006) Role of anemia in progression of chronic kidney disease. Semin Nephrol 26:283–289
84. Roth D, Smith RD, Schulman G et al (1994) Effects of recombinant human erythropoietin on renal function in chronic renal failure predialysis patients. Am J Kidney Dis 24:777–784
85. Kuriyama S, Tomonari H, Yoshida H et al (1997) Reversal of anemia by erythropoietin therapy retards the progression of chronic renal failure, especially in nondiabetic patients. Nephron 77:176–185
86. Jungers P, Choukroun G, Oualim Z et al (2001) Beneficial influence of recombinant human erythropoietin therapy on the rate of progression of chronic renal failure in predialysis patients. Nephrol Dial Transplant 16:307–312
87. Tapolyai M, Kadomatsu S, Perera-Chong M (2003) rhu-erythropoietin (EPO) treatment of pre-ESRD patients slows the rate of progression of renal decline. BMC Nephrol 4:3
88. Gouva C, Nikolopoulos P, Ioannidis JP et al (2004) Treating anemia early in renal failure patients slows the decline of renal function: a randomized controlled trial. Kidney Int 66:753–760
89. Rossert J, Levin A, Roger SD et al (2006) Effect of early correction of anemia on the progression of CKD. Am J Kidney Dis 47:738–750
90. Furumatsu Y, Nagasawa Y, Hamano T et al (2008) Integrated therapies including erythropoietin decrease the incidence of dialysis: lessons from mapping the incidence of end-stage renal disease in Japan. Nephrol Dial Transplant 23:984–990

91. Torffvit O (2009) Effect of epoetin induced normal Hb on progression of nephropathy in insulin treated diabetic patients with renal failure. J Nephrol Urol 1:113–123
92. Furulund H, Linde T, Ahlmen J et al (2003) A randomized controlled trial of haemoglobin normalization with epoetin alpha in pre-dialysis and dialysis patients. Nephrol Dial Transplant 18:353–361
93. Roger SD, McMahon LP, Clarkson A et al (2004) Effects of early and late intervention with epoetin alpha on left ventricular mass among patients with chronic kidney disease: results of a randomized clinical trial. J Am Soc Nephrol 15:148–156
94. Ritz E, Laville M, Bilous RW et al; Anemia Correction in Diabetes Study Investigators (2007) Target level for hemoglobin correction in patients with diabetes and CKD: primary results of the Anemia Correction in Diabetes (ACORD) study. Am J Kidney Dis 49:194–207
95. Drueke TB, Locatelli F, Clyne N et al (2006) Normalization of hemoglobin level in patients with chronic kidney disease and anemia. N Engl J Med 355:2071–2084
96. Singh AK, Szczech L, Tang KL et al (2006) Correction of anemia with epoetin alfa in chronic kidney disease. N Engl J Med 355:2085–2098
97. Szczech LA, Barnhart HX, Inrig JK et al (2008) Secondary analysis of the CHOIR trial epoetin-alpha dose and achieved hemoglobin outcomes. Kidney Int 74:791–798
98. Fliser D, De Groot K, Bahlmann FH et al (2006) Mechanisms of disease: erythropoietin – an old hormone with a new mission? Nat Clin Pract Cardiovasc Med 3:563–572
99. Brines ML, Ghezzi P, Keenan S et al (2000) Erythropoietin crosses the blood–brain barrier to protect against experimental brain injury. Proc Natl Acad Sci U S A 97:10526–10531
100. Celik M, Gokmen N, Erbayraktar S et al (2002) Erythropoietin prevents motor neuron apoptosis and neurologic disability in experimental spinal cord ischemic injury. Proc Natl Acad Sci U S A 99:2258–2263
101. Carlini RG, Alonzo EJ, Dominguez J et al (1999) Effect of recombinant human erythropoietin on endothelial cell apoptosis. Kidney Int 55:546–553
102. Ashley RA, Dubuque SH, Dvorak B et al (2002) Erythropoietin stimulates vasculogenesis in neonatal rat mesenteric microvascular endothelial cells. Pediatr Res 51:472–478
103. Ribatti D, Presta M, Vacca A et al (1999) Human erythropoietin induces a pro-angiogenic phenotype in cultured endothelial cells and stimulates neovascularization in vivo. Blood 93:2627–2636
104. Yang B, Johnson TS, Thomas GL et al (2001) Expression of apoptosis-related genes and proteins in experimental chronic renal scarring. J Am Soc Nephrol 12:275–288
105. Yang B, Johnson TS, Thomas GL et al (2001) Apoptosis and caspase-3 in experimental anti-glomerular basement membrane nephritis. J Am Soc Nephrol 12:485–495
106. Grune T, Sommerburg O, Siems WG (2000) Oxidative stress in anemia. Clin Nephrol 53 [Suppl 1]:S18–S22
107. Digicaylioglu M, Lipton SA (2001) Erythropoietin-mediated neuroprotection involves cross-talk between JAK2 and NF-kappaB signalling cascades. Nature 412:641–647
108. Bernhardt W, Câmpean V, Kany S et al (2006) Preconditional activation of hypoxia-inducible factors ameliorates ischemic acute renal failure. J Am Soc Nephrol 17:1970–1978
109. Bahlman FH, Fliser D (2009) Erythropoetin and renoprotection. Curr Opin Nephrol Hypertens 18:15–20
110. Bahlmann FH, Song R, Boehm SM et al (2004) Low-dose therapy with the long-acting erythropoietin analogue darbepoetin alpha persistently activates endothelial AKT and attenuates progressive organ failure. Circulation 110:1006–1012
111. Fliser D, Bahlmann FH, Haller H (2006) EPO: renoprotection beyond anemia correction. Pediatr Nephrol 21:1785–1789
112. Menne J, Park JK, Shushakova N et al (2007) The continuous erythropoietin receptor activator affects different pathways of diabetic renal injury. J Am Soc Nephrol 18:2046–2053
113. Schiffer M, Tossidou I, Bartels J et al (2008) Erythropoietin-analog as podocyte protective agents in diabetic nephropathy. Kidney Blood Press Res 31:411–415

114. Eto N, Wada T, Inagi R et al (2007) Podocyte protection by darbepoetin: preservation of the cytoskeleton and nephrin expression. Kidney Int 72:455–463
115. Katavetin P, Inagi R, Miyata T et al (2007) Erythropoietin induces heme oxygenase-1 expression and attenuates oxidative stress. Biochem Biophys Res Commun 359:928–934
116. Logar CM, Brinkkoetter PT, Krofft RD et al (2007) Darbepoetin alfa protects podocytes from apoptosis in vitro and in vivo. Kidney Int 72:489–498
117. Coleman TR, Westenfelder C, Togel FE et al (2006) Cytoprotective doses of erythropoietin or carbamylated erythropoietin have markedly different procoagulant and vasoactive activities. Proc Natl Acad Sci U S A 103:5965–5970

Part VI
Novel therapeutic approaches against oxidative stress and hypoxia

Chapter 32
Novel Therapeutic Approaches Against Oxidative Stress and Hypoxia, Targeting Intracellular Sensor Molecules for Oxygen and Oxidative Stress

Toshio Miyata and Charles van Ypersele de Strihou

Abstract Besides hemodynamic and metabolic abnormalities, a broad derangement of oxygen metabolism, such as hypoxia and oxidative stress, has been implicated in the genesis of diabetic kidney lesions. It further impacts on various biological reactions linked to oxygen metabolism (e.g., nitrosative stress, advanced glycation, carbonyl stress, endoplasmic reticulum (ER) stress). The causal role of an impaired oxygen metabolism in diabetic kidney lesions renovates our understanding of current therapeutic benefits accruing from antihypertensive agents, the control of hyperglycemia or hyperinsulinemia, and the dietary correction of obesity. The cellular defense mechanisms against hypoxia and oxidative stress have been recently explored. The hypoxia-inducible factor (HIF) plays a key role to the defense against hypoxia. Its activity is modulated by intracellular oxygen sensors, prolylhydroxylases (PHDs). Small molecular PHD inhibitors improve kidney lesions in several animal models, including diabetic nephropathy. PHD has three isoforms whose respective roles have been delineated in mice by the specific disruption of each PHD gene. Unfortunately, no current inhibitor is specific for a distinct PHD isoform. Nonspecific inhibition of PHDs may induce adverse effects associated with PHD2 inhibition (e.g., angiogenesis). Specific disruption of the *PHD1* gene induces hypoxic tolerance, without angiogenesis and erythrocytosis, through the reprogramming of basal oxygen metabolism and an attendant decreased oxidative stress in hypoxic mitochondria. A specific PHD1 inhibitor might therefore offer a novel therapy against hypoxia. On the other hand, transcriptional factor Nrf2 regulates the basal and inducible expressions of numerous antioxidant stress genes. Disruption of the *Nrf2* gene exacerbates oxidative renal injury. Nrf2 activity is modulated by Keap1, an intracellular sensor for oxidative stress. Inhibitors of Keap1 may thus prove therapeutic against oxidative tissue injury. Altogether, newer approaches targeting intracellular sensor molecules

T. Miyata (✉)
Division of Molecular Medicine and Therapy,
United Centers for Advanced Research and Translational Medicine (ART),
Tohoku University Graduate School of Medicine, 2-1 Seiryo-Machi, Aoba-ku, Sendai,
Miyagi 980-8575, Japan
e-mail: miyata@med.tohoku.ac.jp

T. Miyata et al. (eds.), *Studies on Renal Disorders*, Oxidative Stress in Applied Basic Research and Clinical Practice, DOI 10.1007/978-1-60761-857-7_32,

for oxygen and oxidative stress may offer clinical benefits not only in the diabetic kidney but also in several diseases in which hypoxia or oxidative stress is a final, common pathway.

Keywords Impaired oxygen metabolism · Oxygen sensor · Hypoxia tolerance · Prolylhydroxylase (PHD) · Hypoxia-inducible factor (HIF) · Nrf2 · Keap1

1 Introduction

Several hemodynamic and metabolic derangements have been implicated in the development of renal lesions associated with type 2 diabetes mellitus, for example, hypertension, hyperglycemia, hyperinsulinemis, hyperlipidemia, and obesity (Fig. 1). Prevention of diabetic nephropathy and of its dramatic consequences thus relies on a multipronged approach targeting blood pressure (BP), serum levels of glucose, insulin, lipids, and so forth. Unfortunately, despite significant advances, these therapies do not fully prevent the renal complications of diabetes. Identification of newer culprits and of novel agents that are able to interfere with them should provide additional, dramatically needed benefits.

In this chapter, we propose a central role for oxygen metabolism, especially hypoxia and oxidative stress, in the genesis and progression of diabetic nephropathy, provide a single core hypothesis, including various biological reactions linked to oxygen metabolism (e.g., nitrosative stress, advanced glycation/carbonyl stress, endoplasmic reticulum stress), apply this hypothesis to understand the clinical

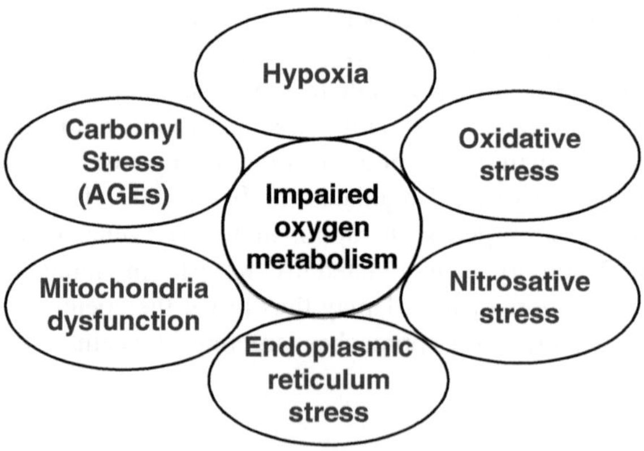

Fig. 1 Sequential events, stemming from hemodynamic and metabolic derangements and leading to hypoxia, oxidative stress, and their eventual consequences in the diabetic kidney. Current therapies targeting hemodynamic or metabolic abnormalities, such as antihypertensive agents (renin-angiotensin system [RAS] inhibitors), control of hyperglycemia/hyperinsulinemia, and dietary correction of obesity, modify the key features of abnormal oxygen metabolism in the diabetic kidney

benefits of current therapies, and, finally, propose more innovative therapeutic approaches, targeting intracellular sensor molecules for oxygen and oxidative stress, not only in the diabetic kidney but also in several diseases in which hypoxia or oxidative stress is a final, common pathway.

2 Abnormal Oxygen Metabolism

2.1 Hypoxia

Oxygen is essential to various biometabolic processes, including oxidative phosphorylation, during mitochondrial respiration. All organs, including the kidney, depend on a sufficient and consistent supply of oxygen. The role of chronic hypoxia in the progression of chronic kidney disease, first proposed in 1998 [1], has been validated in a variety of human and experimental kidney diseases, including diabetic nephropathy (reviewed in [2, 3]). Ries et al. [4] visualized tissue hypoxia in the streptozotocin-induced diabetic kidney by blood oxygen level-dependent (BOLD) imaging, a finding confirmed later by Rosenberger et al. [5] by means of pimonidazole staining (a probe to detect hypoxia) and of hypoxia-inducible factor (HIF). Tissue hypoxia was also documented in a hypertensive, type 2 diabetic rat model [6].

The precise localization of hypoxia in the kidney has proven difficult as few methods are able to identify and quantify tissue oxygenation at the cellular level. Tanaka et al. [7] have used a new hypoxia-responsive reporter vector to generate a novel hypoxia-sensing transgenic rat. In this animal, they identified "diffuse cortical" hypoxia in the puromycin aminonucleoside-induced nephrotic syndrome and "focal and segmental" hypoxia in the remnant kidney. In both models, the degree of hypoxia correlated positively with microscopic tubulointerstitial injury. Localization of tissue hypoxia may thus differ according to the type of renal disease.

Causes of chronic hypoxia in the diabetic kidney are multifactorial. Glomerular efferent arterioles enter the peritubular capillary plexus to offer oxygen to tubular and interstitial cells. Diabetic glomerular and vascular lesions damage efferent arterioles, decrease the number of peritubular capillaries, reduce oxygen diffusion to tubulointerstitial cells, and lead, eventually, to tubular dysfunction and fibrosis [2]. Postglomerular peritubular blood flow is further decreased by vasoactive substances generated in the diabetic kidney, such as angiotensin II and nitric oxide (NO) [8]. As shown by Palm et al. [9] NO regulates oxygen availability, so that its reduction in diabetes causes hypoxia in the renal medulla. Anemia associated with chronic kidney disease also hinders oxygen supply [10, 11].

Together with a significant decrease in oxygen supply, oxygen demand increases in the tubule of the outer medulla of the diabetic kidney: remnant nephrons compensate for tubular loss of nephrons, with an attendant enhanced tubular transport and hence more energy consumption [12].

2.2 Oxidative Stress

Under hypoxia, the cell relies on anaerobic glycolysis to generate adenosine triphosphate (ATP), whereas the residual low oxygen supply still supports some level of oxidative production of ATP through the tricarboxylic acid (TCA) cycle and electron transport chain (ETC). In hypoxic cells, electrons leak from the mitochondrial ETC and generate an excess of reactive oxygen species (ROS), that is, oxidative stress. Although the existence of oxidative stress during hypoxia sounds paradoxical, both an increase and a decrease in oxygen tension lead to oxidative stress. Reoxygenation or high oxygen levels following ischemia further exaggerate ROS generation. This concept opens the way to the clinical use of agents able to scavenge ROS or to prevent their formation in ischemic diseases [13].

Both the existence of oxidative stress and its localization in diabetes have been disputed. Williamson et al. [14] reported that diabetes is characterized by an increased cellular nicotinamide adenine dinucleotide ($NADH/NAD^+$ ratio) and suggested that it is a state of "reductive stress" and "pseudo-hypoxia" rather than oxidative stress. Oxidative stress was postulated in diabetes on the basis of indirect evidence, including increased cellular nicotinamide adenine dinucleotide phosphate ($NADP^+/NADPH$ ratio) and of oxidized to reduced glutathione [15–17]. Such a redox imbalance may, however, rely on nonoxidative mechanisms (e.g., the polyol pathway), so that their presence does not necessarily indicate an oxidative stress. Baynes and his colleagues [18] revisited this issue with a new methodology centered on oxidative protein modifications. They argued against a "generalized" oxidative stress in diabetes: the age-adjusted levels in skin collagen of oxidized amino acids, *ortho*-tyrosine and methionine sulfoxide, proved virtually identical in diabetics and nondiabetics.

In contrast with a generalized oxidative stress, we have demonstrated a local oxidative stress in the human diabetic kidney [19]. Advanced glycation end products (AGEs), generated nonenzymatically with sugars on proteins, include two different classes of structures: oxidative stress-dependent molecules (pentosidine and N^{ε}-(carboxymethyl) lysine [CML]) and oxidative stress-independent molecules (pyrraline). Should tissue AGE formation depend solely on hyperglycemia, all AGE structures would be detected in the diabetic kidney. The identification of individual AGE structures established that this was not the case. Both pentosidine and CML were present in diabetic glomerular lesions, together with other protein modifications derived from the oxidation of lipids (e.g., malondialdehyde-lysine), whereas pyrraline was absent. A local oxidative stress was thus postulated in the diabetic kidney [19, 20]. This contention has been subsequently confirmed in diabetic vascular lesions by Heinecke [21].

Altogether, diabetes is associated not with a generalized but rather with a local oxidative stress. Currently, the presence of local oxidative stress in the diabetic kidney is supported by a large body of evidence gathered in in vitro experiments, in vivo animal, and human studies (reviewed in [22, 23]).

The primary cause of local oxidative stress in the diabetic kidney also remains debated. Numerous enzymatic and nonenzymatic sources produce ROS or affect

ROS production in the diabetic kidney, for example, the renin-angiotensin system (RAS) activation [6, 24], NADPH oxidase activation [25–27], nitric oxide synthase (NOS) and its metabolites (reactive nitrogen species) [28], mitochondrial respiratory chain reaction abnormalities [29, 30], the polyol pathway [23], enhanced advanced glycation and generation of reactive carbonyl compounds (RCOs) [19, 20, 31], auto-oxidation of glucose and lipids, Fenton reaction catalyzed by transition metal ions [32], and glutathione and other sulfhydryl compounds depletion [33].

As already mentioned, hypoxia also induces energy depletion and generates oxidative stress. Aragonés et al. [34] have demonstrated in mice that genetic disruption of the prolylhydroxylase-1 (*PHD1*) gene, an intracellular oxygen sensor, lowers oxygen consumption in mitochondria of skeletal muscle, with an attendant reduction of oxidative stress and an eventual enhanced cellular survival during hypoxia. In agreement with this observation, the activation of HIF-1α reduces ROS generation [35], whereas its inhibition worsens ROS generation [36].

Concurrently, oxidative stress exacerbates the status of hypoxia. In vitro studies in rat proximal tubular cells [37] or in vivo studies in streptozotocin-induced diabetic rats [5] show that high glucose blunts the activation of HIF, an effect fully reversed by treatment with antioxidants, such as α-tocopherol or TEMPOL (4-hydroxy-2,2,6,6-tetramethylpiperidinoxyl). NADPH oxidase activation also aggravates renal hypoxia [38]. Altogether, hypoxia and oxidative stress are closely linked in the diabetic kidney.

2.3 *Broad Derangement of Oxygen Metabolism*

Hypoxia and oxidative stress have an additional impact on various biological reactions linked to oxygen metabolism (Fig. 2). First, oxidative stress modifies proteins either directly through the oxidation of amino acids by ROS or indirectly by an increased generation of RCOs from carbohydrates and lipids (i.e., carbonyl stress) [39], which, in turn, stimulates the production of AGEs. A causal role of oxidative stress in the formation of AGEs is supported by the correlation observed in diabetic and uremic serum between pentosidine (an AGE) and oxidative markers such as dehydroascorbate and advanced oxidation protein products (AOPP) [40, 41] as well as by the colocalization of both oxidation-dependent AGE structures and lipid peroxidation products observed in diabetic glomerular lesions [19, 20]. RCOs also interfere with various cellular functions, independently of protein AGE modifications and influence intracellular signaling by multiple pathways [42], including an interaction with the receptor for AGEs (RAGE) [43].

Second, hypoxia and oxidative stress interfere with the NO system. NO regulates numerous kidney functions, including renal hemodynamics, renin release, and the extracellular fluid volume [44]. It contributes to disease when it is either deficient or in excess. NO deficient animal models indeed develop hypertension, proteinuria, and glomerulosclerosis [45]. NO production and NO bioavailability progressively decline in the diabetic kidney [46], influencing thus both oxygen utilization and

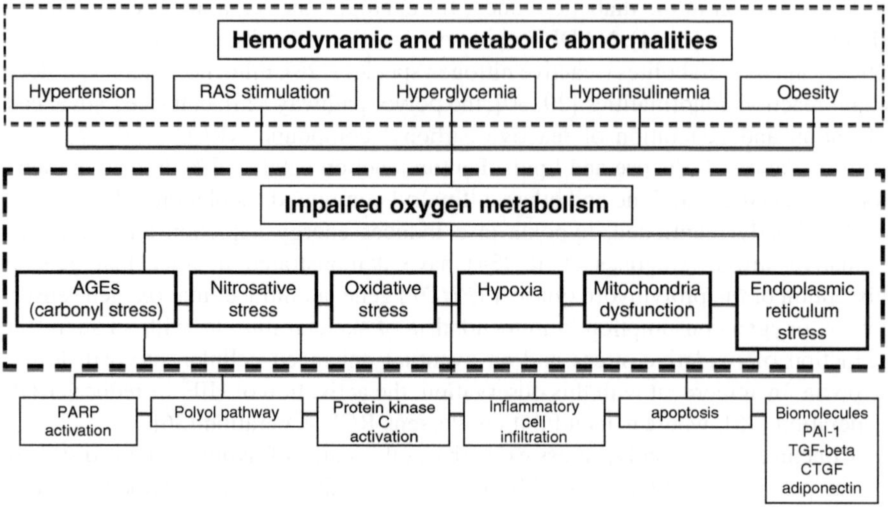

Fig. 2 Abnormal oxygen metabolism in the diabetic kidney. Impaired oxygen metabolism, such as hypoxia and oxidative stress, is a final, common pathway not only in diabetic nephropathy but also in a number of chronic conditions, such as a wide variety of kidney disorders, ischemic heart disease, and stroke. They have an additional impact on various biological reactions linked to oxygen metabolism (e.g., nitrosative stress, advanced glycation, carbonyl stress, and endoplasmic reticulum stress). *AGEs* advanced glycation end products

supply [47]. Overproduction of superoxide and other related ROS, resulting in oxidative stress, blunts the biological effects of NO. Superoxide, combined with NO, forms peroxynitrite, a cytotoxic oxidant [28]. Peroxynitrite activates the nuclear enzyme poly (ADP-ribose) polymerases (PARP) [48], which in turn inhibits the activity of the glycolytic enzyme glyceraldehyde-3-phoshate dehydrogenase (GAPDH) [49], eventually activating the polyol pathway and the formation of AGEs, protein kinase C, and the hexosamine pathway [29], all of which have been implicated in the genesis of diabetic nephropathy. ROS and reactive nitrogen species (i.e., nitrosative stress) thus trigger subsequent cellular dysfunction in diabetes through a multitude of mechanisms [44].

The function of mitochondria is influenced by hypoxia, oxidative stress, and NO [50]. The function of the endoplasmic reticulum (ER) is also modified by hypoxia and oxidative stress [51]. ER, an intracellular compartment, plays a critical role in the processing, folding, and transport of newly synthesized proteins. All cells are able to regulate their ER's capacity and to adapt to an imbalance between protein load and folding capacity, referred to as the ER stress, which has been implicated in the diabetic kidney [52]. Defense against the ER stress includes the unfolded protein response (UPR) with a transient attenuation of new protein synthesis, the degradation of misfolded proteins, and the expression of a variety of anti-ER stress proteins. Excessive ER stress tips the balance beyond the limit of cellular UPR and, occasionally, leads to apoptosis.

The phenomena, stemming from hemodynamic and metabolic derangements and leading to hypoxia, oxidative stress, and their eventual consequences, are tentatively integrated in a hypothetical scheme (see Fig. 1). The interrelationship between these detrimental chain reactions is so complex that a single culprit is probably not sufficient to account for the alterations of the diabetic kidney. Whatever the sequential events, the consequences of an impaired oxygen metabolism play a pivotal role in the genesis and progression of the diabetic kidney.

3 Implications in Current Therapies

Understanding the key features of abnormal oxygen metabolism in the diabetic kidney renovates the interpretation of current therapeutic benefits. Targets of current therapies are numerous and heterogeneous (e.g., blood pressure, glucose, insulin, and obesity). As delineated below, in diabetic animal experiments, renoprotection is not necessarily linked to blood pressure or glycemic control but, interestingly, appears associated with an improved oxygen metabolism.

3.1 Inhibitor of the Renin-Angiotensin System: Blood Pressure Lowering Agent

Several clinical studies, mainly but not only in diabetic patients, have demonstrated that antihypertensive agents that inhibit the RAS, such as angiotensin converting enzyme inhibitors (ACEIs), angiotensin II type 1 receptor blockers (ARBs), or a direct renin inhibitor, achieve better renoprotection than other antihypertensive drugs [53–58]. As a result, ARBs and ACEIs are now part of the standard treatment of patients with diabetic nephropathy, even in the absence of systemic hypertension.

Interestingly, RAS inhibitors provide renoprotection independently of blood pressure (BP) lowering [53–58]. The dissociation between BP and renoprotection has been interpreted within the RAS inhibition hypothesis as a consequence of the substantially higher angiotensin II concentrations within the kidney than in the systemic circulation [59].

Alternatively, recent studies suggest, at least in part, an RAS-independent effect of ARBs (i.e., benefits due to an effect on an impaired oxygen metabolism). In addition to the protective benefits of BP lowering and angiotensin II type 1 receptor (AT_1)-blockade, ARBs (and ACEIs) have unique abilities to correct not only tissue hypoxia (by an increase in postglomerular peritubular blood flow [60] or through AT_2 receptor activation and NO production [61]) but also oxidative stress and nitrosative stress [62, 63], carbonyl stress and advanced glycation [64, 65], redox imbalance [66], and ER stress [67].

In order to dissect the mechanisms of ARBs' protective benefits, we synthesized a novel, nontoxic, ARB derivative, R-147176, characterized by a weak affinity for the AT_1 (6,700 times less effective than olmesartan in AT_1 binding inhibition), but a striking inhibition of oxidative stress and advanced glycation [68]. Despite a minimal effect on blood pressure, it provided a significant renoprotection in two different experimental type 2 diabetic rat models, SHR/NDmcr-cp and Zucker diabetic fatty. The renal benefit of ARB thus partly depends on its potent inhibition of oxidative stress and advanced glycation. Whether or not this compound corrects tissue hypoxia remains undetermined. R-147176, like ARBs, protects not only the kidney but also brain cells in an experimental rat stroke model [69].

3.2 *Glucose Lowering Agent*

In addition to the critical role of hyperglycemia [70], recent studies have incriminated insulin resistance or hyperinsulinemia in the genesis of diabetic renal injury [71]. Insulin sensitizers are therefore recommended for obese, diabetic patients with nephropathy.

Of interest, the renal benefits of insulin and of pioglitazone (an insulin sensitizer) are associated with a reduction of hypoxia [72], oxidative stress [73], nitrosative stress [74], and advanced glycation [73]. Katavetin et al. [37] have shown in a streptozotocin-induced diabetic rat model that hyperglycemia blunts HIF activation, an effect fully reversed by insulin treatment. In a hypertensive type 2 diabetic rat SHR/NDmcr-cp model, we have demonstrated that insulin as well as pioglitazone reduced renal accumulation of AGEs and markers of oxidative stress to the same extent [73]. However, in contrast with insulin, pioglitazone decreased significantly plasma insulin levels and afforded markedly better renoprotection. This puzzling observation is best accounted for by the ability of pioglitazone to reduce the renal expression of transforming growth factor-beta (TGF-β). The latter, together with hyperinsulinemia, might therefore prove a useful therapeutic target [75], independent of glycemic control and impaired oxygen metabolism.

3.3 *Correction of Obesity*

Restriction of energy intake reduces oxidative stress in experimental animals [76, 77]. Recent studies support a link between obesity and hypoxia [78, 79]. Crujeiras et al. [80] demonstrated that energy restriction in obese subjects improves mitochondrial function and reduces oxidative stress.

In an obese type 2 diabetic rat model (SHR/NDmcr-cp), restriction of caloric intake by 30% for 20 weeks corrected obesity. Caloric restriction was associated with a mild, not significant, fall in levels of HbA1c. Nevertheless, despite unchanged BP, hyperglycemia, and hyperinsulinemia [81], proteinuria and histological abnormalities of the kidney were prevented. Renal damage was impressively correlated not only with body weight but also with the renal content of AGEs and the degree of oxidative stress [81]. Renoprotection in this model thus hinges upon the reduction of oxidative stress but remains independent of hypertension and hyperglycemia.

4 Novel Therapeutic Target

4.1 Hypoxia-Inducible Factor

If hypoxia alters oxygen metabolism in diabetic nephropathy, let us review the mechanisms of defense against hypoxia (Fig. 3). Defense against hypoxia hinges on HIF [82, 83], whose activation induces a broad range of genes that participate in erythrocytosis, angiogenesis, glucose metabolism, or cell proliferation/survival, with the eventual protection of hypoxic tissues.

HIF-α is constitutively transcribed and translated. Its level is primarily regulated by its rate of degradation. Oxygen determines its stability through an enzymatic

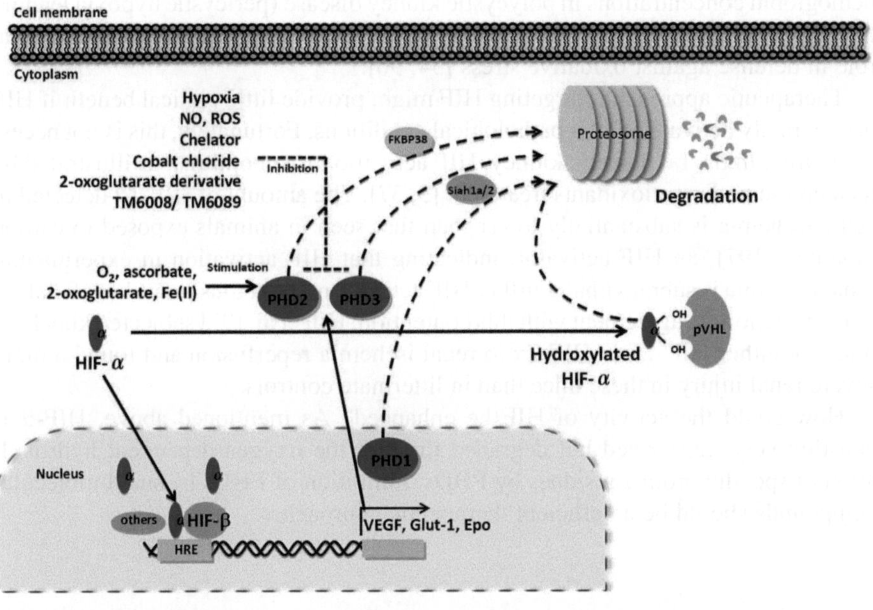

Fig. 3 HIF–PHD pathway under hypoxia. Hypoxia-inducible factor (HIF)-α is constitutively transcribed and translated. Its level is primarily regulated by its rate of degradation. Oxygen determines its stability through its enzymatic hydroxylation by PHDs. The hydroxylated HIF-α is recognized by von Hippel–Lindau tumor suppressor protein (pVHL), which functions as an E3 ubiquitin ligase and is rapidly degraded by the proteasome. Nonhydroxylated HIF-α cannot interact with pVHL and is thus stabilized. It binds to its heterodimeric partner HIF-β mainly in the nucleus and transactivates genes involved in the adaptation to hypoxic-ischemic stress. Expression of PHDs (PHD2 and PHD3) is regulated by HIF. PHDs interact with Siah1a/2 (PHD1 and PHD3) or FKBP38 (PHD2) and are subject to proteasomal degradation. PHD activity is inhibited under hypoxia or by actions with nitric oxide (NO), reactive oxygen species (ROS), transition metal chelators, cobalt chloride, 2-oxoglutarate analogs, or TM6008/TM6089

hydroxylation by prolylhydroxylases (PHDs) [84, 85]. Hydroxylated HIF-α is recognized by Hippel–Lindau tumor suppressor protein (pVHL) [86, 87], which functions as an E3 ubiquitin ligase and is rapidly degraded by the proteasome [88, 89]. Nonhydroxylated HIF-α does not interact with pVHL and is thus stabilized. It binds to its heterodimeric partner HIF-β mainly in the nucleus [90] and transactivates genes involved in the adaptation to hypoxic-ischemic stress.

Three isoforms of the HIF-α subunit have been identified [91] (i.e., HIF-1α, HIF-2α, and HIF-3α). The former two are structurally and functionally similar. In contrast, HIF-3α lacks the structures for transactivation found in the C-termini of HIF-1α and HIF-2α, suggesting an alternative role as a negative regulator of hypoxia-inducible gene expression.

Whereas HIF-1α distribution is rather ubiquitous, that of HIF-2α is localized in certain types of cells [92]. In the kidney, HIF-1α is expressed in tubules, while HIF-2α is confined to endothelial and interstitial cells. Recent studies in mice, utilizing gene disruption of either HIF-1α or HIF-2α, have revealed that HIF-2α functions as a physiological regulator of erythropoietin [93]. The *HIF-2α* gene is responsible for familial erythrocytosis in humans [94] and for comparatively high hemoglobin concentrations in polycystic kidney disease (pericystic hypoxia leading to HIF-2 induction) [95]. In addition, as described below, HIF-2α plays a crucial role in defense against oxidative stress [34, 96].

Therapeutic approaches targeting HIF might provide little clinical benefit if HIF is maximally activated under pathological conditions. Fortunately, this is not necessarily true. In the rat diabetic kidney, HIF activation is suboptimal, as illustrated by its increase under antioxidants treatment [5, 37]. The amount of HIF-1α detected in acute ischemia is substantially lower than that seen in animals exposed to carbon monoxide [97], an HIF activator, indicating that HIF activation in experimental renal ischemia is submaximal. Further HIF activation might thus prove beneficial for renoprotection. In agreement with this contention, Hill et al. [97] subjected knockout mice for either HIF-1α or HIF-2α to renal ischemia reperfusion and found a more severe renal injury in these mice than in littermate controls.

How could the activity of HIF be enhanced? As mentioned above, HIF-α is constitutively transcribed but degraded through the oxygen-dependent hydroxylation of specific proline residues by PHDs. Inhibition of PHDs by small molecular compounds should be an efficient therapeutic approach.

4.2 Oxygen Sensor (PHD)

PHDs belong to the Fe(II) and 2-oxoglutarate-dependent dioxygenase superfamily [91], which incorporates both atoms of molecular oxygen into their substrates: one atom is used in the oxidative decarboxylation of 2-oxoglutarate to yield succinate and carbon dioxide, whereas the other is incorporated directly into the oxidized proline residue of HIF-α. They are called oxygen sensors [98], as their activity rigorously depends on oxygen tension.

Iron is essential for PHD activity, so that transition metal chelators might inhibit PHD activity. Cobalt chloride also inhibits PHD activity through intracellular depletion of ascorbate necessary for iron (reduced) activity [99]. Chemical preconditioning with cobalt chloride also protected the kidney in a variety of experimental models, including ischemic reperfusion [100], progressive uninephrectomized anti-Thy1 nephritis [101], remnant kidney [102], and cisplatin nephropathy [103].

Cobalt chloride also protected the kidney in a hypertensive, type 2 diabetic rat model (SHR/NDmcr-cp) [104]. Given for 20 weeks, it reduced proteinuria as well as histological kidney injury, despite sustained hypertension and metabolic abnormalities. Renal improvement was paralleled by a marked reduction in renal tissue expressions of HIF-regulated genes, including erythropoietin, vascular endothelial growth factor (VEGF), and heme oxygenase 1 (HO-1), and as well as by a reduced renal production of TGF-β and AGEs.

In humans, the erythropoietic effect of cobalt has been established since the 1940s [105, 106]. In the 1970s, cobalt chloride was used in the treatment of anemia associated with chronic renal failure [107]. Unfortunately, its toxicity prevented further clinical use.

Recently, less cumbersome, nontoxic small molecular inhibitors of oxygen sensor (PHDs) have been investigated [98]. Binding of the substrate 2-oxoglutarate to the catalytic domain of PHDs appeared essential for enzymatic PHD activity. Chemical compounds whose structure mimicked 2-oxoglutarate (e.g., *N*-oxalylglycine (dimethyloxalylglycine) [108, 109], *N*-oxalyl-D-phenylalanine [110], L-minosine [97]) were therefore able to inhibit PHD activity.

We utilized a different strategy, docking simulation based on the three-dimensional protein structure of human PHD2, and synthesized two novel inhibitors of PHDs (TM6008 and TM6089) [111]. Both compounds bound to the active site within the PHD2 molecule where HIF binds (Fig. 4). As anticipated, given orally, they stimulated HIF activity in various organs of transgenic rats expressing a hypoxia-responsive reporter vector, and, given locally, they induced angiogenesis in a mouse sponge assay [111].

Another small molecular inhibitor of PHD (FG4487) also offered renal benefits [112]. Given intraperitoneally in a rat ischemic acute renal failure model, it activated both HIF-1α and HIF-2α, induced the expression of HIF target genes, ameliorated tubular injury, and, eventually, improved renal function. A similar renoprotection has been demonstrated for other PHD inhibitors, such as L-minosine and dimethyloxalylglycine [97], both of which increased HIF-1α and HIF-2α expression and protected against renal ischemic injury by decreasing the number of apoptotic cells in the absence of angiogenesis.

4.3 Specific Inhibitor for Oxygen Sensor

Unfortunately, nonspecific inhibition of HIF degradation augments VEGF and erythropoietin production, both of which have proven detrimental in human

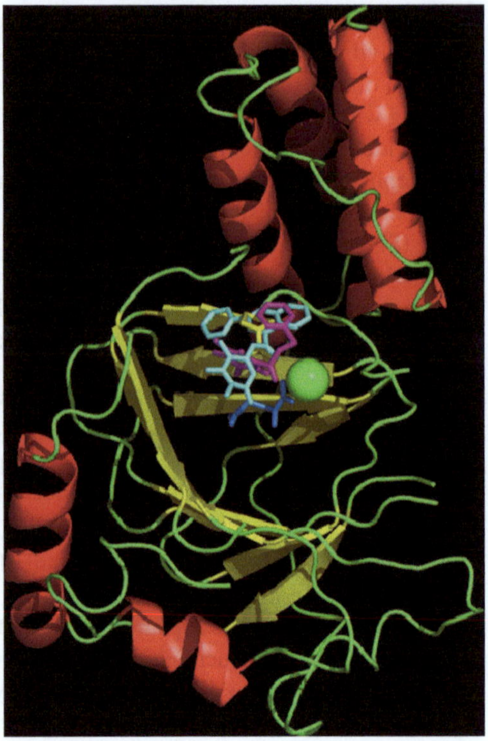

Fig. 4 Predicted binding modes of prolylhydroxylases (PHD) inhibitors in human PHD2. TM6008, TM6089, HIF proline, and 2-oxoglutarate are shown in *light blue*, *magenta*, *yellow*, *blue*, respectively. Fe(II) is shown by a *green sphere*. Figure is drawn by the software Discovery Studio Visualizer 2.0 (Accelrys Software Inc., San Diego, CA)

diabetic retinopathy [113]. Dissociation of the benefits of HIF activation from its effects on VEGF and erythropoietin is therefore needed.

Three different PHD isoforms have been identified [91] (i.e., PHD1, PHD2, PHD3), each of which has its own tissue and subcellular distribution [114]. PHD1 is exclusively nuclear, PHD2 is mainly cytoplasmic (but shuttles between nucleus and cytoplasm [115]), and PHD3 is present in both cytoplasm and nucleus. PHD2 acts as a decisive oxygen sensor in the HIF degradation pathway [116]. In the rat kidney, all three isoforms of PHDs were expressed but PHD2 was the most abundant [117]. PHDs were more strongly expressed in tubular segments of the inner medulla where oxygen tension is physiologically low [118]. Although it decreased PHD activity, hypoxia induced expression of PHD2 and PHD3 through upregulation of HIF-1α [117]. This HIF-induced PHD expression ensures rapid removal of HIF-α after reoxygenation. In addition to hypoxia, NO and ROS also decreased PHD activity [116, 118], but here again both resulted in a feedback upregulation of PHDs' expression due to HIF-α accumulation. Feedback loops may prove very important at different levels of hypoxia signaling [119].

The roles of the three PHD isoforms have recently been delineated by the specific disruption of each PHD gene (Fig. 5). The angiogenic phenotypes of

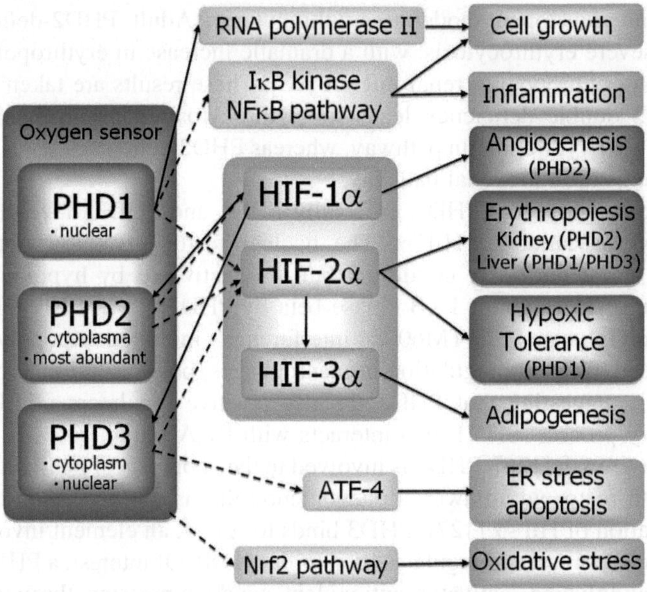

Fig. 5 Respective role of each prolylhydroxylase (PHD) isoform in response to hypoxia. Stimulation and inhibitory interactions are indicated with *solid* and *broken arrows*, respectively. PHD2-HIF-1α pathway regulates angiogenesis and PHD2-HIF-2α dose erythrocytosis in the kidney. PHD1(-HIF-1α) involves hypoxic tolerance through reprogramming basal oxygen metabolism and significant reduction of oxidative stress generated in mitochondria. PHDs also have targets other than the hydroxylation of HIF-α, for example, the nuclear factor κB (NFκB) pathway (inflammation), RNA polymerase II (cell growth), ATF4 (endoplasmic reticulum stress), and the Nrf2 pathway (oxidative stress)

mice with targeted disruptions of these genes [120] revealed that PHD1 and PHD3 gene knockout does not produce apparent angiogenic defects. In contrast, broad-spectrum conditional knockout of PHD2 induced VEGF and an hyperactive angiogenesis, with the formation of mature and perfused blood vessels. In agreement with these observations, TM6008, a novel compound potentially binding human PHD2 in docking simulation studies, induced angiogenesis in mice [111]. PHD3 was also involved in angiogenesis: in mice with hindlimb ischemia, PHD3 gene silencing provided a better therapeutic revascularization than PHD2 gene silencing [121].

Data that demonstrate that upregulation of HIF results in tumor progression [122] might caution against the long-term use of PHD inhibitors. Fortunately, a recent study [123] reported that inhibition of PHD2 prompted endothelial cells to readjust their shape and phenotype to restore oxygen supply, improved tumor perfusion and oxygenation, and inhibited tumor cell invasion, intravasation, and metastasis.

Both PHD1 and PHD3 gene knockout mice had no apparent effect on erythropoietin [93], but double knockout PHD1 and PHD3 mice accumulated HIF-2α in the liver and developed moderate erythrocytosis. Adult PHD2-deficient mice developed severe erythrocytosis, with a dramatic increase in erythropoietin serum levels and in erythropoietin renal mRNA [93]. These results are taken to indicate that PHD1/3 double deficiency leads to erythrocytosis partly by activating the hepatic HIF-2α/erythropoietin pathway, whereas PHD2 deficiency leads to erythrocytosis by activating the renal pathway.

Evidence suggests that PHDs, especially PHD1 and PHD3, have other targets than the hydroxylation of HIF-α. The nuclear factor κB (NFκB) pathway is suppressed under normoxic condition, but it is activated by hypoxia through a modification of its inhibitor I-κB kinase-beta by PHD1 [124]. PHD inhibition by dimethyloxalylglycine or TM6008 interference inhibited lipopolysaccharide (LPS)-induced TNF-α upregulation in macrophages, possibly through NFκB inhibition [125], suggesting that PHDs are also positive regulators of LPS-induced inflammatory process. PHD1 also interacts with RNA polymerase II, which regulates tumor growth [126]. PHD3 is involved in the apoptosis of neuronal cells after nerve growth factor withdrawal, a phenomenon that is not prevented by simultaneous activation of HIF-α [127]. PHD3 binds to ATF4, an element involved in the UPRs under ER stress, and regulates its stability [128]. Of interest, a PHD inhibitor, dimethyloxalylglycine, activates antioxidant genes expression through the Nrf2 pathway [129]. These PHD-dependent but HIF-independent pathways might offer other therapeutic clues for an improved protection against hypoxic or oxidative stress tissue injury.

Dissociation between the benefits of HIF activation and its effects on angiogenesis and erythropoietin has been recently illustrated by Aragonés et al. [34]. The specific disruption of PHD1 unexpectedly induced hypoxic tolerance in muscle cells, without angiogenesis and erythrocytosis induction, at least in part through the activation of HIF-2α. Basal oxygen metabolism was reprogrammed and generation of oxidative stress decreased in hypoxic mitochondria (Fig. 6). Inhibition of PHD1 likely arouses various protective mechanisms, including ATP production, through enhanced glycolysis and restriction of glycolytic intermediates entry into the oxidative phosphorylation of glucose, through induction of pyruvate dehydrogenase kinase, which attenuates entry of electrons into the ETC. Consequently, energy is conserved, oxidative damage is reduced, and cells are protected from hypoxic damage. These findings partly explain why hibernating or hypoxia-tolerant animals are more resistant to ischemic insults [130, 131]. Inhibition of PHD1 also protects cultured rat neuronal cells: PHD1, but not PHD2 and PHD3, knockdown prevented oxidative stress-induced neuronal death [132]. Specific inhibition of PHD1 may thus mediate tissue protection through reduced oxidative stress.

Unfortunately, none of the current, thus far reported, PHD inhibitors are specific for a distinct PHD subtype. A specific PHD1 inhibitor may provide a novel therapy without adverse effects associated with PHD2 inhibition (e.g., polycythemia [93, 133, 134], congestive heart failure [132], placental defects during pregnancy [135]).

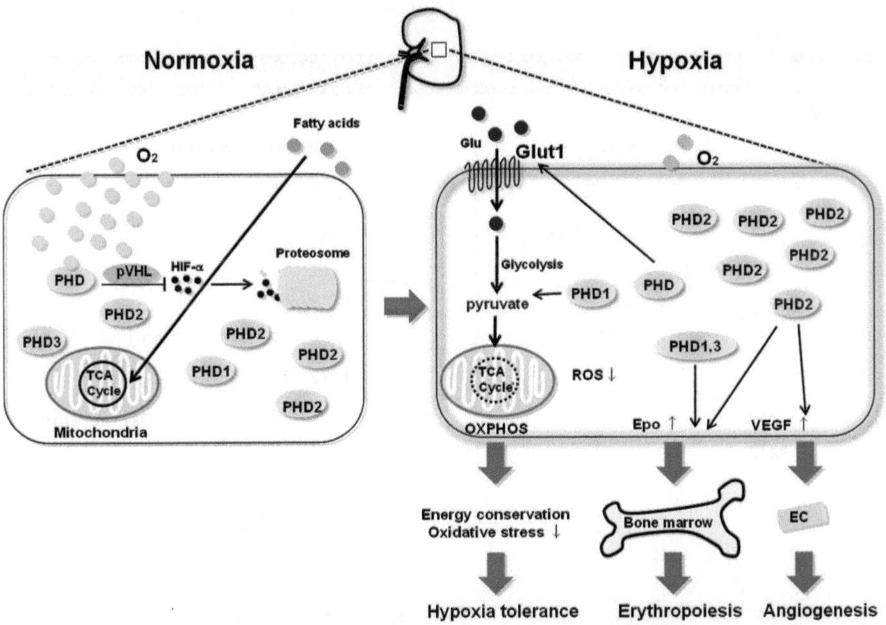

Fig. 6 Inhibition of PHDs, increased oxygen supply, and decreased oxygen demand (hypoxia tolerance). Under hypoxia in the kidney, two different adaptational reactions ensue. First, expressions of erythropoietin, vascular endothelial growth factor (VEGF), and glucose transporter-1 (GLUT-1) are augmented in order to increase the supply of oxygen and nutrient, eventually leading to erythropoietin (PHD2, PHD1/PHD3) and angiogenesis (PHD2). Second, PHD1-induced reactions decrease oxygen demand, conserve energy, reduce oxidative damage, and protect the cell from hypoxic damage (hypoxia tolerance), including stimulation of ATP production through enhanced glycolysis and restriction of the entry of glycolytic intermediates (e.g., pyruvate) into the oxidative phosphorylation (OXPHOS) of glucose

4.4 Nrf2

Transcriptional factor nuclear factor-erythroid 2 p45-related factor 2 (Nrf2) is a basic leucine zipper redox-sensitive transcription factor that regulates the expression of several cellular antioxidant and cytoprotective genes. Upon exposure to oxidative stress or electrophiles, Nrf2 translocates into nuclei, heterodimerizes with a small Maf protein, and eventually binds to the antioxidant/electrophile responsive element (ARE/EpRE), leading to transcriptional activation of antioxidant genes (Fig. 7), including heme oxygenase-1, glutathione peroxidase-2 (GSH-Px2), NAD (P)H-quinone oxidoreductase 1 (NQO1), and glutathione *S*-transferase [136, 137]. Nrf2 thus activates a broad and coordinated set of downstream reactions against oxidative stress.

In a recent study, induction of renal ischemia and reperfusion elevated Nrf2 levels and their downstream target genes in the kidney of wild-type mice [138]. Nrf2-deficiency enhanced susceptibility to both ischemic and nephrotoxic acute

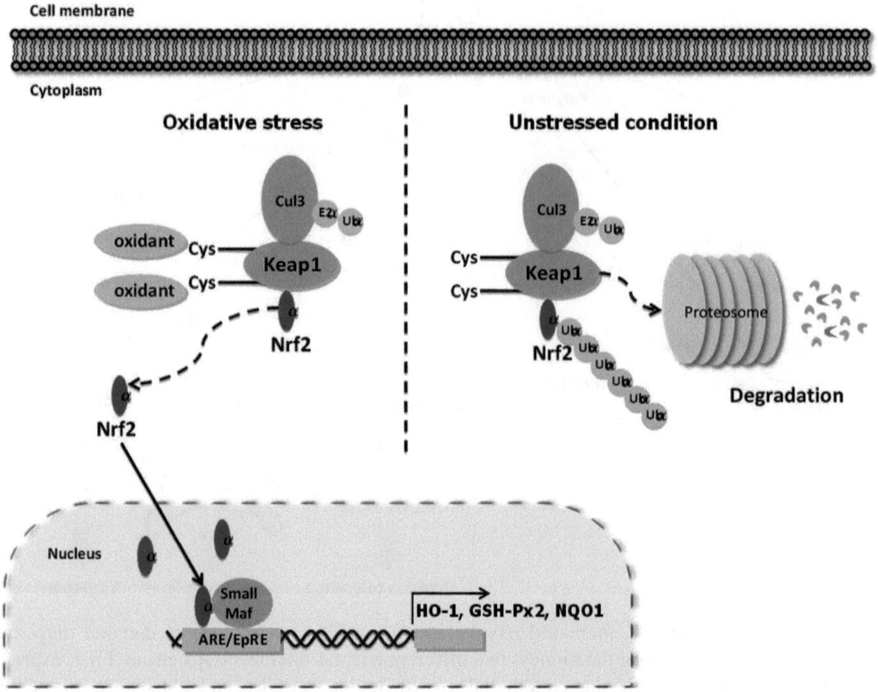

Fig. 7 Nrf2–Keap1 pathway under oxidative stress. Nrf2 is constitutively transcribed and translated. Its level is primarily regulated by its rate of degradation through the Keap1-Cul3 system. Nrf2 is ubiquitinated continuously and degraded within the proteasome. Under oxidative stress, reactive cysteines within the Keap1 moiety are modified by oxidants, which, in turn, induce conformational change, eventually leading to detachment of Nrf2 from Keap1 and to inhibition of its ubiquitination. Oxidative stress thus inhibits degradation of Nrf2 and facilitates nuclear translocation of Nrf2. Nrf2 then heterodimerizes with a small Maf protein, binds to the antioxidant/electrophile responsive element (ARE/EpRE), and transactivates a variety of antioxidant genes

kidney injury [139]: renal function, histology, vascular permeability, and survival were significantly worse in the Nrf2 knockout mice. Treatment of the Nrf2 knockout mice with the antioxidants *N*-acetyl-cysteine or glutathione improved renal function. Furthermore, Nrf2 knockout mice with a streptozotocin-induced diabetes progressively increased their urinary excretion of nitric oxide metabolites and developed renal injury [140]. Nrf2-mediated transcriptional responses are also protective in other experimental diseases, including oxidative lung injury and fibrosis, asthma, and brain ischemia reperfusion injury [141–143].

These data suggest upregulation of Nrf2 as a potential therapeutic target for ameliorating oxidative stress-induced kidney injury.

4.5 Oxidative Stress Sensor (KEAP1)

The regulation of Nrf2 has been recently elucidated. Nrf2 is constitutively transcribed and translated. Its level is primarily regulated by its rate of degradation through the Keap1-Cul3 system [144, 145]. Nrf2 is ubiquitinated continuously and degraded within the proteasome. Keap1 acts as a sensor of oxidative stress and a negative regulator of Nrf2. Under oxidative stress, reactive cysteines within the Keap1 moiety are modified by oxidants and induce conformational changes, eventually leading to the detachment of Nrf2 from Keap1 and inhibition of its ubiquitination (see Fig. 7). Oxidative stress thus inhibits degradation of Nrf2 and facilitates nuclear translocation of Nrf2.

In Keap1 knockdown mice, Nrf2-regulated genes are significantly increased and oxidative injuries in the brain and kidney are markedly ameliorated (Takizawa S, Matsuzaka T, Miyata T, Yamamoto M, unpublished observations). Inhibition of Keap1 could thus result in tissue protection through increased nuclear translocation of Nrf2 and subsequent activation of antioxidant genes.

4.6 *Nrf2 Inducer and Inhibitor of Oxidative Stress Sensor*

Sulforaphane, which is enriched in cruciferous vegetables such as broccoli, enhances in vitro Nrf2 activity and protects oxidative tissue damage in experimental models of ischemia-reperfusion kidney injury [146] and of ischemic stroke [147].

Unfortunately, by contrast to the HIF–PHD pathway, synthetic small molecular compounds able to interfere with the Nrf2-Keap1 system are very few. Effective Keap1 inhibitors are not currently available. However, X-ray crystal structure of Keap1 and the molecular mechanism of interaction between Nrf2 and Keap1 have been clarified [148]. By computer-based virtual screening based on the three-dimensional structure of Keap1, we searched for a compound that binds the active site of Keap1 and inhibits in vitro interaction between Nrf2 and Keap1 (Yamamoto M, Hirayama N, Miyata T, unpublished observation). Its effectiveness should be tested in experimental kidney disease models. If its benefits are confirmed, a specific Keap1 inhibitor may offer another approach to treat oxidative stress injury in kidney disease.

5 Conclusion

Impaired oxygen metabolism, such as hypoxia and oxidative stress, is a final, common pathway not only in diabetic nephropathy but also in a number of chronic conditions, such as a wide variety of kidney disorders, ischemic heart disease, and stroke. Advances in the treatment or prevention of diabetic nephropathy, delineated in the present review, especially those targeting sensor molecules for oxygen and oxidative stress, may thus herald new concepts in the management of a broad spectrum of chronic illnesses linked to an impaired oxygen metabolism.

Acknowledgments We thank Drs. Guo-Hua Fong, Masaomi Nangaku, and Kotaro Takeda for helpful discussion and Dr. Noriaki Hirayama for kindly providing representative pictures of docking simulation.

References

1. Fine LG, Orphanides C, Norman JT. Progressive renal disease: the chronic hypoxia hypothesis. Kidney Int 1998; 53(Suppl 65): S74–S78.
2. Fine LG, Norman JT. Chronic hypoxia as a mechanism of progression of chronic kidney diseases: from hypothesis to novel therapeutics. Kidney Int 2008; 74: 867–872.
3. Singh DK, Winocour P, Farrington K. Mechanisms of disease: the hypoxic tubular hypothesis of diabetic nephropathy. Nat Clin Pract Nephrol 2008; 4: 216–226.
4. Ries M, Basseau F, Tyndal B, et al. Renal diffusion and BOLD MRI in experimental diabetic nephropathy. Blood oxygen level-dependent. J Magn Reson Imaging 2003; 17: 104–113.
5. Rosenberger C, Khamaisi M, Abassi Z, et al. Adaptation to hypoxia in the diabetic rat kidney. Kidney Int 2008; 73: 34–42.
6. Izuhara Y, Nangaku M, Inagi R, et al. Renoprotective properties of angiotensin receptor blockers beyond blood pressure lowering. J Am Soc Nephrol 2005; 16: 3631–3641.
7. Tanaka T, Miyata T, Inagi R, et al. Hypoxia in renal disease with proteinuria and/or glomerular hypertension. Am J Pathol 2004; 165: 1979–1992.
8. Singh P, Deng A, Weir MR, et al. The balance of angiotensin II and nitric oxide in kidney diseases. Curr Opin Nephrol Hypertens 2008; 17: 51–56.
9. Palm F, Friederich M, Carlsson PO, et al. Reduced nitric oxide in diabetic kidneys due to increased hepatic arginine metabolism: implications for renomedullary oxygen availability. Am J Physiol Renal Physiol 2008; 294: F30–F37.
10. Nangaku M, Eckardt KU. Pathogenesis of renal anemia. Semin Nephrol 2006; 26: 261–268.
11. Bahlmann FH, Fliser D. Erythropoietin and renoprotection. Curr Opin Nephrol Hypertens 2009; 18: 15–20.
12. Nath KA, Croat AJ, Hostettor TH. O_2 consumption and oxidant stress in the surviving nephrons. Am J Physiol 1990; 258: F1354–F1362.
13. Kulkarni AC, Kuppusamy P, Parinandi N. Oxygen, the lead actor in the pathophysiologic drama: enactment of the trinity of normoxia, hypoxia, and hyperoxia in disease and therapy. Antioxid Redox Signal 2007; 9: 1717–1730.
14. Williamson JR, Chang K, Frangos M, et al. Hyperglycemic pseudohypoxia and diabetic complications. Diabetes 1993; 42: 801–813.
15. Bravi MC, Pietrangeli P, Laurenti O, et al. Polyol pathway activation and glutathione redox status in non-insulin-dependent diabetic patients. Metabolism 1997; 6: 1194–1198.
16. Jennings PE, Chirico S, Jones AF, et al. Vitamin C metabolites and microangiopathy in diabetes mellitus. Diabetes Res 1987; 6: 151–154.
17. Suzuki E, Yasuda K, Takeda N, et al. Increased oxidized form of human serum albumin in patients with diabetes mellitus. Diabetes Res Clin Pract 1992; 18: 153–158.
18. Wells-Knecht MC, Lyons TJ, McCance DR, et al. Age-dependent increase in *ortho*-tyrosine and methionine sulfoxide in human skin collagen is not accelerated in diabetes. Evidence against a generalized increase in oxidative stress in diabetes. J Clin Invest 1997; 100: 839–846.
19. Horie K, Miyata T, Maeda K, et al. Immunohistochemical colocalization of glycoxidation products and lipid peroxidation products in diabetic renal glomerular lesions. J Clin Invest 1997; 100: 2995–3004.
20. Suzuki D, Miyata T, Saotome N, et al. Immunohistochemical evidence for an increased oxidative stress and carbonyl modification of proteins in diabetic glomerular lesions. J Am Soc Nephrol 1999; 10: 822–832.
21. Pennathur S, Wagner JD, Leeuwenburgh C, et al. A hydroxyl radical-like species oxidizes cynomolgus monkey artery wall proteins in early diabetic vascular disease. J Clin Invest 2001; 107: 853–860.
22. Shah SV, Baliga R, Rajapurkar M, et al. Oxidants in chronic kidney disease. J Am Soc Nephrol 2007; 18: 16–28.

23. Forbes JM, Coughlan MT, Cooper ME. Oxidative stress as a major culprit in kidney disease in diabetes. Diabetes 2008; 57: 1446–1454.
24. Jaimes EA, Galceran JM, Raij L. Angiotensin II induces superoxide anion production by mesangial cells. Kidney Int 1998; 54: 775–784.
25. Gorin Y, Block K, Hernandez J, et al. Nox4 NAD(P)H oxidase mediates hypertrophy and fibronectin expression in the diabetic kidney. J Biol Chem 2005; 280: 39616–39626.
26. Asaba K, Tojo A, Onozato ML, et al. Effects of NADPH oxidase inhibitor in diabetic nephropathy. Kidney Int 2005; 67: 1890–1898.
27. Thallas-Bonke V, Thorpe SR, Coughlan MT, et al. Inhibition of NADPH oxidase prevents advanced glycation end product-mediated damage in diabetic nephropathy through a protein kinase C-alpha-dependent pathway. Diabetes 2008; 57: 460–469.
28. Szabo C. Role of nitrosative stress in the pathogenesis of diabetic vascular dysfunction. Br J Pharmacol 2009; 156: 713–727.
29. Nishikawa T, Edelstein D, Du XL, et al. Normalizing mitochondrial superoxide production blocks three pathways of hyperglycaemic damage. Nature 2000; 404: 787–790.
30. Kiritoshi S, Nishikawa T, Sonoda K, et al. Reactive oxygen species from mitochondria induce cyclooxygenase-2 gene expression in human mesangial cells: potential role in diabetic nephropathy. Diabetes 2003; 52: 2570–2577.
31. Tan AL, Forbes JM, Cooper ME. AGE, RAGE, and ROS in diabetic nephropathy. Semin Nephrol 2007; 27: 130–143.
32. Monnier VM. Transition metals redox: reviving an old plot for diabetic vascular disease. J Clin Invest 2001; 107: 799–801.
33. Baynes JW, Thorpe SR. Role of oxidative stress in diabetic complications: a new perspective on an old paradigm. Diabetes 1999; 48: 1–9.
34. Aragonés J, Schneider M, Van Geyte K, et al. Deficiency or inhibition of oxygen sensor PHD1 induces hypoxia tolerance by reprogramming basal metabolism. Nat Genet 2008; 40: 170–180.
35. Kim JW, Tchernyshyov I, Semenza GL, et al. HIF-1-mediated expression of pyruvate dehydrogenase kinase: a metabolic switch required for cellular adaptation to hypoxia. Cell Metab 2006; 3: 177–185.
36. Bonnet S. Archer SL, Allalunis-Turner J, et al. A mitochondria-K^+ channel axis is suppressed in cancer and its normalization promotes apoptosis and inhibits cancer growth. Cancer Cell 2007; 11: 37–51.
37. Katavetin P, Miyata T, Inagi R, et al. High glucose blunts vascular endothelial growth factor response to hypoxia via the oxidative stress-regulated hypoxia-inducible factor/hypoxia-responsible element pathway. J Am Soc Nephrol 2006; 17: 1405–1413.
38. Yang ZZ, Zhang AY, Yi FX, et al. Redox regulation of HIF-1alpha levels and HO-1 expression in renal medullary interstitial cells. Am J Physiol Renal Physiol 2003; 284: F1207–F1215.
39. Miyata T, Kurokawa K, van Ypersele de Strihou C. From molecular footprints of disease to new therapeutic interventions in diabetic nephropathy: a detective story. Curr Drug Targets Immune Endocr Metabol Disord 2005; 5: 323–329.
40. Miyata T, Wada Y, Cai Z, et al. Implication of an increased oxidative stress in the formation of advanced glycation end products in patients with end-stage renal failure. Kidney Int 1997; 51: 1170–1181.
41. Witko-Sarsat V, Friedlander M, Capeillère-Blandin C, et al. Advanced oxidation protein products as a novel marker of oxidative stress in uremia. Kidney Int 1996; 49: 1304–1313.
42. Miyata, T, Kurokawa K, van Ypersele de Strihou C. Advanced glycation and lipoxidation end products: role of reactive carbonyl compounds generated during carbohydrate and lipid metabolism. J Am Soc Nephrol 2000; 11: 1744–1752.
43. Yan SD, Schmidt AM, Anderson GM, et al. Enhanced cellular oxidant stress by the interaction of advanced glycation end products with their receptors/binding proteins. J Biol Chem 1994; 269: 9889–9897.

44. Modlinger PS, Wilcox CS, Aslam S. Nitric oxide, oxidative stress, and progression of chronic renal failure. Semin Nephrol 2004; 24: 354–365.
45. Wilcox CS. Oxidative stress and nitric oxide deficiency in the kidney: a critical link to hypertension? Am J Physiol Regul Integr Comp Physiol 2005; 289: R913–R935.
46. Prabhakar S, Starnes J, Shi S, et al. Diabetic nephropathy is associated with oxidative stress and decreased renal nitric oxide production. J Am Soc Nephrol 2007; 18: 2945–2952.
47. Palm F, Teerlink T, Hansell P. Nitric oxide and kidney oxygenation. Curr Opin Nephrol Hypertens 2009; 18: 68–73.
48. Garcia SF, Virág L, Jagtap P, et al. Diabetic endothelial dysfunction: the role of poly(ADP-ribose) polymerase activation. Nat Med 2001; 7: 108–113.
49. Du X, Matsumura T, Edelstein D, et al. Inhibition of GAPDH activity by poly(ADP-ribose) polymerase activates three major pathways of hyperglycemic damage in endothelial cells. J Clin Invest 2003; 112: 1049–1057.
50. Friederich M, Hansell P, Palm F. Diabetes, oxidative stress, nitric oxide and mitochondria function. Curr Diabetes Rev 2009; 5: 120–144.
51. Xu C, Bailly-Maitre B, Reed JC. Endoplasmic reticulum stress: cell life and death decisions. J Clin Invest 2005; 115: 2656–2664.
52. Lindenmeyer MT, Rastaldi MP, Ikehata M, et al. Proteinuria and hyperglycemia induce endoplasmic reticulum stress. J Am Soc Nephrol 2008; 19: 2225–2236.
53. Jafar TH, Schmid CH, Landa M, et al. Angiotensin-converting enzyme inhibitors and progression of nondiabetic renal disease. A meta-analysis of patient-level data. Ann Intern Med 2001; 135: 73–87.
54. Lewis EJ, Hunsicker LG, Clarke WR, et al; Collaborative Study Group. Renoprotective effect of the angiotensin-receptor antagonist irbesartan in patients with nephropathy due to type 2 diabetes. N Engl J Med 2001; 345: 851–860.
55. Brenner BM, Cooper ME, de Zeeuw D, et al; RENAAL Study Investigators. Effects of losartan on renal and cardiovascular outcomes in patients with type 2 diabetes and nephropathy. N Engl J Med 2001; 345: 861–869.
56. Parving HH, Lehnert H, Bröchner-Mortensen J, et al; Irbesartan in Patients with type 2 diabetes and Microalbuminuria Study Group. The effect of irbesartan on the development of diabetic nephropathy in patients with type 2 diabetes. N Engl J Med 2001; 345: 870–878.
57. Viberti G, Wheeldon NM; Microalbuminuria reduction with valsartan (MARVAL) Study Investigators. Microalbuminuria reduction with valsartan in patients with type 2 diabetes mellitus. A blood pressure-independent effect. Circulation 2002; 106: 672–678.
58. Parving HH, Persson F, Lewis JB, et al. Aliskiren combined with losartan in type 2 diabetes and nephropathy. N Engl J Med 2008; 358: 2433–2446.
59. Seikaly MG, Arant BS, Seney FD. Endogenous angiotensin concentrations in specific intrarenal fluid compartments of the rat. J Clin Invest 1990; 86: 1352–1357.
60. Manotham K, Tanaka T, Matsumoto M, et al. Evidence of tubular hypoxia in the early phase in the remnant kidney model. J Am Soc Nephrol 2004; 15: 1277–1288.
61. Palm F, Connors SG, Mendonca M, et al. Angiotensin II type 2 receptors and nitric oxide sustain oxygenation in the clipped kidney of early Goldblatt hypertensive rats. Hypertension 2008; 51: 345–351.
62. Onozato ML, Tojo A, Goto A, et al. Oxidative stress and nitric oxide synthase in rat diabetic nephropathy: effects of ACEI and ARB. Kidney Int 2002; 61: 186–194.
63. Benigni A, Corna D, Zoja C, et al. Disruption of the Ang II type 1 receptor promotes longevity in mice. J Clin Invest 2009; 119: 524–530.
64. Miyata T, van Ypersele de Strihou C, Ueda Y, et al. Angiotensin II receptor antagonist and angiotensin converting enzyme (ACE) inhibitor scavenge oxidative radicals and lower the formation of advanced glycation end products. J Am Soc Nephrol 2002; 13: 2478–2487.
65. Nangaku M, Miyata T, Sada T, et al. Anti-hypertensive agents inhibit in vivo the formation of advanced glycation end products and improve renal damage in a type 2 diabetic nephropathy rat model. J Am Soc Nephrol 2003; 14: 1212–1222.

66. Chen S, Ge Y, Si J, et al. Candesartan suppresses chronic renal inflammation by a novel antioxidant action independent of AT1R blockade. Kidney Int 2008; 74: 1128–1138.
67. Sun HL, Sun L, Li YY, et al. ACE-inhibitor suppresses the apoptosis induced by endoplasmic reticulum stress in renal tubular in experimental diabetic rats. Exp Clin Endocrinol Diabetes 2009; 117: 336–344.
68. Izuhara Y, Sada T, Yanagisawa H, et al. A novel sartan derivative with very low angiotensin II type 1 receptor affinity protects the kidney in type 2 diabetic rats. Arterioscler Thromb Vasc Biol 2008; 28: 1767–1773.
69. Takizawa, S. A sartan derivative with a very low angiotensin II receptor affinity ameliorates ischemic cerebral damage. J Cereb Blood Flow Metab 2009; 29: 1665–1672.
70. The Epidemiology of Diabetes Interventions and Complications (EDIC) Study. Sustained effect of intensive treatment of type 1 diabetes mellitus on development and progression of diabetic nephropathy. JAMA 2003; 290: 2159–2167.
71. Sarafidis PA, Ruilope LM. Insulin resistance, hyperinsulinemia, and renal injury: mechanisms and implications. Am J Nephrol 2006; 26: 232–234.
72. Makino H, Miyamoto Y, Sawai K, et al. Altered gene expression related to glomerulogenesis and podocyte structure in early diabetic nephropathy of db/db mice and its restoration by pioglitazone. Diabetes 2006; 55: 2747–2756.
73. Ohtomo S, Izuhara Y, Takizawa S, et al. Thiazolidinediones provide better renoprotection than insulin in an obese, hypertensive type II diabetic rat model. Kidney Int 2007; 72: 1512–1519.
74. Rodriguez WE, Tyagi N, Joshua IG, et al. Pioglitazone mitigates renal glomerular vascular changes in high-fat, high-calorie-induced type 2 diabetes mellitus. Am J Physiol Renal Physiol 2006; 291: F694–F701.
75. Perico N, Remuzzi G. Inhibition of TGF-beta expression: a novel role for thiazolidinediones to implement renoprotection in diabetes. Kidney Int 2007; 72: 1419–1421.
76. Gredilla R, Barja G. Minireview: the role of oxidative stress in relation to caloric restriction and longevity. Endocrinology 2005; 146: 3713–3717.
77. Cai W, He JC, Zhu L, et al. Oral glycotoxins determine the effects of calorie restriction on oxidant stress, age-related diseases, and lifespan. Am J Pathol 2008; 173: 327–336.
78. Trayhurn P, Wang B, Wood IS. Hypoxia in adipose tissue: a basis for the dysregulation of tissue function in obesity? Br J Nutr 2008; 100: 227–235.
79. Ye J. Emerging role of adipose tissue hypoxia in obesity and insulin resistance. Int J Obes 2009; 33: 54–66.
80. Crujeiras AB, Parra D, Goyenechea E, et al. Energy restriction in obese subjects impact differently two mitochondrial function markers. J Physiol Biochem 2008; 64: 211–219.
81. Nangaku M, Izuhara Y, Usuda N, et al. In a type 2 diabetic nephropathy rat model, the improvement of obesity by a low calorie diet reduces oxidative/carbonyl stress and prevents diabetic nephropathy. Nephrol Dial Transplant 2005; 20: 2661–2669.
82. Semenza GL. Targeting HIF-1 for cancer therapy. Nat Rev Cancer 2003; 3: 721–732.
83. Marx J. How cells endure low oxygen. Science 2004; 303: 1454–1456.
84. Epstein AC, Gleadle JM, McNeill LA, et al. Elegans EGL-9 and mammalian homologs define a family of dioxygenases that regulate HIF by prolyl hydroxylation. Cell 2001; 107: 43–54.
85. Schofield CJ, Ratcliffe PJ. Oxygen sensing by HIF hydroxylases. Nat Rev Mol Cell Biol 2004; 5: 343–354.
86. Hon WC, Wilson MI, Harlos K, et al. Structural basis for the recognition of hydroxyproline in HIF-1 alpha by pVHL. Nature 2002; 417: 975–978.
87. Ivan M, Kondo K, Yang H, et al. HIFα targeted for VHL-mediated destruction by proline hydroxylation: implications for O_2 sensing. Science 2001; 292: 464–468.
88. Maxwell PH, Wiesener MS, Chang GW, et al. The tumour suppressor protein VHL targets hypoxia-inducible factors for oxygen-dependent proteolysis. Nature 1999; 399: 271–275.

89. Ohh M, Park CW, Ivan M, et al. Ubiquitination of hypoxia-inducible factor requires direct binding to the beta-domain of the von Hippel-Lindau protein. Nat Cell Biol 2000; 2: 423–427.
90. Eguchi H, Ikuta T, Tachibana T, et al. A nuclear localization signal of human aryl hydrocarbon receptor nuclear translocator/hypoxia-inducible factor 1beta is a novel bipartite type recognized by the two components of nuclear pore-targeting complex. J Biol Chem 1997; 272: 17640–17647.
91. Kaelin WG Jr., Ratcliffe PJ. Oxygen sensing by metazoans: the central role of the HIF hydroxylase pathway. Mol Cell 2008; 30: 393–402.
92. Bernhardt WM, Schmitt R, Rosenberger C, et al. Expression of hypoxia-inducible transcription factors in developing human and rat kidneys. Kidney Int 2006; 69: 114–122.
93. Takeda K, Aguila HL, Parikh NS, et al. Regulation of adult erythropoietin by prolyl hydroxylase domain proteins. Blood 2008; 111: 3229–3235.
94. Percy MJ, Furlow PW, Lucas GS, et al. A gain-of-function mutation in the HIF2α gene in familial erythrocytosis. N Engl J Med 2008; 358: 162–168.
95. Bernhardt WM, Wiesener MS, Weidemann A, et al. Involvement of hypoxia-inducible transcription factors in polycystic kidney disease. Am J Pathol 2007; 170: 830–842.
96. Kojima I, Tanaka T, Inagi R, et al. Protective role of hypoxia-inducible factor-2alpha against ischemic damage and oxidative stress in the kidney. J Am Soc Nephrol 2007; 18: 1218–1226.
97. Hill P, Shukla D, Tran MG, et al. Inhibition of hypoxia inducible factor hydroxylases protects against renal ischemia-reperfusion injury. J Am Soc Nephrol 2008; 19: 39–46.
98. Fraisl P, Aragonés J, Carmeliet P. Inhibition of oxygen sensors as a therapeutic strategy for ischaemic and inflammatory disease. Nat Rev Drug Discov 2009; 8: 139–152.
99. Salnikow K, Donald SP, Bruick RK, et al. Depletion of intracellular ascorbate by the carcinogenic metals nickel and cobalt results in the induction of hypoxic stress. J Biol Chem 2004; 279: 4033–4044.
100. Matsumoto M, Makino Y, Tanaka T, et al. Induction of renoprotective gene expression by cobalt ameliorates ischemic injury of the kidney in rats. J Am Soc Nephrol 2003; 14: 1825–1832.
101. Tanaka T, Matsumoto M, Inagi R, et al. Induction of protective genes by cobalt ameliorates tubulointerstitial injury in the progressive Thy1 nephritis. Kidney Int 2005; 68: 2714–2725.
102. Tanaka T, Kojima I, Ohse T, et al. Cobalt promotes angiogenesis via hypoxia-inducible factor and protects tubulointerstitium in the remnant kidney model. Lab Invest 2005; 85: 1292–1307.
103. Tanaka T, Kojima I, Ohse T, et al. Hypoxia-inducible factor modulates tubular cell survival in cisplatin nephrotoxicity. Am J Physiol Renal Physiol 2005; 289: F1123–F1133.
104. Ohtomo S, Nangaku M, Izuhara Y, et al. Cobalt ameliorates renal injury in an obese, hypertensive type 2 diabetes rat model. Nephrol Dial Transplant 2008; 23: 1166–1172.
105. Smith EL. Presence of cobalt in the anti-pernicious anaemia factor. Nature 1948; 162: 144.
106. Berk L, Burchebal JH, Castle WB. Erythropoietic effect of cobalt in patients with or without anemia. N Engl J Med 1949; 240: 754–761.
107. Edwards MS, Curtis JR. Use of cobaltous chloride in anaemia of maintenance hemodialysis patients. Lancet 1971; 2: 582–583.
108. Ivan M, Haberberger T, Gervasi DC, et al. Biochemical purification and pharmacological inhibition of a mammalian prolyl hydroxylase acting on hypoxia-inducible factor. Proc Natl Acad Sci U S A 2002; 99: 13459–13464.
109. Mole DR, Schlemminger I, McNeill LA, et al. 2-oxoglutarate analogue inhibitors of HIF prolyl hydroxylase. Bioorg Med Chem Lett 2003; 13: 2677–2680.
110. McDonough MA, McNeill LA, Tilliet M, et al. Selective inhibition of factor inhibiting hypoxia-inducible factor. J Am Chem Soc 2005; 127: 7680–7681.
111. Nangaku M, Izuhara Y, Takizawa S, et al. A novel class of prolyl hydroxylase inhibitors induces angiogenesis and exerts organ protection against ischemia. Arterioscler Thromb Vasc Biol 2007; 27: 2548–2554.

112. Bernhardt WM, Câmpean V, Kany S, et al. Preconditional activation of hypoxia-inducible factors ameliorates ischemic acute renal failure. J Am Soc Nephrol 2006; 17: 1970–1978.
113. Watanabe D, Suzuma K, Matsui S, et al. Erythropoietin as a retinal angiogenic factor in proliferative diabetic retinopathy. N Engl J Med 2005; 353: 782–792.
114. Metzen E, Berchner-Pfannschmidt U, Stengel P, et al. Intracellular localisation of human HIF-1 alpha hydroxylases: implications for oxygen sensing. J Cell Sci 2003; 116: 1319–1326.
115. Steinhoff A, Pientka FK, Möckel S, et al. Cellular oxygen sensing: importins and exportins are mediators of intracellular localisation of prolyl-4-hydroxylases PHD1 and PHD2. Biochem Biophys Res Commun 2009; 387: 705–711.
116. Berchner-Pfannschmidt U, Tug S, Trinidad B, et al. Nuclear oxygen sensing: induction of endogenous prolyl-hydroxylase 2 activity by hypoxia and nitric oxide. J Biol Chem 2008; 283: 31745–31753.
117. Fong GH, Takeda K. Role and regulation of prolyl hydroxylase domain proteins. Cell Death Differ 2008; 15: 635–641.
118. Tug S, Delos RB, Fandrey J, et al. Non-hypoxic activation of the negative regulatory feedback loop of prolyl-hydroxylase oxygen sensors. Biochem Biophys Res Commun 2009; 384: 519–523.
119. Berchner-Pfannschmidt U, Yamac H, Trinidad B, et al. Nitric oxide modulates oxygen sensing by hypoxia-inducible factor 1-dependent induction of prolyl hydroxylase 2. J Biol Chem 2007; 282: 1788–1796.
120. Takeda, K, Cowan A, Fong GH. Essential role for prolyl hydroxylase domain protein 2 in oxygen homeostasis of the adult vascular system. Circulation 2007; 116: 774–781.
121. Loinard C, Ginouvès A, Vilar J, et al. Inhibition of prolyl hydroxylase domain proteins promotes therapeutic revascularization. Circulation 2009; 120: 50–59.
122. Pouysségur J, Dayan F, Mazure NM. Hypoxia signalling in cancer and approaches to enforce tumour regression. Nature 2006; 441: 437–443.
123. Mazzone M, Dettori D, Leite de Oliveira R, et al. Heterozygous deficiency of PHD2 restores tumor oxygenation and inhibits metastasis via endothelial normalization. Cell 2009; 136: 839–851.
124. Cummins EP, Berra E, Comerford KM, et al. Prolyl hydroxylase-1 negatively regulates IkappaB kinase-beta, giving insight into hypoxia-induced NFkappaB activity. Proc Natl Acad Sci U S A 2006; 103: 18154–18159.
125. Takeda K, Ichiki T, Narabayashi E, et al. Inhibition of prolyl hydroxylase domain-containing protein suppressed lipopolysaccharide-induced TNF-α expression. Arterioscler Thromb Vasc Biol 2009; 29(12): 2132–2137.
126. Mikhaylova O, Ignacak ML, Barankiewicz TJ, et al. The von Hippel-Lindau tumor suppressor protein and Egl-9-Type proline hydroxylases regulate the large subunit of RNA polymerase II in response to oxidative stress. Mol Cell Biol 2008; 28: 2701–2717.
127. Baranova O, Miranda LF, Pichiule P, et al. Neuron-specific inactivation of the hypoxia inducible factor 1 alpha increases brain injury in a mouse model of transient focal cerebral ischemia. J Neurosci 2007; 27: 6320–6332.
128. Köditz J, Nesper J, Wottawa M, et al. Oxygen-dependent ATF-4 stability is mediated by the PHD3 oxygen sensor. Blood 2007; 110: 3610–3617.
129. Natarajan R, Salloum FN, Fisher BJ, et al. Prolyl hydroxylase inhibition attenuates post-ischemic cardiac injury via induction of endoplasmic reticulum stress genes. Vascul Pharmacol 2009; 51: 110–118.
130. Zhou F, Zhu X, Castellani RJ, et al. Hibernation, a model of neuroprotection. Am J Pathol 2001; 158: 2145–2151.
131. Drew KL, Harris MB, LaManna JC, et al. Hypoxia tolerance in mammalian heterotherms. J Exp Biol 2004; 207: 3155–3162.

132. Siddiq A, Aminova LR, Troy CM, et al. Selective inhibition of hypoxia-inducible factor (HIF) prolyl-hydroxylase 1 mediates neuroprotection against normoxic oxidative death via HIF- and CREB-independent pathways. J Neurosci 2009; 29: 8828–8838.
133. Minamishima YA, Moslehi J, Bardeesy N, et al. Somatic inactivation of the PHD2 prolyl hydroxylase causes polycythemia and congestive heart failure. Blood 2008; 111: 3236–3244.
134. Ladroue C, Carcenac R, Leporrier M, et al. PHD2 mutation and congenital erythrocytosis with paraganglioma. N Engl J Med 2008; 359: 2685–2692.
135. Takeda K, Ho VC, Takeda H, et al. Placental but not heart defects are associated with elevated hypoxia-inducible factor alpha levels in mice lacking prolyl hydroxylase domain protein 2. Mol Cell Biol 2006; 26: 8336–8346.
136. Nguyen T, Sherratt PJ, Pickett CB. Regulatory mechanisms controlling gene expression mediated by the antioxidant response element. Annu Rev Pharmacol Toxicol 2003; 43: 233–260.
137. Kobayashi M, Yamamoto M. Molecular mechanisms activating the Nrf2 Keap1 pathway of antioxidant gene regulation. Antioxid Redox Signal 2005; 7: 385–394.
138. Leonard MO, Kieran NE, Howell K, et al. Reoxygenation-specific activation of the antioxidant transcription factor Nrf2 mediates cytoprotective gene expression in ischemia-reperfusion injury. FASEB J 2006; 20: 2624–2626.
139. Liu M, Grigoryev DN, Crow MT, et al. Transcription factor Nrf2 is protective during ischemic and nephrotoxic acute kidney injury in mice. Kidney Int 2009; 76(3): 277–285.
140. Yoh K, Hirayama A, Ishizaki K, et al. Hyperglycemia induces oxidative and nitrosative stress and increases renal functional impairment in Nrf2-deficient mice. Genes Cells 2008; 13: 1159–1170.
141. Cho HY, Jedlicka AE, Reddy SP, et al. Role of Nrf2 in protection against hyperoxic lung injury in mice. Am J Respir Cell Mol Biol 2002; 26: 175–182.
142. Cho HY, Reddy SP, Yamamoto M, et al. The transcription factor Nrf2 protects against pulmonary fibrosis. FASEB J 2004; 18: 1258–1260.
143. Shih AY, Li P, Murphy TH. A small-molecule-inducible Nrf2-mediated antioxidant response provides effective prophylaxis against cerebral ischemia in vivo. J Neurosci 2005; 25: 10321–10335.
144. Itoh K, Wakabayashi N, Katoh Y, et al. Keap1 represses nuclear activation of antioxidant responsive elements by Nrf2 through binding to the amino-terminal Neh2 domain. Genes Dev 1999; 13: 76–86.
145. Wakabayashi N, Itoh K, Wakabayashi J, et al. Keap1-null mutation leads to postnatal lethality due to constitutive Nrf2 activation. Nat Genet 2003; 35: 238–245.
146. Yoon HY, Kang NI, Lee HK, et al. Sulforaphane protects kidneys against ischemia-reperfusion injury through induction of the Nrf2-dependent phase 2 enzyme. Biochem Pharmacol 2008; 75: 2214–2223.
147. Zhao J, Kobori N, Aronowski J, Dash PK. Sulforaphane reduces infarct volume following focal cerebral ischemia in rodents. Neurosci Lett 2006; 393: 108–112.
148. Tong KI, Kobayashi A, Katsuoka F, et al. Two-site substrate recognition model for the Keap1-Nrf2 system: a hinge and latch mechanism. Biol Chem 2006; 387: 1311–1320.

Chapter 33
Endoplasmic Reticulum Stress as a Target of Therapy Against Oxidative Stress and Hypoxia

Reiko Inagi

Abstract Oxidative stress or hypoxia is the major disturbance that leads to endoplasmic reticulum (ER) dysfunction. The ER maintains protein homeostasis, including the regulation of the concentration, conformation, folding, and trafficking of client proteins. ER dysfunction by the disturbances such as oxidative stress, referred to as ER stress, induces intracellular stress responses, called the unfolded protein response (UPR). The UPR initially serves as an adaptive response, but also induces apoptosis in cells under severe or prolonged ER stress. The linkage of ER stress with oxidative stress or hypoxia, both of which are pathogenic, indicates the potential pathophysiological significance of ER stress across a wide range of diseases. Accumulating evidence indicates that ER stress contributes to glomerular and tubular damages in patients with acute and chronic kidney diseases. In glomeruli, podocyte or mesangial dysfunction tends to induce the adaptive UPR, which involves ER chaperone expression and the attenuation of protein translation to maintain ER homeostasis and ensure cell survival. In tubules, apoptosis, resulting from epithelial cell damage, is caused by the pro-apoptotic UPR, at least in part. These findings emphasize not only the importance of ER stress as a new progression factor but also the interesting future possibility of renoprotective strategy targeting ER stress. These therapeutic approaches may act by breaking the vicious cycle of oxidative stress, hypoxia, and ER stress.

Keywords Chronic kidney disease · Acute kidney injury · Unfolded protein response · Glucose-regulated protein 78 · Inositol-requiring enzyme 1 · Double-stranded RNA-activated-protein kinase-like ER kinase · Activating transcription factor 6 · Hypoxia-inducible factor · Oxidative stress-inducible pathway · Inflammatory response

---

R. Inagi (✉)
Division of Nephrology and Endocrinology,
University of Tokyo School of Medicine, 7-3-1, Hongo Bunkyo-ku,
Tokyo 113-8655, Japan
e-mail: inagi-npr@umin.ac.jp

T. Miyata et al. (eds.), *Studies on Renal Disorders*, Oxidative Stress in Applied Basic Research and Clinical Practice, DOI 10.1007/978-1-60761-857-7_33,

1 Endoplasmic Reticulum Stress and Its Cellular Response

Endoplasmic reticulum (ER) plays an important role in regulation of protein biosynthesis, folding, trafficking, and modification. The ER functions are sensitive to environmental insults such as oxidative stress or hypoxia, which can cause aberrant protein folding. The accumulation of these aberrant unfolded and malfolded proteins in the ER lumen in turn induces a range of ER dysfunctions, collectively referred to as ER stress [1–3]. To ensure the fidelity of protein folding and prevent the accumulation of unfolded or malfolded proteins in the ER, cells experiencing ER stress invoke a well-conserved intracellular signaling known as the unfolded protein response (UPR), which is classified as adaptive or maladaptive. The adaptive UPR is initially aimed at ameliorating the ER functional damage by enhancing protein-folding capacity or attenuating protein translation and thereby ensures cell survival. However, when the ER stress is severe or prolonged, or the adaptive UPR is insufficient for maintaining ER homeostasis, the cells decide to suicide, eventually triggering the maladaptive and pro-apoptotic UPR for cell death.

2 Pathophysiological States of ER Stress: A Link to Oxidative Stress and Hypoxia

ER stress contributes to the pathogenesis of conformational diseases caused by accumulation of malfolded protein in the ER due to mutations associated with protein misfolding [2]. However, recent evidence clearly emphasizes that other factors, including oxidative stress or hypoxia (or nutrient deprivation), induce the impairment of protein-folding capacity and that these lead to pathogenic ER stress in various diseases, including cancer, neurodegenerative and cardiovascular diseases, diabetes, as well as kidney diseases. Most importantly, the crosstalk of stress-induced cellular responses by hypoxia, oxidative stress, and ER stress is highlighted (Fig. 1).

2.1 ER Stress and Hypoxia

Hypoxia and ischemia are well-known ER stressors. Hypoxic conditions associated with the loss of energy initiate an imbalance between protein-folding capacity and protein-folding load, thereby cause the accumulation of misfolded proteins in the ER lumen. These ER functional changes lead to the activation of ER transmembrane signal transducers to initiate the UPR pathway. In tumor cells, the hypoxic microenvironment brought on by poor vascularization brings about physiological ER stress due to the loss of energy for protein folding. Of note, the hypoxia-induced adaptive UPR is crucial to the survival of tumor cells [4]. ER resident chaperone,

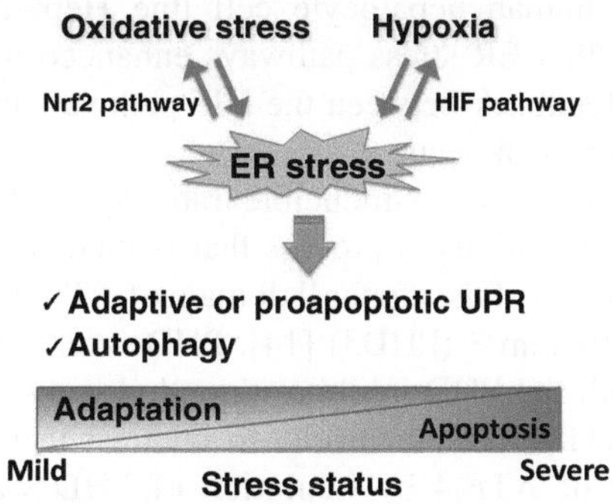

Fig. 1 Pathogenesis of endoplasmic reticulum (ER) stress. Various kinds of pathogenesis are induced by ER stress. Oxidative stress induces ER stress through the accumulation of reactive oxygen species (ROS) and vice versa. Hypoxia, which induces energy consumption for protein folding, accumulates misfolding protein in the ER, thereby causing ER stress. The ER stress induces unfolded protein response (UPR) and autophagy, of which they are cellular protective under mild or short-term ER stress and apoptotic under severe or prolonged ER stress. Importantly, UPR modulates oxidative stress-inducible Nrf2 (nuclear factor-erythroid-derived 2-related factor 2) pathway or hypoxia-inducible factor (HIF) pathway to enhance their target genes, which maintain the cell survival

such as glucose-regulated protein 78 (GRP78), for example, is induced to enhance protein-folding capacity in a wide range of human cancers, and expression levels correlate with tumor progression, metastasis, and drug resistance [5]. While the adaptive UPR promotes the survival of malignant cells, recent evidence suggests the pathogenic involvement of pro-apoptotic UPR, another ER stress response, during hypoxia or ischemia/reperfusion insults in various organs [6, 7]. Ischemia or ischemia/reperfusion of the brain or heart induced several UPR pathways, which are adaptive and organ-protective under brief or moderate stress or maladaptive and apoptotic under prolonged severe stress [8–10]. In coronary artery segments, ER chaperone expression via the adaptive UPR and cell apoptosis via the maladaptive UPR are markedly increased in both smooth muscle cells and macrophages in the fibrous caps of thin-cap atheroma and ruptured plaques, but not in those of thick-cap atheroma and fibrous plaques, suggesting that ER stress plays a pathogenic role in the rupture of atherosclerotic plaques [11].

Most interesting topics on ER stress and hypoxia have emphasized the linkage of the UPR pathway with the hypoxia-inducible signal pathway, namely the hypoxia-inducible factor (HIF) pathway. A proteomic analysis of cultured endothelial cells under hypoxia showed upregulation of the proteins that are responsible for adaptive and pro-apoptotic UPRs, including GRP78 and caspase 12, respectively [12]. Incubation with cobalt chloride, a stabilizer of HIF, increased the expression of ER stress-inducible chaperons such as GRP78 and GRP94. These data reveal that lowering oxygen tensions, probably in part through HIF, alters the expression of a series of proteins involved in UPR. Of note, induction of ER stress by tunicamycin and brefeldin A increased HIF-1α mRNA without altering intracellular calcium concentrations in the human hepatocyte cell line HepG2 exposed to hypoxic conditions, suggesting that ER stress pathways enhanced transcription of HIF-1α mRNA [13]. Positive feedback between the HIF pathway and ER stress responses under hypoxic conditions can occur.

In addition, ATF-4, an ER stress-inducible transcription factor, is translationally induced under anoxic conditions, a process that is mediated by increased protein stability through interaction of the zipper II domain of ATF-4 with the oxygen sensor prolyl-4-hydroxylase domain 3 (PHD3) [14]. PHDs are well-known regulators of HIF, and treatment with the PHD inhibitor dimethyloxalylglycine or proteasomal inhibition increased ATF-4 protein levels. In contrast to the situation with HIF, however, stabilization of ATF-4 by inhibition of PHD was independent of the ubiquitin ligase von Hippel-Lindau protein. These data demonstrate that PHD-dependent oxygen-sensing recruits both the HIF and ATF-4 systems in parallel. To further emphasize the biological significance of interaction between the HIF and ER stress pathways, studies utilizing *Caenorhabditis elegans* showed that HIF-1 deficiency extended lifespan in an UPR transducer IRE-1-dependent manner and was associated with lower levels of ER stress [15].

2.2 ER Stress and Oxidative Stress

Recent accumulated evidence demonstrates that ER stress increases oxidative stress in the stress-damaged cells and vice versa. Reperfusion after ischemia triggers oxidative stress with production of reactive oxygen species (ROS), altering cellular redox-dependent reactions and interfering with protein-folding capacity including protein disulfide bonding, ultimately resulting in protein misfolding in the ER. Studies utilizing overexpression of antioxidative stress enzymes have emphasized the linkage of oxidative stress to the ER stress response. The ischemia-induced UPR was markedly less pronounced in animals overexpressing copper/zinc superoxide dismutase (Cu/Zn-SOD), suggesting that superoxide radicals play a role in this pathological process [16]. Further, cadmium caused generation of ROS with subsequent induction of ER stress in a cultured renal proximal tubular cell line, which in turn led to apoptosis; this cadmium-induced ER stress and apoptosis were significantly attenuated by transfection with manganese SOD [17]. Paradoxically, ER stress also increases intracellular ROS production: increased protein disulfide bonding enhances ROS production in the ER lumen, and alteration of ER Ca^{2+} homeostasis increases cytosolic Ca^{2+}, thereby stimulating mitochondrial ROS production. Of particular note, recent studies have demonstrated that the accumulation of intracellular ROS caused by ER stress is attenuated by the double-stranded RNA-activated protein kinase-like ER kinase (PERK) pathway of the UPR, which simultaneously activates oxidative stress–inducible transcription factor Nrf2 (nuclear factor-erythroid-derived 2-related factor 2) and maintains the redox homeostasis, thereby ensuring cell survival. The antioxidant effects of the PERK pathway are supported by the finding that PERK-deficient cells exposed to tunicamycin, an

ER stress inducer, showed a toxic accumulation of intracellular ROS compared to wild-type cells [18]. These findings are supported by studies demonstrating that translation attenuation through the PERK/eIF2a pathway effectively prevents protein-induced oxidative stress misfolding and maintains the cell [19].

Nitric oxide (NO), produced in excessive levels following ischemia, also contributes to ER stress. Ischemia induces adaptive UPR-mediated translation attenuation (activation of PERK and subsequent phosphorylation of eIF2α). The translation attenuation is completely blocked in endothelial or neuronal NO synthase (NOS) knockout mice with bilateral carotid artery occlusion [20]. Consistent with this, the NO-releasing reagent SNAP activates the PERK pathway [20]. Pretreatment with neuroprotective levels of a NOS inhibitor also ameliorated the ER stress-induced disturbance of ER Ca^{2+} homeostasis by ischemic episodes, which cause Ca^{2+} leakage into the cytosol and subsequent uptake into mitochondria, resulting in mitochondrial ROS generation [21]. The excessive NO production seen in hypoxia, which alters calcium homeostasis in both the ER and mitochondria, may initiate a vicious cycle of ER stress, oxidative stress, and apoptosis. Taken together, these observations indicate that NO plays a role in the ischemia/reperfusion-induced UPR.

Accumulating evidence indicates the involvement of ER stress in several types of inflammation [22]. In inflammation of the central nervous system, for example, interferon-γ, induces ER stress and apoptosis of oligodendrocytes [23], while in lipopolysaccharide-induced inflammation of the lungs, lipopolysaccharide induces ER stress and C/EBP homologous protein (CHOP) expression, leading to the apoptosis of lung cells [7]. At the same time, nuclear factor κB (NFκB), a key transcriptional regulator of genes involved in the inflammatory response, is activated by ER stress, an action caused by translational attenuation and degradation of the inhibitor of NFκB (IκB) [22].

ER stress is also involved in autoimmunity. A microarray analysis of muscle tissue obtained from patients with myositis revealed the induction of UPR proteins such as GRP78, CHOP, and GADD45 in the cells, suggesting that the ER stress response is responsible for the skeletal muscle damage and dysfunction evident in autoimmune myositis [24]. Intestinal inflammation has been shown to result from abnormalities of transcriptional factor of UPR, XBP1, in intestinal epithelial cells, suggesting a link between cell-specific ER stress and organ-specific inflammation [25]. In intestinal epithelial cells, Xbp1 deletion results in spontaneous enteritis, while mutant variants of the gene are associated with both forms of human inflammatory bowel disease (Crohn's disease and ulcerative colitis).

The UPR pathway links not only the oxidative stress-inducible response (Nrf2 pathway) and inflammatory response (NFκB pathways), but also the mammalian target of rapamycin (mTOR) pathway, which regulates cell growth and survival and inflammation through changes in protein synthesis, autophagy, and metabolism. mTOR activation causes ER stress by increasing client protein load in the ER, subsequently causing the activated UPR (IRE1-TRAF2-JNK pathway) to down-regulate mTOR activation [26]. The IRE1-TRAF2-JNK pathway is not only tied to ER stress-mediated apoptosis but also negatively regulates mTOR activity through the inactivation of insulin receptor substrate (IRS1). This may indicate the importance of this negative feedback loop between the mTOR and the UPR pathways. Our recent study also showed a link between hypoxia, UPR, and the mTOR pathway, in which upregulation of metallothionein expression in the hypoxic kidney activates the HIF pathway via the ERK/mTOR pathway [27]. Like the HIF pathway, mTOR and UPR are O_2-sensitive signaling pathways. Thus, the phenotype of hypoxic cells, particularly those associated with inflammation, might be derived from the responses orchestrated by these pathways.

3 Contribution of ER Stress in Kidney Disease

ER stress plays a pathogenic role in diseases associated with the accumulation of malfolded proteins, such as conformational diseases like Alzheimer's, Parkinson's, and Huntington's diseases. Further, ER stress is also associated with a wide range of other conditions, including cancer, cardiac and cerebral ischemia/reperfusion injury, diabetes, and atherosclerosis [2, 28–31]. These associations indicate the importance of ER stress as a novel pathogenic agent and the possibility of its linkage to the pathogenesis and progression of other conditions. Accumulating evidence, including our previous studies, suggests the pathophysiological significance of ER stress in kidney disease [32, 33] (Table 1).

Table 1 Pathogenic contribution of endoplasmic reticulum stress in kidney disease

Site	Species	Disease state
Glomerulus	Human	Membranoproliferative glomerulonephritis
		Crescentic glomerulonephritis
	Rat	Passive Heymann nephritis
		Puromycin nephrosis
		Misfolded protein accumulation in podocytes
		Anti-Thy-1 nephritis
		Type 1 diabetic nephropathy
	Mouse	Focal segmental glomerulosclerosis model
	In vitro	Mislocalization of slit diaphragm components
		Glycated protein-induced apoptosis in podocyte
Tubulointerstitium	Human	Acute kidney injury
		Established diabetic nephropathy
		Minimal change disease
	Rat	Ischemia-reperfusion injury
		Puromycin nephropathy model
		Tubular injury by nephrotoxic drug:
		Antibiotics, immunosuppressants, anticancer drugs
	Mouse	Ischemia/reperfusion injury
		Protein-overload nephropathy

3.1 ER Stress in Glomerular Injury

Podocytes, which are crucial components of the glomerular filtration barrier, play an important role in the development of glomerulosclerosis. Previous reports have identified the induction of ER stress in podocytes in models of membranous nephropathy (Heymann nephritis) [34, 35], focal segmental or minimal-change nephrotic syndrome (puromycin nephrosis) [36], and focal segmental glomerulosclerosis (FSGS) [37]. Among examples, complement-mediated podocyte injury induced an adaptive UPR pathway, as demonstrated by the expression of ER chaperones (GRP78 and GRP94), in proteinuric rats with Heymann nephritis. In puromycin nephrosis rats, nephrotic-range proteinuria increased GRP78 expression in podocytes and led to the mislocalization of nephrin, a key component of the slit diaphragm in the glomerular filtration barrier, to the cytoplasm, demonstrating that proteinuria-induced podocyte damage was associated with ER stress, which in turn affected the biogenesis of slit diaphragm proteins. Further, in vitro studies utilizing human embryonic kidney-293 cells expressing human nephrin have emphasized the pathogenic contribution of ER stress in podocytes. Energy depletion evokes a rapid increase in ER stress, leading to the formation of underglycosylated nephrin and accumulation of unfolded nephrin in the ER, in turn suggesting that ER stress in podocytes may cause an alteration in nephrin folding and trafficking as an underlying factor in the pathomechanism of proteinuria [38]. These observations are consistent with our findings: in podocytes overexpressing a transgene, the excessive product is susceptible to misfolding in the ER, with subsequent accumulation of the misfolded protein in the ER in association with UPR and podocyte dysfunction, including severe proteinuria [39].

With regard to mesangial cells, which maintain the glomerular structure and hemodynamics via their synthesis and assembly of the mesangial matrix, mesangial injury induced the UPR pathway in a rat model of mesangio-proliferative glomerulonephritis, including the induction of ER chaperones (GRP78 and ORP150, 150 kDa oxygen-regulated protein) and translation attenuation via the PERK-eIF2α pathway [40].

The pathogenic contribution of ER stress observed in animal models of glomerular disease is supported by the finding that ER stress is significantly upregulated in renal biopsies from patients with membrano-proliferative glomerulonephritis and crescentic glomerulonephritis [41]. Further, glomerular damage in diabetic nephropathy is also associated with ER stress, which includes GRP78 expression and apoptotic UPR activation through caspase 12 and JNK pathways, strongly suggesting the contribution of ER stress to disease progression [42].

3.2 ER Stress in Tubulointerstitial Injury

A final common pathway to end stage kidney disease is expected to be found in the tubulointerstitium. Tubular cells are particularly sensitive to ER stress inducers such as tunicamycin, which disturbs protein glycosylation in the ER and induces UPR. Evidencing this, mice injected with tunicamycin develop acute renal tubular necrosis [18], while those lacking the antiapoptotic protein BI-1 (Bax Inhibitor-1) in the ER are more sensitive to tunicamycin-induced renal tubule cell death [43]. The effect of massive proteinuria, which mediates tubulointerstitial injury in various chronic kidney diseases, has also been investigated by utilizing puromycin nephropathy in rats [44]. These results demonstrate that protein load in tubular cells invokes ER stress-induced apoptosis via the caspase 12 pathway, suggesting that ER stress contributes to the tubular damage induced by proteinuria.

Chronic hypoxia, another important mediator of tubulointerstitial injury, leads to end stage kidney disease. While the effects of chronic hypoxia on tubular ER stress have not been elucidated, other investigations have successfully shown the induction of ER stress in tubular cells subjected to acute ischemia [45–47]. The major role played by the ER stress response in chronic tubulointerstitial injury was emphasized in a recent study examining knockin mice expressing mutant GRP78, in which heterozygous mutant-GRP78 mice showed significant tubulointerstitial lesions with aging [48]. These data, together with the finding that prolonged proteinuria induced by chronic protein overload accelerates lesion development associated with caspase 12 activation and tubular cell apoptosis, indicate that the ER stress pathway is significantly involved in the pathophysiology of chronic kidney injury with tubulointerstitial damage.

The body of evidence from in vivo and in vitro animal experiments, including our results, is consistent with clinical observations in humans. The adaptive UPR pathways and ER stress-induced apoptosis are invoked in the tubular epithelium of patients suffering from established diabetic nephropathy or minimal change disease, suggesting that hyperglycemia and proteinuria can induce ER stress in tubules in both humans and rodents [49].

Nephrotoxic drugs, such as cisplatin, gentamicin, and cyclosporine, are also known to induce ER stress-mediated apoptosis in tubular cells [50–52]. Of note, in cyclosporine nephropathy, ER stress induces autophagy in the tubules, which is a third type of cell death [53]. We also showed that ER stress induces autophagy in cultured tubular cells [54]. The representative ER stress inducers tunicamycin and brefeldin A significantly increased LC3-II as a marker of autophagy, while immunocytochemistry with LC3 and electron microscopy also showed the activation of autophagy by ER stress. Inhibition by bafilomycin A1 showed that autophagy following ER stress fulfilled its intrinsic function, namely the degradation of cytoplasmic components. We also observed that extracellular-signal-regulated kinase (ERK), a mitogen-activated protein kinase (MAPK) family member, was necessary to the induction of autophagy by ER stress. Further investigation will elucidate the pathogenic role of autophagy in the tubular damage associated with ER stress.

3.3 ER Stress in Aged Kidney

Various pathogenic responses induced by the stress are augmented by aging. They include oxidative stress or ER stress responses. Several studies have demonstrated that the abundance of ER resident chaperons or PDI (protein disulfide isomerase) family enzymes, which act for redox-dependent protein folding and disulfide bond formation, significantly diminishes with age in various organs, including brain, liver, and kidney [55]. In rats, protein expression level of GRP78, a key sensor of ER stress, is lower in aged kidney (>18 months old) than in young kidney (1 month old), a change that is associated with an imbalance in UPR activation (diminished adaptive UPR and enhanced pro-apoptotic UPR). Heterozygous mutant-GRP78 knockin mice, in which GRP78 function is diminished, show progressive tubulointerstitial lesions with aging as compared to wild-type mice [48]. Further, given that ER-resident proteins are highly susceptible to oxidative stress, the age-related accumulation of oxidatively modified (carbonylated) GRP78 or PDI causes ER dysfunction with age [56]. Taken together, these findings indicate that the insufficient ER function and imbalance in UPR activation (diminished adaptive UPR and enhanced pro-apoptotic UPR) with age may contribute to the induction of the aging-related senescence phenotype, including renal fibrosis.

4 Therapeutic Approaches Targeting ER Stress

From the findings that hypoxic or oxidative stress response links ER stress response, therapeutic approaches targeting ER stress may be beneficial due to their ability to break the vicious cycle of oxidative stress, hypoxia, and ER stress. Indeed, several studies emphasize the idea that artificial modulation of ER stress may provide protection to the cell. Cells overexpressing GRP78 are resistant to conditions associated with ER stress [57]. Further, systemic overexpression of ORP150 protects the kidney against ischemia/reperfusion injury [46]. However, while gene transfer may be promising for future therapy, other therapeutic modalities are needed to address immediate concerns (Table 2).

Preconditioning was first identified by the observation that in ischemic diseases brief ischemic treatment before the subsequent insult induced a state of resistance to blood loss by initiating a cascade of biochemical events, allowing for the upregulation of the cellular protective genes in the tissues. In our previous study, we observed beneficial effects of preconditioning with ER stress by utilizing a subnephritogenic dose of the ER stress inducer (tunicamycin or thapsigargin) in a model of mesangio-proliferative glomerulonephritis [40]. Given that overwhelming ER stress leads to eventual cell death via pro-apoptotic UPR, including activation of caspase 12, induction of ER stress as a therapeutic approach may seem paradoxical. However, ER stress also induces the adaptive UPR, shutting down translation and expressing ER chaperones to aid in protein folding, which improves the efficiency

Table 2 Therapeutic modality targeting endoplasmic reticulum stress

Modality	Function(s)
ER chaperone overexpression	
GRP78	Translation shutdown↑
ORP150	Protein folding↑
ER stress preconditioning	
ER stress inducers	Basal adaptive UPR↑
Chemical compounds	
DTTox	GRP78 expression↑
BIX	GRP78 expression↑
Salubrinal	Adaptive UPR (PERK-eIF2α)↑
Benzodiazepinones	IRE1/TRAF2/ASK1↓
Chemical chaperones	
4-PBA	Protein folding↑, ERAD↑, caspase 12↓
TUDCA	Adaptive UPR↑, proapoptotic UPR↓
Others	
Anti-oxidative stress compounds	
TM2002	Oxidative stress↓
Butylated hydroxyanisole	Protein-folding capacity↑
Anti-inflammatory drug	
Mizoribine	Intracellular energy for protein folding↑
Anti-hypertensive drug	
Angiotensin II type 1 receptor blocker	Proapoptotic UPR (CHOP)↓

of energy consumption. Cytoprotection by preconditioned targeting of ER stress has also been confirmed in cultured renal tubular cells treated with clinically relevant nephrotoxic drugs and in tubular injury in mice with renal ischemia/reperfusion [58, 59]. Thus, preconditioned targeting of ER stress enhances the adaptive UPR, thereby protecting the cell more effectively against the pathogenic environment via maintenance of ER function.

Chemical compounds that target the ER stress pathways have been shown effective across a range of situations, building enthusiasm for the therapeutic potential latent in ER stress response augmentation. One study found that trans-4,5-dihydroxy-1,2-dithiane (DTTox) stimulated an increase in GRP78 expression and protected the proximal tubular epithelium against a nephrotoxic chemical [60]. A GRP78 reporter assay system identified the chemical compound BIX (BiP inducer X, 1-(3,4-dihydroxyphenyl)-2-thiocyanato-ethanone) as an inducer of GRP78 mRNA. In gerbils subjected to forebrain ischemia, prior treatment with BIX protected against disease manifestation [9]. Further, screening for inhibitors of ER stress–induced neuronal death identified the chemical compound salubrinal, which suppresses protein phosphatases responsible for the dephosphorylation of eIF2α, thereby increasing the accumulation of phosphorylated eIF2α and providing protection from the apoptosis induced by several ER stress inducers [61]. Kim et al. used a cell-based high throughput screening assay to identify components that halted ER stress-induced apoptosis [62]. Results showed that benzodiazepinones selectively enhanced ASK1 phosphorylation at serine 967, which reduces but does not induce ASK1 activity, resulting in inhibition of IRE1-ASK1 pathway. These findings suggest the usefulness of a modulator of ASK1 function as an ER stress-induced apoptosis inhibitor.

Alternatively, chemical chaperones present another possible treatment method. Chemical chaperones such as sodium 4-phenylbutyrate (4-PBA) are a group of compounds known to improve ER folding capacity and facilitate the trafficking of unfolded or malfolded proteins by stabilizing their conformation. In one study, pre- or posttreatment with 4-PBA at therapeutic dose attenuated infarction volume, hemispheric swelling, and apoptosis and also improved neurological status in a mouse model of brain ischemia, possibly due to a decrease in the protein load retained by the ER [63]. Mutated nephrin associated with misfolding and mislocalization is rescued by 4-PBA, suggesting the beneficial effect of ER stress modulators in nephrosis induced by nephrin mutation [64]. Endogenous bile acid derivatives, such as tauroursodeoxycholic acid (TUDCA), can also modulate ER function, protecting liver cells against ER stress–induced apoptosis or restoring glucose homeostasis in type 2 diabetic mice [65, 66]. Unlike 4-PBA, however, TUDCA has not been shown to act as a chaperone that promotes the folding and trafficking of unfolded or malfolded proteins. Instead, TUDCA is likely to increase the stability of unfolded or malfolded proteins, possibly as a result of its influence on UPR activation with subsequent enhancement of the ERAD pathway. ER stress plays a role in advanced glycation end products (AGE)-induced apoptosis in murine podocytes, and TUDCA prevents apoptosis by blocking an ER stress-mediated apoptotic pathway [67].

From the evidence of crosstalk of oxidative stress, hypoxia, inflammation, and ER stress, antioxidative stress, antihypoxia, or anti-inflammation drugs may also be promising to reduce pathogenic ER stress. As an example, effective attenuation of the ER stress response has also been observed in compounds that suppress oxidative stress. TM2002, an inhibitor of oxidative protein glycation, shows renoprotective effects associated with a reduction in ER stress in ischemic/reperfusion rats [68]. Malhotra et al. demonstrate that antioxidants such as butylated hydroxyanisole reduce ER stress-inducible apoptosis and improve protein-folding capacity by reducing intracellular ROS level, suggesting the effectiveness of intervention to reduce ROS for cell survival through improvement of ER homeostasis [69]. A chemical HIF stabilizer prolyl hydroxylase inhibitor, which attenuates ischemic cardiac injury, selectively activates adaptive UPR and lowers proapoptotic UPR [70]. Mizoribine, an clinically used immunosuppressant, inhibits purine nucleotide biosynthesis and subsequently restores the intracellular energy balance during ER stress by salvaging adeosine triphosphate (ATP) levels. Mizoribine treatment reduced ER stress and rescued the mislocalization of nephrin to the cytoplasm in podocytes with glucose starvation, suggesting that the remission of proteinuria by mizoribine is mediated by the reduction of ER stress [36]. This evidence indicates the possibility that combination therapy of anti-ER stress drug and antioxidative stress, antihypoxic, or anti-inflammatory drug shows the beneficial effects synergistically. Further, an angiotensin II type 1 receptor blocker also shows the ER stress regulating effects. Angiotensin II type 1 receptor is one of the ER stress inducers, while the molecular mechanisms of angiotensin II induce ER stress. Therefore, it is reasonable that hypertrophic and failing heart after aortic constriction induces ER stress, which contributes to cardiac myocyte apoptosis during the disease progression, and angiotensin II type 1 receptor blocker prevents upregulation of ER stress, thereby damaging the heart [71].

5 Conclusion

ER stress stimulates adaptive UPR to restore ER homeostasis and the pro-apoptotic UPR to eliminate cells under prolonged stress. The balance of these two contrasting UPRs appears to depend on either or both ER stress status (severity or duration) or cell type. In acute and chronic kidney disease, ER stress contributes to the development and progression of glomerular and tubular disease, and accumulating evidence suggests the possibility of a novel therapeutic approach targeting ER stress. The finding that ER stress links hypoxia, oxidative stress, and inflammation further highlights the potential benefit of maintaining ER homeostasis by modulating ER stress status as a means of protecting the kidney against various pathogenic environments.

Acknowledgments This work was supported by grants-in-aid for scientific research from the Japan Society for the Promotion of Science (19590939) and a grant from the Kidney Foundation, Japan (JKFB09-2).

References

1. Malhotra JD, Kaufman RJ: Endoplasmic reticulum stress and oxidative stress: a vicious cycle or a double-edged sword? Antioxid Redox Signal 2007;9:2277–2293. Review.
2. Yoshida H: ER stress and diseases. FEBS J 2007;274:630–658.
3. Ron D, Walter P: Signal integration in the endoplasmic reticulum unfolded protein response. Nat Rev Mol Cell Biol 2007;8:519–529.
4. Koumenis C: ER stress, hypoxia tolerance and tumor progression. Curr Mol Med 2006;6:55–69.
5. Lee AS: GRP78 induction in cancer: therapeutic and prognostic implications. Cancer Res 2007;67:3496–3499.
6. Orrenius S, Zhivotovsky B, Nicotera P: Regulation of cell death: the calcium-apoptosis link. Nat Rev Mol Cell Biol 2003;4:552–565.
7. Endo M, Mori M, Akira S, Gotoh T: C/EBP homologous protein (CHOP) is crucial for the induction of caspase-11 and the pathogenesis of lipopolysaccharide-induced inflammation. J Immunol 2006;176:6245–6253.
8. Paschen W, Mengesdorf T: Cellular abnormalities linked to endoplasmic reticulum dysfunction in cerebrovascular disease – therapeutic potential. Pharmacol Ther 2005;108:362–375.
9. Oida Y, Izuta H, Oyagi A, Shimazawa M, Kudo T, Imaizumi K, Hara H: Induction of BiP, an ER-resident protein, prevents the neuronal death induced by transient forebrain ischemia in gerbil. Brain Res 2008;1208:217–224.

10. Glembotski CC: The role of the unfolded protein response in the heart [review]. J Mol Cell Cardiol 2008;44:453–459.
11. Myoishi M, Hao H, Minamino T, Watanabe K, Nishihira K, Hatakeyama K, Asada Y, Okada K, Ishibashi-Ueda H, Gabbiani G, Bochaton-Piallat ML, Mochizuki N, Kitakaze M: Increased endoplasmic reticulum stress in atherosclerotic plaques associated with acute coronary syndrome. Circulation 2007;116:1226–1233.
12. Ostergaard L, Simonsen U, Eskildsen-Helmond Y, Vorum H, Uldbjerg N, Honoré B, Mulvany MJ: Proteomics reveals lowering oxygen alters cytoskeletal and endoplasmatic stress proteins in human endothelial cells. Proteomics 2009;19:4457–4467.
13. Werno C, Zhou J, Brüne B: A23187, ionomycin and thapsigargin upregulate mRNA of HIF-1alpha via endoplasmic reticulum stress rather than a rise in intracellular calcium. J Cell Physiol 2008;215:708–714.
14. Köditz J, Nesper J, Wottawa M, Stiehl DP, Camenisch G, Franke C, Myllyharju J, Wenger RH, Katschinski DM: Oxygen-dependent ATF-4 stability is mediated by the PHD3 oxygen sensor. Blood 2007;110:3610–3617.
15. Chen D, Thomas EL, Kapahi P: HIF-1 modulates dietary restriction-mediated lifespan extension via IRE-1 in *Caenorhabditis elegans*. PLoS Genet 2009; 5:e1000486.
16. Hayashi T, Saito A, Okuno S, Ferrand-Drake M, Dodd RL, Nishi T, Maier CM, Kinouchi H, Chan PH: Oxidative damage to the endoplasmic reticulum is implicated in ischemic neuronal death. J Cereb Blood Flow Metabol 2003;23:1117–1128.
17. Yokouchi M, Hiramatsu N, Hayakawa K, Okamura M, Du S, Kasai A, Takano Y, Shitamura A, Shimada T, Yao J, Kitamura M: Involvement of selective reactive oxygen species upstream of proapoptotic branches of unfolded protein response. J Biol Chem 2008;283:4252–4260.
18. Zinszner H, Kuroda M, Wang X, Batchvarova N, Lightfoot RT, Remotti H, Stevens JL, Ron D: CHOP is implicated in programmed cell death in response to impaired function of the endoplasmic reticulum. Genes Dev 1998;12:982–995.
19. Back SH, Scheuner D, Han J, Song B, Ribick M, Wang J, Gildersleeve RD, Pennathur S, Kaufman RJ: Translation attenuation through eIF2alpha phosphorylation prevents oxidative stress and maintains the differentiated state in beta cells. Cell Metab 2009;10:13–26.
20. De Gracia DJ, Montie HL: Cerebral ischemia and the unfolded protein response. J Neurochem 2004;91:1–8.
21. Kohno K, Higuchi T, Ohta S, Kohno K, Kumon Y, Sakaki S: Neuroprotective nitric oxide synthase inhibitor reduces intracellular calcium accumulation following transient global ischemia in the gerbil. Neurosci Lett 1997;224:17–20.
22. Zhang K, Kaufman RJ: From endoplasmic-reticulum stress to the inflammatory response [review]. Nature 2008;454:455–462.
23. Lin W, Harding HP, Ron D, Popko B: Endoplasmic reticulum stress modulates the response of myelinating oligodendrocytes to the immune cytokine interferon-γ. J Cell Biol 2005;169:603–612.
24. Nagaraju K, Casciola-Rosen L, Lundberg I, Rawat R, Cutting S, Thapliyal R, Chang J, Dwivedi S, Mitsak M, Chen YW, Plotz P, Rosen A, Hoffman E, Raben N: Activation of the endoplasmic reticulum stress response in autoimmune myositis: potential role in muscle fiber damage and dysfunction. Arthritis Rheum 2005:52:1824–1835.
25. Kaser A, Lee AH, Franke A, Glickman JN, Zeissig S, Tilg H, Nieuwenhuis EE, Higgins DE, Schreiber S, Glimcher LH, Blumberg RS: XBP1 links ER stress to intestinal inflammation and confers genetic risk for human inflammatory bowel disease. Cell 2008;134:743–756.
26. Wouters BG, Koritzinsky M: Hypoxia signalling through mTOR and the unfolded protein response in cancer. Nat Rev Cancer 2008;8:851–864.
27. Kojima I, Tanaka T, Inagi R, Nishi H, Aburatani H, Kato H, Miyata T, Fujita T, Nangaku M: Metallothionein is upregulated by hypoxia and stabilizes hypoxia-inducible factor in the kidney. Kidney Int 2009;75:268–277.
28. Zhang K, Kaufman RJ: The unfolded protein response: a stress signaling pathway critical for health and disease. Neurology 2006;66:S102–S109.

29. Oyadomari S, Mori M: Roles of CHOP/GADD153 in endoplasmic reticulum stress. Cell Death Differ 2004;11:381–389.
30. Xu C, Bailly-Maitre B, Reed JC: Endoplasmic reticulum stress: Cell life and death decisions. J Clin Invest 2005;115:2656–2664.
31. Ogawa S, Kitao Y, Hori O: Ischemia-induced neuronal cell death and stress response. Antioxid Redox Signal 2007;9:573–587.
32. Inagi R: Endoplasmic reticulum stress as a progression factor for kidney injury [review]. Curr Opin Pharmacol 2010;10(2):156–165.
33. Kitamura M: Endoplasmic reticulum stress in glomerulonephritis: the bad guy turns good? J Am Soc Nephrol 2009;20:1871–1873.
34. Cybulsky AV, Takano T, Papillon J, Khadir A, Liu J, Peng H: Complement C5b-9 membrane attack complex increases expression of endoplasmic reticulum stress proteins in glomerular epithelial cells. J Biol Chem 2002;277:41342–41351.
35. Cybulsky AV, Takano T, Papillon J, Bijian K: Role of the endoplasmic reticulum unfolded protein response in glomerular epithelial cell injury. J Biol Chem 2005;280:24396–24403.
36. Nakajo A, Khoshnoodi J, Takenaka H, Hagiwara E, Watanabe T, Kawakami H, Kurayama R, Sekine Y, Bessho F, Takahashi S, Swiatecka-Urban A, Tryggvason K, Yan K: Mizoribine corrects defective nephrin biogenesis by restoring intracellular energy balance. J Am Soc Nephrol 2007;18:2554–2564.
37. Cybulsky AV, Takano T, Papillon J, Bijian K, Guillemette J, Kennedy CR: Glomerular epithelial cell injury associated with mutant α-actinin-4. Am J Physiol Renal Physiol 2009:297:F987–F995.
38. Fujii Y, Khoshnoodi J, Takenaka H, Hosoyamada M, Nakajo A, Bessho F, Kudo A, Takahashi S, Arimura Y, Yamada A, Nagasawa T, Ruotsalainen V, Tryggvason K, Lee AS, Yan K: The effect of dexamethasone on defective nephrin transport caused by ER stress: a potential mechanism for the therapeutic action of glucocorticoids in the acquired glomerular diseases. Kidney Int 2006;69:1350–1359.
39. Inagi R, Nangaku M, Onogi H, Ueyama H, Kitao Y, Nakazato K, Ogawa S, Kurokawa K, Couser WG, Miyata T: Involvement of endoplasmic reticulum (ER) stress in podocyte injury induced by excessive protein accumulation. Kidney Int 2005;68:2639–2650.
40. Inagi R, Kumagai T, Nishi H, Kawakami T, Miyata T, Fujita T, Nangaku M: Preconditioning with endoplasmic reticulum stress ameliorates mesangioproliferative glomerulonephritis. J Am Soc Nephrol 2008;19:915–922.
41. Markan S, Kohli HS, Joshi K, Minz RW, Sud K, Ahuja M, Anand S, Khullar M: Up regulation of the GRP-78 and GADD-153 and down regulation of Bcl-2 proteins in primary glomerular diseases: a possible involvement of the ER stress pathway in glomerulonephritis. Mol Cell Biochem 2009;324:131–138.
42. Liu G, Sun Y, Li Z, Song T, Wang H, Zhang Y, Ge Z: Apoptosis induced by endoplasmic reticulum stress involved in diabetic kidney disease. Biochem Biophys Res Commun 2008;370:651–656.
43. Xu Q, Reed JC: Bax inhibitor-1, a mammalian apoptosis suppressor identified by functional screening in yeast. Mol Cell 1998;1:337–346.
44. Ohse T, Inagi R, Tanaka T, Ota T, Miyata T, Kojima I, Ingelfinger JR, Ogawa S, Fujita T, Nangaku M: Albumin induces endoplasmic reticulum stress and apoptosis in renal proximal tubular cells. Kidney Int 2006;70:1447–1455.
45. Kuznetsov G, Bush KT, Zhang PL, Nigam SK: Perturbations in maturation of secretory proteins and their association with endoplasmic reticulum chaperones in a cell culture model for epithelial ischemia. Proc Natl Acad Sci USA 1996;93:8584–8589.
46. Bando Y, Tsukamoto Y, Katayama T, Ozawa K, Kitao Y, Hori O, Stern DM, Yamauchi A, Ogawa S: ORP150/HSP12A protects renal tubular epithelium from ischemia-induced cell death. FASEB J 2004;18:1401–1403.
47. Bailly-Maitre B, Fondevila C, Kaldas F, Droin N, Luciano F, Ricci JE, Croxton R, Krajewska M, Zapata JM, Kupiec-Weglinski JW, Farmer D, Reed JC: Cytoprotective gene bi-1 is required for

intrinsic protection from endoplasmic reticulum stress and ischemia-reperfusion injury. Proc Natl Acad Sci U S A 2006;103:2809–2814.
48. Kimura K, Jin H, Ogawa M, Aoe T: Dysfunction of the ER chaperone BiP accelerates the renal tubular injury. Biochem Biophys Res Commun 2008;366:1048–1053.
49. Lindenmeyer MT, Rastaldi MP, Ikehata M, Neusser MA, Kretzler M, Cohen CD, Schlöndorff D: Proteinuria and hyperglycemia induce endoplasmic reticulum stress. J Am Soc Nephrol 2008;19:2225–2236.
50. Lorz C, Justo P, Sanz A, Subira D, Egido J, Ortiz A: Paracetamol-induced renal tubular injury: a role for ER stress. J Am Soc Nephrol 2004;15:380–389.
51. Liu H, Baliga R: Endoplasmic reticulum stress-associated caspase 12 mediates cisplatin-induced LLC-PK1 cell apoptosis. J Am Soc Nephrol 2005;16:1985–1992.
52. Peyrou M, Hanna PE, Cribb AE: Cisplatin, gentamicin, and p-aminophenol induce markers of endoplasmic reticulum stress in the rat kidneys. Toxicol Sci 2007;99:346–353.
53. Pallet N, Bouvier N, Legendre C, Gilleron J, Codogno P, Beaune P, Thervet E, Anglicheau D: Autophagy protects renal tubular cells against cyclosporine toxicity. Autophagy 2008;4:783–791.
54. Kawakami T, Inagi R, Takano H, Sato S, Ingelfinger JR, Fujita T, Nangaku M: Endoplasmic reticulum stress induces autophagy in renal proximal tubular cells. Nephrol Dial Transplant 2009;24:2665–2672.
55. Naidoo N: The endoplasmic reticulum stress response and aging [review]. Rev Neurosci 2009;20:23–37.
56. Nuss JE, Choksi KB, DeFord JH, Papaconstantinou J: Decreased enzyme activities of chaperones PDI and BiP in aged mouse livers. Biochem Biophys Res Commun 2008;365:355–361.
57. Morris JA, Dorner AJ, Edwards CA, Hendershot LM, Kaufman RJ: Immunoglobulin binding protein (BiP) function is required to protect cells from endoplasmic reticulum stress but is not required for the secretion of selective proteins. J Biol Chem 1997;272:4327–4334.
58. Peyrou M, Cribb AE: Effect of endoplasmic reticulum stress preconditioning on cytotoxicity of clinically relevant nephrotoxins in renal cell lines. Toxicol In Vitro 2007;21:878–886.
59. Prachasilchai W, Sonoda H, Yokota-Ikeda N, Oshikawa S, Aikawa C, Uchida K, Ito K, Kudo T, Imaizumi K, Ikeda M: A protective role of unfolded protein response in mouse ischemic acute kidney injury. Eur J Pharmacol 2008;592:138–145.
60. Asmellash S, Stevens JL, Ichimura T: Modulating the endoplasmic reticulum stress response with trans-4,5-dihydroxy-1,2-dithiane prevents chemically induced renal injury in vivo. Toxicol Sci 2005;88:576–584.
61. Boyce M, Bryant KF, Jousse C, Long K, Harding HP, Scheuner D, Kaufman RJ, Ma D, Coen DM, Ron D, Yuan J: A selective inhibitor of eIF2α dephosphorylation protects cells from ER stress. Science 2005;307:935–939.
62. Kim I, Shu CW, Xu W, Shiau CW, Grant D, Vasile S, Cosford ND, Reed JC: Chemical biology investigation of cell death pathways activated by endoplasmic reticulum stress reveals cytoprotective modulators of ASK1. J Biol Chem 2009;284:1593–1603.
63. Qi X, Hosoi T, Okuma Y, Kaneko M, Nomura Y: Sodium 4-phenylbutyrate protects against cerebral ischemic injury. Mol Pharmacol 2004;66:899–908.
64. Liu XL, Doné SC, Yan K, Kilpeläinen P, Pikkarainen T, Tryggvason K: Defective trafficking of nephrin missense mutants rescued by a chemical chaperone. J Am Soc Nephrol 2004;15:1731–1738.
65. Xie Q, Khaoustov VI, Chung CC, Sohn J, Krishnan B, Lewis DE, Yoffe B: Effect of tauroursodeoxycholic acid on endoplasmic reticulum stress-induced caspase-12 activation. Hepatology 2002;36:592–601.
66. Ozcan U, Yilmaz E, Ozcan L, Furuhashi M, Vaillancourt E, Smith RO, Gorgun CZ, Hotamisligil GS: Chemical chaperones reduce ER stress and restore glucose homeostasis in a mouse model of type 2 diabetes. Science 2006;313:1137–1140.

67. Chen Y, Liu CP, Xu KF, Mao XD, Lu YB, Fang L, Yang JW, Liu C: Effect of taurine-conjugated ursodeoxycholic acid on endoplasmic reticulum stress and apoptosis induced by advanced glycation end products in cultured mouse podocytes. Am J Nephrol 2008;28:1014–1022.
68. Izuhara Y, Nangaku M, Takizawa S, Takahashi S, Shao J, Oishi H, Kobayashi H, van Ypersele de Strihou C, Miyata T: A novel class of advanced glycation inhibitors ameliorates renal and cardiovascular damage in experimental rat models. Nephrol Dial Transplant 2008;23:497–509.
69. Malhotra JD, Miao H, Zhang K, Wolfson A, Pennathur S, Pipe SW, Kaufman RJ: Antioxidants reduce endoplasmic reticulum stress and improve protein secretion. Proc Natl Acad Sci USA 2008;105:18525–18530.
70. Natarajan R, Salloum FN, Fisher BJ, Smithson L, Almenara J, Fowler AA 3rd: Prolyl hydroxylase inhibition attenuates post-ischemic cardiac injury via induction of endoplasmic reticulum stress genes. Vascul Pharmacol 2009;51:110–118.
71. Okada K, Minamino T, Tsukamoto Y, Liao Y, Tsukamoto O, Takashima S, Hirata A, Fujita M, Nagamachi Y, Nakatani T, Yutani C, Ozawa K, Ogawa S, Tomoike H, Hori M, Kitakaze M: Prolonged endoplasmic reticulum stress in hypertrophic and failing heart after aortic constriction: possible contribution of endoplasmic reticulum stress to cardiac myocyte apoptosis. Circulation 2004;110:705–712.
72. Cullinan SB, Diehl JA: Coordination of ER and oxidative stress signaling: the PERK/Nrf2 signaling pathway. Int J Biochem Cell Biol 2006;38:317–332.

Chapter 34
Stem Cell Therapy Against Oxidative Stress and Hypoxia

Takashi Yokoo and Motoko Yanagita

Abstract Stem cell therapy has the potential to treat renal diseases, and a number of different approaches have been reported. Although these therapies hold enormous promise, enthusiasm for their use in medical practice is tempered by the numerous disappointments of the past decade. However, the efforts of a large number of researchers are removing obstacles step by step. This chapter reviews many of the recent advances in research on stem cell therapy, with a particular focus on oxidative stress-induced renal disease. Progress to date is discussed under four categories or approaches: (a) gene delivery of angiogenic factors using stem cells as a vehicle for accessing the damaged kidney; (b) inducing mobilization of stem cells from respective niches; (c) supplementation of stem cells by transplantation; and (d) de novo generation of renal tissue from stem cells. In addition, we discuss the optimal stem cell sources and types for these purposes.

Keywords Endothelial progenitor cells · Hematopoietic stem cells · Mesenchymal stem cells · Embryonic stem cell · Induced pluripotent stem cells

1 Introduction

As discussed in other chapters, oxidative stress and hypoxia play a critical role in the progression of kidney injury and, thus, understanding how to control such factors offers new possibilities for treating end stage renal disease. Rationales for novel therapeutic effects include increasing oxygen production from neovascularization and hematopoiesis, and several experimental trials of chemicals and growth factors to achieve such an outcome have been conducted [1]. Recent advances in

T. Yokoo (✉)
Division of Nephrology, Jikei Medical University, 3-25-8,
Nishi-Shimbashi, Minato-ku, Tokyo 105-8461, Japan
e-mail: tyokoo@jikei.ac.jp

T. Miyata et al. (eds.), *Studies on Renal Disorders*, Oxidative Stress in Applied Basic Research and Clinical Practice, DOI 10.1007/978-1-60761-857-7_34,

stem cell research have also suggested redox management using stem cells as a clinical application, and the scope for such an approach is extensive. First, stem cells recruited to an ischemic lesion could be used as a vehicle for angiogenic factors to target a site of ischemia and induce angiogenesis. Such applications of stem cells as a drug delivery system (DDS) seem feasible and practical. Second, supplementation of stem cells studied in experimental models of ischemic renal disease showed that transplantation and mobilization of stem or progenitor cells could improve neovascularization and recovery of ischemic tissue and organ [2]. The underlying mechanism of this therapeutic effect was attributed, at least in part, to (a) differentiation of the administered stem cells into kidney resident cells or endothelial cells; (b) therapeutic fusion with the existing host cells; and (c) release of paracrine signals from the administered stem cells [3]. Third, establishing erythropoietin (EPO)-producing tissue from stem cells could provide an alternative strategy in stem cell–based medicine.

This chapter reviews the recent advances and remaining challenges in stem cell research and discusses the potential clinical application of stem cells for treating oxidative stress in kidney.

2 Stem Cell Sources for Therapy Against Oxidative Stress

Several types of stem cells might be used in therapeutic strategies against oxidative stress. This section reviews the possible sources of stem cells for this purpose.

2.1 Endothelial Progenitor Cells

Human peripheral blood contains bone marrow–derived endothelial progenitor cells (EPCs), which may contribute to renal repair by maintaining vascular integrity [4, 5]. EPCs are characterized by the expression of early hematopoietic stem cell markers CD34, CD133, and the vascular endothelial-cell growth factor receptor-2, and have been differentiated in vitro into cells with endothelial characteristics [6]. Recently, populations of resident mural cells were found to participate in re-endothelialization and might therefore act in a salvage mechanism to restore the renal microcirculation [7]. In several forms of renal disease, sloughing of endothelial cells is mediated by oxidative stress, and inefficient subsequent recruitment of EPCs to hypoxic tissues can lead to chronic renal failure [8]. Therefore, transfer or mobilization of EPCs is a potential therapeutic strategy for oxidative stress–induced renal diseases. In fact, infusion of EPCs was shown to restore damaged endothelial surfaces and stimulate angiogenesis [9]. Recently, it was also shown that these re-endothelializing cells derive from a common myeloid precursor, and that risk factors such as diabetes could unfavorably alter the immunogenic properties of these myeloid-derived endothelial cells [10]. In addition, EPCs reportedly showed reduced migratory capacity in response to stromal-derived factor (SDF)-1, impaired angiogenic function, and enhanced cell senescence in several oxidative stress–related disorders, including atherosclerosis, smoking, and diabetes [11–13]. These findings underpin the need for rational design of EPCs for therapy against oxidative stress.

2.2 *Hematopoietic Stem Cells*

Hematopoietic stem cells (HSCs) are perhaps the most extensively studied and "traditional" form of stem cells. HSCs can differentiate into many hematopoietic cell lineages, as well as other cell types, including hepatocytes and myocardium [14]. It was previously believed that bone marrow–derived HSCs could differentiate into renal-resident cells after renal ischemia or reperfusion injury [15, 16]; however, recent studies suggested that the levels of bone marrow–derived cells that engraft injured tubules and develop into functional renal tissue are very low, and thus their overall contribution to renal repair would be minor [17, 18]. Therefore, current studies are trying to enhance HSCs' migration into injured kidney to ascertain their true therapeutic potential [19].

HSCs may also have potential as a DDS to target angiogenic factors into ischemic kidney. For example, inflammatory cells such as macrophages and neutrophils can be recruited to and activated at sites of ischemia. Since these cells are derived from HSCs, genetic manipulation of bone marrow–derived cells or bone marrow per se using transplantation-based technology might be exploited to deliver angiogenic factors and modify the oxidative insult. Such an approach involves isolating the HSCs, genetic manipulation to insert an angiogenic gene, and administering them back into the affected subject.

2.3 *Mesenchymal Stem Cells*

It has been suggested that mesenchymal stem cells (MSCs) have the capacity for site-specific differentiation into various cell types, including chondrocytes, adipocytes, myocytes, cardiomyocytes, bone marrow stromal cells, and thymic stromal cells [20, 21], and that injection of MSCs could preserve renal function based on the induction of various experimental models of oxidative stress–induced renal failure [22]. However, Kunter et al. [23] reported recently that MSCs injected intraglomerularly maldifferentiated into adipocytes in vivo, thus offsetting the early beneficial effect of the stems cells in preserving damaged glomeruli and maintaining renal function. The precise glomerular microenvironment necessary for proper MSC differentiation thus requires evaluation to enable efficient and successful cell therapy for oxidative stress–induced renal failure using MSCs.

Our studies into the generation of EPO-producing tissue used primary human MSCs (hMSCs) obtained from the bone marrow of healthy volunteers. As discussed later, embryonic stem (ES) cells are the ideal candidate source of cells for therapeutic intervention [24]. Unlike ES cells, however, hMSCs injected into established metanephroi might not integrate into renal structures during organ culture, which is a distinct disadvantage for the potential clinical feasibility of forming functional renal structures. Our finding that hMSCs do not express WT1 or Pax2 (unpublished data) also suggested that hMSCs do not possess a complete set of nephrogenic molecular features. Nevertheless, one advantage of hMSCs over ES cells as a cell source for therapeutic use is the ease with which adult MSCs can be isolated from autologous bone marrow and used therapeutically without serious ethical issues or the requirement for immunosuppressants.

2.4 Embryonic Stem Cells

ES cells are undifferentiated pluripotent stem cells isolated from the inner cell mass of blastocysts [25]. ES cells can differentiate into several cell types of mesodermal, endodermal, and ectodermal lineages, depending on culture conditions, and are a potential source of cells for angiogenesis as well as EPO-producing tissue. The application of ES cells for therapeutic intervention has been approved in many disease models, including Parkinson's disease and diabetes [26, 27]. Since human ES cells can differentiate into kidney structures when injected into immunosuppressed mice [28, 29], studies have focused on identifying the precise culture conditions that allow the differentiation of ES cells into renal cells in vitro. To this end, Schuldiner et al. showed that human ES cells cultured with eight growth factors, including hepatocyte growth factor (HGF) and activin A, differentiated into cells expressing WT-1 and renin [30]. More recently, mouse ES cells stably transfected with Wnt4 (Wnt4-ES cells) were differentiated into tubular-like structures expressing AQP-2 in the presence of HGF and activin A [31]. Using such in vitro techniques, it may eventually be possible to identify the key molecules that determine the fate of ES cells, although establishing a whole, functional kidney in vitro for clinical use with this technology might remain difficult. An ex vivo culture system, in which ES cells (or ES-derived cells) are cultured in the developing metanephros, was also investigated to determine the capacity of ES cells to differentiate into kidney cells integrated into the kidney structure [32]. ROSA26 ES cells were stimulated with developmental signals in the microenvironment of a developing kidney following injection into a metanephros cultured in vitro. ES cell-derived, β-galactosidase-positive cells were identified in epithelial structures resembling renal tubules with an efficiency approaching 50% [32]. Based on these results, Kim and Dressler then attempted to identify the nephrogenic growth factors needed to induce differentiation of ES cells into renal epithelial cells [33]. When injected into a developing metanephros, ES cells treated with retinoic acid, activin A, and BMP7 contributed to tubular epithelia with near 100% efficiency. Furthermore, Vignewu et al. showed that ES cells that expressed brachyury, a marker of mesoderm specification, became a renal progenitor population in the presence of activin A [34]. After injection into a developing metanephros, these cells might be incorporated into the blastimal cells of the nephrogenic zone. In addition, a single injection of the same cells into developing live newborn mouse kidneys showed them stably integrating into proximal tubules with normal morphology and polarization for 7 months without teratoma formation [34]. Taken together, these data highlight ES cells as a potential source of renal stem cells to be differentiated into renal-resident cells including EPO-producing populations.

2.5 Induced Pluripotent Stem Cells

Major obstacles to using ES cells include the use of donated eggs and the possible immune response to non-self cells. Therefore, an ideal cell source may be cells with all the properties of ES cells but are derived from an adult source, such as the skin. The first attempt to make such patient-specific stem cells used somatic cell nuclear transfer or cloning, which involved reprogramming DNA from an adult cell by transplanting it into the cytoplasmic environment of an unfertilized egg [35] or, more recently, of a newly fertilized egg [36]. ES cells derived through nuclear transfer have been generated in mice, albeit at a fairly low rate. The first reported successful cloning in humans by Hwang et al. [37] was probably a parthenogenetic ES cell derived from a blastocyst; however, later claims by this group were found to be fraudulent. Theoretically, it should be possible to produce pluripotent stem cells from an adult human source and, indeed, nonhuman primate pluripotent stem cells have been produced from adult skin fibroblasts [38]. However, there remains no reported successful development of a human ES cell line using nuclear transfer. In this context, Takahashi and Yamanaka [39] produced pluripotent ES-like cells from cultured somatic cells by retroviral transfer with Ocr3/4, Sox2, c-Myc, and Klf4, which are transcription factors associated with pluripotency. These cells were termed induced pluripotent stem (iPS) cells [40]. Rate-reprogrammed iPS cells were subsequently derived by the reactivation of Fbx15 [40], Oct4 [40], or Nanog [41], all of which carry a drug-resistance marker inserted into the respective endogenous locus by homologous recombination or a transgene containing the Nanog promoter. Furthermore, iPS cells were also isolated based on their ES-like morphology, without the use of transgenic donor cells [42], and the therapeutic potential of autologous iPS cells in a mouse model of a hereditary disease has been reported [43]. Recently, human iPS cells were successfully induced from adult skin fibroblasts using the same four factors as detailed above [44]. iPS cells are epigenetically and biologically indistinguishable from normal ES cells and, therefore, are another potential source of patient-specific renal stem cells for forming EPO-producing tissue.

3 DDS against Oxidative Stress Using Stem Cells

Stem cells as a DDS is another potential means of therapeutic intervention against oxidative stress, enabling the continuous delivery of several known regulators of angiogenesis into the temporally and spatially optimal site for combating the effects of oxidative stress. HGF is an example of an angiogenic candidate for delivery since it can exert anti-inflammatory actions in endothelial cells and renal epithelial cells and was shown to be protective in ischemic renal injury [45]. Vascular endothelial growth factor (VEGF) is another candidate for therapeutic angiogenesis. Expressed by podocytes, tubular epithelial cells, and endothelial cells [46–49], VEGF could contribute to maintaining glomerular capillary integrity, as evidenced by the effects of decreasing intraglomerular VEGF levels by genomic manipulation or antibodies (i.e., capillary endothelial cells swell, capillary loops collapse, and proteinuria develops) [50]. Therefore, administration of these factors might help endothelial cells to recover from oxidative stress–induced injury.

We recently established an in vivo differentiation system of transplantation-based gene therapy [51] to enable the continuous and site-specific delivery needed to get these angiogenic agents to the sites of renal injury. The system was initially tested for efficiency of gene delivery by targeting an anti-inflammatory cytokine to an inflamed site [52, 53]. Before differentiation and transplantation, HSCs were genetically modified to express an anti-inflammatory cytokine via retroviral transfection, such that anti-inflammatory mononuclear cells would be continuously supplied from the reconstituted bone marrow and long lasting suppression of local inflammation was achieved. The bone marrow of female recipient mice was reconstituted with that of male mice that were genetically modified to express the interleukin-1 receptor antagonist (IL-1Ra) gene or a mock gene [54]. Glomerulonephritis was induced in the IL-1Ra and mock chimeras using antiglomerular basement membrane serum 8 weeks after the primary transplantation. Serum creatinine and urea nitrogen levels as well as urine albumin excretion levels rose progressively in the mock chimera, whereas these increases were suppressed significantly in the IL-1Ra chimera. This therapeutic effect persisted for 4 months after the primary bone marrow reconstitution, demonstrating that the donor cells secreting anti-inflammatory cytokine were continuously supplied from the reconstituted bone marrow [54]. This result indicated that bone marrow reconstitution with anti-inflammatory stem cells confers long-lasting resistance against glomerular inflammation and confirmed the potential application of this system for delivering angiogenic factors into the kidney.

We next attempted to modify the system using another source of HSCs, since taking stem cells from the bone marrow is highly inconvenient for clinical use, especially for treating nonlethal and symptom-free diseases such as chronic renal failure. Umbilical cord blood cells were chosen as a viable and useful alternative since (a) a substantial amount of blood (about 100 mL from each delivery [55]) may be collected without pain or risk to the mother or infant, (b) they contain a significantly higher number of hematopoietic progenitor cells compared with adult peripheral blood [56], (c) a low incidence and severity of graft-vs.-host disease after transplantation is encountered due to the immature immune system [56], and (d) recently, large-scale banks of cord blood have been set up or are being considered throughout the world. $CD34^{+}$ cells from human cord blood have already been transduced with a foreign gene using a retroviral vector [57]. Therefore, these cells were the first alternative HSC source selected for our gene delivery system. After retroviral transfection with the human β-glucuronidase (HBG) gene as a reporter, human cord blood–derived $CD34^{+}$ cells were transplanted into nonobese diabetic/severe combined immunodeficiency mice since this strain is characterized by a functional deficit in natural killer cells, absence of circulating complement, and defects in the differentiation and function of antigen-presenting cells, as well as an absence of T- and B-cell function, all of which facilitate reconstitution with human hematopoietic cells [58, 59]. Flow cytometric analysis revealed that $24.1 \pm 14.5\%$ of bone marrow cells in these chimera mice expressed human HLA 8 weeks after transplantation. Further, a clonogenic assay showed sustained engraftment of human hematopoietic cells expressing HBG. CD14-positive cells were also recruited into the glomeruli upon immune activation using Lipopolysaccharide (LPS) treatment and they secreted bioactive HBG, suggesting that cord blood–derived $CD34^{+}$cells could efficiently differentiate into monocytic cells while maintaining transgene expression [60]. These data demonstrated that human cord blood cells present an alternative source of HSCs for transplantation-based gene delivery systems. This option would allow the next step aimed at clinical application of stem cell gene therapy for oxidative renal diseases to proceed.

4 Supplementing Stem Cells by Mobilization and Transplantation

There are two strategies by which stem cell or EPCs could be supplied at the local site of injury. The most practical way may be to induce mobilization from an established stem cell niche (i.e., bone marrow). It was recently demonstrated that circulating stem/progenitor cells are preferentially recruited to ischemic sites through interactions between the chemokine, SDF-1, and its receptor CXCR4, both of which are HIF-target genes [61–63]. The expression of SDF-1 in endothelial cells indicates the presence of tissue ischemia and activates the recruitment of stem/progenitor cells expressing CXCR4. Tissue oxygen tension and hypoxia also influence the proliferation and differentiation of recruited progenitor cells. Interestingly, SDF-1 is expressed in the bone marrow niche, where low oxygen tension promotes stem cell expansion. In this context, the ischemic tissue may serve as an "acquired stem cell niche," thus recruiting the circulating stem cells and promoting their expansion.

Mobilization of stem cells and EPCs has been extensively documented in acute renal ischemia [50]. That these cells are consistently found at sites of oxidative injury alludes to the existence of intrinsic factors generated during oxidative stress that may be responsible for their mobilization. Identification of such factors could be exploited in therapeutic tools for oxidative stress. EPCs can be mobilized pharmacologically using various factors, including statins, VEGF, EPO, angiotensin-converting enzyme inhibitor, and estrogens [64–68]. Since these agents are multifunctional and not available simply for the purpose of stem cell mobilization, more therapy-specific molecules are required. Uric acid exhibits a short-lived surge after ischemia/reperfusion injury and thus it could act as a stress-signaling molecule when discharged from ischemic tissue to activate the downstream mobilization and recruitment of stem cells and EPCs [69]. In fact, exogenous uric acid leads to a rapid mobilization of EPCs and HSCs and protection of the kidney against ischemic injury. In contrast, Togel and coworkers [70] reported that cyclophosphamid- and granulocyte colony-stimulating factor (G-CSF)–induced mobilization of endogenous HSCs in the mouse renal ischemia model worsened renal failure, allegedly because of the induction of granulocytosis. Such findings restate the need for more highly specific inducers of stem cell mobilization for clinical use.

The other strategy is direct supplementation of isolated stem cells via systemic administration, which may improve neovascularization and recovery of ischemic tissue. Many studies have attempted to clarify the therapeutic potential of such an approach, using bone marrow transplantation of marked donor cells to trace their progeny. Among them, Lin et al. [71] were the first to report that bone marrow stem cells ($Rh^{lo}Lin^{-}Sca\text{-}1^{+}c\text{-}kit^{+}$ cells) isolated from male Rosa26 mice, which ubiquitously express the LacZ gene, could differentiate into renal proximal tubular cells and contribute to renal tubular regeneration when transplanted into female mice with renal ischemia/reperfusion injury. Kale and coworkers [72] subsequently reported the therapeutic potential of bone marrow stem cell infusion for the treatment of ischemic renal injury. They introduced an ischemia/reperfusion injury in mice, with and without bone marrow ablation, to control the contribution of bone marrow–derived cells. Renal damage in the mice with bone marrow ablation was much worse than observed in normal mice, and this exacerbation was reversed by stem cell transfusion. This report provided the conceptual basis for therapeutic strategies involving exogenous renal stem cells to enhance recovery from oxidative stress. Since then, a numbers of papers have reported using different bone marrow fractions or different experimental models to investigate these potential therapies. These studies generally involved the transplantation of bone marrow cells marked with LacZ, enhanced green fluorescent protein (EGFP), or a genetic marker (Y chromosome) and their detection after induction of renal damage. Progeny of these donor cells were then detected using X-gal staining, fluorescence microscopy, or fluorescent in situ hybridization for the Y chromosome, respectively. Renal tubular cells bearing these markers were detected, indicating that some fraction of the transplanted bone marrow cells (e.g., bulk fraction, hematopoietic stem cells, or mesenchymal stroma) contributed to the renal regeneration following ischemic injury, confirming the therapeutic potential of such modalities. More recently, however, Duffield et al. precisely demonstrated that all of the detection systems used in these earlier studies, although well established, were capable of producing false positives that could overestimate the contribution of bone marrow–derived cells to the renal repair [18]. Care is clearly required when evaluating therapeutic effects of stem cell administration.

5 De Novo Establishment of EPO Producers from Stem Cells

EPO is a glycoprotein hormone produced mainly by the kidney in an inverse relationship to blood oxygen content. EPO is a major regulator of red blood cells in the circulation by stimulating erythropoiesis and erythrocyte differentiation. Anagnostou et al. [73] showed that EPO receptors are also expressed in mature endothelial cells. Since angiogenesis is the endogenous response to hypoxia, it is feasible that EPO also stimulates angiogenesis and proliferation of endothelial cells during kidney development. Therefore, supplementation of EPO by establishing EPO-producing cells could provide a therapeutic tool against oxidative stress.

The gene therapy approach [74], whereby the EPO gene would be transferred directly to a recipient in vivo, and the cell therapy approach [75], whereby the EPO gene would be transferred into isolated cells followed by their transplantation into the recipient organism, have both been successfully applied in combination with an artificial regulatory system [76–79]. However, both these methods require viral vector-mediated delivery, which presents potential ethical consequences and could prove unable to mimic physiological oxygen-dependent control of EPO production due to the complexity and sensitivity of EPO regulation [80]. The ultimate therapeutic approach, then, is the establishment of functional tissue derived from autologous stem cells, capable of generating endogenous levels of EPO and retaining physiological regulatory pathways to conditions such as anemia.

We attempted to establish an EPO-producing tissue from MSCs. Initially, we built an organized and functional kidney structure using the developing heterozoic embryo as an "organ factory." Human MSCs expressing neurotrophin GDNF were microinjected at the site of budding. The recipient embryo was grown in a whole embryo culture system, and the formed metanephros was developed in organ culture. With this combination (termed relay culture), donor cells can be stimulated with numerous factors in a spatially and temporally identical manner to the endogenous developmental process of nephron formation [81]. Using a xenobiotic developmental process for growing embryos thus allows endogenous hMSCs to undergo epithelial conversion and be transformed into an orchestrated nephron comprising glomerular epithelial cells (podocytes) and tubular epithelial cells that are linked, as well as stromal cells that probably contain the EPO-producing cells. The hMSC-derived kidney primordia were then transplanted into host omentum to allow growth and differentiation into a functional renal unit with integration of vessels [82]. The resultant hMSC-derived neokidney was equivalent to a human nephron [83]. We subsequently confirmed that the neokidney vasculature in the omentum

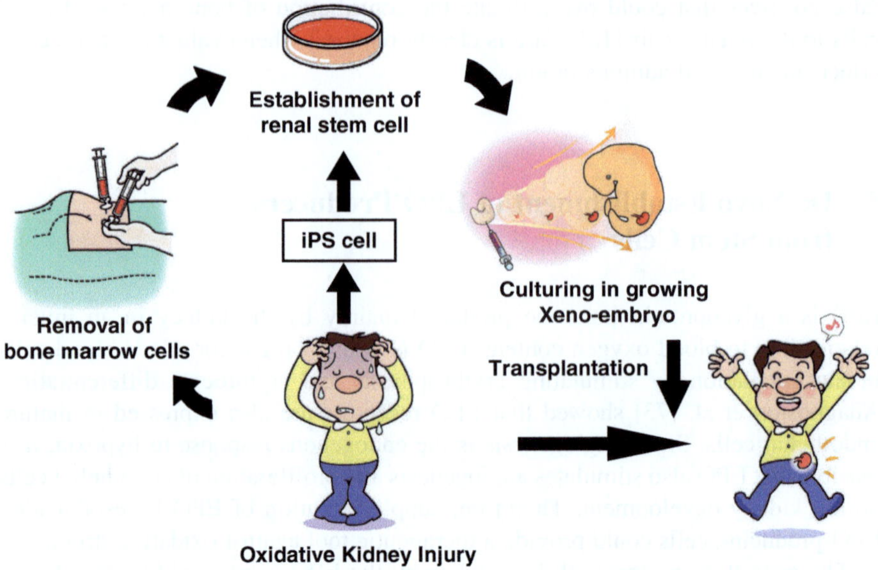

Fig. 1 Putative scenario for the application of stem cell therapy to oxidative kidney injury. Renal stem cells derived from bone marrow cell or skin fibroblast of patient with oxidative kidney injury are cultured in growing xeno-embryo for a given time to develop into kidney primordia, followed by autologous implantation into the omentum of the same patients. Kidney primordia eventually become self-organ that produce erythropoietin and angiogenic factors. The patient might be cured from oxidative stress

originated from the host and communicated with the host circulation. In fact, analysis of liquid secreted from the expanded ureter showed higher urea nitrogen and creatinine concentrations than in the recipient sera, with similar concentrations to native urine [84]. This suggested that the neokidney was capable of producing urine by filtering the recipient's blood. Furthermore, the levels of EPO increased in response to anemia induction and the presence of neokidney in rats suppressed the recipient (rat) EPO production, with enhanced recovery from anemia compared to control animals lacking the neokidney [84]. These data indicate that autologous EPO-producing tissue could be established from MSCs by our methods and suggest a powerful therapeutic tool to combat oxidative stress-induced renal disease (Fig. 1).

6 Conclusion

In this chapter we reviewed recent research into using renal stem cells to treat kidney diseases and proposed their possible therapeutic application against oxidative stress. We know that prior to the total loss of renal structure, kidney function can be restored by reactivating quiescent renal stem cells or supplying renal stem cells expanded sufficiently in vitro. Such stem cells may contribute to neovascularization or hematopoiesis, leading to recovery from hypoxic shock.

We believe that emerging knowledge of kidney stem cell biology and developmental biology will enable the development of new therapeutic strategies for the treatment of oxidative stress–induced renal diseases that aim to regain damaged components in the kidney or restore kidney function.

References

1. Nangaku M. Chronic hypoxia and tubulointerstitial injury: A final common pathway to end-stage renal failure. J Am Soc Nephrol 17: 17–25, 2006
2. Kocher AA, Schuster MD, Szabolcs MJ et al. Neovascularization of ischemic myocardium by human bone marrow-derived angioblasts prevents cardiomyocyte apoptosis, reduces remodeling and improves cardiac function. Nat Med 7: 430–436, 2001
3. Chen J, Parrk H-C, Addabbo F, Ni J, Pelger E, Li H, Lotkin M, Goligorsky MS. Kidney-derived mesenchymal stem cells contribute to vasculogenesis, angiogenesis and endothelial repair. Kidney Int 74: 879–889, 2008
4. Rookmaaker MB, Smits AM, Tolboom H, Van't EK, Martens AC, Goldschmeding R, Joles JA, van Zonneveld AJ, Grone HJ, Rabelink TJ, Verhaar MC. Bone-marrow derived cells contribute to endothelial repair in experimental glomerulonephritis. Am J Pathol 163: 553–562, 2003
5. Rookmaaker MB, Tolboom H, Goldschmeding R, Zwaginga JJ, Rabelink TJ, Verhaar MC. Bone-marrow-derived cells contribute to endothelial repair after thrombotic mictoangiopathy. Blood 99: 1095, 2002
6. Asahara T, Murohara T, Sullivan A et al. Isolation of putative progenitor endothelial cells for angiogenesis. Science 275: 964–967, 1997
7. Hillebrands JL, Klatter FA, van Dijk WD, et al. Bone marrow does not contribute substantially to endothelial cell replacement in transplant arteriosclerosis. Nat Med 8: 194–195, 2002
8. Reinders MEJ, Rabelink TJ, Briscoe DM. Angiogenesis and endothelial cell repair in renal disease and allograft rejection. J Am Soc Nephrol 17: 932–942, 2006
9. Rabelink TJ, de Boer HC, de Koning EJ, van Zonneveld AJ. Endothelial progenitor cells: more than an inflammatory response? Arterioscler Thromb Vasc Biol 24: 834–838, 2004
10. Loomans CJM, Wan R, de Crom R, van Haperen R, de Boer HC, Leenen PJM, Drexhage HA, Rabelink TJ, van Zonneveld AJ, Staarl FJT. Angiogenic murine endothelial progenitor cells are derived from a myeloid bone marrow fraction and can be identified by endothelial NO synthase expression. Atherioscler Thromb Vasc Biol 26: 1760–1767, 2006
11. Rauscher FM, Goldschmidt-Clermont PJ, Davis BH, Wang T, Gregg D, Ramaswami P, Pippen AM, Annex BH, Dong C, Taylor DA. Aging, progenitor cell exhaustion, and atherosclerosis. Circulation 108: 457–463, 2003
12. Hill JM, Zalos G, Halcox JP, Schenke WH, Waclawiw MA, Wuyyumi AA, Finkel T. Circulating endothelial progenitor cells, vascular function, and cardiovascular risk. N Eng J Med 348: 593–600, 2003
13. Loomans CJ, de Koning EJ, Staal FJ, Rookmaaker MB, Cerseyden C, deBoer HC, Verhaar MC, Braam B, Rabelink TJ, van Zonneveld AJ. Endothelial progenitor cell dysfunction: a novel concept in the pathogenesis of vascular complications of type 1 diabetes. Diabetes 53: 195–199, 2004
14. Vieyra DS, Jackson KA, Goodell MA. Plasticity and tissue regenerative potential of bone marrow-derived cells. Stem Cell Rev 1: 65–69, 2005

15. Kale S, Karihaloo A, Clark PR et al. Bone marrow stem cells contribute to repair of the ischemically injured renal tubule. J Clin Invest 112: 42–49, 2003
16. Lin F, Cordes K, Li L, et al. Hematopoietic stem cells contribute to the regeneration of renal tubules after renal ischemia-reperfusion injury in mice. J Am Soc Nephrol 14: 1188–1199, 2003
17. Lin F, Moran A, Igarashi P. Intrarenal cells, not bone marrow derived cells, are the major source for regeneration in postischemic kidney. J Clin Invest 115: 1756–1764, 2005
18. Duffield JS, Park KM, Hsiao L-L, Kelley VR, Scadden DT, Ichimura T, Bonventre JV. Restoration of tubular epithelial cells during repair of the postischemic kidney occurs independently of bone marrow-derived stem cells. J Clin Invest 115: 1743–1755, 2005
19. Stroo I, Stokman G, Taske GJD, Glorquin S, Leemans JC. Haematopoietic stem cell migration to the ischemic damaged kidney is not altered by manipulating the SDF-1/CXCR4-axis. Nephrol Dial Transplant 24: 2082–2088, 2009
20. Liechty KW, MacKenzie TC, Shaaban AF, et al. Human mesenchymal stem cells engraft and demonstrate site-specific differentiation after in utero transplantation in sheep. Nat Med 6: 1282–1286, 2000
21. Prockop DJ: Marrow stromal cells as stem cells for nonhematopoietic tissues. Science 276: 71–74, 1997
22. Bussolati B, Camussi G. Stem cells in acute kidney injury. Contrib Nephrol 156: 250–258, 2007
23. Kunter U, Rong S, Boor P, et al. Mesenchymal stem cells prevent progressive experimental renal failure but maldifferentiate into glomerular adipocytes. J Am Soc Nephrol 18: 1754–1764, 2007
24. Yamamoto M, Cui L, Johkura K, et al. Branching ducts similar to mesonephric ducts or ureteric buds in teratomas originating from mouse embryonic stem cells. Am J Physiol Renal Physiol 290: F52–F60, 2005
25. Martin GR: Isolation of a pluripotent cell line from early mouse embryos cultured in medium conditioned by teratocarcinoma stem cell. Proc Natl Acad Sci USA 78: 7634–7638, 1981
26. Bjorklund LM, Sanchez-Pernaute R, Chung S, et al. Embryonic stem cells develop into functional dopaminergic neurons after transplantation in a Parkinson rat model. Proc Natl Acad Sci U S A 99: 2344–2349, 2002
27. Blyszczuk P, Czyz J, Kania G, et al. Expression of Pax4 in embryonic stem cells promotes differentiation of nestin-positive progenitor and insulin-producing cells. Proc Natl Acad Sci USA 100: 998–1003, 2003
28. Reubinoff BE, Pera MF, Fong CY, Trounson A, Bongso A. Embryonic stem cell lines from human blastocysts: Somatic differentiation in vitro. Nat Biotechnol 18: 399–404, 2000
29. Thomson JA, Etskovitz-Eldor J, Shapiro SS, et al. Embryonic stem cell lines from human blastocysts. Science 282: 1145–1147, 1998
30. Schuldiner M, Yanuka O, Itskovitz-Ekdor J, Melton DA, Benvenisty N. Effects of eight growth factors on the differentiation of cells derived from human embryonic stem cells. Proc Natl Acad Sci USA 97: 11307–11312, 2000
31. Kobayashi T, Tanaka H, Kuwana H, et al. Wnt4-transformed mouse embryonic stem cells differentiate into renal tubular cells. Biochem Biophys Res Commun 336: 585–595, 2005
32. Steenhard BM, Isom KS, Cazcarro P, et al. Integration of embryonic stem cells in metanephric kidney organ culture. J Am Soc Nephrol 16: 1623–1631, 2005
33. Kim D, Dressler G. Nephrogenic factors promote differentiation of mouse embryonic stem cells into renal epithelia. J Am Soc Nephrol 16: 3527–3534, 2005
34. Vignewu C, Polgar K, Striker G, et al. Mouse embryonic stem cell-derived embryoid bodies generate progenitors that integrate long term into renal proximal tubules in vivo. J Am Soc Nephrol 18: 1709–1720, 2007
35. Wilmut I, Schnieke AE, McWhir J, Kind AJ, Cambell KH. Viable offspring derived from fetal and adult mammalian cells. Nature 385: 810–813, 1997

36. Egli D, Rosains J, Birkhoff G, Eggan K. Developmental reprogramming after chromosome transfer into mitotic mouse zygotes. Nature 447: 679–685, 2007
37. Hwang WS, Ryu YJ, Park JH, et al. Evidence of a pluripotent human embryonic stem cell line derived from a cloned blastocyst. Science 303: 1669–1674, 2004 (Retraction in Kennedy D. Science 2006;311:335)
38. Byrne JA, Pedersen DA, Clepper LL, et al. Producing primate embryonic stem cells by somatic cell nuclear transfer. Nature 450: 497–502, 2007
39. Takahashi K, Yamanaka S. Induction of pluripotent stem cells from mouse embryonic and adult fibroblast cultures by defined factors. Cell 126: 663–676, 2006
40. Okita K, Ichisaka T, Yamanaka S. Generation of germline-competent induced pluripotent stem cells. Nature 448: 313–317, 2007
41. Wernig M, Meissner A, Foreman R, et al. In vitro reprogramming of fibroblasts into a pluripotent ES-cell-like state. Nature 448: 318–324, 2007
42. Meissner A, Wernig M, Jaenisch R. Direct reprogramming of genetically unmodified fibroblasts into pluripotent stem cells. Nat Biotechnol 25: 1177–1181, 2007
43. Hanna J, Wernig M, Markoulaki S, et al. Treatment of sickle cell anemia mouse model with iPS cells generated from autologous skin. Science 318: 1920–1923, 2007
44. Takahashi K, Tanabe K, Ohnuki M, et al. Induction of pluripotent stem cells from adult human fibroblasts by defined factors. Cell 131:1–12, 2007
45. Mizuno S, Nakamura T. Prevention of neutrophil extravasation by hepatocyte growth factor leads to attenuations of tubular apoptosis and renal dysfunction in mouse ischemic kidneys. Am J Pathol 166: 1895–1905, 2005
46. Shulman K, Rosen S, Tognazzi K, Manseau EJ, Brown LF. Expression of vascular permeability factor (VPF/VEGF) is altered in many glomerular diseases. J Am Soc Nephrol 7: 661–666, 1996
47. Brown LF, Berse B, Tognazzi K, Manseau EJ, Van de Water L, Senger DR, Dvorak HF, Rosen S. Vascular permeability factor mRNA and protein expression in human kidney. Kidney Int 42: 1457–1461, 1992
48. Haltia A, Solin ML, Jalanko H, Holmberg C, Miettinen A, Holthofer H, Mechanisms of proteinuria: vascular permeability factor in congenital nephritic syndrome of the Finnish type. Pediatr Res 40: 652–657, 1996
49. Eremina V, Quaggin SE. The role of VEGF-A in glomerular development and function. Curr Opin Nephrol Hypertens 13: 9–15, 2004
50. Patschan D, Krupincza K, Patschan S, Hang Z, Hamby C, Goligorsky MS. Dynamics of mobilization and homing of endothelial progenitor cells after acute renal ischemia: Modulation by ischemic preconditioning. Am J Physiol Renal Physiol 291: F176–F185, 2006
51. Yokoo T, Kawamura T. Gene therapy for glomerulonephritis using bone marrow stem cells. Clin Exp Nephrol 6: 190–194, 2002.
52. Yokoo T, Ohashi T, Utsunomiya Y, Kojima H, Imasawa T, Kogure T, Hisada Y, Okabe M, Eto Y, Kawamura T, Hosoya T. Prophylaxis of antibody-induced acute glomerulonephritis with genetically modified bone marrow-derived vehicle cells. Hum Gene Ther 10: 2673–2678, 1999
53. Yokoo T, Ohashi T, Utsunomiya Y, Shiba H, Shen J-S, Hisada Y, Eto Y, Hosoya T. Inflamed glomeruli-specific gene activation using recombinant adenovirus with the Cre/loxP system. J Am Soc Nephrol 11: 2330–2337, 2001
54. Yokoo T, Ohashi T, Utsunomiya Y, Shen J-S, Hisada Y, Eto Y, Kawamura T, Hosoya T. Genetically modified bone marrow continuously supplies anti-inflammatory cells and suppresses renal injury in mouse Goodpasture syndrome. Blood 98: 57–64, 2001
55. Broxmeyer HE, Gluckman E, Auerbach A, Douglas GW, Friedman H, Cooper S, Hangoc G, Kurtzberg J, Bard J, Boyse EA. Human umbilical cord blood: A clinically useful source of transplantable hematopoietic stem/progenitor cells. Int J Cell Cloning 8: 76–91, 1990
56. Cairo MS, Wagner JE. Placental and/or umbilical cord blood: An alternative source of hematopoietic stem cells for transplantation. Blood 90: 4665–4678, 1997

57. Ohashi T, Iizuka S, Sly WS, Machiki Y, Eto Y. Efficient and persistent expression of b-glucuronidase gene in $CD34^+$ cells from human umbilical cord blood by retroviral vector. Eur J Haematol 61: 235–239, 1998
58. Shultz LD, Scweitzer PA, Christianson SW, Gott B, Schweitzer TB, Tennent B, McKenna S, Mobraaten L, Rajan TV, Greiner DL, Leiter EH. Multiple defects in innate and adaptive immunologic function in NOD/LtSz-scid mice. J Immunol 154: 180–191, 1995
59. Larochelle A, Vormoor J, Hanenberg H, Wang JC, Bhatia M, Lapidot T, Moritz T, Murdoch B, Xiao XL, Kato I, Williams DA, Dick JE. Identification of primitive human hematopoietic cells capable of repopulating NOD/SCID mouse bone marrow: implication for gene therapy. Nat Med 2: 189–233, 1996
60. Yokoo T, Ohashi T, Utsunomiya Y, Okamoto A, Suzuki T, Shen J-S, Tanaka T, Kawamura T, Hosoya T. Gene delivery using human cord blood-derived $CD34^+$cells into inflamed glomeruli in NOD/SCID mice. Kidney Int 64: 102–109, 2003
61. Ceradini DJ, Gurtner GC. Homing to hypoxia: HIF-1 as a mediator of progenitor cell recruitment to injured tissue. Trends Cardiovasc Med 15: 57–63, 2005
62. Ceradini DJ, Kulkarni AR, Callaghan MJ, Tepper OM, Bastidas N, Kleinman ME, Capla JM, Galiano RD, Levine JP, Gurtner GC. Progenitor cell trafficking is regulated by hypoxic gradients through HIF-1 induction of SDF-1. Nat Med 10: 858–864, 2004
63. Fine LG, Norman JT. Chronic hypoxia as a mechanism of progression of chronic kidney diseases: from hypothesis to novel therapeutics. Kidney Int 74: 867–872, 2008
64. Dimmeler S, Aicher A, Vasa M et al. HMG-CoA reductase inhibitors (statins) increase endothelial progenitor cells via the PI3-kinase/Akt pathway. J Clin Invest 108: 391–397, 2001
65. Asahara T, Takahashi T, Masuda H, et al. VEGF contributes to postnatal neovascularization by mobilizing bone marrow–derived endothelial progenitor cells. EMBO J 18: 3964–3972, 1999
66. Bahlmann FH, De Groot K, Spandau JM, et al. Erythropoietin regulates endothelial progenitor cells. Blood 103: 921–926, 2004
67. Min TQ, Zhu CJ, Xiang WX, Hui ZJ, Peng SY. Improvement in endothelial progenitor cells from peripheral blood by ramipril therapy in patients with stable coronary artery disease. Cardiovasc Drugs Ther 18: 203–209, 2004
68. Iwakura A, Luedemann C, Shastry S, et al. Estrogen-mediated, endothelial nitric oxide synthase-dependent mobilization of bone marrow–derived endothelial progenitor cells contributes to reendothelialization after arterial injury. Circulation 108: 3115–3121, 2003
69. Kuo M-C, Patschan D, Patschan S, Cohen-Gould L, Park H-C, Ni J, Addabbo F, Goligorsky MS. Ischemia-induced exocytosis of Weibel-Palade bodies mobilizes stem cells. J Am Soc Nephrol 19: 2321–2330, 2008
70. Togel F, Isaac J, Westenfelder C. Hematopoietic stem cell mobilization-associated granulocytosis severely worsens acute renal failure. J Am Soc Nephrol 15: 1261–1267, 2004
71. Lin F, Cordes K, Li L, Hood L, Couser WG, Shankland SJ, Igarashi P. Hematopoietic stem cells contribute to the regeneration of renal tubules after renal ischemia-reperfusion injury in mice. J Am Soc Nephrol 14: 1188–1199, 2003
72. Kale SKA, Clark PR, Kashigarian M, Krause DS, Cantley LG. Bone marrow stem cells contribute to repair of the ischemically injured renal tubule. J Clin Invest 112: 42–49, 2003
73. Anagnostou A, Liu Z, Steiner M, Chin K, Lee ES, Kessimian N, Noguchi CT. Erythropoietin receptor mRNA expression in human endothelial cells. Proc Natl Acad Sci USA 91: 3974–3948, 1994
74. Hamamori Y, Samal B, Tian J, et al. Myoblast transfer of human erythropoietin gene in a mouse model of renal failure. J Clin Invest 95: 1808–1813, 1995
75. Svensson EC, Black HB, Dugger DL et al. Long-term erythropoietin expression in rodents and non-human primates following intramuscular injection of a replication-defective adenovirus vector. Hum Gene Ther 8: 1797–1806, 1997
76. Ye X, Rivera VM, Zoltick P et al. Regulated delivery of therapeutic proteins after in vivo somatic cell gene transfer. Science 283: 88–91, 1999

77. Abruzzese RV, Godin D, Mehta V et al. Ligand-dependent regulation of vascular endothelial growth factor and erythropoietin expression by a plasmid-based autoinducible GeneSwith system. Mol Ther 2: 276–287, 2000
78. Bohl D, Salvetti A, Moullier PH, Heard JM. Control of erythropoietin delivery by doxycycline in mice after intramuscular injection of adeno-associated vector. Blood 92: 1512–1517, 1998
79. Rinsch C, Regulier E, Deglon N et al. A gene therapy approach to regulated delivery of erythropoietin as a function of oxygen tension. Hum Gene Ther 8: 1881–1889, 1997
80. Eckardt K-U. After 15 years of success – perspectives of erythropoietin therapy. Nephrol Dial Transpl 16: 1745–1749, 2001
81. Yokoo T, Ohashi T, Shen J-S, Sakurai K, Miyazaki Y, Utsunomiya Y, Takahashi M, Terada Y, Eto Y, Kawamura T, Osumi N, Hosoya T. Human mesenchymal stem cells in rodent whole-embryo culture are reprogrammed to contribute to kidney tissue. Proc Natl Acad Sci USA 102: 3296–3300, 2005
82. Rogers S, Lowell JA, Hammerman NA, Hammerman MR. Transplantation of developing metanephroi into adult rats. Kidney Int 54: 27–37, 1998
83. Yokoo T, Fukui A, Ohashi T, Miyazaki Y, Utsunomiya Y, Kawamura T, Hosoya T, Okabe M, Kobayashi E. Xenobiotic kidney organogenesis from human mesenchymal stem cells using a growing rodent embryo. J Am Soc Nephrol 17: 1026–1034, 2006
84. Yokoo T, Fukui A, Matsumoto K, Ohashi T, Sado Y, Suzuki H, Kawamura T, Okabe M, Hosoya T, Kobayashi E. Generation of a transplantable erythropoietin-producer derived from human mesenchymal stem cells. Transplantation 85: 1654–1658, 2008

Name Index

T. Miyata et al. (eds.), *Studies on Renal Disorders*, Oxidative Stress in Applied Basic Research and Clinical Practice, DOI 10.1007/978-1-60761-857-7,

C

N

S

Subject Index

MIX
Papier aus verantwortungsvollen Quellen
Paper from responsible sources
FSC® C105338

If you have any concerns about our products,
you can contact us on
ProductSafety@springernature.com

In case Publisher is established outside the EU,
the EU authorized representative is:
Springer Nature Customer Service Center GmbH
Europaplatz 3, 69115 Heidelberg, Germany

Printed by Libri Plureos GmbH
in Hamburg, Germany